THE PENGUIN DESK ENCYCLOPEDIA OF

SCIENCE

AND

MATHEMATICS

THE PENGUIN DESK
ENCYCLOPEDIA OF
SCIENCE
AND
MATHEMATICS

Bryan Bunch
and
Jenny Tesar

**PENGUIN
REFERENCE**

BOOK DESIGN: Gerry Burstein
ART: Steve Virkus
COMPOSITION: G & H SOHO, Ltd.
COPYEDITING: Felice Levy
EDITORIAL ASSISTANT: Mary Bunch
INDEX: Mary Bunch, Jenny Tesar, Bryan Bunch, and Felice Levy

PENGUIN REFERENCE
Published by the Penguin Group
Penguin Putnam Inc., 375 Hudson Street, New York, New York 10014 U.S.A.
Penguin Books Ltd, 27 Wrights Lane, London W8 5TZ, England
Penguin Books, Australia Ltd, Ringwood, Victoria, Australia
Penguin Books Canada Ltd, 10 Alcorn Avenue, Toronto, Ontario, Canada M4V 3B2
Penguin Books (N.Z.) Ltd, 182–190 Wairau Road, Auckland 10, New Zealand

Penguin Books Ltd, Registered Offices: Harmondsworth, Middlesex, England

First published in 2000 by Penguin Reference, a member of Penguin Putnam Inc.

10 9 8 7 6 5 4 3 2 1

Library of Congress Cataloging-in-Publication Data
Bunch, Bryan H.
The Penguin desk encyclopedia of science and mathematics / Bryan Bunch and Jenny Tesar.
p. cm.
Includes index
ISBN 0-670-88528-2
1. Science—Encyclopedias. 2. Mathematics—Encyclopedias. I. Tesar, Jenny E. II. Title

Q121 .B86 2000
503—dc21   00-039970

This book is printed on acid-free paper. 8
Printed in the United States of America

For Will, Sally, and Jim

For Lesley and Jennifer

*For all the kids and grownups
who wonder why—*

How to Use This Book

The Penguin Desk Encyclopedia of Science and Mathematics has two parts. Several thousand alphabetical entries are followed by a detailed index. The entries are useful by themselves, but the index makes this volume into a complete reference tool.

Using the Index to Find Concepts in Context. Important topics, key words, and significant scientists and mathematicians have their own entries, alphabetized using the letter-by-letter system (e.g., **blueberry, blue-green algae, blue jet**).

The entries also contain significant information about topics that do not have their own separate entries. This information, including biographical information about scientists and mathematicians, alternative names for given entities, histories of ideas, details concerning closely related topics, taxonomic information about organisms, and so forth can best be accessed through the index. Here are some examples: lactation and prolactin are discussed in the entry **mammary gland**; scalar product and inner products of vectors are other names for **dot product**; copper sulfate is discussed as an example of a **hydrate**; biographical details of Edward Suess are in **Gondwanaland**; Swiss chard is in **beet**; electric induction and the Van de Graaf generator are described in **static electricity**. This arrangement places concepts in context instead of isolating them as short, individual words, the way a science dictionary does.

Thus, if you don't find the topic you seek among the entry words, the location of a discussion of that topic can nearly always be found in the index.

Using Cross-References to Expand Entries. Within an entry, scientific or mathematical terms that also have entries of their own may be employed. The existence of useful cross references from one entry to another is marked by setting terms a special light boldface type, like this. Thus in the entry **kinkajou** the animal is identified as part of the raccoon family; and in the entry on **multiplication**, separate rules are given for natural numbers and irrational numbers. The use of the special type signals that for more information, you can turn to the entries **raccoon, natural number,** and **irrational number** for expansion of ideas or to clear up any questions about specific meanings.

Entry words are not treated as cross-references wherever they are used but only where they can provide important or useful information. Frequently the important information is an illustration or a table. When the word star is used as a cross-reference with a given star, such as **Sirius**, or solar system is marked in the **Mercury** entry, it means that there is more information in tables accompanying entries **star** and **solar system**. Similarly the cross-reference to skin in **sweat gland** leads to an illustration at **skin** that depicts the gland. If you need more information about a particular topic, the typographical clue for a cross-reference will lead you to related entries.

A

aardvark (*animals*)   Looking rather like a pig with big ears, a long snout, and a long tail, the aardvark (*Orycteropus afer*) lives in sub-Saharan Africa. It gets its name from Afrikaans words meaning "earth pig." Unlike other mammals, it has tubular teeth that lack enamel. Primarily nocturnal, the aardvark uses its clawed feet to tear apart ant and termite nests and its long, sticky tongue to lap up insects.

abacus (*arithmetic*)   Several mechanical calculating devices work by moving counters from one place to another by hand. The abacus, widely used in various formats, arranges the counters on wires in a frame. The basic operation involves moving a counter from one end of a wire to another. The wires mirror a decimal system of numeration, so there is a ones wire (a counter is worth 1), a tens wire, (a counter is worth 10), a hundreds, thousands, etc., for as many wires as are likely to be needed. Moving 2 counters up on the tens wire and 5 on the ones wire indicates 25 in the simplest system.

abalone (*animals*)   A gastropod mollusk that lives in shallow coastal waters, the abalone has a flattened spiral shell with a colorful, highly iridescent interior. A row of holes in the shell is for an outgoing water current that carries respiratory and excretory wastes. On the abalone's underside are a large muscular foot and many sensory projections.

Abbe, Cleveland (*astronomer/meteorologist*)   [American: 1838–1916] After being educated as an astronomer, Abbe achieved lasting fame as a meteorologist, founding the division of the U.S. government now known as the Weather Service. He was also among the first and most influential scientists to push for standard time zones across the United States.

abdomen (*anatomy*)   In arthropods and vertebrates, the abdomen is the part of the body posterior to the thorax; that is, behind or below the part of the body that in mammals includes the chest. In mammals, the abdomen is separated from the thorax by the muscular diaphragm. The mammalian abdomen contains the stomach, intestines, liver, pancreas, kidneys, bladder, and other organs (but not the heart and lungs, which are in the thorax); in females, it also contains the ovaries and uterus. The arrangement is somewhat different in other animals. For example, a spider's heart is in its abdomen but its stomach is in its cephalothorax (joined head and thorax).

Abel, Niels Henrik (*mathematician*)   [Norwegian: 1802–29] Abel is best known for his proof of the impossibility of solving the general equation of degree 5 (the quintic) by arithmetic operations or **radicals**; however, he was anticipated in this by Paolo Ruffini [Italian: 1765–1822]. Abel was among the creators of **group** theory (groups with the **commutative property** are called abelian in his honor) and made major advances in the theory of **functions**.

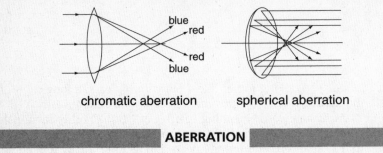

chromatic aberration   spherical aberration

ABERRATION

aberration (*waves*)   Aberration is distortion in **lenses** and mirrors caused by behavior of light.

In a lens chromatic aberration occurs because different colors of light travel through materials at different speeds. Light passing through a curved lens is bent because it takes longer to pass through the thicker part of the lens. As in a rainbow, white light spreads into colors from red to violet, in a process called dispersion. Violet light travels 1% faster through glass than red, so red and violet head in different directions after passing through the lens. A rainbow halo appears around the image and light is spread out, interfering with sharpness. Dispersion also reduces intensity of light at any given point. Chromatic aberration was a problem for early astronomers using refracting telescopes. Instead of a single point of white light, a star's image became a line of dimmer points of different colors. As early as 1732, Chester Moor Hall [English: 1703–71] developed a system in which a second lens counteracts chromatic aberration by bending the colors the other way.

Reflecting telescopes do not produce chromatic aberration the way a lens does. In a reflector, light is directed by a curved mirror toward a single point. Early reflecting mirrors were sections of spheres that caused light from the edges of the mirror to focus at a slightly different place from light reflected by the center. For an image to focus properly with a spherical mirror, the light would have to originate at the center of the sphere. Spherical aberration can be resolved by changing the shape of the mirror or by adding a correcting lens. Spherical aberration became famous briefly when the Hubble Space Telescope was put into use in 1990 and early images were fuzzy. The Hubble's spherical aberration was caused by an error in shaping of the lens. Astronauts installed corrective lenses and the Hubble's spherical aberration was eliminated.

aberration of light (*astronomy*)   The aberration of light (also called the aberration of starlight or stellar aberration) causes stars to appear to shift position as a result of Earth's speed of revolution about the Sun. This effect was announced by James Bradley [English: 1693–1762] in 1729.

Think of light from a distant star as a beam of photons. If an observer moves at an angle to the beam, the photons appear to fall somewhat toward the direction of motion. Earth moves through space in several ways. Bradley discovered that the apparent position of a star changes with the seasons, but not in the way that can be explained by parallax (a much smaller shift). He calculated that the observed change in star position combines Earth's movement and a finite speed for light. Aberration provided the first observational proof that Earth indeed revolves about the Sun; it also allowed Bradley to determine the best estimate for the speed of light at that time.

abrasion (*physics*)   Whenever two solids touch and one moves even the slightest distance with respect to the other, some looser particles are knocked off or transfer from one surface to the other, a process called abrasion. Each surface is worn away. Abrasion can be avoided only by preventing contact.

A fluid cannot abrade, but moving fluid carries away some of a solid past which it flows, abrading the solid. Usually, the fluid causes less abrasion than another solid would, so a fluid between two solids tends to reduce abrasion of both the solids, the principle behind lubrication.

The most dramatic forms of abrasion occur when a fluid made from small solid particles, such as sand or dust, abrades the more ordinary form of solids. Each small particle moving against a surface can carry a bit of the surface away. Sand blasting is often used to remove paint or dirt. The toadstool-shaped rocks seen in deserts are caused by low-blowing sand, which abrades the lower portions of a rock. The upper parts are exposed to the less effective erosion by wind and so the tops of the toadstools form over the narrow stems.

abscission layer (*botany*)   In a process called abscission, plants regularly shed their leaves, either as part of normal aging of the leaves or—in the case of deciduous plants—at the end of a growing season. An abscission layer of weak, thin-walled cells forms at the point where the leaf grows out of the stem. Ethylene produced in the abscission layer promotes the production of enzymes that break down the cell walls, weakening the leaf's attachment to the stem and causing it to fall off. After the leaf is shed, a layer of protective cells, the leaf scar, forms on the stem.

absolute value (*algebra*)   The distance of a number from 0 is its absolute value, indicated by vertical lines, as in $|{-6}|$. Distance is always a positive number or 0 so absolute value is always nonnegative: $|{-6}| = 6$, $|3/4| = 3/4$, $|0| = 0$, and $|{-\pi}| = \pi$. For a complex number such as $x + iy$, the absolute value (also called amplitude, modulus, or norm) is the square root of $x^2 + y^2$, so $|3 - 4i|$ (often written $\|3 - 4i\|$) is $\sqrt{3^2 + 4^2} = 5$.

absolute zero (*thermodynamics*)   No temperature can drop below the temperature designated as absolute zero (0), which is $-459.7°F$ ($-273.15°C$). This temperature is exact because the temperature scale is defined in terms of absolute 0—if you know that absolute 0 is $-459.7°F$ and that ice melts at $32°F$, then you have set the scale. Absolute 0 exists because heat and temperature are both based on the movement of particles. Temperature is a measure of the average momentum of the particles (usually atoms or molecules) that make up matter. Momentum depends on particle speed and mass. For most purposes, the mass of particles does not change, so the important factor in temperature is particle speed. When average speed falls to zero, it can fall no farther, thus absolute 0 is absolute.

absorption (*biology*)   Cells take in water and dissolved substances from their surroundings in a process called absorption. For example, water from soil enters cells of plant roots and then moves through the roots and up the stem to the leaves. Similarly, digested food in the human small intestine is absorbed into the bloodstream through the walls of the small intestine.

absorption and adsorption (*chemistry and physics*)   When a substance is *ab*sorbed, it is taken into the body of a solid, just as a sponge that soaks up water takes the water into its pores. When a substance is *ad*sorbed, a surface reaction on a solid accepts a part of the substance and holds it, as fat or oil is adsorbed on the surface of an iron skillet, seasoning it. A solid can *ab*sorb a fluid if the solid has a structure with holes in it, but a solid can *ad*sorb a liquid, gas, or even another solid if there is an attraction that holds the particles (atoms or molecules) of each material closely together in an outer layer. Adsorption of one element or molecule onto another surface is an important way to change surface characteristics of solids, which is useful in electronic devices and in providing protective coats on tools.

abyssal plain (*oceanography*)   Following the invention of sonar in 1915, surveys revealed very large, flat regions, called abyssal plains, lining the Atlantic Ocean on each side of the mid-oceanic ridge system. These deep (*abyss* originally meant "bottomless") plains formed where tiny particles of soil, rock, or organic matter settled, creating the flattest places on Earth. Where deep ocean trenches surround a basin, however, as in the Pacific, abyssal plains are fewer and smaller.

abyssal zone (*ecology*)   That part of the ocean that receives no sunlight and thus is always in total darkness is called the abyssal zone. It begins at a depth of about 6500 ft (2000 m) and extends downward to the ocean's deepest reaches. Food is scarce in this zone, and inhabitants must be able to withstand extreme pressure and cold temperatures. Nonetheless, there is great diversity of life, including organisms with bioluminescence, huge mouths, and many other interesting adaptations to the environment. Abyssal animals include viper fish, spiny eels, squid, bivalve mollusks, bristle worms, sea anemones, and sea cucumbers.

acacia (*plants*)   Closely related to mimosas, acacias (*Acacia*) are shrubs and trees native mainly to warm, arid regions of Africa and Australia. Most species have highly divided compound leaves but some develop flat leaflike structures called phyllodes, which actually are modified stems. The small, often fragrant flowers are borne in clusters. The fruit is a pod. The best-quality gum arabic adhesive comes from *A. senegal,* a small tree of tropical Africa. Many Australian species are commonly called wattles, including the golden wattle (*A. pycnantha*), which forms part of Australia's coat of arms. The koa (*A. koa*) of Hawaii, prized for its beautiful wood, grows to heights of 50 ft (15 m) or more.

acceleration (*mechanics)*   Acceleration combines changes in speed and direction; that is, acceleration is the rate of change of velocity. Going faster is one kind of acceleration but also any turn from a straight path is acceleration, even while maintaining the same speed. Slowing down is called deceleration in ordinary language but is negative acceleration in physics. Acceleration is measured as a distance per the square of a time, such as mi per sq sec or m/s^2.

You can feel acceleration, but you cannot feel speed. Einstein recognized that there is no physical difference between forces caused by acceleration and those caused by gravity. A person in a closed spaceship who is pulled toward one wall cannot tell whether the cause is a change in speed or direction or an approach to a great mass, such as a star or planet.

acclimatization (*ecology*) The process by which an organism adjusts to changed living conditions is called acclimatization or acclimation. The adjustments, which are reversible, may be physiological or behavioral. For example, the number of red blood cells, which transport oxygen from the lungs to other parts of the body, gradually increases in humans who move from sea level to a mountain community, where air is thinner.

accuracy (*measurement*) Accuracy in measurement refers to the amount of relative error or the number of significant digits, not to whether a measurement has been made correctly. The smaller the relative error or the more significant digits in a measurement, the more accurate it is.

acetone (*compounds*) Acetone (CH_3COCH_3) is the simplest ketone. It is widely used as a solvent. It is unusual for an organic solvent in that it dissolves easily in water.

acetylcholine (*biochemistry*) Acetylcholine is the neurotransmitter responsible for moving all skeletal muscles of vertebrates. When this chemical is released from the tips of motor neurons, it transmits an impulse to muscle fibers to begin the process of contraction. After acetylcholine transmits the impulse, it is quickly broken down by the enzyme cholinesterase. Acetylcholine also is released by neurons in parts of the brain and by neurons that control the sweat glands and heartbeat.

Acheulean (*paleoanthropology*) From about 1,500,000 to about 200,000 years ago ancestors of humans made hand axes and other roughly flaked stone tools called cleavers or picks. This stone-tool industry was named Acheulean after the St. Acheul site in France. The distribution in space and time suggests that *Homo ergaster*, *Homo heidelbergensis*, and *Homo erectus* made these tools, which gradually became more refined, especially in flaking and chipping techniques, over hundreds of thousands of years. Not all human ancestors during this time were involved in the Acheulean tradition, however; groups on the edge of Acheulean territory—England, northern Europe, Java, and China—used stone tools but not hand axes.

acid (*compounds*) An acid is a chemical with a sour taste, such as vinegar (acetic acid) or lemon (citric acid), although tasting a strong acid is dangerous and should not be attempted. The chemical definition of an acid is that it is a compound that when dissolved in water liberates a proton. The freed proton can then enter into reactions with other materials that the water touches. Such chemical reactions often produce compounds called salts, especially when the other reactant is a base. Acids are often detected by chemical reactions that produce color changes, especially by changing an indicator chemical called litmus to red. Acids are measured in strength in terms of pH; the lower the pH, the more acid in a solution or the stronger the acid.

acid (*geology*) Igneous rock that is high in quartz and feldspar but low in magnesium and iron is called acid because of an earlier incorrect belief that quartz and feldspar (and similar minerals) derive from an acid. A typical acid rock, such as granite or obsidian, is less dense and often lighter in color than rocks classed as basic.

acid rain (*environment*) The burning of fossil fuels, particularly coal and oil, releases sulfur dioxide and nitrogen oxides into the atmosphere, where the gases combine with oxygen and water to form solutions of sulfuric acid and nitric acid. When rain, snow, or other precipitation falls, it carries the acids to the ground. If the precipitation has an acidity below pH 5, it is called acid rain. The acids seep into the ground and enter bodies of water, changing soil and water chemistry. For example, the acids can deplete soil of calcium and magnesium needed by plants. Acid rain has been linked to the decline and death of forests and the disappearance of fish from many lakes.

acoel (*animals*) Marine invertebrates only about 0.1 in. (0.25 cm) long, acoels are the oldest living ancestors of all animals with bilateral symmetry. They have long been classified with the flatworms but this classification is a matter of dispute because of the acoels' primitive structure.

acorn worm (*animals*) Looking like a long worm with a head shaped like an acorn, acorn worms are members of Phylum Hemichordata, a group considered to be closely related to the chordates. Hemichordates typically have gill slits and a supporting structure similar to a notochord. Acorn worms range in length from 1 to 80 in. (2.5 to 200 cm). They spend most of their time in tunnels they dig in mud and sand in shallow marine waters, mainly in the tropics and subtropics. Acorn worms swallow the sand or mud, digest organic matter found therein, then excrete the wastes outside their tunnels.

acoustics (*sciences*) Acoustics, from a Greek word for hearing, is the science of sound, especially its generation and wave patterns; its propagation through various substances, called mediums; and its perception by humans. Sound waves are the archetype of all longitudinal waves, so acoustics also applies to seismic (earthquake) waves and to similar waves in stars, including the Sun.

acquired characteristic (*evolution*) A change in an organism as a result of its adaptation to its environment is called an acquired characteristic. A weightlifter's bulging muscles are an example. Jean-Baptiste Lamarck proposed that acquired characteristics could be passed on to offspring. This theory was disproved when Mendel's work on genetics was rediscovered in 1900.

acrylic (*compounds*) Acrylic is a fiber formed from a compound called acrylonitrile (C_3H_3N) and used in fabrics such as Orlon or artificial turf. As a class of compounds, however, acrylics include methyl methacrylate ($C_5H_8O_2$), the basis of acrylic paints and also of transparent plastics such as Lucite. Superglue is another acrylic compound.

actin (*biochemistry*)   A protein that forms thin filaments, actin is an important component of the internal skeleton—the cytoskeleton—of eukaryotic cells. Together with other protein filaments, actin helps maintain cell shape, plays a role in cell movement, and is involved in cell division. Actin also is 1 of the 2 major proteins involved in contraction of muscles.

actinide (*elements*)   The 14 elements that range from atomic mass 89 to 103 are known as the actinide series after element 89, actinium. These elements are radioactive because of their large and unstable nuclei and are alike chemically because of a peculiarity of the arrangement of their electrons. As atoms become more massive, the allowable positions for electrons increase. Between actinium and lawrencium (103), the additional electron for each higher atomic number slips into a position closer to the nucleus, leaving the outer electrons, which control chemical reactions, the same for each of the 14 actinide elements. A similar process occurs with the rare earths, which are also known as the lanthanide series.

actinium (*elements*)   Actinium is a radioactive metal, atomic number 89, discovered among the decay products of uranium by André-Louis Debierne [French: 1874–1949] in 1899 while working with Marie Sklodowka Curie and Pierre Curie. The element's name comes from the Greek *aktis*, meaning "ray," because it is about 150 times as radioactive as radium. It is the first element of the actinide series in the periodic table. Chemical symbol: Ac.

actinomycetes (*monera*)   Actinomycetes are referred to as funguslike bacteria. The cells remain together in branching filaments like those of fungi. Also like fungi, many species produce spores. Most actinomycetes are decomposers, playing an important role in decay in soils. Certain species of *Streptomyces* produce antibiotics such as streptomycin and tetracycline. Pathogens include *Actinomyces israelii*, which is part of the normal bacterial population in the human mouth and tonsils, but can invade other tissues and cause abscesses; a similar disease in cattle and other animals is caused by *Actinomyces bovis*.

action (*mechanics*)   Action is an entity that combines several familiar physics measures. Planck's constant, the smallest unit of action, is measured as an amount of energy per frequency of a wave. Replacing frequency with the related idea of period makes action into energy × time (the reciprocal of frequency is period, a measure of time). Similarly, recognizing energy as work (force times distance) leads to interpreting action as force × distance × time.

Nearly all of physics can be expressed in terms of action. Newton's laws of motion become those paths that maximize or minimize action. Basic formulas of quantum theory similarly can be derived from constraints on action—in Niels Bohr's theory of the atom, electrons must have orbits for which the action is a natural number, for instance.

active transport (*physiology*)   When molecules pass through a cell membrane by diffusion, they move from a region of high concentration to a region of low concentration; no cell energy is needed for this process. In contrast, when molecules pass through a cell membrane by active transport, they move from a region of low concentration to a region of high concentration—a process that requires energy, which comes from the breakdown of ATP. Examples of active transport include the removal of iodine from the blood for concentration in the thyroid gland and the absorption of sodium and potassium ions from the soil by root cells.

Adams, John Couch (*astronomer*) [English: 1819–92] At 24 Adams correctly calculated the location of the previously unknown planet Neptune using deviations of Uranus from its predicted orbit. Adams's results were first ignored and then poorly applied, so credit for Neptune went to Urbain Leverrier [French: 1811–77], who calculated the position a few months after Adams, and Johann Galle [German: 1812–1910], who based a search on Leverrier's calculations and first recognized the planet.

adaptation (*evolution*) Any change in an organism that makes it better fitted for life in a certain environment is called an adaptation. Some adaptations are acquired during the organism's lifetime. Others occur through evolution and are acquired genetically, by offspring from their parents. There are 3 basic types of adaptations: structural, based on how the organism is built; physiological, based on the ways in which the parts of the organism operate; and behavioral, based on all the things the organism does. Examples of these types and how they interact are seen in termites. To grow, a termite sheds its exoskeleton and forms a new, larger exoskeleton—a structural adaptation. The termite cannot digest the wood it eats but depends on microorganisms living in its digestive tract for digestion—a physiological adaptation. When the termite sheds its exoskeleton it also loses the part of its digestive system that contains the microorganisms. It retrieves them by eating the discarded exoskeleton—a behavioral adaptation.

Some organisms are highly adapted to a particular niche, making it very difficult for them to cope if their environment changes. This group includes endangered species. Other organisms are much more tolerant and able to thrive in a variety of niches. This latter category includes weeds, Norway rats, and other species that humans consider pests.

adaptive optics (*astronomy*) Adaptive optics has been called "untwinkling the stars," since it removes twinkling caused by Earth's atmosphere. Air is filled with dust, and different parts of the atmosphere vary in density. Consequently, starlight bounces about during its passage through the atmosphere and twinkles. Knowing how the light has been altered at a given instant makes it possible to compensate for the twinkling, improving the resolution of astronomical telescopes. Laser light is reflected by sodium ions abundant in a layer about 60 mi (95 km) high. The resulting artificial star is monitored for changes in its image caused by atmospheric conditions. A computer makes corresponding corrections in the image of a real star.

adaptive radiation (*evolution*) As populations of a species spread out and adapt to different habitats as a result of natural selection, they gradually become genetically distinct from one another and from their common ancestor. Over time this process of adaptive radiation, or divergent evolution, leads to the formation of different subspecies and, later, species that cannot interbreed with one another. The process is well illustrated by the Galapagos Islands' finches, first studied by Darwin. Some 10,000 years ago, small South American finches arrived on the islands. Facing little competition, the finches began to occupy a variety of niches. Thirteen species of finches have evolved from the original species. While all 13 species share the same basic characteristics, there are obvious differences among them, particularly in beak structure and feeding habits. Each species has a bill adapted to its particular diet: seeds, nectar, fruit, leaves, insects, ticks, or even the blood of seabirds.

adder (*animals*)   Native to Europe and Asia as far north as the Arctic Circle, the adder (*Vipera berus*) is a viper up to 2 ft (0.6 m) long. It lives in many kinds of habitats, using its venom to kill prey such as lizards, small mammals, and birds. The puff adder (*Bitis arietans*), native to drier areas of Africa, is a viper up to 5 ft (1.5 m) long. When frightened it inflates itself and makes a loud hissing sound.

Addison, Thomas (*endocrinologist*)   [Engish: 1793–1860] In addition to important research on pneumonia, appendicitis, pernicious anemia, and other ailments, Addison in 1855 described the illness now called Addison's disease and linked it to the cortex (outer layer) of the adrenal glands. Characterized by weakness and weight loss, Addison's disease develops when the adrenal cortex does not secrete sufficient amounts of hormones.

addition (*arithmetic*)   Addition is an operation on 2 numbers called addends that produces a third called the sum. For natural numbers addition can be accomplished by starting with the first addend and counting as many more numbers as the second addend; to add 5 to 3, think 3 4_1 5_2 6_3 7_4 8_5 making the sum 8. This counting-on definition corresponds to the sum being the number of members in the union of 2 sets that have no common members. Different definitions are required for numbers other than natural numbers. Real numbers such as 2.4 or π can be added by combining lengths of line segments. For complex numbers, such as $2 - 3i$, where i is $\sqrt{-1}$, the real and imaginary parts add separately.

additive inverse (*numbers*)   For a given number, its additive inverse is another number such that the sum of the 2 is the identity element for addition, or 0. In group theory the additive inverse is an element that combines with a given element to produce identity—for example, for rotations of a hexagon, the additive inverse of a turn of 60° counterclockwise is a turn of 60° clockwise, the same as no turn.

adenovirus (*viruses*)   Members of the virus family Adenoviridae are widespread in nature and a common cause of acute respiratory disease, including pneumonia. Each virus particle is a 20-sided polyhedron. Its protein coat has protruding glycoprotein "fibers"; no envelope encloses the particle. The genetic material consists of double-stranded linear DNA.

adhesion (*materials science*)   Adhesion occurs when 2 substances, usually 2 different materials or 2 separate bodies of the same material, become attached to each other so that a force is needed to separate them. Most adhesion is thought to be caused by electromagnetic forces between molecules, which, for example, cause a liquid such as water to adhere to glass. Some adhesion may result from small regions of vacuum that result when a force begins to separate the bodies.

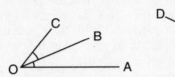

Angles AOB and BOC are adjacent but not AOB and AOC. OB is not between them.

AOC and DOB are vertical but AOC and AOD are adjacent.

ADJACENT ANGLE

adjacent angle (*geometry*)   Two angles are adjacent if they have a common endpoint and if a half line that is 1 side of each angle is between the other sides.

adrenal gland (*anatomy*)   Most 4-legged vertebrates have 2 adrenal glands, 1 atop each kidney. An adrenal gland has 2 parts—an inner medulla and an outer cortex—each of which secretes hormones that control vital body processes.

advection (*meteorology*)   Meteorologists call effects of horizontal movement of air advection ("carrying to"), especially when the air is transporting heat away from an environment or carrying moisture into an environment. An advection fog is caused when warm, moist air moves over a cool surface, lowering the dew point enough for water vapor to form small drops of liquid. Advection's horizontal motion is the bottom current of a large, flattened convection cell.

adventitious root (*botany*)   Roots that sprout from stems or leaves are called adventitious roots. They include prop, aerial, and air roots. The main function of prop roots is to help support the plant. Corn plants develop prop roots near the base of the stem, and banyan trees produce long prop roots from their branches. Aerial roots may function to anchor the plant and absorb water and minerals, as in epiphytes such as orchids, or they may function solely as support organs, as in English ivy and other vines. Air roots are grown by mangroves and other plants that live in swampy places where the soil is low in oxygen; these roots absorb oxygen and help anchor the plant.

aerobe (*biology*)   An organism that requires oxygen for respiration and therefore lives only in the presence of air or free oxygen is an aerobe. All plants and most animals are aerobes, as contrasted with anaerobes.

aflatoxin (*biochemistry*)   Aflatoxin is a toxin produced by the mold fungus *Aspergillus flavus*, which sometimes contaminates grains and nuts, especially peanuts. It is highly toxic to animals, including humans, and can cause severe liver damage, including liver cancer.

African Plate (*geology*)   A large section of Earth's lithosphere stretching from the mid-Atlantic ridge on the west to beyond Madagascar on the east, and from the Mediterranean on the north to the Atlantic-Indian ridge on the south, is named the African Plate after its main continent. The eastern, southern, and western boundaries are all spreading centers, pushing the plate slowly northward. This plate shows signs of breaking in two along Africa's rift valley, although the incipient rupture seems to have been arrested.

aftershock (*geology*)   Any earthquake produced by further movement along the same fault after a large earthquake is termed an aftershock of that quake; movements along the same fault before the big quake are foreshocks (recognized only after the major earthquake in most cases). Most large earthquakes have smaller aftershocks hours, days, or even years afterward; in some instances, the aftershocks are nearly as large as the main earthquake. An earthquake sometimes triggers the rupture of other nearby faults, producing secondary earthquakes that are not considered aftershocks.

agar (*biochemistry*)   A mixture of carbohydrates extracted from certain red algae, notably *Gelidium*, agar is well-known as a laboratory medium for growing bacteria. Small amounts of agar will form a gel with water at 100°F (37.5°C). The gel will not dissolve at temperatures below 185°F (85°C). This allows incubation of bacteria at comparatively high temperatures. In addition, agar is nontoxic and resists attack by most microorganisms. An even greater use of agar, however, is as a thickener and emulsifier in food, pharmaceutical, and other industries.

Agassiz, Louis (*geologist*)   [Swiss-American: 1807–73] After producing a major description of known fossil fishes from 1833–44, Agassiz turned to the question of the effects of glaciers. He was the first to prove that glaciers move slowly, carrying rock and changing the landscape. Before he came to Harvard University in the United States, he had concluded that much of northern Europe had once been covered with glaciers. From this base he studied glaciation in North America, establishing and popularizing the concept of a worldwide Ice Age in which glaciers occupied much of the Northern Hemisphere.

agave (*plants*)   Members of the genus *Agave* are succulents native to warm, arid regions of the Americas. The long, stiff leaves are arranged in rosettes at ground level. Most species require 5 to 20 years or even longer to mature and produce a tall flowering stem; in the century plant (*A. americana*)—which typically blooms when it is about 10 years old—this stem may reach a height of 40 ft (12 m). After flowering once, agaves die. Agaves are widely cultivated as ornamentals and for fiber (sisal rope comes from *A. sisalana*). In Mexico the sap of certain species is fermented for pulque or distilled for tequila and mescal.

The agave family, Agavaceae, comprises at least 700 species of tropical and subtropical plants native mainly to arid regions. Like agave, many are succulents with fleshy leaves borne in rosettes. The family includes dracaenas, yuccas, and snake plants (*Sansevieria*).

aggregate (*rocks and minerals*)   Clumps of minerals that combine various poorly formed crystals are called aggregates. Aggregates are classed by their forms—botryoidal (like a bunch of grapes), dendritic (branching like trees or river systems), fibrous (in strings), foliated (in sheets), lamellar (in flat planes, thicker than sheets), and mamillated (with bumpy protrusions).

aggression (*ethology*)   Aggression—behavior used by an animal to threaten or injure another animal—is most common in situations in which animals are competing for a resource, such as food, a mate, or nesting sites. Aggression is not the same as predation, which is hunting and killing other animals (prey) for food.

Animals generally try to avoid fighting that might result in serious injury or death. Thus aggressive behavior often begins with threat displays designed to scare off opponents. For example, a male red-winged blackbird (*Agelaius phoeniceus*) faced with a rival will spread his tail and wings, displaying his bright red shoulder patches. The meaning of such a display is understood by other members of the species.

agouti (*animals*)   Rodents native to tropical forests of Central and South America, agoutis (*Dasyprocta*) are about 2 ft (0.6 m) long. They have small ears, a very short tail, and long legs with broad flat claws on the toes. They are swift runners and can jump vertically into the air to a height of 6.5 ft (2 m).

Agricola, Georgius (*geologist*)   [German: 1494–1555] Although best known by the Latinized form of his name, Agricola was born Georg Bauer. Both names mean "farmer," but Agricola was a physician who specialized in diseases of miners in the main mining centers of Saxony. His principle contribution to science is his posthumously published (1556) book *De Re Metallica*, a clear, well-illustrated summary of what German miners had learned and how they operated. His earlier writings on minerals and geology are also valuable resources.

Agricultural Revolution (*archaeology*)   The Agricultural Revolution is the period in which humans in a particular region first domesticated animals and planted and harvested crops. It occurred independently in the Middle East, Southeast Asia, China, Mexico, and South America. The earliest dates in the Middle East are about 11,000 years ago and most crops and animals in all regions were domesticated by about 6000 years ago.

Early humans gathered wild plant materials, just as chimpanzees do today. Although humans killed animals for food, in some cases animals probably were wounded or trapped and kept alive for a time. Most anthropologists also believe that people kept young animals as pets. These practices led to domestication.

There are various theories concerning why the Agricultural Revolution occurred. Some change encouraged people to apply what they knew about plants and animals to the production of larger amounts of food. Domestication may have resulted from urbanization and specialization or perhaps a climatic change reduced resources locally.

Ahmes (*mathematician*)   [Egyptian: c1650 BC] The name Ahmes (or Ahmose) is attached to a document called the Rhind papyrus, first discovered in 1858, when a Scot named Henry Rhind purchased it in Egypt. Ahmes notes that he is copying material that is about 200 years old. The manuscript, which opens with the dramatic words "Directions for attaining knowledge of all dark things," contains 85 problems and their solution, most involving arithmetic but a few concerning geometry.

ailanthus (*plants*) Native to Asia and Australia, ailanthus (*Ailanthus*) are deciduous trees that grow to 70 ft (21 m) in height. They have large compound leaves that resemble fern fronds. The small greenish flowers are unisexual, with male and female flowers usually on separate trees. The male flowers have an unattractive scent. The female flowers develop into winged fruits called samaras.

air (*atmospherics*) Air is often described as a mixture of gases but it is more like a solution of oxygen (20.05% by volume in dry air) in nitrogen (78.08%) with a few other gases. Water vapor near the surface changes these ratios, since at 80°F (26.5°C) on a damp day, 4% of air may be water vapor. Water vapor vanishes almost completely above the troposphere. Air in the troposphere and stratosphere also includes argon (0.93%), carbon dioxide (0.035% and rising), and even smaller amounts of neon, helium, krypton, and xenon. Ozone in the stratosphere reaches only about 0.001%. The lightest elements rise into the higher layers of the atmosphere, notably hydrogen and helium in the thermosphere and beyond.

airglow (*atmospherics*) Air molecules in the thermosphere, raised to a higher energy level by incoming ultraviolet radiation from the Sun, give back that energy at night as light called airglow. Airglow on a clear, moonless night provides more illumination than all the stars together.

air mass (*meteorology*) An air mass is a large region of the troposphere in which, for a given elevation, temperature and humidity are nearly uniform. If atmospheric currents are slow, an air mass develops that reflects the land or water below it. Air rising from oceans is humid while that from land tends to be drier. Air masses migrate from regions of their formation and change the weather along the way. Meteorologists classify air masses into 5 types: continental tropical (**cT**) with warm, dry air; maritime tropical (**mT**) with warm, wet air; continental polar (**cP**) with cool, dry air; maritime polar (**mP**) with cool, damp air; and continental arctic (**cA**), very cold and very dry.

air pollution (*environment*) Contamination of the atmosphere by toxic substances is called air pollution. Air pollutants include solids (particulate matter), gases such as sulfur dioxide and nitric oxide, and radioactive compounds. Some pollutants are emitted directly from a source, such as dust from volcanoes or lead from smelters. Other pollutants form in the atmosphere through chemical interactions—for example, acid rain and the ozone in photochemical smog.

air pressure (*meteorology*) A column of air reaching to the top of the atmosphere rests on every surface on Earth. This weight for a given area is one common measure of air pressure. For example, an area of 1 sq ft at sea level is covered by a column of air that weighs about 14.7 lb (a column 1 m² has a mass of 101,325 N), often called a standard atmosphere. More frequently air pressure is measured by how high the pressure will lift a column of mercury into a vacuum—29.92 in. (76 cm) is a standard atmosphere. Meteorologists prefer a unit called the millibar, which is exactly 100 N per m², so 1000 millibars, or 1 bar, is slightly less than a standard atmosphere. Air pressure is noticeably reduced at higher elevations and increased at lower ones. The difference is approximately 1 in. (2.5 cm) of mercury per 1000 ft (300 m) of elevation. Air pressure also varies with humidity and temperature, since warm or humid air is less dense than cold or dry air.

Airy, Sir George Biddel (*astronomer/geophysicist*)   [English: 1801–92] Airy was a famously unlucky astronomer, best known for ignoring the calculations that John Couch Adams presented to him that would have located the previously unknown planet Neptune and for failing to determine the precise size of the solar system through observations of the transit of Venus. Nevertheless, he was Astronomer Royal from 1835 to 1881, directing the Greenwich observatory during his lengthy tenure. In addition to developing many improved tools for astronomical observations, Airy correctly suggested the principle of isostasy to describe the way that mountain ranges maintain gravitational equilibrium with the crust.

alarm call (*ethology*)   Many birds and mammals that are preyed upon have evolved alarm calls to communicate real or potential danger—and to let the predator know that it has been seen. The prairie dog acting as a sentinel for its colony utters a sharp "yap," alerting other prairie dogs to dash into burrows. Species may have a variety of alarm sounds, such as different calls to indicate whether a predator is in the air or on the ground.

albatross (*animals*)   The albatrosses, Family Diomedeidae, are large, graceful seabirds with stout bodies, long, narrow wings, and webbed feet. They sleep and feed on the water's surface, sometimes diving a short distance to grab their favorite food, squid. Albatrosses spend almost their entire lives at sea, coming ashore only to breed and raise their young. They typically mate for life, and may live for more than 50 years.

albedo (*astronomy*)   Albedo is a number used to describe the amount of light an object reflects, ranging from 0 for a pure black that absorbs all light to 1 for a totally reflective surface. Venus with its white cloud cover has an albedo of 0.65, while the Moon's is only 0.12.

albinism (*zoology*)   The lack of normal pigmentation, or coloring, in an animal's skin, hair, feathers, or eyes is called albinism. Albinism generally is a hereditary characteristic owed to a recessive gene (that is, the individual must inherit the gene from both parents). In rare cases albinism results from a genetic mutation and is called leucism. Albinism usually makes an animal too visible to predators for it to survive in the wild. Races of domestic albino animals, such as white mice, rabbits, and pigeons, are the result of experimental breeding.

alchemy (*chemistry*)   Alchemy, whose Arabic-sounding name is said to derive from the Greek for "transmutation of elements as practiced in Egypt," was a complex mixture of spiritual and scientific principles used to control matter. Geber [Arabian: c721–c815] based his popular alchemical theories on the 4 elements of the Greeks and claimed that a suitable elixir (also known as philosopher's stone) could alter the proportions of the elements in a metal such as lead to change it to gold. Alchemists in the Middle Ages in Europe added to Geber's ideas such notions as the panacea, or universal cure, for illness.

Although they failed in their lofty goals, alchemists such as Geber laid the groundwork for chemistry. Geber himself carefully described how to make certain compounds and refine metals. Newton was very interested in alchemy and devoted many years to its study.

alcohol (*compounds*)   An alcohol is a hydrocarbon that contains a hydroxide ion.

Grain alcohol, or ethanol, produced by fermenting grains, fruits, or potatoes, is the alcohol of beer, wine, and spirits. Its chemical formula is C_2H_6O. Ethanol is also added to gasoline as an oxidizer or even burned instead of gasoline.

Methanol (CH_4O) is the simplest alcohol. Distilling wood in the absence of air produces wood alcohol, or methanol (roughly "intoxicating timber"). After imbibing methanol, the initial reaction is similar to that with ethanol, but then the body converts methanol to highly toxic formic acid and formaldehyde. Methanol today is manufactured mostly from carbon monoxide and hydrogen or natural gas.

A third familiar alcohol is isopropyl, known as "rubbing alcohol." It is one isomer of propanol, C_3H_9O.

alder (*plants*)   Members of the birch family, alders (genus *Alnus*) comprise about 35 species of deciduous trees and shrubs native mainly to moist soils in temperate regions of the Northern Hemisphere. Flowers are borne in catkins that generally appear before leaves in the spring. Male catkins are long; female catkins are short and develop into fruit that resemble small, dark cones. Some species, such as the black alder (*A. glutinosa*), are important sources of timber.

alfalfa (*plants*)   A member of the legume family, alfalfa or lucerne (*Medicago sativa*) is believed to be native to central Asia, though cultivated forms were known in China and Greece more than 2000 years ago. A bushy perennial that generally grows to about 3 ft (90 cm) tall, it produces clusters of purplish flowers followed by curled seed pods. Its long roots penetrate more than 20 ft (6 m) deep, enabling it to exploit underground water supplies and making it a valued forage plant and soil enricher in dry climates.

algae (*protists*)   Algae are the only organisms other than plants that possess chlorophyll and can make their own food through the process of photosynthesis. They range from one-celled forms to giant kelp. Most algae live in aquatic environments, where they form the basis of almost all food chains, filling a role comparable to that of grass and trees on land. Other algae live in damp land environments, including on soil, trees, and animal hairs.

Structurally, algae are comparatively simple. They lack vascular tissue and thus, unlike plants, do not have true roots, stems, and leaves. In addition to chlorophyll, algal cells may contain other pigments. For example, in red algae the chlorophyll is masked by red phycoerythrins. Different groups of algae also differ on the composition of their cell walls and stored foods. For example, green algae have cell walls composed mainly of cellulose and store food reserves as starch; golden algae usually have silica as well as cellulose in their cell walls and store most food reserves as oils.

In addition to the protists, one group of organisms commonly called blue-green algae are classified as monera.

algebra (*mathematics*)   Traditional algebra uses variables to write equations along with general methods for solving the equations. Diophantus [Hellenic: c210–90] is termed the father of algebra because his *Arithmetica* of about AD 250 uses letters as variables in describing problems, although without signs of operation or equality. About 300 years later, Muhammad ibn-Musa al-Khwarizmi wrote *Al-jabr wa'l muqabalah* ("Restoration and Balancing"), which gives general rules for solving equations—the word algebra derives from *Al-jabr*. Algebra as we know it today started with François Vièta [French: 1540–1603], the first to write and solve general equations of the type we might now write as $ax^2 + bx + c = 0$. Vièta stated that arithmetic deals with specific problems, while algebra provides rules for solving all problems of a given type.

In the 19th century mathematicians carried this idea further, developing several algebras that are mathematical structures obeying a few specific rules but that are not otherwise specified. Such abstract algebras form an active branch of modern mathematics, while the traditional algebra of numbers, variables, and equations remains a necessary tool for all mathematics.

algebraic number (*numbers*)   Algebraic numbers are the solutions to equations in 1 variable for which all numbers used in the equation are rational numbers—including exponents as well as coefficients of the variable. Most numbers commonly encountered are algebraic, including all rational numbers as well as square roots and all other roots of rational numbers; but there are an infinity of transcendental numbers, real numbers that are not algebraic.

Algol (*stars*)   Algol means "the ghoul," after the evil demon that feeds on corpses. The star Algol's bad reputation comes from its noticeable change in magnitude, from 2.2 to 3.5 and back every 69 hours. In 1782 John Goodricke [English: 1764–87] correctly guessed that Algol is an eclipsing binary star—the less-bright companion orbits between the brighter star and our line of vision. Algol is actually a triple-star system, but the orbit of the third star does not take it in front of the others. The system, officially Beta Persei, is 82 light-years away.

algorithm (*mathematics*)   Any sequence of steps that, if followed correctly, will always produce a correct result is called an algorithm (sometimes algorism) after al-Khwarizmi, who described the rules needed to calculate in the Hindu-Arabic numeration system. A familiar example is the long-division algorithm. Algorithms occur in all branches of mathematics, but their general nature is a part of advanced logic called meta-mathematics. Logicians have proved that there are true results of mathematics that cannot be calculated with any algorithm.

Alhazen (Abu al-Hassan ibn al Haytham) (*physicist*)   [Egyptian: c965–1038]. Alhazen correctly recognized that light is produced by some objects and reflected or refracted by others; many previous writers thought light originates in the eye. Alhazen's *The Treasury of Optics* was published in Latin in 1572, becoming a major influence on Kepler. His book discusses lenses, mirrors, both flat and curved, the rainbow as an effect of light, and how pinholes focus light.

alkali metal (*elements*)   The alkali metals are those elements with exactly 1 electron in their outer shell, making them chemically active. Hydrogen fits the definition but has not been solidified successfully on Earth (metallic hydrogen may exist on Jupiter or other gas-giant planets). The metals are lithium, sodium, potassium, rubidium, cesium, and probably francium(not enough francium has been synthesized to be certain of its properties).

alkaline earth metal (*elements*) The alkaline earth metals are so named because their mostly insoluble compounds, known long before any of the metals themselves, include such minerals as limestone and dolomite, which form much of Earth's crust. Each alkaline earth metal has 2 electrons in its outer shell, so the metals are nearly as reactive as the alkali metals to which they bear considerable similarity—for example, beryllium, magnesium, calcium, strontium, barium, and radium are each white or silvery metals.

alkaloid (*biochemistry*) Alkaloids are basic nitrogenous organic compounds produced by several groups of flowering plants, including the nightshade family (Solanaceae) and poppy family (Papaveraceae). Most alkaloids have a bitter taste and are poisonous to animals, but some are valuable medicines. Atropine, caffeine, cocaine, morphine, nicotine, and quinine are all alkaloids.

alkane (*compounds*) Alkanes are saturated hydrocarbons (every carbon atom has as many hydrogen atoms connected by single covalent bonds as possible). The simplest alkane is methane, 1 carbon atom bonded to 4 hydrogens. Adding more carbon atoms produces such hydrocarbons as ethane, propane, and butane. Related types of hydrocarbons are the alkenes and alkynes, which are unsaturated, permitting double bonds between carbon atoms (alkenes, such as ethylene) and triple bonds (alkynes, such as acetylene).

allele (*genetics*) A particular form of a gene, different from other forms of the same gene, is called an allele (from the Greek word meaning "other"). Thus the gene for brown eyes in humans and the gene for blue eyes are alleles of the gene that controls eye color. An organism has 2 alleles for a characteristic, one on each of a pair of chromosomes. How that characteristic is expressed depends on the dominance relationship between the 2 alleles.

allelopathy (*ecology*) The production by a plant species of chemical compounds that enter the environment and inhibit growth of other plant species is called allelopathy. This type of "chemical warfare" has been found in many species and many kinds of ecosystems. For example, decaying leaves of black walnut trees (*Juglans nigra*) release toxins that inhibit the growth of seedlings of many other tree species. As a result, these species are not likely to grow under black walnut trees and thus not likely to compete with black walnuts for soil, water, and other resources.

allergy (*physiology*) An allergy is characterized by the production of antibodies that react with specific or closely related antigens (foreign substances); the function of the antibodies is to help destroy the antigens. Pollen, bee stings, dust, and various foods and medicines are among the substances that cause allergic reactions in some people. Symptoms of an allergy, such as sneezing, rashes, itching, and swelling, generally are caused by the release of histamine. Antihistamines are drugs used to counteract the effects of histamine.

Allosaurus (*paleontology*)   A large meat-eating dinosaur of the late Jurassic to early Cretaceous, *Allosaurus* ("other lizard") attained lengths of 40 ft (12 m), including its long, heavy tail. *Allosaurus* may have weighed as much as 4 tons. Like all 2-footed dinosaurs, it had powerful legs. Its small arms were fortified with long, curved claws and its large head had long jaws filled with teeth up to 4 in. (10 cm) in length.

allotrope (*elements*)   Two different forms of the same phase of a single element are called allotropes. Familiar examples include diamond, graphite, buckminster-fullerene—all forms of solid carbon—and atmospheric oxygen and ozone—forms of gaseous oxygen. Less well known are the allotropes of sulfur, which occur in liquid sulfur as well as in solids and gases. Allotropes may be different crystal structures (diamond and graphite) or different molecular structures (oxygen and ozone). Properties are often different—white phosphorus bursts into flame when exposed to air, while red and black phosphorus are less reactive.

alloy (*materials science*)   An alloy is a solid solution in which the solvent is a metal. While many alloys, such as brass or bronze, consist of 2 or more metals, other common alloys such as cast iron and steel also include nonmetals. Cast iron, for example, is iron alloyed with nonmetals carbon, silicon, sulfur, and phosphorus as well as manganese metal.

Alloys often have properties that are more desirable than the materials in the solution. They may be stronger, harder, softer, brighter, more magnetic, or more easily melted, for example.

allspice (*plants*)   Native to Central America, the allspice tree (*Pimenta dioica*) is a tropical evergreen with small white flowers that grows to about 40 ft (12 m). All parts of the tree are aromatic. The dried berries are the source of allspice, a spice so-named because its flavor resembles a combination of cinnamon, nutmeg, and cloves. The name allspice also is given to several unrelated but fragrant shrubs, including Carolina allspice (*Calycanthus floridus*) and Japanese allspice (*Chimonanthus praecox*).

alluvial feature (*geology*)   The word "alluvial" refers to sediment deposited by running water, so an alluvial feature is part of the landscape formed from such sediment. The features include deltas, but the term is usually used for small piles of sediment formed by runoff and occasional streams down hills or mountains in normally arid regions. The sediments are alluvial fans or, if steeper, alluvial cones. The flat region caused by occasional floods of a stream is an alluvial plain or a floodplain.

almond (*plants*)   A member of the rose family native to the Mediterranean region, the almond (*Prunus amygdalus*, or *P. dulcis*) is a deciduous shrub or tree up to 40 ft (12 m) tall. It has white or pink flowers that develop into a fleshy fruit similar to a peach. The seed, also called an almond, is incorrectly referred to as a nut.

aloe (*plants*)   Members of the genus *Aloe* are succulents native to Africa and the Middle East. They include perennial herbs, shrubs, and trees and have thick, pointed leaves, often with toothed margins, arranged in rosettes. The colorful flowers are borne on spikes. *Aloe vera* is a popular pot plant; its sap has been used medicinally since the days of ancient Egypt.

Alpha Centauri (*stars*))   The common name of this bright binary star is Rigil Kentaurus, or Rigil Kent ("the foot of the Centaur"), but nearly everyone uses its astronomer's name, Alpha Centauri, which means brightest star in the constellation Centaur. There is a third star in the system, Proxima Centauri, but it is far from the two bright stars and very dim, invisible to the naked eye. Often Alpha Centauri means the whole 3-star system.

alpha particle (*particles*)   The most massive particles that stream from radioactive substances are alpha particles. After natural radioactivity was discovered in 1896, physicists rushed to learn the nature of the new energy. By 1898 Rutherford had shown that one part of the radiation from uranium passes through a barrier that stops another. Rutherford named the less penetrating part alpha radiation and the other beta radiation. By 1905 Rutherford correctly concluded that alpha radiation consists of ionized helium. Today physicists describe alpha particles as helium-4 nuclei—2 protons joined to 2 neutrons.

alpine tundra (*ecology*)   High mountain regions, above the tree line but below an area of perpetual snow, support an ecosystem referred to as alpine tundra. The soil is thin and stony; unlike arctic tundra, there is no permafrost. Temperatures are never very high during daytime and usually drop below freezing at night. The land is subject to strong winds and high light intensity, including high levels of ultraviolet radiation. Many of the organisms that live here are endemic species well adapted to the unique environmental conditions. Lichens and plants such as grasses and mosses dominate the vegetation. Larger animals may arrive in spring to feed but with the arrival of winter they generally migrate to lower altitudes.

Altamira (*paleoanthropology*)   The first cave recognized to have remarkable paintings from the Paleolithic Age is Altamira in northwestern Spain. The dramatic red-and-black painted bison were discovered in 1879 by a 5-year-old girl, who saw them on the ceiling while visiting the cave with her father. The bison are Magdalenian from about 15,500 BC, but the earliest art in the cave consists of engravings of animals that date from about 7000 years before that.

alternate angle (*geometry*)   Pairs of angles formed when 2 lines are both intersected by 1 other line, called the transversal, are alternate if the interior of each angle is on a different side of the transversal and either both interiors are between the 2 lines (interior alternate angles) or else both are outside the 2 lines (exterior alternate angles). (See art on page 20.)

alternating current (*electricity*)   Alternating current is electric current in which an electric field flows first in 1 direction and then the opposite way (typically reversing 60 times a second). It is produced by switching current direction back and forth at the source. Current supplied over power lines is alternating current partly because its potential can be easily raised and lowered with a transformer. When potential differences are high, current travels through wires farther with less loss of power.

 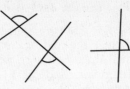

interior alternate
angles

exterior alternate
angles

corresponding
angles

ALTERNATE ANGLE

alternation of generations (*biology*)   All plants plus some algae and animals (particularly cnidarians) have life cycles that involves the alternation of 2 distinctly different generations: an asexual generation and a sexual one. In plants, including ferns, the 2 generations are the **sporophyte and gametophyte**. The cells of the sporophyte have the normal, or diploid, number ($2n$) of chromosomes. The sporophytes produce spores that contain only 1 of each pair of chromosomes (n). These germinate into gametophytes (n), which produce eggs (n) and sperm (n) in their reproductive organs. The union of an egg and sperm leads to the growth of a new sporophyte ($2n$).

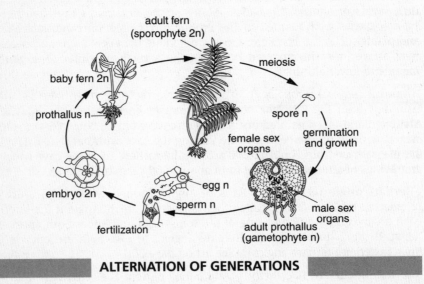

adult fern
(sporophyte 2n)

meiosis

baby fern 2n

prothallus n

spore n

germination
and growth

female sex
organs

embryo 2n

egg n

sperm n

male sex
organs

fertilization

adult prothallus
(gametophyte n)

ALTERNATION OF GENERATIONS

altitude (*geometry*)   An altitude is a number that is the height of a **polygon** or solid and also a line segment whose length provides that number. The essential feature of all altitudes is that they are **perpendicular** to a side of a polygon or base of a solid. For some polygons the altitude varies with the choice of base; the altitude of a triangle is a segment from a vertex that is perpendicular to the opposite side taken as a base, so each triangle has 3 altitudes, which may not be equal in length. When 2 sides or bases are parallel, as in a parallelogram, rectangle, trapezoid, cylinder, or prism, an altitude is perpendicular to both sides and may be placed anywhere that this can be accomplished. For a cone or pyramid, however, there is only a single altitude from the vertex to the base.

altruism (*ethology*)   An altruistic act benefits other individuals at some risk or cost to the individual that performs the act. Some altruistic behavior is believed to have evolved because it offers a genetic reward. For example, parental care of the young—often involving risks to the health or life of the older animal— helps ensure that the parent's genes are passed on to succeeding generations. Another form of altruism has been termed reciprocal altruism. Individuals aid other members of the group with the instinctual understanding that such aid will be reciprocated at some time in the future. For example, vampire bats that are successful in a night's hunt for blood will, on returning to their communal home, regurgitate some of the blood to feed unsuccessful hunters.

alum (*compounds*)   The familiar alum used to stop bleeding in small cuts and famous for its bitter taste is potassium aluminum sulfate. It is a representative of a group of similar alums used in industrial processes, such as fixing dyes. In other alums, sodium or lithium might substitute in the sulfate for potassium, and iron or chromium for aluminum.

aluminum (*elements*)   Since it is strong and light and conducts electricity easily, aluminum is one of the most commonly used elements. Before 1886, however, when an inexpensive manufacturing method was developed, aluminum metal was rare and expensive. The familiar white metal, atomic number 13, was discovered in 1825 by Hans Christian Oersted. Its name, spelled aluminium everywhere but in the United States, comes from the bitter aluminum compound alum. Chemical symbol: Al.

Alvarez, Luis (*physicist*)   [American: 1911–88] Alvarez contributed to experimental nuclear and particle physics in many ways; his 1968 Nobel Prize was for the discovery of various high-energy particle resonance states. Alvarez was a prolific inventor, working on both radar systems for navigation and the nuclear bomb during World War II, as well as the variable-focus lens for eyeglasses and the stabilizer used to keep hand-held cameras and binoculars steady. His contribution to the work of his son Walter Alvarez in identifying the probable cause of the K/T event added to his reputation outside physics.

Alvarez, Walter (*geologist*)   [American: 1940–] In 1980 Alvarez and coworkers located in Italy a thin layer of sediment associated with the boundary between the Cretaceous and Tertiary periods, a time of mass extinctions now known as the K/T event. Alvarez sent some of the sediment to his father, Luis Alvarez , a physicist, who found the sample enriched with iridium, a rare element in Earth's crust. Subsequent research showed the iridium layer to be worldwide. The Alvarezes proposed a theory that the mass extinctions were caused when a large body enriched in iridium struck Earth and produced great clouds of dust that obscured the Sun for several years.

alveolus (*anatomy*)   The functional unit of the vertebrate lung is a microscopic sac called an alveolus. Clusters of alveoli are at the end of the tiny tubes that carry air into and out of the lung. The walls of an alveolus are only about 1 cell thick, and blood capillaries surround them. By the process of diffusion, oxygen moves from the alveolus into the blood and carbon dioxide moves from the blood into the alveoli.

Some estimates suggest that there are about 700,000,000 alveoli in 1 human lung, creating a total alveolar surface across which gas exchange can occur of about 645 sq ft (60 sq m).

amalgam (*materials science*)   An amalgam is alloy in which the solvent metal is mercury. The most familiar amalgam contains copper, silver, and zinc, which, because it is pliable when formed but soon hardens, is used in filling teeth. Gold and silver miners often use mercury to dissolve those metals from ores and then boil off the mercury from the resulting amalgam.

amaranth (*plants*)   Members of the genus *Amaranthus* are annual herbs native to temperate and tropical regions. The name is derived from the Greek *amarantos*, meaning "unfading," a reference to the plants' long-lasting flowers. Many species have colorful leaves and are cultivated as ornamentals. Some species are grown as herbs and vegetables. Others, including tumbleweeds and pigweeds, are weeds.

amaryllis (*plants*)   Native to tropical America, amaryllises (*Hippeastrum*) are widely cultivated for their large funnel-shaped flowers. Numerous varieties, or cultivars, with different colored blossoms have been developed. An amaryllis has an underground stem called a bulb, a basal cluster of leaves, and a flower stalk up to 3 ft (0.9 m) high. Close relatives include daffodils, onions, kaffir lilies (*Clivia*), spider lilies (*Hymenocallis*), and snowdrops (*Galanthus*). All are classified either in the lily family or in a separate family, Amaryllidaceae.

amber (*paleontology*)   Some trees, particularly certain pines, produce a sticky resin that under proper conditions hardens and becomes a fossilized material called amber. Tiny organisms trapped in the soft resin may be preserved intact, providing a wealth of information about species that lived millions of years ago. Scientists have even extracted DNA from ancient bacteria and insects preserved in amber.

Transparent or translucent, amber usually is yellowish in color. It has been carved and used in jewelry since ancient times. Because amber produces a negative electric charge when rubbed with a cloth, its Greek name, *elektron*, became the basis of the word "electricity."

ambergris (*biochemistry*)   Ambergris is a substance produced in the intestines of sperm whales, probably in response to irritation caused by the sharp, indigestible beaks of squid, the sperm whales' principal food. Ambergris freshly vomited up or removed from the intestine of a sperm whale is a blackish, greasy, foul-smelling liquid. On exposure to air and light it turns gray, hardens to a waxy mass, and develops a pleasant aroma. From ancient times until fairly recently, ambergris was extremely valuable, for it was used by perfumers as a fixative, to help fragrances retain their scent. Today trade in ambergris is banned, and the perfume industry uses synthetic substitutes.

americium (*elements*) In 1944 Glenn T. Seaborg and coworkers synthesized the radioactive metal with atomic number 95 and named it americium for the Americas. Americium is used sometimes as a commercial source of gamma rays, although it is quite dangerous to humans. One use is to ionize air in smoke detectors. Chemical symbol: Am.

Ames, Bruce (*biochemist/molecular biologist*)   [American: 1928–] In the 1970s Ames developed a simple, low-cost technique for determining the mutagenicity (mutation-causing potential) of chemicals. Called the Ames test, it introduces a test chemical into a culture of a particular strain of *Salmonella typhimurium* bacteria and then measures the rate at which the bacteria mutate. Because mutations play an important role in the development of cancer, the test has been widely used to help identify cancer-causing chemicals. However, not all substances that cause cancer in animals—dioxin, for example—give a positive Ames test, and a positive Ames test is not proof of carcinogenicity in animals.

Common Amino Acids			
alanine	glutamic acid	leucine	serine
arginine	glutamine	lysine	threonine
asparagine	glycine	methionine	tryptophan
aspartic acid	histidine	phenylalanine	tyrosine
cysteine	isoleucine	proline	valine

amino acid (*biochemistry*)   The building blocks of proteins (large polypeptides) are organic compounds called amino acids. Organisms can synthesize some or all of the amino acids they need; those that an organism must obtain from the environment are called essential amino acids. Humans, for example, must obtain at least 9 amino acids from the foods they eat.

An amino acid molecule is characterized by the presence of at least 1 amino group (NH_2), which is basic, and 1 carboxyl group (COOH), which is acidic. Amino acids are linked together by peptide bonds, which join the amino group of 1 amino acid with the carboxyl group of another. In this manner, peptides of varying lengths are formed. There are 20 common amino acids; modifying common amino acids after they become linked together produces rarer ones. For example, lysine and proline may be changed to hydroxylysine and hydroxyproline after they become part of the protein collagen.

amino-acid dating (*paleoanthropology*)   In living creatures most of the amino acids that build proteins possess handedness, rotating polarized light counterclockwise. Over long periods of time, random motions of atoms change some of these left-handed molecules in fossils to right-handed versions in a process called racemization. In a closed environment or in hard eggshells, the rate of racemization stays the same, so a suitable sample can be dated from the ratio of right-handed to left-handed amino acids. This method is useful for fossils too old for radiocarbon dating (good to about 40,000 years before the present) and not old enough for such methods as potassium-argon dating (useful for samples older than 200,000 years).

ammonia (*compounds*) Ammonia is hydrogen nitride (NH_3). In household applications it is dissolved in water, where it reacts to produce a complex solution of ammonium ions and hydronium ions in chemical balance with ammonia and water. Ammonia used as fertilizer is called anhydrous, since it is not dissolved in water when applied. Ammonia is important not only as fertilizer, but as a source of nitrogen in many industrial applications. Nitrogen in air and hydrogen gas are combined to produce ammonia in a process invented in 1908 by Fritz Haber [German: 1868–1934], fixing atmospheric nitrogen so that it can be used in other chemicals such as explosives and pesticides.

ammonite (*paleontology*) Ammonites were cephalopod mollusks with a shell, usually coiled, that was internally divided into a series of chambers by thin walls. The living animal occupied the last—and largest—chamber. A tube called a siphuncle extended from the ammonite's body into the other chambers, which were filled with a mixture of gases. By altering the density of this mixture, the ammonite could control buoyancy, enabling it to move toward or away from the water's surface. Ammonites first appeared in seas more than 400 million years ago. They became widespread before dying out 65 million years ago. There were hundreds of species, ranging in diameter from less than 1 in. (2.5 cm) to almost 6 ft (1.8 m).

ammonium (*compounds*) Ammonium is a radical, NH_4, that is a part of many industrially important chemicals such as the salt ammonium chloride or such fertilizers as ammonium phosphate, sulfate, or nitrate. Ammonium nitrate, which has a higher percentage of nitrogen than the other compounds mentioned, can also be converted into a powerful explosive.

amniote (*zoology*) Reptiles, birds, and mammals are sometimes referred to as amniotes because the embryos that develop from fertilized eggs are surrounded by 4 membranes: the amnion, allantois, chorion, and yolk sac. The amnion forms a sac around the embryo; it holds amniotic fluid, a watery solution that keeps the embryo moist and acts as a shock absorber. In reptiles and birds the allantois and chorion serve as embryonic respiratory and excretory organs; the yolk sac provides food for the embryo. In most mammals the chorion fuses with the allantois to form part of the placenta and the yolk sac is small and insignificant.

amoeba (*protists*) The many species of amoeba (from the Greek *amoibe*, meaning "change") are one-celled protists with irregular, ever-changing body shapes. Amoebas form extensions called pseudopods, which are used for locomotion and food gathering.

Depending on the species, amoebas range in size from 0.0004 in. (0.01 mm) to 0.04 in. (1 mm). They occur throughout the world in fresh and salt water, in moist soil, and as animal parasites. The species *Entamoeba histolytica* causes amoebic dysentery, an intestinal disorder of humans. Some amoebas form thick protective cysts and become inactive during droughts and other unfavorable environmental conditions.

amorphous substance (*materials science*)   A form of an element or compound that does not form crystals is amorphous. While glass is an amorphous substance, "amorphous" is more commonly used with forms of elements that lack crystal structure, such as amorphous antimony, carbon, silicon, selenium, or sulfur.

Ampère, André-Marie (*physicist*)   [French: 1775–1836] Ampère's name is a household word because the ampere, the fundamental unit of electric current, is named in his honor. Ampère learned of the connection between electricity and magnetism soon after its discovery in 1820 by Hans Christian Oersted. He immediately began a program to measure this relationship in various circumstances. Ampère developed methods for measuring current numerically and discovered the general principle, now Ampère's law, that the magnetic force generated between 2 current-carrying wires varies directly with the product of their currents and inversely with the square of the distance between the wires.

amphibian (*animals*)   The name amphibian comes from Greek words meaning "dual life." It refers to the fact that most of these vertebrates live in water and use gills for respiration in the first (larval) stage of life; then, following metamorphosis into adulthood, they live at least partly on land, using lungs to breathe air. Their thin nonscaly skin, kept moist by secretions from numerous mucous glands, is an important respiratory organ. Amphibians are ectothermic (cold-blooded), absorbing heat from the environment.

The more than 4000 species of amphibians, the majority of which live in the tropics, are divided into 3 orders: caecilians, salamanders, and frogs and toads. The smallest species is probably the Cuban pygmy frog (*Sminthillus limbatus*), which is less than 0.5 in. (1.2 cm) long. The largest is the giant salamander of Japan (*Andrias japonicus*), which grows to lengths of over 5 ft (1.5 m).

amplitude (*numbers*)   Each complex number is associated with a positive real number variously called its amplitude, modulus, or absolute value. The amplitude corresponds to the distance from the origin to the point for the complex number. For a complex number expressed as $x + iy$, the amplitude, written either as $\| x + iy \|$ or $| x + iy |$, is the square root of $x^2 + y^2$; for the number expressed in polar form as (r, θ) the amplitude is $|r|$, the real absolute value of r.

amplitude (*waves*)   Amplitude is a numerical measure of intensity used for energy traveling as waves. The amplitude is half the distance from the highest point of a wave (crest) to the lowest (trough). For longitudinal waves amplitude is measured from the region of greatest density to the lowest. Amplitude of light is brightness and of sound, loudness. The amplitude of both the sine and cosine functions is 1.

anabolism and catabolism (*physiology*)   Synthesis and breakdown of substances occur constantly in living cells. Processes by which simple substances are combined to form more complex substances are known collectively as anabolism. The synthesis of food storage compounds such as fats and the production of new cellular materials are examples. Conversely, processes in which complex substances split into simpler compounds are known as catabolism. The breakdown of foods to release energy involves a series of catabolic reactions.

anaconda (*animals*)   The largest boa is the heavy-bodied anaconda, which lives in tropical forests of South America. It attains lengths up to 30 ft (9 m). Unlike other boas it is semiaquatic, often dropping from trees into the water to prey on waterfowl and other animals.

anadromous fish (*animals*)   Some fish migrate great distances to spawn in waters best suited to the development of their young. Fish such as salmon that ascend rivers to spawn in shallow freshwater are said to be anadromous. In contrast, freshwater eels and other fish that migrate downstream into the ocean to spawn are catadromous.

anaerobe (*biology*)   An organism that lives in an oxygen-free environment is an anaerobe. Instead of obtaining energy by cell respiration, as aerobes do, anaerobes obtain energy by fermentation (sometimes called anaerobic respiration).

Obligate anaerobes, such as certain bacteria, are poisoned by oxygen and must live in an oxygen-free environment. Yeasts and many internal parasites of animals are examples of facultative anaerobes (and facultative aerobes); they can live with or without oxygen and can obtain energy from either respiration or fermentation.

analytic geometry (*mathematics*)   Analytic geometry, also called coordinate geometry, began as employing coordinate systems to develop proofs in geometry—using algebra to do geometry. The subject was invented in this form independently by Descartes and Fermat around 1630. The marriage of algebra and geometry proved to be far more useful than that, however, for almost every equation in 2 variables describes a curve in a coordinate system (and equations in more variables describe surfaces in higher dimensions). A modern definition holds that analytic geometry is the use of equations to describe curves or surfaces, and the study of how the properties of curves appear in equations. Solutions to systems of equations occur where curves or surfaces intersect. Properties of curves can be recognized from equations—for example, all conic sections are curves associated with equations of the second degree—and vice versa.

Anaxagoras (*astronomer*)   [Ionian: c500–c428 BC] The last of the Ionian philosophers, whose tradition started with Thales, Anaxagoras established a rational view of the universe, explaining correctly the phases of the Moon and eclipses and identifying the Sun and Moon with rocky meteorites. In Athens he was tried for impiety and barely escaped death.

Anaximander (*geography*)   [Ionian: 611–547 BC] Anaximander prepared the first map that attempted to show the whole Earth. He recognized that Earth must be curved and not flat—he had used a sundial to find the solstices and equinoxes, and believed Earth to be suspended in space within a sphere of stars—so his map showed Earth as a cylinder.

anchovy (*animals*)   Anchovies, Family Engraulidae, are small fish that live in large schools, mainly in shallow tropical and warm temperate marine waters.

They are 4 to 10 in. (10 to 25 cm) long, with a long snout, big eyes, and a mostly silvery body. Their fins lack spines.

and (*logic*)   The connective and, symbolized by $\wedge$ or & and also called conjunction, produces a combined sentence that is true if and only if both sentences it connects are true. For example, "The sky is blue and the Sun is shining" is true if there is a blue sky and it is now sunny, but false if the sky is not blue or the Sun is not shining or both. Similarly, the and gate in an electronic device passes a signal of two 1s, but registers 0 for a signal with a 1 and 0 or with two 0s.

Anderson, Carl David (*physicist*)   [American: 1905–91] Anderson is the only particle physicist credited with the discovery of 2 elementary subatomic particles, the positron in 1932 and the muon in 1936, although neither is fundamental by itself. (The positron is the antiparticle of the electron and the muon is the "fat electron" of the second family of elementary particles.) The positron had been predicted 4 years prior to its discovery, but few had believed the prediction of Paul Adrien Maurice Dirac. The muon was thought, when Anderson first located it in cosmic-ray tracks, to be another predicted particle (the pion of Hideki Yukawa, predicted in 1932). Anderson soon demonstrated that his new particle had different properties and was totally unexpected.

Andrews, Roy Chapman (*zoologist*)   [American: 1884–1960] Andrews became world famous as a fossil hunter in the 1920s, when he led 4 expeditions to Mongolia's Gobi Desert. His discoveries included *Protoceratops* bones and eggs (the first dinosaur eggs ever found); the first fossils of several dinosaur genera, including *Oviraptor* and *Velociraptor*; and evidence that Stone Age people lived in central Asia. Also a leading authority on whales, Andrews spent his entire professional career at the American Museum of Natural History in New York, where he served as director from 1935 until his retirement in 1941.

Andromeda galaxy (*galaxies*)   A giant spiral collection of stars in constellation Andromeda, the Andromeda galaxy is also called M31 (Messier 31), NGC 224 (New Galactic Catalog 224), or the Great Nebula (without use of a powerful telescope the galaxy appears as a patch of light, or nebula). The Andromeda galaxy contains some 300 billion stars that form a spiral shape about twice the size of the Milky Way.

angiosperm (*plants*)   Seed-producing plants that bear flowers and produce their seeds within protective fruits are angiosperms ("enclosed seeds"). They are the dominant form of plant life on Earth, ranging from duckweed not much bigger than a pinhead to the banyan tree, which may spread outward over 1 a (0.4 ha) or more. Angiosperms are found worldwide in almost every kind of habitat, but their numbers and variety are greatest in the tropics. The approximately 230,000 known species make up Phylum Magnoliophyta. The phylum is divided into 2 classes, commonly known as monocots and dicots.

angle (*geometry*)   A geometric angle is the formal version of the sharp corners on certain geometric figures. The word "triangle" means 3 sharp corners, for example, not 3 sides. The earliest ideas of angle were the parts of a polygon where 2 sides come together or else the amount of opening between an adja-

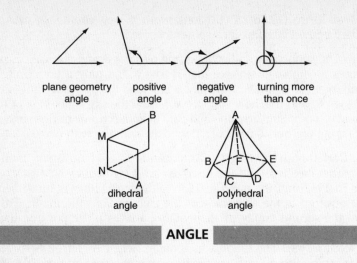

plane geometry angle — positive angle — negative angle — turning more than once

dihedral angle — polyhedral angle

ANGLE

cent pair of sides. Angles today, however, are conceived of as 2 adjacent sides extended infinitely. This angle is 2 half lines with a common endpoint. The amount of opening is a measurement and not the angle itself. This definition applies only to angles in the plane for geometry based on sets of points.

Away from plane geometry, a different definition is employed, one based on the idea of turning a half line about a point, as one turns a crank handle. Turning the half line counterclockwise produces a positive angle, while turning in the opposite direction gives a negative angle. One can keep turning and turning the half line through complete circles, producing angles of any size whatever. A turn through 1 complete circle plus a quarter turn more is a different angle from a quarter turn alone, for example. The definition based on turning is used in trigonometry and most higher mathematics.

Replacing the half line with a half plane in these definitions results in one kind of angle in 3 dimensions, while another kind of 3-dimensional angle is formed from parts of planes that meet at a common point.

anglerfish (*animals*) Anglerfish, constituting the order Lophiiformes, are bizarre-looking marine fish named for the modified spine atop the head. This spine acts as a fishing rod, its fleshy tip used as "bait" to lure prey to the mouth. Anglers include goosefish, frogfish, and batfish, as well as deep-sea anglers. Among the latter, only the females, which grow to 4 ft (1.2 m) long, have a fishing rod; the males are tiny and live as parasites on their mates.

angular momentum (*mechanics*) Angular momentum is a quantity that develops when a body spins about an axis. It is more complex than ordinary momentum, which is mass times velocity.

The distance of each part of a spinning body from the center affects acceleration and force, so that distance, r, needs to be considered for rotating bodies. The equivalent of ordinary mass, m, for a spinning body is mr^2, where r represents either the distance to the center of an object circling a point outside itself or an average distance for an object rotating about a point within it. This is the moment of inertia, or I.

Velocity in a circle is conveniently measured in radians. Change in an angle in radians times the distance r from the center gives angular velocity, traditionally written with a small omega (ω). Angular momentum then becomes $I\omega$, corresponding to the mv of ordinary momentum. Angular momentum in a closed system, like ordinary momentum, obeys a conservation law.

anhinga (*animals*)   Looking somewhat like small, slim cormorants, anhingas (*Anhinga*)—also called darters or snakebirds—are about 3 ft (0.9 m) long. They have a very long pointed bill, very long neck, and long narrow tail. Anhingas live along lakes, rivers, and swamps in warm temperate and tropical regions. They are expert swimmers and hunt for fish and invertebrates underwater.

animal (*zoology*)   Animals differ from the other main group of multicellular organisms, plants, in several basic ways. Animals cannot make their own food. Their cells do not have walls. They commonly store carbohydrates as glycogen. And they usually react quickly to stimuli. An animal's form is rather invariable, and growth generally ends when the animal reaches maturity. Most animals have the ability to move about from one place to another.

There are more than 2,000,000 known species of living animals divided into 2 groups: invertebrates and vertebrates. They range from microscopic rotifers to blue whales that top the scales at 260,000 lb (118,000 kg).

animal communication (*ethology*)   Animals communicate information in many ways. The use of sound is widespread: a cricket chirps to attract a mate, a starling utters an alarm call to warn other starlings that a predator is near, a hyena "laughs" to tell its pack that food is available. Visual communication includes the use of coloring, such as the different plumages of male and female cardinals; postures and displays; and markings, such as the scars made on tree trunks by bears. Chemical communication may involve taste or smell; substances called pheromones are used to attract mates, mark trails or territory, distinguish members of a group from strangers, and so on. Touch is important among animals with poor vision and those living in dark environments. Electrical impulses are used for communication by electric eels and other fish living in muddy waters.

ankylosaur (*paleontology*)   The ankylosaurs were a group of plant-eating armored dinosaurs that lived during the Cretaceous. The 4 short legs supported a barrel-shaped body that was protected by numerous bony plates and spikes. In at least some species, the tail had a thickened, bony end that was probably used in battles with predators. The largest member of the group, *Ankylosaurus*, was more than 33 ft (10 m) long and weighed 5 tons.

annelid (*animals*)   The phylum Annelida comprises approximately 9000 species of segmented worms. They are found worldwide in all kinds of environments. There are 3 major classes: bristle worms, earthworms and their relatives, and leeches. Most are less than 5 in. (12.5 cm) long, but the Australian earthworm *Megascolides australis* grows to lengths of more than 10 ft (3 m).

An elongated body that consists of a series of segments characterizes annelids. Except for the head and the last segment, all segments are alike externally, generally armed with bristles (setae) for movement. Internally, the segments also show repetition, with similar muscles, a nerve ganglion, a pair of excretory organs, and so on. Only the digestive system of an annelid species shows significant differentiation into a number of differently formed regions between the mouth at the head end and the anus at the rear end.

annual plant (*botany*) Plants that complete their entire life cycle—from seed germination to seed production—in a single growing season are called annuals. The plants die at the end of the season; the seeds survive to start a new cycle at the beginning of the next growing season. Annuals typically are herbaceous, with soft green stems. Beans, marigolds, and tobacco are examples.

Some plants are either annual or perennial plants, depending on the climate. For example, sweet pepper (*Capsicum annuum*) is an annual in northern temperate areas but a perennial in warmer places.

annual ring (*botany*) Each growing season, new xylem is added to the wood of a tree stem. In many trees, such as oaks, there is a distinct difference between xylem that forms in the spring, when soil water is plentiful, and xylem that forms in the dryer summer months. Spring xylem has numerous wide vessels and looks porous under a microscope; summer xylem is much denser and darker in color. The number of these dark annual rings equals the tree's age. Examination of the rings may also provide data about weather in years past. Very narrow annual rings, for example, suggest drought conditions.

annulus (*geometry*) An annulus—also called a ring—is the region between 2 circles in the same plane that have the same center. It is the set of points outside the smaller circle but inside the larger one; the same region including both circles is also called an annulus.

anode (*electricity*) An anode is the place that a positive charge builds on a battery or in a diode. It is called the anode ("ascending path") because Michael Faraday, who named it, pictured an "electric fluid" moving out of the anode to a location called the cathode. In diodes and similar tubes or transistors, electrons do move in this way, but in the other direction (from cathode to anode). In batteries, negative ions move from the anode to the cathode or positive ions from the cathode to the anode, so that the anode develops its positive charge. Current then moves along a circuit between the cathode and anode.

anole (*animals*) Native to warm areas of the Americas, anoles are small, agile lizards. They spend most of their time in trees, hunting during daylight for insects. Males have a large flap of skin called a throat fan, which often is brightly colored and which can be flared out to frighten an enemy or impress a female. Anoles can change their color, an ability that has led people incorrectly to call them chameleons.

ant (*animals*) Ants (Family Formicidae) are social insects that live in colonies containing from 50 to tens of millions of individuals. A caste system exists in a colony, generally including winged queens, wingless female workers and soldiers, and winged males. Ants have club-shaped antennae, a constricted "waist" in the abdomen, and a crop that is used to store food that can be regurgitated to feed both larvae and other adults. Ants are found in almost every ecological niche on land, from polar regions to the tropics and from seashores to mountaintops. Depending on the species, they feed on living and dead animals, vegetation, nectar, fungi, and honeydew of aphids and other insects. Some species are parasites of other ants.

Ants belong to Order Hymenoptera ("membrane wings"), which also includes bees, wasps, ichneumons, sawflies, and others. Hymenopterans have 2 pairs of thin, veined wings. Small hooks on the front margin of the hind wings interlock with the forewings.

antagonist (*biochemistry*)   A substance—usually a hormone or drug—that counteracts the effects of another substance (the agonist) is called an antagonist. Antagonists act in various ways. For instance, an antagonist may compete for the same receptor sites on molecules as the agonist, or it may produce chemical effects opposite to those of the agonist, thereby neutralizing the agonist's effects.

Antarctic Plate (*geology*)   The section of Earth's lithosphere called the Antarctic Plate appears to have surrounded the south pole for millions of years, while other crustal plates moved away from it. Mid-oceanic ridges form spreading centers that border the plate nearly all the way around—the Southeast Indian rise in the Indian and Pacific oceans, the Pacific-Antarctic ridge in the Pacific, the Chile rise in the Pacific, and the Atlantic-Indian ridge in the Atlantic and Indian oceans.

Antares (*stars*)   Antares—whose name, "rival of Mars," refers to its red color—is a supergiant star orbited by a small companion in the constellation Scorpius. Antares is about 285 times as wide as the Sun. Its companion star, just lower than the limit of naked-eye visibility from Earth, is losing gas that streams toward Antares, producing a detectable radio emission.

anteater (*animals*)   Toothless mammals that feed mainly on ants and termites, anteaters constitute the family Myrmecophagidae. They are well adapted for breaking into insect nests and capturing prey. They have long limbs armed with claws; a long cylindrical snout; and a sticky, extensible tongue. They are native to Central and South America. The giant anteater, or ant bear (*Myrmecophaga tridactyla*), lives on the ground but other species spend most of their time in trees.

antelope (*animals*)   Antelopes—including the addax, bongo, gazelles, gemsbok, hartebeest, impala, kudus, nilgai, wildebeests (gnus), and others—are hoofed mammals in the bovid family. They are a varied group, with markedly different appearance and behavior, though most are quick-moving gregarious creatures with slender legs, powerful rear-leg muscles, and excellent sight, smell, and hearing. Horns may be straight, curved, or spiral. The royal antelope (*Neotragus pygmaeus*) of western Africa stands about 10 in. (25 cm) tall at the shoulders and weighs about 15 lb (6.8 kg). In contrast, the giant eland (*Taurotragus derbianus*), also of western Africa, may attain a shoulder height of more than 6 ft (1.8 m) and weigh more than 1200 lb (545 kg). Most antelopes are native to Africa but a few species survive in Asia. Antelopes generally inhabit grasslands and forests, but some species live in swamps, deserts, and high mountain regions.

antenna (*anatomy*)   Projecting from the head of certain arthropods, particularly insects and crustaceans, are a pair of many-jointed antennae. The antennae are laden with small sense organs. Ants communicate by tapping their antennae, bees taste with their antennae, silk moths use their antennae to scent mates, and blood-sucking insects depend on their antennae to sense temperature changes that indicate the proximity of a warm-blooded vertebrate. In some crustaceans, antennae are modified for other purposes; for example, the antennae of male brine shrimp are large claspers used during mating.

anthropology (*sciences*)   Anthropology is the study of human beings, but certain aspects of human life are treated as biology. Anthropologists investigate group processes while biologists study human development and the functioning of individuals. Physical anthropology is concerned with the origin of the species and with physical variations from one group of humans to another. Cultural anthropology concerns the development of social structures and variations in patterns of behavior from one group to another. Often the 2 aspects are combined for the study of early humans and their hominid relatives; this combination is termed paleoanthropology (*paleo-* means "ancient"). Paleoanthropology as a discipline is replaced by archaeology after the Neolithic Revolution.

antibiotic (*biochemistry*)   Produced by microorganisms such as molds and bacteria, antibiotics ("against life") are organic compounds that kill or inhibit the growth of other organisms, usually other microorganisms. Well-known antibiotics include penicillin, produced by *Penicillium* mold, and streptomycin and actinomycin, produced by *Streptomyces* bacteria. Antibiotics work in various ways: some interfere with the formation of bacterial cell walls; others inhibit synthesis of proteins, DNA, and RNA. Some antibiotics target a few species whereas others are active against a broad range of organisms. Since the 1940s people have used antibiotics to treat a wide variety of human and animal diseases.

antibody (*biochemistry*)   Also known as immunoglobulins, antibodies are large, highly specific proteins that play an vital role in the immune response. Antibodies are secreted by vertebrate white blood cells called B lymphocytes in response to foreign chemicals called antigens, such as protein molecules on the surface of bacterial cells. When a B lymphocyte comes in contact with an antigen, it engulfs the antigen. The B cell then produces antibodies shaped to combine with other antigen molecules of exactly the same design. A single B lymphocyte can produce up to 10,000,000 antibodies an hour. These are released into the bloodstream to hunt down, bind to, and inactivate the antigens.

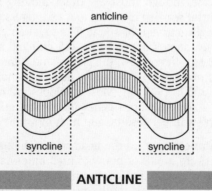

ANTICLINE

anticline (*geology*)   A fold in rock strata that curves up and then down to form an arch is called an anticline. Anticlines a few feet high can often be seen in road cuts with dips, called synclines, between them. Large anticlines produced where colliding tectonic plates have folded the rock are the framework of such mountain ranges as the Alps and Rockies.

anticyclone (*meteorology*)   Winds in the Northern Hemisphere are turned east by the Coriolis effect if flowing north and south if flowing east. Thus northern winds always turn to the right. The result is that large-scale movements of wind (and ocean currents as well) flow in a clockwise direction in the Northern Hemisphere. Such a rotating air mass is called an anticyclone because it moves in the opposite direction from the cyclones of the Southern Hemisphere. Air moves into a region of low pressure or outward from one of high pressure, but either pattern causes an anticyclone in the Northern Hemisphere. A hurricane is the most extreme variety of anticyclone.

antiderivative (*calculus*)   The antiderivative is essentially the same as the indefinite integral, although defined differently. Nearly all useful functions have a derivative function whose value is the slope of the original function. The antiderivative of a given function $f(x)$ is another function $F(x) + C$ whose derivative is $f(x)$. Since the derivative of a constant is always 0, the antiderivative includes an arbitrary constant, C, that may be assigned a particular value to match a problem situation.

antigen (*biochemistry*)   Short for "antibody-generating substance," an antigen is a foreign protein or other chemical that can cause an immune response by a host's body. This response, which involves the production of antibodies, is designed to destroy the antigen. Antigens include toxins (snake venom, for example) as well as molecules on the surface of invading bacteria and other cells.

antimatter (*particles*)   Any collection of antiparticles can be described as antimatter, but the term usually means atoms made from antiprotons, antineutrons, and antielectrons. When antimatter meets ordinary matter, both substances vanish and their energy becomes high-energy gamma radiation. Thus, antimatter cannot exist for long on Earth. In 1995 physicists succeeded in the production of a few antiatoms of antihydrogen that promptly vanished as predicted.

antimony (*elements*)   The brittle, hard metal antimony, atomic number 51, is thought to have been first produced in metallic form by Rhazes [Arabic: c845–c930], but had long been recognized in compounds. Its chemical symbol, Sb, is from the Latin name for an antimony compound, *stibium*. The origin of the name antimony is unclear, although it may be a corruption of Arabic. The metal has many uses, combining with lead or other metals to make alloys for bearing housings and type and with other elements to make compounds used in paints and ceramics.

antioxidant (*biochemistry*)   Any substance that inhibits or prevents the process of oxidation is an antioxidant. The term is often applied to compounds that neutralize free radicals (molecules with unpaired electrons). For example, the enzyme superoxide dismutase absorbs certain radicals and breaks them down into hydrogen peroxide; vitamin E sponges up free radicals that would otherwise damage cell membranes.

antiparticle (*particles*) Every subatomic particle has a near twin called the antiparticle. When Paul A.M. Dirac developed a mathematical version of electron theory in 1930, he observed that one solution to his main equation predicted a particle that would be exactly the same as the electron but positively rather than negatively charged. The predicted particle, discovered 2 years later, was named the positron. The same equation predicts near twins for all subatomic particles. The twins are called antiparticles, so another name for positron is antielectron. The photon is its own antiparticle, but other neutral particles can have antiparticles that differ in specific ways. Neutrons, neutral particles composed of quarks, are paired with equally neutral antineutrons formed from antiquarks. Neutrinos and antineutrinos differ in a subtle characteristic called intrinsic spin angular momentum.

A particle and its antiparticle that meet annihilate each other, producing a burst of electromagnetic energy. For example, a stream of antielectrons found jetting from the Milky Way in 1997 was recognized by the characteristic of energy resulting from annihilation by electrons.

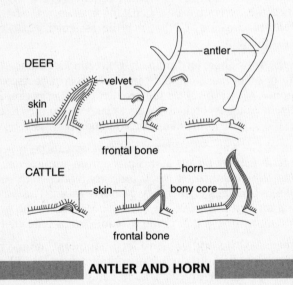

ANTLER AND HORN

antler and horn (*anatomy*) Many ruminant mammals have bony antlers or horns on their heads. These structures, which vary greatly in size and shape, are used for defense and for fighting between males at mating time. Male members of the deer family, Cervidae, have antlers (only in the caribou genus do females have them as well). In some species the antlers are branched, with each branch ending in a "point." Antlers are annual growths, shed after each mating season. As antlers grow they are covered with a layer of soft, finely haired skin called velvet. When antlers are fully grown the blood supply is cut off and the velvet dies. Deer rub against trees to remove the velvet, exposing the bone of the antlers.

Cattle, sheep, and goats are among the animals with horns. Depending on the species, the horns, typically curved and pointed, may be present on both sexes or on males only. Unlike antlers, horns are permanent, with continuous growth from the base to compensate for wear. Horns on the snouts of rhinoceroses are different; they are composed of tightly packed strands of hair.

Apatosaurus (*paleontology*)   Popularly known by the name brontosaurus, *Apatosaurus* was one of the largest dinosaurs that ever walked on Earth. It measured up to 82 ft (25 m) in length and weighed some 30 tons. It had a small, elongated head; a long neck; an enormous body supported by 4 heavy legs; and a long, heavy tail. A plant eater that was preyed on by smaller, more agile carnivorous dinosaurs, it lived more than 140 million years ago.

ape (*animals*)   The term "ape" is applied to large, tailless primates with complex brains. They include gibbons and the great apes: chimpanzees, gorillas, and orangutans. All are members of the hominoid superfamily. (Some tailless monkeys are known as apes—such as the Barbary ape (*Macaca sylvanus*)—but these are not apes in the biological sense.)

aphelion (*solar system*)   The point in an orbit about the Sun that is farthest from the Sun is the aphelion; the nearest is the perihelion. The aphelion is often confused with the apogee, the point in an orbit about Earth farthest from Earth.

aphid (*animals*)   Also called plant lice, aphids are small pear-shaped insects classified in the same order as cicadas. Some species have wings, others are wingless. Aphids secrete a sugary fluid called honeydew that is highly prized by ants. Indeed, many species of ants care for and protect aphids, stroking the aphids with their antennae until the aphids excrete a drop of honeydew. Aphids feed on plant sap and cause much damage to cultivated plants.

Apian, Peter (*astronomer*)   [German: 1495–1552] In 1540 Apian reported on 5 comets he had observed, including the one we now know as comet Halley. He was the first to note that the tails of comets always point away from the Sun, an effect caused by the solar wind.

apogee (*solar system*)   The highest point of an orbit about Earth, whether the orbiting body is the Moon or an artificial satellite, is its apogee. By extension, any highest point is called an apogee, even when no orbit is involved.

Apollo, Amor, and Atens asteroids (*solar system*)   More than 60 asteroids are known to cross Earth's orbit. They are called the Apollo asteroids after the asteroid Apollo, the first orbit crosser to be discovered (by Karl Reinmuth [German: 1892–1979] in 1932). In addition to the Apollos, more than 100 known asteroids have orbits near Earth's. The Atens asteroids actually travel inside Earth's orbit, while the Amor group is just outside it. After formal search operations for such asteroids began in the 1990s, astronomers observed small asteroids passing close to Earth, one at about 60,000 miles (100,000 km). There is little doubt that asteroids have collided with Earth in the past; by definition these are Apollo asteroids.

It was once thought that all the asteroids were in orbit between Mars and Jupiter, so any largish object with an orbit beyond Jupiter was thought to be a comet. Now we know that asteroids are found in various parts of the solar system, although most do occur between Mars and Jupiter.

Apollonius (*mathematician*)   [Hellenic: c262–c190 BC] Apollonius, also called Apollonius of Perga, is considered the greatest of the Hellenic geometers. Much of his work extends Euclid's *Elements*. Although many of his writings are lost, his books were described by some of their early readers, allowing mathematicians to reconstruct the missing books. His famous *Conics* was in 8 books, of which the first 7 survive. *Conics* introduced the definition of conic sections in terms of sections of a plane through a double cone and proceeded to establish many properties of those curves. *Conics* comes very close to inventing analytic geometry in its techniques.

apothem (*geometry*)   The apothem is the short radius of a plane regular figure; that is, either the distance from the center of a regular polygon to a side or a line segment from the center perpendicular to the side. The apothem bisects the side. The area of a regular polygon is half the product of the apothem and the perimeter.

appendix (*anatomy*)   At the junction of the small and large intestines in mammals is a pouch called the cecum. Projecting from the cecum is a fingerlike tube, the vermiform appendix. In plant-eating mammals, the cecum and appendix are well developed and play a role in digestion, particularly the digestion of cellulose. In humans, the appendix is about 4 in. (10 cm) long; as in other non-herbivore mammals, it has no known function.

apple (*plants*)   Members of the rose family, apples (*Malus*) are deciduous shrubs and trees that grow to a maximum height of about 40 ft (12 m).They require a dormancy, or period of relatively cold weather. Apples are widely cultivated for their fleshy, edible fruits, and thousands of varieties have been developed. The fruits, called pomes, usually are red, yellow, or green and average 2 to 4 in. (5 to 10 cm) in diameter. The edible portion is actually a greatly enlarged receptacle (top portion of the flower stalk).

approximation (*arithmetic*)   An approximation is a simplified substitute number, used especially in place of large natural numbers or measurements. Exact measurement is theoretically impossible, and irrational numbers cannot be written exactly as decimals since such decimals always have an infinite sequence of varying digits.

The most common form of approximation is rounding, which can be done by measurement rules or by statistical rules. A unit is chosen for the approximation, say thousands for a crowd of people. The estimate is then made to the next smaller unit, hundreds. If there are fewer than 500 in the last interval, round down to the smaller number of thousands—6300 is approximated as 6000. By measurement rules, if there are 5 or more hundreds, round up to the next higher number of thousands—6500 is rounded to 7000. In statistical rounding, if there are exactly 5 smaller units, round to the even number of larger units; so 6500 is rounded to 6000 and 7500 is rounded to 8000.

aquifer (*geology*)   Any water that fills small openings (pores) in rock and soil below the water table is an aquifer, but the word is most often associated with a large region where a connected body of water exists at some depth in soil or rock. Such an aquifer can exist even when the region above it gets little water. In that case, the water in the aquifer has moved downhill along the top of a sinking impermeable layer or has flowed along the layer because all the pores available have been filled and additional water has entered the aquifer.

Arabian Plate (*geology*)   The Arabian Plate, extending from the Red Sea to the Persian Gulf and from the Indian Ocean to Anatolia, is a small part of Earth's lithosphere that is pushing into the Iranian region of the Eurasian Plate and raising the plateau of Iran.

arachnid (*animals*)   The arthropod class Arachnida includes spiders, scorpions, daddy longlegs, mites, ticks, and others. The segmented arachnid body has 2 regions: a cephalothorax (fused head and thorax) and an abdomen. The cephalothorax bears 6 pairs of appendages, including 4 pairs of walking legs. Unlike other arthropods, arachnids lack antennae; instead, the first pair of appendages are chelicerae, which are modified into pincers in some orders and fangs in spiders. Most arachnids breathe by means of air-filled sacs in the abdomen called book lungs.

Arachnids are predators. Instead of grinding and chewing their food, they secrete liquefying chemicals into the food, then suck up the "soup." There are more than 60,000 described species, inhabiting every type of land environment. The smallest are mites about 0.01 in. (0.025 cm) long. The longest is probably the South American rock scorpion (*Hadogenes troglodytes*), which can exceed 8 in. (20 cm).

aralia (*plants*)   Members of the genus *Aralia* have prickly stems, large compound leaves, and clusters of tiny creamy flowers that develop into berries. A well-known North American species is the angelica tree, also known as Hercules' club (*A. spinosa*), which grows along streams and in moist woods and attains a height of 30 ft (9 m) or more.

The aralia family, Araliaceae, comprises some 700 species of mostly tropical and subtropical shrubs, trees, vines, and herbs. Other members of the family include ginseng, ivy, and schefflera (*Schefflera*).

arborvitae (*plants*)   Members of the cypress family, arborvitaes (*Thuja*) are evergreen conifer shrubs and trees native to North America and eastern Asia. Young leaves are needle-shaped but adult leaves are scalelike and overlapping. The small seed-bearing cones grow erect at the tips of branches. The giant arborvitae or western red cedar (*T. plicata*) of western North America grows to 200 ft (60 m).

arbovirus (*viruses*)   Short for "arthropod-borne viruses," arboviruses are transmitted primarily by blood-sucking arthropods, such as mosquitoes, fleas, lice, and ticks. The arthropods pick up the virus while feeding on blood from an infected individual. The virus replicates in the arthropod and spreads to its salivary glands. The arthropod spreads the disease to new hosts when it bites them. Diseases caused by arboviruses—which are members of several virus families—include yellow fever, dengue, sandfly fever, and equine encephalitis.

arc (*geometry*)   An arc is any continuous part of a circle or other curve; an arc can be deformed (straightened) to form a line segment. For a circle, 2 points on the circle define 2 arcs. Unless the points are endpoints of the diameter, the longer arc is the major arc and the shorter one the minor arc.

Archaea (*biology*)   In recent years it has become apparent that certain organisms originally considered to be bacteria—and looking very much like bacteria under a microscope—are biochemically and genetically very different. These organisms are now placed in an entirely separate group, Archaea ("ancient ones"). Several types are adapted to life in extreme environments, but archaeans also are found among marine plankton.

Thermophiles ("heat lovers") live in hot springs, near deep-sea vents, and in other environments where temperatures may reach 212°F (100°C) or more. Methanogens ("methane makers") are anaerobic organisms that cannot tolerate oxygen; they obtain the energy they need for life by converting hydrogen and carbon dioxide into methane. Halophiles ("salt lovers") live in saline ponds and other places with extremely high salt concentrations.

The origin and evolutionary history of archaeans is still unclear. Like bacteria, archaeans lack a clearly defined nucleus. But in many ways they resemble animals and other eukaryotes. For example, the archaean *Halobacterium* has a red light-sensitive pigment that is remarkably similar to the light-detecting pigment in the retina of vertebrate eyes.

archaeology (*sciences*)   Archaeology is the scientific study of human culture and history from the dawn of village life to modern times, based on the retrieval of artifacts, including early texts. It is often grouped with the social sciences. The case for archaeology as a natural science begins by noting that its discoveries are reported in the science pages of newspapers and even in science magazines that contain new research in other sciences. Modern archaeologists increasingly employ scientific tools, such as radiocarbon dating or other physical means of dating artifacts, satellite surveys, and electromagnetic methods of finding objects. Archaeology is closely allied to the science of paleoanthropology, merging seamlessly with it about 10,000 to 40,000 years before the present.

Archaeopteris (*paleontology*)   The earliest known modern tree, *Archaeopteris* made up an estimated 90% of Earth's forests 360,000,000 to 345,000,000 years ago. Its photosynthetic activities altered Earth's atmosphere, helping increase oxygen levels from 5% to 20% and decreasing carbon dioxide levels from 10% to 1%. Although *Archaeopteris* is believed not to be a direct ancestor of today's trees, it had a comparable structure. It was the first known plant with branching trunks and an extensive root system, and the first long-lived perennial.

Archaeopteryx (*paleontology*)   Living some 140 million years ago, *Archaeopteryx* ("ancient wing") was one of the precursors of modern birds. The size of a large pigeon and weighing 1 lb (500 g) or less, it had a wishbone and feathers on its wings and tail. It was probably capable of some flight, such as gliding from high tree branches to lower ones. It also had many reptilian features, including a long, flexible tail, clawed fingers, and teeth in both jaws. Based on the original fossil discovery of *Archaeopteryx* in 1861, Thomas Huxley [English:

1825–95] proposed that dinosaurs and birds are related—a theory that continues to be widely held and supported by recent discoveries.

archaic *Homo sapiens* (*paleoanthropology*)   Certain very old fossils with some primitive features but otherwise similar to modern humans, are classed as archaic *Homo sapiens*. These fragmentary fossils, mostly from the southern tip of Africa, appear to represent *H. sapiens* but are dated at 120,000 or more years in the past (Cro Magnon humans—the first recognized as fully modern—appeared about 35,000 years ago). The archaic *H. sapiens* are sometimes referred to as anatomically modern humans because, even though their fossil bones are similar to those of present-day human populations, the living hominids might well have been very different from modern humans in behavior or other ways not apparent from bones alone.

Archimedes (*mathematician/physicist/inventor*)   [Hellenic: c287–212 BC] Archimedes was the greatest of Greek mathematicians. He also contributed to physics and developed inventions employing physical concepts.

In mathematics Archimedes computed close limits on the value of π by comparing polygons inscribed in and circumscribed about a circle. He discovered how to calculate the volume of a sphere and used techniques similar to limits to find centers of gravity and areas of shapes bordered by a parabola and a line. Archimedes created a notation that could express any natural number, no matter how large, impossible with ordinary Greek numerals.

Archimedes' discovery of the hydrostatic principle while in his bath is well known. He observed that a body immersed partly in a fluid displaces a weight of the fluid equal to the weight of the body. He invented the Archimedian screw, a helix in a tube that when turned raises a fluid to its top. Archimedes also worked out the mathematics of simple machines and probably discovered the compound pulley. According to the story, after declaring "Give me a place to stand on and with a lever I will move the whole world," Archimedes demonstrated the power of simple machines by using compound pulleys to launch a large ship single-handedly.

Archimedes' principle (*physics*)   A solid floats in a liquid when the part of the solid below the liquid exactly displaces the amount of liquid that has the same mass as the entire solid. The rest of the solid rises above the surface. This is a consequence of Archimedes' principle. According to old accounts, Archimedes discovered this concept while taking a bath (and then ran naked through the streets shouting *"Eureka,"* Greek for "I have found it"). At the time, Archimedes had been set the task of discovering whether a crown was pure gold or adulterated. Although the metal crown sank in water, its weight was reduced by the weight of the volume of water displaced, the essential idea behind Archimedes' principle.

Arctic and Antarctic Circle (*geography*)   As it orbits the Sun, Earth maintains a tilt of 23.45° in the same direction so the regions around the poles are sometimes in total darkness. The portion of Earth's surface near the north or south pole that has at least one 24-hour period of darkness is within the Arctic or the Antarctic Circle. The Arctic and Antarctic Circles are the latitudes of 66.55° N and 66.55° S. Since the poles are each at 90°, the tilt of 23.45° results in the circles being 90 − 23.45 = 66.55 degrees of latitude.

Arcturus (*stars*)   Arcturus is a red supergiant that is the brightest star in the constellation Boötes, the herdsman. It is 46 times as wide as the Sun with a volume 12,167 times that of the Sun.

Ardipithecus ramidus (*paleoanthropology*)   *Ardipithecus ramidus* was a small bipedal creature halfway between chimpanzees and humans. It roamed woodlands along the banks of the Aramis River in the Afar region of Ethiopia some 4,400,000 years ago. The name *ramidus* (based on the Afar word for "root") reflects its status as the earliest known hominid.

A. *ramidus* is almost as close to the chimpanzee line as it is to the australopithecines. Its teeth were more those of a chimpanzee than those of *Australopithecus afarensis,* a species that appeared in the same region about 800,000 years after *ramidus.* But *ramidus* walked bipedally, not on knuckles like a chimpanzee.

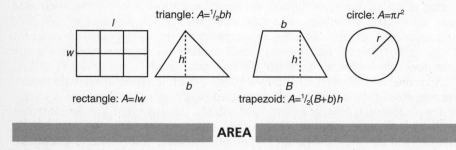

triangle: $A = \frac{1}{2}bh$ circle: $A = \pi r^2$

rectangle: $A = lw$ trapezoid: $A = \frac{1}{2}(B+b)h$

AREA

area (*measurement*)   A number that describes the extent of a surface is called its area. The concept of area is simplest for rectangles with sides that can be measured in integral amounts of a single linear unit. A square with the linear unit used for each side is termed a square unit. The number of square units that will cover the rectangle is the product of the numbers of linear units in 2 adjacent sides. The ancient Greek geometers showed how this product of the length and width can be considered as area even when the numbers are not integral; how areas of triangles relate to rectangles; and how to break any polygon into triangles. Curved figures in the plane, such as circles, can be approximated by polygons as closely as is needed for any practical purpose and curved surfaces can similarly be approximated by parts of polyhedrons.

Argand diagram (*numbers*)   An Argand diagram shows a complex number on a plane, locating the number $x + iy$ at the point (x, y). Jean Robert Argand [Swiss: 1768–1822] published this method in 1806, but the same approach was also described in 1798 in a little-read paper by Casper Wessel [Danish: 1745–1818]. The Argand diagram is also called the Gaussian plane since Gauss described it in 1832; however, he probably independently invented the method much earlier.

argon (*elements*)   An inert gas, atomic number 18, argon was discovered in 1894 by William Ramsay [Scottish: 1852–1916]. The name, from the Greek *argus*, neutral or inactive, is especially appropriate since it is the only element not known to form any compounds. After nitrogen and oxygen, argon is the most common gas in the atmosphere, about 1% of dry air at sea level. Chemical symbol: Ar.

Aristarchus of Samos (*astronomer*)   [Greek: c320–c230 BC] Aristarchus noted that when the Moon is half illuminated, it is at one vertex of a right triangle with Earth and the Sun. He used trigonometry to determine relative distances of the Sun and Moon, but his angle measurements were quite rough, causing him to underestimate the ratio by a factor of 20. He also used the size of Earth 's shadow on the Moon during an eclipse to obtain a nearly correct ratio of the size of the Moon to Earth. These measurements helped Aristarchus to recognize that the Sun is much larger than Earth. Aristarchus proposed that Earth revolves about the Sun and that stars are at a practically infinite distance.

Aristotle (*scientist/philosopher*)   [Greek: 384–322 BC] Influenced by his father, the physician Nicomachus, Aristotle developed an early interest in science. As a student of Plato he formed a love of philosophy and logic. His later association with Alexander the Great expanded his interest in plants and animals. Ancient scholars attributed more than 400 treatises to Aristotle, encompassing all Grecian knowledge about the world at that time. Few treatises have survived, and there is much debate over their authenticity.

In mechanics Aristotle explained the functioning of levers. In chemistry he defined elements as materials that cannot be decomposed into other substances. In biology he classified the 500 known kinds of animals into 8 classes, described development of the chicken embryo, and explained the structure of ruminant stomachs.

arithmetic, elementary (*mathematics*)   Traditionally, elementary arithmetic consists of counting, adding, subtracting, multiplying, and dividing natural numbers and fractions, including fractions expressed in decimal form. Elementary school mathematics usually includes such additional topics as geometry and measurement, ratio and proportion, and even simple probability and algebra. The word *arithmetic* is also used in mathematics to mean number theory, so the adjective *elementary* is needed.

arithmetic sequence and series (*calculus*)   A sequence for which each term is the same distance from the one preceding is called arithmetic (air-rihth-MET-ic—in older books, an arithmetic progression). An arithmetic sequence can be formed from an initial term, a, and a common difference, d, added to each preceding term, giving a sequence $a, a + d, a + 2d, a + 3d, \ldots$. The nth term is $a + (n - 2)d$. For example, if $a = 7$ and $d = 4$, the sequence is 7, 11, 15, 19, 23, ...

An arithmetic series is the indicated sum of an arithmetic sequence, such as $7 + 11 + 15 + 19 + 23 + \ldots$. The sum of an arithmetic series is half the product formed by multiplying the number of terms by the sum of the first and last term, or $S = 1/2\ n[2a + (n - 1)d]$. For the 5-term series $7 + 11 + 15 + 19 + 23$, the sum is $1/2\ 5(7 + 23) = 75$.

armadillo (*animals*)   Named for their armorlike covering, which consists of a series of bony plates within the skin, armadillos ("little armored ones") constitute the mammal Family Dasypodidae. The largest of the approximately 20 species is the giant armadillo (*Priodontes giganteus*), which may be more than 40 in. (100 cm) long and weigh over 120 lb (54 kg). All species are native to the Americas. Armadillos are nocturnal, burrowing creatures that feed mainly on insects. They are surprisingly fast runners and can hold their breath for up to 10 minutes underwater. If attacked, most armadillos roll up into a ball so that only the armored surfaces are exposed.

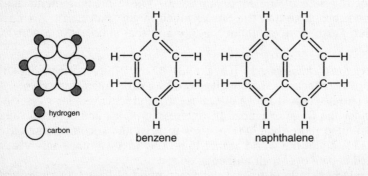

hydrogen
carbon
benzene   naphthalene

AROMATIC

aromatic (*compounds*)   Hydrocarbons that are based on a ring or rings of 6 carbon atoms are called aromatics because many of them have a pronounced smell—naphthalene, the basis of moth balls, is an example. Benzene, the simplest aromatic, was known for 40 years as a major chemical mystery: no one could unravel its structure although it was known to have 6 carbon atoms and 6 hydrogen atoms. In 1865, however, Friedrich Kekulé recognized in a dream that the carbon atoms in benzene link to form a ring with 1 hydrogen atom attached to each carbon atom. Petroleum contains aromatic hydrocarbons, but the chief source has been coal tar derived from coal distillation. Aromatics are important sources of dyes and other industrial chemicals.

array (*algebra*)   The arrangement of entities into rows and columns is a rectangular array. Such arrays of numbers or variables are the fundamental structures of a matrix. Rarely an array takes some other geometric form, such as a triangle. In statistics, a set of numbers arranged by size is also called an array.

Arrhenius, Svante (*chemist*)   [Swedish: 1859–1927] Arrhenius was the first to explain (in 1884) why salty water conducts an electric current but distilled water does not. The salt in solution disassociates into positive ions (called cations) and negative ions (anions). An electrical potential difference makes ions move toward the poles, producing the current. Arrhenius recognized that the same process occurs in all electrolytes (fluids that conduct electricity) and later researchers extended his work to explain ionic crystals.

Arrhenius also contributed to many other fields and is best known today for being the first to recognize (in 1896) the greenhouse effect caused by carbon dioxide in the atmosphere.

arrowworm (*animals*)   Constituting the invertebrate phylum Chaetognatha ("bristle jaw"), arrowworms are slender torpedo-shaped animals 0.5 to 4 in. (1 to 8 cm) long. Using lateral fins and a tail fin to dart around like tiny arrows, they are part of the marine plankton worldwide, though particularly plentiful in shallow tropical waters. Arrowworms are primarily carnivores, with sicklelike bristles around the mouth to capture prey. A fold of tissue, called the hood, can be pulled down over the bristles and mouth to streamline the head while the arrowworm swims.

arsenic (*elements*)   Although the earliest bronze of the Bronze Age was copper alloyed with arsenic, this nonmetal, atomic number 33, was not isolated until 1250, by Albertus Magnus [German: 1193–1280]. Its name is from the Latin *arsenicum* for the yellow compound orpiment. Arsenic is best known as a poison, but some of the most advanced computer chips are based on a compound of arsenic with gallium. Chemical symbol: As.

arthropod (*animals*)   More than 75% of all known living and fossil animals are members of Phylum Arthropoda ("joint feet"), comprising over 1,000,000 species. They include crustaceans, arachnids, centipedes, millipedes, and insects. Biologists believe that another 9,000,000 species of arthropods, especially in tropical rain forests, remain to be discovered and described. Ranging in size from near-microscopic insects to the Japanese spider crab (*Macrocheira kaempferi*), arthropods inhabit almost every imaginable ecological niche, from the deep sea to mountain peaks and the air.

Arthropods have bilateral symmetry and segmented bodies with 3 regions—head, thorax, and abdomen—that may be fused together. At least some of the segments bear paired, jointed appendages, which are modified in various ways to form antennae, mouthparts, legs, and so on. An exoskeleton composed largely of chitin and proteins covers the body and is molted periodically; at joints the exoskeleton is flexible to permit movement.

arum (*plants*)   The arum family, Araceae, comprises some 1800 species, most native to ponds, swamps, and other watery habitats in tropical and subtropical lands. They are characterized by an inflorescence called a spadix. The spadix is a thick fleshy column that bears numerous small flowers. In many species the spadix is enclosed in a leaflike, funnel-shaped pouch. The family includes arums (*Arum*), anthuriums (*Anthurium*), calla lilies (*Zantedeschia*), philodendrons (*Philodendron*), taro (*Colocasia esculenta*), skunk cabbage (*Symplocarpus foetidus*), and jack-in-the-pulpit (*Arisaema triphyllum*). Perhaps the most fascinating species is the krubi (*Amorphophallus titanum*) of Sumatra, which produces an inflorescence as much as 7 ft (2.1 m) tall—the largest known floral cluster.

asbestos (*rocks and minerals*)   Asbestos (Greek for "unquenchable") was originally the trade name used for mineral fibers separated from rock, spun, and processed into sheets or other forms. Asbestos helps prevent fire from spreading, since it does not burn easily, and is useful for insulation or resistance to acids. Asbestos is also the name of several fibrous magnesium-based silicates, including fibrous forms of actinolite, crocidolite, and serpentine (known as fibrous chrysotile) that provide asbestos fibers.

By 1935 physicians had begun to suspect that asbestos exposure causes lung disease, including cancer. After a definite link between asbestos and cancer was established, the use of asbestos was phased out in a process that is ongoing.

Ascaris (*animals*)   The **roundworm** *Ascaris lumbricoides* is a parasite of humans. The adult worms, growing to 12 in. (30 cm) or more in length, live in the small intestine. Females lay thousands of eggs daily, which are excreted in feces. When embryonic worms are swallowed by a new host—perhaps on inadequately washed vegetables grown in soil contaminated with human feces—they pass to the intestine and hatch into larvae. The larvae burrow into blood vessels and travel throughout the body, causing serious illness.

asexual reproduction (*biology*)   Asexual reproduction involves only 1 parent and results in offspring that are genetically identical to that parent. One-celled organisms, fungi, many invertebrates, and plants reproduce asexually, at least during part of their life cycles. The simplest type of asexual reproduction is fission, found among bacteria and other 1-celled organisms. In this process, the parent cell divides in half, producing 2 small daughter cells, which separate and grow to adult size. Fission can quickly result in a large population of descendants. For example, a bacterium that can divide when it is 30 minutes old can give rise—in favorable environmental conditions—to billions of bacteria in less than a day. Other types of asexual reproduction include budding, fragmentation, spore formation, and vegetative propagation.

ash (*plants*)   Ashes (*Fraxinus*) are deciduous shrubs and trees native mostly to northern temperate regions. Species typically have opposite, compound leaves and inconspicuous flowers borne in dense clusters called panicles. Well-known species include the white ash (*F. americana*) of eastern North America and the European ash (*F. excelsior*) of Europe and western Asia, both of which can attain heights of 120 ft (37 m). The flowering ashes, such as the manna ash (*F. ornus*) of Europe and western Asia, have showy, fragrant flowers.

ash, volcanic (*geology*)   Gases that escape from an erupting volcano often carry bits of rock, molten rock, and even very large rocks. A common emission consists of rock pieces too large to be called dust but smaller than gravel, usually less than about 0.25 in. (0.6 cm) in diameter. Such an emission is called "ash" from its resemblance to the residue left after burning wood or coal, although volcanic ash is not itself produced by burning. Ash from an active volcano can be a nuisance hundreds of miles from the eruption, but its great weight (about 3 times that of water) can collapse buildings nearer the volcano. Massive deposits of ash may bury buildings, as in the famous eruption of Vesuvius in AD 79 that buried Pompeii.

asparagus (*plants*)   Native to warm regions of Europe, Africa, and North Africa, the genus *Asparagus* includes perennial herbs, climbers, and shrubs. All have fernlike foliage that consists of small scalelike structures that are not true leaves. Common asparagus (*A. officinalis*) is cultivated for its edible upright shoots, which arise from horizontal stems called rhizomes. The shoots are picked when they are about 8 in. (20 cm) tall. Unpicked, they develop into branching stems up to 4 ft (1.2 m) or more in height that bear tiny flowers and berry fruits. The houseplant called asparagus fern (*A. setaceus*) is related.

asphalt (*rocks and minerals*)   Asphalt is a naturally occurring residue of petroleum, found in large deposits called asphalt lakes or in rock as an embedded mineral. Essentially the same material occurs as a residue of petroleum distillation. The solid portion may be called bitumen. Tar and pitch, similar in nature to asphalt, are fluids produced by distilling coal, wood, or petroleum (natural conifer resins are also called pitch). Asphalt is familiar as the blacktop of highway construction; it is also used in shingles and roofing materials. Natural asphalt or bitumen was used as mortar in Mesopotamia some 5000 years ago and as part of the embalming process for Egyptian mummies.

ass (*animals*)   Asses, hoofed mammals native to Africa and Asia, resemble and can interbreed with their close relatives, horses. The largest species is the kiang (*Equus kiang*) of Tibet, which may be 4 ft (1.2 m) tall at the shoulders. The Egyptians are believed to have domesticated the African wild ass (*E. asinus*) about 4000 BC. Breeding a male domesticated ass, or donkey, with a female horse produces a sterile animal called a mule (a cross between a female donkey and male horse is a hinny). Mules combine some of the best traits of both parents; they are exceptionally strong and surefooted, have tremendous stamina, and are long-lived.

associative property (*algebra*)   A binary operation ($\bullet$) on a set of elements is associative if the result of sequential operations does not depend on how elements are paired: $(a \bullet b) \bullet c = a \bullet (b \bullet c)$. For example, addition and multiplication of numbers exhibit the associative property, as demonstrated by $(12 + 2) + 3 = 12 + (2 + 3) = 17$ or $(12 \times 2) \times 3 = 12 \times (2 \times 3) = 72$. Subtraction and division are not associative: $(12 - 2) - 3 = 7$ while $12 - (2 - 3) = 13$, and $(12 \div 2) \div 3 = 2$ but $12 \div (2 \div 3) = 18$. One of the main requirements for a group is that the operation be associative.

astatine (*elements*)   Astatine is a radioactive halogen, atomic number 85, discovered in 1940 by Emilio Segrè and coworkers. Its name comes from the Greek *astatos*, "unstable," since its longest lived isotope has a half-life of about 8 hours. Despite this, about 1 oz (35 g) may exist naturally on Earth, since astatine is in the decay chain of uranium and thorium. Chemical symbol: At.

asteroid (*solar system*)   Many rocky or metallic bodies orbit the Sun. Those that are from dozens of feet to hundreds of miles in diameter, but much smaller than planets, are asteroids, or minor planets. *Asteroid* means "starlike." Although asteroids are not much like stars, they appear as points of light through a telescope, as stars do, not as disks, as planets do. Most of the nearly 20,000 known asteroids travel in a wide belt between Mars and Jupiter, but some are found near Earth or beyond Jupiter. About 5000 asteroids have had precise orbits determined. These are designated with a number, signifying the order in which the orbits were calculated, and a name, as in 1 Ceres or 433 Eros. Theory suggests that there are about a million asteroids with a diameter greater than 0.6 mile (1 km) yet to be discovered.

Ceres was the first asteroid to be discovered—in 1801 by Giuseppe Piazzi [Italian: 1746–1826], who considered it to be a small planet. Gauss calculated the orbit. Ceres is still the largest asteroid known, with a diameter approaching 600 miles (1000 km).

asthenosphere (*geology*)   The plates that make up Earth's lithosphere move slowly with respect to each other and with respect to the mantle. Geologists think that this is possible because there is a partly fluid layer called the asthenosphere ("sphere of weakness") beneath the plates. This layer is not directly beneath the crust, as once supposed, but near the top of the mantle some 45 mi (70 km) to 160 mi (260 km) below the surface; thus roots of the plates extend to the base of the lithosphere.

astronomical unit (*measurement*)   The average distance from Earth to the Sun, 92,943,721.11 mi (149,578,706.91 km) is the astronomical unit (A.U.), a convenient measure of distances within the solar system, which extends approximately 100,000 A.U. from one edge of the Oort cloud to the other. Pluto at its farthest point is slightly less than 50 A.U. from the Sun.

astronomy (*sciences*)   Astronomy, which began with the naked-eye study of the Sun, Moon, and stars, is the oldest science. Today astronomers study the universe in many ways. The astronomical telescope of 1609 marked the beginning of modern astronomy. Optical telescopes collect visible light, but some modern "telescopes" collect gamma rays, X rays, infrared and ultraviolet light, radio waves and microwaves, neutrinos, and cosmic rays. Astronomers also study meteorites for clues to the solar system. Space probes from Earth have reported on our Moon, other planets in the solar system, asteroids and comets, and the Sun. A few such probes are traveling out of the solar system, although still far from the region around even the nearest stars.

astrophysics (*sciences*)   Astrophysics ("star physics") began as the analysis of light from stars using the spectroscope. Later studies added the nuclear fusion processes that produce the energy that stars radiate. The study of fusion led to the analysis of the life cycle of stars. Astrophysics today goes beyond stars to encompass such physical phenomena as neutron stars, black holes, and gamma-ray bursts. The physics of collections of stars, such as binary stars, clusters of stars, galaxies, and clusters of galaxies is sometimes included in astrophysics, but the study of the origin and evolution of the universe is called cosmology.

asymptote (*analytic geometry*)   A line that in one direction approaches closer and closer to a curve, but that does not meet that curve, is an asymptote. For example, the exponential function $y = 2^x$ has the x axis as an asymptote (it is asymptotic to the negative x axis); a hyperbola such as $x^2 - y^2 = 1$ closes in on the lines $y = x$ and $y = -x$, asymptotes to both branches; and the tangent curve is asymptotic to the line $x = \pi/2$ and to an infinity of related parallel lines.

atavism (*genetics*)   The reappearance of a trait possessed by an ancestor after an absence of several generations is termed atavism. Usually a recessive gene causes atavism. The individual possessing the characteristic may be said to be a throwback to an earlier time.

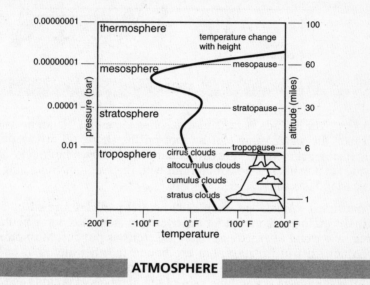

atmosphere (*atmospherics*)   The atmosphere consists of gases called air that form a sphere around Earth (*atmo-* means "vapor" or "gas"). The atmosphere originated from volcanoes and objects from space, but living organisms completely reworked the air over time. Gases compress, so lower layers are much denser than upper ones. The mass of the lower 2 mi (5 km) equals that of everything above, which extends hundreds of miles.

Layers differ by temperature. In the lowest 6 mi (10 km) of air, the troposphere, temperatures fall relatively quickly with height. At the top of the troposphere, or tropopause, temperatures begin to rise, marking the stratosphere. The upper stratosphere is enriched with ozone produced by ultraviolet radiation. The next layers are the mesosphere and thermosphere. The thermosphere is almost empty space—low altitude Earth satellites, the space shuttle, and space station all stay within the thermosphere. The thermosphere is named for its fast-moving molecules, but there are so few molecules that a thermometer shaded from the sun would register a temperature near absolute zero.

atoll (*geology*)   Tropical seas, especially the Pacific, contain rings of low islands of coral called atolls (Bikini and Eniwetok are familiar examples). Coral animals cannot grow in waters deeper than about 180 ft (50 m) because they depend on symbiosis with photosynthesizing algae. Since atolls are found in deep water, their origin was a long-time mystery. Darwin visited several atolls in 1836. He also observed volcanic islands fringed with coral reefs as well as islands with coral barrier reefs offshore. In 1837 Darwin proposed the correct

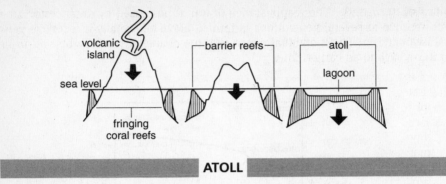

volcanic island

sea level

fringing coral reefs

barrier reefs

atoll

lagoon

ATOLL

explanation. The 3 kinds of reefs form a series. A central island, he said, "gradually sinks, the water gains foot by foot on the shore, till at last the highest peak is finally submerged." Erosion continually removes both land and coral, but living coral maintains an equilibrium with the sea, remaining long after the volcanic island has sunk.

atom (*particles*)   Atoms are electrically neutral particles formed by a nucleus of at least 1 proton and, except for the hydrogen atom, 1 or more neutrons, surrounded by a cloud of exactly as many electrons as protons. Most atoms, however, do not exist in isolation, but are bound to other atoms in one way or another. Each atom is identified by two numbers, the number of protons (called the **atomic number**) and the sum of the numbers of protons and neutrons (approximately the **atomic mass**). These two numbers are conventionally indicated by chemists as a superscript before (atomic mass) and a subscript before (atomic number) the chemical symbol, as in $^{14}_{6}C$, or by physicists with a superscript before (atomic number) and one after (atomic mass), as in $^{6}C^{14}$. Each configuration stands for carbon of atomic mass 14 and an atomic number of 6. Since all atoms with the same atomic number are considered to be a single element, another way to name atoms of a specific element consists of hyphenating the atomic mass number after the name of the element, as in carbon-14.

The idea of the atom goes back to ancient Greek philosophers, who thought that matter is composed of tiny indivisible particles. The concept was put on a scientific basis by Dalton in 1803 and became the foundation of chemistry. In 1899 J.J. Thomson demonstrated that atoms are *not* indivisible after all but contain electrons that can be separated from a nucleus. Atomic nuclei lacking 1 or more electrons or with extra electrons are called **ions**. Like all particles, atoms can also be understood as waves of matter.

atomic clock (*particles*)   A number of different devices that keep accurate time with high precision are known as atomic clocks. Modern physics can be said to have begun when Galileo used his own pulse to time pendulum swings in 1581. Today's physics is built on the measurement of time to tiny parts of a second with atomic clocks. The first such clocks measured the natural vibration of a nitrogen atom in ammonia, and so were properly atomic clocks. More recent clocks rely on measuring the frequency of electromagnetic radiation emitted from or absorbed by specific atoms. The fundamental basis of time in physics today is the frequency with which cesium-133 absorbs or emits radiation; 9,192,631,770 cycles of that wave is taken as the amount of 1 second, defining the second independently of Earth's movements.

atomic mass (*physics*)   The combined number of protons and neutrons in an atom is approximately its atomic mass. This combined number, which identifies the isotope, is hyphenated after the element name, as in carbon-12, or is a superscript to the symbol (C^{12}). Some of the mass of individual nuclear particles is replaced by energy when the particles combine to form the nucleus—the same energy that is released during nuclear fusion. The actual atomic mass is based on a small measure (1.660×10^{-27} kg), exactly 1/12 of the mass of carbon-12, the atomic mass unit (amu). The most common isotope of oxygen, oxygen-16, has an atomic mass of 15.99491 amu, while an atom with a larger nucleus, such as iron-56, loses more energy to fusion, having an atomic mass of 55.9349 amu. Most elements are mixtures of several different isotopes and the number reported for the element, often called the atomic weight, is the average atomic mass for the isotopes. Only 91.66% of natural iron is iron-56, for example, while the rest is iron-54, iron-57, and iron-58, giving an atomic weight of 55.847 for the naturally occurring element.

atomic number (*physics*)   The number of protons in the nucleus of an atom or ion is its atomic number. For a nonionized atom, the atomic number is also the number of electrons present. Each element is identified by its atomic number, often written as a subscript or superscript in front of the element's chemical symbol. Thus, potassium with 19 protons is indicated as $_{19}K$ or ^{19}K

ATP (*biochemistry*)   Adenosine triphosphate (ATP) is the principal source of energy in living cells. When ATP is formed, energy is stored in its molecular bonds. When these bonds are broken, energy is released.

An ATP molecule is made up of the nitrogen base adenine, the sugar ribose, and 3 phosphate groups. High-energy, relatively weak bonds link together the phosphate groups. When 1 of these bonds is broken, a phosphate group is removed, energy is released, and ADP (adenosine diphosphate) is formed. Removal of a second phosphate group releases more energy and creates AMP (adenosine monophosphate). Usually the last phosphate group is transferred to another molecule, in a process called phosphorylation, instead of being removed. This process transfers some energy to the phosphorylated compound for use in a later reaction.

Audubon, John James (*naturalist/painter*)   [American: 1785–1851] Before Audubon, painters of birds typically used stuffed birds as subjects and created stylized pictures. Audubon drew intricately detailed pictures of birds in their natural habitats. His ambitious 4-volume *Birds of America* was published in London between 1827 and 1838. Audubon also was one of the earliest proponents of conservation.

auk (*animals*)   The auk family, Alcidae, includes auks, guillemots, murres, murrelets, and puffins. All are seabirds of the Arctic and sub-Arctic, with a large head, stout body, short wings, webbed feet, and short tail. The plumage usually is black on top and white on the underparts. Adroit swimmers and divers, alcids feed mainly on fish, which they capture with the beak. Alcids are clumsy on land and go ashore only in spring to breed, forming large colonies on cliffs and rocky islands. The largest living alcid is the razorbill (*Alca toda*), up to 18 in. (45 cm) long.

Aurignacian (*paleoanthropology*)   Between 35,000 and 30,000 years ago, the Aurignacian tool tradition of the Cro Magnon peoples appeared in western and southern Europe. It was characterized by points made of bone and stone, blades, and burins. The culture was heavily dependent on late Ice Age herds of reindeer for food, bone, and horn.

aurochs (*paleontology*)   An ancestor of modern domestic cattle, the aurochs (*Bos primigenius)* was a giant ox that lived in Europe, North Africa, and Asia beginning about 1 million years ago. It was first domesticated around 6000 BC in the Near East. The last known aurochs died in 1627. The bull weighed over 2200 lb (1000 kg) and had a shoulder height that exceeded 7 ft (2.1 m); the female was smaller.

aurora (*atmospherics*)   The night sky in regions near Earth's poles is often filled with bright colors that form moving curtains of light across the sky. These are the aurora borealis (named after gods of dawn and the north wind) and the aurora australis, or southern aurora. The auroras are also called the northern and southern lights. Frequent at high latitudes, they sometimes extend as far as 40° from the equator.

The auroras are caused by charged particles of the solar wind interacting with the upper atmosphere. Auroras occur near the poles because Earth's magnetic field emerges from the north and south magnetic poles. This magnetosphere deflects most particles but captures some that then spiral along magnetic field lines toward the poles. Where the solar particles slam into air particles, some of the energy produced is released as light, creating the colors of the auroras. Variations in the strength of the solar wind produce the movement of the curtains of light. Solar flares, which release powerful blasts of the solar wind, overpower the magnetic field, and push the auroras down to lower latitudes.

Australian Plate (*geology*)   Until recently the Australian Plate, a large region of Earth's lithosphere surrounding Australia, was thought to be the eastern portion of a larger Indian-Australian Plate, but seismic studies have shown a partial fracture between the Indian Plate and the Australian. Until about 60,000,000 years ago, the Australian Plate was joined to the Antarctic Plate instead of the Indian, but the Australian Plate, whose northern boundary includes New Guinea, has pushed its way into the Pacific Plate and almost to the Philippine Plate.

australopithecine (*paleoanthropology*)   Australopithecines were a diverse group of hominids that evolved in Africa; they are thought to be the immediate ancestors of the first human species. Unlike their presumed ancestor, *Ardipithecus*, australopithecines were more like humans than like chimpanzees, although the australopithecine brain was much smaller for body size than the human brain and more within the range of larger chimpanzee brains. From their anatomy and from preserved footprints it is clear that they were bipedal. Aside from being smaller, australopithecine anatomy below the neck is essentially human. Above the neck, not only brain size but also teeth are similar to those of chimpanzees, reflecting a common vegetarian diet. Like modern chimpanzees, australopithecines probably made and used simple tools. Although walking upright, australopithecines were more adapted physically to climbing trees than humans are. Like modern gorillas (but unlike chimpanzees), male australopithecines at about 100 lb (45 kg) were much larger than females, who weighed about 66 lb (30 kg).

Australopithecus afarensis (*paleoanthropology*) *Australopithecus afarensis* is an early australopithecine species from shortly before 3,000,000 years ago. It is famous for the fossil known as Lucy, about 40% of the skeleton of a small woman who lived in Ethiopia about 3,180,000 years ago. In addition to Lucy and other fossils from Ethiopia, a trail of footprints left by 3 members of *A. afarensis* in volcanic ash at Laetoli, Tanzania, was discovered in 1978. These footprints demonstrated that *A. afarensis* walked upright like modern humans, unlike any of the great apes.

Australopithecus africanus (*paleoanthropology*) *Australopithecus africanus* is a small australopithecine that was the first nonhuman hominid to be discovered. It was found throughout much of southern and eastern Africa, with fossils dating from between 3,000,000 and 2,500,000 years ago. Although recognized as a hominid by Raymond Dart [Australian-South African: 1893–1988] in 1924, very few anthropologists accepted Dart's classification until the second half of the 1900s. Many thought Dart's proposed species was simply an early ape. But *A. africanus* walked upright and was large-brained for an ape, although much smaller brained than members of *Homo* of the same general size. *A. africanus* is thought by many to be the immediate ancestor of humans.

Australopithecus anamensis (*paleoanthropology*) Some 21 fossils of a human ancestor from about 4,200,000 to 3,900,000 years ago were found at a site near Lake Turkana in Northern Kenya in 1994. They have been identified as a previously unknown bipedal hominid now called *Australopithecus anamensis.*

Australopithecus bahrelghazalia (*paleoanthropology*) The first hominid remains from central Africa, discovered in 1995, have been named *Australopithecus bahrelghazalia*. This australopithecine flourished some 3,000,000 years ago in what is now Chad. The specific name *bahrelghazalia* refers to the province where the fossil jaw was found.

autotroph (*biology*) Autotrophs are organisms that synthesize their own energy-containing organic food from inorganic substances. Most autotrophs, including chlorophyll-containing protists and plants, make sugar from carbon dioxide and water, using the sun's light energy in a process called photosynthesis. Certain autotrophic bacteria derive energy not from the Sun but from simple mineral substances. For example, methanogenic bacteria that live in the absence of oxygen (in habitats such as marshes and rice paddies) produce methane and water from carbon dioxide and water; in the process, energy needed to carry out life processes is released.

auxin (*biochemistry*) One of the main classes of plant growth regulators, auxins influence many aspects of plant growth, including root formation, development of vascular tissue, and certain responses (tropisms) to external stimuli. Auxins are produced mainly at the growing tip of a shoot, or stem, and then transported to other parts of the plant.

Some effects result from the combined action of auxins and other growth regulators. For example, auxins and gibberellins work together to elongate stems. A combination of auxins and cytokinins results in much faster cell division than with either chemical working alone.

avalanche (*geology*) The sudden flow of snow, often bringing ice and rock with it, down the side of a mountain is an avalanche. Snow on steep slopes is lightly held. Even a small disturbance, such as a loud sound or melting at the base, can set snow sliding downhill, picking up more snow and accelerating as it travels. Soon the avalanche gains great power and size, knocking down trees and structures and burying any villages in its path.

average (*statistics*) An average can be any measure of central tendency, but commonly the arithmetic mean, usually called the mean, is used as the average. The mean of n numbers is the sum of the numbers divided by n. An average may also be weighted by assigning a factor called a weight to each of the n numbers before dividing by n. For example, the mean of scores 3, 5, 7, 8, 8, and 9 is 40/6 = $6^2/_3$. If the scores less than 6 are weighted by a factor of 1.2, the weighted average would be (3.6 + 6 + 7 + 8 + 8 + 9)/6 or 41.6/6 = 6 14/15.

Avery, Oswald T. (*bacteriologist*) [American: 1877–1955] The genetic revolution can be said to have started with the 1944 publication of a paper by Avery and his colleagues in which they proved that DNA—not protein, as many believed at the time—is the agent of heredity. Their work built on that of Frederick Griffith [English: 1881–1941], who discovered that a "transforming substance" can be passed from one bacterium to another and cause a permanent change in the latter. Working with the same type of bacteria, Avery and his coworkers purified and tested different bacterial chemicals, eliminating all except DNA as the transforming, or genetic, material.

avocado (*plants*) A member of the laurel family, the avocado or alligator pear (*Persea americana*) is native to the American tropics but is widely cultivated for its edible fruit. The avocado is an evergreen tree that grows to heights of about 60 ft (18 m). The large pear-shaped fruit is fleshy and contains a single, inedible seed.

avocet (*animals*) The wading birds that make up Family Recurvirostridae, found in temperate and tropical regions worldwide, include avocets, stilts, and the ibisbill. All are characterized by a long, slender, upward-curving bill and by extremely long legs. They feed by sweeping the bill from side to side in shallow water, stirring up crustaceans, aquatic insects, seeds, and other food. They range in length from 11 to 20 in. (28 to 50 cm).

Avogadro, Amedeo (*chemist*) [Italian: 1776–1856] In 1811 Avogadro proposed that equal volumes of gas at the same temperature and pressure contain the same number of molecules. Avogadro himself coined the word "molecule" to mean the smallest part of a compound. On the basis of his law Avogadro became the first to show that water is H_2O—that is, composed of molecules of 2 atoms of hydrogen and 1 of oxygen. The related concept, called Avogadro's number, is that when the mass of a compound in grams is equal to the molecular weight, the total number of molecules is always the same, equal to 1 mole.

axiom (*logic*)   In modern logic a statement accepted as true without proof and used for structuring a body of knowledge is an axiom or, synonymously, a postulate. In older usage axiom was reserved for statements accepted as true for all entities, such as "2 things each equal to a 3rd are equal to each other." Thus axioms were generally accepted rules of reasoning while postulates applied to specific subject matter.

axiomatic system (*logic*)   An axiomatic system is the arrangement of a body of knowledge into a logical structure, in which each proposition is proved on the basis of a limited number of axioms or postulates about a few undefined terms and in which all other terms are defined on the basis of those undefined ones. The model axiomatic system is Euclid's *Elements*, in which 13 books of theorems about geometry and numbers derive from 5 postulates about points, lines, circles, and angles, along with 4 axioms about equality and 1 axiom stating that "the whole is greater than the part." Euclid attempted to define *point* and *line*, but modern axiomatic systems leave these terms undefined. Two recent examples of axiomatic systems are Giuseppe Peano's number structure, based on 5 postulates with 2 undefined terms, and axiomatic set theory, based on 8 axioms with 3 undefined terms.

axis (*geometry*)   An axis is a line with a special location or a special purpose.

An axis of symmetry for a curve or other figure in a plane is a line for which the part on one side of the line is the mirror image of the part on the other. The conic sections each have axes of symmetry; the parabola has 1, the ellipse and hyperbola each have 2, and the circle has an infinite number. When a 2-dimensional figure, such as a triangle, is rotated in 3-dimensional space about an axis of symmetry (forming a cone from a triangle), the line becomes an axis of revolution.

In coordinate geometry, axes are usually mutually perpendicular lines used to locate points—for 2 dimensions, the x and y axes, with a z axis added for 3 dimensions. Complex numbers are located using a horizontal real axis and a vertical imaginary axis.

aye-aye (*animals*)   The sole species in the primate family Daubentoniidae, the aye-aye (*Daubentonia madagascariensis*) is about 16 in. (40 cm) long excluding its bushy tail, which may be 24 in. (60 cm) long. The aye-aye has long wooly fur except on the ears, which are hairless. The fingers are long and thin, especially the middle one. All digits bear claws except the big toes, which have nails. Aye-ayes live in forests and thickets on Madagascar. They are nocturnal and feed on fruits, eggs, bamboo shoots, and wood-boring insect larvae, locating the latter by tapping on wood with the long middle finger and listening for movement.

Babbage, Charles (*mathematician*)   [English: 1791–1871] Babbage designed and partially built the first mechanical computers. In 1832 he built a demonstration model of his first advanced calculator, the Difference Engine, designed to compute logarithms and other functions. This model works, and Babbage's plans have been used to create fully functioning versions. Babbage also designed a device he called the Analytical Engine. This was supposed to use punched cards as input for problem-solving programs and have the equivalent of memory and a central processing system, but it was never built. Babbage was the first to use mathematics to study a complex system (the English postal system); his ideas led to the flat-rate postage stamp. He was also an inventor credited with the first ophthalmoscope, speedometer, skeleton key, and cowcatcher for locomotives.

baboon (*animals*)   Large, intelligent African monkeys that live in groups called troops, baboons inhabit much of the nonforested areas south of the Sahara Desert. Omnivores, they eat a wide variety of foods, storing it temporarily in large cheek pouches. Baboons have heavy brow ridges over the eyes, strong limbs, and a prominent bare pad on each buttock. The largest species, the chacma (*Papio ursinus*), weighs up to 90 lb (41 kg). The mandrill (*Papio sphinx*), a less gregarious forest-dwelling baboon, is particularly colorful, with a red nose surrounded by light blue ridges and buttock pads that are bright red or light blue.

bacillus (*monera*)   Rod-shaped bacteria are called bacilli. They occur singly or in chains, and may have cigar shapes or pointed ends. Many are decomposers, including the nitrogen-fixing bacteria of genus *Rhizobium,* and *Lactobacillus acidophilus*, which sours milk. Others are disease-causing parasites, including *Mycobacterium tuberculosis*, which causes tuberculosis, and *Agrobacterium tumefaciens*, which causes crown gall in plants.

backbone (*anatomy*)   There is, of course, no single backbone in any vertebrate, or "animal with a backbone." Instead there are a number of small bones, called vertebrae, linked together by ligaments and separated by disks of cartilage. This backbone is also known as the vertebral column or spinal column. It surrounds and protects the spinal cord.

background radiation (*geology*)   Existing radioactivity at any location is called background radiation. After radioactivity was discovered in 1896, physicists soon recognized that Earth itself is slightly radioactive—mostly as a result of thorium and uranium and radioactive isotopes like potassium-40, which are widely distributed in the crust. Earth's radioactivity produces most of the background radiation at a given point, which must be subtracted in most scientific work involving radiation measurement. The same radioactivity is thought to account for most of Earth's inner heat. A small contribution to background radiation comes from cosmic rays.

Bacon, Francis (*philosopher*)   [English: 1561–1626] Bacon's main influence on science was his emphasis on developing general laws that apply to observations, laws formed by thoughtfully collecting as many instances as possible and abstracting the laws in a process called induction. As a scientist himself, Bacon was nearly correct in explaining heat, although he reversed cause and effect by suggesting that the expansion of heated materials causes movements of their particles. Despite his advocacy of experiment, Bacon performed few experiments himself. However, one indirectly led to his death—he caught cold trying to find out if snow could preserve food.

Bacon, Roger (*philosopher/alchemist*)   [English: c1214–92] Bacon compiled a great deal of scientific information in one of the first attempts at a general encyclopedia, including the first account of spectacles for improved vision and a description of gunpowder. He explained to a new audience such ancient Greek discoveries as the roundness of Earth and the great size of the universe (although Bacon, like the Greeks, greatly underestimated it). Bacon proposed that knowledge depends on experimentation and mathematics, even asking the Pope to use his influence to encourage experiments.

bacteria (*monera*)   First seen by Leeuwenhoek around 1675, bacteria are tiny 1-celled prokaryotes classified in Kingdom Monera. There are 3 main bacterial shapes: coccus (spherical or ovoid), bacillus (rod-shaped), and spirillum (spiral). Numerous variations in these shapes occur; for example, some bacilli have pointed ends. Also, certain bacteria have structural features such as protective layers of slime, stalks to attach to the substrate, and whiplike flagella for movement. Members of certain species link together in pairs, chains, or clusters. In unfavorable conditions, some species form endospores and become dormant.

More than 2000 species of bacteria have been identified. They seldom exceed 0.0002 in. (0.005 mm) in their greatest dimension, and until recently it was believed that all bacteria are microscopic. But in 1993 it was confirmed that *Epulopiscium fishelsoni*, discovered in the intestine of surgeonfish, is a bacterium visible to the unaided eye. Even bigger is *Thiomargarita namibiensis*, discovered in 1997 in sediment obtained off the coast of Namibia; it reaches a diameter of up to 0.03 in. (0.76 mm).

Bacteria live in every habitat on Earth, including rock hundreds of feet below the surface. Aquatic species are important food sources at the base of many food chains. Soil decomposers and nitrogen fixers recycle nutrients needed by plants. Many bacteria flourish in other organisms, some beneficially or harmlessly, others as agents of disease. In addition, people use bacteria to make foods, particularly yogurt and other dairy products, and medicines, including antibiotics and hormones.

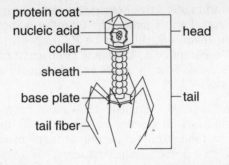

protein coat
nucleic acid
collar
sheath
base plate
tail fiber

head

tail

BACTERIOPHAGE

bacteriophage (*viruses*) Any virus that attacks bacteria is known as a bacteriophage, or phage. Bacteriophages vary greatly in size and shape, and there are both DNA and RNA types. Among the most widely studied are the T phages—DNA viruses that infect the common bacterium *Escherichia coli*. A T phage has a "head" in the shape of a 20-sided polyhedron, a tubular "tail," and leglike "tail fibers" that are used to attach the phage to specific receptor sites on the surface of a bacterial cell. Following attachment, the tail touches down onto the cell and then contracts, thereby injecting the phage's DNA into the bacterium.

badger (*animals*) Members of the weasel family, badgers have a small head, broad body, short legs, strong clawed toes adapted to digging, and scent glands. They have dense fur with distinctive markings; the European badger (*Meles meles*), for example, has a striking black and white face. Badgers are mostly nocturnal, walking quietly with their nose close to the ground, sniffing for worms and other food. Most badgers are Eurasian. The American badger (*Taxidea taxus*) is native to western North America.

Baekeland, Leo Hendrik (*chemist*) [Belgian-American: 1863–1944] In the 1890s Baekeland came to the United States, where he invented a paper used for printing photographs. After selling his paper business to Eastman Kodak, Baekeland invented and manufactured the first plastic that became solid and insoluble with the application of heat (previous plastics burned easily); he named this plastic Bakelite.

Baer, Karl Ernst von (*embryologist*) [Estonian: 1792–1876] Famed for his discovery of the mammalian egg cell, Baer also discovered the membranes around the embryo and described their function. He formulated basic principles of embryonic development.

The purpose of early development of a zygote (fertilized egg) is to form 3 layers of tissue, which then form all the various organs of the body. An embryo develops from simple to complex; general characteristics appear before special characteristics. In their early stages, young embryos of various species resemble one another, but as each species develops it moves progressively away from this resemblance.

Baeyer, Adolf von (*chemist*)   [German: 1835–1917] Baeyer synthesized his first chemical at age 12 and then bought himself a lump of indigo dye for his 13th birthday, but he was nearly 30 before becoming the first to analyze the dye and synthesize it. Baeyer's other famous achievement was the 1863 creation of the first barbiturate (named for his then girlfriend Barbara). Baeyer was the preeminent organic chemist of his day, working with many compounds and developing a still effective theory of chemical stability.

Baily's beads (*solar system*)   During an eclipse of the Sun, both before and after the Sun passes behind the Moon a number of bright spots appear along the edge of the Moon's dark disk. These are Baily's beads, caused by sunlight passing through valleys between the mountains of the Moon. Astronomer Francis Baily [English: 1774–1844] first observed the beads in 1836.

bald cypress (*plants*)   The 3 species of bald cypresses (*Taxodium*) are distant relatives of true cypresses. They are deciduous conifer trees, reaching heights of 120 ft (35 m) or more, native to wet, swampy habitats in Mexico and the southeastern United States. Their roots develop cone-shaped protuberances called knees that rise above the water. It is theorized that the knees, which consist of light, spongy wood, obtain air for underwater parts of the tree. Bald cypresses that grow in dry locations seldom develop knees.

The bald cypress family, Taxodiaceae, comprises about 15 species—remnants of a much larger group that was widespread in the Northern Hemisphere some 60,000,000 years ago. The family includes redwoods and sequoias.

baleen (*anatomy*)   Hanging down in fringed sheets from the upper jaws of certain whales, baleen is a horny outgrowth of the whale's skin. (It is sometimes called whalebone, though it isn't bone.) A whale uses its baleen to strain tiny animals, usually krill, from seawater. The whale engulfs a huge mouthful of water, then closes its mouth and pushes its tongue against the roof of the mouth. This forces water out of the mouth between the baleen sheets, leaving behind the food organisms. The whale then swallows its food.

ball (*geometry*)   Two closely related objects are considered balls in mathematics. An open ball has no skin; it is the set of all points in 3-dimensional space less than a specified distance from a single point. A closed ball includes its own skin; it is the set of all points in 3-dimensional space that are less than or equal to a specified distance from a single point.

The common term for a perfectly round 3-dimensional object is sphere. A sphere is the set of all points in 3-dimensional space equidistant from a single point—that is, the skin of a ball.

ball lightning (*meteorology*)   Although not accepted by science until recently, observers have long reported that lightning sometimes causes glowing spheres that appear both outside and inside buildings. This phenomenon is called ball lightning. A typical appearance occurred in 1938 when a pilot of an early airliner opened a window to get a better view in a rainstorm. A ball of lightning entered the cockpit, singed his eyebrows, rolled into the passenger cabin, and exploded. A recent theory is that the balls are glowing silicon fibers induced by a conventional lightning strike on the ground.

ballooning (*ethology*)   In certain species of spiders and moths, the young scatter in various directions by a method called ballooning. They crawl to some height—the tops of trees, for example—where they spin a long thread of silk or in some other way increase body exposure until they are dislodged and carried away by air currents. Ballooning can carry the young animals for long distances; dispersal distances of 20 mi (32 km) have been measured for newly hatched *Lymantria dispar* caterpillars, explaining why these pests are called gypsy moths.

balsa (*plants*)   Native to the American tropics, the balsa (*Ochroma lagopus*) is a tall, fast-growing evergreen tree known for its strong, extremely light wood. Its bell-shaped flowers give rise to long pods that contain numerous tiny seeds surrounded by brown hairs.

bamboo (*plants*)   Members of the grass family native to warm regions around the world, bamboos are extremely diverse in color, shape, and size; they range from a few feet to 120 ft (37 m) tall. All have a woody culm, or stem, which in most species is hollow and divided by walled nodes. Bamboos are rapid growers; however, they generally flower and produce seeds only at intervals of 12 to 120 years. After flowering, the culms die; underground stems, called rhizomes, live on to produce new growth. It is common for an entire forest of bamboo to flower simultaneously and then die back in the same season. This can cause ecological havoc; pandas, for example, which feed exclusively on bamboo, have died of starvation when their forests flowered and died back.

banana (*plants*)   Among Earth's largest plants without a woody stem are the banana "trees" of the genus *Musa*, of which there are many species and hybrids. Like palms, bananas have trunks that actually consist of the bases of encircling leaves pressed closely together. Native to tropical Asia, bananas grow to 30 ft (9 m) tall and have long, wide leaves. The elongated fruits of some species are edible.

band (*anthropology*)   For humans or other primates band is the level of organization above family. A band usually has several families living in a specific territory that they recognize as belonging to the band but not in enough proximity to be called a village. Among humans only hunter-gatherers form bands.

bandicoot (*animals*)   Constituting the family Peramelidae, bandicoots ("pig rats") are marsupials native to Australia. They have a pointed muzzle for rooting in soil and front feet adapted for digging. The hind limbs are comparatively long. Bandicoots average 10 to 20 in. (25 to 50 cm) in length, depending on the species. Mostly nocturnal, they feed on insects and plant matter.

Banks, Sir Joseph (*botanist*)   [English: 1743–1820] Interested in plants from childhood, Banks helped make botany an academic discipline, developed an extensive botanical collection (donated to the British Museum), and helped establish Kew Gardens in London. He undertook several major collecting trips, the most famous of which was his around-the-world voyage aboard the *Endeavour* on the 1768–71 expedition led by James Cook [English: 1728–79]. This expedition made known to Europeans such wonders as the marsupial animals of Australia.

Banting, Frederick (*biochemist*) [Canadian: 1891–1941] The hormone insulin, which controls blood sugar levels, was first extracted from the pancreas by Banting and Charles H. Best [Canadian: 1899–1978] in 1921, using a technique conceived by Banting. In 1922 a 14-year-old boy suffering from type I diabetes mellitus (a disease resulting from insufficient insulin) showed rapid improvement after becoming the first person to be treated with insulin. Banting and the head of the laboratory in which he and Best worked, John Macleod [Canadian: 1876–1935], were awarded the 1923 Nobel Prize in physiology or medicine.

banyan (*plants*) Native to India, the banyan (*Ficus benghalensis*) is an evergreen fig that begins life as an epiphyte on the branch of a host tree (which it eventually kills). The banyan grows aerial roots that reach the ground, enter the soil, and develop into trunks. As large horizontal branches grow from a trunk, they too send down roots. Dozens or even hundreds of "accessory" trunks form in this manner and the group of banyan trunks gets broader and broader, reaching as much as 2000 feet (600 m) in circumference.

baobab (*plants*) A distinctive feature of the African savannas, the baobab (*Adansonia digitata*) is a deciduous tree that has a life span of 2000 years or more. It has an enormous barrel-like trunk that functions as a water reservoir for the tree; the trunk may be 60 ft (18 m) tall and 30 ft (9 m) in diameter. The large compound leaves have 3 to 7 glossy leaflets. The sweet-scented white flowers hang from long stalks; they give rise to large gourdlike fruits with an edible pulp. A related species called the bottle tree (*A. gregori*) is native to Australia.

The baobab is classified in the bombax family, Bombacaceae, which is comprised of often very large tropical trees. Other members of the family include balsa, durian, and kapok (*Ceiba pentandra*).

bar (*atmospherics*) The bar is a unit of measurement close to the standard pressure of the atmosphere at sea level but easily convertible to pascals, the official pressure unit in SI measurement. Standard air pressure may be given as 14.7 lb per sq in. (101,325 pascals) or as the pressure that raises a column of mercury in a barometer to 29.92 in. (76 cm). A bar is 0.9867 of that standard, but exactly 100,000 pascals. Meteorologists often use the millibar, 0.001 bar, to report air pressure.

barberry (*plants*) Native to temperate regions in the Northern Hemisphere, barberries (*Berberis*) are shrubs with bright yellow wood and branches armed with sharp thorns. The common barberry (*B. vulgaris*), native to Europe but introduced elsewhere, is the alternate host of the wheat rust fungus (*Puccinia graminis*).

Bardeen, John (*physicist*) [American: 1908–91] Few persons have won more than a single Nobel Prize and Bardeen is the only one so far to have won twice for physics—in 1956 for his contribution to the theory behind the transistor and in 1972 for helping to explain superconductivity. Bardeen recognized that in superconducting metals free electrons are interacting with the electrons bound into the metallic structure. This concept is not only part of the explanation of superconducting metals, but it helps explain many other interactions. Bardeen's insight was combined in 1957 with the concept of electrons pairing with holes developed by Leon Cooper [American: 1930–] and with calculations applying the ideas on a macroscopic scale by Bob Schrieffer [American: 1931–] to form the BCS theory of low-temperature superconduction.

barium (*elements*)   Most people know barium only as the key ingredient in a chalky liquid used to make the digestive system visible on X rays. Barium compounds are used for this purpose because they are dense—the name is based on the Greek word *baros*, meaning heavy. A metal with atomic number 56, barium was discovered in 1808 by Humphry Davy. Chemical symbol: Ba.

bark (*botany*)   A woody stem more than 1 or 2 years old has an outer section commonly called bark. It consists of all the tissues external to the vascular cambium—that is, it includes a thin layer of phloem, a thin layer of cork cambium, and a layer of cork. The thickness of the cork and its pattern of splitting and sloughing off are often characteristic of a species.

barley (*plants*)   A member of the grass family, common barley (*Hordeum vulgare*) is widely cultivated in temperate regions for its grains, or fruits. It grows to about 2.5 ft (75 cm). At the base of each leaf is an ear-shaped appendage that envelops the stem. The small flowers form in groups of 3 at the top of a stalk and produce 6 rows of grains.

barnacle (*animals*)   Though sometimes mistaken for mollusks, barnacles actually are small crustaceans that as adults live permanently attached to rocks, ship bottoms, whales, and other marine surfaces. The young larva is free-swimming and resembles larvae of other crustaceans. But as a larva metamorphoses into an adult, it attaches by the head end to a solid object, using an adhesive substance secreted by glands at the base of rudimentary antennae. A shell of calcified plates surrounds the body. The 6 pairs of thoracic appendages are modified into long feathery cirri, which extend from the shell and trap food particles.

Barnard's star (*stars*)   With an apparent magnitude of 9.5, Barnard's star is invisible to the naked eye. It is famous for having the largest proper motion of any star, moving across the sky the apparent distance of the full Moon every 180 years. The rapid proper motion occurs because Barnard's star is close (5.9 light-years away) and because it actually is moving quickly. Since the main component of motion is toward the Sun, in about 10,000 years Barnard's star will replace Proxima Centauri as the nearest star. Barnard's star is also the first star other than the Sun thought to have a planetary system, although evidence of planets is no longer accepted.

barometer (*atmospherics*)   In 1643 Evangelista Torricelli, following a suggestion from Galileo, invented the barometer. Galileo had asked why vacuum pumps failed to work for wells deeper than about 30 ft (10 m). Torricelli guessed that air pressure lifted the water in such pumps rather than the vacuum pulling the water. He used a glass tube, stoppered at one end and filled with mercury, to find how high mercury, heavier than water, would rise. It rose 30 in. (76 cm). In 1648 Pascal demonstrated that the level of mercury falls at higher altitudes, confirming the theory that the barometer measures air pressure.

Barometers called altimeters have been used since Pascal's time to measure height above sea level. Some barometers are still made with mercury, but most use changes in the volume of low-pressure air enclosed in a flexible metal box; these are called aneroid barometers. Torricelli recognized that slight changes in the height of a column of mercury reflect changes in air pressure. Such changes became a cornerstone of weather forecasting. Torricelli's barometer also produced the first artificial vacuum. The region above the mercury, aside from a few atoms of mercury vapor, is a moderately good vacuum.

Barosaurus (*paleontology*)   A giant plant-eating dinosaur of the late Jurassic, *Barosaurus* was a close relative of another huge dinosaur, *Diplodocus*. It grew to lengths of 75 ft (23 m) and weights of 88,000 lb (40,000 kg). Its small head was held aloft by a neck that was 2.7 to 3 ft (9 to 10 m) in length and composed of exceptionally long vertebrae; indeed, the neck made up half the animal's total length.

barracuda (*animals*)   Barracudas (*Sphyraena*) are moderate to large marine fish found in warm and tropical waters. The largest species is the great barracuda (*S. barracuda*), up to 6 ft (1.8 m) long. Barracudas are swift, powerful predators, with a streamlined body and jaws armed with many sharp teeth. There are 2 dorsal fins; the tail is forked.

barrier island (*oceanography*)   A barrier island (or barrier beach) is a low island of sand, clay, and pebbles that develops near shore in places where the ocean floor slopes gently. Ocean waves are circular movements, with the tops of the wave advancing while the bottoms retreat. As waves approach shore, the retreating lower portion moves loose material backward and deposits it below the motion of the waves. The deposits trap sediment flowing away from the main beach and stop sediment coming from the open ocean. Gradually an island rises with a region of calm water called a lagoon between the barrier island and the main shore. A coral reef around an island forms by a different mechanism. (*See also* **atoll.**)

Barringer crater (*solar system*)   Barringer crater in Arizona is one of the few obvious meteorite impact sites on Earth, but when mining geologist Daniel Barringer [American: 1860–1929] began to investigate it, impact theories were unfashionable. Late 19th century geologists had assumed that volcanoes caused all craters on the Moon and on Earth. In 1905 Barringer published his conclusion that the Arizona crater—nearly a mile (1.2 km) across and 600 ft (180 m) deep—was caused by a meteorite. Few agreed at first, but geological studies supported Barringer, causing scientists also to revise their opinions on lunar craters. Barringer crater was formed about 50,000 years ago when a mass of approximately 300,000 tons of iron and nickel struck.

baryon (*particles*)   Subatomic particles that are made from 3 quarks are called baryons. These particles are the proton, neutron, and hyperons. In the standard model of particle physics, the number of baryons in any isolated interaction is constant, so the total number of baryons in the universe is a very large constant. Most proposed revisions of the standard model, however, include a small probability that a baryon can be lost in certain interactions.

basalt (*rocks and minerals*)   Basalt is a dense black, gray, or greenish igneous rock that forms when lava cools. It is the main rock found in the ocean floor, produced by lava flowing from mid-oceanic ridges. Basalt is characterized by a high-percentage of iron and magnesium (it is mafic, from "magnesium and ferrous"). Gabbro is a similar mafic rock, but basalt cooled into rock on the surface while gabbro cooled within Earth's interior. Both are rather low in silica for igneous rock, with about half the mineral being comprised of it.

base (*compounds*)   A base (sometimes called an alkali) is a chemical with a bitter taste, such as ammonia, milk of magnesia, or lye; tasting a strong base is dangerous and should not be attempted. The chemical definition of a base is that it is a compound that when dissolved in water will take away a proton from other chemicals when one is available. The addition of a base to water frees an ion of hydroxide—a combination of 1 oxygen and 1 hydrogen atom that carries a single negative charge, making it receptive to an available proton, which carries a single positive charge. Bases react with acids to produce salts, and with the indicator litmus to make its color blue. The strength of a base is measured in terms of pH; the higher the pH, the more of a base in a solution or the stronger the base.

base (*mathematics*)   The word *base* is used throughout mathematics with different meanings. A base may be the amount on which interest is calculated or the open sets whose union forms an abstract entity called a topological space.

In algebra, a base means either the number used with an exponent to create a power (in $2^3 = 8$, for example, 2 is the base) or equivalently the same number written as a subscript to a logarithm. For $\log_2 8 = 3$ the 2 is the base of the logarithm. Usually if no base appears with the abbreviation log, the base is 10, while ln for the natural logarithm implies a base of e.

For numeration systems a base is a natural number whose powers are added to produce a name for a number. Using 10 as a base, 1234.05 means $(1 \times 10^3) + (2 \times 10^2) + (1 \times 10^1) + (4 \times 10^0) + (0 \times 10^{-1}) + (5 \times 10^{-2})$.

In geometry the base is the side of a polygon or a polyhedron taken to be its bottom. For an isosceles triangle, however, the base is the side that differs in length, so the base angles are those that include that base as one side.

basement (*geology*)   In most places Earth's solid crust has sediment as its top layers, whether soil, sand, or sedimentary rock. In a few places solidified lava or magma overlies a layer of sediment; in even fewer places there is no sedimentary layer at all. Everywhere that a set of sedimentary layers exists, the bottom layer in the sequence lies on either igneous rock or metamorphic rock, the basement. Exposed igneous or metamorphic rock is basement rock also. Note that the basement is not the same as bedrock.

basic (*geology*)   Geologists call igneous rock basic (or mafic) when it contains considerable iron and magnesium and some feldspar but no quartz. (*Mafic* is made from *ma* for magnesium and *f* for ferrous plus *-ic*.) The word basic is applied because of an early incorrect belief that silicate minerals, such as quartz and feldspar, derive from acids; basic rocks contain smaller amounts of silicate minerals. Rocks that have no silicate minerals at all are ultrabasic or ultramafic.

Typically the more basic a rock is, the darker its color. Basalt and gabbro are typical basic rocks, while dunite and peridotite are ultrabasic.

basin (*geology*)   Geologists call several different kinds of formations basins—the lowlands between mountain ranges; a large depressed region of sedimentary rock; and the part of a region drained by a particular stream system. Southwestern Texas and Arizona through southern New Mexico and California, where rows of mountains separate wide basins, is called the Basin and Range Province by geologists.

bat (*animals*)   The only mammals capable of true flight, bats have wings consisting of a double layer of skin stretched over thin, greatly elongated arm and finger bones; the skin also extends backward to attach to the sides of the body and the legs. Strong, sharp claws on the thumbs and toes enable bats to grasp objects and to hang upside down from cave ceilings. Bats are mainly nocturnal; they can see but depend primarily on echolocation to find their way. Diet depends on the species: most bats are insect-eaters but some feed on fruit, nectar, small animals, or the blood of birds and mammals. Bats live clustered in large colonies. Those inhabiting cooler regions may hibernate or migrate to avoid low winter temperatures. The smallest species are the bumblebee bats of Thailand, which weigh less than a penny. The largest are the flying foxes, also of Southeast Asia, which are up to 15 in. (38 cm) long, with a wingspan up to 5 ft (1.5 m).

Batesian mimicry (*zoology*)   First described by Henry W. Bates [English: 1825–92] after observing butterflies in the Amazon basin, Batesian mimicry occurs with 2 species living in the same area that look similar. The model species is distasteful or dangerous to predators; the mimic species is harmless but is avoided by predators who mistake it for the model. For example, the harmless Arizona mountain king snake (*Lampropeltis pyromelana*) has coloration that mimics that of the poisonous western coral snake (*Microides euryxanthus*).

Bateson, William (*biologist*)   [English: 1861–1926] After graduating from Cambridge University, Bateson did research in the United States on the worm-like marine animal *Balanoglossus*, then classified as an echinoderm. Bateson recognized that the animal had a notochord, correctly identified it as a chordate, and hypothesized that chordates may have evolved from echinoderms (today a widely accepted theory). Following his return to Cambridge in 1885, Bateson theorized that variations within a species are not related to environmental conditions. The discovery of Mendel's work in 1900 supported Bateson's theory. Bateson also showed that, unlike the characteristics studied by Mendel, some characteristics are governed by more than 1 gene. He coined the name "genetics" for the new science of heredity.

batholith (*geology*)   Giant masses of igneous rock or metamorphic rock called batholiths form many mountain ranges. These have pushed up as magma through the other rock present and then cooled to form such typical continental rocks as granite. By agreement among geologists, the batholith must have a top surface of at least 40 sq mi (100 km²). A similar rock formation, but too small for this criterion, is called a stock.

In plate tectonics theory the batholith is caused by an ocean plate moving below a continent, pushing up and melting the continental edge.

batôn de commandment (*paleoanthropology*)   A characteristic artifact of the Magdalenian period, an engraved reindeer horn pierced with a large hole, is called a batôn de commandment ("stick of commând") because paleoanthropologists thought it might be a ceremonial sign of a chief or priest. However, no one knows whether the artifact had some useful purpose, such as straightening spear shafts, or had religious significance, or was simply ornamental.

bayberry (*plants*)   Species of 2 different families are commonly called bayberries. Bayberries (*Myrica*) of Family Myricaceae are shrubs or small trees with aromatic foliage and berrylike fruits often coated with a wax that is used to make candles. Jamaica bayberry (*Pimenta acris*), also called the bay rum tree, of Family Myrtaceae, is the source of an aromatic oil used in perfumes.

Bayes' theorem (*statistics*)   People often need to know the probability an event will occur on the condition that some other specified event has occurred. For example, the probability of drawing an ace depends on whether the hand held already contains 1 or more aces; this is called conditional probability. Bayes' theorem, first derived by Thomas Bayes [English: 1702–61], generalizes this concept in a way useful for situations where probabilities cannot be directly calculated. Using estimates for all possible outcomes, Bayes' theorem provides the probability that a given event has caused an observed outcome.

beach (*oceanography*)   To geologists an ocean beach is the region between high and low tides, but in ordinary language a beach requires gently sloping land rather than a steep cliff; flat regions adjacent to lakes or even rivers are also beaches. Beaches occur because waves and ocean currents carry small particles forward toward land, but slow on passing over land, leaving the particles behind. The size of the particles carried depends both on water speed and available material. Where currents carry sand (usually currents nearly parallel to the shoreline), sand beaches form; but in places with slow currents, beaches tend to be rocky. The size of the particles deposited also reflects the steepness of the beach. A steep beach is usually covered with pebbles or smooth rock because the water returns more quickly to the sea, carrying sand with it.

Beadle, George Wells (*geneticist*)   [American: 1903–89] In the 1940s Beadle and Edward L. Tatum [American: 1909–75], experimenting with the red bread mold *Neurospora crassa*, demonstrated that all biochemical reactions in an organism are controlled by genes. Each gene controls a specific step in a reaction by producing a specific enzyme, or catalyst. A change, or mutation, in a gene alters the enzyme, preventing it from doing its job properly. This 1-gene, 1-enzyme hypothesis was refined in the 1950s to the more accurate 1-gene, 1-polypeptide hypothesis after other researchers showed that genes control the synthesis of all proteins, not just enzymes, and that each gene controls synthesis of a polypeptide (many proteins are composed of more than 1 polypeptide chain). Beadle and Tatum shared the 1958 Nobel Prize in medicine or physiology with Joshua Lederberg.

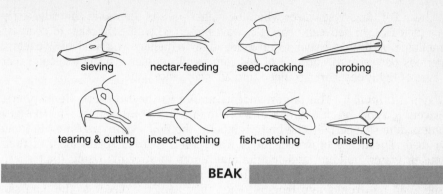

sieving nectar-feeding seed-cracking probing

tearing & cutting insect-catching fish-catching chiseling

BEAK

beak (*anatomy*) The jaws of a bird are extended to form a pointed bill, or beak. This structure has evolved into a wide variety of designs, each suited for obtaining a certain kind of food. In addition birds use their beaks to break up food, clean their feathers, gather nesting materials, build nests, and peck at enemies. Some other animals, including turtles and octopuses, also have jaws called beaks.

bean (*plants*) Various bushy and climbing members of the legume family produce edible seeds called beans; the plants themselves are called beans too. Hundreds of varieties of the common bean (*Phaseolus vulgaris*), native to Central America, have been developed, including kidney, green, navy, pinto, and snap beans. The lima bean (*P. limensis*) is native to South America. Mung beans (*P. aureus*) are native to Asia. The soybean (*Glycine max*), created through cultivation, is believed to be a descendant of an Asian species. The broad bean (*Vicia faba*) is native to northern Africa and southwestern Asia.

bear (*animals*) Constituting Family Ursidae, bears are heavy, powerfully built carnivorous mammals native mainly to the Northern Hemisphere. They range in size from the lesser panda to polar bears, which weigh as much as 1800 lb (800 kg) and are the world's largest land carnivores. A bear has small ears and eyes, a very short tail, and plantigrade feet (both the heel and sole touch the ground, as in humans). The molar teeth are flattened and suitable for grinding a variety of plant and animal food. Most species are solitary and nocturnal, though polar bears often are active during the day. Some species gain a lot of weight in autumn and sleep in dens through most of the winter, awakening and wandering about during periods of mild weather.

bearberry (*plants*) Members of the heath family, bearberries (*Arctostaphylos*) are evergreen shrubs and small trees native to the Northern Hemisphere. The smooth bark often peels or flakes and the simple leaves are leathery. The urn-shaped flowers, white to pink, form in clusters at the ends of branches. The fruit is a fleshy berrylike drupe. Species native to western North America are known as manzanita.

beat (*waves*) A beat occurs when 2 waves of different frequencies are superimposed. For sounds, 2 high-pressure regions of the waves come together to form the beat. For transverse waves when the higher parts of the waves coincide, amplitude increases. In between beats, the waves interfere with each other to decrease amplitude. A beat is easier to measure than frequency. A sound of 256 hertz (periods per second) and one of 244 hertz will produce 4 beats a second. For waves of light, beats are arranged into an interference pattern.

beaver (*animals*)   The largest rodents native to the Northern Hemisphere, beavers (*Castor*) reach a length of more than 3 ft (0.9 m), plus a 1-ft- (0.3-m-) long paddle-shaped tail. The body is stocky, the limbs short, and the hind feet webbed. The dense brownish fur is waterproof. Beavers spend most of their time in water, building and maintaining dams in streams. Using their sharp teeth, a pair of beavers cuts down trees along a stream's banks. They gnaw off limbs and cut trunks into lengths. Then they haul the pieces to the dam site and use mud and rocks to hold the pieces in place. Once the dam is finished, the beavers construct a lodge, either on the dam or nearby.

Becquerel, Henri (*physicist*)   [French: 1852–1908] Four generations of Becquerel men, of which Henri was the third, were prominent physicists. In 1896 Henri Becquerel, reacting to the recent discovery of X rays, continued his father's and grandfather's research on phosphorescence by starting an experiment aimed at looking for X rays from phosphorescent uranium salts. In the process he accidentally became the first to observe radioactivity. He soon identified the uranium as the source of the radiation and recognized that the radiation is not X rays, although it shares some properties of X rays such as the ability to penetrate opaque objects.

bedrock (*geology*)   Bedrock was originally a miner's term for the rock (or even clay) layer that was below the bottom layer of ore. It has come to mean that part of Earth's crust that is attached to the crust below it, nearly always covering a region as a layer of solid rock. Bedrock, unlike basement rock, can be sedimentary rock, igneous rock, or metamorphic rock. One factor in determining the soil of a region is the composition of the bedrock, although other factors are usually more important.

bee (*animals*)   Members of the same order of insects as ants, bees have mouthparts designed for sucking and a constricted "waist" in the abdomen. Females have an ovipositor modified as a stinger. Bees sip nectar, which is turned into honey in the bees' crop. They collect pollen on the hairs that cover their body and legs and in a specialized structure on the hind legs called a pollen basket. Pollen and honey are fed to the bees' larvae. Most species are solitary but the best-known species, such as honeybees and bumblebees, are social insects, with a caste system that includes a queen, worker females, and males called drones.

beech (*plants*)   Native to northern temperate regions, beeches (*Fagus*) are deciduous trees that grow to heights of 100 ft (30 m) or more. They have smooth gray bark, simple dark green leaves, and tiny unisexual flowers. The female flowers give rise to small edible nuts.

The beech family, Fagaceae, comprises some 900 species of trees and shrubs. All produce a nonfleshy fruit that is at least partially surrounded by a husk called a cupule. The family includes chestnuts, oaks, chinquapins (*Castanopsis*), and tan oaks (*Lithocarpus*).

beet (*plants*)   Cultivated as food for humans and livestock, the common beet (*Beta vulgaris*) generally has a large, fleshy root and fleshy leaves that are green, red, or yellow. One variety, Swiss chard or spinach beet, has a small, branched root.

The beet or goosefoot family, Chenopodiaceae, comprises approximately 1400 species of herbs and shrubs with inconspicuous flowers that lack petals. Some species, such as saltbushes (*Atriplex*), are halophytes adapted to soils with high salt content. The family also includes spinach (*Spinacia oleracea*), Russian thistle (*Salsola kali*), and quinoa.

beetle (*animals*)   The most numerous order in the animal kingdom, with some 500,000 described species, is Coleoptera ("sheath wings")—the beetles. The front wings are tough, leathery sheaths that meet in a straight line down the back. The membranous back wings are of primary importance in flight. Beetles have chewing mouthparts both as larvae (grubs) and adults. The great majority are vegetarians. Many, including the Japanese beetle (*Popillia japonica*) and Colorado potato beetle (*Leptinotarsa decemlineata*), cause serious damage to crops, trees, and other plants. Some beetles are predators of other insects. The ladybird beetles, or ladybugs (Family Coccinellidae), for example, feed voraciously on aphids and other plant pests.

Behring, Emil von (*biochemist*)   [German: 1854–1917] In 1901 Behring, a founder of the field of immunology, was awarded the first Nobel Prize in physiology or medicine for his work with diphtheria. He demonstrated that injecting dead or weakened diphtheria bacteria into an animal causes the animal's blood to produce a chemical—which he called an antitoxin—that neutralizes diphtheria toxin, providing immunity against the disease. Together with Shibasaburo Kitasato, Behring similarly showed that an animal's blood produces tetanus antitoxin when exposed to tetanus bacteria.

Bell, Charles (*anatomist*)   [Scottish: 1774–1842] A pioneer in the study of the human nervous system, Bell discovered that nerves are actually bundles of fibers, each with its own origin. He showed that some nerve fibers have sensory functions and others have motor functions. Bell's palsy, a paralysis or weakness of muscles controlled by the facial nerve, is named after him.

Bell Burnell, Jocelyn (*astronomer*)   [English: 1943–] While surveying quasars with a radio telescope built by Antony Hewish [English: 1924–] at Cambridge University in England, Bell recognized in 1967 that a periodic radio signal had too short a period and was too regular to come from any then known astronomical phenomenon. After a brief period during which Bell and Hewish suspected extraterrestrial life as a source and Bell discovered a second similarly periodic signal, the sources were named pulsars. They were later recognized as certain neutron stars. Hewish was given the 1974 Nobel Prize in physics for the discovery.

belladonna (*plants*)   A member of the nightshade family native to the Mediterranean region, belladonna or deadly nightshade (*Atropa belladonna*) is an extremely toxic perennial herb that can be deadly if eaten. It grows to about 3 ft (0.9 m) tall and has pointed leaves, reddish bell-shaped flowers, and black berries. The leaves and roots of the belladona plant are the source of atropine and other alkaloids with medicinal uses.

bellflower (*plants*)   Popular ornamentals with bell-shaped flowers, bellflowers (*Campanula*) are herbs widely distributed throughout the Northern Hemisphere. The flowers usually are bluish purple; the fruits are dry capsules. Well-known species include the Scottish bluebell, or harebell (*C. rotundifolia*), a perennial that grows to about 18 in. (45 cm) tall.

Benedict, Ruth (*anthropologist*)   [American: 1887–1948] Benedict studied Native Americans in the southwestern United States in the 1920s and 1930s, especially the Zuni, Cochiti, and Pima. Later she wrote against racism (1940) and, at the request of the U.S. military, described the main characteristics of traditional Japanese culture (1946).

benthic zone (*ecology*)   The bottom, or floor, of an ocean is termed the benthic zone. It is divided by the continental shelf into a littoral zone and a deep-sea zone.

Benthic communities close to shore, in the littoral zone, are usually extremely productive. The waters are enriched by organic nutrients from the land and sunlight penetrates to the bottom, enabling photosynthesis by algae and rooted plants.

No photosynthesizing organisms can live in the dark deep-sea zone. The organisms that live here feed mainly on dead plankton, fish, and other organisms that sink to the bottom.

benzene (*compounds*)   Benzene is the simplest of the aromatics, consisting of the basic ring of 6 carbon atoms, each with a single hydrogen atom attached (C_6H_6). Benzene was discovered by Michael Faraday in 1825 in illuminating gas made from whale oil then used in London, but its ring-shaped structure was not uncovered until 1865, by Friedrich Kekulé. Benzene is used as a starting material for plastics and in airplane fuel.

berkelium (*elements*)   In 1949 Glenn T. Seaborg and coworkers synthesized a radioactive metal, atomic number 97, at the University of California at Berkeley, naming it berkelium for Berkeley. Although one isotope has a half-life of over a year, there is no known use for the metal or its compounds. Chemical symbol: Bk.

Bernoulli family (*mathematicians/physicists*)   [Swiss: Jacques (Jakob), 1654–1705; Jean (Johann), 1667–1748; Daniel, 1700–82] A dozen members of this family, all descended from Nicolaus Bernoulli [Swiss: 1623–1708], achieved distinction in mathematics or science, especially Jacques, Jean, and Daniel.

Jacques learned calculus from Leibniz and applied the new techniques in the investigation of various curves and infinite series. He was the first to publish work using polar coordinates, although Newton had used them earlier. He also wrote the first account of probability theory, including permutations and combinations.

Jacques' brother Jean is credited with inventing in 1696 the calculus of variations, the technique for finding a function that produces a maximum or minimum.

Daniel is best known as the physicist who discovered the Bernoulli principle: the speed of a fluid, such as air, varies inversely with pressure, which creates lift for heavier-than-air flight.

beryllium (*elements*)   In 1798 Louis-Nicolas Vauquelin [French: 1763–1829] recognized the presence of a metal in the gems beryl and emerald, naming it beryllium from the Greek *beryllos*, gem. The metal, atomic number 4, is light, stiff, and stable, qualities it contributes to its alloys. Despite a sweet taste, beryllium and most of its compounds are extremely poisonous. Chemical symbol: Be.

Berzelius, Jöns Jackob (*chemist*)   [Swedish: 1779–1848] Berzelius was the dominant chemist of the first part of the 19th century. He introduced the system of symbols we now use, divided chemistry into organic and inorganic branches, and introduced much of the basic vocabulary of chemistry—including such words as allotrope, catalysis, halogen, isomer, polymer, and protein. He was a careful experimenter and in 1828 created the first really good table of atomic masses, covering 28 elements, an important step toward the development of the periodic table. Berzelius's early endorsement of Dalton's atomic theory led to its general acceptance. As the author of a widely used textbook and the editor of a yearly review of chemical progress, however, Berzelius upheld many conservative ideas that we now know to be incorrect.

Bessel, Friedrich Wilhelm (*astronomer/mathematician*)   [German: 1784–1846] Mathematicians and physicists often use Bessel functions, developed by Bessel to analyze the motions of planets and stars. In 1838 Bessel was the first to measure the distance to a star (61 Cygni) using parallax and a special instrument he invented known as the heliometer. With the heliometer Bessel also discovered that Sirius has an unseen companion that causes its position to shift slightly as the companion orbits the larger star.

beta decay and rays (*particles*)   Ordinary beta decay is the change of a neutron into a proton with the emission of an electron and an antineutrino. Its name comes from an old term for free electrons. After natural radioactivity was discovered in 1896, physicists rushed to learn the nature of the new energy. By 1898 Rutherford had shown that one part of the radiation from uranium passes through a barrier that stops another. Rutherford named the penetrating part beta radiation and the other alpha radiation. Even after beta rays were shown to be electrons, the name continued in use in the context of natural radioactivity.

Although ordinary beta decay occurs most often in the atomic nucleus, it also is the decay mode of an isolated neutron. Two other forms of radioactivity are also called beta decay. In one, an atomic nucleus captures an electron (ordinary beta decay in reverse); in the other a proton in the nucleus emits a positron and a neutrino

Betelgeuse (*stars*)   All stars but 2 have only been seen as points of light from Earth. One is our Sun and the other is the nearby red supergiant Betelgeuse, whose disk was first photographed in 1996. Its width varies from between 300 to 420 times that of the Sun. Betelgeuse is so large that, at maximum size, if placed at the center of the solar system, it would engulf all the planets through Mars.

Bethe, Hans (*physicist*) [German-American: 1906–] Bethe's work has permeated modern physics and astrophysics, but he is known in part for something that actually was a joke. In 1948, at the request of George Gamow [Russian-American: 1904–68], Bethe added his name to an important paper by Gamow and Ralph Alpher [American: 1921–] that explained how elements are created in stars; the joke was that the paper's authors were Alpher, Bethe, Gamow (pronounced alphuhr, beta, gammov, similar to alpha, beta, and gamma, the beginning of the Greek alphabet). Bethe's own earlier work (1939) explained how stars produce energy by nuclear fusion, so adding his name to the Alpher and Gamow paper was not totally inappropriate. In 1947 Bethe was among the several founders of quantum electrodynamics.

betweenness (*geometry*) Euclid's *Elements* has been criticized as an axiomatic system for ideas used in proofs without inclusion in postulates or definitions. One omission is a definition for betweenness, an idea crucial to many early proofs. A modern definition is that a point C is between points A and B if all 3 are on the same line and if $AC + CB = AB$. Note that a point not on the same line with A and B cannot be between A and B.

biennial plant (*botany*) Plants that live 2 growing seasons are called biennials. Usually a biennial grows vigorously during the first season, producing roots, stems, and leaves. In the second season it develops flowers, fruits, and seeds. Beets, carrots, and turnips are examples of biennials.

big bang (*cosmology*) The accepted explanation of how the universe began is known as the big bang theory. The theory, a consequence of Einstein's general relativity, postulates that the universe originated with a sort of explosion some 12,000,000,000 to 20,000,000,000 years ago. The explosion, or big bang, caused all parts of the universe to rush away from one another. Such an expansion is observed, along with other evidence for the big bang theory, such as the discovery in 1965 of cosmic background radiation, long electromagnetic waves that arrive equally from all directions. This radiation has the characteristics expected from a universe that resulted from the explosion of a small, dense region.

bile (*biochemistry*) Bile is a secretion produced in the liver of vertebrates. It is usually stored in the gallbladder before passing through a duct into the small intestine. Bile contains a mixture of organic salts that emulsify fats, breaking them into tiny droplets. This greatly increases the surface area of the fats, so they can be broken down more easily by the enzyme lipase. Bile also contains pigments released during the breakdown of hemoglobin. These bile pigments give feces its characteristic color.

billfish (*animals*) Constituting Family Istiophoridae, billfish have an upper jaw that extends into a long round bill. There are 2 dorsal fins, the front one much bigger than the second. Found in tropical marine waters, generally near the surface, billfish include marlins, sailfish, and spearfish. Billfish are fast swimmers and many migrate long distances. They feed on fish, squid, and other animals. The largest billfish, the black marlin (*Makaira indica*) of the Indo-Pacific, can exceed 14 ft (4.3 m) in length and weigh over 1500 lb (680 kg).

billion (*numeration*) The word billion has different meanings in American and British English. In the United States a billion is 1,000,000,000 while in Britain (and similarly in the German equivalent) it is 1,000,000,000,000—America's trillion. The British word for 1,000,000,000 is milliard. All intervals higher than a billion also differ—for example, an American quintillion, or 1,000,000,000,000,000,000, is a British trillion.

binary numeral (*numeration*) Writing all numbers as combinations of 2 different symbols is binary notation. Binary numerals are used for systems with 2 choices, such as true or false in logic or on or off in switches. Thus, binary numerals are used for data and programs in computers. Commonly 0 and 1 with their usual meanings are the 2 symbols. The first few natural numbers in binary numerals are 1, 10, 11, 100, 101, 110, 111, 1000. Decimal equivalents are 1, 2, 3, 4, 5, 6, 7, 8.

binary operation (*algebra*) A way to combine 2 entities to produce a third is called a binary operation; addition and multiplication of numbers and union and intersection of sets are examples. Binary operations on a set need not exhibit closure; for example, subtraction is a binary operation on natural numbers that is not closed.

binary star (*stars*) At least half the stars in the universe are multiple star systems in orbit about each other; a pair orbiting each other is a binary star. Astronomers may see both stars of a binary through a telescope, but often they recognize a binary star because an invisible companion's gravitational pull causes periodic motions in the visible star. This can be detected as a Doppler effect shift or when one star periodically eclipses the other. Binaries and multiple stars are common because creation of orbiting stars is a way to conserve angular momentum as a rotating cloud of gas condenses. Speeding up the cloud as it condenses tends to break it into clumps. Interactions between members of binaries cause novas as well as unusual electromagnetic radiation.

binocular vision (*physiology*) Human eyes are located on the front of the head and focus on the same object at the same time, though from slightly different angles. This binocular ("2-eyed") vision gives the impression of distance and depth, enabling the brain to determine accurately the position and speed of objects as much as 200 ft (60 m) distant. In contrast, some animals—particularly prey such as rabbits—have eyes located on the sides of the head. The 2 eyes focus largely on 2 different views, which increases the animal's ability to detect approaching predators.

binomial coefficient (*algebra*) The power of a binomial (an expression with 2 parts joined by addition) for any exponent is written as the sum of a series of products of powers of the 2 parts, with each product multiplied by a number called the binomial coefficient. When the power $(x + y)^4$ is written as $x^4 + 4x^3y + 6x^2y^2 + 4xy^3 + y^4$, the binomial coefficients are 1, 4, 6, 4, and 1. Binomial coefficients appear as the rows of Pascal's triangle. They can be calculated using formulas based on factorials. For the rth term of the nth power of a binomial, the coefficient is $n!$ divided by the product $(r - 1)!(n - r + 1)!$. The $(r + 1)$th binomial coefficient is the same as the number of combinations of n things taken r at a time, often symbolized by writing n above r in parenthesis. This number, found by dividing $n!$ by the product $r!(n - r)!$, is often called the binomial coefficient.

binomial name (*taxonomy*) Each kind, or species, of organism identified by scientists is given a 2-word, or binomial, name. This name, usually Latin, consists of the organism's genus and species, and is underlined or written in italics. For example, the binomial name of the mountain lion is *Felis concolor*. The first word is the genus, the second the species. This binomial name is the same the world over, unlike common names, which often differ from one place to another. Mountain lions, for example, are also called pumas and cougars—and that's just in English! Another source of confusion is the use of one common name for different species. In Scotland, a bluebell is *Campanula rotundifolia*, a member of the bellflower family; in the Rocky Mountains of North America, "bluebell" refers to *Mertensia ciliata* of the borage family.

biochemical oxygen demand (*environment*) The amount of dissolved oxygen consumed by microorganisms as they decompose organic matter in a water sample at a specific temperature in a given length of time is called the biological or biochemical oxygen demand (BOD). The greater the amount of oxygen consumed, the greater the amount of organic matter in the water. BOD is used to approximate the amount of organic pollution in sewage and other effluents.

biochemistry (*sciences*) Biochemistry is the study of chemical compounds and processes occurring in organisms. Like the world around them, organisms are made up of atoms and molecules that can take part in a vast variety of chemical reactions. Every life process—digestion, respiration, reproduction, etc.—involves a series of chemical reactions; anything that inhibits or otherwise interferes with these reactions may cause illness or death.

biodegradable substance (*environment*) Materials that can be broken down, or degraded, into simpler substances by living organisms—mainly bacteria and fungi—are said to be biodegradable. Materials that are not biodegradable can cause serious problems. For example, nonbiodegradable plastic wastes create problems of disposal; nonbiodegradable pesticides present problems as a result of biological magnification.

biodiversity (*ecology*) Earth's variety of life, or biodiversity, can be measured on 3 levels: the variety of genes within a species; the variety of genes among the estimated 3,000,000 to 100,000,000 species, of which about 1,750,000 have been identified by scientists; and the variety of ecosystems that provide homes for these species. This biodiversity makes Earth unique, maintains ecological stability, and provides a wealth of resources to humans. But due to habitat loss, pollution, introduction of exotic species, and other human activities, Earth is experiencing what is believed to be the most rapid loss of biodiversity in the planet's history. Scientists estimate that more than 20,000 species now become extinct each year.

biological clock (*biology*) Almost all organisms have been shown to have mechanisms known as biological clocks; these control rhythmic changes in body metabolism. These changes are related to rhythmic environmental changes. Circadian rhythms—such as the human wake-sleep cycle—follow the

Sun, with a period of about 24 hours. Other common rhythms are related to tides, the lunar cycle, and an annual cycle.

Scientists have identified clock genes in everything from bread molds to fruit flies to mice. These clocks keep time by producing proteins that act as on-off switches. For example, a fruit fly (*Drosophila*) clock gene directs production of the protein PER. When PER levels are low, the gene switches on to make more. When sufficient PER has accumulated, the gene shuts down. This ebb and flow of PER repeats itself every 24 hours. It occurs in tissues throughout the fruit fly's body, suggesting that each clock has its own schedule, tailoring protein production to meet the needs of, say, eyes or muscles.

biological control (*environment*)   Instead of applying environmentally harmful chemical pesticides, many undesirable organisms can be effectively controlled by employing their natural enemies. For example, Japanese beetles (*Popillia japonica*) can be controlled by inoculating lawns with the bacterium *Bacillus popilliae*, which causes a disease fatal to beetles. Aphids, spider mites, corn earworms, and many other destructive insects can be kept in check by various species of ladybird beetles (ladybugs). Other types of biological control include the use of sex pheromones to trap pests and the release of large numbers of sterile males to mate with females (causing the females to produce infertile eggs).

biological magnification (*environment*)   As pesticides and heavy metals move up food chains, they become concentrated in the tissues of animals. The effects of such biological magnification, or bioaccumulation, can be seen in osprey (*Pandion haliaetus*) populations along the Atlantic coast of North America. Beginning in the 1940s DDT was heavily used in the area to combat mosquitoes and other pests. Some of the DDT entered aquatic food chains and was passed from small organisms to larger organisms. As ospreys ate contaminated fish, the level of DDT in their bodies increased. Eventually DDT concentrations were high enough to interfere with the ospreys' ability to use calcium. Female ospreys began laying thin-shelled eggs that were easily crushed by a nesting parent. By the 1970s, osprey populations had declined sharply. After DDT was banned in 1972, populations slowly began to recover.

biology (*sciences*)   Biology is the study of living things. It includes all the facts, laws, and theories that relate to organisms both living and extinct. There are numerous approaches to the study of life, and thus many ways in which biological knowledge can be organized. Some subdivisions of biology focus on types of organisms, such as the study of plants (botany), animals (zoology), and fungi (mycology). Other fields deal with specific life processes (digestion, reproduction, etc.). Still others study aspects of individuals organisms, such as structure (anatomy), development and growth (developmental biology), and heredity and variation (genetics). Additional fields look at the classification of organisms (taxonomy), relations of organisms to their environments (ecology), and origin and differentiation of organisms (evolution).

Certain disciplines combine aspects of biology and other sciences. For example, biomechanics applies mechanical principles to the study of organisms while biochemistry deals with chemical compounds and processes occurring in organisms.

bioluminescence (*physiology*)   Various organisms, including species of bacteria, protists, fungi, invertebrates, and fish, emit cold light—that is, light accompanied by almost no heat. The process, called bioluminescence, involves oxidation of luciferin, controlled by the enzyme luciferase. The greatest variety of bioluminescent species are marine. But the best-loved light producers are fireflies, which have abdominal organs containing luciferin. When oxygen reacts with the luciferin, a flash of light is produced.

Many deep-sea fish do not produce their own light but instead have symbiotic relationships with light-producing bacteria. The bacteria live in specialized organs in the fish and receive a constant source of food via the fish's blood; the fish are supplied with light that attracts mates and prey.

biomass (*ecology*)   The dry weight of all the living material in an area at a given time is called its biomass. Calculations of biomass enable ecologists to learn about an ecosystem's productivity and stability, and to compare one ecosystem with another. Temperate forests have a high biomass, while the open ocean has a very low biomass. Biomass also varies according to trophic level; in any ecosystem the vegetation, or food producers, represent the largest portion of biomass while predators at the top of food chains represent the smallest portion. Burning of biomass—wood, crop residues, and dried animal dung—is a significant cause of air pollution in many parts of the world.

biome (*ecology*)   A group of communities and ecosystems that occupies a large area of land and has a characteristic climate, soil, and mixture of plants and animals is called a biome. Major biomes include tundra, taiga, temperate deciduous forest, temperate grassland, tropical rain forest, savanna, chaparral, and desert. Each biome merges gradually into its neighbors in transition zones called ecotones.

biophysics (*sciences*)   Biophysics is defined as the physics of living organisms. It applies principles of mechanics and thermodynamics to such biological processes as locomotion, energy use, respiration, and circulation.

biosphere (*ecology*)   All the living organisms on Earth make up the biosphere. The term biosphere also is used as a synonym for ecosphere, an ecosystem that encompasses all living things and the totality of habitats in which they live: the upper layer of Earth's crust, its surface and underground waters, and the lower region of the atmosphere.

Biot, Jean Baptiste (*physicist*)   [French: 1774–1862] Biot was a physicist whose main discoveries affected astronomy, earth science, and chemistry. His 1803 report on a meteorite fall convinced scientists for the first time that rocks fall from the sky. His balloon flights with Joseph-Louis Gay-Lussac [French: 1778–1850] established that Earth's magnetic field extends into the atmosphere. In 1815 Biot showed that some organic compounds have 2 chemically identical forms that rotate polarized light in different directions, correctly speculating that the effect is caused by differences in the shape of the molecules.

biotic and abiotic factors (*ecology*)   Some features of an ecosystem result from the presence and activities of living things; these are called biotic factors. Competition, predation, and photosynthesis are biotic factors. Chemical and physical features such as temperature, precipitation, and light intensity are abiotic factors. There are constant interactions between biotic and abiotic factors. Plants influence the movement of water; drought affects plant growth; fungi secrete acids that weaken the structure of rocks; and so on.

birch (*plants*)   Birches (genus *Betula*) are deciduous trees and shrubs native to the Northern Hemisphere. They are distinguished by their white or grayish bark, which often peels off in papery sheets or scales. Male and female flowers are borne on hanging catkins in springtime; female flowers develop into small conelike fruit. Many species are valued as ornamentals or sources of timber. Sap of the sweet birch (*B. lenta*) is used to make birch beer. Native Americans used the bark of the paper birch (*B. papyrifera*), which is impervious to water, to make canoes and wigwams. The birch family, Betulaceae, also includes alders, hazels (*Corylus*), hornbeams (*Carpinus*), and hop hornbeams (*Ostrya*).

bird (*animals*)   Birds are air-breathing vertebrates designed for flying (flightless birds, such as the ostrich, evolved from flying species). Birds are characterized by a pair of wings, a body covering of feathers, a beak, scales on legs and feet, and hollow bones. They are endothermic, maintaining a high body temperature by internal metabolic processes, and lay eggs.

The approximately 10,000 known species of birds are divided into some 28 orders (the number varies slightly depending on the classification scheme). The smallest species is the bee hummingbird (*Mellisuga helenae*) of Cuba, only 2.25 in. (5.7 cm) long, with a wingspan of 4 in. (10 cm). The heaviest and tallest is the ostrich (*Struthio camelus*), with weights to 345 lb (156 kg) and heights to more than 8 ft (2.4 m). The largest wingspread belongs to the wandering albatross (*Diomedea exulans*), at more than 11 ft (3.4 m).

bird of prey (*animals*)   Members of the orders Falconiformes and Strigiformes are commonly referred to as birds of prey, or raptors. Falconiformes, such as falcons and hawks, are daytime hunters, while Strigiformes—owls—are active at night. All have superb eyesight, hooked beaks, sharp claws, and other adaptations that make them formidable predators.

bisect (*geometry*)   To bisect ("cut in two") is to divide into halves, referring usually to line segments or angles. A line segment is bisected by locating 2 points equally distant from the segment's endpoints; the line joining those points bisects (and is perpendicular to) the segment. An angle is bisected by first locating a point on each ray at equal distances from the vertex and then constructing a third point equally distant from each of the first 2; the line through that third point and the vertex bisects the angle.

In analytic geometry, the bisector of the line segment joining (x_1, y_1) and (x_2, y_2) is the point whose coordinates are the average of the coordinates of the endpoints, or $((x_1 + x_2)/2, (y_1 + y_2)/2)$.

bismuth (*elements*)   Bismuth is a metal, atomic number 83, with unusual magnetic properties—it is the element most repelled by a magnetic field and increases the most in electrical resistance in a magnetic field. In 1753 Claude J. Geoffrey the Younger [French: fl1750] was the first to recognize that bismuth is not just an odd form of tin or lead. Its name is from the German *weisse masse*, white mass—the pure metal is white tinged with pink—changed to *bismat* as part of the same consonant shift that changed the name wisent to bison. Chemical symbol: Bi.

bison (*animals*) The genus *Bison*, of the bovid family of mammals, comprises 2 species. The American bison is commonly called the buffalo. The European bison is sometimes called the wisent. Bison have a broad head with sharp horns, a short neck, and a hump in the shoulder region. The American species is the larger, attaining a shoulder height of 6.5 ft (2 m) and a weight of 2000 lb (900 kg) or more. Intensive hunting almost caused the extinction of both species.

bivalve (*animals*) Clams, oysters, mussels, and scallops are members of the mollusk class Bivalvia. The soft body of a bivalve is compressed between a shell formed of 2 pieces, called valves, held together by a hinge ligament. Bivalves lack a head, have a bladelike foot, and take in oxygen via a pair of gills. They inhabit every type of aquatic environment and are generally rather sedentary. Many live in burrows in sand or mud. Most feed on small organic particles filtered from the water.

Black, Joseph (*chemist*) [English: 1728–99] Black's contributions to chemistry and materials science were few, but fundamental. In his doctoral thesis he introduced quantitative methods to chemistry and showed that carbon dioxide, previously known only from respiration and fermentation, can be produced by heating calcium carbonate and also will recombine with the resulting calcium oxide. His experiments detected carbon dioxide in air and showed that it forms an acid in water.

Black's other work is now part of thermodynamics. He discovered latent heat and also observed, but failed to understand, differences in specific heat between materials.

Blackett, Patrick M. S. (*physicist*) [English: 1897–1974] Blackett was awarded the 1948 Nobel Prize in physics for his development of methods for observing particle interactions. He was the first to photograph the transmutation of one element (nitrogen) into another (oxygen, after bombardment by an alpha particle) and the first to show matter (electrons and positrons) arising out of energy (gamma radiation). During World War II Blackett created the methods now known as operational research, which he used to improve greatly the ability of planes to sink submarines and to make research on radar and the atomic bomb more efficient.

black hole (*astronomy*) When a body becomes so massive for its size that not even light can escape the powerful gravitational pull it exerts, it is called a black hole. Black holes were predicted by John Michell [English: c1724–93] as early as 1784 but became understandable only with the application of relativity theory. When a body collapses to less than its minimum size, the matter shrinks to a point surrounded by a region in which any matter or energy is forever trapped; the edge of this region is called the event horizon. For a mass the size of Earth, the radius of the event horizon is about 0.4 in. (1 cm); for the mass of the Sun it is about 2 mi (3 km). Black holes occur whenever a star exceeding 3 times the mass of the Sun collapses after fusion has stopped.

Evidence for black holes has been observed at the centers of galaxies, including our own Milky Way. Quasars may be radiation caused by matter falling into a galactic black hole. Smaller black holes, thought to be remains of supernova explosions of very large stars, have also been detected.

bladder (*biology*)　Any thin-walled sac that contains fluid or gas is a bladder. Many brown algae have air-filled bladders along their fronds to help the fronds float in sunlit water. Bladderworts have bladders that aid in capturing tiny animals. Bony fish have a swim bladder to adjust specific gravity. Mammals and some other vertebrates have a urinary bladder, where urine is stored until it is excreted.

bladderwort (*plants*)　Bladderworts (*Utricularia*) are carnivorous plants with small bladders on their leaves. Each bladder opens via a flaplike trapdoor, with sensitive trigger hairs near the door's free-hanging edge. An empty bladder has concave walls and contains a partial vacuum. When an animal touches the trigger hairs, the door opens inward, the victim is sucked into the bladder, and the door closes—all in less than a second. Digestive enzymes are then released into the bladder to break down the victim's tissues.

blade (*paleoanthropology*)　In Europe in the later part of the Paleolithic Age people made some stone tools with parallel edges. Such tools are called blades when they are at least twice as long as their other dimensions. The working edges, however, were usually at the end, making such blades more like modern chisels. Often blades were reshaped by further flaking into other tools. If there is 1 working edge along 1 of the parallel sides, the blade becomes a backed knife, while 2 such edges mean the tool is simply a knife.

bleach (*compounds*)　Bleach is any substance used to neutralize colors of dyes or pigments. Common household bleach is based on chlorine in water. Chlorine reacts with water to make both hydrochloric acid (HCl) and free atomic oxygen (O), each of which can combine with dye molecules. A solution of sulfur dioxide (SO_2) in water also bleaches. Oxygen alone, especially when the molecules are broken by sunlight, can bleach fabrics, so white towels dried in sunlight are whiter than those dried indoors.

blood (*anatomy*)　Most animals, including all vertebrates, have a fluid tissue called blood that functions as a transport system within the body. Blood delivers nutrients and oxygen to cells and removes waste materials. It transports many other materials, too, including hormones and mineral salts; helps maintain and regulate the body's water content and temperature; fights infection; and forms clots at wound sites.

　　Vertebrate blood has a liquid portion called plasma in which are suspended red blood cells, white blood cells, and platelets. An adult human has approximately 5 qt (5 L) of blood. Plasma accounts for about 55% of this volume; cells and platelets make up the rest. One cubic millimeter of human blood generally has about 5,000,000 red blood cells, 6000 to 8000 white blood cells (a number that jumps if there is an infection), and 300,000 to 400,000 platelets.

blood group (*physiology*)　The blood of all individuals within a single mammal species is not exactly identical. Various antigens that stimulate the production of antibodies may or may not be present. This results in different blood groups within the species.

Among humans the 4 major blood groups are A, B, AB, and O. These are based on antigens A and B in red blood cells and 2 corresponding antibodies in plasma, *a* and *b*. Type A blood has antigen A and antibody *b*. Type B has antigen B and antibody *a*. Type AB has antigens A and B but neither antibody. Type O has neither antigen but both antibodies. Type O blood can be transfused into people with other blood types without ill effect. (Antibodies in donor blood become so diluted in the recipient's bloodstream that their effect is insignificant.) But if, say, type A blood is transfused into a type O person, the recipient's *a* antibodies will clump together, or agglutinate, the introduced A antigens.

bloom (*ecology*) Occasionally, 1-celled algae reproduce exceptionally fast, creating a dense growth, or bloom, on the water's surface. The heavy concentrations of algae—as many as 1,000,000,000 cells per quart (liter)—discolor the water, turning it green, red, yellow, or brown depending on the type of algae causing the bloom. Often, formation of a bloom is strictly a natural phenomenon, as when currents carry nutrient-rich water from deep in the ocean to the surface. Sometimes, however, pollution causes or enhances blooms. For example, blue-green algae often form blooms in lakes that receive a rapid increase in nutrients, perhaps in sewage or runoff of fertilizers. A few dozen species of algae are commonly associated with harmful blooms. Some of these blooms are misleadingly referred to as red tides.

blubber (*anatomy*) Whales, seals, and some other aquatic mammals have beneath the skin a thick layer of fatty tissue called blubber. The blubber acts as an insulator, protecting the body from loss of heat. Blubber also reduces a whale's specific gravity, helping the animal float, and serves as a reserve food supply.

blueberry (*plants*) Several species of shrubs of the genus *Vaccinium*, generally classified in the heath family, are cultivated for their edible berries. These include varieties, or cultivars, of blueberries (*V. ashei* and *V. corymbosum*), cranberries (*V. macrocarpon*), and lingonberries (*V. vitis-idaea*).

blue-green algae (*monera*) Formerly considered algae, blue-green algae are now classified with bacteria in Phylum Monera and are more properly called cyanobacteria. Like bacteria, they are prokaryotes, lacking nuclei and membrane-bound organelles. Some are single-celled but most form clusters or filaments. They differ from other bacteria in possessing the green pigment chlorophyll, which they use to produce food and oxygen in photosynthesis. Many species also have the bluish pigment phycocyanin, but some species are reddish due to the pigment phycoerythrin. Blooms of a reddish species of *Oscillatoria* give the Red Sea its name, and eating *Spirulina* gives African flamingos their pink color.

Fossils of cyanobacteria 3,500,000,000 years old are the oldest evidence of life on Earth. Cyanobacteria were the dominant form of life for more than 2,000,000,000 years. Their photosynthetic activity is believed to be responsible for Earth's oxygen-rich atmosphere. Also, it has been suggested that the first eukaryotes originated when cyanobacteria took up residence within other cells. The cyanobacteria made food, which they shared with their hosts, and in return had a place to live. Over time, the cyanobacteria evolved into chloroplasts.

Most cyanobacteria live in fresh water and moist soil, or on dead logs and tree bark. Some species are components of lichens. Many species, such as members of *Anabaena*, carry out nitrogen fixation.

blue jet (*meteorology*)   Blue jets are visible blue cones that briefly appear above electrically active regions of thunderstorms. Unlike lightning, they proceed upward from the cloud and widen quickly. They differ from sprites in color, size, and interactions with other electromagnetic phenomena.

B meson (*particles*)   Any meson containing a bottom quark or antiquark is called a B meson. The other quark in a B meson determines its type, such as the neutral B meson (bottom and antidown), the positive B meson (antibottom and up), the neutral strange B meson (antibottom and strange), and the negative charmed B meson (antibottom and anticharm). B mesons are not part of ordinary matter, but are created in certain particle accelerators for experiments testing basic theories.

boa (*animals*)   Nonpoisonous, thick-bodied snakes that kill by constricting prey, boas live in a variety of tropical and temperate habitats. Some, such as the rosy boa (*Lichanura trivirgata*) of Mexico and the U.S. Southwest, seldom exceed 3 ft (0.9 m) in length. Others, such as the boa constrictor (*Boa constrictor*) and anaconda, grow to more than 10 ft (3 m) long.

Boas, Frank (*anthropologist*)   [American: 1858–1942] Boas specialized in various Native American groups of northern and western Canada, although his base of operations was the United States. He had a great influence on American anthropology, developing rigorous methods and pioneering in the use of statistics. In addition, Boas contributed to the introduction of stratiographic methods to archaeology.

Bode's law (*solar system*)   In 1776 astronomer Johann Elert Bode [German: 1747–1826] published a discovery made in 1772 by Johann Daniel Titius [German: 1729–96] that the distance from the Sun of each of the then known planets is determined by a simple number sequence, now called Bode's law. The sequence is formed by adding 4 to 0, 3, 6, 12, 24, 48, 96, 192, 384, and 768 producing 4, 7, 10, 16, 28, 52, 100, 196, 388, and 772. In astronomical units (A.U.) the distances of the planets are 0.4 (Mercury), 0.7 (Venus), 1.0 (Earth), 1.5 (Mars—near 1.6), 5.2 (Jupiter), and 9.5 (Saturn—near 10). In 1781 William Herschel discovered Uranus, which fit the rule, being 19.1 (near 19.6). Then in 1801 the asteroid Ceres was found near 2.8; soon other large asteroids were also found about 2.8 A.U. from the Sun. Astronomers came to believe (incorrectly, it was later shown) that a large planet had occupied the orbit at 2.8 that was decreed by Bode's law. But the fit of Bode's law to Uranus and Ceres was just a coincidence. When Neptune was found, it was at about 30.0 instead of

38.8, the next distance described by the "law." Pluto at 39.4 A.U. is even farther away from Bode's tenth number.

bog (*ecology*)   Bogs are a type of wetland characterized by thick peat deposits, mats of floating vegetation, and brown, acidic water. Bogs form in depressions with poor drainage, often in places formerly scoured by glaciers, and receive their water only from precipitation. Low levels of oxygen slow decomposition of dead plant matter, which gradually accumulates as layers of peat on the bottom of the bog. The most common bog plant is *Sphagnum* moss, which grows on the undulating mats. Animal populations tend to be limited because of the acidic conditions.

Bohr, Niels (*physicist*)   [Danish: 1885–1962] Bohr's greatest contribution to modern physics may have been leading the Copenhagen Institute of Theoretical Physics from 1918 to 1943, the principal incubator of quantum theory. Bohr made his reputation in 1913 with the Bohr model of the hydrogen atom, which explained the spectrum of glowing hydrogen gas in terms of sudden passages of an electron from one orbit to another. In 1939 Bohr also explained the nucleus of heavy elements, this time in analogy with a water drop; Bohr's model predicted the properties of uranium's isotope 235, used in early atomic bombs. Bohr also proposed the correspondence principle and complementarity as philosophical bases for quantum theory.

bohrium (*elements*)   Bohrium is a radioactive metal, atomic number 107, synthesized in 1977 by Russian scientists at Dubna and named for Danish physicist Niels Bohr, who developed the first quantum-mechanical model of an atom. Chemical symbol: Bh.

boiling point (*physics*)   Boiling is not the same as evaporation; evaporation is a surface process while boiling occurs within a liquid. The boiling point of a liquid is the temperature at which gas bubbles begin to form within the liquid. It varies with pressure, since the gas must overcome the pressure on the liquid to form bubbles that do not collapse and become reabsorbed. At 1 bar the boiling point of water is 212°F (100°C), but this falls rapidly with lower pressures. As liquid boils, it normally cools at the same rate as it is being heated; since temperature remains constant, cooking in a boiling liquid depends mainly on time.

Boltzmann, Ludwig (*physicist*)   [Austrian: 1844–1906] Boltzmann established the mathematical basis of thermodynamics, especially the definitions of heat and entropy in terms of the movement and energy of particles in gases. He completed a statistical analysis of how molecules behave in gases (the Maxwell-Boltzmann statistics), 1 of 3 ways assemblies of particles can distribute themselves (the others are the Fermi-Dirac statistics for particles that behave like electrons and the Bose-Einstein statistics for particles that behave like photons).

bomb, volcanic (*geology*)   A rock fragment ejected from a volcano is called a bomb if it is larger than an inch or two in diameter, but volcanic bombs may be enormous boulders weighing as much as 100 tons and tossed as far as 6 mi (10 km) from the crater. Some bombs remain hot enough to be somewhat soft on landing.

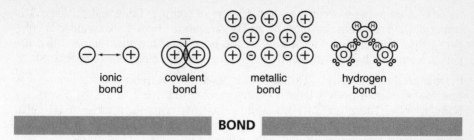

ionic bond   covalent bond   metallic bond   hydrogen bond

BOND

bond (*compounds*) Whenever 2 or more atoms form a chemical relationship, the connection between the atoms is called a bond. The basic bonds are ionic, covalent, metallic, and hydrogen—other varieties are variations. All bonds use charge to hold atoms together.

In an ionic bond, an electron leaves one atom for another, giving the abandoned atom a positive charge and the atom with the electron a negative charge. Charged atoms are called ions and the attraction between them is an ionic bond.

In a covalent bond, electrons stay with their atoms but are shared. The probability of an electron being in the shared region is greater than the probability of it being elsewhere. Positively charged atomic nuclei are attracted to the shared region, which forms the bond. Combinations of atoms connected by covalent bonds are called molecules.

In a metal, some electrons are free to move and the loose electrons form a negatively charged sea to which the nuclei are attracted.

A hydrogen bond occurs when hydrogen's lone electron leaves its atom and enters a covalent bond, exposing hydrogen's proton to attract electrons of a third atom.

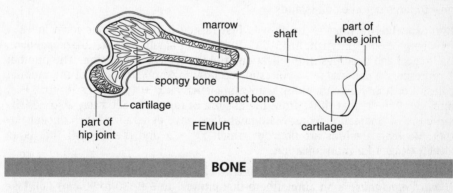

marrow   shaft   part of knee joint

spongy bone

cartilage

compact bone

part of hip joint   FEMUR   cartilage

BONE

bone (*anatomy*) A hard connective tissue, bone is the primary component of the skeleton of most vertebrates. Bone consists mainly of an extracellular matrix in which living bone cells are embedded. There are 2 types of bone tissue: compact bone, which is very dense and hard, and spongy bone, which is softer and has numerous spaces filled with bone marrow.

A typical long bone, such as the femur in the upper leg, has 2 enlarged ends connected by a shaft. The ends, which meet other bones at movable joints, are made of spongy bone covered with thin layers of compact bone and cartilage. The center of the shaft contains marrow and is surrounded by a thick cylinder of compact bone. Anchored to the outside of the bone is a thin membrane called the periosteum, which contains blood vessels and bone-forming cells (osteoblasts).

bone marrow (*anatomy*) Found in cavities within vertebrate bones, marrow is one of the human body's largest organs, accounting for 2 to 5% of a person's body weight. There are 2 types. Yellow marrow consists mostly of fat cells, which function as protective tissue and store energy in the form of lipids. Red marrow manufactures red blood cells, white blood cells, and platelets.

bony fish (*animals*) Almost all fish have bony skeletons and true jaws and are classified as Osteichthyes. There are 2 subclasses: lobe-finned fish, represented by coelacanths and lungfish; and ray-finned fish, an extremely diverse group that includes bass, salmon, tuna, carp, seahorses, eels, herring, and flatfish. The body of bony fish is typically protected by scales, which vary in composition and structure according to species. Some species are toothless while others have teeth, which may be located on the jaw, in the mouth wall, or on the bone of the throat. Fertilization usually is external, with females releasing eggs into the water and males releasing sperm over the floating eggs.

booby (*animals*) Large seabirds with long wings, wedge-shaped tails, and webbed feet, boobies (*Sula*) fly comparatively high over the ocean, then plunge to the water to grab fish with a pointed bill. Their name comes from the Spanish *bobo*, meaning "stupid fellow," and refers to their lack of fear of humans.

The booby family, Sulidae, also includes 3 species of gannets, which are similar in appearance to boobies. However, boobies live in the tropics and subtropics and gannets live in temperate areas. Outside the breeding season, boobies typically roost on land at night while gannets roost on water. Both types nest in large colonies on islands.

Boolean algebra (*logic*) Logic was first turned into a symbolic form in 1854 by George Boole [English: 1815–64]. In Boolean algebra letters are propositions, the logical and becomes multiplication, and the logical or addition. The number 1 represents a universal set. Boole showed how logic obeys many of the familiar rules of ordinary algebra along with some, such as $x + (yz) = (x + y)(x + z)$ and $1 + 1 = 1$, that differ from the algebra of numbers. Set theory also has the structure of Boolean algebra. Mathematicians have generalized the concept further: Boolean algebras are rings for which $xx = x$ and there is a 1 that is an identity element for multiplication.

borage (*plants*) Native to the Mediterranean region and western Asia, borage (*Borago officinalis*) is an annual herb that grows 1 to 3 ft (30 to 90 cm) tall. Like all members of the borage family, Boraginaceae, its leaves are covered with bristly hairs. The small blue flowers are borne in clusters. The family also includes comfreys (*Symphytum*), forget-me-nots (*Myosotis*), and heliotropes (*Heliotropium*).

borer (*animals*) Various species of worms, mollusks, and insects are among invertebrate animals that bore into other animals or plants, often causing serious damage. Hookworms bore through the skin of people who walk barefoot on contaminated ground. Some piddocks and the teredo are mollusks that bore into rock and wood; the oyster drill (*Urosalpinx cinerea*) is a mollusk that bores through oyster shells and sucks out the soft animals. Many beetles bore into plants during the larval stage, including the wood-boring apple tree borer (*Saperda candida*) and locust borer (*Megacyllene robiniae*). The European corn borer, the larva of the maize moth (*Pyrausta nubilalis*), mainly bores into the stems of corn plants.

Born, Max (*physicist*)   [German: 1882–1970] Born originally trained to be a mathematician, and his higher mathematics contributed greatly to quantum theory. He suggested using matrices to Werner Heisenberg, an approach that produced the first successful quantum mechanics. Born conceived of using the wave equation of quantum mechanics as the basis for determining the probability that a particle is in a particular location; this made the otherwise inexplicable behavior of subatomic particles somewhat easier to understand.

boron (*elements*)   Boron is a nonmetal, atomic number 5, that occurs in compounds concentrated in ancient lake beds that have become desert, often as borax (sodium tetraborate), from which the name derives. Borax is used in soaps and cleaning powders and to make boric acid. Both borax and boric acid are used in special glasses, while other boron compounds are among the hardest substances known. The element, chemical symbol B, was discovered in 1808 by Joseph-Louis Gay-Lussac [French: 1778–1850] and Louis-Jacques Thénard [French: 1777–1857].

borrow (*arithmetic*)   In the most common decimal system algorithm for subtraction, the steps proceed from right to left by decimal places, and negative numbers are not permitted. To avoid negatives when subtracting a larger digit from a smaller, borrow 10 units by renaming 1 of the units in the next higher place. To subtract 27 from 53, since 3 – 7 would be negative, rename 53 (5 tens 3 ones) as 4 tens 13 ones.

Bose Einstein condensate (*materials science*)   Atoms, like subatomic particles, possess spin, which can either be an odd multiple of the 1/2 spin common to fermions or an even multiple, common to bosons. Bosonic behavior was calculated by Satyendranath Bose [Indian: 1894–1974] and Einstein. In contrast to fermions, which cannot be in the same place at the same time, bosons overlap easily with each other.

Spin properties of atoms are usually not noticeable because their normal motion causes spin states to become mixed. Einstein predicted that very cold atoms with even spins could condense into a new form of matter, now called a Bose Einstein condensate (BEC). The first Bose Einstein condensates were created in 1995. Because atoms in a BEC behave like photons, devices similar to lasers that use atoms instead of light are possible.

boson (*particles*)   Particles with integral spin, which allows collective action, are called bosons. Collective action is behavior similar to that of light, which consists of bosons called photons. Any number of photons can be in the same state at once, for example, in a laser. The observed bosons are the photon, pions and other mesons, gluons, W particles, and Z particles. Bosons that are predicted, but that have not been observed, include the Higgs particle and the graviton.

Matter particles, which have nonintegral spin, relate to each other through forces that exist as a result of the exchange of other bosons. For example, charged particles such as electrons and protons attract or repel through exchanging photons; and protons and neutrons in atomic nuclei attract each other by exchanging pions.

Bosons are named for Satyendranath Bose [Indian: 1894–1974], who first explored the mathematics of their behavior in 1924. The mathematics was put into final form by Einstein. Thus, a gas that consists of atoms behaving as bosons is called a Bose Einstein condensation.

botany (*sciences*)   Botany is the branch of biology that deals with the study of plants. It can be divided into many overlapping fields. Some botanists specialize in fields such as plant physiology, anatomy, or ecology. Others focus on specific types of plants, such as mosses and liverworts (bryology). Still others devote their efforts to applied plant sciences such as the study of fruits (pomology) or ornamental plants (floriculture).

bottom quark (*particles*)   Of the 6 known quarks, bottom—once called beauty—is the second most massive, a giant with almost 10,000 times the mass of the electron. Like all quarks, bottom is never observed in isolation, but is always accompanied by at least one other quark. Such pairs of quarks constitute B mesons, first observed in the early 1980s.

bougainvillea (*plants*)   Named after the explorer Louis Antoine de Bougainville [French: 1729–1811], bougainvilleas (*Bougainvillea*) are woody shrubs and vines native to tropical South America. They are widely cultivated for their showy red or purple petal-like bracts, which actually are modified leaves that surround the small, inconspicuous flowers.

bound (*calculus*)   A bound is a number that is either greater than or equal to every other number in a set of numbers (upper bound) or less than or equal to every other number (lower bound). When a set of numbers increases or decreases "without bound," the symbol ∞ is used (often read as "to infinity"). In calculus the bounds of special interest are the greatest lower bound and the least upper bound, numbers that may or may not be in the set. For example, 1/3 is the least upper bound of the set of decimal approximations to it, which are 0.3, 0.33, 0.333, 0.3333, and so on.

bovid (*animals*)   Hoofed mammals of the family Bovidae are a diverse group of some 125 species of ruminants, comprising antelopes, goats, sheep, cattle, and others. Bovids have a pair of hollow, unbranched, permanent horns; in most species both males and females are horned, though male horns are generally better developed. The premolar and molar teeth have ridges that aid in grinding grass and other vegetation. Many bovids inhabit grasslands but mountain goats live on rocky mountains and musk oxen are found on northern tundra.

bowerbird (*animals*)   Perching birds of Family Ptilinorhynchidae are named for the elaborate bowers of twigs and other materials built by males during the mating season. Bowerbirds live in forests of Australia and New Guinea, mostly on the ground. They are about 9 to 15 in. (23 to 38 cm) long, with short wings and legs. Some species have brilliant plumage.

boxwood (*plants*) Members of the genus *Buxus* are evergreen shrubs and small trees widely distributed in tropical and temperate regions. The English, or common boxwood (*B. sempervirens*) grows to heights of 20 ft (6 m). It has long been a favored plant for topiary and hedging, and its extremely hard wood was often used to make boxes. The boxwood family, Buxaceae, also includes *Pachysandra*, of which *P. terminalis* is a particularly popular ground cover, and jojoba (*Simmondsia chinensis*), with seeds that yield a liquid wax used in cosmetics and other products.

Boyle, Robert (*chemist-physicist*) [Anglo-Irish: 1627–91] Boyle's name is most closely associated with his contribution to the gas laws (the volume of a gas varies inversely with pressure), but his greatest influence was on chemistry. He was the first to develop modern concepts of element and compound; to distinguish between acids, bases, and neutral substances; and to conduct and publish experiments along the lines now called the scientific method. In his work with a vacuum he demonstrated that all masses fall toward Earth at the same speed and that the vacuum does not conduct sound but does conduct the electric force.

brachiopod (*animals*) Commonly known as lampshells because they look like ancient Roman oil lamps, members of the phylum Brachiopoda have soft bodies enclosed in a pair of hinged, usually oval-shaped, shells called valves. They are found in seas worldwide. Following a free-swimming larval stage, members of most species settle to the bottom and attach themselves by a muscular stalk to the seafloor. A feeding organ, the lophophore, generates currents that bring in oxygenated water and food particles and sweep away wastes. Brachiopods were much more common during the Paleozoic than today; their fossils are valuable for dating rocks.

bracket fungus (*fungi*) Members of the club fungi, bracket fungi form tough, leathery shelves, or brackets, that grow horizontally from the bark of trees, sometimes singly, sometimes in groups. Some species, saprophytes on dead trees, are important agents of decay; others are parasites that kill the trees they attack.

brain (*anatomy*) The vertebrate brain is a large mass of nerve tissue contained within a hard skull and further protected by membranes called meninges. Together with the rest of the central nervous system, the brain controls and coordinates body activities. It has 3 basic regions. The forebrain consists mainly of the cerebrum; in fish the cerebrum is concerned primarily with the sense of smell, whereas in mammals it is also the center of intelligence, memory, and emotions. The midbrain includes the optic lobes and is mainly involved with sight and hearing. The hindbrain includes the medulla, which is continuous with the spinal cord and which controls breathing and many other involuntary activities; and the cerebellum, which helps maintain balance.

Many invertebrates have ganglia (concentrations of nerve cell bodies) in the head that function as primitive brains. For example, an insect has a brain consisting of several consolidated ganglia; nerves connect it to the antennae, eyes, and other organs of the head and there are connections to a nerve cord for coordination of other ganglia in the body.

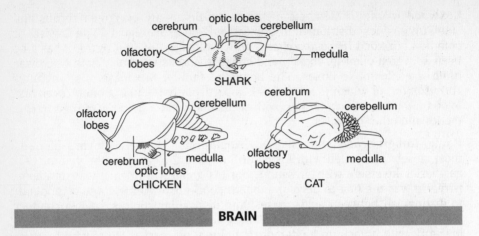

BRAIN

Brains differ greatly in size and complexity. An insect's brain may be less than ⅟₂₅ in. (1 mm) across. An elephant's brain weighs about 11 lb (5 kg). The most complex and highly developed brain is that of humans, of which about ⅔ is the cerebrum.

bramble (*plants*)   Raspberries, blackberries, and dewberries make up a group of shrubs in the rose family called brambles (*Rubus*). They have perennial underground parts and biennial prickly stems called canes. The flowers and fruits develop on the canes in their second year. Raspberry fruits form a cup-shaped cluster that separates from the receptacle (flower stalk) at maturity; blackberry and dewberry fruits remain attached to the receptacle.

brass (*materials science*)   Brass is an alloy of copper and zinc with perhaps some other metals (aluminum, iron, lead, manganese, nickel, or tin) included for special uses. Brass is rather brittle and stiff, but has a brilliant appearance and is easily plated with chromium or nickel—qualities that make it suitable for the coiled horns on trumpets, trombones, and other metal horns, known collectively as brasses.

Brazil nut (*plants*)   Native to tropical rain forests of South America, the Brazil nut tree (*Bertholletia excelsa*) grows to 160 ft (49 m). It has long leathery leaves and creamy yellow flowers. Each fruit—a hard capsule—contains 12 to 24 large 3-sided edible seeds.

All members of the Brazil nut family, Lecythidaceae, are tropical shrubs or trees with large undivided leaves and capsule fruits. Other tropical American species in the family include the cannonball tree (*Couroupita guianensis*) and sapucaias (*Lecythis*).

breadfruit (*plants*)   Native to tropical islands of the South Pacific, the bread-fruit tree (*Artocarpus altilis*) attains heights of 60 ft (18 m). It has large shiny leaves and unisexual flowers borne in separate male and female clusters. The tree is cultivated for its large edible fruits, each of which is actually a multiple fruit composed of ripened ovaries and other parts of all the female flowers in a cluster. A related tree, the jackfruit (*A. heterophyllus*), produces similar fruits.

brick (*archaeology*) The tools used by early humans were made by shaping natural materials, sometimes treating wood or stone with heat to improve hardness. Brick started the same way, as clay shaped into a rectangular solid and dried by the Sun. Clay brick was most popular where both stone and wood were hard to find, as in the centers of early civilization in Mesopotamia, but Sun-dried brick was used everywhere. Early people learned that Sun-dried brick holds its shape better as a composite, mixed with straw. By about 3500 BC the practice of hardening brick with high heat in kilns was in use, but this more expensive brick was saved for outer layers or for important buildings. Adobe, still in use, is essentially the same as Sun-dried brick, and fired brick also is popular around the world.

bristle worm (*animals*) Also called polychaetes, bristle worms have fleshy lobes (parapodia) armed with stiff bristles (setae) extending from the sides of each body segment. These annelid worms generally live in the sea, from shallow coastal waters to great depths. Some bristle worms swim through the water or crawl about. Others live in tubes they build or burrow into sand.

Broca, Paul (*neurologist/anthropologist*) [French: 1824–80] A noted medical researcher who made important contributions on tumors and aneurysms (pathological dilations of blood vessels), Broca is best remembered for his discovery of the part of the human brain that controls speech, now called Broca's area. As a result of his studies of skeletons exhumed from an ancient cemetery in Paris, Broca also developed a strong interest in physical anthropology. He established the Society of Anthropology of Paris and was a founder of the French Association for the Advancement of Science.

Broglie, Prince Louis-Victor de (*physicist*) [French: 1892–1987] Broglie's doctoral thesis in 1924 explained that wave equations also apply to particles. Broglie showed that every material object is associated with a wave, most observable for small bits of matter, such as electrons. By 1926 Broglie's waves, showing the predicted properties, had been found. Erwin Schrödinger used Broglie's idea to create the wave equation for quantum mechanics and Max Born interpreted the waves as probabilities.

bromeliad (*plants*) Members of Family Bromeliaceae are a diverse group of comparatively small tropical and semitropical American plants. Many species are characterized by long stiff leaves with spiny margins arranged in a circular cluster close to the ground. Flower stalks up to 4 ft (1.2 m) tall grow from the center of the cluster. Among the best-known bromeliads are pineapples. Other bromeliads, such as Spanish moss (*Tillandsia usneoides*), which hangs from trees throughout the southern United States, are epiphytes.

bromine (*elements*) The halogen bromine, atomic number 35 and chemical symbol Br, is the only nonmetal liquid at room temperature. It was recognized in 1826 by Antoine-Jérôme Balard, [French: 1802–76] and named from the Greek *bromos*, "stench," because of its smelly vapors. Since its properties are midway between those of chlorine and iodine bromine was first thought to be "iodine chloride." Potassium bromide, called bromide, is a sedative, whose properties led to the use of the word "bromide" to mean a banal saying that might put one to sleep.

bronze (*materials science*)   Bronze is copper alloyed with tin (and perhaps other elements, especially antimony or phosphorus). Copper was the first metal regularly used by humans, but it is hard to cast and rather weak. Metalworkers soon learned that copper mixed with arsenic or tin melts easier and can become stronger or harder. Bronze became the metal of choice until easy methods of smelting iron became available, so the period from about 3000 BC to about 1500 BC is the Bronze Age. Today the main use for bronze is where corrosion resistance is important (as in ships' propellers) or where large objects must be cast (as in church bells), since bronze is easier to cast than iron or steel.

brood (*ethology*)   Young animals—particularly the young of birds and insects—that are born and cared for at the same time form a brood. The term brood also is used as a verb, to describe the provision of warmth and protection to the young. Among herring gulls (*Larus argentatus*), for example, the parents care for and feed their chicks as long as the chicks remain in the nest. Their brooding behavior includes responding to the sight of an open mouth by dropping food into it.

broom (*plants*)   Shrubs of 2 genera of legumes, *Cytisus* and *Genista*, are commonly called brooms because of their stiff broomlike stems. They are native to dry, mild areas of Europe, western Asia, and northern Africa. Species exhibit a wide variety of shapes and sizes, from low-growing mats to shrubs 10 ft (3 m) or more in height. Among the best-known species is Scotch broom (*C. scoparius*), an erect deciduous shrub with yellow pealike flowers borne singly or in pairs.

brown dwarf (*stars*)   Brown dwarfs are bodies too small to be stars but too large to be planets. They glow dimly as a result of energy released by gravitational contraction. A brown dwarf must be between 10 and 80 times the size of Jupiter. The first definite brown dwarf was located orbiting the star Gliese 229 late in 1995. Although about 50 times Jupiter's mass, its diameter is about the same as Jupiter's. Since then a dozen or so brown dwarfs have been located, most orbiting close to ordinary stars.

brownfield (*environment*)   A brownfield is an abandoned or underutilized site, usually in an urban area, that is contaminated with hazardous substances and must be cleaned up before it can be put back into productive use. Closed industrial facilities, empty warehouses, razed gas stations, abandoned railway lines, and idled ports are examples of brownfields. Though remediation of brownfields can be extremely costly, the process offers distinct advantages over development of open lands (greenfields).

Brownian motion (*particles*)   When tiny particles suspended in liquid are observed under a microscope, they move randomly as if they were alive. Robert Brown [English: 1773–1858] first observed this motion in pollen particles in 1827, and soon found that suspended nonliving small particles also move. In 1905 Einstein established that the particles move because they are pushed randomly by unseen molecules. Einstein's work offered the first clear proof of the Greek idea that matter is made from unseen particles and of the theory that heat is the motion of such particles.

bryophyte (*plants*)   The bryophytes comprise more than 20,000 species of mosses, liverworts, and hornworts. They were first recognized as a distinct division of plants in 1821 by Samuel Gray [English: 1766–1828], although the name Bryophyta ("moss plants") was not proposed until later in the 1800s. Bryophytes are small plants, generally less than 9 in. (23 cm) long, that grow in moist habitats. They lack true roots, stems, and leaves; that is, they do not have vascular tissue but they are composed of structures that look like roots, stems, and leaves. They do not produce flowers or seeds. There is a distinct alternation of generations, with the green gamete-producing generation dominant and the nongreen spore-producing generation growing from and dependent on it.

bryozoan (*animals*)   Popularly called moss animals, members of the phylum Bryozoa are common marine invertebrates. Individually microscopic, they live together in large colonies, some of which resemble mossy expanses. Each animal has a circular crown of tentacles (lophophore) around its mouth that draws bacteria and protists toward the mouth. The mouth has a muscular pharynx that sucks in the food.

The bryozoan fossil record dates back more than 500,000,000 years. Because certain fossil species are closely associated with specific rock formations, they are widely used for determining the age of rock specimens.

bubble chamber (*particles*)   When an energetic subatomic particle passes through a liquid cooled below its normal freezing point, it creates a trail of bubbles. While observing bubbles in a glass of beer early in the 1950s, Donald Glaser [American: 1926-] recognized that this effect could be used to observe the paths of subatomic particles. The bubble chamber detectors based on Glaser's idea became the main particle detectors of his time. Glaser received the 1960 Nobel Prize in physics for his invention.

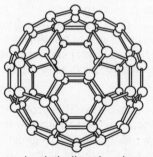

buckyball molecule

BUCKMINSTERFULLERENE

buckminsterfullerene (*materials science*)   Carbon can link its atoms to each other in several ways, producing such allotropes as graphite and buckminsterfullerene; the latter resembles a section of a graphite sheet folded into a closed surface. Some of the hexagons that are the normal arrangement of atoms in graphite are replaced by pentagons. The laws of geometry dictate the smallest way in which the closed surface will form—each molecule contains 12 pentagons made from 60 carbon atoms, but the sides of the pentagons create 20 hexagons. This is the same arrangement as that of the geodesic dome designed by Buckminster Fuller [American: 1895–1983], so this allotrope of carbon is called buckminsterfullerene. Individual molecules are known as buckyballs. Fullerenes are buckminsterfullerene and carbon allotropes that have similar ball-like molecules with more than 60 carbon atoms. (*See also* **nanotube**.)

buckthorn (*plants*)   Members of the genus *Rhamnus* are shrubs and small trees native mainly to northern temperate regions. They generally have thorn-bearing branches, small greenish flowers, and small berrylike fruits. The fruits and bark contain anthraquinones, pigmented compounds long used as laxatives.

The buckthorn family, Rhamnaceae, also includes the jujube tree (*Zizyphus jujuba*), a native of China that has been cultivated for thousands of years for its fleshy fruits, called Chinese dates.

buckwheat (*plants*)   Native to central Asia, buckwheat (*Fagopyrum sagittatum*) is an annual herb that grows to about 2 ft (0.6 m) tall. Its clusters of small white flowers produce fruits that look like miniature beechnuts. Buckwheat has been cultivated for more than 1000 years for its starchy seeds. It is not related to wheat but is classified in the family Polygonaceae, which also includes rhubarb and sorrel (*Rumex*).

bud (*botany*)   Among the most prominent features of a growing woody stem are its buds. A bud consists mainly of growth tissue. In most cases, tough bud scales (actually modified leaves) surround the bud, protecting it from injury and drying out. Some buds give rise to leaves and stems, some produce flowers, and some produce twigs bearing both leaves and flowers. When a bud opens, the scales fall off, leaving a scar around the stem.

At the tip of a stem usually is a terminal bud. The distance between 2 terminal bud scale scars equals the length of growth in 1 growing season. Buds that form at the juncture of a leaf stalk and the stem are lateral, or axillary, buds. Occasionally buds arise elsewhere, such as at a point of injury; these are called adventitious buds.

budding (*biology*)   A form of asexual reproduction, budding involves the development of an outgrowth, or bud, from the body of a mature organism. This bud grows into a new individual that is genetically identical to its parent. The offspring—which eventually is as big as the parent—may become detached from the parent or remain attached as part of a colony. Organisms that reproduce by budding include yeasts, sponges, and hydras.

buffalo (*animals*)   Several mammals of the bovid family that are native to Asia and Africa are commonly called buffaloes. (The North American buffalo is actually a bison.) The largest is the widely domesticated Asian water buffalo, or carabao (*Bubalus bubalis*), which stands up to 6 ft (1.8 m) tall at the shoulder and has thick horns that sweep out and backward.

buffer (*chemistry*)   A buffer is a mixture of an acid and a corresponding base that together resist changes in pH caused by the addition to the solution of more of any acid or base (not just the ones used as the buffer). Loosely, people refer to an acid or a base by itself as a buffer when it is added to an existing solution to improve the buffering capacity of the solution—for example, when limestone (a base) is added to lakes affected by acid rain.

bug (*animals*)   True bugs constitute the insect order Heteroptera ("half wings"). Their front wings are thick and leathery on the half near the base and thin and membranous on the outer half. At rest, the wings overlap on the back of the abdomen. Bugs have piercing and sucking mouthparts that are enclosed in a stout beak. The great majority feed on plants.

The order includes assassin bugs, bedbugs, stinkbugs, water striders, water bugs, and other species. Many are serious pests, including the chinch bug (*Blissus leucopterus*), which feeds on cereal crops, and assassin bugs of the genus *Triatoma*, which suck human blood, spreading parasitic protozoa that cause Chagas' disease. The largest is the giant water bug (*Lethocerus grandis*) of South America, which exceeds 4 in. (10 cm) in length.

bulb (*botany*)   Bulbs are short underground stems that are surrounded by thick, fleshy leaf scales. The leaf scales store food and protect the plant bud inside. In addition to growing into a complete plant, a bulb can produce small bulbs that generate new plants—a type of vegetative propagation. Onions, lilies, and tulips are among plants that produce bulbs.

Bunsen, Robert (*chemist*)   [German: 1811–99] Bunsen's name is familiar from the simple gas heater used in chemistry laboratories (the Bunsen burner), although his only connection with the burner is that an assistant in his laboratory made and sold the devices. However, Bunsen's main contribution to science did involve heated gases: in 1859 he and Gustav Robert Kirchhoff developed the spectroscope, a tool for recognizing elements from the light they produce when heated. Bunsen and Kirchhoff used their spectroscope to discover the elements cesium and rubidium. Bunsen also devised various types of laboratory equipment for better electrolysis, for measurement of heat and radiation, and for filtering chemicals.

buoyancy (*mechanics*)   Buoyancy is the property a material must have to rise or float in a fluid; the buoyant material is any body with a lower density than the fluid. The force that causes the buoyant material to rise or to find equilibrium is gravity. A volume of the fluid equal to that of the buoyant body has greater mass than the body, so the fluid flows under the buoyant body, pushing it up in accordance with Archimedes' principle. A hot air balloon, for example, rises because cold, dense air flows below it, not because hot air itself lifts the balloon.

Burgess shale (*paleontology*)   Located on a saddle between two peaks in the Canadian Rockies, the Burgess shale is a treasure trove of fossils of soft-bodied marine animals preserved in great detail and including many previously unknown species. Discovered by paleontologist Charles D. Walcott [American: 1850–1927] in 1909, it dates from about 520 million years ago. At that time, this part of Earth's crust formed the continental margin of North America and was actually near the equator; later crustal movements brought it to its current location. While most of the Burgess shale animals were obviously related to modern groups—especially arthropods, but also sponges, mollusks, worms, etc.—others were unlike any living animals.

burin (*paleoanthropology*)   Chunky stone tools with sharp points are called burins ("engravers"), although burins are thought to have been general-purpose tools used for many tasks beyond engraving. If a pointed tool is long and slender, however, it is an awl. Symmetrical thin tools with points have various other names.

buttercup (*plants*)   The genus name for buttercups, *Ranunculus*, is derived from Latin words meaning "little frog," a reference to the moist habitats in which these herbs are often found. Buttercups generally have glossy, bright yellow flowers. Many species are less than 6 in. (15 cm) tall but some grow to heights of 3 ft (0.9 m).

The buttercup family, Ranunculaceae, comprises approximately 2000 species of herbs and shrubs. Most are native to northern temperate areas and favor damp habitats. The family includes aconites (*Aconitum*), anemones (*Anemone*), clematis (*Clematis*), columbines (*Aquilegia*), and delphiniums.

butterfly and moth (*animals*)   Classifying insects of Order Lepidoptera as butterflies or moths is somewhat subjective. Traditionally, the word "butterfly" refers to species that are active during daytime and have antennae with an enlarged club at the tip; "moth" refers to species that are nocturnal and lack clubbed antennae. But a group of tropical American moths are believed to be more closely related to butterflies. And skippers, traditionally considered to be thick-bodied butterflies, are now thought to share more characteristics with moths.

cabbage (*plants*)   A single member of the mustard family, *Brassica oleracea* is the genesis of not only cabbage but also broccoli, cauliflower, kale, kohlrabi, and Brussels sprouts. This resulted from selection by plant breeders for specific traits: leaves (kale), stem (kohlrabi), stem and flowers (broccoli), flower clusters (cauliflower), lateral buds (Brussels sprouts), and terminal buds (cabbage). The wild ancestor is believed to have been a kalelike plant native to coastal Europe.

cacao (*plants*)   The cacao or cocoa tree (*Theobroma cacao*) is an evergreen native to lowland tropical forests of Central and South America, where it often grows in the shade of taller trees. Averaging heights of about 26 ft (8 m), it has large leathery leaves and small flowers that grow directly from branches and the trunk. A flower gives rise to a pod that contains 20 to 40 seeds, or cacao beans. These are fermented and roasted to produce cocoa and chocolate.

The cacao family, Sterculiaceae, is named for its foul-smelling flowers designed to attract pollinating flies; the Latin *stercus* means "dung." The family comprises about 1000 species, including the kola nut trees (*Cola*) native to tropical Africa. Fresh kola nuts are chewed as a stimulant; dried nuts are used in the production of cola beverages.

cactus (*plants*)   Native to the Americas, the approximately 2000 species of Family Cactaceae live chiefly in warm, dry habitats. They exhibit a startling diversity, ranging from small "buttons" to plump globes to the multibranched saguaro (*Carnegiea gigantea*), which can exceed 50 ft (15 m) in height. Although they evolved from broad-leafed plants, most cacti lack green leaves; rather, photosynthesis is carried out by the green stem and branches, which have a thick cuticle, often covered with wax, that limits water loss.

Leaves, shoots, and flowers arise from pads of tissue called areoles, which are arranged in a distinctive pattern useful in species identification. Typically, the leaves have been modified into spines or hairs. Roots lie close to the surface and spread widely—a saguaro's roots, for example, may cover an area 100 ft (30 m) in diameter. In addition, many cacti have a swollen taproot, which may be larger than the rest of the plant. Flowers depend on color and scent to attract animal pollinators. Fruits are fleshy and juicy in some species, hard and dry in others.

caddis fly (*animals*)   Constituting the order Trichoptera, caddis flies are small, dull-colored mothlike insects. They have very long antennae and 2 pairs of membranous wings that are held tentlike when at rest. The flies' larvae, called caddisworms, are aquatic; some construct portable cases of sand, pebbles, twigs, and other materials, bound together with silk spun from modified salivary glands. The adults live in foliage near the streams of their birth.

cadmium (*elements*) Cadmium, atomic number 48, is between zinc and mercury in the periodic table. While cadmium metal looks like zinc, it is as toxic and as useful as mercury. Among cadmium's common applications are solders, bearing alloys, electroplating materials, batteries, and control rods for nuclear reactors. As a result of its usefulness, cadmium contamination has become a major problem at abandoned industrial sites and in waste disposal. Cadmium was first recognized in 1817 by Friedrich Strohmeyer [German: 1776–1835], who named it from *cadmia*, a zinc ore. Chemical symbol: Cd.

caecilian (*animals*) Amphibians often mistaken for worms, caecilians are long, slender, legless creatures. They have small, poorly developed eyes and depend on 2 retractable tentacles—1 on either side of the head—to sense odors and tastes. Most species are less than 2 ft (0.6 m) long, but *Caecilia thompsoni* of Colombia can reach lengths of over 4 ft (1.2 m). Caecilians spend their adult life burrowing through damp soil in tropical forests, feeding on worms, insects, and other small animals.

calabash (*plants*) Native to the American tropics, the calabash tree (*Crescentia cujete*) grows to about 40 ft (12 m) tall. Its bell-shaped yellow-purple flowers give rise to hard-shelled fruits (technically berries) that may be 12 in. (30 cm) in diameter. The shells often are used to make bowls and other containers. The hard shell of fruits of the bottle or calabash gourd (*Lagernaria siceraria*), a vine of the Old World tropics, is used in a similar manner.

calcite (*rocks and minerals*) Calcite ("lime rock") is the crystal form of $CaCO_3$, the main mineral in limestone and marble. Calcite has many crystal shapes and often forms a clear or translucent large crystal. The transparent form, Iceland spar, is known for double refraction of light. Calcite also occurs as tiny white crystals in pure marble and stalactites and stalagmites. Many animal shells are calcite, fossilizing as limestone beds.

calcium (*elements*) The alkaline earth metal calcium, atomic number 20, was discovered in 1808 by Humphry Davy through electrolysis. Its name comes from the Latin *calx*, "lime" or "limestone," and its symbol is Ca. Heating limestone, $CaCO_3$, drives off carbon dioxide, leaving calcium oxide, CaO, which combines with water to make lime, used in plaster. A calcium compound is also the basis of cement. Calcium is the most abundant metal in our bodies, important both structurally as the metallic element in bone and biochemically as a messenger used by cells.

calculus (*mathematics*) Calculus (also known as analysis, especially when the most general approaches are used) is a method for handling quantities that can change value continuously, such as velocity and acceleration, or that are defined by very general curves, such as the area of a figure bounded by curves. As arithmetic deals with numbers and algebra with variables, calculus handles functions. Operations to obtain one function from another are called the differentiation (finding a **derivative**) and integration (obtaining the **indefinite integral**). Both are applications of another operation, finding the **limit** of a function.

In 1665 and 1666 **Newton** expanded on methods of approximation developed by earlier mathematicians, creating the first version of calculus. About 10 years later **Leibniz** developed the same mathematical tools, using different symbols and words with somewhat different meanings. Leibniz published his first account in 1684 and Newton his first public account in 1687, although Newton had circulated manuscripts much earlier. Most of the symbolism and language used today derive from Leibniz.

caldera (*geology*)   A large basin that may contain several volcanic vents or even separate **volcanoes** is called a caldera. The most common cause of caldera formation is collapse of the **crust** after a volcano's **magma** reservoir has emptied; this may occur rapidly during a larger eruption or slowly as magma moves out of the reservoir over many years. Calderas include the dormant Crater Lake in Oregon; Long Valley in California, which shows signs of new magma entering the reservoir; and Rabaul on New Guinea, where the caldera is over 1000 years old but volcanoes within it continue to erupt frequently.

calendar (*solar system*)   All plants and animals respond to changes caused by the rotation and revolution of Earth about the Sun; some also respond to the revolution of the Moon about Earth. Humans have employed astronomy to convert these changes into a system known as the calendar, after a Roman word, *calends,* for the first day of a new moon. Not only do different groups of humans use different starting points for their calendars, but they also emphasize different events. Some, such as Jews and Muslims, base their calendars on the 19-year Metonic cycle that relates lunar months to solar years. Others, especially in the West, use the official Gregorian calendar, which is closely tied to the annual 365.256-day earthly rotations. All calendars approximate the lunar month of about 29 days and most add "leap" days to the year from time to time to keep the calendar year about the same as the astronomical year.

californium (*elements*)   Californium is a radioactive **actinide** metal, atomic number 98, first made in 1949 by **Glenn T. Seaborg** and coworkers at the University of California, for which it is named. Californium has an isotope with a **half-life** of about 800 years that is a powerful neutron emitter; it is used for a steady supply of neutrons. Chemical symbol: Cf.

Callisto (*solar system*)   One of Jupiter's 16 moons, Callisto is second in size—2986 mi (4806 km) in diameter—to Ganymede. It orbits Jupiter beyond Ganymede at about 1 million mi (2 million km). The heavily cratered satellite, discovered by **Galileo** in 1610, is made largely from water ice.

Camarasaurus (*paleontology*)   A plant-eating dinosaur of the late Jurassic, *Camarasaurus* grew to lengths of 57 ft (17.5 m) and weights of 68,000 lb (31,000 kg). Its small head (with a tiny brain) sat atop a long neck. Its front legs were well-developed, and the animal walked on all fours. *Camarasaurus* apparently spent much of its time browsing on trees and large bushes in swamps and along rivers.

cambium (*botany*)   A growth tissue called cambium enables plant stems and roots to grow in diameter. There are 2 types of cambium: vascular and cork. Vascular cambium lies between the xylem and phloem (the vascular tissues). Cell division on the inner side of the vascular cambium produces new xylem; cell division on the outer side produces new phloem. Cork cambium forms near the surface of stems and roots. It produces cork.

camel (*animals*)   Hoofed ruminant mammals of the family Camelidae include 2 species of camels (*Camelus*) as well as llamas and their relatives. The Arabian camel, or dromedary (*C. dromedarius*), is about 7 ft (2.1 m) tall at the shoulder and has 1 hump. The slightly smaller Bactrian camel (*C. bactrianus*) has 2 humps. The humps are masses of fat that nourish the camels when food is scarce in their desert and semidesert habitats. Other adaptations to the environment include the camels' ability to shut their nostrils to keep out sand, long eyelashes to catch sand, broad bone ridges above their eyes to shield them from the strong noonday sun, and the ability to go without water for prolonged periods of time.

Camptosaurus (*paleontology*)   A relatively small plant-eating dinosaur of the late Jurassic, *Camptosaurus* grew to lengths of 20 ft (6 m) and weights of 2200 lb (1000 kg). Its hind legs were significantly longer than its forelimbs, but it moved mainly on all fours. Its toothless mouth ended in a horny beak that may have been an adaptation for browsing on shrubs.

cancel (*arithmetic/algebra*)   Simplifying computation either by removing sums of 0 or common factors is called canceling although the meaning of *cancel* is different in the 2 forms. For addition, simplify a sum such as $3 + 4 - 3 + 7$ by putting a stroke (/) called a cancel through both 3 and –3 to show the sum is 0. In algebra, use the same concept to simplify an equation such as $x + 3 = 2x + 3$, since the threes on each side "cancel" each other. For fractions, canceling is used to remove common factors between a numerator and denominator. In $5x/5y$ the fives can be canceled, for example.

cancer (*physiology*)   Cancer is a disease characterized by the uncontrolled growth of abnormal cells. The result is a large mass of cells called a malignant tumor. Cells can potentially break free from the tumor and migrate to other parts of the body—a process called metastasis—with life-threatening results. Cancer is a complex process that involves the presence of oncogenes coupled with damage to the tumor suppressor genes that regulate cell division.

candle, standard (*measurement*)   Astronomers call stars or supernovas whose approximate brightness is thought to be known standard candles. The traditional measure for brightness was the wax candle; the apparent brightness of the candle was used to estimate distance, also the purpose of astronomical standard candles. (The SI measurement for luminosity is the candela.) The original standard candles are Cepheid variables, whose brightness is proportional to their period. Even when a Cepheid is in a far-off galaxy, its period is easy to observe, telling the brightness and therefore providing an estimate for distance.

canna (*plants*)   Native to the American tropics and subtropics, cannas (*Canna*) are tall, large-leafed herbs with showy flowers. The flowers, typically yellow or red, form in clusters at the tip of stout stalks up to 8 ft (2.4 m) tall. The 3 true petals of each flower are small; the showy part of the flower actually consists of petal-like stamens.

cannibal (*anthropology*)   A person who knowingly consumes part of the flesh of another human is a cannibal. Although there have long been reports of tribes that regularly practice cannibalism, mainly for ceremonial purposes—especially in Central and South America, Africa, and the Pacific Ocean islands—many anthropologists claim that there is no hard evidence of cannibalism as a cultural practice; others disagree, especially with regard to the Aztec and other Native Americans. Some paleoanthropologists have found fossils that suggest cannibalism was practiced by early humans such as the Neandertals.

cannibalism (*ethology*)   Eating other members of an individual's own species, called cannibalism, is surprisingly widespread in the animal world, and can offer nutritional, reproductive, and other benefits. Gulls often eat the nourishing eggs of their neighbors. In times of food scarcity, parents of burying beetles eat some of their newly hatched offspring so that the family remains only as big as the food supply can support.

Sexual cannibalism, in which a female eats her male partner, occurs among mantises and many spiders. It is not uncommon, for example, for a female praying mantis to start eating her mate while they are still copulating.

Canopus (*stars*)   This bright star, second after Sirius, has often been used by spacecraft as a beacon to orient themselves—appropriately, its name means "helmsman." It is a yellow supergiant about 25 times as wide as the Sun.

canopy (*ecology*)   The top layer of a forest, formed by the crowns of the tallest trees, is the canopy. It receives the greatest amount of sunlight, and therefore most of a forest's photosynthesis occurs here. The canopy also controls the forest's temperature and humidity and absorbs the impact of heavy rains and winds. In some forests the canopy is dense and continuous, limiting growth of shorter plants. Elsewhere, treetops are more widely spaced, allowing sunlight to reach lower layers. Many species call the canopy home, including epiphytes, leaf-eating insects, insect-eating birds, and birds of prey.

Cantor, Georg (*mathematician*)   [German: 1845–1918] Cantor was the first mathematician to cope directly with infinity, becoming the founder of the theory of infinite sets with a series of discoveries starting in 1874. (Bernhard Bolzano [Bohemian: 1781–1848] preceded Cantor in accepting infinity and using sets, but his work had little influence.) By 1895 Cantor had determined how to compare infinite sets. He proved that the rational numbers have the same number of members as the natural numbers, but the real numbers and all points in 3-dimensional space are larger sets, equal in size to each other. Cantor also found an infinity of infinities. His work solved problems of infinite series but also introduced paradoxes not yet completely resolved.

Capella (*stars*)   Capella is a binary system of two similar stars, each a yellow main-sequence star about 3 times the mass of the Sun. The stars orbit each other with a period of 104 days. The combination of 2 moderately bright stars not very far from the Sun puts Capella on the brightest stars list.

caper (*plants*)   The caper bush (*Capparis spinosa*) is a spiny evergreen shrub native to the Mediterranean region. It grows to about 3 ft (0.9 m) and has simple leaves and clusters of large white flowers. The flower buds are picked before opening, pickled or dry-salted, and sold as condiments.

capillary action (*physics*)   If molecules in a liquid are attracted more to a solid than they are to each other, the liquid flows along the solid, even against the force of gravity. This flow is called capillary action. The adhesion of water to fibers in a washcloth or kerosene to the material in a wick carries the liquid out of a basin into which the washcloth or wick is dipped. Since water is attracted to glass, water flows upward in a glass tube, an effect more noticeable in a tube of small diameter because the flow continues until the weight of the water is greater than the force of adhesion.

captive breeding (*environment*)   Breeding wild animals in captivity in order to reintroduce them to their native habitats is a technique used to try to save species from extinction. It has helped return Père David's deer to China, Arabian oryxes to Jordan, golden lion tamarins to Brazil, and condors to the mountains of southern California.

capuchin (*animals*)   Named for the peak of thick hair atop the head that resembles the hood worn by Capuchin monks, capuchins (*Cebus*) are lively, mischievous monkeys once used by organ grinders to collect money from onlookers. They also are called ring-tailed monkeys because they usually carry their tail coiled at the tip. Capuchins are among the most intelligent members of the New World monkey family Cebidae. The 58 species in this family also include howlers, wooly monkeys, spider monkeys, squirrel monkeys, and ukaris. They live high in trees of tropical forests of Central and South America. All are excellent climbers, with long limbs and long furry tails that are prehensile in some species. Mostly diurnal, they feed mainly on fruits and insects.

capybara (*animals*)   The largest rodent, the capybara (*Hydrochaeris hydrochaeris*) is about the size of a young pig, growing to more than 4 ft (1.2 m) long and weighing over 100 lb (45 kg). It is native to forests of tropical South America, where it lives in dense vegetation along the banks of rivers and lakes. The capybara is a strong swimmer.

carbohydrate (*biochemistry*)   Carbohydrates are organic compounds composed of carbon, hydrogen, and oxygen, typically with a 2 to 1 proportion of hydrogen atoms to oxygen atoms. The simple sugar glucose, for example, has the formula $C_6H_{12}O_6$. More complex carbohydrates, called polysaccharides, are huge molecules formed by linking together hundreds or thousands of simple sugar molecules.

The carbohydrates sugar, starch, and glycogen are major sources of energy for most organisms. Another carbohydrate, cellulose, is the main structural component of plants.

carbon (*elements*)   A nonmetal (atomic number 6, chemical symbol C), carbon has been known since prehistoric times. Native carbon occurs as diamond or graphite crystals and as the amorphous substance that is the principal element in coal (the name is from the Latin *carbo*, "coal"). Since 1985 carbon has been recognized in the form of large, hollow molecules called fullerenes, especially the 60-atom form buckminsterfullerene. In 1991 a different kind of large carbon molecule was found, the nanotube.

Carbon is also part of millions of compounds, ranging from simple carbon dioxide in the air and calcium carbonate (limestone) to complex organic compounds. It is essential to all forms of life and the main direct source of energy through burning of wood, coal, and hydrocarbons. There are 7 known isotopes. The most common form is carbon-12, used as the basis of the atomic mass unit, while carbon-14 is used in radiocarbon dating.

carbonate (*compounds*)   The carbonate radical is CO_3, which has a valence of −2. It is familiar in combinations with light metals such as calcium and sodium. Calcium carbonate ($CaCO_3$) is an abundant mineral, best known in limestone and marble. Sodium carbonate (washing soda) is a hydrate of Na_2CO_3 used in making soap and in other industrial processes; sodium bicarbonate, $NaHCO_3$, is used as a source of carbon dioxide in baking. In modern chemistry the word "bicarbonate" (misleading if it suggests 2 carbonates, since it is actually the radical HCO_3) has been replaced with "hydrogen carbonate," so baking soda today is sodium hydrogen carbonate.

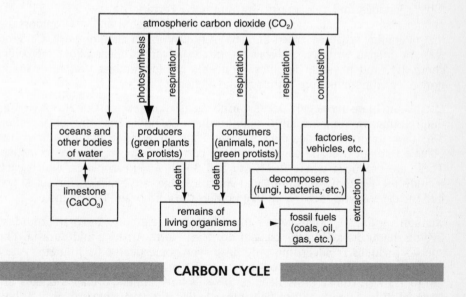

CARBON CYCLE

carbon cycle (*biochemistry*)   Carbon continually circulates through the environment, largely due to 3 life processes: photosynthesis by chlorophyll-containing organisms, food intake by animals and other non-photosynthesizing organisms, and respiration by all living things. Carbon leaves the cycle for storage in such reservoirs as the atmosphere, the oceans, fossil fuels, and rocks such as limestone. The time of storage ranges from instants to eons. The total amount of carbon in the environment remains steady, but there can be shifts in its distribution. For example, the great increase in the burning of fossil fuels that has occurred since the beginning of the Industrial Revolution is considered to be responsible for the marked increase in atmospheric carbon.

carbon dioxide and monoxide (*compounds*)   When carbon is oxidized by fire or respiration, the product is usually carbon dioxide (CO_2), since a carbon atom seeks 4 additional electrons to fill its outer shell, and each oxygen atom provides 2. Carbon dioxide is necessary to most forms of life, since it is one basis of photosynthesis. Carbon dioxide is also the greenhouse effect gas most involved in global warming.

When there is limited oxygen, carbon will accept 4 electrons from oxygen's outer shell, leaving 2 hanging out. Carbon monoxide (CO) is the result. Carbon monoxide is stable but reactive—it burns in air if ignited. Carbon monoxide is a poison that reacts with iron in hemoglobin, destroying blood's oxygen-carrying capacity.

Cardano, Girolamo (*mathematician*)   [Italian: 1501–76] Cardano's *Ars magna* ("Great art") of 1545 is often taken as the beginning of modern mathematics, although Cardano himself made few original contributions. The book is most noted for the first publication of the solutions of the general polynomial equation of degree 3 (the cubic) and degree 4 (the quartic). But the cubic was solved by Niccolò Tartaglia and the quartic by Ludovico Ferrari [Italian: 1522–65]. *Ars magna* also contained many general rules about algebra and was among the first to employ negative numbers and imaginary numbers. In another work, Cardano was the first to write on probability. Cardano advanced ideas about medicine, chemistry, and physics, but these concepts are lost among Cardano's many now-rejected notions from astrology and alchemy.

Caribbean Plate (*geology*)   Most Caribbean Islands (except Cuba), as well as Central America from Yucatán though northern Colombia and Venezuela, rest on a small tectonic plate called the Caribbean Plate. An island arc that runs from Hispaniola through Grenada consists in large part of volcanoes, dormant or active. These are produced where the North and South American plates, which each include the western Atlantic to mid-Atlantic ridge, are experiencing subduction into the deep Puerto Rico trench that borders the Caribbean Plate on the east.

caribou (*animals*)   Members of the deer family, caribou (*Rangifer tarandus*) live on the tundra and in the taiga of northern North America and Eurasia. The species includes reindeer, the only deer ever domesticated by humans. Adult caribou average lengths of 5 to 7 ft (1.5 to 2.1 m) and weights of 175 to 250 lb (80 to 113 kg). Unlike other deer, both sexes have antlers; a male's antlers are large, used in defense and to fight other males during mating season. Caribou feet have broad hoofs designed for walking on soft ground and digging in snow. The animals live in herds that may number tens of thousands of individuals. The herds move almost constantly, drifting from one feeding ground to another. In spring pregnant cows migrate to calving grounds; at the onset of winter both sexes migrate to wintering grounds. Lichens, including so-called reindeer moss (*Cladonia rangiferina*), are their primary food.

carnivore (*ecology*)   An animal that feeds mainly on the flesh of other animals is called a carnivore. Starfish, spiders, owls, and lions are examples. Each carnivore has special adaptations for its chosen diet. Many centipedes, for example, hunt small arthropods by means of sense organs on their antennae, then inject the victims with a lethal poison.

The term carnivore also refers specifically to animals of the mammalian order Carnivora. These mammals (dogs, cats, foxes, raccoons, bears, etc.) have enlarged canine teeth, or fangs, designed for cutting flesh.

carnivorous plant (*botany*)   Most plants produce all the food they need in a series of processes that begins with photosynthesis. But nearly 600 plant species and subspecies obtain part or even all of their food by trapping and digesting insects and other small animals. These carnivorous (meat-eating) plants are small and typically grow in wet, highly acidic soils. The acidity severely limits the availability of nitrogen and other essential plant nutrients. Animal-trapping adaptations enable carnivorous plants to compensate for the soils' deficiencies.

Two types of traps are found among carnivorous plants. Active traps use movement to catch prey. Examples include Venus's-flytrap, bladderworts, and the waterwheel plant (*Aldrovanda*). Passive traps use sticky surfaces or one-way traps rather than movement. Sundews (*Drosophyllum*, etc.), pitcher plants, and cobra lilies (*Dorlingtonia*) are examples.

Carnot, Nicolas "Sadi" (*physicist*)   [French: 1796–1832] Carnot was interested in the improvement of steam engines, asking such questions as whether steam is the most efficient medium for obtaining motion from heat. His 1824 paper on the subject is the founding document of thermodynamics. Carnot used a few reasonable assumptions, logic, and mathematics to establish that efficiency depends entirely on the difference in heat between the source and the heat sink (where heat is removed at the end of a cycle). Thus the question of which substance is most efficient became irrelevant. Carnot also determined the maximum possible amount of work to be obtained from a particular combination of heat source and sink, and offered the first modern definition of work.

carob (*plants*)   A legume native to the Mediterranean region, carob (*Ceratonia siliqua*) is an evergreen tree that grows to about 50 ft (15 m) tall. It is cultivated for its edible dark brown pods, which are about 1 ft (30 cm) long.

carotenoid (*biochemistry*)   A group of yellow, orange, and red fat-soluble pigments, carotenoids are found in many organisms. The most common carotenoids are carotenes, which are hydrocarbons, and xanthophylls, which are oxygenated derivatives of carotenes.

Carotenoids are present in plant and algal cells that carry out photosynthesis. There they help capture light energy, which is then transferred to the green pigment chlorophyll. Because carotenoids do not break down as quickly as chlorophyll, their presence in leaves becomes evident in autumn. Carotenoids give color to carrots, tomatoes, crustacean cells, and many corals. Beta carotene is the source of vitamin A in animals.

carp (*animals*)   The biggest member of the large freshwater fish family Cyprinidae is the carp (*Cyprinus carpio*), which exists in many varieties and grows to about 4 ft (1.2 m) long and a weight of 80 lb (32 kg). The family also includes breams, chubs, dace, goldfish, minnows, and shiners. All have teeth only in the throat and fins that lack spines. The species vary greatly in habits and habitats: some feed mainly on algae while others eat mostly small insects and crustaceans; some prefer shallow muddy pools while others gravitate to clear streams.

carrot (*plants*)   The approximately 3000 species of the carrot family, Umbelliferae (also called Apiaceae), generally are herbaceous plants with hollow stems and compound leaves. The flat-topped flower cluster, or inflorescence, is called an umbel. Several species are cultivated for their edible roots, including the carrot (*Daucus carota*) and parsnip (*Pastinaca sativa*). Celery (*Apium graveolens*) and parsley (*Petroselinum crispum*) are grown for their leaves. Anise (*Pimpinella anisum*), caraway (*Carum carvi*), coriander (*Coriandrum sativum*), cumin (*Cuminum cyminum*), dill (*Anethum graveolens*), and fennel (*Foeniculum vulgare*) provide seeds and leaves used to flavor foods. Some species are poisonous, including poison hemlock (*Conium maculatum*), which contains alkaloids that paralyze respiratory nerves. (The ancient Greek philosopher Socrates, condemned to death, died by hemlock poisoning.)

carry (*arithmetic*)   In the decimal system algorithm for addition, the steps proceed from right to left by decimal places. If a sum in one place has 2 or more digits, carry all digits but the last to the next place to the left, where the digits represent units of that place. For example, in adding 37 + 28 + 16, the sum in the ones place is 21. Carry the 2 to the tens place, where it means 2 tens, and complete.

carrying capacity (*ecology*)   The maximum number of individuals of a species that can be supported by the resources of a particular area is known as that area's carrying capacity. Food, water, shelter, and other limiting factors determine the carrying capacity.

Carson, Rachel (*marine biologist*)   [American: 1907–64] The 1962 publication of Carson's *Silent Spring* prompted worldwide interest in the environment and concern about the damage caused by chemical pesticides. In the book Carson linked pesticides such as DDT to the alarming decline in populations of eagles and other wildlife, which ingested the pesticides as they ate contaminated food. Carson pointed out that the pesticides accumulated in human tissues, too, and were even passed from pregnant women to their fetuses. Carson's work led to the 1972 ban on DDT in the United States.

Cartesian coordinates (*analytic geometry*)   Location of a point using distances from perpendicular axes creates ordered pairs or triples called Cartesian coordinates (after Descartes, an inventor of the method and first to publish a detailed account). In a plane a point is identified by the directed distance from the vertical or y axis (called the abscissa and indicated as x) followed by the directed distance from the horizontal or x axis (called the ordinate and indicated as y) combined as the ordered pair (x, y). Points in 3 dimensions are similarly identified by ordered triples of the form (x, y, z).

cartilage (*anatomy*)   Cartilage is a tough, elastic material that makes up part or all of the skeleton of vertebrates. A shark's skeleton consists entirely of cartilage, while in humans cartilage is found at the ends of long bones; between segments of the spine; and in structures such as the ear, nose, epiglottis, and larynx.

cartilaginous fish (*animals*)   Sharks, skates, rays, and chimaeras—members of the fish class Chondrichthyes—are characterized by skeletons made of cartilage. Their body is covered with toothlike placoid scales, creating a rough, sandpapery texture. There may be as many as 5 rows of teeth, which usually are not fused to the jaw; if a tooth in the front row is lost, another tooth moves forward and takes its place. Males have a copulatory organ called the clasper, and fertilization is internal. Some species lay eggs; others bear live young.

cashew (*plants*)   Native to tropical South America, the cashew (*Anacardium occidentale*) is an evergreen shrub or tree that grows to 40 ft (12 m) in height. It has leathery leaves and broad clusters of small, fragrant flowers. What appears to be a fleshy, pear-shaped fruit is actually the enlarged tip of the flower stalk. The actual cashew fruit, at the end of this stalk, is a kidney-shaped organ consisting of a double shell enclosing a single edible seed.

The cashew family, Anacardiaceae, comprises about 600 species, most of them tropical shrubs and trees. Many of the plants produce irritating resins that can cause skin rashes and blisters. The family includes mangoes, pistachios, poison ivy, and sumacs.

Cassini, Giovanni (*astronomer*)   [Italian-French: 1625–1712] Cassini's work established much of what we know of the solar system, including size, rotational periods of nearby planets, exact orbits of the moons of Jupiter, satellites of Saturn, and the gap in Saturn's rings named after him. He used parallax with a second observer on the other side of Earth to find the distance to Mars, from which other solar system distances can be computed.

caste system (*ethology*)   All ants and termites and many species of bees and wasps are social insects that live in colonies. These colonies maintain division of labor among groups called castes. Members of one caste differ anatomically or by age from members of other castes. A honeybee (genus *Apis*) colony, for example, consists of some 30,000 worker bees and 1 adult queen bee, who is much larger than the workers. Some workers are foragers that gather pollen. Other workers feed the queen and larvae. Others are "undertakers," removing dead larvae and bees from the hive.

casuarina (*plants*)   Native mainly to Australia and Southeast Asia, the genus *Casuarina* is comprised of small to large flowering trees characterized by wiry drooping branchlets and leaves reduced to small scales—hence the common if misleading name, Australian pine. The best-known species, the common casuarina (*C. equisetifolia*), is typically 50 to 100 ft (15 to 30 m) tall. It is also called ironwood or beefwood because of its extremely hard wood, which was used by Polynesians to make fishhooks and weapons.

cat (*animals*)   The cat family, Felidae, comprises about 35 species, including cheetahs, jaguars, leopards, lions, lynxes, ocelots, pumas, tigers, and domestic cats. The cats inhabit a variety of habitats, from South American rain forests to the rocky Himalayas. All are great hunters and, except for lions, usually hunt alone. They have powerful leg muscles and are extremely agile and well coordinated. Except for the cheetah, cats have retractile claws, which can be pulled into a sheath and thus protected against becoming blunt. Their sharp, pointed canine teeth are designed for stabbing. Cats typically stalk prey until they are close enough to rush and jump onto a victim, aiming for the neck and breaking the backbone. Most rely largely on sight but also have well-developed senses of smell and hearing; their whiskers are extremely sensitive to touch.

catalyst (*chemistry*)   A substance that speeds up a chemical reaction between 2 materials without itself being changed when the reaction is complete is called a catalyst. For biochemical processes, protein catalysts are called enzymes. A catalyst often works by briefly attaching itself to 2 molecules, bringing them physically closer together. When this proximity causes the molecules to react with each other, the catalyst releases its hold. Common catalysts are metals, such as platinum or copper, and metal compounds, such as nickel iodide. The catalytic converters used in automobiles, for example, use platinum and other catalysts to help exhaust gases convert from carbon monoxide to carbon dioxide or nitric oxides to become nitrogen and oxygen.

catastrophism (*geology*)   The theory that sudden events in the past, such as a worldwide flood, created existing geological forms, is catastrophism. Catastrophism starts with a belief, based on interpretations of the Bible, that Earth's history extends back a few thousand years only. In such a short time, great catastrophes seem the only possible explanation for observed geological formations. Catastrophism in geological science was replaced in the 19th century by uniformitarianism, based upon the theory of a very ancient Earth. In recent years a partial catastrophism has been revived in the sense that some major changes are thought to have occurred rapidly, such as the K/T event.

catenary (*geometry*)   A commonly seen curve is the catenary, formed when a flexible cable or chain is suspended from 2 ends. This curve looks rather like a parabola, although not really the same. The surface formed by revolving a catenary about its axis (a catenoid) is the surface of minimum area, which means that catenaries also appear in hanging soap bubbles. Despite its ubiquity, the catenary was not recognized as a separate curve until 1690, when the Bernoulli family investigated it.

caterpillar (*zoology*)   The larvae of butterflies and moths are called caterpillars. Their bodies are somewhat wormlike, with 3 pairs of jointed legs and several pairs of fleshy legs. Most caterpillars are equipped with chewing mouthparts. They eat enormous amounts—hence the great damage they cause to plants. To accommodate their continuous increase in size they periodically molt, or shed the exoskeleton. Following its last molt a caterpillar stops eating, finds a sheltered spot, secretes silk to enclose itself in a cocoon, and becomes a chrysalis.

catfish (*animals*)   Distributed around the world, mainly in freshwater, catfish are named for the whiskerlike barbels (feelers) around the mouth. The barbels help locate food at night and on cloudy days—the time when the fish is most active. Catfish typically lack scales, but some have an outer covering of bony armor. The fins contain spines; in some species poison glands surround the spines. Catfish range in length from a few inches to more than 10 ft (3.3 m).

cathode (*electricity*)   A cathode is a region where a negative charge builds on a battery or is emitted from a diode or similar device. It is called the cathode ("descending path") because Michael Faraday, who named it, pictured an "electric fluid" reentering an electric cell at the cathode; but electrons flow the other way. That is, a circuit between cathode and anode carries a current that moves a few electrons away from the cathode. In batteries negatively charged ions move toward or positively charged ones away from the cathode, giving it negative charge. In a cathode tube, such as the main tube of a television set, electrons from the cathode cause fluorescence when they strike the face of the tube.

cattail (*plants*)   Reedlike perennial herbs that grow to 8 ft (2.4 m) tall, cattails (*Typha*) inhabit swamps, the edges of ponds, and other wet habitats. They have erect, unbranched stems and erect, flat leaves. The minute petalless flowers are densely clustered in a cylindrical spike. The male flowers are borne uppermost; the female flowers—which form the brown "cat's tail"—are below.

cattle (*animals*)   The term cattle generally refers to certain domesticated mammals of the genus *Bos* in the bovid family. The genus also includes the gaur and yak, which have domesticated forms. Cattle were first domesticated as long as 8000 years ago. Most modern cattle breeds are classified as *B. taurus* and descended from a European species; Brahman or zebu cattle, *B. indicus*, characterized by a shoulder hump, originated from an Indian species.

cave and cavern (*geology*)   A cave is any natural tunnel or roofed opening in rock or soil large enough for a human to enter, often extending to dark regions. The term cavern is given to large caves usually found in limestone, gypsum, dolomite, or sandstone. Caves and caverns occur because water that contains dissolved carbon dioxide, such as rainwater, dissolves some minerals. Over millions of years, most rock is dissolved and carried away; however, other minerals that are slowly deposited from dripping water in caves result in formations such as stalactites and stalagmites or dramatic waves in rock walls. Some limestone caverns extend for hundreds of miles underground. Such caves may contain underground rivers and usually have organisms specialized for living in the dark.

Rarely caves form by other mechanisms. Lava tubes, for example, occur where fluid lava has drained away from the inner part of a stream of lava while outer walls have already solidified.

cave art (*paleoanthropology*)   Paintings, engravings, and sometimes sculpture from the latest Ice Age have survived for 10,000 to nearly 30,000 years in natural caverns, mostly in southeastern Europe. More than 200 painted or engraved caves from the Ice Age have been found, most with just a few rude sketches. However, sets of spectacular paintings and engravings of animals have also been found in caves such as Altamira in Spain and Lascaux, Cosquer, and Chauvet in France. The earliest cave art consists of outlines of human hands formed by blowing pigment around a hand placed on a cave wall. Later drawings show animals such as auks, cave bears, wild cattle, deer, horses, and mammoths as well as occasional humans dressed as shamans.

cave life (*ecology*)   A surprising variety of organisms are adapted for life in caves, which typically are dark, cold, and damp. No plants live here, except in the twilight zone at the entrance, for they require light for photosynthesis. Bacteria and fungi are plentiful, living on dead organisms, animal droppings, and organic material that enters from the outside. Cave animals are divided into 3 groups. Bats, flies, packrats, and other trogloxenes ("cave visitors") spend only part of their lives in caves, mostly near an entrance; they must leave to find food or for other purposes. Certain segmented worms, snails, spiders, and crickets are troglophiles ("cave guests"); they can pass their entire lives in caves but also can live elsewhere. Certain flatworms, insects, crayfish, fish, and salamanders are troglobites ("cave dwellers"); they can survive only in caves. Many troglobites have an excellent sense of touch but are blind and lack pigmentation.

Cavendish, Henry (*chemist/physicist*)   [English: 1731–1810] A wealthy recluse, Cavendish worked on scientific problems for his entire long life. He was the first to show that water is a compound and that air is a mixture—previously both were thought to be elements. In the process he discovered hydrogen and developed methods for weighing and measuring gases. But he is best known for his experiment determining the gravitational constant G. This experiment is often called "weighing the Earth," since knowing G permits one to calculate Earth's mass. Cavendish used Earth's mass to find its density, which is greater than that of stone, implying a heavy core. He also made advances in the study of heat and electricity; but these results were not published until long after his death and after their independent discovery by others.

cavy (*animals*)   Native to South America, cavies (Family Caviidae) are small plump rodents with short legs; some have a short tail, others are tailless. The largest are the Patagonian cavies, or maras, which can grow to 30 in. (76 cm) long. The best known is the guinea pig (*Cavia porcellus*), originally domesticated for meat by South American Indians some 3000 years ago and now also a popular pet and laboratory animal.

Cayley, Sir George (*physicist*)   [English: 1773–1857] As a child Cayley became interested in the possibility of heavier-than-air flight. Throughout his life he created model helicopters and gliders that flew, working out the main principles of heavier-than-air flight for the first time. Although one of his gliders carried a human pilot aloft, the only available engines of his time were heavy steam devices, too massive for flight. Cayley is also credited with inventing the tractor that runs on treads instead of wheels and a number of other useful devices.

cedar (*plants*)   The name cedar is commonly given to a variety of unrelated trees, but true cedars comprise the genus *Cedrus* in the pine family. These cedars are large evergreen trees native to mountains of North Africa and Asia. They grow to about 125 ft (38 m) and have wide-spreading branches with stiff needlelike leaves clustered on short spurs. The cones are erect. The best-known species is cedar of Lebanon (*C. libani*).

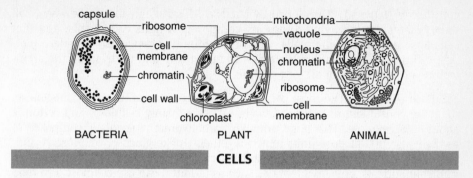

capsule
ribosome
cell membrane
chromatin
cell wall

mitochondria
vacuole
nucleus
chromatin
ribosome
cell membrane

chloroplast

BACTERIA   PLANT   ANIMAL

CELLS

cell (*cells*)   When **Robert Hooke** looked at cork under a microscope, he saw it was composed of many tiny spaces surrounded by walls. He named the spaces cells. Almost 200 years passed before biologists recognized that cells are the basic building blocks of all living things. Unlike the spaces in cork, however, living cells are anything but empty. Enclosed by a cell membrane, they contain a fluid (**cytoplasm**) in which highly specialized **organelles** and other structures carry out basic life functions. Plant, fungal, bacterial, and some protist cells have rigid cell walls outside the membrane.

Cells exist alone as complete organisms, as parts of a colony, or as units of a multicellular organism. Most cells are microscopic; the smallest are certain bacteria. The largest cells are the yolks of birds, led by the yolk cell of an ostrich egg, which is about the size of a baseball. Though cells tend to be spherical, many have unique shapes related to function. The nerve cell that extends from your spinal cord to your toe, for example, is incredibly long and has many branches. This shape aids in the speedy transmission of messages between the spinal cord and toe.

cell death (*physiology*)   Scientists have recognized 2 forms of cell death, necrosis and apoptosis. Necrosis is the pathological death of cells; it is induced by illness or injury. Necrosis can occur as the result of an inadequate blood supply to the cells, infection by a disease-causing organism, or damage by exposure to temperature extremes, radiation, or toxic chemicals.

In contrast, apoptosis is a programmed cell death that follows instructions resident in the cell's DNA. It is deliberate and orderly, and occurs both in developmental and adult stages of an organism. For example, apoptosis occurs as a tadpole loses its tail and metamorphoses into an adult frog. It is common in adult tissues that replace themselves frequently, such as human white blood cells.

cell division (*cells*)   All cells arise from the division of preexisting cells. There are 2 types of cell division. Most cells divide by **mitosis** but the cells that produce gametes (egg and sperm cells) divide by **meiosis**.

cell membrane (*cells*)   Every living cell is enclosed by a thin cell membrane (also called a plasma membrane). The membrane separates the cell from its environment, provides shape and some protection to the cell, and plays a critical role in the transport of materials in and out of the cell.

cell theory (*biology*)   Developed by Matthias Schleiden and Theodor Schwann, cell theory states that all living things are composed of units called cells, and that cells arise only from other cells. One of the founding concepts of modern biology, cell theory has enabled biologists to investigate and understand the architecture and processes of all forms of life.

cellulose (*botany*)   The most common organic compound on Earth is cellulose, the main constituent of plant cell walls. Wood is mainly cellulose, and cotton is about 90% cellulose. This large, complex carbohydrate molecule is composed of as many as 3000 glucose units joined together. Unlike glucose, cellulose is difficult to digest. Only a few microorganisms can break it down. Cattle and other ruminant animals, plus invertebrates such as termites, can use cellulose as a food source because cellulose-digesting microorganisms live in their digestive tracts.

center of mass (*mechanics*)   Newton discovered that the laws of motion and gravity describe objects behaving as if all their mass were at a point, called the center of mass. For a uniform object with symmetry in several directions, such as a ball or rod, the center of mass is the geometric center of the object. The center of mass is not necessarily part of the material body; the center of mass of a hollow sphere is the center of the sphere, for example. For objects in a uniform gravitational field, such as near Earth, the center of mass is also the center of gravity.

centipede (*animals*)   Centipedes ("hundred legs") are arthropods with 1 pair of legs on each body segment. The number of segments ranges from fewer than 20 to more than 150, depending on the species. One pair of legs is modified as poison claws. The centipede uses these claws to seize prey and inject it with poison. Centipedes live in moist land habitats, hiding under rocks and decaying logs during the day and hunting at night.

central angle (*geometry*)   The central angle of a circle is an angle in the plane of the circle that has its vertex at the circle's center. The measure of any angle in either degrees or radians is based on the proportion of a whole circle enclosed by a central angle of that circle. One-fourth of a whole circle, for example, has as the central angle 90°, or $\pi/2$ in radian measure.

centrifugal and centripetal force (*mechanics*)   Newton's first law of motion is that unless a force acts on a body, the body moves in a straight line. A force that causes a body to move in a circle is centripetal ("center seeking"); its direction is toward the center of the circle. Inertia, nevertheless, tends to keep the body moving straight, a fictitious force that must be overcome by the real centripetal force. If the mass is m, the speed v, and the radius of the circle r, inertia acts as mv^2/r. The inertia is often called centrifugal force, although some would banish the term. Centrifugal force is like the "force" a passenger in an automobile feels when the car undergoes acceleration—nothing pushes or pulls the passenger, whose body stays at rest or moves straight until a force from the car acts on it.

cephalopod (*animals*)   Octopuses, squids, cuttlefish, and nautiluses are members of the mollusk class Cephalopoda (from Greek words meaning "head foot"). In octopuses and squids, the foot is divided into a number of fleshy arms that surround the large head. The nautilus mouth is surrounded by numerous tentacles. An external shell is present in nautiluses; squids have internal remnants of a shell; octopuses have no shell. All cephalopods are marine carnivores that feed on invertebrates and fish.

Cepheid variables (*stars*) Some young, yellow supergiant stars undergo regular changes in size that result in changes in magnitude. Such variable stars are called Cepheid after the first one discovered, Delta Cephei, described in 1784 by John Goodricke [English: 1764–87]. In 1912 Henrietta Leavitt recognized that the absolute magnitude of all Cepheid variables can be determined from the period of variation. Since then, the Cepheids have been used as one of the main ways to determine actual distances to nearby galaxies, for if both absolute magnitude and apparent magnitude are known, it is relatively easy to calculate distance.

ceramic (*materials science*) The first materials known to have been manufactured by humans were ceramics, lumps of loess, sometimes shaped into small statues, heated in hot ovens in central Europe about 30,000 years ago. Most ceramics since then have been made by firing soils, especially clay. Such ceramics include bricks, pottery, terra-cotta, tile, and porcelain. Glass, made from sand, is sometimes considered a ceramic. All of these ceramics are made at high heat. Cement and plaster are ceramics produced by the action of water on powdered and previously heated rock.

The definition of ceramic is loose. All ceramics are hard, brittle, heat-resistant materials, mostly made from metal-oxygen compounds (oxides) held together with ionic bonds. Modern ceramics include ferrites, iron-based ceramics that were discovered in the 1940s and are now essential in modern electronics. Most modern ceramics, such as the ones used for lining furnaces, are made from pure compounds, not soils.

Cerenkov radiation (*particles*) Nothing can travel faster than the speed of light in a vacuum, but a physical particle can exceed that speed in a substance. Any charged particle traveling faster than light in a liquid or solid produces a wake of electromagnetic radiation called Cerenkov radiation or Cerenkov light. The wake forms a cone of polarized light for which the vertex angle in a given medium depends entirely on the momentum of the particle. If the rest mass of the particle is known, measuring the angle enables a physicist to deduce its speed; or, if the speed is measured, mass can be identified. Cerenkov radiation is also used to detect uncharged particles, such as neutrinos, by the light produced when a particle knocks out an electron or muon from an atom. The phenomenon was discovered in 1934 by Pavel A. Cerenkov [Russian: 1904–90; also spelled Cherenkov].

cerium (*elements*) Cerium, atomic number 58, is among the least rare of the rare earths, slightly less abundant than neodymium, but more easily recovered from several common ores. It was recognized in 1803 by Martin Klaproth [German: 1743–1817], who named it for the asteroid Ceres, discovered 2 years earlier. The highly reactive metal was not isolated until 1875. Like water, solid cerium is less dense than liquid. Among its many uses is as the basis of self-cleaning ovens, where it catalyzes various hydrocarbon residues. Chemical symbol: Ce.

cesium (*elements*) The alkali metal atomic number 55 was discovered in 1860 by Gustav Robert Kirchhoff and Robert Bunsen as a blue line in a spectrogram; thus it was named cesium from the Latin *caesius*, "sky blue" (chemical symbol: Cs). Cesium was the first element to be discovered spectroscopically. Liquid at room temperature, cesium is used in photoelectric cells, since light releases electrons from its surface, and in atomic clocks.

cetacean (*animals*) Technically all members of Order Cetacea are whales, though smaller species are commonly called dolphins and porpoises. Cetaceans are aquatic mammals with a hairless torpedo-shaped body. The front limbs are modified into flippers; through the course of evolution the hind limbs were lost. The powerful tail has horizontal flukes. Beneath the skin is a fatty layer called blubber. Cetaceans lack external ears. The eyes are protected by a greasy secretion. The nostrils are on top of the head, opening through a blowhole.

Chadwick, James (*physicist*) [English: 1891–1974] Chadwick's early experiments established that atomic number is determined by the number of protons in an atom (now used as the definition). In 1932 he discovered the fourth known subatomic particle, the neutron (following the electron, photon, and proton); its existence had been suspected since 1924. Chadwick's later experiments with particle accelerators became the basis of his contribution to the invention of the nuclear fission bomb.

chameleon (*animals*) Lizards of the family Chamaeleontidae, native to warmer parts of Africa, southern Europe, and Asia, are noted for their ability to change color in response to temperature, available light, fright, and other stimuli. Chameleons range in length from 2 in. (5 cm) to almost 2 ft (60 cm). They have a large head, large bulging eyes, a body flattened from side to side, grasping feet, and a tail that typically is long and capable of grasping. A chameleon's main weapon is its muscular tongue, which can shoot forward to lengths greater than the animal's body to capture insect prey. (Certain anoles of the Americas are incorrectly called chameleons.)

chamois (*animals*) An exceptionally agile, surefooted climber found in the high mountains of central and southern Europe and southwestern Asia, the chamois (*Rupicapra rupicapra*) is a hoofed mammal of the bovid family. It stands about 31 in. (80 cm) tall at the shoulder and weighs up to 110 lb (50 kg). It is goatlike in appearance, with black horns that curve backward at the tip to form a hook.

chance (*probability*) Chance is usually thought of in terms of events that have not occurred; a small chance means the event is unlikely to happen. When an event can be repeated, the chance relates to how often a particular outcome occurs. Chance in that case is a ratio of the selected outcomes (called successes) to the total number of possibilities. If an outcome never occurs, then its chance is 0, but if it always occurs, the ratio becomes 1. For many events in ordinary life, such as weather, chance is based on analysis of circumstances of past events. For example, if a successful outcome is rain, a meteorologist determines that a certain weather pattern has previously led to rain, say, about 2 times in 10. The ratio is 2 to 10, and the chance is expressed as 20% chance of rain. Probability theory is a mathematical formulation of this idea.

Chandrasekhar, Subrahmanyan (*astronomer*) [Indian-American: 1910–95] Chandrasekhar developed the theory of white dwarf stars and in 1930 showed that they occur only when a star with a mass less than about 1.4 times that of the Sun collapses after fusion ends (the Chandrasekhar limit). Above the Chandrasekhar limit a collapsing star forms a **neutron star** or, for masses greater than 3 Suns, a black hole. Chandrasekhar also developed mathematical treatments of stars and galaxies that included relativistic effects.

chaos theory (*mathematics*)   Recognition that extremely small changes can lead to unpredictable results is credited to Edward Lorenz [American: 1917–], who in 1961 observed this effect in equations that describe weather. His "butterfly effect" theory states that even the flutter of a butterfly's wings changes weather unpredictably. Lorenz's later work with other systems and that of other mathematicians on similar problems became known as chaos theory. Chaos theory investigates not only unpredictable effects but also similar patterns that emerge from differing initial conditions. Many patterns in chaos theory are best described as **fractals**.

chaparral (*ecology*)   In areas with a so-called Mediterranean climate—characterized by mild, wet winters and hot, dry summers—vegetation consists mainly of low-growing shrubs with leathery evergreen leaves that resist water loss. In winter and spring, masses of annual plants create a colorful display. Birds and small mammals such as mice are common. This biome has various local names, including chaparral in California and maquis around the Mediterranean. Rainfall averages 10 to 20 in. (25 to 50 cm) annually.

characin (*animals*)   Members of the family Characidae are freshwater tropical fish. They are related to and look somewhat like **carps** and minnows, but typically have an adipose (fatty) fin behind the dorsal fin and teeth in the jaw rather than the throat. Most characins are small; many are brightly colored. The group includes tetras and piranhas, popular aquarium fish native to South America.

charge (*particles*)   Particles have a property called charge, which is either positive, negative, or 0. The charge of one electron is –1. Other charges for fundamental subatomic particles are either –1, +1, $-1/3$, $+1/3$, $-2/3$, or $+2/3$. Extensive experiments have revealed no other possible values. Particles whose charge is a nonzero multiple of $1/3$ never travel alone, but always combine to give charges of –1, 0, or +1.

Charge produces the electromagnetic force through exchange of photons between charged particles. When particles with like charges—both negative or both positive—exchange photons, electromagnetic force pushes them apart, while particles with unlike positive or negative charges are attracted to each other. Particles with 0 charge are unaffected. Photons are the same as electromagnetic waves. Thus, charge is the basis of electricity, magnetism, and light. Most chemical reactions are caused by charge interactions of atomic electrons.

charge-coupled device (*astronomy*)   Charge-coupled devices (CCDs) are used with telescopes to collect and record astronomical images, replacing photographic film. They use a **photoelectric phenomenon** to turn photons of light into trapped electrons. The number of electrons in each small region called a pixel ("picture element") is counted electronically to determine the energy of the photon. A typical CCD is an array of 2048 by 2048 pixels and is more sensitive than the best film.

Charles, Jacques-Alexandre (*chemist*) [French: 1746–1823] Less than 3 months after the Montgolfier brothers made the first hot-air balloon flight in 1783, Charles flew the first hydrogen-lifted balloon, on August 27. He also investigated the relationship of gas volume to temperature and independently reached the same conclusion as Guillaume Amontons [French: 1663–1705] in 1699 and Joseph-Louis Gay-Lussac [French: 1778–1850] in 1802: all gases expand approximately equal amounts for the same change in temperature, a proposition now often called Charles's law or Gay-Lussac's law.

charm quark (*particles*) Although the original quark theory was based on 3 quarks, Sheldon Glashow [American: 1932–] predicted in 1964 that a fourth quark, which he called charm, was needed to match the 4 leptons known at that time. The charm quark was established in 1974 when the J/psi particle, which is a meson that includes a charm quark, was discovered.

Charon (*solar system*) Pluto has one known moon, Charon, discovered in 1978. Charon—728 mi (1172 km) in diameter—is about half the size of Pluto. Pluto and Charon rotate and revolve synchronously: the same side of each always faces the other, like a double planet system.

Chatelperronian (*paleoanthropology*) A tool tradition contemporary with the earliest Aurignacian in Spain and southern France (36,000 to 33,000 years ago) is called Chatelperronian. Like the Aurignacian, the Chatelperronian includes bone points, stone blades, and burins, but techniques were less well developed and many tools resemble those of the earlier Mousterian. Some paleoanthropologists believe that the Chatelperronian industry was produced by Neandertals, not Cro Magnons.

Chauvet cave (*paleoanthropology*) Chauvet cave, near the Ardèche River and Vallon Pont-d'Arc, France, was discovered in 1994. It contains more than 300 paintings as well as human footprints, hearths, tools, and fossils from about 20,000 years ago—the height of the recent Ice Age. Among its dramatic relics is a cave bear skull set facing a painting of a group of bears. Also notable are paintings of predators, including a panther, a hyena, and several owls.

chelate (*chemistry*) Certain compounds have the ability to bind strongly to metal atoms in solution and to maintain a stable complex while not becoming a new compound. Such complexes are termed chelates; the compounds that form chelates with metals are called chelating agents. Chelating agents are used to remove unwanted metals, including heavy metals that can damage living tissues.

chemiluminescence (*chemistry*) Light emitted from a chemical reaction is called chemiluminescence. For example, the reaction between one form of phosphorus and oxygen when exposed to air is a blue-green glow.

chemistry (*sciences*) Chemistry is concerned with the ways that matter interacts at the atomic and molecular levels, interactions that mainly occur between electrons in the outer shells of atoms. The shapes of molecules strongly influence these interactions, especially molecules that are the subjects of organic chemistry. Organic chemistry concerns the properties, structures, and interactions of complex molecules based on carbon, whether or not produced by living organisms. **Biochemistry** deals with the behavior of **compounds** within organisms. The chemistry of molecules without carbon or of simple carbon-based molecules is called inorganic. The intersection of chemistry and physics in the study of **crystals** and other bulk properties of matter is called **materials science**.

cherry (*plants*) Members of the **rose** family, cherries are trees native to Europe, Asia, and North America. Some, such as the chokecherry (*Prunus virginiana*), seldom reach more than 25 ft (7.6 m) in height; in contrast, the wild black cherry (*P. serotina*) grows to 100 ft (30 m) tall. Some species, and varieties thereof, are raised for their edible fruits or as ornamentals.

The genus *Prunus* comprises some 430 species of shrubs and trees native mainly to northern temperate regions. Like cherries, all produce fleshy drupe fruits that contain a single seed called a stone. Many are cultivated as ornamentals or for their fruits or seeds, including the **almond**, apricot (*P. armeniaca*), peach (*P. persica*), and plum (*P. domestica*).

chestnut (*plants*) Members of the **beech** family, chestnuts (*Castanea*) are deciduous shrubs and trees native to temperate areas of the Northern Hemisphere. They grow to heights of 100 ft (30 m) and have furrowed bark, glossy green leaves, and unisexual flowers. The female flowers give rise to spiny fruits with brown, edible seeds.

Horse chestnuts (*Aesculus*) are deciduous shrubs and trees classified in the family Hippocastanaceae. Native to northern temperate regions, they grow to heights of 100 ft (30 m) or more. They have compound leaves and showy clusters of fragrant flowers. The large fruit—a spiny capsule—contains 1 or 2 brown, inedible seeds popularly called conkers or buckeyes (because of their resemblance to deer eyes).

The Chinese water chestnut (*Eleocharis dulcis*) is the edible underground stem, or corm, of a **sedge** that grows in water. The corm's round shape, chestnut-brown skin, and chestnutlike flavor and texture gave rise to its name.

Chicxulub crater (*geology*) The theory that the K/T event of mass extinction was produced by the impact of a large object, such as an **asteroid**, some 65,000,000 years ago is based on the 1980 discovery of a worldwide layer of sediment enriched in **iridium**. In 1990 evidence of a large impact crater in Yucatán, Mexico, known as Chicxulub (CHIK-zoo-lub) crater, became generally known. The crater dates from the time of the extinction event, and evidence links it to the iridium layer. Many think the impact at Chicxulub was the cause of the K/T mass extinction.

chigger (*animals*) Small 6-legged larvae of **mites** of the family Trombiculidae, chiggers are skin parasites that suck the blood of birds and mammals. Their saliva contains substances that cause localized itching and inflammation. Some chiggers, particularly certain species in Southeast Asia, transmit human diseases, including scrub typhus (tsutsugamushi fever).

chimpanzee (*animals*) Humans' closest living relatives, the 2 species of chimpanzees (genus *Pan*) are **hominid primates** of tropical Africa. They grow to 5.5 ft (1.7 m) tall and may weigh more than 150 lb (68 kg). Chimpanzees have a highly mobile, hairless face, with a low forehead and large brow ridges over the eyes. Their arms are long. Chimpanzees are active both in the trees and on the ground, and feed on a variety of plant and animal matter.

chinchilla (*animals*)   Native to South America, chinchillas (*Chinchilla*) are rodents famed for their dense silky fur. A chinchilla grows to about 15 in. (38 cm) long, not counting its long tail. It is gregarious, living in large colonies in underground burrows, from which it emerges at dusk to forage for plant food.

chitin (*biochemistry*)   Tough but flexible, chitin is a nitrogen-containing polysaccharide derived from the sugar glucose; it is insoluble in water. The name, from the Greek word for tunic, recognizes chitin's role as the main component of the cell walls of fungi and the exoskeletons of insects and other arthropods. Chitin also is present in other organisms, including certain sponges and annelid worms.

chiton (*animals*)   Chitons are mollusks with a dorsal shell consisting of 8 broad overlapping plates. This arrangement allows the animal to curl up, exposing only the shell to a predator. Most species live in shallow marine waters, usually clinging to rocks or other solid substrates. They feed on algae. The largest species, the gumboot chiton (*Cryptochiton stelleri*) of the North Pacific, grows to lengths of more than 1 ft (0.3 m).

chlorine (*elements*)   Chlorine is a halogen gas, atomic number 17, recognized as an element in 1810 by Humphry Davy and named from the Greek *chloros*, "yellow-green," its color. It has been used as a weapon because it is both poisonous and heavier than air but is more familiar today as the life-saving antiseptic in drinking water and swimming pools. Although its compounds have many important uses, the long-lived chlorine-based pesticides cause environmental damage. Chlorine is also the catalyst linked to destruction of ozone in the upper atmosphere. Chemical symbol: Cl.

chlorofluorocarbon (*compounds*)   A chlorofluorocarbon begins with a compound such as ethane (C_2H_6) or methane (CH_4); then chlorine and fluorine replace some or all of the hydrogen. An example is dichlorodifluoromethane (CCl_2F_2), also known as Freon-12 (Freon is a trade name for fluorocarbons). Because both fluorine and chlorine are highly reactive, they bind tightly to carbon, creating the especially inert compounds once widely used as refrigerants and spray propellants. However, ultraviolet radiation in the ozone layer of the atmosphere breaks chlorofluorocarbon molecules apart, permitting free chlorine to act as a catalyst that turns ozone to oxygen. Since the ozone layer prevents harmful radiation from reaching Earth, the use of chlorofluorocarbons has largely been discontinued.

chlorophyll (*biochemistry*)   The pigment that makes plant leaves and algae green, chlorophyll captures the light energy needed for the food-making process photosynthesis. It is usually contained in cell structures called chloroplasts. Several kinds of chlorophyll exist, differing slightly in their chemical structure and able to absorb light at different wavelengths.

chloroplast (*cells*)   The chlorophyll-containing structures, or plastids, found in plant and algae cells are called chloroplasts. They are the site of the food-making process photosynthesis. Chloroplasts come in various shapes and sizes, ranging from comparatively large spirals in the algae *Spirogyra* to small disk shapes in most plants. The latter, however, can change their shape and position within a cell in response to changes in light intensity. Many biologists believe chloroplasts were originally free-living bacteria that developed symbiotic relationships with eukaryotes.

choice, axiom of (*logic*) Many proofs in mathematics require the ability to choose a set, such as all even numbers, from a collection of sets. Agreement that such choices can be accomplished for sets with an infinity of elements is the axiom of choice, first stated by Ernst Zermelo [German: 1871–1956] in 1904. Specifically, the axiom states that you can choose exactly one element from each of an infinite number of sets that have no elements in common. In 1963 Paul Cohen [American: 1934–] demonstrated that the axiom is independent of set theory.

cholesterol (*biochemistry*) A white fatty alcohol (sterol), cholesterol occurs in all animal cells but not in plants. It is synthesized in the liver, with excess production excreted in bile. It is an essential component of cell membranes. In addition, cholesterol is needed for the production of many important steroids, including the sex hormones estrogen and androgen.

chord (*geometry*) A chord of a circle is a line segment with both endpoints on a circle. This idea has been generalized to include line segments that have both endpoints on any curve or even on surfaces, such as a chord of a sphere. (*See also* **tangent**.)

chordate (*animals*) The phylum Chordata includes 2 groups of relatively primitive animals (tunicates and lancelets) as well as the most highly evolved of all animals, the vertebrates. Three characteristics distinguish chordates from other animals: a dorsal, hollow nerve cord known as the spinal cord; a stiff dorsal rod called the notochord; and paired gill slits in the throat region. These structures are present during at least part of a chordate's life cycle. For example, fish have gill slits throughout life but humans and other mammals have gill slits only in the embryonic stage.

chromatin (*biochemistry*) The nucleus of a nondividing cell contains a tangle of long, thin threads called chromatin; these consist of nucleic acids and proteins. During cell division the chromatin condenses, becoming rod-shaped bodies called chromosomes.

chromatography (*chemistry*) Any method of separating molecules that is based on structure or composition is called chromatography. The original version of this method was used to separate an ink into pigments by using osmosis to draw the ink through filter paper (paper chromatography), thus accounting for the name ("color writing"). More generally, any fluid that contains several different kinds of molecules can often be separated by flowing it through a permeable solid that has a different degree of interaction with each of the molecules, such as thin films of silica or columns of plastic beads. The concept has been extended to include separation based on the sequence of evaporation of volatile gases (gas chromatography) or the sequence in which solutions precipitate (liquid chromatography).

chromatophore (*cells*) Chameleons are famed for their ability to quickly change color in order to better blend in with the background. This ability is owed to the presence in the skin of pigment-containing cells called chromatophores (from Greek words meaning "color carrier"). In some chromatophores, the pigment moves in response to a stimulus: when all the pigment is concentrated in the center of the cells, the animal is pale; as the pigment disperses throughout the cells, the animal darkens. In other cases, an animal's color changes as chromatophores expand or contract. An animal may have several different types of chromatophores, each containing different color pigments.

chromium (*elements*) Chromium is familiar as a highly reflective coating on metal parts, but its name, from the Greek *chroma* ("color"), was given because all of its compounds are brightly colored. Chrome yellow (lead chromate) is a well-known pigment, and chrome glass is emerald green. The metal, atomic number 24, was discovered in 1797 by Louis-Nicolas Vauquelin [French: 1763–1829]. Chemical symbol: Cr.

chromosome (*genetics*) A chromosome is a cell structure composed of DNA. It carries genetic information in the form of genes. Prokaryotes (bacteria and blue-green algae) have a single, circular chromosome to which a few protein molecules are loosely attached. Eukaryotes (all higher organisms) have threadlike chromosomes, with a complex of tightly bound proteins—called histones—that play a role in condensing the chromosomes during cell division.

Most cells in higher organisms contain 2 copies of each chromosome. For example, each human cell has 46 chromosomes, or 23 pairs. Of these, 22 pairs consist of identical chromosomes; the 23rd pair are the sex chromosomes, which are identical in a female but different in a male. The gametes, or sex cells, contain only 1 chromosome of each pair.

chromosphere (*solar system*) During an eclipse, a pink zone can be seen surrounding the Sun. It is the chromosphere, which corresponds in some ways to the lower atmosphere, or troposphere, on Earth. The chromosphere is very active. It is formed from spikes that rise from the Sun and fall back into it. Although the spikes may rise to 4000 mi (7000 km), that is only about 1% of the solar radius.

chrysalis (*zoology*) The pupa of a butterfly or moth is known as a chrysalis. The term chrysalis also is used as a synonym for the silken cocoon that encloses the butterfly or moth pupa. At first this case is pale and soft. Soon it hardens and darkens, however, often matching the twig or other structure to which it is attached.

cicada (*animals*) Members of Family Cicadidae are winged, stout-bodied insects known for their shrill songs produced by a pair of vibratory organs in the abdomen. Only males sing—to attract females—and their songs are distinctive for each species. Among the most interesting species are the periodical cicadas (*Magicicada*), which remain in the nymph stage for up to 17 years. Thus in some years great numbers of adults mature while in other years there are no adults.

Cicadas belong to Order Homoptera ("uniform wings"), a diverse group of insects that also includes aphids, hoppers, scale insects, mealy bugs, whiteflies and spittlebugs. Homopterans have piercing-sucking mouthparts and feed on the sap of plants. There are both winged and wingless forms. In winged forms, the front wings have a uniform texture; when the insects are at rest they are held tentlike above the body.

cichlid (*animals*)   Members of the family Cichlidae are the only freshwater fish that have 2 sets of jaws. Those in the mouth scrape, bite, or suck up food; those in the throat—actually remodeled gill arches—crush, chew, or slice food. Another unique characteristic of cichlids is the effort they put into raising their young. Parents guard their fertilized eggs and young hatchlings, often by holding them in their mouth. The hundreds of species vary greatly in color, shape, and habit and occupy many different ecological niches. Cichlids live in warm rivers and lakes in the Americas (as far north as Texas), Africa, Madagascar, India, and Sri Lanka.

cilium (*cells*)   Many kinds of cells have fine hairlike projections from their surface. Called cilia, these organelles are used for movement; as they beat in unison they move either the cell or fluid surrounding the cell. Paramecia and related protozoa—collectively called ciliates—are covered with cilia; some of the cilia are used to propel the protozoa through water while others set up a current that sweeps food toward the gullet. The larvae of many aquatic animals are covered with ciliated cells, enabling them to swim about.

Ciliated cells line the trachea of humans and other air-breathing vertebrates. They trap dust and other particles in mucus, which can then be expelled. Ciliated cells move water through the gills of some mollusks, sperm through the sperm ducts of earthworms, and food through the digestive tract of ribbon worms.

cinder cone (*geology*)   The classic tall, symmetrical volcano with steep sides and a crater at the top is most often a cinder cone, formed entirely from ash and bombs ejected from the crater. Such a volcano is not as stable as a composite volcano, which in addition to ash and bombs also produces lava flows. The solidified lava forms layers with the ash, making composite volcanoes both less symmetrical and stronger than cinder cone volcanoes.

cinnamon (*plants*)   Members of the laurel family, cinnamons (*Cinnamomum*) comprise some 250 species of evergreen trees and shrubs native to Asia. They grow to heights of about 30 ft (9 m). Cinnamon spice is derived from the dried bark of several species, including the cinnamon tree (*C. zeylanicum*) and the cassia (*C. cassia*). Camphor oil is extracted from the camphor tree (*C. camphora*) and used in medicines, mothballs, and other products; camphor wood is fragrant, insect-repellent, and prized for making chests and other furniture.

circle (*geometry*)   The set of points in a plane that have an equal distance from a single point in the plane, called the center, is a circle. Notice that the circle is a curve, not a region of the plane; the region inside a circle is sometimes called a disk. All circles are similar figures, so the parts of any circle have the same ratio no matter what the circle's size. Just as all squares have a ratio of the perimeter to a side of 4 and of the diagonal to a side of $\sqrt{2}$, the ratio of the distance around a circle, or circumference, to the diameter is π, a number approximately 3.14156. The number π appears also in the formula for the area A of a circle in terms of its radius r, which is $A = \pi r^2$.

circuit (*electricity*)   For an electric field to move through a conductor, the conductor must form a closed path from a region of negative charge (negative terminal) to one of positive charge (positive terminal). Such a closed path is an electric circuit. The simplest kind of circuit is called a series circuit; there is only a single path, which may have several devices connected to it one after another. If 1 device stops transmitting current, the whole circuit does also. More often a parallel circuit is used. Each device to be powered is connected separately to the 2 terminals, usually by adding loops to the basic circuit. If 1 device fails, the others continue to work.

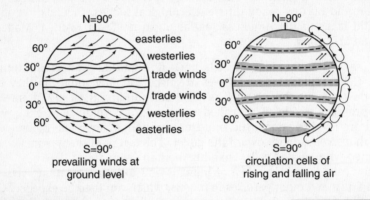

prevailing winds at
ground level

circulation cells of
rising and falling air

CIRCULATION OF ATMOSPHERE

circulation of atmosphere (*atmospherics*)   Sailors from early times noticed that winds tend to come from a particular direction at a given place. These prevailing winds mainly arise from cold air from polar regions sinking below warm air from tropical regions. However, this exchange, called the circulation of the atmosphere, is made more complex by several factors. If Earth did not rotate, the prevailing winds would be from the north in the Northern Hemisphere and from the south in the Southern Hemisphere. Earth's rotation induces a Coriolis effect that acts like a force turning winds north of the equator to the right and south of the equator to the left. After traveling about 30° of latitude either to the north or south, a wind pattern turns through a right angle and begins to head the other way. This process causes the winds from about 30° to 60° either north or south latitude to flow in a reverse swirl from the ones from 0° to 30° and from 60° to 90°. Note that winds at ground level are the opposite of those high in the atmosphere (see illustration).

circulatory system (*anatomy*)   The principal function of an animal's circulatory system is to transport food, gases, hormones, and other materials throughout the body. All vertebrates and most invertebrates have a circulatory system (exceptions include sponges, cnidarians, and flatworms). In some invertebrates, such as most arthropods, the system is open: the heart pumps the blood into spaces between the internal organs and the blood slowly seeps back to the heart. Certain worms and all vertebrates have closed circulatory systems, with the heart pumping blood through blood vessels.

circumference (*geometry*)   The circumference of a circle is the length of the curve, equivalent to the perimeter of a polygon. The circumference, *C*, is always in the same ratio to the diameter, *d*, for a circle of any size, a number called π (approximately 3.14156), so the formula for the distance around a circle is $C = \pi d$, or, since the radius *r* is half the diameter, $C = 2\pi r$.

circumpolar star (*stars*)   As the name suggests, circumpolar stars appear from Earth to rotate the line extending through north and south poles. At any given latitude except 0° (the equator), some stars never rise or set, and are therefore circumpolar. At the poles (90°) all visible stars are circumpolar. In the Northern Hemisphere, circumpolar stars today appear to rotate about Polaris, but there is no star at the center of rotation in the Southern Hemisphere.

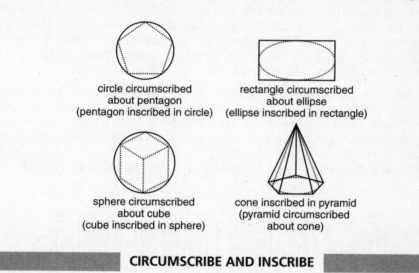

circle circumscribed
about pentagon
(pentagon inscribed in circle)

rectangle circumscribed
about ellipse
(ellipse inscribed in rectangle)

sphere circumscribed
about cube
(cube inscribed in sphere)

cone inscribed in pyramid
(pyramid circumscribed
about cone)

CIRCUMSCRIBE AND INSCRIBE

circumscribe and inscribe (*geometry*)   A plane geometric figure, such as a circle, is circumscribed about a polygon if all the vertices of the polygon fall on the figure and no sides of the polygon intersect the figure. The polygon is also inscribed in the figure. To circumscribe a polygon about a plane figure with curved sides, make all sides of the polygon tangent to the curve and let no sides intersect (the curve is then inscribed in the polygon). In 3 dimensions, different pairs of closed surface require separate formal definitions that use the same basic idea. For example, a polyhedron is inscribed in a sphere when all the vertices are on the sphere and those are the only points of intersection of the 2 figures (the sphere is circumscribed about the polyhedron); but a cone is inscribed in a pyramid when the base of the cone is inscribed in the base of the pyramid and the vertices coincide.

cirque (*geology*)   Erosion from a glacier that forms atop a mountain leaves a characteristic rounded basin near the peak that is called a cirque. A pair of cirques on opposite sides of the peak make a sharp ridge called an arête, while several cirques can combine to form a sharp horn-shaped peak called a matterhorn, similar to the peak on Matterhorn mountain in Switzerland.

citrus (*plants*) Members of the rue family, citruses (*Citrus*) are evergreen shrubs and small trees native to subtropical Asia. Their branches usually are armed with spines, and they have large leathery leaves and clusters of highly fragrant white flowers. Most species have long been cultivated for their edible fruits, which are leathery-skinned berries called hesperidiums. The best-known species include orange (*C. sinensis*), lemon (*C. limon*), grapefruit (*C. paradisi*), lime (*C. aurantifolia*), and tangerine (*C. reticulata*). Horticulturists have created hundreds of citrus varieties as well as hybrids such as the tangelo, which is a cross between the tangerine and grapefruit.

civet (*animals*) About the size of a domestic cat, civets are carnivorous mammals found mostly in forests in warm regions of Asia and Africa. They are nocturnal, eating a variety of plant and animal food. Civets are classified in Family Viverridae together with genets, linsangs, fossas, and the binturong. Most family members have a relatively small head with a pointed muzzle with well-developed senses of sight, smell, and hearing; they have perianal glands that produce a thick musky substance, also called civet.

Clactonian (*paleoanthropology*) Clactonian refers to a very early Paleolithic Age stone-tool industry without hand axes that existed in England and northern France about 250,000 years ago. The industry is named for Clacton-on-Sea, a site in England. In addition to Clactonian stone tools, in 1911 a wooden lance tip from about 300,000 years ago was found at Clacton, suggesting that early humans hunted game or defended themselves from predators (or other humans) with spears.

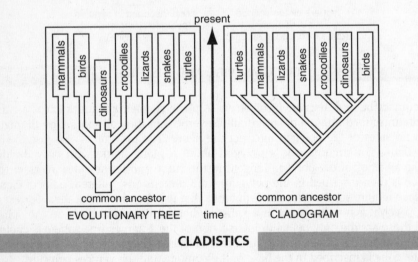

CLADISTICS

cladistics (*taxonomy*) Two methods are commonly used to classify organisms: evolutionary taxonomy and cladistics. Both are based on the principle of evolution, or descent from a common ancestor. Evolutionary taxonomy, established by Darwin considers branching of the evolutionary tree plus the degree of divergence or change that has occurred in a branch since it split off from the ancestral group—that is, a combination of genealogy and genetics. Cladistics, established by Willy Hennig [German: 1913–76], is concerned solely with how and when an evolutionary tree branched—that is, genealogy. Thus, evolutionary taxonomists classify birds in one group and crocodiles in a group with lizards and snakes, based on similarity of inherited features. Cladists consider crocodiles and birds to be more closely related to each other than to lizards and snakes; they would place crocodiles and birds in one group, lizards and snakes in another to reflect the more recent origin from a common ancestor of crocodiles and birds.

clam (*animals*) Clams are bivalve mollusks with a 2-piece shell hinged at the top and closed by a pair of powerful muscles. The muscular foot is used to burrow in sand or mud. Siphons formed by the margins of the mantle extend upward into the water—one drawing water to the clam, the other expelling it. Most clams are only a few inches long but the giant clam (*Tridacna gigas*) of the Pacific and Indian oceans can attain lengths of 4 ft (1.2 m) and weights approaching 600 lb (272 kg).

clan (*anthropology*) A clan of humans is a cultural group within a tribe that follows a particular set of spiritual practices, such as maintaining an animal totem or marrying only into certain clans. If a tribe has clan structure, everyone in the tribe is born into one of the clans. Often the clan has its own legendary ancestor. Sometimes there is a hereditary clan leader.

class (*logic*) A class is a set of elements defined by some proposition. The proposition is true for elements in the class and false for those in an implied larger set called the range or domain. The proposition "Chairs have more than 2 legs and a back" defines the class of chairs within the domain furniture; for a domain of, say, animals, the proposition defines a class with different elements. Often, however, class is taken to be synonymous with set. (*See also* **set theory.**)

classical physics (*physics*) The physics developed before the 20th century and any results that do not depend on relativity or quantum theory are called classical; the physics that employs relativity (which dates from 1905) or quantum concepts (from 1900 onward) is modern. Much of classical physics is not exactly correct, but for objects larger than atoms and smaller than stars, the deviation from modern physics is too small to make a difference. Many relativistic effects become noticeable only near the speed of light, a speed achieved only by subatomic particles. Thus engineers nearly always use classical physics, for which the mathematics is much easier than for most of modern physics.

classification (*taxonomy*) Scientists have identified more than 2 million kinds of organisms—from various dinosaurs to some 300,000 different kinds of beetles. Modern scientific classification of these organisms began around 1735, when Linnaeus invented a system based on structural similarities. He grouped all plants into Kingdom Plantae and all animals into Kingdom Animalia. He divided each kingdom into smaller and smaller groups, in what is known as a taxonomic hierarchy. Linnaeus also introduced binomial names for each kind of organism. As our knowledge of living things has grown, the classification scheme begun by Linnaeus has evolved and expanded. Today most scientists use schemes that divide all organisms into one of 6 kingdoms. Because relationships within some kingdoms are still unclear, there is much variety among schemes. For example, some place all flagellated protists in a single phylum, while others place nonphotosynthetic flagellates in Phylum Mastigophora and photosynthetic flagellates in Phylum Euglenophyta. Names of groups also vary. For example, the phylum of flowering plants is Magnoliophyta in some schemes, Anthophyta in others. (*See also* **taxonomy**.)

Clausius, Rudolf (*physicist*) [German: 1822–88] Clausius continued the work on thermodynamics originated by Nicolas "Sadi" Carnot and in 1850 became the first to state the second law of thermodynamics, which resolved some difficulties in Carnot's theory. In 1865 Clausius introduced the term entropy to describe the distribution of heat in a system, restating the second law as "in a closed system, entropy always increases." Clausius also contributed to chemistry by correctly proposing in 1851 that an electric current passes through a solution because of ionization, although few believed him at the time.

clay (*soils*) When feldspar or similar silica minerals are broken apart by weathering, very small particles tend to be produced. Particles smaller than 1/6400 in. (0.004 mm), mostly silica minerals, are called clay. Pure deposits of clay are the basis of many ceramics, ranging from bricks to porcelain, but clay is also an important component of most soils.

clear-cutting (*environment*) Trees in a forest can be cut selectively, leaving a good part of the forest intact. But most commercial operations use a technique called clear-cutting, in which large wooded areas are completely stripped of vegetation. Wildlife habitats are lost and often much of the soil is eroded away. The areas may be replanted with seedling trees but many years must pass before they once again support productive, diverse communities.

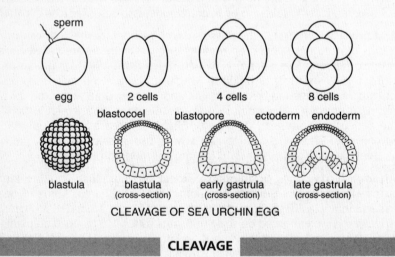

CLEAVAGE OF SEA URCHIN EGG

CLEAVAGE

cleavage (*biology*) Sexual reproduction results in the creation of a fertilized egg, or zygote, which soon goes through rapid, repeated divisions in a process called cleavage. In animals the single fertilized cell divides into 2. Each of the 2 cells divides again, producing 4 cells, then 8 cells, 16 cells, etc. After several hundred divisions, the developing organism, or embryo, resembles a hollow ball of cells and is called a blastula. Cleavage is followed by gastrulation. During this stage, 1 side of the blastula folds inward and the ball becomes a double-layered cup called the gastrula. A third layer of cells then grows between the 2 layers. At first the cells of the 3 layers are more or less alike. But soon they begin the process of differentiation. The outermost cell layer, the ectoderm, forms skin and nerve tissue. The innermost layer, the endoderm, forms the digestive and respiratory systems. The middle layer, the mesoderm, gives rise to muscles, most internal organs, and—in vertebrates—the skeleton.

climate zone (*geography*)   Annual patterns of precipitation and temperature for a given locality form its climate. Although climate varies from one place to the next, within a broad geographic region the climate is usually roughly similar.

Climate begins with average temperature within a band defined by latitude—surrounding the equator in the band between the Tropic of Cancer and Capricorn warm temperatures prevail (tropical climate). North and south of the Arctic and Antarctic Circle are cold regions (frigid climate). In between the climate varies more from season to season (temperate climate).

Classification into tropical, frigid, or temperate does not take precipitation into account. Winds, seas, and altitude also affect patterns of climate. One classification scheme for climate zones consists of subpolar; desert; cool, dry; cool, humid; subtropical; tropical; tropical, dry; and Mediterranean. Mediterranean climate has moderate temperatures, some rain in winter, and a dry summer.

cline (*evolution*)   A cline is a gradual change over a geographic area in a characteristic of a species. It usually is associated with a gradual environmental change, such as a change of temperature or humidity. For example, house sparrows that live in colder parts of the species' range are larger than house sparrows that live in warmer parts of the range.

cloaca (*anatomy*)   In insects and some other invertebrates, the cloaca (from the Latin for "sewer") is the final section of the digestive tract. In most vertebrates—including fish, amphibians, reptiles, and birds—the cloaca is a common terminal chamber for the digestive, urinary, and reproductive tracts. From the cloaca materials move to the outside through a small opening called the anus.

clone (*genetics*)   Genetically identical cells or organisms arising by asexual reproduction from a single ancestor are known as clones. Plants produced by vegetative propagation, such as strawberry plants produced by means of runners (horizontal stems) from a parent plant, are clones. Species of snails and aphids are among animals that create clones of themselves. Scientists have artificially produced clones of a variety of animals using body cells that contain a full set of chromosomes. For example, in 1996 Ian Wilmut [Scottish: 1944–] created a clone named Dolly from the DNA in 1 udder cell of a 6-year-old female sheep. Dolly's genetic makeup was exactly the same as her mother's.

closed curve (*geometry*)   A closed curve is a curve in a plane that has no endpoints. It is a simple closed curve if it separates the plane into an inside and an outside, as in a circle or a polygon; otherwise it may have 2 or more loops, regions that by themselves completely enclose part of a plane.

closure (*algebra*) An operation on a set *S* possesses closure, or is closed, if applying the operation to an element or to elements of *S* results in another element of *S*. For example, the set of natural numbers is closed for the unary operation of squaring or for the binary operation of addition (for example 2^2 and $2 + 3$ are natural numbers), but not closed for taking the positive square root or for subtraction (for example, $\sqrt{2}$ is not a natural number, nor is $2 - 3$).

clotting (*physiology*) When a blood vessel is cut or broken, the escape of blood is soon stopped as some of the blood forms a solid clot. Blood clotting, or coagulation, starts when cells of the injured blood vessel release a chemical called thromboplastin. Blood platelets, which begin to stick to the injured vessel, release additional thromboplastin. Calcium ions in the blood plasma stimulate the thromboplastin to start a chain reaction that leads to the production of long tangled strands of an insoluble protein called fibrin. The fibrin and platelets trap red blood cells, forming the clot.

cloud (*atmospherics*) Clouds are floating masses of tiny water droplets or ice particles condensed on microscopic particles of dust. Water vapor is invisible, but near microscopic amounts of liquid or ice dispersed through air appear white. Black clouds occur where sunlight is completely blocked by a cloud; red clouds occur near sunset and dawn because scattering of sunlight by air lets only red light through.

The system of cloud names was devised in 1802 by Luke Howard [English: 1772–1864]. Lowest are foglike stratus and puffy cumulus clouds, only ½ mile (1 km) above ground. Much higher are thin wisps of cirrus clouds, at about 6 mi (10 km). Howard's system combines each of these 3 types into low-lying stratocumulus and much higher—about 5 mi (8 km)—cirrostratus and cirrocumulus clouds. "Nimbo" indicates a cloud that produces precipitation, as in cumulonimbus ("thunderhead") and nimbostratus, low, dark clouds that produce steady rain or snow. Between lower and upper levels are altostratus and altocumulus clouds, about 2 to 3 mi (3 to 5 km) high.

cloud chamber (*particles*) The first device developed to observe the passage of charged subatomic particles in 3 dimensions was a small windowed box; looking through the window, a physicist could observe and photograph trails of vapor similar to airplane contrails. The cloud chamber, invented by Charles T. R. Wilson [English: 1869–1959] in 1895 simply to produce clouds in the laboratory was adapted by Wilson in 1911 to study the then new phenomenon of radioactivity. A sudden vacuum induced in the chamber causes the air to become supersaturated. Any charged particle—or X ray or gamma ray—ionizes molecules during passage, and water condensing on the ions appears as a visible streak. An ingenious device using a Leyden jar and an early mercury-vapor lamp produced light at the right moment to record the path on film.

clover (*plants*) Annual and perennial herbs in the legume family, clovers grow to heights of 8 to 24 in. (20 to 60 cm). They make up the genus *Trifolium* ("3 leaves"); most clovers have compound leaves composed of 3 leaflets. Small flowers, in shades from white to pinkish purple, are clustered in dense heads. The fruit is a pod.

club fungus (*fungi*) Forming the phylum Basidiomycota, club fungi derive their name from the club-shaped, spore-producing structure called the basidium, discovered in 1837 by Joseph Léveillé [French: 1796–1870]. Each basidium typically bears 4 spores. The group includes mushrooms, bracket fungi, puffballs, stinkhorns, and the parasitic rusts and smuts.

club moss (*plants*) Plants of the genus *Lycopodium* are commonly called club mosses. They are not true mosses but many *Lycopodium* superficially resemble mosses. They are small, mostly less than 1 ft (30 cm) tall, with small narrow leaves usually arranged spirally on the stem. Most species grow in moist, shady soil but some tropical species are epiphytes that grow on tree bark. Spores are produced on specialized leaves called sporophylls. In some species sporophylls occur in compact clusters at the tips of branches, creating a clublike structure called a cone or strobilus.

cluster of atoms (*materials science*) Small aggregates of atoms of single metals, called clusters, have properties that differ from the bulk materials. Although clusters are not bonded together the way molecules are, they tend to form in characteristic numbers. For example, sodium atoms that are heated and then allowed to congregate form clusters of 8, 20, 40, or 58 atoms rather than other numbers of atoms. Each cluster acts somewhat as a molecule.

cnidarian (*animals*) Also known as coelenterates, members of the invertebrate phylum Cnidaria include more than 9000 species of hydroids (such as hydra), jellyfish, corals, and sea anemones. Species range from microscopic in size to the jellyfish *Cyanea arctica*, which may have a diameter of more than 6.5 ft (2 m) and tentacles over 130 ft (40 m) long. Cnidarians are almost exclusively marine, with only a few known freshwater species, notably hydroids. Cnidarians have radial symmetry, a mouth surrounded by tentacles, and stinging cells called nematocysts that aid in capturing prey. Most species have 2 different stages in their life cycle: a free-swimming medusa stage that reproduces sexually and a sessile polyp stage that reproduces asexually.

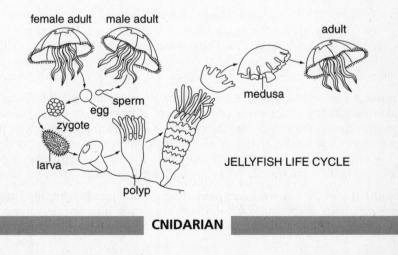

JELLYFISH LIFE CYCLE

CNIDARIAN

coal (*rocks and minerals*) Coal is technically a rock that is largely the mineral carbon, consisting of fossilized plants, just as some limestone consists of fossilized shells. Several different stages of coal are used for fuel. Peat, decaying plant remains submerged in water, is the raw material of coal of the future but it can be burned after drying. Lignite (brown coal) and bituminous (soft) coal are sedimentary or slightly metamorphosed rocks with more potential energy than peat. Anthracite (hard) coal is a strongly metamorphic form. It has even more energy when burned, but is so difficult to light and slow burning that it is mainly used in industry.

cobalt (*elements*) Cobalt is a metal, atomic number 27, discovered in 1735 by Georg Brandt [Swedish: 1694–1768] in his analysis of an ore well known to miners. The deep blue ore was known in German as *Kobalt*, "goblin," since the miners claimed it was copper ore from which a goblin had removed the copper. Cobalt is a transition element and part of the iron triad. It is an important part of many iron-based alloys as well as the basis of bright-blue dyes. Radioactive cobalt-60 is commonly used in medicine and the element is essential for human health as part of vitamin B_{12}. Chemical symbol: Co.

cobra (*animals*) Cobras are venomous snakes native to Africa, southern Asia, and the Philippines. They range in length from 5 to 15 ft (1.5 to 4.5 m), with the largest being the king cobra (*Ophiophagus hannah*). When excited or frightened, a cobra expands the ribs below the neck; this forces outward loose skin surrounding the neck to form a so-called hood. The cobra hisses and may even "spit" venom; the ringhals cobra (*Hemachatus haemachatus*) can spray venom from its fangs to a distance of more than 7 ft (2.1 m). Cobras belong to the family Elapidae, which also includes coral snakes, kraits, and mambas. All are highly poisonous. They feed on various small vertebrates.

coca (*plants*) Native to the Andes of South America, the coca plant (*Erythroxylum coca*) is an evergreen shrub that grows to about 15 ft (4.6 m) tall. It has spirally arranged leaves, clusters of small white flowers, and orange-red berries. The leaves contain a mixture of alkaloids, including cocaine.

coccus (*monera*) Spherical and egg-shaped bacteria are called cocci. They occur singly or in pairs, chains, or clusters. For example, *Streptococcus* bacteria link together to form long chains. Many cocci are harmless or ecologically valuable; for example, *Nitrosococcus* are soil bacteria that change ammonia to nitrates. Other groups, such as streptococci and staphylococci, cause disease.

cockroach (*animals*) The species of cockroaches (Family Blattidae) that infest human dwellings are notorious pests. Almost all species, however, particularly in warm climates, live outdoors; they favor dark, hidden places such as piles of decaying vegetation, the underside of tree bark, and caves. Active at night, they are big eaters, feeding on both plant and animal matter. Cockroaches have flattened oval bodies, are usually winged, and have legs designed for running.

coconut (*plants*) The coconut palm (*Cocos nucifera*) is a tropical tree that grows mainly in sandy soil along and near seashores. It attains heights up to 100 ft (30 m) and is topped by a crown of dark green leaves, each up to 20 ft (6 m) long. The clusters of cream flowers develop into fruits that have a large seed—the coconut—surrounded by a fibrous husk. The fruits float and can be dispersed great distances by the sea.

cocoon (*zoology*)   A protective case or covering around a cluster of eggs or an animal is called a cocoon. Earthworms and spiders form cocoons for their eggs. The larvae of insects that undergo complete metamorphosis spin cocoons in which the pupae develop. Among these are the cocoons spun by the silkworm, the caterpillar of the moth *Bombyx mori*. Parrotfish (Family Scaridae), which live in tropical reefs, rest at night in mucus cocoons they form around themselves.

Cocos Plate (*geology*)   Off the west coast of Mexico and Central America, the Cocos Plate is a small part of oceanic lithosphere named after Costa Rica's Cocos Island, which projects from the Cocos ridge. The ridge, a spreading center that is the lower boundary of the Cocos Plate, extends into the Pacific west of the border between Costa Rica and Panama. The movement of the Cocos Plate eastward has raised Central America and continues to produce active volcanoes and severe earthquakes in the region.

cod (*animals*)   The cod family, Gadidae, consists of mostly bottom-living fish, including cods, hakes, haddocks, and pollocks. Except for the freshwater burbot (*Lota lota*), all live in cool marine waters of the Northern Hemisphere. The fish have an elongated body and a large head. The fins lack spines; the dorsal and anal fins are long. Most species have a fleshy barbel hanging from the lower jaw. The largest species, the Atlantic cod (*Gadus morhua*), grows to 6 ft (1.8 m) and over 200 lb (90 kg).

code (*mathematics*)   A code is any way of symbolizing information. One branch of mathematics concerned with codes is called information theory. Computer languages are codes, with the computer coding information into strings of on and off switches; the strings are coded in machine language as binary numerals, which themselves are coded through use of a programming language.

Another sense of code involves hiding information from all but one possessing a special method, called a key, for turning coded messages into readable form. This kind of coding is heavily influenced by mathematical ideas. The most common code of commerce, the public-key code, for example, is based on finding factors of very large numbers and other concepts from number theory.

codon (*genetics*)   The basic unit of the genetic code is the codon, a sequence of 3 nucleotides on a DNA or messenger RNA (mRNA) molecule that codes for a specific amino acid. Both DNA and RNA have 4 bases, which means there are 64 (4 $\times$ 4 $\times$ 4) possible combinations, or codons—more than enough to code for the 20 amino acids found in proteins. In fact, there are 2 or more codons for each amino acid. For example, mRNA codons for glutamic acid are GAA and GAG. In addition, 1 codon initiates synthesis and 3 codons function as stop signals.

coefficient (*algebra*)   For a term formed by multiplication only, all factors except those under consideration form the coefficient of that part. For example, in the term $3x^2y$ the coefficient of x^2y is 3, of y is $3x^2$ and of x is $3xy$. Almost always, however, the intended coefficient is the numerical part, such as 3 for

$3x^2y$. When there is no indication of a special consideration, the coefficient is taken to be the numerical coefficient.

coelacanth (*animals*)   In 1938 a strange 4.5-ft- (1.4-m-) long fish was caught in the waters between South Africa and Madagascar. Identified as a coelacanth and named *Latimeria chalumnae*, it was the first living specimen of a family of lobe-finned fish known only from fossils at least 70,000,000 years old. Since 1938 additional coelacanths have been found in the same area and in 1998 2 coelacanths were discovered in Indonesian waters, nearly 6000 mi (9650 km) away. In addition to fleshy lobes at the base of fins, coelacanths have a 3-lobed paddlelike tail.

coelom (*anatomy*)   The main body cavity of many animals is a fluid-filled space called the coelom (from the Greek word meaning "hollow"). The coelom contains organs such as the heart, lungs, and digestive tract. In some animals, the coelom is divided into separate spaces. In mammals, for example, the largest coelomic spaces are the thoracic and abdominal cavities. In mollusks and arthropods, the coelom is greatly reduced; the main body cavity is the haemocoele, which is filled with circulating blood.

Coelophysis (*paleontology*)   One of the earliest and most primitive dinosaurs, *Coelophysis* lived some 220 million years ago in the late Triassic. A graceful, swift-running predator, it probably ate insects and small invertebrates, though fossils of a juvenile in the stomach of an adult indicate that cannibalism also occurred. *Coelophysis* grew to about 6 ft (1.8 m) in length, half of which was tail. The name, "hollow-boned beast," refers to the vertebrae and limb bones, which were hollow.

coelurosaur (*paleontology*)   The meat-eating dinosaurs known as coelurosaurs were generally small and lightly built, with long, flexible necks and small heads. Well-known species included *Hesperornis*, *Oviraptor*, and *Velociraptor*. At least two species, *Caudipteryx zoui* and *Protarchaeopteryx robusta*, which lived an estimated 135 million to 122 million years ago, had feathers, providing strong evidence that coelurosaurs may have been the ancestors of modern birds. The short forearms of these two species indicate that neither could fly; paleontologists who discovered the fossils hypothesized that the feathers may have evolved to provide insulation or to attract mates.

coenzyme (*biochemistry*)   A coenzyme is an organic molecule that bonds weakly with a specific enzyme during a reaction. Like the enzyme, it acts as a catalyst; it is neither altered nor used up during the reaction. Common coenzymes include coenzyme A, flavin adenine dinucleotide (FAD), and nicotinamide adenine dinucleotide (NAD$^+$), which have roles in respiration. Vitamins often act as coenzymes, which helps explain why they are essential to an organism's well-being.

coevolution (*evolution*)   Sometimes different species interact so closely that they influence one another's evolution. Such coevolution is well represented among flowers and their animal pollinators. Particularly striking is the mutualism between yuccas and yucca moths. These plants and insects are totally dependent on one another. A female yucca moth has specially adapted mouthparts for collecting pollen from a yucca flower. She carries this pollen to another yucca flower. After drilling a hole and laying her eggs in the ovary of the second flower, she pollinates the flower. Pollination leads to fertilization and development of seeds in the yucca's ovary. The moth's offspring eat some of the yucca seeds before boring out of the ovary, but they leave behind enough for propagation of the plant.

coffee (*plants*)   The source of a popular beverage and food flavoring, coffees (*Coffea*) are tropical evergreen shrubs with glossy leaves and dense clusters of small, fragrant flowers. The fruit is a berry with 2 seeds ("beans") that are dried and roasted to produce coffee. The most widely grown species is *C. arabica*, which generally grows to 23 ft (7 m).

collagen (*biochemistry*)   A fibrous protein consisting of 3 long polypeptide chains twisted around one another and connected by hydrogen bonds, collagen is strong but not elastic. It forms the white fibers found in vertebrate connective tissue. An important structural component of cartilage and tendons, collagen is the most abundant protein in the extracellular matrix that holds cells together. It also is part of a layer under the skin that connects the skin to the surface muscles.

collinear (*geometry*)   Points lying on the same line are called collinear. Any 2 points are collinear, but 3 points A, B, and C are collinear if and only if the sum of the distances between 2 pairs of the points is equal to the distance between the third pair (for example, if AB + BC = AC). Planes having a common line are also called collinear. Any 2 planes that are not parallel are collinear, but 3 or more collinear planes must share the same line as the first 2.

colloid (*materials science*)   A combination of 2 materials in which one is in a gas, liquid, or solid phase (called the medium) and the other is dispersed through the first in tiny clusters of atoms or molecules or in very large molecules is a colloid. If the particles are large enough to see, however, the combination is called a suspension. The tiny clusters or large molecules of a colloid are called particles. If light passes through a colloid, the particles induce scattering, unlike a solution or other homogenous material. A colloid in which liquid or solid particles are dispersed in a gas is also called an aerosol; one with gas in a liquid or solid is a foam; one with liquid particles in another liquid is an emulsion; and solid particles in a liquid or a solid is a gel or a sol. Some examples include the aerosols fog (water in air) and smoke (soot in air); the foams whipped cream (air in milk) and Styrofoam (air in styrene); the emulsion mayonnaise (oil in egg); the gel gelatin (protein in water); and the sol ruby glass (metal in glass).

Colloids with water as the medium are among the most useful of substances and also are important in all living organisms, forming most of the mass of a preponderance of organisms. Some such colloids are called hydrophilic because the particles are attracted to water; if there is no attraction, the colloids are hydrophobic and eventually the particles will clump and fall out of the medium. Chemical reactions can also cause particles to clump together, such as the influence of acid on milk (curdling) or salt water on muddy water, which forms deltas where rivers flow into the sea.

colony (*biology*)   Several types of colonies exist among living things. Some single-celled organisms are attached together as a group of more or less similar cells. For example, in *Anabaena* algae individuals of the same species are linked together in chainlike filaments. A more complex colony is the Portuguese man-of-war (*Physalia physalia*), which looks like a large jellyfish but actually consists of a variety of specialized individuals—some catch and digest prey, some contract their bodies to propel the colony through the water, some are concerned with reproduction, etc.

A group of separate organisms may live together, either temporarily or permanently. Many birds form breeding colonies, generally with their own kind but in some cases with other species. Social insects such as honeybees live permanently in colonies of tens of thousands of specialized individuals.

color (*optics*) Color is based on the wavelength or **frequency** of light waves. Colors of light range from about 400 nanometers (nm) for violet through 500 nm for green to about 650 nm for red (1 nm = 0.00003937 in.).

Color of materials is based on **reflection**. Sulfur, for example, absorbs all colors but yellow and reflects yellow; that is, light of the yellow wavelength raises electrons in sulfur atoms to a higher state from which they fall back emitting yellow light, while other wavelengths simply raise the temperature of the material. Color vision involves much more than simple wavelengths, however, since the eye and brain use inputs of contrast and intensity as well as wavelengths in color perception.

color force (*particles*) The color force binds quarks and quark combinations together. Quarks are said to exist in 3 different colors, most often called red, blue, and green on the analogy of the 3 primary colors for light, which when mixed in equal strength produce white light. A fundamental rule for quark combinations is that they can exist only in white forms; a **baryon** can combine red, blue, and green quarks or a **meson** can link red and antired. Antiquarks have anticolors—red and antired also produce white. Another rule is that quarks of different colors are attracted to each other by a powerful, if short-range, force: the color force.

colugo (*animals*) Constituting the **mammal** order Dermoptera, the 2 species of colugos (*Cynocephalus*) are native to forested areas of Southwest Asia and the Indo-Pacific. About the size of a small cat, a colugo has a large gliding membrane on each side of the body that extends from the neck to the tail. A colugo spends its days sleeping on a roost tree, hanging from a branch by powerful claws. At dusk and dawn it glides to its food trees to feed on fruits and other plant matter. Colugos are sometimes called flying lemurs, but they cannot fly and are not related to lemurs.

coma (*solar system*) Chinese astronomers called **comets** "hairy stars" after the coma, which is the ball of gas and dust that surrounds the comet's solid body (called the nucleus). The well-known tail of a comet is simply the coma blown away from the nucleus by the solar wind. The coma shines partly because it reflects sunlight, but it also produces its own light by fluorescence, glowing with energy that is also supplied by sunlight. The coma can be as large as 60,000 mi (100,000 km) in diameter.

comb and wattle (*anatomy*) The bare parts around the head of certain birds have skin outgrowths that play a role in courtship and in recognition between sexes.

A comb is an erect fleshy appendage on the crown of the head, best seen in domestic chickens and their ancestors, the junglefowl of Southeast Asia. Males have larger combs than females, and studies indicate that females use comb size and color in choosing mates.

A wattle is a floppy piece of skin that hangs from the throat, as in chickens and turkeys. In wild turkeys (*Meleagris gallopavo*), the male's wattle is infused with blood and bright red when he is trying to attract a female, but the blood drains and the wattle turns pale blue if he becomes frightened.

combination (*mathematics*)   The technique of counting how many different ways elements from a given set can be combined, a part of combinatrics, is called combinations. A combination is an arrangement of the elements of a set without considering order—the arrangement ABC is considered the same as BCA. The combinations of the 4 letters ABCD taken 3 at a time are ABC, ABD, ACD, and BCD. The number of combinations of n elements taken r at a time is n factorial divided by the product of r factorial with $(n - r)$ factorial, or $n!/r!(n - r)!$.

comb jelly (*animals*)   Also known as sea gooseberries and Venus's girdles, comb jellies are small, transparent marine invertebrates constituting Phylum Ctenophora ("comb-bearers"). Comb jellies have 8 "comb rows" of fused cilia radiating from top to bottom like lines of longitude on a globe. The beating of the cilia propel the comb jelly through the water. However, comb jellies are weak swimmers and are mainly carried about by currents and tides. They feed on fish eggs and tiny animals. Most species are armed with 2 long tentacles, which help capture prey.

combustion (*chemistry*)   Combustion is the process in which oxygen combines with other atoms to form new compounds, releasing energy in the process as heat. Often combustion also produces light, either by causing fluorescence in reaction particles that are gases or by incandescence of small particles. Although fire, in which oxygen from air combines with carbon compounds, is the most familiar combustion reaction, rusting of iron is sometimes considered combustion as well. Explosions may be caused by rapid combustion of small particles, such as flour in air, or gases, such as gasoline vapor in air. But many explosive compounds have oxygen built into them (nitroglycerine, for example); such explosive reactions are usually not considered combustion.

comet (*solar system*)   Comets are striking but irregular visitors in the sky, sometimes appearing as "stars" joined to long streaks called tails. Today we recognize comets as bodies from the far reaches of the solar system that have

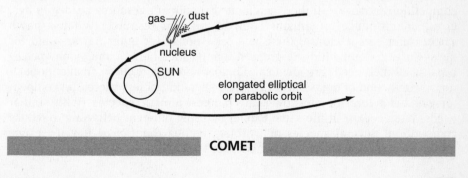

gas — dust
nucleus
SUN
elongated elliptical or parabolic orbit

COMET

drifted into a fall toward the Sun. They are lit by sunlight when they approach the Sun. Dust and gas, which flow out of the comet as it gets closer to the Sun's heat, form the tail, propelled by the solar wind.

Aristotle believed comets to be atmospheric phenomena like lightning or rainbows. In 1577, however, Tycho Brahe established that a comet was at least 4 times farther than the Moon. In 1705 Edmond Halley showed that some comets can orbit the Sun, although some plunge into the Sun. In 1949 astronomers began to describe comets as mixtures of dust and frozen gases—the "dirty snowball" theory. This model was verified by satellite visits to the vicinity of comet Halley in 1985. Today many astronomers believe that comets are remnants of the gas and dust that condensed to form the solar system.

commensalism (*ecology*) The type of symbiosis in which one species—the commensal—benefits from the relationship while the other species—the host—is neither helped nor harmed is called commensalism. An example is the relationship between suckerfish (*Remora*) and other fish, particularly sharks. A remora has a flat suction disk (actually a modified fin) behind its head, which it uses to attach itself to the underside of a host. It feeds on pieces of food that drop from the host's mouth. It also may gain protection; a predator isn't likely to get close to a remora attached to a shark.

community (*ecology*) Populations of various kinds of organisms that live together in a particular area, or habitat, and interact with one another make up a community. Communities range greatly in size, from a small pond to a coral reef to a vast expanse of coniferous forest; they usually are named after a dominant feature (lake community, sagebrush community, and so on). Interactions within a community may involve predation, symbiosis, and competition for resources. These interactions are a critical force in population size and stability, as well as in natural selection.

commutative property (*algebra*) A binary operation that yields the same answer when the elements are reversed exhibits the commutative property. If • represents an arbitrary operation, $a • b = b • a$ is true for commutative operations. Operations such as addition and multiplication for numbers and intersection and union for sets are commutative, but subtraction and division of numbers are not (for example, $4 - 2$ does not equal $2 - 4$). A group with a commutative operation is called abelian after Niels Henrik Abel.

competition (*ecology*) If there is a limited supply of a resource needed to live, grow, or reproduce, organisms compete for that resource. When a hawk catches and eats a mouse, there is 1 less mouse for other hawks—and for owls—to eat. Competition is greatest among members of the same species because all their needs are identical. Competition may result in smaller populations, elimination of organisms or species, migration of less successful competitors, or behavioral or physiological changes. Similar species (with similar needs) often coexist in the same habitat by using different behaviors to reduce competition. Several species of warblers (genus *Dendroica*) live in spruce forests in Maine; each typically searches for food at a different level in a tree.

complement (*sets*)   Generally all the elements of all the sets in a given discussion form a single set, the universal set for that discussion. Then for any set S under discussion, the complement is a set that consists of all the elements of the universal set that are not part of S. The complement may be symbolized with a bar over the name or with a prime sign after it. If the universal set is the colors of the rainbow, then the complement of the primary additive colors {red, green, blue} is the set {orange, yellow, indigo, violet}.

complementarity (*particles*)   In the 1930s Niels Bohr proposed a principle of complementarity in response to wave-particle duality and the uncertainty principle: A measurement that shows that an entity is a wave cannot also show that it is a particle, and vice versa; and the act of observation of a subatomic particle influences it so that if you measure position precisely, you will not be able to pin down momentum, and vice versa. Bohr suggested that complementarity is a basic principle of physics and even of life in general, applying it to philosophical situations such as a complementary duality between free will and determinism.

complementary angles (*geometry*)   When the sum of the measures of 2 angles is a right angle, the angles are complementary. They do not need to be adjacent angles; for example, the 2 acute angles in a right triangle are always complementary.

complete (*logic*)   An axiomatic system is complete if every true proposition about the undefined and defined terms of the system can be proved from the axioms. Only some true propositions are actually proved, but the possibility of proving them is established. The axiomatic version of logic based on true and false propositions connected by and, or, and not is complete, as is the version that also includes quantifiers; but more complex systems, such as arithmetic or set theory, are not.

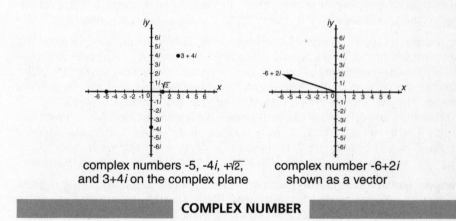

complex numbers -5, -4*i*, +√2, and 3+4*i* on the complex plane

complex number -6+2*i* shown as a vector

COMPLEX NUMBER

complex number (*numbers*)   A complex number is of the form $x + iy$ where i is the square root of negative 1. Although a complex number is the sum of a real number and an imaginary number (like a complex sentence, which has an independent and a dependent clause), all numbers are included in this set. For real numbers, y is 0; for integers, x is an integer and y is 0; etc. While real numbers are shown on a line, such as the x axis, complex numbers require a plane. The usual y axis is replaced by an iy axis. It is often useful to picture complex numbers as vectors, arrows from the origin to a point on the complex plane. Despite their seemingly abstract nature, complex numbers are the basis of theoretical electronics, fluid flow including aerodynamics, and quantum theory.

component (*algebra*)   A vector with n dimensions can be shown as a set of numbers arranged in order, called an ordered n-tuple, such as the ordered triple (0, 2, –4). Each number in the n-tuple is a component, so the components of the vector (0, 2, –4) are 0, 2, and –4. The components are the same as the "shadows," called projections, of an arrow vector on the axes of the space. For example an arrow from the origin to the point (0, 2, –4) has projections of a vector from 0 to 2 on the y axis, and from 0 to –4 on the z axis, and a 0-length projection on the x axis. The projection vectors are also called components. Components that are projections are vectors while components that form an ordered n-tuple are scalars.

composite (*materials science*)   A composite is a solid in which 2 materials are intermingled while maintaining their separate identities. Typically one material is a ceramic, plastic, or adhesive that is used to bind together fibers of a material such as steel, glass, carbon, boron, or asbestos. The bulk material provides protection and generally has better flexibility than the stiff fibers.

That thin fibers have high resistance to stretching forces (high tensile strength) compared with the same materials in bulk was discovered by A.A. Griffith [English: 1893–1963] in 1920. In 1942 fiberglass, the combination of thin glass fibers embedded in polyesters, was introduced in airborne radar equipment. Fiberglass remains a popular material. Graphite fibers were developed in 1963 and were being used in composites by the 1970s, replacing fiberglass as the composite of choice where flexibility and strength are important. Specialized composites may someday replace metal parts in engines.

composite (*plants*)   Plants of the huge aster family Asteraceae (also called Compositae) have a characteristic "blossom" that actually is an inflorescence— the capitulum—consisting of numerous flowers of 2 types on a single base. The daisy is a typical example. The outer flowers, wrongly called petals, are ray flowers. The tiny inner flowers, forming a central disk, are disk flowers.

More than 20,000 species of composites have been identified. Most have non-woody (herbaceous) stems. They are most abundant in temperate regions. Asters (*Aster*), coneflowers (*Echinacea*), dandelions, ragweeds, marigolds (*Tagetes*), sunflowers (*Helianthus*), thistles, and yarrow are well-known examples.

compost (*environment*)   Food scraps, grass clippings, paper, and other organic wastes can be decomposed by microorganisms—mostly bacteria and fungi—to form a nutrient-rich material called compost, which can be used as a soil fertilizer. One composting method involves placing the wastes in long piles and turning the material periodically; this aerates the wastes and hastens composting. Another method uses mechanical mixing and aerating. By returning useful materials to the environment, composting reduces polluting wastes.

compound (*chemistry*)   A compound is any material made from 2 or more different substances combined by chemical bonds. A given compound always maintains the same proportion of substances in any size sample since it is formed from elements that are linked together at the atomic level. Compounds are usually formed from molecules linked by covalent bonds or from crystals based on ionic bonds. Water is a compound with 2 hydrogen atoms linked to each oxygen molecule, while sodium chloride (table salt) is an ionic compound with 1 sodium ion for each chlorine ion, but salt water is a mixture, not a compound, since the amount of salt to water can be varied.

compound eye (*anatomy*)   The 2 prominent eyes of a housefly—like those of many arthropods, particularly insects and crustaceans—are actually compound eyes composed of many individual units called ommatidia. Each ommatidium has a light-sensitive retinula that connects through nerve fibers to the optic nerve, which carries impulses to the brain. There is no focusing mechanism and no structure (such as the iris of the human eye) to adapt to different light intensities.

Since each ommatidium forms its own image, the overall result is a mosaic. The slightest shift in an object being viewed will change the mosaic. Thus, compound eyes are very efficient in detecting movement. The greater the number of ommatidia, the sharper the vision. Each dragonfly eye has about 28,000 ommatidia, whereas the eye of a housefly has about 4000.

Compton, Arthur Holly (*physicist*)   [American: 1892–1962] Compton's experimental work established that Einstein had been correct in 1905 when he explained that light, then known only as waves, also behaves as particles. In 1923 Compton discovered what we now call the Compton effect, scattering of light waves by molecules that can be explained if light behaves as particles. Compton was able to observe changes in the momentum of electrons that result from the impact of a particle of light, or photon. In another series of experiments in the 1930s Compton demonstrated that cosmic rays must be charged particles that probably originate outside the Milky Way galaxy.

computable (*logic*)   If a function is defined so that for any given value of the variable or variables, the function's value can be determined by applying a finite number of operations a finite number of times, the function is computable. One technical definition of computable is based on a concept called a Turing machine after Alan M. Turing, who developed it in 1937. A Turing machine has strictly limited capabilities. It can find the successor of a natural number; add, subtract, and multiply 2 numbers; recognize whether or not a number is a member of a set; and pick one of a list of numbers. If a Turing machine can find the value of a function, the function is computable. Turing also showed in 1937 that functions exist that are not computable.

conclusion (*logic*)   For an if . . . then statement, the part following "then" is the conclusion (the part between "if" and "then" is the antecedent); similarly, in a statement cast in the form p implies q, p is the antecedent and q the conclusion or consequent. If both p and "if p then q" are true, then the conclusion q is true. This rule, called *modus ponens* or the rule of detachment, forms the basis of most proofs; consequently the final statement of a proof is the proof's conclusion.

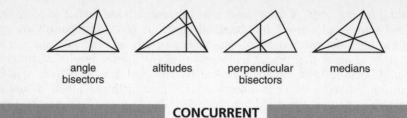

angle
bisectors

altitudes

perpendicular
bisectors

medians

CONCURRENT

concurrent (*geometry*)   When 2 or more lines or line segments have a point in common, they are concurrent. Examples of special interest involve 3 lines in most cases. For a given triangle the 3 angle bisectors, 3 altitudes, 3 perpendicular bisectors of the sides, and 3 medians are concurrent (but not necessarily at the same point for each type of line segment). The word concurrent is also used for 3 or more planes with a point in common.

condensation (*atmospherics*)   Water vapor, a gas nearly always encountered mixed with air, can suddenly change to liquid water or even ice, giving up its latent heat in the process. This change is called condensation. Condensation of water from air does not necessarily occur at the temperatures we think of as boiling or freezing points. On a warm, damp summer day, condensation occurs as liquid on any cool surface below a temperature known as the dew point, which varies with the humidity. If the surface is cold enough—below freezing, or 32°F (0°C)—condensation appears directly as ice and is known as frost. Condensation in air, instead of on a surface, does not necessary occur at the dew point or even below it. Microscopic particles of materials that attract water, such as tiny salt crystals or sulfur trioxide ash from burning, are needed to serve as condensation nuclei.

condensed matter (*sciences*)   The study of the properties of condensed matter, in which atoms or molecules remain close to each other, principally in the solid or liquid phases as opposed to as gases and plasmas, was formerly called solid-state physics. The principal emphasis for condensed matter physics is on the behavior of electrons and magnetic properties, including semiconductors—the basis of modern electronics—and superconductivity.

conditioned response (*ethology*)   A conditioned response is a learned response to a specific stimulus. In experiments begun in 1889, Ivan Pavlov showed that dogs can be conditioned to salivate at the sound of a bell. Each time Pavlov presented dogs with food, he rang a bell. After repeated trials, the dogs salivated whenever the bell was rung, even if no food was presented.

Many animals are trained using conditioned responses, with positive reinforcement for the desired behavior (a bone to a dog that sits, for example). An example of negative conditioning in nature involves blue jays and the larvae of sawflies. Sawfly larvae have conspicuous colors and an awful taste. A blue jay quickly vomits after eating sawfly larvae. One or two such experiences and the blue jay is conditioned to look elsewhere for a meal.

condor (*animals*)   Large, graceful flyers that are particularly adept at soaring, condors are members of the American vulture family, Charartidae. There are 2 species. The California condor (*Gymnogyps californianus*) appears to be coming back from the brink of extinction thanks to **captive breeding** programs. The Andean condor (*Vultur gryphus*) of South America is the largest **bird of prey** and among the largest of flying birds, with a wingspread of about 10 ft (3 m).

conduction (*materials science*)   Conduction is transport of energy through material caused by one particle pushing those particles around it. Essentially the same process occurs when sound, electricity, or heat is conducted. For sound the particles are molecules or atoms. For electricity the particles are electrons, **ions** in fluids, or "holes" (places where there is no electron) in **semiconductors**. For conduction of heat the particles include both molecules and electrons.

Electrons that are weakly bound to atoms or molecules can move away from home locations, conducting electricity and contributing to heat conduction. Metals contain a sea of such free electrons. Applying an electric force frees some electrons from that sea. In a completed circuit, the free electrons jostle their neighbors to form an electric current that flows through the circuit. Metals and other materials that conduct electricity in this way also tend to be good conductors of heat since the free electrons jostle molecules as well as each other.

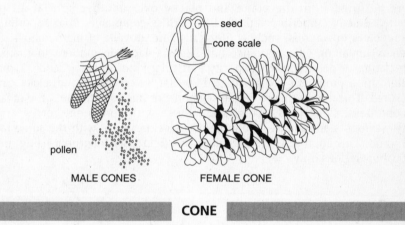

MALE CONES   FEMALE CONE

CONE

cone (*botany*)   Cones are reproductive structures of most conifers and other gymnosperms, as well as of simpler vascular plants such as club mosses and horsetails. A cone is actually a specialized branch with scalelike leaves called sporophylls arranged around a central axis.

Pines and other conifers have two kinds of cones, male and female. These usually are borne on the same tree, but in some species (junipers, for example) male and female cones are on separate trees. Male cones, which produce pollen, tend to be long and narrow. Female cones are small and fleshy when young. Following pollination and fertilization, the female cones grow and become woody; when the seeds are mature, the cone scales open so that the seeds can be shed.

cone (*geometry*)   A cone may be a geometric **solid** or a **surface**. The surface is the **set of points** that includes all the points on every **line** that contains 1 point on a closed **curve** and that passes through a point not in the plane of the curve.

conglomerate

Picture keeping 1 point of a line fixed, while rotating another point of the line in a closed curve, especially a circle. The cone on which the conic sections is based is such a circular cone, for which the fixed point is on a line perpendicular to the circle's center.

The solid called a cone combines the conical surface from the fixed point to the curve with the region inside the curve. Commonly one thinks of a right circular cone, which has a circular base and an altitude perpendicular to the base. The volume V of a cone is $\frac{1}{3}$ the product of the area of the base (B) and the altitude h, or $V = \frac{1}{3} Bh$.

conglomerate (*rocks and minerals*) Conglomerate is sedimentary rock formed when rounded pebbles from seashores or streams are buried and become cemented together by sand along with silica or calcite. Some pebbles may be quite large, even boulders. Pebbles often consist of tough rock called quartzite, a metamorphosed quartz sandstone. When embedded rocks are sharp edged instead of rounded, the formation is called breccia.

congruence (*geometry*) Congruence in Euclid's *Elements* required that one figure be lifted from the plane and placed over another so that all parts matched exactly (superimposition). Later critics complained that superimposition involves operations, such as lifting a figure, outside of mathematics. Such a general meaning of congruence has been replaced in modern mathematics with axioms or definitions that specify particular kinds of congruence. Typically a ruler and a protractor are postulated so that line segments and angles can be measured. This allows one to define segments or angles with the same measure as congruent. Congruent triangles are those with congruent corresponding parts (segments and angles); congruent circles are those with the same radius; and so on. This approach classifies the same figures as congruent using mathematical operations only.

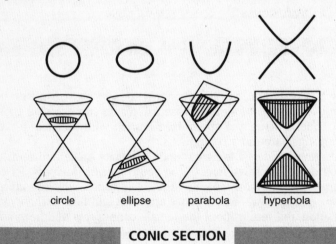

circle   ellipse   parabola   hyperbola

CONIC SECTION

conic section (*geometry*) The conic sections are certain curves in a plane, including the circle and ellipse. These can be recognized as the same curve as seen from 2 different angles. Ancient Greeks noticed this relationship. They pictured light rays between the eye and circle as forming a cone, with the intersection of a plane and the cone as the circle or ellipse. When the plane intersects the cone of vision at large angles, however, the curves do not close, forming a parabola or a hyperbola. A plane cutting any 3-dimensional figure selects a 2-dimensional figure called a section, so curves formed with a cone are the conic sections, or conics. All graphs of equations of the second degree are conic sections. Modern definitions use a double cone, which gives the hyperbola 2 parts. A line, crossed pairs of lines, and even a single point are degenerate conics.

conifer (*plants*)   The conifers ("cone bearers") were first recognized as a distinct division of plants in 1907 by Charles Edwin Bessey [American: 1845–1915]. The approximately 550 known species are mainly trees and shrubs with small evergreen leaves usually in the shape of narrow needles or flat scales. Many species have specialized cells and ducts in their stem tissues that secrete and transport resin. Conifers reproduce by seeds, which usually are borne on woody cones. The group includes cypresses, pines, redwoods, and yews. Conifers occur worldwide but are most common in temperate regions, especially in the Northern Hemisphere, where they form extensive forests.

conjugate (*numbers*)   Conjugate has many meanings in mathematics. For numbers conjugate describes 2 binomial expressions whose only difference is the sign of the second term, such as $3 + \sqrt{2}$ and $3 - \sqrt{2}$ or, for complex numbers, $4 - 3i$ and $4 + 3i$. The product of the conjugates is the difference of 2 squares, so for radical expressions such as $3 + \sqrt{2}$ and $3 - \sqrt{2}$, the product eliminates the radical, needed for rationalization of a denominator: $(3 + \sqrt{2})(3 - \sqrt{2}) = 9 - 2 = 7$. For complex conjugates the product eliminates the imaginary part: $(4 - 3i)(4 + 3i) = 16 + 9 = 25$. The product of 2 complex conjugates, $(x + iy)(x - iy) = x^2 + y^2$, is always the square of the amplitude of either.

conjugation (*biology*)   Certain protists, fungi, and bacteria reproduce sexually by a process called conjugation. Two cells of individuals of the same species fuse together and either exchange or donate genetic material. In *Paramecium* and other ciliated protists, for example, 2 organisms fuse, exchange genetic material, separate, and then reproduce asexually by splitting in half. In the green algae *Spirogyra*, cells on 2 adjacent filaments form a connecting tube; then the contents of 1 cell migrate into the other cell and unite with its contents, forming a zygote. The zygote develops into a zygospore that, under favorable conditions, develops into a new *Spirogyra*.

connective tissue (*anatomy*)   The connective tissues of vertebrates serve to bind together to protect other tissues and organs and to support the body. They include bone, cartilage, ligaments, and tendons; adipose (fat) tissue; membranes such as those lining the abdominal cavity and covering the brain; and the fluid tissues blood and lymph. Unlike other types of tissue, connective tissue consists largely of extracellular material; the cells are comparatively inconspicuous.

conodont (*paleontology*)   In the 1800s scientists began to find cone-shaped, toothlike microfossils. These came to be called conodonts from the Greek for cone, *konos*, and tooth, *odonotos*). Existing in a wide variety of shapes, these marine fossils date from 515 million to 210 million years ago. However, it wasn't until 1982 that the first complete conodont fossils were found, in Scotland. The eel-like animal has been described as a "tail with teeth at the front end." The presence of a notochord and eye muscles, previously unique to vertebrates, suggests that conodonts may be ancestors of vertebrates.

conservation (*environment*)   Conservation comprises actions taken to save, protect, and improve Earth's resources so that these resources will be available for future generations to use and enjoy. It involves a broad variety of efforts, such as preserving wildlife habitats, reducing overexploitation of resources, protecting farmland against soil erosion, and preventing pollution.

conservation laws (*physics*)   Many theories in physics state that a certain quantity does not change unless there is addition or subtraction to the quantity from outside. These theories are known as conservation laws; the quantity is conserved in a closed system. The first conservation law was Newton's third law of motion of 1687, equivalent to the statement that momentum is conserved in a closed system. Similarly angular momentum is conserved. A hundred years later Lavoisier introduced conservation of mass as an important tool in chemistry. The 19th century added conservation of energy, which became the first law of thermodynamics. In 1905 Einstein introduced the famous equation $E = mc^2$, which combines conservation laws for mass and energy into a single law. Particle physicists in the 20th century added conservation of charge, spin, and other quantities. In 1918 Emmy Noether proved mathematically that every conservation law is a consequence of a form of symmetry.

consistent (*logic*)   A set of propositions is consistent if all the propositions are true when any one of them is true. In contrast, a set is inconsistent if it contains 2 or more propositions that cannot both be true. An axiomatic system is consistent if there is no possible proof of both a proposition (p) and its negation (not p).

constellation (*stars*)   Constellations originally consisted of several bright stars treated as a group. Before recorded history, people began naming these groups. By Sumerian times (3000–2500 BC), stories were being told about particular constellations. Most of our present knowledge of such stories comes from Greek sources.

A group of stars that outlines a particular shape is called an asterism. Most asterisms are part of constellations; for example, the asterism Big Dipper is part of the constellation Ursa Major.

Today astronomers use constellations for mapping the sky. Each part of the sky is named by a particular constellation. These constellations, especially in the Southern Hemisphere, may not be traditional ones, but rather, groups of stars astronomers have named so that for reference purposes all of the sky is labeled. The International Astronomical Union has decreed that each such constellation be bounded by straight north-south and east-west lines. As a result, many of the larger traditional constellations extend beyond the boundaries of modern astronomical constellations.

The Constellations

NAME	ABBREVIATION	TRANSLATION	REMARKS
Andromeda	And	Andromeda	daughter of Cepheus in Greek myth; lies next to square of Pegasus and contains M31, the Andromeda galaxy
Antlia	Ant	air pump	named by Nicolas Lacaille in 1750; in Southern Hemisphere, south of Hydra
Apus	Aps	swift, or bird of paradise	named by Johann Bayer in 1603; in Southern Hemisphere, near pole
Aquarius	Aqr	Water Bearer	in zodiac—Sun passes through in late February and early March
Aquila	Aql	eagle	visible in summer in Northern Hemisphere; contains Altair
Ara	Ara	altar	part of Centaurus group; in Southern Hemisphere, south of Scorpius
Aries	Ari	Ram	in zodiac—visible in Northern Hemisphere in autumn
Auriga	Aur	charioteer	visible in Northern Hemisphere in winter between Gemini and Taurus; contains Capella
Boötes	Boo	herdsman	kite-shaped constellation at end of handle of Big Dipper; best seen in spring in Northern Hemisphere; contains Arcturus
Caelum	Cae	chisel, or graving tool	faint constellation in Southern Hemisphere; named by Lacaille in 1750
Camelopardalis	Cam	giraffe	large, faint constellation near north pole; named by Johannes Bartsch in 1661
Cancer	Cnc	Crab	faintest constellation in zodiac—Sun passes through in late July and early August
Canes Venatici	CVn	hunting dogs	lies between Boötes and Ursa Major; named by Johannes Hevelius in 1687
Canis Major	CMa	big dog	in Southern Hemisphere most of the year, but visible in north in winter; contains Sirius and Adhara
Canis Minor	CMi	little dog	with Canis Major, 1 of 2 dogs of Orion, lying near Orion's feet; contains Procyon
Capricornus	Cap	Goat (or Sea Goat)	in zodiac—Sun passes through in late January, early February

(continued)

The Constellations (continued)

NAME	ABBREVIATION	TRANSLATION	REMARKS
Carina	Car	ship's keel	in Southern Hemisphere; contains Canopus (named for helmsman of Argonaut's ship) and Eta Carinae, a star that was almost as bright as Sirius for a few years in the 19th century, but now barely visible; formerly part of Argo Navis; renamed by Lacaille in 1750
Cassiopeia	Cas	Cassiopeia	named for queen in Greek myth; a prominent W-shape near the north pole and pointing toward Polaris
Centaurus	Cen	Centaur	named for half-man, half-horse in Greek myth; in Southern Hemisphere; contains Alpha Centauri (Rigil Kentaurus) and Proxima Centauri, star closest to Earth
Cepheus	Cep	Cepheus	named for king in Greek myth; near north pole
Cetus	Cet	whale	named after sea monster slain by Perseus; in southern sky; best seen in autumn in Northern Hemisphere
Chamaeleon	Cha	chameleon	near south pole; named by Bayer in 1603
Circinus	Cir	compass	in Southern Hemisphere; named by Lacaille in 1750
Columba	Col	dove	between Lepus and Pictor in Southern Hemisphere; named by Bayer in 1603
Coma Berenices	Com	Berenice's hair	between Boötes and Leo; named for 3rd century BC Egyptian queen
Corona Australis	CrA	southern crown	in Southern Hemisphere near Sagittarius; also known as Sagittarius's Crown
Corona Borealis	CrB	northern crown	between Boötes and Hercules; best seen in northern spring; also known as Ariadne's Crown
Corvus	Crv	crow	in Southern Hemisphere below Virgo; named for bird companion of Orpheus
Crater	Crt	cup	in Southern Hemisphere between Leo and Hydra
Crux	Cru	southern cross	named by Augustine Royer in 1679; contains Beta Crucis and Acrux; smallest constellation
Cygnus	Cyg	swan	sometimes known as Northern Cross because of its shape; best seen in northern summer; contains Deneb

NAME	ABBREVIATION	TRANSLATION	REMARKS
Delphinus	Del	dolphin	between Pegasus and Aquila; 4 of its stars form an asterism called Job's coffin
Dorado	Dor	goldfish (swordfish)	in Southern Hemisphere; contains Large Magellanic Cloud; named by Bayer in 1603
Draco	Dra	dragon	large constellation near north pole; named for dragon slain by Hercules
Equuleus	Equ	little horse	dim constellation near Pegasus
Eridanus	Eri	River Eridanus	extends from equator far into Southern Hemisphere; contains Achernar
Fornax	For	furnace	in Southern Hemisphere next to Eridanus; named by Lacaille in 1750
Gemini	Gem	Twins	in zodiac—Sun passes through in late June to late July; contains Pollux and Castor, named for twins in Greek myth
Grus	Gru	crane	in Southern Hemisphere below Piscis Austrinus; named by Bayer in 1603
Hercules	Her	Hercules	best seen in summer in Northern Hemisphere; named for well-known hero from Greek myth
Horologium	Hor	pendulum clock	in Southern Hemisphere near Eridanus; named by Lacaille in 1750
Hydra	Hya	water monster	largest constellation; named for many-headed monster from Greek myth
Hydrus	Hyi	sea serpent	near south pole; named by Bayer in 1603
Indus	Ind	Indian	near south pole; named by Bayer in 1603
Lacerta	Lac	lizard	between Andromeda and Cygnus; named by Hevelius in 1687
Leo	Leo	Lion	in zodiac—Sun passes through between mid-August and mid-September; contains Regulus; visual source of Leonid meteor shower in November
Leo Minor	LMi	little lion	north of Leo; named by Hevelius in 1690
Lepus	Lep	hare	prey of Orion since it is at Orion's foot; best seen in northern winter

(continued)

constellation

The Constellations *(continued)*

NAME	ABBREVIATION	TRANSLATION	REMARKS
Libra	Lib	Scales	in zodiac between Virgo and Scorpius—Sun passes through during November
Lupus	Lup	wolf	in Southern Hemisphere between Scorpius and Centaurus
Lynx	Lyn	lynx	near Ursa Major; named by Hevelius in 1687
Lyra	Lyr	lyre	best seen in northern summer; contains Vega
Mensa	Men	table mountain	near south pole; contains part of Large Magellanic Cloud; named by Lacaille in 1750
Microscopium	Mic	microscope	in Southern Hemisphere next to Sagittarius; named by Lacaille in 1750
Monoceros	Mon	unicorn	between Canis Major and Orion, best seen in winter in Northern Hemisphere; named by Bartsch in 1624
Musca	Mus	fly	in Southern Hemisphere near Crux; named by Bayer in 1603
Norma	Nor	carpenter's square	in Southern Hemisphere near Lupus and Ara; named by Lacaille in 1750
Octans	Oct	octant	faint constellation near south pole; named by Lacaille in 1750
Ophiuchus	Oph	Ophiuchus (serpent bearer)	near Hercules and Scorpius; most visible in summer in Northern Hemisphere; giant in Greek myth
Orion	Ori	hunter	character in Greek myth; visible to the south in winter in Northern Hemisphere; contains Rigel, Betelgeuse, and Bellatrix; 3 other bright stars make Orion's belt while a glowing nebula is his sword hanging from a belt
Pavo	Pav	peacock	in Southern Hemisphere; named by Bayer in 1603
Pegasus	Peg	Pegasus	best seen during autumn in Northern Hemisphere; prominent Square of Pegasus is easily recognized asterism; named after winged horse in Greek myth
Perseus	Per	Perseus	named after important hero in Greek myth; best seen during autumn in Northern Hemisphere; appears to be source of Perseid meteor shower in late July and early August

NAME	ABBREVIATION	TRANSLATION	REMARKS
Phoenix	Phe	Phoenix	in Southern Hemisphere near mouth of Eridanus; named for mythical bird by Bayer in 1603
Pictor	Pic	painter	faint constellation in Southern Hemisphere; named by Lacaille in 1750
Pisces	Psc	Fish	in zodiac—Sun passes through from mid-March to mid-April; contains imaginary line for vernal equinox used as starting point when measuring right ascension of celestial bodies
Piscis Austrinus	PsA	southern fish	in Southern Hemisphere below Cetus; contains Fomalhaut
Puppis	Pup	ship's stern	in Southern Hemisphere next to Canis Major; renamed by Lacaille in 1750 (formerly part of Argo Navis)
Pyxis	Pyx	ship's compass	in Southern Hemisphere below Hydra; renamed by Lacaille in 1750 (formerly part of Argo Navis)
Reticulum	Ret	net	faint constellation in Southern Hemisphere; named by Lacaille in 1750
Sagitta	Sge	arrow	although faint, one of 48 constellations listed by Ptolemy; near Aquila, it is most visible in summer in Northern Hemisphere
Sagittarius	Sgr	Archer	in zodiac—Sun passes through between mid-December and mid-January; from Earth the center of the Milky Way galaxy appears to be in Sagittarius
Scorpius	Sco	Scorpion	in zodiac—Sun passes through in last week of November, but in Northern Hemisphere constellation is most visible in summer; contains Antares and Shaula
Sculptor	Scl	sculptor	faint constellation in Southern Hemisphere below Cetus; named by Lacaille in 1750
Scutum	Sct	shield	originally Sobieski's shield, after a patron of Hevelius, who named Southern Hemisphere constellation in 1687
Serpens	Ser	serpent	snake held by, or wound around, giant Ophiuchus; sometimes considered as 2 parts, the head (Serpens Caput) and the tail (Serpens Cauda); best seen in Northern Hemisphere in spring and summer

(continued)

The Constellations (continued)

NAME	ABBREVIATION	TRANSLATION	REMARKS
Sextans	Sex	sextant	faint constellation between Leo and Hydra named by Hevelius in 1690
Taurus	Tau	Bull	in zodiac—Sun passes through from mid-May to mid-June, although best seen in Northern Hemisphere in winter; contains Aldebran and Elnath, as well as Pleiades star cluster (sometimes thought of as a tiny constellation itself)
Telescopium	Tel	telescope	faint constellation in Southern Hemisphere below Sagittarius; named by Lacaille in 1750
Triangulum	Tri	triangle	small constellation between Andromeda and Perseus; contains spiral galaxy M33
Triangulum Australe	TrA	southern triangle	in Southern Hemisphere next to Centaurus; named by Bayer in 1603
Tucana	Tuc	toucan	near south pole; contains Small Magellanic Cloud; named by Bayer in 1603
Ursa Major	UMa	big bear	circumpolar about north pole, but best seen in winter; contains asterism Big Dipper (also called the Plow)
Ursa Minor	UMi	little bear	at the north pole; contains asterism Little Dipper and Polaris, the current north star
Vela	Vel	ship's sails	in Southern Hemisphere next to Centaurus; renamed by Lacaille in 1750 (formerly part of Argo Navis)
Virgo	Vir	Virgin	in zodiac—Sun passes through from mid-September to early November; contains Spica
Volans	Vol	flying fish	faint constellation in Southern Hemisphere next to Carina; named by Bayer in 1603

consumer (*ecology*) Organisms that obtain energy by eating organic matter are called consumers. All animals are consumers; so are bacteria, fungi, and many protists. Consumers include herbivores, carnivores, omnivores, scavengers, and decomposers. Either directly or indirectly, they depend on producers for their nourishment.

continent (*geography*)   A continent is a large mass of lighter rock rising above the seafloor. Geographers count only the part above the sea surface, but geologists include the continental shelf. Continents are not permanent features of Earth but last longer than oceans by far. The oldest parts of continents are about

4,000,000,000 year old. Plate movements have pushed continents together and pulled them apart throughout Earth's history. Today the 6 or 7 continents are North America, South America, Africa, Europe and Asia, Australia, and Antarctica. Europe and Asia are often treated as a single continent, Eurasia.

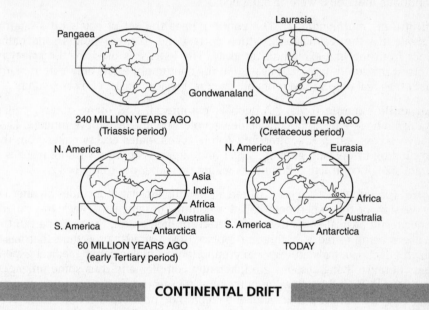

240 MILLION YEARS AGO
(Triassic period)

120 MILLION YEARS AGO
(Cretaceous period)

60 MILLION YEARS AGO
(early Tertiary period)

TODAY

CONTINENTAL DRIFT

continental drift (*geology*) Continents slowly change their relative positions over long periods of time in movement known as continental drift. The motion of continents results from plate tectonics, and the word "drift" is not taken literally. The current explanation is that the movement of tectonic plates results largely from lithosphere under the oceans being created and destroyed. The concept of continental drift was popularized by Alfred Lothar Wegener starting in 1912. Wegener proposed that about 200,000,000 years ago all the continents had "drifted" together to form a single landmass, which he named Pangaea. Pangaea broke apart and the pieces moved to form today's continents.

continental shelf (*oceanography*) Continental rock does not stop at the ocean edge. Ocean levels rise and fall (mainly because of ice ages), changing the position of the edge of the land. Continental rock currently extends 100 mi (160 km) or more beneath the sea, forming what is called a continental shelf. A continental shelf ends where a steeper drop-off called a continental slope begins; such a slope terminates in an abyssal plain or a trench.

continuity (*calculus*) A curve is continuous if there are no gaps. It can be traced with a pencil keeping the pencil point on a point of the curve at all times. These informal characteristics describe continuity, but 19th-century mathematicians encountered odd curves that they called pathological, showing the need for formal definitions. A modern definition is for a function *f*, which is continuous for a number *a* in its domain provided that *f* is defined for every domain value in the **neighborhood** of a and the **limit** of $f(x)$ as x goes to a is $f(a)$; this means that the curve must close in on the point $(a, f(a))$ and pass through it. Even this definition was not sufficient to deal with all pathological curves and stronger types of continuity, called absolutely continuous and uniformly continuous functions, were also defined.

continuum (*mathematics*)   The continuum is the set of **points** on a **line** or, equivalently, the **real numbers**, which correspond to those points. The name comes from the line exhibiting **continuity**. The letter c is used for the **infinity** of points in the continuum. It appears that the continuum exists only within mathematics, as real objects and probably even real space consist of discrete units.

contractile vacuole (*cells*)   Amoebas, paramecia, and many other 1-celled organisms have 1 or more contractile vacuoles, organelles whose function is to remove excess water. A contractile vacuole accumulates excess water from the surrounding cytoplasm, growing larger and larger. When the vacuole reaches a certain size, it contracts suddenly, expelling the water out of the cell.

control (*philosophy of science*)   Often a scientific **experiment** includes an alternative version of the experiment called a control, designed to check the experiment's result against some unknown defect. In the control, the main part of the **hypothesis** being tested is omitted or replaced by a similar procedure that has a known effect. Controls are especially important in biological or medical experiments because living systems are inherently complex and thus some unknown variable may be influencing the outcome. Often tests of a new drug or medical procedure are double-blind: neither patients nor physicians know which subjects are in the control group (those receiving a placebo) and which are in the experimental group (those receiving the drug or undergoing the procedure).

convection (*thermodynamics*)   Convection is one of the 3 ways (with conduction and radiation) that **heat** is transported.

Responding to gravitational force, parts of a fluid that are denser than surrounding parts sink. The sinking fluids displace less dense material and cause it to rise. From the point of view of someone surrounded by the fluid, as we are by air, it is more obvious that the less dense fluid is rising than that the denser fluid is sinking. Thus people say "hot air rises," although that occurs only because cold air sinks.

When a fluid is heated from the bottom, the unheated fluid sinks and the heated fluid is pushed away—but then the previously unheated fluid becomes heated. The cycle continues to repeat and a current is set up, a process called convection. This current carries faster moving particles from one place to another, thus transporting heat.

convergence (*calculus*)   Convergence is the general name for one mathematical entity getting very close to another. The entities may be **sequences**, **series**, **curves**, or **functions**. When sequences, series, or functions converge, they approach a value called the **limit**. When a curve converges, it can approach a line that is its **asymptote** or a point that is not on the curve itself—a spiral can become very close to a point in its center without the center being part of the spiral. As with other mathematical ideas concerned with the infinitely small, careful definitions of strong forms of convergence, such as absolute convergence and uniform convergence, are needed to prevent errors based on vague ideas of convergence.

convergent evolution (*evolution*)   Organisms that are not closely related may independently adapt to similar environments by evolving similar structures. Examples of such convergent evolution, or parallelism, are the shark and the dolphin, both of which evolved streamlined shapes, fins, and tail flukes as adaptations to life in the sea. Similarly, members of the cactus and euphorbia families evolved such closely related characteristics as numerous spines, thick waxy surfaces, and extensive water storage tissues to adapt to desert life.

conveyor belt (*oceanography*)   Transport of heat by ocean currents is known as the conveyor belt. Cold, salty water from near the poles sinks and flows along the bottom of Earth's oceans in various currents. Warm surface water flows in to replace it. The cool water rises far from where it sank, as in the Peru Current near the equator, bringing up nutrients from the ocean bottom. The conveyor belt moves heat from the tropics to polar regions, cooling the first and warming the second. Because of the Coriolis effect, the warm upper current of the Atlantic warms Europe much more than it does Canada.

coordinate (*analytic geometry*)   The numbers used to locate points in analytic geometry are called coordinates, whether Cartesian coordinates, polar coordinates, or parts of another coordinate system. A point is located by different numbers depending on the coordinate system. In general the number of coordinates needed for a point depends on the dimension of the space—1 coordinate for a line, 2 coordinates for a plane, 3 for ordinary space, and so forth—rather than on the system used.

coordinate system (*analytic geometry*)   In analytic geometry numbers are correlated with points in spaces. The connection between numbers and points is a coordinate system. The most familiar consists of 2 perpendicular lines, called axes, defining the 2-dimensional plane. Any point is a specific distance and direction from each axis; the directed distances are the 2 numbers for the point. A point 3 units to the left of the vertical axis and 4 units below the horizontal correlates with $(+3, -4)$. In 1 dimension there is only 1 axis, and any number correlates with distance of the point to the right $(+)$ or left $(-)$ of 0.

A coordinate system in 3-dimensional space uses 3 axes, usually called x, y, and z, that define 3 planes. The choice of which side of the xy plane to put the $+$ part of the z axis determines whether the coordinate system is right-handed or left-handed. In a right-handed system, if the x and y axes are in the normal orientation, the positive z axis points away from the viewer; in the left-handed version, the positive z axis points toward the viewer. In algebra, however, the positive x axis is usually shown pointing toward the viewer, the positive y axis to the right, and the z axis up—a right-handed system.

Many coordinate systems are based on different geometric ways to represent numbers. Polar coordinates are one example, but mathematicians also locate numbers on curved surfaces or by systems of intersecting conic sections.

copepod (*animals*)   Believed to be the most numerous animals on Earth, copepods are tiny crustaceans found in both fresh and salt water, where they are a major component of plankton. The first pair of antennae are very long, there are no compound eyes, and the body is shaped somewhat like a bottle. Most species are free-living, acting as primary or secondary consumers in aquatic food chains. Other species, usually with greatly modified bodies, are parasitic on fish and invertebrates.

Copernicus, Nicolaus (*astronomer*)   [Polish: 1473–1543] Copernicus was a scholarly church canon who succeeded in convincing almost all later astronomers that Earth rotates on an axis and revolves about the Sun. An early Greek, Aristarchus, had proposed similar ideas, but few accepted them. Before Copernicus's book *De revolutionibus orbium coelestium* was published in 1543, nearly everyone endorsed the Ptolemaic system, based on a stationary Earth. The ideas expressed by Copernicus are not exactly what we believe today—for example, Copernicus used circles moving on circles to make the system work.

coplanar (*geometry*)   Points, lines, or other geometric figures that lie in the same plane are coplanar. Any 3 points or any 2 intersecting or parallel lines are necessarily coplanar, often stated as 3 points or 2 intersecting or 2 parallel lines determine a plane.

copper (*elements*)   Copper, or possibly gold, was the first element to be recognized as a separate substance. Each metal occurs in a relatively pure form called native. The main source of copper in the ancient world was Cyprus, *cuprum* in Latin, hence the chemical symbol Cu. Copper, atomic number 29, is in the same column of the periodic table as gold and silver. Like them, it reacts weakly with other chemicals and is very malleable, making it useful in many ways. Copper conducts heat and electricity better than other common metals. The principal use of pure or nearly pure copper is for electrical connections. Although copper is soft, its alloys with tin (bronze) and zinc (brass) are strong and used where corrosion is a problem.

coprolite (*paleontology*)   Fossilized feces, or coprolites, offer valuable insights into the diets and habitats of prehistoric animals. For example, a 65-million-year-old coprolite found in western Canada in 1998 was 17 in. (44 cm) long and 5 in. (13 cm) wide. It contained pieces of bone from a plant-eating dinosaur, and is assumed to be the first *Tyrannosaurus rex* coprolite ever found, since *T. rex* was the only known meat-eating dinosaur in the area at that time that could produce such large feces. The coprolite indicates that *T. rex* crushed bones before it swallowed them.

coral (*animals*)   Classified as cnidarians, corals are small marine animals, most common in warm seas. They look like miniature sea anemones, with an important difference: a coral animal lives permanently attached to a hard cup of calcium carbonate, sometimes hardened with other minerals, that it secretes around itself. Some species are solitary but most live in colonies of thousands of individuals, their cups fused together into intricately folded, branched, or spherical masses. These species are the principal builders of coral reefs.

coral bleaching (*environment*)   Living inside the tissues of the coral animals that build coral reefs are microscopic algae. The animals provide shelter and nutrients for the algae, and the algae provide sugar and other foods for the animals. If coral animals become stressed by pollution, temperature changes, or other environmental factors, they lose the algae from their tissues and turn white, as if they had been bleached. If the stress is short term and not extreme, the coral animals usually recover. But if the stress is high and persists, the corals may die.

coral reef (*oceanography*)   Certain species of corals in shallow tropical and subtropical seas contribute to the construction of enormous structures called coral reefs. Over a period of centuries, the calcium carbonate deposited by tiny coral animals, a material also called coral, remains behind as the animals die. New generations build on the surface of the mass, gradually enlarging it until a long ridge under the sea, a reef, results. If waves break up large sections of a reef and pile the coral above sea level, coral islands are created. The Great Barrier Reef, a coral reef extending some 1250 mi (2000 km) along the eastern coast of Australia, is thought to be the largest structure ever built by organisms.

core (*geology*)   The innermost part of Earth, an irregular ball of iron alloy covering the region 2000 mi (3500 km) from Earth's center, is known as the core. It has 2 layers, a liquid outer core and a solid inner core. From the surface the distance to the outer core is about 1800 mi (2900 km); the edge of the inner core is about 3000 mi (5000 km) deep. High temperatures, which may reach 9000°F (5000°C), keep the outer core liquid, while the pressure, about 3,000,000 times air pressure at sea level, solidifies the inner core.

Iron is the major constituent of both cores, but density studies show the iron is alloyed with a lighter element, such as silicon, sulfur, carbon, or oxygen. Because iron **meteoroids** contain a high percentage of nickel, it is commonly believed that Earth's core also contains nickel. Other **terrestrial planets** are also thought to have metal cores, while **gas giants** may feature layers of rock surrounding metal, much like an entire terrestrial planet, as the core.

Coriolis effect (*meteorology*)   Any object that moves over the surface of a rotating sphere is affected by a **fictitious force** called the Coriolis effect, first explained by Gustave Coriolis [French: 1792–1843] in 1835. There are 2 components.

Because Earth is nearly a **ball** rotating about its **axis** from west to east, the motion of its surface at the equator is rapid—a point there moves 24,900 mi (40,000 km) in 24 hours. Speed becomes less the closer the point is to one of the poles, where velocity approaches 0. Thus a body moving north from the equator with a steady speed relative to the surface, but not traveling on it (an airplane, an ocean current, or a wind, for example), finds that the Earth below it slows down, causing the body's path to veer eastward with respect to the surface. The faster the body moves directly north, the more it shifts east along the surface.

Centrifugal force is also a fictitious force that increases with velocity. On Earth centrifugal force acts **perpendicular** to the axis of rotation. But a body moving east or west and unconnected to the surface has a different velocity than the surface below it, resulting in a higher or lower centrifugal force. Eastward motion causes the body to curve toward the equator, while westward movement results in curving toward the nearer pole.

cork (*botany*)   As the stems and roots of a young woody plant grow in diameter, the outer layer, or epidermis, is stretched and eventually tears. A growth tissue called cork cambium forms just below the epidermis. In some plants only a single cork cambium develops; in other species many layers may form over time. The cork cambium produces cork—the outer bark of trees and shrubs. Cork protects the inner tissues against excessive evaporation, mechanical injury, and insects and other pests.

Commercial cork comes from the cork oak, *Quercus suber,* which is native to southern Europe and North Africa. This tree produces a prodigious amount of cork, and it is possible to strip the outer layers every 9 or 10 years without injuring the tree.

corm (*botany*)   Superficially resembling a bulb, a corm is a thickened underground stem surrounded by thin, dry leaf scales. Primarily a food storage unit, a corm also can be used for vegetative propagation; that is, clumps of new corms can grow from nodes. The gladiolus and crocus are examples of plants that produce corms.

cormorant (*animals*)   Long-necked, long-bodied birds that make up Family Phalacrocoracidae, cormorants are famous for their skill in catching fish. Most cormorants are strong fliers and excel at diving and swimming underwater. They live mainly in tropical and temperate areas, near either oceans or inland waters. Cormorants are gregarious, nesting in sizable colonies. One species, the Galapagos cormorant (*Nannopterum harrisi*), has lost the ability to fly.

corn (*plants*)   Native to tropical America, corn or maize (*Zea mays*) is an annual grass growing up to 15 ft (4.5 m) tall. It is widely cultivated for its grains (fruits) and hundreds of varieties have been developed. Corn has a hard, jointed stem and large leaves. Male flowers develop at the top of the stem in a cluster called a tassel. Female flowers form in clusters at joints lower on the stem. Following pollination, the female flowers develop into spikes called ears. Each ear has a central pith, called a cob, on which seeds—usually yellow—are arranged in rows. The ear is enveloped in a husk formed by several leaves.

corona (*solar system*)   The outer layer of the Sun, similar to the upper atmosphere of Earth in some ways, is the corona; the name means "crown," as the corona appears as a crown of streamers surrounding the black disk of the Moon during a solar eclipse. The corona extends tens of millions of miles into space and gradually merges with the solar wind.

correspondence principle (*particles*)   Quantum theory, developed in the first half of the 20th century, has quite different laws from those of the earlier physics, now called "classical." Niels Bohr proposed the correspondence principle partly in an attempt to unify physics and partly as a method of discovery. It states that a collection of enough subatomic particles obeying quantum laws will approximate the behavior of large objects obeying corresponding classical laws.

cosecant (*trigonometry*)   One of the 6 functions of trigonometry, the cosecant (csc) is defined as the reciprocal of the sine (sin); that is, $\csc \theta = 1/\sin \theta$. In a right triangle $\csc \theta$ is equal to the ratio of the length of the hypotenuse (side opposite the right angle) to the length of the angle opposite θ.

cosine (*trigonometry*) The cosine (cos) is one of the 2 basic functions in trigonometry. It is the complement of the sine (sin); that is, for 2 complementary angles *A* and *B*, cos *A* = sin *B* and sin *A* = cos *B*. In terms of a unit circle, cos θ equals the *x* coordinate of the point where the terminal side of θ meets the circle. In a right triangle cos θ is the ratio of the length of the side adjacent to θ to the length of the hypotenuse (side opposite the right angle).

cosmic background radiation (*cosmology*) Cosmic background radiation is a pervasive bath of electromagnetic waves that originated during the big bang. In 1965 Arno Penzias [German-American: 1933–] and Robert Wilson [American: 1936–] discovered this microwave signal from space. Theoretical calculations had predicted such a remnant from the big bang. After its origin as high-energy gamma rays, the radiation lengthened in response to expansion of the universe. Its energy today corresponds to temperature of –454.747°F (–270.415°C, or 2.735°C above absolute zero).

Microwaves from any direction measure almost the same as radiation from any other direction. The Doppler effect—as we move toward waves they reach us more frequently—however, raises the energy slightly and reveals the direction and speed with which we travel through the universe.

cosmic ray (*astronomy/physics*) Cosmic rays are fast-moving nuclei of hydrogen (protons), helium (alpha particles), and heavier elements along with energetic electrons—thus, they are not technically rays, but particles. The particles are produced by the Sun and other stars, with the most energetic ones thought to come from supernovas. Such primary cosmic rays produce secondary cosmic rays when they collide with atoms or each other; the secondary rays include gamma radiation (high-energy photons), positrons, mesons, and heavy hadrons known as strange particles. Cosmic rays provide limited information about their origin because their paths have been altered by interstellar and intergalactic magnetic fields, but the secondary cosmic rays were one of the main tools of particle physicists in the 1940s and 1950s. Some cosmic rays provide higher energies than any particle accelerator, but these do not occur frequently enough for useful scientific studies.

cosmology (*sciences*) Cosmology is that part of astronomy concerned with the origin and evolution of the universe. Early cosmological theories were restricted to the solar system and visible stars. Modern theoretical cosmology began in 1915 with Einstein's general relativity theory. Observational cosmology followed in the 1920s with Hubble's work establishing galaxies outside the Milky Way and demonstrating the expansion of the universe. In 1948 George Gamow [Russian-American: 1904–68] and several collaborators formulated the big bang theory to explain these observations, a theory that has been accepted since the 1965 discovery of cosmic background radiation. However, a small minority of cosmologists cling to a steady-state theory as an alternative explanation.

Cosquer cave (*paleoanthropology*) Cosquer cave in France has only 1 opening, which has long been covered by the Mediterranean Sea. Its cave art was found in 1991 by diver Henri Cosquer. The earliest paintings, hand stencils, date from 28,000 years ago. About 100 animal paintings are more recent (c16,000 BC). Sea animals, such as seals and auks, occur along with such game animals as horses, chamois, ibex, and bison. Water has destroyed lower parts of some animals, and no paintings appear below the waterline.

cotangent (*trigonometry*)　One of the 6 functions of trigonometry, the cotangent (cot; sometimes ctn) is defined as the reciprocal of the tangent (tan); that is, cot θ = 1/tan θ. In a right triangle cot θ is equal to the ratio of the length of the side adjacent to θ to the length of the angle opposite θ.

cotinga (*animals*)　Cotingas (Family Cotingidae) are perching birds native to forests of the American tropics and subtropics. They range in length from 3.5 to 18 in. (9 to 46 cm). Some species are dull-colored but others are extremely colorful and ornate. For example, the male cock-of-the-rock (*Rupicola*) is bright orange or red, with a fan-shaped crest; the female is smaller and brown, with only a rudimentary crest. Cotingas eat mainly fruit.

cotton (*plants*)　Cotton (*Gossypium*) is a shrubby warm-weather perennial that grows to heights of about 6.5 ft (2 m). After the flower's cream-colored petals fall, the ovary enlarges into a capsule called the cotton boll, which contains numerous seeds covered with short white hairs, or fibers. For some 5000 years people have woven the fibers into fabric. Cotton seeds are pressed to obtain an oil used in foods, cosmetics, soaps, and other products.

Coulomb, Charles (*physicist*)　[French: 1736–1806] Although Coulomb is best remembered for his work with electric charges, he also was the first to show that the force of friction is always proportional to the pressure exerted at 90° to the surface. His experiments with electricity, using a sensitive apparatus called the torsion balance, showed that the force between 2 charges displays inverse variation with respect to the square of the distance between the charges and direct variation with respect to the product of their size (mathematically the same as the law of gravity).

countable (*sets*)　A set is countable (also called denumerable) if its members can be matched in one-to-one correspondence with the natural numbers or with a subset of the natural numbers. For example, the set of even numbers is countable because for every natural number there is an even number that is its double and for every even number there is a natural number that is half the even number. Not only are all finite sets countable, but so are all subsets of the natural numbers and of the integers. Surprisingly, it can be proved that the set of rational numbers is countable, but the sets of real numbers, points on a line, complex numbers, points in space, and many other sets of interest are not countable. Several proofs of this were among the significant contributions of Georg Cantor at the end of the 19th century.

counting (*arithmetic*)　Matching the members of a set with the first natural numbers in order is counting; the number matched with the last member of the set is the count: 1 apple, 2 apples, 3 apples, no more, so the count is 3 apples. Sets with the same count can be put into one-to-one correspondence and vice versa; this ensures that recounting a set in a different order will produce the same result. Counting is the basic operation of natural number arithmetic from which all others derive.

courtship (*ethology*)   In many animal species mating is preceded by courtship rituals designed to assess the quality of potential mates and to ensure that mating occurs with a sexually mature member of the opposite sex of the right species. These rituals take many forms. A male greater bird-of-paradise (*Paradisaea apoda*) shows off his colorful plumage as he performs an elaborate courtship display. Among elephant seals (*Mirounga angustirotris*), males fight for dominance, with the winner having the right to mate with females in the area. Male satin bowerbirds (*Ptilonorhynchus violaceus*) attract females by offering elaborately built and decorated structures in which to mate.

Cousteau, Jacques-Yves (*oceanographer*) [French: 1910–97] An explorer with a lifelong fascination for the sea, Cousteau coinvented the Aqua-Lung, or scuba (self-contained underwater breathing apparatus), the first device that allowed sustained, unencumbered diving. A strong advocate of marine conservation, Cousteau wrote many books and produced nearly 100 films, fostering appreciation of the previously unseen undersea environment.

cowrie (*animals*)   Snails that live in warm coastal waters of the Pacific and Indian oceans, cowries (genus *Cypraea*) have glossy, often beautifully colored and patterned, flat-bottomed shells. Most snails have the mantle inside the shell. But the cowrie's mantle covers the shell and produces new shell layers on the outside.

coyote (*animals*)   A member of the dog family widespread in North and Central America, the coyote (*Canis latrans*) has adapted to a variety of habitats and diets, even expanding its range into suburban areas. It has a body length of up to 3.3 ft (1 m), with a thick coat and bushy tail. Coyotes typically live in family groups. They are primarily nocturnal, resting in dens during the day.

crab (*animals*)   So-called true crabs are crustaceans with a broad, flattened shell, or carapace. The abdomen, greatly reduced, is folded under the thorax. There are 4 pairs of walking legs (crabs typically walk sideways in a scuttling gait). Most species are marine but some inhabit freshwater or land. Most are scavengers and predators. Many live in burrows. Certain species camouflage themselves by attaching seaweed and small invertebrates to their back.

Crab Nebula (*astronomy*)   The Crab Nebula (also called M1 or NGC 1952) is the most famous example of the remnant of a supernova explosion. Chinese records report a bright new star appearing in July 1054—a supernova—that was visible for 23 days even in daylight. The nebula was observed by telescope in the 18th century. By 1937, after its expansion rate had been measured, the nebula was recognized as the debris from the supernova of 1054. In 1969 astronomers located the remains of the original star, now a pulsar with the mass of the Sun condensed into a sphere a few miles across.

crane (*animals*)   Tall birds with a long straight bill, a long neck, and long legs, cranes (Family Gruidae) live in temperate and tropical regions on all continents except South America. They inhabit grasslands and marshes, feeding on a variety of plant and animal matter. Among the largest is the whooping crane (*Grus* *americana*), the tallest bird in North America, up to 5 ft (1.5 m) tall with a wingspan of 7 ft (2.1 m). Cranes perform elaborate courtship dances and are believed to mate for life.

Cranial Nerves

NUMBER AND NAME	EXTERNAL CONNECTIONS	PRIMARY FUNCTION(S)
I olfactory	nasal cavity	sensory (smell)
II optic	retina of eye	sensory (sight)
III oculomotor	muscles of eyeball, iris, lens, upper eyelid	motor (movement of eye, eyelid)
IV trochlear	muscle of eyeball	motor (rotation of eyeball)
V trigeminal	head, face, jaw	sensory (scalp, face, teeth); motor (chewing, movement of tongue)
VI abducens	muscle of eyeball	motor (rotation of eyeball)
VII facial	scalp, face, tongue, neck	sensory (taste); motor (facial expression, chewing, salivation, movement of neck)
VIII auditory	inner ear	sensory (hearing, balance)
IX glossopharyngeal	tongue, pharynx	sensory (taste, touch); motor (swallowing)
X vagus	most organs in thorax and abdomen	sensory (larynx, lungs, esophagus, stomach, etc.); motor (constricts bronchi, slows heartbeat, stimulates stomach and gallbladder, etc.)
XI accessory	larynx, pharynx	motor (larynx, pharynx)
XII hypoglossal	tongue	motor (movement of tongue)

cranial nerve (*anatomy*)   Pairs of cranial nerves originate from the base of the vertebrate brain and connect mainly to parts of the head and neck. Some consist of sensory fibers that carry information from sense organs to the brain, some of motor fibers that transfer information from the brain to muscles, and some of both sensory and motor fibers. All vertebrates have the cranial nerves numbered I through X; fish and amphibians lack nerves XI and XII.

crater (*geology*)   Two kinds of large natural pits are called craters. The word originally applied to large circular openings on volcanoes caused by gases from below pushing away rock and soil during eruptions. Similarly shaped pits first recognized on the Moon and later on other rocky bodies in the solar system, including Earth, were originally thought to be volcanic craters. Further study showed that most of the Moon's craters had been caused by the impact of large meteorites or planetesimals, but the name *crater* was retained. Impact craters such as Arizona's Barringer crater tend to be more circular than volcanic craters and to have raised rims.

crayfish (*animals*)   Closely related to lobsters and similar in structure, crayfish are freshwater crustaceans up to 16 in. (40 cm) long. They hide under stones during the day and are active at night, feeding on a variety of plant and animal matter.

creeper (*animals*)   Members of several families of perching birds, native to various parts of the world, are referred to as creepers. The name alludes to their crawling gait up and down tree trunks in search of insects. Honeycreepers, also perching birds, behave similarly but primarily feed on nectar.

creodont (*paleontology*)   The creodonts ("flesh teeth") were the dominant group of carnivorous mammals in the early Tertiary. They were once believed to be ancestors of modern carnivores, but scientists now place them in a separate order. The creodonts ranged from the size of rats to species the size of wolves and bears. They had powerful jaws, with teeth designed to tear, slice, and grind meat. One of the last creodonts was *Hyaenodon*, a large hyenalike animal that became extinct about 25 million years ago.

Crick, Francis H.C. (*biophysicist/geneticist*)   [English: 1916–] In 1953 Crick and James Watson showed that the DNA molecule is a double helix consisting of 2 complementary chains wound around each other and attached by their nucleotide bases. Crick shared the 1962 Nobel Prize in physiology or medicine with Watson and Maurice H.F. Wilkins [English: 1916–].

cricket (*animals*)   Closely related to long-horned grasshoppers and similar in appearance, crickets are famed for the male's ability to produce sounds in a process called stridulation. The sounds, unique for each species, are used to attract mates and defend territory against other males. The hearing mechanism is located on the front legs. Crickets, usually colored green or brown, eat plants and dead animal matter.

crinoid (*animals*)   The oldest and most primitive of living echinoderms, crinoids (Class Crinoidea) have a small body with feathery, flowerlike arms surrounding the mouth. The waving arms collect food particles and direct them toward the mouth—a simple opening that never closes. All crinoids are marine, with some species found at depths greater than 10,000 ft (3000 m). Stalked crinoids that are permanently anchored to the seafloor are called sea lilies; free-moving species without stalks are called feather stars. Crinoids are comparatively small, though the stalks of some species may reach lengths of almost 2 ft (0.6 m). In some extinct species, stalks grew more than 75 feet (23 m) long.

crocodilian (*animals*)   Crocodiles, alligators, caimans, and the gavial constitute the reptile order Crocodylia. Predators all, they have a strong body and a long powerful tail, flattened from side to side and used for swimming, defense, and stunning prey. The toes on the 4 strong legs are webbed. If need be crocodilians can move fast on land, but they are more at home in water. They live primarily in tropical and subtropical habitats, their dark greenish black color blending with the background vegetation. Following mating, which takes place in water, the female lays eggs in a hole she digs in the ground or in a nest made of piled-up vegetation.

Cro Magnon (*paleoanthropology*)   Skeletons of early Europeans who used stone tools were first found in 1868 at a rock shelter known as Cro Magnon near Les Eyzies, France; thus, the earliest modern humans in Europe are popularly called Cro Magnons. These humans completely displaced the Neandertals in Europe about 36,000 years ago. Modern Europeans are physically the same as the Cro Magnons, although not quite as robust on average. The Cro Magnon culture is the last phase of the Paleolithic Age, as bone and antler tools began to displace stone for many uses.

Crookes, Sir William (*physicist*)   [English: 1832–1919] William Crookes trained as a chemist, but became interested in spectrometry and electrical discharges into near vacuums; his major discoveries emerged from these interests. In 1861 he recognized that a green line on a spectrogram must be a new element; he isolated the element and named it thallium. In 1875 he invented the radiometer, an evacuated transparent bulb that has an arrangement of vanes, black on one side and silvered on the other, that spins when placed in a light (the cause is unequal heating of the gases left in the bulb near the black and the silver vanes). Crookes's name is preserved in his vacuum tubes. He did not invent the vacuum tube with a cathode and anode to produce an electric current, but his Crookes tubes were superior. With them he discovered the basic properties of the electric discharge, or cathode ray, from which J.J. Thomson was to discover the electron, confirming Crookes's own views on the nature of cathode rays.

crop (*anatomy*)   Several kinds of animals have an organ called a crop where food is either digested or stored. In birds the crop is an expansion of the lower esophagus. Food is moistened and stored here temporarily before passing into the stomach. Invertebrates that have crops include earthworms, snails, and insects.

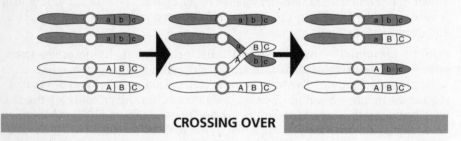

CROSSING OVER

crossing over (*genetics*)   Sometimes during cell division, while 2 homologous chromosomes lie alongside one another, they may become tangled and exchange corresponding segments. This is one way in which chromosomes can change as they pass from generation to generation, and it plays an important role in evolution. The phenomenon was discovered by Thomas Hunt Morgan, who called it crossing over.

cross pollination (*botany*)   The transfer of pollen from a male reproductive organ of one seed plant to a female reproductive organ of another plant of the same species is called cross pollination. This depends on intermediary agents such as wind, water, and animals. For example, when a bee visits a flower in search of nectar, it brushes against the flower's reproductive organs. Pollen picked up at a flower visited earlier falls off the bee's body onto the sticky female stigma.

cross product (*mechanics*)   For 3-dimensional vectors a product called the cross product **a** × **b**, also known as the vector product, exists. The cross product **a** × **b** is a vector whose length is the product of the lengths of **a** and **b** multiplied by the sine of the angle between them. The product **a** × **b** is perpendicular to the plane containing **a** and **b** so that the 3 vectors **a, b,** and **a** × **b** form a right-handed system (like a right-handed coordinate system). This form of multiplication is not commutative.

crow (*animals*)   Perching birds of the crow family, Corvidae, are found worldwide in temperate and tropical habitats. They include crows, jackdaws, jays, magpies, and ravens. These birds are aggressive; have stout, heavy bills; and eat almost anything. Though considered songbirds, their voices are harsh, often mimicking the sounds of other animals. They are probably the most intelligent of all birds.

crust (*geology*)   For terrestrial planets and large moons, the crust consists of lighter rock that forms a thin layer around the planet rather like the skin on a pudding. Denser solid rock called the mantle lies below the crust. The boundary between crust and mantle on Earth is the Mohorovicic discontinuity (Moho). Crust beneath continents varies from 15 to 30 mi (25 to 50 km) in depth, while under the oceans it is generally from 3 to 5 mi (5 to 8 km) thick. Continental crust is mostly lighter rock, such as granite, while the ocean floor, produced by rising magma at mid-oceanic ridges, is basalt.

crustacean (*animals*)   Lobsters, shrimp, barnacles, and their kin constitute the arthropod class Crustacea ("hard shell"). Crustaceans are characterized by a hard, jointed exoskeleton that must periodically be replaced as the animal grows. On the head are 2 pairs of sensory antennae, a pair of stalked compound eyes, and various mouthparts used to cut food and carry it into the mouth. On the thorax are up to 8 pairs of walking legs. The claws on the first pair of legs typically are of unequal size, with one greatly enlarged and ending in a heavy pincer. Crustaceans vary in size from minute species such as copepods to the Japanese spider crab (*Macrocheira kaempferi*), which can measure 12 ft (3.7 m) from claw to claw. Most species are marine but many live in freshwater and a few, notably pill bugs, inhabit moist places on land.

cryogenics (*thermodynamics*)   Cryogenics is the study of the behavior of materials at very low temperatures and the production of such temperatures. Temperatures near absolute zero are called cryogenic ("producing icy cold").

Near absolute 0 the noise caused by moving atoms and molecules is greatly reduced, allowing materials to have properties, such as superconductivity, that cannot exist at higher temperatures, and permitting measurements of subtle effects. Some materials, such as Bose Einstein condensations, exist only at very low temperatures.

cryptic coloration (*zoology*)   Often referred to as camouflage, cryptic coloration is any coloration that enables an animal—or its eggs—to blend in with its surroundings, thereby helping it to avoid detection by predators. In some cases the animal's color matches the background color: the caterpillars of certain moths are the same green as the leaves on which they feed. Surface-swimming fish have a cryptic coloration called countershading; they have dark tops and light bellies to counteract natural shadows—except for the Nile catfish (*Synodontis batensoda*), which is colored the opposite way because it swims belly up! Irregular patterns, such as the mottled, sandy coloration on a skate's back, break up the outline of an animal's body. Chameleons and flounders can change their color or color pattern to resemble the background. Other animals assume postures that reinforce the color effect. Many insects, for example, look like decaying leaves, dead twigs, or bits of lichen. As long as they do not move they are almost impossible to see.

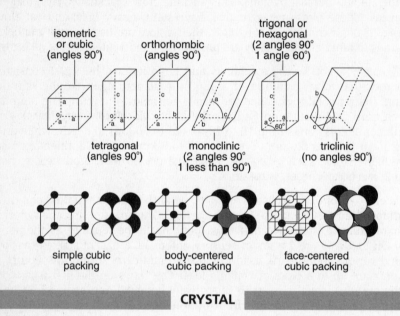

CRYSTAL

crystal (*materials science*)   Mineral formations with regular and symmetric shapes, flat sides, and sharp edges are crystals, especially if the mineral is translucent or transparent. Most solid materials and some liquids have a regular crystal structure, although usually on a microscopic scale. Large mineral crystals amplify this regular structure. Solids form crystals when allowed to cool slowly or to come out of solution because atoms settle into arrangements that have a lower energy than irregular structures. A solid that forms quickly is not crystalline. The shapes of atoms or molecules and distribution of charge determine whether they will pack in cubic forms, tetrahedral forms, or some other arrangement, although some materials, such as carbon, have 2 or more crystal structures. Otherwise all crystals of a given substance have the same angle between faces. There are 6 main crystal structures, 3 with all right angles, 2 with 2 right angles, and 1 with no right angles.

cube and cube root (*numbers*)   Multiplying a number by itself twice produces the cube of that number, which can be indicated with the exponent ^{3}; for example, $4^3 = 4 \times 4 \times 4 = 64$. For a given number p, the cube root, r, is a number that, when multiplied by itself twice, has the product p, or $r^3 = p$. For example, the cube root of 64 is $+4$ and of -64 is -4.

cuckoo (*animals*)   Named for the sound of the male's song, cuckoos are forest birds found worldwide. They are primarily insect-eaters. Species known as true cuckoos are native to the Old World. Most are brood parasites, laying their eggs in other birds' nests and letting the host species incubate the eggs and care for the young. American species include anis and the roadrunner; only a couple of American cuckoos are occasional brood parasites.

Curie, Marie Sklodowska (*physicist*)   [Polish-French: 1867–1934] Curie began working with radioactivity in 1896, the year it was discovered. Curie coined the word *radioactivity* and developed a device to measure it. Curie was the first to discover that thorium is radioactive; she also found that some uranium ore is more radioactive than uranium itself. Marie Curie and her husband Pierre Curie demonstrated in 1898 that some of the excess radioactivity was produced by previously undiscovered elements, polonium and radium.

Curie, Pierre (*physicist*)   [French: 1859–1906] In 1880 Curie and his brother Jacques [French: 1855–1941] observed that certain crystals respond to pressure by separating electric charge on opposing faces. An electric current applied to such crystals results in a change in length of the crystals. The Curies named the phenomenon piezoelectricity ("pressure electricity"). Pierre Curie then investigated magnetism; he was the first to determine that at a certain temperature, specific to each substance, the magnetic properties of the substance change (the Curie point). When his wife, Marie Sklodowska Curie, studied radioactivity, Pierre Curie joined her research effort and became a codiscoverer of polonium and radium.

curium (*elements*)   Curium, atomic number 96, is an artificial radioactive actinide element discovered in 1944 by Glenn T. Seaborg and coworkers. It is named after Marie Sklodowska Curie and Pierre Curie. Produced originally by bombarding plutonium with alpha particles, curium is extremely toxic. It has been considered as a power source for space probes, but so far has not been used for that or any other purpose. Chemical symbol: Cm.

current, electric (*electricity*)   An electric current is the movement of an electric field through a material that permits electric conduction. During passage of the field, electrons are nudged, either in 1 direction (direct current) or in back and forth waves (alternating current). The conducting material remains neutral (no net charge), and few electrons flow into or out of the wire. Electrical effects, such as filament heating in lamps or rotor motion in engines, are induced by the moving electric fields. For example, a filament heats when electron motion caused by the field increases vibration of atoms and molecules.

curvature (*geometry*)   The idea of curvature grows out of the notion that a small circle bends more than a large one. Think of turning a vehicle on a curve that is an arc of a circle; the larger the circle, the less you need to turn. A very large circle approaches a line, which has the curvature 0. For circles the curvature is defined as the reciprocal of the radius. For curves other than circles or the line, curvature varies from point to point and must be found as a limit.

A surface also has a curvature that requires advanced mathematics to calculate. For surfaces curvature can be a positive number (a sphere) or a negative number (the inner part of a torus). A surface with positive curvature lies entirely on one side of a plane tangent to the surface, while one with negative curvature lies on both sides of the tangent plane.

curve (*geometry*) Broadly speaking, a curve is any figure that can be created by moving a point, just as a curve can be sketched on paper by a pencil point. Notice that a line or a triangle is a curve in mathematics, although in ordinary speech straight figures are not called curves. Any figure represented by an equation in 2 variables is a curve in 2 dimensions (unless the figure is a point). Specifying curves in higher dimensions requires mathematics beyond the scope of this book, but the intuitive idea of a figure formed by a moving point is still accurate.

cusp (*geometry*) The point where a curve exactly and smoothly reverses direction is a cusp. At the cusp the changing slope of the curve approaching from one side becomes the same as the slope coming from the other. Thus a line can be drawn through the cusp tangent to both parts of the curve.

cuticle (*biology*) A cuticle is a thin, noncellular layer secreted by and covering the epidermis of many plants and invertebrate animals. In plants the cuticle is waxy and extremely efficient in preventing water loss; it typically is thicker on the upper surfaces of leaves, which are more exposed to the drying action of sunlight, than on the lower surfaces. Many parasitic worms have thick cuticles that protect the animals against being digested by their hosts. In insects and other arthropods the cuticle forms a protective exoskeleton of rigid chitin plates connected at joints by flexible chitin.

cutworm (*animals*) The caterpillars of many species of owlet moths (Family Noctuidae) are called cutworms because they cut through plants at or near ground level. Many are serious crop pests.

Cuvier, Georges (*zoologist/paleontologist*) [French: 1769–1832] Cuvier played a major role in establishing the fields of comparative anatomy and paleontology. He made the first important studies of fossils, applying his knowledge of comparative anatomy to the reconstruction of skeletons of extinct animals found in rocks quarried near Paris. Cuvier noted the similarity of fossil and living forms but did not believe in evolution. His theory of catastrophism proposed that a succession of disasters through time killed all life, after which new life-forms were created.

cycad (*plants*) When dinosaurs roamed Earth some 200,000,000 years ago, the dominant plants, particularly in tropical lands, were cycads, which look like palm trees with fernlike leaves. Today there are fewer than 150 species of cycads, forming the phylum Cycadophyta. Cycads are the most primitive living seed-bearing plants. Some have underground stems, with the leaves rising from the soil's surface. Others have trunklike stems—seldom more than several feet high—covered with the remains of old leaf stalks. Male and female cones form on separate plants. Seeds are produced on the female cones following pollination. The largest cone of all plants, weighing 90 lb (40 kg) or more, is produced by the African cycad *Encephalartos caffer*.

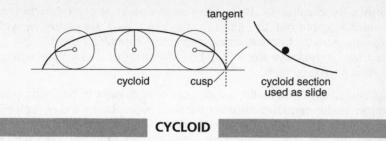

tangent

cycloid   cusp   cycloid section
used as slide

CYCLOID

cycloid (*geometry*) The curved path that a point on the rim of a wheel takes as the wheel rolls on a flat surface is a cycloid, nicknamed "the Helen of geometers." Problems posed in 1654 by Pascal started so many arguments that the Trojan War comparison seems appropriate. Many properties of the curve are unusual. Huygens used cycloids to create an accurate pendulum clock. An object will slide down a cycloid from any point on the curve to the bottom in the same amount of time and slide faster down a cycloid than any curve except a straight drop.

cyclone (*meteorology*) In the mid-19th century a former British sea captain recognized that storms in the Southern Hemisphere follow a large counterclockwise pattern that he named a cyclone ("coil of a snake"). This pattern is caused by the Coriolis effect, which deflects winds moving south from the equator to the west and winds moving west toward the south—in every case, the winds veer toward the left. Meteorologist generally refer to all circular movements around areas of low pressure as cyclones, although technically such storms in the Northern Hemisphere are anticyclones. Great storms that occur in the Indian Ocean, similar to the hurricanes of the North Atlantic Ocean, are also called cyclones.

cylinder (*geometry*) A cylinder is both a surface and a solid. The most familiar solid form is a right circular cylinder, made by rotating a rectangle about an axis through the midpoints of opposite sides. Both bases are congruent circles. Other solid cylinders also have congruent bases in 2 parallel planes that are closed curves. The volume V of a solid cylinder is the product of the altitude h and the area of one base, B ($V = Bh$).

    A cylindrical surface, also called a cylinder, is the set of points on a line through a curve combined with the points on every line parallel to the original one and passing through the curve. The curves do not have to be closed like a circle, but can be parabolas or other curves.

cypress (*plants*) True cypresses (*Cupressus*) are evergreen conifers native to warm temperate and subtropical regions. They range from shrubs to the Himalayan cypress (*C. torulosa*), native to Nepal, which can be 150 ft (45 m) tall. Cypresses have aromatic scalelike leaves. The small round cones grow erect. The cypress family, Cupressaceae, also includes arborvitaes, junipers, and *Fitzroya cupressoides*, a native of southern South America that grows to 240 ft (73 m) and has an age span of at least 1000 years.

cytokinin (*biochemistry*)   One of the main classes of plant growth regulators, cytokinins promote cell division. In many ways, they work together with auxins to promote growth. In other cases, cytokinins counteract the effects of auxins. For example, cytokinins promote the development of lateral buds whereas auxins inhibit such growth.

cytoplasm (*cells*)   Originally, a living cell was thought to be filled mainly with a simple, undifferentiated substance called protoplasm. Today, however, it is apparent that the contents of a cell are varied and complex. These contents make up the cytoplasm. Cytoplasm consists of a fluid portion called cytosol, plus specialized structures called organelles. The organelles are held in place within the cytosol by a network of protein filaments, which also help maintain the cell's shape. Cytosol contains enzymes, nutrients, and other dissolved molecules; it helps transport substances from one part of the cell to another.

D

daffodil (*plants*) Native to the Mediterranean region, herbs of the genus *Narcissus* are variously known as daffodils, narcissuses, and jonquils. They grow from a bulb (underground stem) that gives rise to a basal cluster of long, grasslike leaves and a flower stalk that bears 1 or more flowers. The flowers, usually white or yellow, have a central trumpet- or cup-shaped appendage called a corona.

Dalton, John (*chemist*) [English: 1766–1844] Dalton is best known for his 1803 revival on a scientific basis of the ancient Greek idea that all materials are formed from tiny particles (atoms). In part Dalton had prepared for this idea with his development of 2 gas laws—a rule for the pressure of a mixture of gases (the law of partial pressures) and one for the way volume changes with temperature (called Charles's law, although first stated by Dalton). Dalton also developed the theory of red-green color blindness (he himself was color blind); hence color blindness is called daltonism. Dalton was the first to demonstrate that water in springs originates as precipitation.

dandelion (*plants*) Taking its name from the French *dent de lion*, meaning "lion's tooth," the dandelion (genus *Taraxacum*) has a rosette of leaves with deeply toothed margins. Its distinctive yellow flowering heads—on stalks up to 12 in. (30 cm) tall—are followed by fluffy balls of seeds adapted for wind dispersal. A hardy perennial of the composite family, it is common throughout the Northern Hemisphere.

Daphnia (*animals*) Often called water fleas—a name given by Jan Swammerdam in 1669—members of the genus *Daphnia* are tiny crustaceans, less than 0.2 in. (5 mm) long. They make up an important part of the plankton in lakes and other bodies of fresh water, serving as the basic food of many fish. *Daphnia* themselves feed on microscopic algae and particles of organic waste. Although *Daphnia* can reproduce sexually, males are seldom found because during much of the year the females reproduce by parthenogenesis.

dark matter (*cosmology*) Motions of galaxies indicate that they are responding to mass that we cannot otherwise detect. Astronomers call this mass dark matter, but it is not simply black; the missing mass neither reflects nor blocks electromagnetic radiation at any wavelength. Astronomers have suggested that dark matter may be cold or hot, referring to the speed of bodies. Cold dark matter could consist of moderately large bodies of ordinary matter, such as planets or stars, nicknamed MACHOS (massive astrophysical compact halo objects), that are too cool to radiate much energy. Theorists have also proposed slow-moving subatomic particles known as WIMPS (weakly interacting massive particles) as dark matter, although no WIMPS have ever been observed. Hot dark matter must be fast-moving subatomic particles, such as speedier WIMPS or massive neutrinos.

Darwin, Charles (*naturalist*)   [English: 1809–82] The 1859 publication of *On the Origin of Species*, in which Darwin demonstrated that all living things evolved from earlier forms of life by the process of natural selection, revolutionized human thought. Although initially controversial, Darwin's theory of evolution, which also was set forth by Alfred Russel Wallace, became accepted as one of the foundations of modern biology.

Darwin originally believed in the orthodox theory of his time: that each species had been created individually, and had remained unchanged since Earth's beginning. But his observations of fossils and living organisms during a 5-year voyage around the world aboard H.M.S. *Beagle* (1831–36) led to his conclusion that new species arose as existing species gradually changed in response to environmental conditions. Among his best-known evidence was the adaptive radiation of finches he studied on the Galapagos Islands.

Davisson, Clinton (*physicist*)   [American: 1881–1958] In 1925 while investigating how electrons are reflected from different surfaces, Davisson accidentally created large crystals of nickel for some of his samples. The following year he recalled the unusual nickel crystal's electron reflections after learning of Prince Louis-Victor de Broglie's theory that electrons behave like waves. Davisson realized that he had observed wave behavior in electrons and conducted more experiments verifying this and, of course, Broglie's theory.

Davy, Sir Humphry (*chemist*)   [English: 1778–1829] In 1807 Davy constructed the most powerful electric battery to that time and began to perform electrolysis on available materials. Metallic potassium emerged easily from molten potash and metallic sodium from lye. The next year Davy used improved methods to obtain the first samples of barium, calcium, magnesium, and strontium. He also independently produced boron a few days after its first discoverers and was the first to recognize chlorine as an element. His invention of the miner's safety lamp in 1815 saved many lives and his electric arc lamp of 1809 was the start of electric lighting.

day (*astronomy*)   Normally a day in astronomy is the period from one sunrise to the next, including both the sunlit period and night. On Earth this is 24 hours.

People often think that 1 day is the same as 1 rotation of a planet, but a day also is influenced by the revolution of a planet in its orbit. For Earth, 1 rotation is 4 minutes short of a day—23 hours 56 minutes. The other 4 minutes in the day are contributed by the planet's movements in relation to the Sun. Mercury, close to the Sun and with a slow rotation period, has a day that is dramatically different from its 59-Earth-day period of rotation. Because of Mercury's swift passage of 88 Earth days around the Sun, the Sun only rises at the same place on the planet once each 176 Earth days. For Mercury the day is twice as long as the year.

deamination (*biochemistry*)   The process of deamination involves the removal of an amino group (NH_2) from a molecule. In mammals, for example, the metabolism of proteins produces amino acids that cannot be stored in the body. Enzymes in the liver remove the amino group from the amino acids. In a series of additional reactions in the liver, the amino group is converted first to ammonia, NH_3, and then to urea, $CO(NH_2)_2$.

death (*physiology*) Death, the end of life, results from irreversible changes in essential life functions. Death can involve the entire organism (somatic death) or be limited to organs (organ death) or individual cells (cell death). All living things eventually die, or at least cease to exist. It can be argued that an amoeba, however, can live forever, since instead of dying it can divide to form 2 new amoebas.

decay (*particles*) In physics, the term decay refers to a spontaneous change in an atom or subatomic particle that results in a different atom or in 2 or more subatomic particles. Atomic decay of one atom into another is a direct result of particle decay, and excess particles released from the particle decay are the cause of radioactivity.

decibel (*measurement*) The bel (B) measures change of intensity of sound (it is also used for power and voltage), but a tenth of a bel, or decibel (dB), is commonly used. Because the bel and decibel are defined as logarithms, an increase of 10 dB (1B) in loudness above 0 dB—the threshold of hearing—is $10^1 = 10$ times as loud while an increase of 20 dB (2 B) above 0 is $10^2 = 100$ times as loud. A sound of x dB is $10^x/10$ times as intense as the smallest audible sound.

deciduous plant (*botany*) Woody plants that shed their leaves at the end of the growing season, leaving the branches temporarily bare, are said to be deciduous. They are typically found in temperate regions and other places with pronounced seasonal variations in the amount of water available to plant roots. By dropping their leaves, the plants reduce the amount of water lost by transpiration during periods when water is scarce.

decimal fraction (*numeration*) The place-value system of numeration can be extended to places that are smaller than 1. For the decimal system, these places begin tenths, hundredths, thousandths. A dot, called a decimal point, is used to separate the ones place and the tenths place. A decimal numeral with any place less than 1 is a decimal fraction. Many rational numbers and all irrational numbers require an infinite number of decimals places to be expressed exactly.

decimal system (*numeration*) Many numeration systems are based on 10, probably because humans have 10 fingers. Systems based on 10 are called decimal (from the Latin for 10). Our decimal Hindu-Arabic numeration system uses 10 digits and decimal place value to show numbers. For whole numbers the place farthest right shows ones, while the next one left shows tens, and so on, with each place indicating the next higher power of 10. A decimal point at the right of a whole number marks the beginning of a decimal fraction with places showing negative powers. For example, the meaning of 3029.481 is $(3 \times 10^3) + (0 \times 10^2) + (2 \times 10) + (9 \times 1) + (4 \times 10^{-1}) + (8 \times 10^{-2}) + (1 \times 10^{-3})$.

decision problem (*logic*) The question of whether an algorithm exists for deciding the truth or falsity of a whole class of statements is a decision problem. Consider, for example, polynomial equations in 1 variable with coefficients that are integers, such as $3x^4 + 5x^3 - x^2 - x - 82 = 0$. The question of whether there exists an algorithm for solving all such equations so that x is also an integer has been established to have a negative solution (even though one might solve this particular example by showing that $x = 2$ is a solution). The similar question of whether there exists an algorithm for solving all polynomial equations of the fourth degree or less so that x is a complex number, however, has a positive solution, since such an algorithm exists.

declination (*astronomy*)   The height above (+) or below (–) the extension of Earth's equator into space, measured in degrees, is called declination. It is used with right ascension to describe the location of any celestial body. Declination is a direct translation of latitude to the sky. For example, Polaris is nearly above the north pole (90°N) and has a declination of + 89°11′.

Declination in mathematics refers to an angle below the horizon—above is inclination. (*See also* **magnetic pole**.)

decomposer (*ecology*)   Decomposers, or decay organisms, obtain energy by breaking down the wastes and remains of other organisms. Bacteria and fungi such as mushrooms are decomposers. Decomposers are essential members of an ecosystem: by breaking down organic matter they release nitrogen, phosphates, and other substances that can then be reused by food-producing organisms.

Dedekind cut (*numbers*)   In 1872 Richard Dedekind [German: 1831–1916] gave the first sound definition of the real number, one that made no appeals to intuition but that was based on clearly defined entities known to exist. Dedekind assumed only that we can recognize any 2 different rational numbers and determine which is greater. He then defined a real number as an entity, now called a Dedekind cut, that separates the rational numbers into 2 sets—all numbers greater than the cut and all numbers less than it.

deduction (*logic*)   A deduction in formal logic consists of a sequence of propositions that begins with an assumed proposition (one taken as given) followed by axioms, definitions, or conclusions of if . . . then implications for which the if clause is one of the items previously listed (the given proposition, an axiom, or a definition). The last proposition in the list, or conclusion, is that which has been proved by deduction. An important theorem called the deduction theorem states that if a conclusion is proved this way, then the proposition "If the given proposition, then the conclusion" is true. Deduction in formal logic is the underlying process behind direct proof in mathematics.

deer (*animals*)   Constituting Family Cervidae, deer are hoofed, ruminant mammals native to the Americas, Eurasia, and northern Africa. They are unique in that they have antlers. Deer live in herds, can run swiftly on long powerful legs, and have keen senses of sight and smell. The species are quite similar in appearance but vary markedly in size. Pudus of South America may be only 12 in. (30 cm) high at the shoulder and weigh as little as 15 lb (6.8 kg). The largest deer is the moose, which may be 6.5 ft (2 m) high at the shoulder and weigh over 1600 lb (725 kg).

definite integral (*calculus*)   The definite integral for a continuous function is the area between 2 specified vertical lines and also bounded by the function's graph and the x axis. Thus the definite integral is a number while the indefinite integral is a function. If vertical lines through a and b on the x axis are the left and right bounds of a function $f(x)$, then the area between the curve and the x axis is symbolized $\int_a^b f(x)dx$, where the area above the axis is taken as positive and area below is negative. Replacing b with the variable x produces an indefinite integral. If $F(x)$ is the antiderivative of $f(x)$, then $\int_a^b f(x)dx = F(b) - F(a)$.

deforestation (*environment*)   As much as 50% of the world's tropical rain forests were destroyed by humans during the 20th century; temperate forests also suffered significant deforestation during this period. A number of harmful environmental effects can ensue from deforestation. Loss of habitats results in the decline of wildlife populations and of biodiversity. Erosion and flooding increase because there are no tree roots to hold the soil; the fertile soil is blown off the bare ground or washed into rivers. Large-scale deforestation can also change rainfall patterns, intensify global warming, and cause other climatic changes.

degenerate conic (*analytic geometry*)   The equation for a degenerate conic may have the form of the equation for a circle, ellipse, parabola, or hyperbola, but the graph is a point, line, or pair of lines. A degenerate circle has a radius of 0, becoming a point; virtually the same happens to a degenerate ellipse. When the distance between the 2 branches of a hyperbola is 0, intersecting lines result. The degenerate curve of a parabola can be either 1 line or a pair of parallel lines.

degree (*algebra*)   If an equation uses 1 variable only, its degree is the same as the greatest exponent used with that variable; for example, $2x^2 + 3x - 7 = 0$ is a second-degree equation. If there is more than a single variable, the degree is the largest of the sums of all the exponents in each term. For example, the equation $2x^3y + x^2y^3 - 3xy + 5y = 0$ has a degree of 5 since the term x^2y^3 has the greatest exponent sum.

degrees of freedom (*mechanics*)   The number of variables needed to describe a physical state is called its degrees of freedom. Consider rigid bodies in motion. A wheel rotating on an axle has 1 degree of freedom; a checker sliding on a board has 2; a domino on the board 3, since the domino may be at different angles. A pair of scissors that can open and close as it slides has 4 degrees of freedom, since 2 locate the scissors and 1 each describes the angle of a blade. Systems of atoms or molecules are also described this way. A molecule of 2 atoms of hydrogen gas has 5 degrees of freedom—3 to locate it and 2 to describe rotation about the molecule's center—pitch and yaw, as pilots call such rotations. (Roll is not counted as it does not affect total kinetic energy.)

Deimos (*solar system*)   Deimos is the smaller—7 mi (12 km) in diameter—of the two irregularly shaped satellites of Mars. Named after one of the god Mars's chariot horses ("terror"), Deimos was discovered in 1877 by Asoph Hall [American: 1829–1907].

Delbrück, Max (*molecular biologist*)   [American: 1906–81] One of the founders of molecular biology, Delbrück focused much of his research on bacteriophages (viruses that infect bacteria). Delbrück frequently collaborated with Salvador Luria and Alfred Hershey [American: 1908–97]. In 1943 he and Luria developed a test for measuring mutation rates, which created the field of bacterial genetics. In 1946 Delbrück and Hershey independently discovered genetic recombination. The 3 men shared the 1969 Nobel Prize in physiology or medicine.

delphinium (*plants*)   Comprising some 250 species of biennial and perennial herbs native mainly to northern temperate regions, the genus *Delphinium* gets its name from the Greek *delphis*, meaning "dolphin," which the flower buds were thought to resemble. Delphiniums have stout stalks, up to 7 ft (2 m) tall, topped by dense clusters of bluish flowers. Some species are called larkspurs and, in the western United States, locoweed. The latter species contain a poison, aconite, that can be fatal to livestock and humans.

delta (*geography*)   A delta is a flat plain formed of sediments carried by a stream and deposited at the mouth of the stream. The name "delta" comes from the plain at the mouth of the Nile, which is triangular like the Greek capital delta, Δ. A delta forms only where the mouth of a stream enters a calm part of the sea or a lake. Otherwise currents disperse the sediments.

dendrochronology (*archaeology*)   Dendrochronology ("study of tree time") is the science of dating archaeological remains with tree rings. Patterns of warm and cool years, wet and dry years, affect the size and color of annual rings in the trunks of trees. Patterns of rings have been matched, with the inner rings of today's trees linked to outer rings from trees cut down many years ago. Continued matching with older and older remains has made it possible to date wood from thousands of years ago. By 1984 researchers had linked a series of Irish oak trees that could be used for dating as far back as 7272 years.

denitrification (*biochemistry*)   Certain kinds of soil bacteria, particularly species of the genera *Bacillus* and *Pseudomonas*, obtain energy by breaking down nitrogen-containing compounds in animal wastes and decaying organic matter. This process, called denitrification, is an important part of the nitrogen cycle because one of the end products is nitrogen gas, which is returned to the atmosphere.

density (*mechanics*)   Density is the ratio of mass to volume measured in units such as lb per cu in. (kg/m³). It is commonly confused with weight; lead is denser than water, not heavier. Density is often compared to that of water. Such a comparison, made with the densest water—at 39.2°F (4°C)—is called relative density (formerly specific gravity). Relative density of lead is about 11 (it is 11 times as dense as water); the lightest metal, lithium, has a relative density of about 0.5, so it floats on water.

dentition (*anatomy*)   The number, kind, and arrangement of an animal's teeth constitute its dentition. In mammals basic dentition is expressed in a dental formula that indicates the number of incisors, canines, premolars, and molars on one side of the upper and lower jaws. The dental formula for the domestic cat:

i	*c*	*pm*	*m*
3	1	3	1
3	1	2	1

indicates that the cat has 8 teeth on one side of the upper jaw and 7 teeth on one side of the lower jaw, or 30 teeth in all.

denumerable infinity (*numbers*) In 1874 Georg Cantor published the first proof that not all infinities are the same. The smallest infinity, which is the size of the set of natural numbers, is called a denumerable, or countable, infinity, since it can in principle be counted. The integers and, surprisingly, the rational numbers are denumerable, but the real numbers and the set of points in space are not.

deposition (*geology*) Soil or rock removed by erosion often collects in localized regions, a process called deposition. Soil particles removed by a river may be deposited at a delta; rock removed by a glacier forms moraines and other structures; sand may be piled into dunes by the wind; and dissolved limestone creates stalactites and stalagmites.

derivative (*calculus*) The derivative gets its name because it is a function derived from another function. The derivative is the rate of change, or slope of the graph, of the original function. It is symbolized in several ways: if the original function is $y = f(x)$, the derivative is $f'(x)$ or y' or $Df(x)$ or very often dy/dx. Finding the derivative employs the operation called differentiation.

Since the derivative is the rate of change, if $f(x)$ describes distance traveled in time x, then $f'(x) = dy/dx$ is velocity. As a function the derivative also may have its own derivative, the second derivative, symbolized as $f''(x) = d^2y/dx^2$; the second derivative of change in distance, for example, is acceleration. A derivative does not necessarily exist for all values of a function; for example, if the graph has a sharp corner, there is no definite slope or derivative.

Descartes, René (*mathematician*) [French: 1596–1650] Descartes was the first to publish a detailed account of how to use coordinates for locating points in space, although Fermat developed the same ideas around the same time. In Descartes's appendix to *Discours de la méthode* he introduced analytic geometry and used the new technique to solve problems in geometry. His ideas made calculus and other advances in mathematics possible.

desert (*ecology*) One of Earth's major biomes, deserts are characterized by extreme dryness, with usually less than 10 in. (25 cm) of precipitation annually—not enough to support grasslands or forests. Temperatures vary greatly, perhaps nearing 100°F (38°C) during the day and then dropping close to freezing (32°F, or 0°C) at night. Another common feature is high wind, which can produce violent sand and dust storms. Desert organisms have various adaptations for water conservation. For example, cacti store water in stout stems, ephemerals complete their life cycles in a matter of days following a rainstorm, and rats excrete metabolic wastes in almost solid form. Adaptations for other aspects of desert life include the camel's long eyelashes, which keep sand out of eyes, and the tendency of desert birds to lay eggs in caves, crevices, and other spots protected from the harsh sun.

desertification (*environment*) The creation of desertlike conditions in semiarid regions, induced by climatic changes, human activities, or some combination of the two, is called desertification. A prolonged drought, deforestation, overgrazing, irrigation practices that cause salinization of soils, and farming methods that exhaust soil nutrients or cause erosion can lead to desertification. Desertification occurs worldwide, with millions of acres of productive land destroyed annually.

detritus feeder (*ecology*)   Detritus, or organic waste matter, includes dead plants and animals, leaves, feces, and discarded snakeskins and insect exoskeletons. Organisms that feed on detritus include both scavengers and decomposers. These organisms are essential to a clean and healthy environment, since detritus makes up by far the majority of organic material in any ecosystem.

deuterium (*elements*)   The isotope hydrogen-2 is called deuterium ("second substance") or heavy hydrogen (chemical symbol D or ^{2}H). The "heavy" part is the nucleus, which has a neutron in addition to the proton that characterizes hydrogen. The ion, or bare nucleus, is a deuteron. About 0.016% of the hydrogen in ocean water is hydrogen-2, which is enough to make deuterium abundant. Water with a high percentage of deuterium is called heavy water.

developmental biology (*sciences*)   The progressive changes that occur during the growth and development of organisms are the focus of developmental biology. A major division of this science is embryology, the study of developing embryos of plants and animals from the 1-cell stage to the moment of birth. In recent years growing attention has focused on the relationships between developmental processes and the deteriorative changes of aging.

deviation (*statistics*)   The difference between the average, or arithmetic mean, of a set of numbers and each of the numbers is a deviation from the mean, or simply a deviation. Subtracting the mean from the number makes some deviations positive and some negative, but the absolute value of such differences, called the absolute deviation, is more useful. The average of the absolute deviations, called the mean absolute deviation, is a number that tells how variable the numbers in the set are. (*See also* **standard deviation.**)

dew (*meteorology*)   Water vapor in air below the dew point (but above freezing temperatures) becomes liquid when it loses heat to a cold surface. The drops of water produced on the surface are called dew. Dew tends to form during clear nights, when temperatures drop rapidly, or near bodies of water, where humidity is high. Often some objects are cooler than others, so dew may appear on grass but not on the ground or, when conditions are reversed, on the ground but not on grass. If temperatures are below freezing, the same conditions produce a thin coat of ice known as frost.

Dewar flask (*thermodynamics*)   The cryogenic experiments of James Dewar [Scottish: 1842–1923] were considerably aided by his 1892 invention of the Dewar flask, which has silvered double walls with a vacuum between them, effectively insulating the interior from transmission of heat by conduction or convection and lowering transmission through radiation. With his flasks and various means for lowering temperatures, Dewar became the first to liquefy hydrogen (1898). A version of the flask known as a thermos bottle is used to keep foods hot or cold.

dew point (*atmospherics*)   The temperature at which water vapor in air begins condensation is called the dew point. It varies with humidity—when relative humidity reaches 100%, or saturation, the dew point is reached. At a temperature of 50°F (10°C) air that contains 0.8% water by weight is at its dew point. Water will condense from such air on a surface cooler than 50°F. At night most surfaces lose heat more quickly than air cools, so water vapor from air near saturation condenses on them as dew or, if it is colder than freezing, as frost.

diabase (*rocks and minerals*)   Diabase is similar to basalt, but the parent molten rock has failed to reach the surface and become lava. Buried diabase cools more slowly than basalt, so crystals of diabase are larger and there may be separation of minerals. Thus diabase often looks like grainy basalt flecked with a lighter mineral.

diagonal (*geometry*)   A diagonal is any line segment connecting 2 vertices of a polygon that is not a side of the polygon, or a segment connecting 2 vertices of a polyhedron that does not lie within a side. The most commonly encountered are diagonals of squares or rectangles, which join opposite vertices. For a polygon with n sides, the total number of diagonals is half the product of n and $n - 3$; for example, a pentagon possesses $1/2 \times 5 \times (5 - 3) = 5$ diagonals while a rectangle has $1/2 \times 4 \times (4 - 3) = 2$. Diagonals of polygons often have special properties proved in elementary geometry: diagonals of rectangles are congruent; diagonals of parallelograms bisect each other; diagonals of kites are perpendicular.

dialysis (*chemistry*)   A thin sheet, or membrane, may have a structure that permits ions or small molecules in solution to pass from one side to another while blocking larger molecules, a process called dialysis. Such a membrane is called semipermeable. When a solution is on one side of the membrane and pure solvent on the other, dialysis can reduce the concentration of small molecules and ions while maintaining that of large molecules, such as proteins. For this reason dialysis of blood is used to remove wastes without losing important proteins or blood cells.

diameter (*geometry*)   The diameter of a circle is its longest chord, a line segment that passes through the circle's center, or the number that is the length of that segment. The diameter is twice the radius. (*See also* **tangent**.)

diamond (*rocks and minerals*)   Diamond is a mineral form of carbon. Carbon can occur as a native element (a mineral not combined into a compound). Most native carbon is graphite, usually found as soft, greasy, black inclusions in many kinds of rock. In contrast diamond is extremely hard and transparent, and usually occurs as crystals shaped as octahedrons. Diamonds come from unusual structures similar to volcanoes called pipes; the diamonds are embedded in a mineral called kimberlite. Diamonds form at great depths under high pressure and temperature, then are carried to the surface in a kimberlite pipe. Because diamonds are hard and resist wear, they also survive passage from original locations into gravel deposits in rivers and on beaches.

diaphragm (*anatomy*)   Any partition that separates one area from another is a diaphragm (from Greek words meaning "through fence"). However, the term usually refers to the dome-shaped muscular layer that separates the thoracic and abdominal cavities of mammals. This diaphragm plays an important role in breathing. As the diaphragm contracts, it flattens; the thoracic cavity is enlarged, creating a partial vacuum so that air rushes into the lungs. As the diaphragm relaxes, the volume of the thoracic cavity is reduced and air is pushed out of the lungs.

diatom (*protists*)   The most abundant food producers in aquatic ecosystems are the diatoms, one-celled and colonial algae that have unique cell walls made largely of silica. These transparent cell walls consist of a top valve that overlaps a bottom valve, as in a pillbox. Numerous pores decorate the cell wall in a pattern characteristic of the species. The pores reduce the diatom's weight, helping it to float at levels where water is sunlit. Much of a diatom's food reserves are stored as oil, which also provides buoyancy.

Most diatoms are free-floating members of the plankton. There are two basic types. Pennate diatoms have an elongated shape and are most common in fresh water and in shallow parts of the sea. Centric diatoms are circular, triangular, or irregular in shape and are most common in the open ocean.

Scientists have identified some 10,000 species of diatoms. They are grouped with other golden algae in Phylum Chrysophyta.

diatomaceous earth (*rocks and minerals*)   When diatoms die, their glassy cell walls remain intact and sink to the bottom of a body of water. Over long periods of time, layers thousands of feet thick can accumulate on ocean floors. Beds of such diatomaceous earth occur in various parts of the world, many now on land lifted upward by geological changes. It has been estimated that the cell walls of as many as 40,000,000 diatoms were needed to produce 1 in.^{3} (16.4 ml) of diatomaceous earth.

Diatryma (*paleontology*)   A flightless, ground-dwelling bird that lived during the Eocene, *Diatryma* stood about 7 ft (2.1 m) tall. It had a stocky body, sturdy legs more than 4 ft (1.2 m) long, and the largest head and bill of any known bird. Current thinking is that *Diatryma* was comparatively slow moving, and used its formidable parrotlike bill to eat plant matter.

Dicke, Robert (*physicist*)   [American: 1916–97] In 1961 Dicke conducted experiments that verified the equivalence principle for inertia and gravity to at least 1 part in 100,000,000,000. He proposed a plausible alternative theory of gravity to Einstein's general theory of relativity; however, it was not established because the differences between the theories are too slight to be tested. In 1964 Dicke independently formulated the theory that cosmic background radiation was created by the big bang. As he and coworkers were developing an experiment to look for evidence, Robert Wilson [American: 1936–] and Arno Penzias [American: 1933–] found the radiation accidentally and Dicke identified its cause.

dicynodont (*paleontology*)   Dicynodonts ("two doglike teeth"), a group of extinct therapsid reptiles, were either toothless or endowed with a pair of fangs. In either case, they used the horn-covered edges of their jaws to cut up plant food. Their wide diversity of species ranged in length from 1 to 10 ft (0.3 to 3 m). Most species were wiped out at the end of the Permian (the P/Tr event), but some, such as *Lystrosaurus,* recovered; herds of *Lystrosaurus* dominated many regions of the world during the early Triassic.

dielectric (*materials science*) A dielectric is an insulator for electricity but the term dielectric ("through electric") would seem to convey the opposite. Dielectrics insulate because an electric field is trapped by them—in an electric field more lines of force pass through a dielectric than through a vacuum, hence the name. The dielectric is used in a device called a capacitor, which consists of 2 parallel conducting plates of opposite charge. A dielectric between the plates stores the charge, which can be released by touching one of the charged plates with a conductor.

diet (*physiology*) The kinds and amounts of foods normally consumed by an animal make up its diet. The aardvark's (*Orycteropus afer*) diet consists mainly of ants and termites; the sea otter (*Enhydra lutris*) eats fish and marine invertebrates. Diet is able to provide all the nutrients needed by an animal to remain healthy. For example, the human diet must provide 6 kinds of nutrients: carbohydrates, proteins, fats, vitamins, minerals, and water.

differentiation (*biology*) All the cells of an organism resulting from sexual reproduction contain the same genes or DNA as the original fertilized egg cell. Initially the fertilized egg cell passes through a stage of rapid cell division called cleavage, producing hundreds of simple, unspecialized cells. The process by which the unspecialized cells of the young embryo develop into the many different kinds of cells found in the adult organism is called differentiation. Differentiation depends on gene switching—the activation of some genes and the deactivation of others. That is, different combinations of genes are expressed in different cells, and at different times within a cell. There are many mechanisms for regulating gene expression.

diffraction (*optics*) When a wave encounters either an object or a hole, some of the wave near the edges of the object or hole is bent. This effect, called diffraction, is most noticeable for waves of single wavelength and objects or holes near to or smaller than that wavelength. Photographs taken in monochromatic (single-wavelength) light show strange patterns caused by the bent waves interfering with each other, since the amount of bending varies with closeness to the edge of the object or hole. White light is spread into a rainbow by interference as it passes through closely spaced slits in a diffraction grating. Since atoms are about the size of the wavelength of an X ray, their positions can be determined from X-ray diffraction patterns. (See art on page 176.)

diffusion (*physics*) Diffusion is movement of molecules of a fluid from one place to another as a result of random motion. Two different fluids at the same temperature have the same average kinetic energy of their molecules, so the average speed of molecules of the lighter fluid is faster than the average speed of the more massive molecules. For example, at the same temperature and pressure, hydrogen molecules move faster than molecules of carbon dioxide. If a mixture of the 2 different gases is contained so that molecules can diffuse through long (compared to a molecule's size) narrow tubes passing through the container (unglazed pottery provides such tubes, for example), more of the faster gas

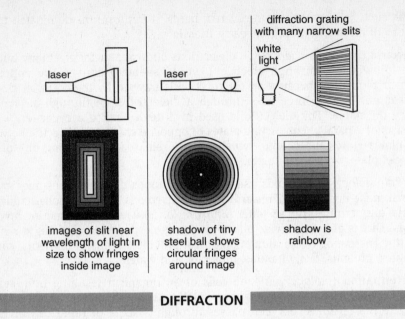

images of slit near
wavelength of light in
size to show fringes
inside image

shadow of tiny
steel ball shows
circular fringes
around image

shadow is
rainbow

DIFFRACTION

molecules escape from the container over time. Diffusion can be used to separate isotopes and is a principal means for separating uranium-235 from uranium-238 for use in nuclear weapons and reactors.

digestion (*physiology*)   The food of animals typically consists of large complex organic molecules. During the process of digestion those molecules are converted into smaller, simpler molecules that can enter individual cells in the animal's body. Carbohydrates are broken into simple sugars, proteins into amino acids, and fats into fatty acids and glycerol. Digestion may include mechanical actions, such as chewing and grinding. But digestion is mainly a chemical process, dependent on the action of enzymes. Except for the simplest invertebrates, digestion takes place in specialized digestive systems.

digestive system (*anatomy*)   The function of an animal's digestive system is to break down complex food molecules into smaller molecules that can reach the cells of the body. The simplest digestive systems, seen in flatworms such as *Planaria*, have a single opening; food is taken in through the mouth and digested in an intestine, with undigested materials eliminated through the mouth. Advanced digestive systems, as seen in vertebrates, have a long gastrointestinal tract plus several glands and organs, such as the pancreas and liver, that aid in the digestive process.

digit (*numeration*)   The name digit ("finger") for each of the numbers 0, 1, 2, 3, 4, 5, 6, 7, 8, and 9 and for their single-character numerals is carried over from finger-based numeration systems that were widely used in classical and medieval times.

dihedral angle (*geometry*)   A dihedral ("2 bases") angle is 1 of the 4 angles formed when 2 nonparallel planes meet. The measure of a dihedral angle is that of a plane angle with a ray in each plane, each ray perpendicular to the line of intersection of the planes.

dike (*geology*)   An intrusion of magma through existing rock is called a dike after it cools to form rock, typically basalt or diabase. Dikes are vertical, or nearly so, and are more like walls than columns. When existing rock is in layers, the dike pushes its way through layer after layer, following vertical cracks.

dilatation (*geometry*)   A transformation that produces similar figures is called a dilatation. If the identity transformation is excluded, dilatation of a polygon results in all the line segments of the transformed polygon being parallel to the original; dilatation of a circle produces another circle with the same center (concentric) to the first.

dimension (*geometry*)   In elementary geometry the 3 dimensions are length, width, and height. Figures on a plane extend in 2 dimensions, length and width. A line has only 1, length. To locate a point on a line, 1 number is needed, while a pair of coordinates are needed for a plane; 3 coordinates are required in the space used in elementary geometry. A space has as many dimensions as the number of coordinates needed to describe a point. The space of elementary solid geometry is Euclidean space.

Mathematicians have generalized the idea of dimension in several ways. Some spaces have an infinite number of dimensions. A fractal figure exists in a space for which the dimension is not a natural number, typically, the ratio of 2 logarithms.

dimension (*physics*)   The number of variables needed to locate a particle in space is the space's dimension. Before Einstein's 1905 special theory of relativity, physics was based on the idea that 3 space variables (x, y, and z) are sufficient. Einstein showed that since space and time are inextricably linked, the location of a particle is (x, y, z, t), where the time variable t has a special role. This space is different from 4-dimensional Euclidean space, which has complete symmetry among the 4 variables.

Einstein's success prompted Theodor Kaluza [Polish: 1885–1944] to develop a theory in 1919 combining gravity and electromagnetism in a space of 5 dimensions. Kaluza's theory was clarified in 1926 by Oskar Klein [American: 1895–1977], who showed why the fifth dimension is present but unobservable. In recent years the Kaluza-Klein idea has been taken up by specialists in string theory, who propose that space has 10 or 11 or even 26 dimensions with all but the first 4 unobservable because they are curled into tiny curves smaller than the nucleus of an atom.

dimensional analysis (*mathematics*)   Most measurements use the dimensions of length (L), mass (M), or time (T). Force is ML/T^2, meaning "the product of mass and length per the square of time"; acceleration is L/T^2, work is L^2M/T^2, angular momentum is L^2M/T, and so forth. Dimensional analysis recognizes that equations are correct only if both sides have exactly the same dimensional structure. For example, the equation $F = ma$ for force = mass × acceleration is analyzed dimensionally as $ML/T^2 = ML/T^2$, so the equation can be true.

Dimetrodon (*paleontology*)   *Dimetrodon* was a carnivorous pelycosaur reptile, common in North America during the Permian. The most notable feature of this 11.5-ft- (3.5-m-) long predator was the huge sail on its back, formed by spines joined together by a membrane. It is believed that the sail functioned as a temperature regulator, absorbing the Sun's heat when the animal was cold and giving off heat when it was too warm.

dinoflagellate (*protists*)   Members of Phylum Pyrrhophyta, dinoflagellates are single-celled aquatic protists with two flagella for movement—one flagellum encircling the body, causing it to spin, the second trailing behind like a tail. The cell walls typically consist of cellulose plates that fit together like pieces of armor; their distinctive patterns are used by taxonomists for classification.

Many species contain chlorophyll; they are second only to diatoms in their importance as food producers in aquatic ecosystems. Other dinoflagellates acquire some or all of their energy by eating other organisms. Some dinoflagellates produce potent nerve toxins capable of poisoning fish and humans; in large numbers, or blooms, they can cause widespread damage.

dinosaur (*paleontology*)   During the Jurassic and Cretaceous, reptiles called dinosaurs were the dominant form of land vertebrates, filling ecological niches from Alaska to Antarctica. Believed to have evolved from small, crocodilelike reptiles more than 240 million years ago, dinosaurs became extinct during the K/T event about 65 million years ago. The name, from the Greek for "terrifying lizards," is a misnomer coined in 1842 by Richard Owen [English: 1804–92]; dinosaurs weren't lizards and not all were terrifying.

Approximately 775 genera of dinosaurs have been identified. They range in size from the 6-lb (3-kg) *Compsognathus* to the 165,000-lb (75,000-kg) *Brachiosaurus* and include plant eaters and meat eaters. Some walked on all fours and others were basically 2-footed. Fossils of some dinosaurs uncovered in the late 1990s showed clear evidence of feathers, strengthening the theory that birds are direct descendants of dinosaurs.

There is much controversy over the taxonomic classification of dinosaurs. Many systems simply place dinosaurs in 2 orders—Ornithischia and Saurischia—in the reptile subclass Archosauria. The Ornithischia had pelvic bones arranged like those of a bird hip. They included the duck-billed hadrosaurs, the armored ankylosaurs, the horned dinosaurs such as *Triceratops*, and the plated dinosaurs such as *Stegosaurus*. The Saurischia had pelvic bones arranged like those of a lizard hip. A much more heterogeneous group than the ornithischians, the saurischians are divided into 3 suborders: prosauropods, theropods, and sauropods.

diode (*electricity*)   The original diode is an evacuated chamber (vacuum tube) that contains a cathode and an anode. The cathode is a metal device that receives an electric current, which heats the metal, causing it to emit electrons (the Edison effect). Because of the vacuum, electrons reach the anode, which connects to a circuit, causing current to flow. A solid-state diode joins 2 regions of a semiconductor, one prepared to act as an emitter and one as a collector of electrons.

By itself, the main use of such a diode is to change alternating current to direct current. Adding an electronic barrier to a diode allows control of the amount of current that flows from the anode. A small current in the barrier, which can be turned on or off or varied continuously, repels electrons. The large current through the diode also turns on or off, or variations are amplified, from changes in the current in the barrier. A vacuum tube with such a barrier is called a triode; the solid-state device is a transistor.

diophantine equation (*algebra*)   An equation in 2 or more variables with integral coefficients and variables defined only for integers is called diophantine, after Diophantus [Hellenic: c210–c290]. Diophantus is known as the father of algebra because of his use of equations and symbolism in *Arithmetica* and his acceptance of positive rational numbers as solutions (although modern diophantine equations require integral solutions). Hilbert's Tenth Problem, posed in 1900—to find a general algorithm for solving diophantine equations—has been shown to have no solution.

dioxin (*compounds*)   Dioxin is the general name for a group of related chemicals. The most common of the dioxins is 2, 3, 7, 8 tetrachlorodibenzo dioxin (TCDD), a contaminant in the herbicide Agent Orange used as a defoliant in the Vietnam War. Other dioxins have been found in ash from garbage incinerators and in residue from manufacturing processes such as papermaking. Dioxins are poisonous to many animals. In humans they have been proven to cause skin disease and there is some evidence linking them with cancer.

Diplodocus (*paleontology*)   A giant plant-eating dinosaur of the late Jurassic, *Diplodocus* grew to lengths of 75 ft (23 m) and weights of 35,000 lb (16,000 kg). It had a small head, long neck, heavy body, and long tail. When standing on its hind legs to feed, its head could be 40 ft (12 m) or more above the ground. The name, meaning "double beam," refers to certain tailbones that had both forward and backward projections—rather like rails of a sled—that helped support the body's weight.

Dirac, Paul Adrien Maurice (*physicist/mathematician*)   [English-American: 1902–84] Dirac studied both engineering and mathematics before turning to particle physics. In 1928 Dirac set out to rework the basic wave equation of quantum mechanics so that Einstein's special relativity is taken into account. By 1930 he had created a new wave equation. Among other virtues, Dirac's equation accounts for spin, discovered experimentally in 1925 but without a theoretical basis at that time. The symmetry of Dirac's equation suggested that there must be a particle exactly like the electron, except with a positive instead of negative charge. An energetic photon would produce from its energy alone both the electron and its positive counterpart, the antiparticle. The predicted antiparticle, the positron, was observed in 1932 by Carl David Anderson. Dirac's theory also predicted an antiparticle for every particle, all of which have since been found.

direct current (*electricity*)   Direct current is an electric current for which an electric field travels through a conductor in 1 direction only, nudging electrons in the conductor to move in that direction. The passage of static electricity from 1 object to another produces a surge of direct current. Batteries, solar cells, and similar devices, however, create steady direct currents. Alternating current can be changed to direct current, needed for many applications, by a rectifier, such as a diode.

directed distance (*analytic geometry*)   Distance, ordinarily a nonnegative number, is assigned a + or − to indicate direction from the axis in Cartesian coordinates (direction from the pole in polar coordinates). A positive number indicates a directed distance above, to the right, or toward the viewer, while a negative number shows directed distances below, on the left, or away from the viewer.

direct proof (*logic*)   Direct proof is based on rules that produce 1 true proposition from 2. In direct proof a proposition is shown to be the conclusion of a chain constructed from such rules applied to given (accepted as true) propositions along with axioms, definitions, and theorems. For example, given that 2 sides of a triangle have the same length, a definition and a theorem show that a line bisecting their vertex produces 2 congruent triangles. This provides a direct proof that the angles at the other 2 vertices have the same size. (*See also* **indirect proof**.)

direct variation (*analytic geometry*)   A relation between 2 variables such that 1 of them is a constant multiple of the other or, equivalently, such that their ratio is a constant, is called direct variation; for example, $y = cx$ where c is a number. (*See also* **inverse variation**.)

discontinuity (*analytic geometry*)   A gap in the graph of a function is a discontinuity. Such a gap may occur where the function is defined (as in a step function) or where the function is not defined, as in $y = 1/x$, not defined when $x = 0$. When only 1 point is missing, the discontinuity can often be removed—for example, $y = x^0$ is not defined at $x = 0$, but $x^0 = 1$ for all values in the neighborhood of 0, so the graph may be made continuous by inserting the single point (0, 1).

discriminant (*algebra*)   For a quadratic equation of the form $ax^2 + by^2 + c = 0$, the number $b^2 - 4ac$ is the discriminant. If $b^2 - 4ac = 0$, the equation has 1 solution. If $b^2 - 4ac$ is greater than 0 there are 2 real solutions. When $b^2 - 4ac$ is less than 0 there are no real solutions, but 2 complex solutions. A discriminant that is a perfect square indicates a pair of rational solutions. Other types of equations also have easily found discriminants (based on the constants) that describe types of solutions.

disease (*biology*)   Any abnormality other than injury that interferes with the proper functioning of an organism is a disease. The disease may affect the entire organism or only certain organs or parts of organs. Many diseases, such as distemper in cats, are caused by viruses, bacteria, and other pathogens. These diseases are infectious—that is, they can spread from one organism to another. Some diseases, such as hemophilia (a blood-clotting disorder) in humans, are genetic. They are transferred from parent to child by abnormal DNA. Other diseases are caused by a deficiency of vitamins or other essential nutrients. For example, a lack of nitrogen causes stunted growth in plants. Still other diseases are caused by adverse environmental conditions. For example, inhaling tobacco smoke causes respiratory illnesses in both smokers and nonsmokers.

dislocation (*materials science*)   In solids composed of crystals, atoms seldom line up perfectly. A dislocation is a place in a crystal where a small section of a plane of atoms fails to align. In 1934 Sir Geoffrey Taylor [English: 1886–1975] recognized that dislocations in metals account for ductility and other properties. Traditional techniques, such as hammering, heating, and rapidly cooling metals, change properties because they increase or decrease the number of dislocations.

dispersal (*ecology*)   The movement of organisms away from their place of birth or from centers of population is called dispersal. Dispersal techniques are numerous and varied. Fish swim from one place to another, birds fly, mammals walk. Parasites are carried to new places by their hosts, where they may fall off or be excreted in the hosts' wastes. Seed dispersal depends on wind, water, and other mechanisms. Barriers limit dispersal. For example, an ocean is a barrier to freshwater fish and desert soil is a barrier to an orchid seed.

dispersion of light (*optics*)   Separation of light into colors by a prism or in a rainbow is called dispersion of light. As light passes through transparent and translucent materials, different wavelengths of light travel at different speeds. For most transparent materials higher frequencies interact more often with atoms, so blue and violet light travel slower than red. Combined with refraction, this effect causes dispersion.

display (*ethology*)   Any ritualized behavior used to communicate with other organisms is termed a display. Displays can involve postures, movements, and vocalizations. Courtship displays are designed to attract a mate; for example, a male frigatebird inflates his red throat sac. Appeasement displays, such as the slinking of a dog and other postures of submission, are meant to pacify attackers. Distraction displays distract predators. Threat displays scare off opponents.

disproof (*logic*)   In logic the existence of a single instance that contradicts a proposition is a disproof of the proposition. For example, the disproof of the proposition "All primes are odd" is the true statement "2 is a prime and not odd." If a disproof exists for a proposition, the proposition is false.

dissociation (*chemistry*)   When some compounds go into solution, they may dissociate, or break into separate ions. For example, sodium chloride (table salt) dissociates into positive sodium ions and negative chlorine ions. Acids and bases partly dissociate water itself, acids producing positive hydronium ions and bases negative hydroxide ions. Many substances, sugar for example, dissolve without dissociating. If dissociation has occurred, the solution will conduct electricity.

distance (*geometry*)   Distance is a nonnegative real number based on the unit from a number line. The distance between any 2 points associated with the real numbers a and b on the number line is the absolute value of $b - a$, or $|b - a|$. Distance is also the length of a segment between a and b.

If points are located with Cartesian coordinates as (x_1, y_1) and (x_2, y_2), the distance is the square root of $(x_2 - x_1)^2 + (y_2 - y_1)^2$, or for points (x_1, y_1, z_1) and (x_2, y_2, z_2) in 3 dimensions, the square root of $(x_2 - x_1)^2 + (y_2 - y_1)^2 + (z_2 - z_1)^2$.

The distance between 2 figures is the least of the distances between a point in one figure and a point in the other. The distance between a point and a line (or plane) is the length of the line segment from the point perpendicular to the line (or plane). For **parallel** lines (or planes), this distance is the length of the line segment perpendicular to both lines (or planes).

distillation (*chemistry*)   One way to separate components of a **solution** is by distillation, accomplished by heating until the component with the lower boiling point begins to vaporize. The vapor can be captured and cooled to obtain, for example, liquid enriched in ethyl **alcohol** from a solution of alcohol and water. A solution with several components, such as liquid air or crude petroleum, can be separated by removing each gas as it boils in a process called fractional distillation. Wood or coal can be distilled in a vacuum to obtain liquids or gases by destructive distillation; this process also produces nearly pure carbon in the form of charcoal or coke.

distraction display (*ethology*)   A ritualized behavior used as a diversionary tactic is termed a distraction display. It usually occurs in response to a predator. For example, if a nesting killdeer (*Charadrius vociferus*) is threatened by a fox, it pretends that it is injured, flopping about and dragging a wing as it moves away from the nest. The fox follows, but before it reaches the killdeer, the bird flies off.

distributive properties (*algebra*)   A distributive property connects 2 binary operations in an algebraic structure. One operation applied to the result of the second has the same outcome as the first operation applied separately to each of the entities combined by the second. The most familiar example is the distributive property of multiplication over addition: if a, b, and c are numbers, $a \times (b + c) = (a \times b) + (a \times c)$. Note that addition is not distributive over multiplication: $a + (b \times c) \neq (a + b) \times (a + c)$. Sometimes either operation can distribute over the other, as happens with **union of sets** and **intersection of sets**; if A, B, and C are sets then not only is $A \cap (B \cup C) = (A \cap B) \cup (A \cap C)$ true, but so is $A \cup (B \cap C) = (A \cup B) \cap (A \cup C)$.

division (*arithmetic*)   The operation of division for **natural numbers** can be understood as repeated subtraction. Unlike addition or multiplication, division of 2 natural numbers may fail to produce an answer that is also a natural number (natural numbers are not closed for division). To divide a number such as 13 (the dividend) by 4 (the divisor) using repeated subtraction, keep track of how many times you subtract 4 (the number of times is the quotient) and what is left over (the remainder); e.g., $13 - 4_1 = 9$; $9 - 4_2 = 5$; $5 - 4_3 = 1$; at this point you cannot subtract again in the natural numbers, so the quotient is 3 and remainder 1. For numbers in general, repeated subtraction is replaced by defining division as the **inverse** of multiplication: $13 \div 4 = 3.25$ because $3.25 \times 4 = 13$.

divisor (*numbers*)   In number theory a divisor is a **natural number** greater than 1 that when divided into another natural number produces a 0 remainder. Another way to say this is that the smaller number divides the larger one. The word divisor can often be used interchangeably with factor—only the point of view (multiplication for factor, division for divisor) is different.

DNA (*genetics*)   One of the 2 kinds of **nucleic acid**, deoxyribonucleic acid, or DNA, is the hereditary material found in the **chromosomes** of almost all organisms (only in certain viruses is **RNA** the hereditary material). DNA also is found else-

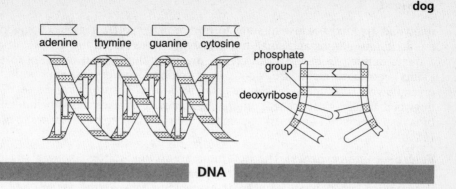

adenine   thymine   guanine   cytosine

phosphate group

deoxyribose

DNA

where in cells, including plant and animal mitochondria and plant chloroplasts. A DNA molecule resembles a long, twisted ladder; this shape, called a double helix, was determined by James Watson and Francis Crick in 1953. The ladder's sides are composed of alternating deoxyribose sugar and phosphate groups. The "rungs" consist of pairs of nitrogen bases, of which there are 4 types: adenine, guanine, thymine, and cytosine. Adenine always pairs with thymine and cytosine always pairs with guanine. The highly specific sequence of bases provides the blueprint needed by cells for protein synthesis. An alteration, or mutation, in the DNA may have fatal consequences to the cell or even to the entire organism.

dodecahedron (*geometry*) Any polyhedron with 12 faces is a dodecahedron ("12 bases"), but nearly always the word means a regular dodecahedron, 1 of the 5 Platonic solids. All 12 faces are regular pentagons, so there are 60 vertices. If adjacent faces are painted black and white, a dodecahedron is like a soccer ball with corners. The 60 vertices are where the 60 carbon atoms in a molecule of buckminsterfullerene lie.

dodo (*animals*) Related to the pigeon, the dodo (*Raphus cucullatus*) was a stout flightless bird native to Mauritius in the Indian Ocean. It had a large head with an enormous hooked bill, strong feet, and rudimentary wings. The dodo became extinct in the late 17th century following the arrival of European colonizers, who hunted dodos extensively and imported predators such as pigs, which ate the birds' eggs. Two related species, called solitaires, lived on nearby islands and became extinct for the same reasons.

dog (*animals*) The dog family, Canidae, comprises 34 species of carnivorous mammals, including coyotes, foxes, jackals, and wolves. Species are found on all continents except Antarctica. Dogs are built for endurance rather than speed. Their senses of smell and hearing are very acute, their sight less so. Their diet is varied, including plant as well as animal matter. Dogs typically mark their territory by repeatedly urinating on objects within or along the territory's edges. Some species form packs with strict dominance hierarchies. The fennec, or desert fox (*Fennecus zerda*), of the Sahara is the smallest species, about 14 in. (35 cm) long, including tail, and 11 lb (5 kg) in weight. The gray wolf (*Canis lupus*) of the Northern Hemisphere is the largest, up to 6.5 ft (2 m) long, including tail, and weighing as much as 175 lb (80 kg).

dogwood (*plants*)   Native mainly to northern temperate regions, dogwoods are deciduous shrubs and small trees. They are classified in the genus *Cornus*, a name derived from the Latin *cornu*, meaning "horn" and referring to the plants' very hard wood. Dogwoods bear compact clusters of small white or green flowers surrounded by showy petal-like bracts. The fruits are berries. The dogwood family, Cornaceae, comprising some 110 species, also includes the broadleaf (*Griselinia littoralis*), a common New Zealand tree that grows to 60 ft (18 m).

doldrums (*atmospherics*)   The doldrums are a wind-current zone along the equator where breezes are very light or nonexistent. The circulation of atmosphere is controlled largely by differences in temperature between the equator and the poles (and the Coriolis effect, which changes wind directions). At the equator, especially where it crosses an ocean, temperatures are warm all year and air rises from the resulting perpetual high pressure. Thus there are almost no horizontal winds.

dolomite (*rocks and minerals*)   Dolomite is both a mineral and a rock. It was named after Déodat Dolomieu [French: 1750–1801]. The mineral is a white glassy or pearly crystal form of the compound $CaMg(CO_3)_2$ probably derived from limestone. The rock is a white or creamy form of recrystallized limestone with a large percentage of the mineral.

dolphin and porpoise (*animals*)   Small cetaceans with beaklike snouts and slender bodies are generally called dolphins. Those with blunter snouts and stockier bodies are called porpoises. Dolphins and porpoises range in length from 4 to 14 ft (1.2 to 4.2 m). They inhabit all oceans plus the estuaries of many large rivers. Some species, called freshwater or river dolphins, live in rivers and lakes. Dolphins and porpoises feed mainly on fish, locating prey by echolocation.

The name dolphin also is given to fast-moving fish of the genus *Coryphaena*. Found in tropical seas, these dolphins grow to almost 6 ft (1.8 m).

domain (*algebra*)   A function is a rule that connects one set, the domain, to another, the range. If the function connects the radius of a circle to its area, for example, the domain is the radius' length, always a positive real number. But for a function that connects a number of eggs to total cost, the domain is the natural numbers. A variable is an implied function associated with specific domains. In the examples, the variable r for radius has the domain positive real numbers while the variable n for number of eggs has the domain natural numbers. It is the custom (but not the rule) to use n for domains of natural numbers or integers; x for domains of real numbers; and z for domains of complex numbers.

An unrelated meaning of domain in algebra describes certain algebraic structures, especially the structure of the integers.

domestication (*biology*)   Practiced since ancient times, domestication is the adaptation of plants and animals to live and breed under human control. Domesticated organisms provide humans with most of their food as well as clothing fibers, medicines, and many other products. Organisms also are domesticated for other purposes—for example, as ornamentals (roses, peacocks), pets (dogs, cats), or beasts of burden (horses, camels).

A wild species can gradually be changed by selective breeding, reproducing only those individuals with the most desirable traits. Mating closely related individuals—inbreeding—can increase certain traits in a population, while mating individuals that are not closely related—outbreeding—can introduce new genes into a population. Over time domesticated organisms may become so altered that they show little if any resemblance to their wild ancestors. For example, there is little similarity between Rhode Island Reds, White Leghorns, and the many other domestic breeds of poultry and their ancestor, the jungle fowl of India (*Gallus bankiva*).

dominance (*genetics*)   Mendel discovered that certain plant characteristics appear in 1 of 2 contrasting forms, such as tallness and shortness in pea plants. Mendel also found that a short pea plant that self-pollinated produces offspring (called the F_1 generation) that are short. But if a short plant is mated with a tall plant, all the offspring are tall. He called traits expressed in the F_1 generation dominant and traits that are hidden recessive. Today we know that characteristics such as pea plant height are determined by a pair of genes—one inherited from the male parent, the other from the female parent—and that genes can exist in different forms (alleles). Mendel's original plants were pure for the height trait; that is, they had genotypes of TT (tall) and tt (short). The F_1 generation were hybrids, with a genotype of Tt.

In some situations there is only partial dominance of one allele over another. For example, when a red (RR) snapdragon is crossed with a white (WW) snapdragon, all the offspring are pink (RW). When pink (RW) snapdragons are crossed, their offspring are 25% red (RR), 50% pink (RW), and 25% white (WW).

dominance hierarchy (*ethology*)   Many animals that live in groups have established dominance hierarchies, with individuals at the top of the hierarchy dominant to those lower down. A well-known example is the pecking order among domestic chickens. One hen usually is dominant to all the others; she can peck any of the other hens without being pecked in return. A second hen is subordinate to the top hen but dominant to all others. A third hen is dominant to all except the top 2, and so on. Roosters have their own pecking order and do not peck hens. Animals that rank high in a dominance hierarchy get first crack at food, mates, nesting sites, and so on. They generally are stronger and have the most offspring.

dopamine (*biochemistry*)   Dopamine is a neurotransmitter produced in the vertebrate brain. This chemical is believed to play a role in controlling the movement of muscles. In humans, low levels of dopamine produce symptoms of Parkinson's disease, which often is characterized by muscle tremors.

Doppler effect (*waves*)   Waves of any kind, including sound and light waves, appear to increase or decrease in frequency when produced by a moving source; this is known as the Doppler effect after physicist Christian Doppler [Austrian: 1803–53]. As a result, sound from an approaching vehicle, such as a siren on a moving car, has a higher pitch, with the pitch falling as the vehicle

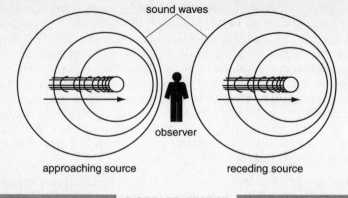

sound waves

observer

approaching source   receding source

DOPPLER EFFECT

passes and moves away. Light from distant galaxies is affected by a related phenomenon. As the universe expands, electromagnetic waves emitted in the distant past become longer in wavelength, shifting frequencies lower, which for visible light is toward the red.

dormancy (*biology*) During unfavorable environmental conditions many organisms enter a state of relative inactivity called dormancy (from the French *dormir*, meaning "to sleep"). Metabolism slows down and growth ceases until conditions improve. Pea seeds remain dormant until they are moistened. Oak trees are dormant in winter, then burst into bloom when spring arrives. Mud turtles go through estivation during hot, dry summers and pocket mice undergo hibernation through cold winters.

dormouse (*animals*) Native to Europe and Asia, dormice (Family Gliridae) are small rodents famed for their long periods of hibernation, which can last 6 months at a time; their name comes from the French *dormir*, meaning "to sleep." Dormice resemble small squirrels, with well-developed eyes, round ears, soft fur, and a bushy tail almost as long as the body. They are nocturnal, sleeping during the day in tree holes, among rocks, or in deserted burrows of other animals.

dot product (*algebra*) If 2 vectors **a** and **b** of the same dimension are shown as ordered n-tuples, the dot product **a · b**, also known as the scalar product or inner product, is the scalar created by multiplying each component in **a** with the corresponding component in **b** and adding the products. For example, if **a** = (1, 2, 3) and **b** = (4, 5, 6), then **a · b** = $1 \cdot 4 + 2 \cdot 5 + 3 \cdot 6 = 32$. When vectors are pictured as arrows, **a · b** is the product of the lengths of **a** and **b** with the cosine of the angle between the vectors.

down quark (*particles*) The 2 light quarks that are the basis of ordinary matter are called down and up, named for a kind of spin characteristic. The down quark with a charge of $-\frac{1}{3}$ is needed twice to balance out an up ($+\frac{2}{3}$) in the neutron but only once to reduce the excess positive charge of 2 ups in a proton. The antidown quark has a charge of $+\frac{1}{3}$.

dracaena (*plants*)   Generally classified in the agave family, members of the genus *Dracaena* are tropical and subtropical plants with sword-shaped leaves that have parallel veins. The small flowers are borne in various kinds of inflorescences. Some species, such as the corn plant (*D. fragrans*), are grown as ornamentals. The biggest species is the dragon tree (*D. draco*), native to the Canary Islands, which grows to a height of 60 ft (18 m) or more. It produces a dark red resin ("dragon's blood") that once was used as a varnish on violins.

dragonfly (*animals*)   The approximately 5000 species of insects in the order Odonata are brightly colored, dexterous fliers with elongated wings and an extremely long abdomen. The larger species, with wingspans up to 5.5 in. (14 cm), are dragonflies, with differently shaped fore and hind wings. Smaller, more delicate species are damselflies, with similarly shaped fore and hind wings. Odonata depend on their large compound eyes to detect insect prey, which they snatch in flight. In most species the immature stage, called a nymph, is an aquatic predator.

drift, glacial (*geology*)   A jumble of rocks unrelated to the local bedrock, often including large erratics, is left behind when a glacier melts. This assembly is termed glacial drift. The name *drift* was originally given to these deposits in the mistaken belief that the rocks had been carried to their present locations by a large flood.

Drosophila (*animals*)   Generally called fruit flies or pomace flies, *Drosophila* are very small insects often seen on decaying fruits and vegetation. The best-known species, *D. melanogaster*, has been extremely useful in genetics research because of its 10-day life cycle and very large chromosomes. Some species of *Drosophila* produce the world's longest sperm; for example, the sperm of *D. bifurca* is 2 in. (5 cm) long—20 times longer than the fly's body, and about 1000 times longer than a human sperm.

drought (*meteorology*)   A drought is a period of at least 3 weeks when considerably less water falls as precipitation than is needed for plants and animals and for maintaining the water table. The actual amount of precipitation needed varies from place to place and from season to season. In the United States the government issues the Drought Severity Index and the Crop Moisture Index; they take into account supply and demand for water in particular regions. A Drought Severity Index below −2 represents drought conditions.

drumlin (*geology*)   A drumlin is a narrow, smooth small hill made of glacial till or smoothed bedrock that aligns with the direction of glacial movement that has formed the drumlin. Typically a drumlin is about 1200 ft (400 m) long, 300 ft (100 m) wide, and 80 ft (25 m) high. Drumlins occur in swarms of hundreds that are only about as far apart as they are long. The exact cause of the streamlining by glaciers is not well understood.

dual theorem (*geometry*)   In projective geometry and a few other parts of mathematics, theorems occur in pairs where 2 elements, such as points and lines, can be interchanged in one theorem to produce the other. This occurs in projective geometry because the axioms for the system are themselves duals, existing in pairs with points and lines interchanged from one member to the other. Proving any theorem in projective geometry also proves the dual theorem.

dubnium (*elements*)   Dubnium is a radioactive metal, atomic number 105, whose first few atoms were created by Russian scientists at Dubna in 1967. Larger amounts were created at Dubna in 1970 and independently that year by a U.S. team led by Albert Ghiorso [American: 1915–]. Dubnium is named for the Joint Institute for Nuclear Research at Dubna, Russia. Chemical symbol: Db.

duckweed (*plants*)   The smallest flowering plants are the duckweeds, Family Lemnaceae, so named because they are a favorite food of ducks (as well as other animals). A single *Wolffia borealis* duckweed is less than 0.04 in. (1 mm) across and weighs no more than 2 grains of table salt. Duckweeds are aquatic, found in ponds, ditches, and other quiet freshwater habitats. They do not have distinct stems or leaves, but consist of a flattened green "frond" that floats on the water's surface and carries out photosynthesis. Some species have 1 or 2 dangling roots; others lack roots. The miniature flower gives rise to a tiny 1-seeded fruit called a utricle.

ductility (*materials science*)   Ductility is the ability of a material to be stretched without breaking; it is the opposite of hardness. A ductile metal dents when struck a strong blow, but hard metals resist. Ductile materials can be pulled into wires. Ductility is the result of dislocations in crystal structure. Repeated hammering produces more dislocations, making material more ductile. Heating to just below melting and then cooling reduces dislocations, hardening metals.

dugong (*animals*)   A marine mammal that lives in shallow tropical coastal waters in the Pacific and Indian oceans, the dugong (*Dugong dugon*) can be 13 ft (4m) long and weigh 2000 lb (900 kg). The flippers (which, unlike manatee flippers, lack nails) are used for propulsion by young dugongs; adults use the 2-lobed tail fin for propulsion and the flippers for steering. The dugong diet consists mainly of sea grasses.

Steller's sea cow (*Hydrodamalis gigas*) was a dugong of the northern Pacific and Bering Sea that grew to almost 26 ft (8 m) and weights of 11,000 lb (5000 kg). Hunting caused its extinction in the 18th century.

dune (*geology*)   A dune is a long mound made from sand that has been created by winds or, if underwater, by currents. Underwater dunes are much smaller than those on the surface, appearing as ripples a few inches high in the sand. Dunes that form on beaches may be 100 ft (30 m) high and some dunes in sandy deserts can reach 300 ft (100 m). The characteristic shape of a dune is a gentle slope on the side facing the wind or current and a steep drop-off on the other side. Sand particles are bounced along the windward side and roll down the trailing edge, causing the whole dune to move slowly in the direction of the wind or current.

durian (*plants*)   Named for its fruit, the durian (*Durio zibethinus*) is an evergreen tree native to tropical rain forests of Southeast Asia. It grows to heights of 100 ft (30 m) or more. The fruit, which may weigh up to 100 lb (45 kg), has a thick green or yellowish rind thickly covered with sharp spines. The custardlike pulp has a foul odor but a delicious flavor.

dust (*geology/astronomy*)   Many different kinds of small particles are called dust, including tiny particles that fall so slowly in still air that they seem suspended (some produced by volcano eruptions); ground rock worn away by glaciers and transported away in meltwater; small particles of regolith formed by collisions that cover the surface of the Moon and other rocky bodies in space; and the still tinier particles of compounds found in nebulas and even in intergalactic space. Dust that accumulates on Earth may be lifted by winds from arid soils, produced by volcanoes, or even be from meteors and dust particles from space.

dust cloud (*astronomy*)   Dark clouds obscuring the stars, some so noticeable that early South Americans such as the Inca named them, are called dust clouds or dust nebulas. They consist mostly of fine grains of materials such as silicon dioxide (quartz) and carbon (diamond). Because of the small size of the grains—about 0.00004 in. (1 micrometer) across—individual grains are invisible, but concentrations produce scattering of light that blots out the light of stars behind them.

dust devil (*atmospherics*)   Small-scale rotating winds form for poorly understood reasons. Over land such winds, like miniature tornadoes, pick up dust, sand, dead leaves, or other debris, swirling the material around and lifting it above the ground. This visible manifestation of the wind is called a dust devil.

dwarf star (*stars*)   Dwarf is the official classification given to an ordinary star. All stars on the main sequence of the Hertzsprung-Russell diagram are dwarfs, though many are much larger than the Sun, itself a typical yellow dwarf. Astronomers do not use the term dwarf often, however; but they do write of brown dwarfs and white dwarfs, neither of which are stars on the main sequence at all.

dye (*compounds*)   A dye is a chemical used to color materials. Paints color by covering and pigments provide color in various ways, but dyes enter all parts of a solid or liquid. The best dyes, called fast, chemically combine with molecules, making the color fade-resistant.

Early dyes were natural materials, such as madder and woad from plants or royal purple and cochineal from animals. In 1856 William Henry Perkin [English: 1838–1907] synthesized the dye mauve and started a fashion as well as an industry. Following Perkin, dye development and manufacture became the driving force in both German and English chemical industries.

The ability of fast dyes to attach themselves chemically has made them useful in biology. Dyes stain tissues for examination, differentiate between types of bacteria, and treat disease.

dysprosium (*elements*)   A rare earth, atomic number 66, dysprosium was recognized in 1886 by Paul-Emile Lecoq de Boisbaudran [French: 1838–1912] but not prepared as a metal until 1950 (its name, from the Greek *dysprositos*, "difficult of access," is appropriate). So far there are few uses for dysprosium. Chemical symbol: Dy.

E

e (*numbers*)   The number *e* is a transcendental number that is approximately 2.718281828459045 . . . . It is nicknamed Euler's constant because of extensive use by Euler and is often identified as the base of the natural logarithms; however, that is only 1 of many uses for this fundamental constant. There are various ways to define *e*, including as the limit of $(1 + 1/x)^x$ as x increases to infinity and as the infinite sum $e = 1 + 1/1! + 1/2! + 1/3! + \cdots$. One common application of *e* occurs in the basic equation for growth and decay, $A = A_0 e^{ct}$, where A is the final amount that began with A_0 and that grows or decays at a rate determined by the constant c for a time t.

eagle (*animals*)   Large members of the hawk family are commonly called eagles. All are superb fliers with excellent eyesight, a strongly hooked beak, and powerful feet with sharp claws. The golden eagle (*Aquila chrysaetos*), found in mountainous regions throughout the Northern Hemisphere, was a symbol of imperial power in ancient Rome and later European empires. Sea eagles, or ernes, which feed mainly on fish, include the bald eagle (*Haliaeetus leucocephalus*), the national bird of the United States. The harpy eagle (*Harpia harpyja*) of rain forests of Central and South America preys on sloths, monkeys, and other tree-dwelling mammals.

ear (*anatomy*)   The vertebrate ear is a sense organ concerned with hearing and balance, or equilibrium. In humans the ear has 3 major sections. The external ear directs sound waves to the eardrum, or tympanic membrane, causing it to vibrate. Three tiny bones in the middle ear transmit the vibrations from the eardrum to a membrane at the entrance to the inner ear. The vibrations are then transmitted to fluid in the cochlea, a snail-shaped structure in the inner ear. As the vibrations travel through the cochlea, they set up vibrations in receptors called hair cells, which initiate impulses to the brain. Also in the inner ear are semicircular canals and associated organs that detect position and movements of the head.

earth (*chemistry*)   Chemists before the 19th century called some minerals that were difficult to break down, and therefore believed to be elements, earths. Aluminia is an example. With better techniques of analysis, the earths were found to be oxides of previously unknown metals, such as aluminum. The terminology is preserved in the name of the rare earth group of elements.

Earth (*solar system*)   The third planet from the Sun is the only one in the solar system known to harbor life. From out in space, Earth appears as a blue-and-

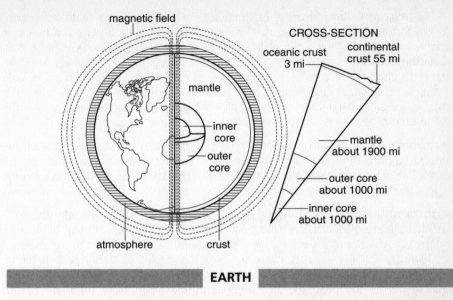

magnetic field

CROSS-SECTION

oceanic crust 3 mi

continental crust 55 mi

mantle

inner core

outer core

mantle about 1900 mi

outer core about 1000 mi

inner core about 1000 mi

atmosphere   crust

EARTH

white sphere—some 70% of the surface is water and clouds cover about half the sky. Earth's substantial atmosphere and strong magnetic field shield living organisms from most harmful radiation from space.

Earth itself consists of three main layers. The rocky crust varies from 55 mi (89 km) deep under continents to 3 mi (5 km) under oceans. Below the crust, the mantle of silicate rock rich in iron extends to about 1900 mi (3060 km) below the surface. The top of the mantle is semiliquid to about 150 mi (250 km). The crust is broken into large plates that move slowly on this layer. Beneath the mantle lies Earth's iron-and-nickel core, with a lumpy outer boundary. The temperature at the center may be 7200°F (4000°C). The outer core is liquid due to this great heat, while an inner core is solid as a result of enormous pressure.

Earth Day (*environment*)   Conceived by Gaylord Nelson [American: 1916–], Earth Day was first held on April 22, 1970, to demonstrate nationwide concern for the environment, making the issue part of the political dialogue in the United States. Today Earth Day is celebrated internationally with a great variety of events designed to heighten public awareness and understanding of environmental issues.

earthquake (*geology*)   When solid rock moves rapidly at a fault that has broken under stress, waves spread through rock in all directions. These seismic waves are felt at the surface as an earthquake—motion of both bedrock and soil. Earthquakes are often weak and barely noticeable as a slight movement or rumble, but several times each year very large earthquakes strike somewhere on Earth. These can knock down walls, produce cracks in the soil, or move one part of the surface several feet from its original position. In populated regions earthquake damage can kill large numbers of people. An earthquake that moves the seafloor may cause a tsunami that can be even more destructive.

The most common sources of earthquakes are the great faults that are the borders of tectonic plates, but rarer earthquakes at faults within a plate can be more destructive because they can travel farther in all directions through the plate. Not all earthquakes are caused by breakage at faults. Moving magma beneath a volcano induces a characteristic series of medium-sized earthquakes that often aid in predicting an eruption. A large landslide or avalanche can be detected some distance away in a seismogram because of the small earthquake produced. Pressure of artificial reservoirs as they grow upstream of a large dam also triggers earthquakes.

earthshine (*solar system*) Sunlight reflects from Earth into space. Some of the reflected light is bounced back to Earth from the Moon, where it is called earthshine. Earthshine is weak, but it makes the dark part of the Moon's disk barely visible near or even at the new moon.

earthworm (*animals*) Segmented annelid worms that burrow through the soil, earthworms are of enormous ecological importance. Their burrowing helps aerate the soil, distribute water to greater depths, and bring nutrients to the surface. Earthworms actually eat soil, digesting decaying organic material and excreting the remainder in nutrient-rich fecal matter called castings. Most earthworms are reddish due to the presence of a hemoglobinlike respiratory pigment. Earthworms are hermaphrodites, with 2 pairs of male testes and 1 pair of female ovaries.

earwig (*animals*) Found mostly in warmer climates, members of the insect order Dermaptera probably got their common name from an old superstition that these small, nocturnal insects crawl inside the ears of sleeping people. Earwigs typically have a small head, 4 wings (the forewings short and leathery, the hind wings large and membranous), and a long abdomen that terminates in a pair of pincers. Some species are parasites but most feed on plants and small animals.

ebony (*plants*) Blackish woods from a variety of trees are commonly called ebonies, but true ebony trees are classified in the genus *Diospyros*. The finest ebony wood comes from the macassar ebony (*D. ebenum*), a native of India and Sri Lanka that grows to about 60 ft (18 m). So-called American ebony comes from the persimmon (*D. virginiana*), a tree of eastern North America better known for its large edible fruits.

eccentricity (*astronomy*) The amount of deviation of a conic section from a circle is eccentricity. The Greeks and subsequent astronomers up to Kepler had thought that orbits must be circles, but Kepler discovered that planetary orbits are ellipses with the Sun at one focus of curve. Earth's nearly circular orbit has an eccentricity of 0.0167, but Pluto's highly elliptical orbit is 0.2448. When eccentricity reaches 1, as it does for some comets, the orbit is a parabola; above 1, it is a hyperbola.

echidna (*animals*) Also called spiny anteaters, echidnas are monotreme mammals of Australia and New Guinea. They grow to 2.5 ft (7.6 m) long and weights of 15 lb (6.8 kg). Echidnas are covered from pointed snout to stubby tail with sharp spines. They use large claws on the feet to dig up ants, termites, and worms, which they then pick up by a long tongue. The female lays an egg that she places in a pouch that forms on her belly during mating season. After the egg hatches, the young echidna feeds on milk flowing from the mother's mammary glands.

echinoderm (*animals*)   Crinoids, starfish, sea urchins, sand dollars, and sea cucumbers are members of Phylum Echinodermata ("hedgehog-skinned"). All of the approximately 6000 described species are marine, inhabiting ecological niches from shores to great depths. Echinoderms have a thin skin that covers an internal skeleton of calcium carbonate plates, from which project numerous tiny spines. They have radial symmetry, typically with 5 parts arranged around an axis that runs through the mouth. A unique water vascular system with tube feet for locomotion allows water to enter and circulate through the body.

echo (*waves*)   Return of a sound, seismic wave, microwave, or radio wave to its source after bouncing from a reflecting surface is an echo. For light or infrared radiation, the echo is called a reflection. Echoes can be used to locate a reflecting surface (echolocation). Bats and dolphins use audio echolocation to locate prey. Radar is echolocation with microwaves.

echolocation (*ethology*)   Many bats plus certain species of dolphins, whales, shrews, and nocturnal birds use sound echoes to track prey and detect obstacles, a process called echolocation. The animal emits sounds that bounce off nearby objects and return to the animal as echoes. Bats use pulses of ultrasound far above the limit of human hearing but other echolocators—oilbirds (*Steatornis caripensis*), for example—use sounds audible to humans. The process can be extremely accurate: some bats can distinguish insects 15 ft (4.6 m) away.

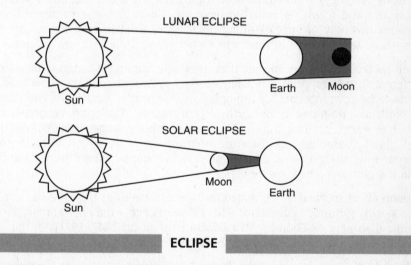

ECLIPSE

eclipse (*astronomy*)   When astronomers observe one heavenly body passing between one viewing point and another in such a way as to cut off all or most of the light from the farther body, they call the event an eclipse. The most noticeable eclipses on Earth come when the Moon passes in front of the Sun (a solar eclipse). When Earth passes between the Sun and Moon, so that Earth's shadow darkens the Moon, it is called a lunar eclipse even though the cause is somewhat different from a solar eclipse. The eclipsing of one star by the less bright member of a binary star is one of the causes of variable stars.

ecliptic (*solar system*)   The ecliptic is the plane defined by Earth's orbit about the Sun. It is named after the solar and lunar eclipses, all of which occur only along that plane. All the planets except Pluto orbit the Sun in nearly the same plane, with Mercury deviating the most—other than Pluto at about 17°—at about 7°.

ecology (*sciences*)   Derived from the Greek *oikos* (house) and *logos* (discussion), ecology is the study of the complex interactions of organisms and their environments. Ecologists are interested in such topics as the diversity of organisms in a community, niches occupied by each type of organism, changes in predator-prey populations, consequences of the introduction of exotic species, and flow of energy and nutrients through the community.

ecosystem (*ecology*)   A community of organisms and the abiotic (physical) environment in which they live make up an ecosystem. Grasslands, forests, deserts, ponds, rivers, and oceans are ecosystems. Even Earth as a whole can be considered an ecosystem. Energy and nutrients continuously flow through an ecosystem in sequences called food chains. Disruptions in food chains, such as loss of plants due to fire or disappearance of a predator due to hunting, can drastically alter an ecosystem. Research indicates that the greater the diversity of species living in an ecosystem, the greater its productivity and stability.

ecotone (*ecology*)   The transition zone between 2 communities, such as a grassland and forest, is called an ecotone. There is some evidence that the greatest diversity of species—long believed to exist in rain forest communities—may actually exist in the ecotones between rain forests and savannas.

ecotype (*evolution*)   A species that lives in a variety of habitats may exhibit different forms, or ecotypes. Each ecotype represents certain genetic adaptations to its environment. For example, there appear to be at least two ecotypes of bottlenose dolphins in the northwestern Atlantic. The coastal ecotype, which inhabits warm, shallow waters, has a smaller body, larger flippers, and different blood characteristics than the offshore ecotype, which lives in cooler, deeper waters. Over time, an ecotype may become so altered from its ancestor that it represents a new species.

Edison effect (*particles*)   A heated metal emits electrons that in a vacuum flow to a nearby grounded conductor. This Edison effect is the only known pure scientific discovery of Thomas Alva Edison [American: 1847–1931]. In the hands of later inventors the Edison effect became the basis of the diode vacuum tube.

eel (*animals*)   Members of the order Anguilliformes, eels are long snakelike fish with smooth slimy skin. They lack pelvic and ventral fins; the long dorsal and anal fins usually connect to the tail fins. One family (Anguillidae) lives in freshwater but spawns in the sea. All other eels live in salt water. Eels are mainly predators, though freshwater eels also scavenge. Most species are smaller than 3 ft (0.9 m) in length but the moray eel (*Thyrsoidea macrurus*) grows as long as 10 ft (3 m).

The electric eel (*Electrophorus electricus*), which lives in the Amazon and other South American rivers, is not a true eel but a relative of carp.

eelgrass (*plants*)   The name eelgrass usually refers to *Zostera marina* of the family Zosteraceae. This perennial herb has eel-shaped leaves that arise from rhizomes (underground stems). It is one of very few flowering plants adapted to life in the sea. Eelgrass lives on tidal mud flats and submerged in brackish and salt waters. Eelgrass communities are extremely diverse and productive; many fish and shellfish spend all or part of their lives there.

efficiency (*thermodynamics*)   Efficiency is a measure that applies to any transformation of energy from one form to another. It is the ratio of the amount of work done to the amount of energy used to produce the work. Since both work and energy are measured in the same units, this ratio is a pure number, most commonly expressed as a percent. No transformations in the real world have 100% efficiency. All machines or transformations of chemical to electrical or kinetic energy, or even transformations of matter to energy in nuclear reactions lose some energy, nearly all as unusable heat.

effluent (*environment*)   Wastewater that flows out of a treatment plant, sewer, or industrial plant into a river or other body of water is called effluent. It may be untreated or partially or completely treated prior to discharge. The effects of an effluent on the environment depend on the amounts and types of pollutants it carries.

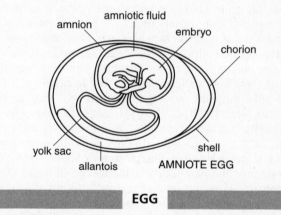

AMNIOTE EGG

EGG

egg (*cells*)   Female reproductive cells, produced in ovaries of female organisms, are called eggs, or ova. The union of an egg with a sperm from a male (a process called fertilization) is the beginning of a new organism. Though each is only a single cell, unfertilized eggs vary greatly in size. Mammalian ova range from 0.002 to 0.01 in. (0.05 to 0.25 mm) in diameter while chicken ova are 1.2 in. (30 mm) in diameter.

The term "egg" also refers to the structures enclosing the developing embryos of insects, reptiles, and birds. These eggs have protective cases or shells that prevent drying out on land. The amniote eggs of reptiles and birds—so named because they have a tissue called the amnion—also contain yolk and membranes important to the embryos' survival. Ostriches lay the largest of all eggs—1 ostrich egg is as big as 2 dozen domestic hens' eggs.

Ehrlich, Paul (*bacteriologist/immunologist*)   [German: 1854–1915] A pioneer in the fields of hematology, immunology, and chemotherapy, Ehrlich made his first important contribution while still a student. He discovered that cells and structures within cells have an affinity for specific dyes, allowing them to be stained different colors. Using dyes, Ehrlich discovered the various types of white blood cells. In 1890 he joined the laboratory of Koch, where he developed a method for staining the newly discovered tuberculosis bacterium; found that poisons induce the blood to produce antibodies; and standardized diphtheria antitoxin so that it could be used by doctors. Ehrlich suggested that chemicals could be used to kill bacteria in the human body without harming the body itself—a technique he termed chemotherapy. He discovered salvarsan, a compound that proved to be effective in treating syphilis. Ehrlich shared the 1908 Nobel Prize in physiology or medicine with Elie Metchnikoff [Russian: 1845–1916], who discovered the role of white blood cells in fighting infection.

Einstein, Albert (*physicist*)   [German-Swiss-American: 1879–1955] Einstein was the best-known scientist of the 20th century, a man whose name was virtually synonymous with genius. His contributions to physics began in 1905 with 3 major results, any of which would have assured him a place in the history of science. The first explained Brownian motion in terms of the movements of molecules, making it possible to establish the physical existence and size of molecules. The second concerned one of the photoelectric phenomena and became the foundation of quantum theory; it explained that light, viewed at the time entirely in terms of electromagnetic waves, sometimes behaves as particles that transfer energy on the basis of the quantum. The third was the special theory of relativity that links time to space and energy to matter.

From 1907 to 1915 Einstein developed general relativity, a theory of gravity more accurate than Newton's. This theory became the basis of cosmology and describes how gravity interacts with electromagnetic waves. In 1924 Einstein collaborated with Satyendranath Bose [Indian: 1894–1974] to develop a mathematical treatment of particles that, like photons, lose their individuality in collections (now known as the Bose-Einstein statistics). In failed efforts in 1935 to show that the interpretation of quantum theory in terms of probability is incorrect, Einstein was instrumental in developing the theoretical basis for what is sometimes called teleportation of photons (and what Einstein called "spooky action at a distance"), achieved in the 1990s. His last major effort was an attempt to unify electromagnetism and gravity into a single unified field theory. Although unsuccessful in this attempt, others continue working on the problem in terms of string theory and supersymmetry.

einsteinium (*elements*)   Einsteinium, atomic number 99, is a radioactive actinide metal in the same family as uranium and plutonium. It was discovered by a team led by Albert Ghiorso [American: 1915–] in the United States in 1952 in the debris from an H-bomb explosion, but classified as a military secret for several years. At the time its existence was revealed, the leading 20th-century scientist, Albert Einstein, had recently died, so the discoverers named the element after him. Chemical symbol: Es.

elasticity (*materials science*) Elasticity is the property of returning to an original size after being stretched or compressed. Metals, including alloys, and solid elements are all somewhat elastic. The change in length is called strain and the imposed force is stress; the ratio of stress to strain is Young's modulus, measured in pounds per square inch (pascals). Barely elastic lead has a modulus of 2,000,000 lb per sq in. (0.16×10^{11} Pa), but very elastic tungsten has a modulus of 50,000,000 lb per sq in. (3.6×10^{11}). Metals and similar materials are elastic because pressure or stretching causes spaces between atoms in the crystalline structure to change. Rubber is elastic for a different reason. It is a polymer whose lengthy molecule is coiled like a spring. Stretchable proteins, such as silk or wool, are elastic for the same reason.

elastin (*biochemistry*) A protein that stretches easily, elastin is found in vertebrate connective tissues where the ability to extend or be elastic is important. Elastic fibers are common in artery walls; certain ligaments; and the lungs and tubes leading to the lungs, including the larynx.

electric fish (*animals*) Several types of freshwater fish have evolved an electrical organ, located in the tail, that can emit electric pulses unique to each species. The pulses create an electric field around a fish, which is used to perceive surrounding objects, help navigate, stun prey, or communicate with other members of the species. The strongest discharges—up to 650 volts—are produced by the electric eel (*Electrophorus electricus*) of tropical South America, which grows to almost 10 ft (3 m); it is not a true eel but is related to carp.

electricity (*physics*) Although electricity is a commonly used term, it is too general to be useful in science. Electricity encompasses everything from charge, which is a fundamental force, to the electromagnetic spectrum, waves often thought of as an effect of electricity rather than as a part of it. Most of the time electricity refers either to electric current or to static electricity, 2 different manifestations of charge reflecting movement of an electric field or else movement or placement of large number of electrons or ions. Electronic devices, also based on movement or placement of electrons, work with a few electrons at a time rather than large-scale currents.

electrode potential, standard (*chemistry*) The standard electrode potential is a negative or positive number that describes how powerfully an element acts in oxidation and reduction. For example the alkali metals are reducing agents (potassium's standard electrode potential is –2.925) while halogens are oxidizing agents (chlorine's is +1.360). The arrangement of elements into a numerical order based on standard electrode potential is called the electromotive, or electrochemical, series. Elements higher in the series can displace lower elements in chemical reactions. Hydrogen is taken as the 0 point on the scale.

electrolysis (*chemistry*) The word electrolysis means destruction by an electric current (as in electrolysis of hair) but in chemistry the word refers to production of any chemical change from an electric current passing through materials. An electric current in metals causes no chemical change because only the electrons move. An ionic fluid, however, conducts electricity via moving ions. When the ions reach a terminal, electric charges are neutralized and the ions become the elemental form. Thus when molten salt (NaCl) undergoes electrolysis, sodium appears at the **cathode** and chlorine gas bubbles from the **anode**. More complex reactions occur in **solutions** since the solvent may also be involved. Electrolysis of a dilute solution of NaCl in water releases hydrogen and oxygen instead of sodium and chlorine, although a stronger solution produces chlorine instead of oxygen.

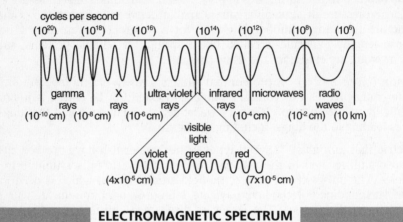

cycles per second
(10^{20}) (10^{18}) (10^{16}) (10^{14}) (10^{12}) (10^8) (10^6)

gamma rays $(10^{-10}\,cm)$ — X rays $(10^{-8}\,cm)$ — ultra-violet rays $(10^{-6}\,cm)$ — infrared rays $(10^{-4}\,cm)$ — microwaves $(10^{-2}\,cm)$ — radio waves $(10\,km)$

visible light
violet green red
$(4\times10^{-5}\,cm)$ $(7\times10^{-5}\,cm)$

ELECTROMAGNETIC SPECTRUM

electromagnetic spectrum (*waves*) The electromagnetic spectrum is the full range of **light** in all forms. In this extended sense, light ranges from long, low-energy **radio waves** to short, high-energy **gamma radiation**. Visible light is a tiny strip near the middle.

The original **spectrum** was the rainbow. The rainbow colors also appear when white light is separated. After the **wave** nature of light was established by **Thomas Young** in 1801, scientists determined that color is wavelength—long waves are red and short ones violet. At that time both infrared ("below red") and ultraviolet ("beyond violet") radiation were discovered as unseen energy striking on either side of light's spectrum, hence invisible light.

In 1873 **James Clerk Maxwell** reasoned that light waves are produced by rapid vibration of a tiny electric charge (this was a quarter century before the discovery of the **electron**, which provides that tiny charge). A slower oscillation of charge would excite waves longer than infrared. In 1888 an experiment to test the theory produced radio waves. Unexpectedly, in 1895 **Wilhelm Roentgen** found that bombarding glass with an electric discharge induces a more energetic radiation. Roentgen called the radiation X rays but correctly suspected X rays to be short electromagnetic waves. Even shorter waves, gamma radiation, were found in natural radioactivity 5 years later. Because the whole electromagnetic spectrum consists of different frequencies of the same entity, scientists know that there will be no further unsuspected invisible rays of this type.

electron (*particles*) All of electronics and most events in materials science result from interactions of electrons. The electron—the first subatomic particle to be identified, by J.J. Thomson in 1897—is a low-mass elementary particle found in the outer reaches of every atom but also able to move easily from one atom to another or even through empty space.

One property of the electron accounts for its fundamental importance: charge, the quality that produces both electric and magnetic fields. When a charged particle moves, it creates electromagnetic waves—light, radio waves, X rays, etc. A large number of electrons moving in the same direction is an electric current. An electron in an atom that changes from one energy level to another produces or absorbs electromagnetic waves, accounting for light at some wavelengths and heat at other wavelengths. Controlled movements of electrons in semiconductors are the basis of electronics.

The electron and its antiparticle the positron are the lightest particles to possess charge, so both are stable, although the common electron quickly annihilates any rare positrons that appear.

element (*materials science*) A chemical element is a substance made entirely from atoms or ions that have the same number of protons. Atoms with the same atomic number are not necessarily identical, but they behave the same chemically and so form a single element. Small clusters of atoms have different properties from elements in bulk, and even large amounts of the same atom can exist in several allotropes. For example, atoms in a sample of the element carbon all have 6 protons, but may have 6, 7, or even 8 neutrons (carbon-12, carbon-13, and carbon-14) and be arranged as hard, transparent diamond or as soft, black graphite.

elementary particle (*particles*) The ancient Greeks conceived of the idea of elements from which all else is constructed, with the smallest part of an element called an atom. By the beginning of the 19th century scientists also concluded that matter is made from indivisible atoms. Within the century, however, a particle smaller than the smallest atom, the electron, was discovered. Soon other subatomic particles were discovered and thought to be fundamental. In the 1960s the discovery of quarks showed that some entities thought to be fundamental are made from smaller particles. Today in the standard model of particle physics, 6 quarks and 6 leptons are the elementary particles of matter—but a few experiments suggest that even these may be made from smaller particles.

elements 110–118 (*elements*) A few atoms of elements 110, 111, 112, 114, 116, and 118 have been created in laboratories, where these very heavy elements soon decayed into lighter ones. Elements 110, 111, and 112 were synthesized by the Society for Heavy Ion Research in Darmstadt, Germany, in 1994 (110 and 111) and 1996. One atom of element 114 was created in 1998 by the Joint Institute for Nuclear Research in Dubna, Russia, followed by atoms of different isotopes of element 114 there and at the Lawrence Berkeley National Laboratory (LBNL) in Berkeley, California, in 1999. The LBNL also produced 3 atoms of element 118 in 1999, which decayed to element 116 and then to element 114. The half-life of the 1998 isotope of element 114 was about 30 seconds. It was thought that the isotope of element 114 with atomic mass 298 would be very long lived, but efforts to create it have so far fallen short.

elephant (*animals*)   The largest land mammal is the African elephant (*Loxodonta africana*), which may stand more than 11 ft (3 m) tall and weigh 14,500 lb (6577 kg). The other living species of elephant, the Asian elephant (*Elephas maximus*), is slightly smaller. An elephant's most distinguishing feature is its long trunk, or proboscis. This extremely versatile organ is used for breathing, smell, touch, sound production, spraying water or dust on its back, and such delicate functions as picking up and cracking open a peanut. An elephant has thick skin, big ears, small eyes, thick legs, and broad feet. Males have incisor teeth that curve outward, forming large tusks; female African elephants generally have small tusks but female Asian elephants are tuskless. Elephants are vegetarians; they live either on grasslands or in forests, depending on the subspecies.

elk (*animals*)   Americans and Europeans apply the name elk to 2 different species of deer. The North American elk, or wapiti, is classified either as *Cervus canadensis* or together with the red deer of Eurasia and northern Africa as *C. elephus*. The European elk (*Alces alces*) is known as the moose in North America.

ellipse (*geometry*)   An ellipse is the conic section shaped like an elongated circle. Every point on the ellipse is located so that the sum of the distances from 2 distinct points called the foci (FOE-sigh; sing.: focus) is a constant. Halfway between the foci is the center of the ellipse. A line segment with endpoints on the ellipse that passes through the center and foci is an axis of the ellipse, while another segment perpendicular to that axis at the center is the other axis. The endpoints of the axes are the vertices of the ellipse.

   The equation of an ellipse is a second-degree equation in 2 variables set equal to 0 for which the coefficients of the 2 squared terms have the same sign but different values, such as $9x^2 + 4y^2 + 18x + 24y - 99 = 0$. An ellipse centered at the origin with axes whose lengths are $2a$ and $2b$ can also be put into the form $x^2/a^2 + y^2/b^2 = 1$.

elm (*plants*)   Native to northern temperate regions, elms (*Ulmus*) are deciduous shrubs and trees with toothed leaves, clusters of inconspicuous flowers, and winged fruits called samaras. Among the tallest and most graceful elms are the American elm (*U. americana*), English elm (*U. procera*), and Japanese elm (*U. japonica*), all of which attain heights of more than 100 ft (30 m). The elm family, Ulmaceae, which comprises more than 150 species of shrubs and trees, also includes hackberries (*Celtis*).

El Niño (*oceanography*)   El Niño is a change in conditions off the west coast of South America during which normally dry inland regions receive torrential rains while warm ocean temperatures kill or drive away fish, causing starvation in ocean birds that feed on them. It is now recognized that this is part of a worldwide weather phenomenon that begins in the central Pacific Ocean. A large mass of warm water develops and eventually spreads in all directions, covering the colder currents, such as the one along Peru and Ecuador. El Niño is blamed for droughts in India and Australia, for rain in the southwestern United States, and in general for changing the climate around the world. It even affects Earth's rotation, slowing it slightly.

Meteorologists now recognize that a similar cooling of the Pacific, called La Niña, also influences worldwide weather patterns, strengthening hurricanes in the Atlantic, for example.

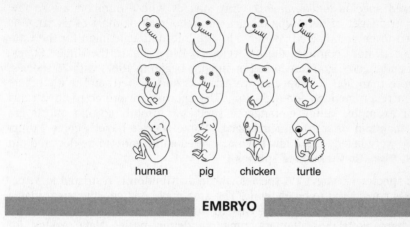

human   pig   chicken   turtle

EMBRYO

embryo (*biology*)   The early developing stage of an organism created as a result of sexual reproduction is called the embryo. In animals such as humans, this is the stage from the beginning of cleavage—which starts soon after fertilization of an egg—up to the time of birth. During the initial part of embryonic development, related organisms greatly resemble each other. Later in the development process they develop the characteristics that distinguish them from one another.

emigration (*ethology*)   The permanent movement of animals out of an area is called emigration. Emigrations often result from large population increases, which create shortages of food and other resources. The phenomenon is common among small rodents, particularly lemmings, and certain birds and insects.

empty set (*sets*)   The set with no members is the empty set, symbolized sometimes as {} and sometimes as ∅. The empty set, sometimes called the null set, is a subset of every set (even of itself). Not only is its presence needed to make the calculations of numbers of subsets correct, but also many important axiomatic systems begin with the assumption that the empty set exists. Note that the empty set is not the same as 0, nor is it nothing.

emu (*animals*)   A flightless bird with a long neck and coarse hairlike feathers, the emu (*Dromaius novaehollandiae*) is the largest bird of Australia, standing about 5 ft (1.5 m) tall. It has muscular legs and can run at speeds up to 30 mph (50 km/h). The emu also is a good swimmer. Emus inhabit shrubland and forests, feeding mainly on plant matter but also eating many insects.

The closely related cassowaries (genus *Casuarius*) live in rain forests of northern Australia and New Guinea. They, too, are big flightless birds with hairlike feathers. Atop their head is a large bony "helmet" called a casque.

Enceladus (*solar system*)   One of Saturn's 18 moons, Enceladus enjoys large regions apparently unmarked by meteor bombardment. Tidal forces probably heated the interior and made its surface soft enough to smooth over craters formed by meteor impacts. Enceladus, discovered by William Herschel in 1789, is 309.6 mi (498.2 km) in diameter.

endangered species (*environment*) Any species whose numbers are so few that it is in danger of becoming extinct throughout all or most of its range is considered to be endangered. Worldwide, the International Union for the Conservation of Nature compiles these species on Red Lists. In the United States, listing of endangered and threatened organisms began in 1967, with 78 species. Today more than 900 U.S. species are listed as endangered, and many others have been recommended for inclusion. The species are protected in various ways; for example, killing endangered species is prohibited and efforts are made to maintain or restore their habitats. Some species have become extinct since they were listed, and a few have recovered sufficiently to be declared out of danger. (*See also* **threatened species**.)

endemic species (*ecology*) A species whose distribution is restricted to a specific area is considered to be endemic. The area may be large—the sugar maple (*Acer saccharum*) is endemic to the eastern United States—or only a few square miles in diameter. Islands often are home to endemic species. New Zealand, for example, has many species found nowhere else in the world, including tuataras, kiwis, and wattlebirds such as the kokako (*Callaeas cinerea*).

endocrine gland (*anatomy*) Glands that have no ducts connecting them to specific body parts are endocrine glands. They work closely with the nervous system to coordinate and regulate the body's many activities. Endocrine glands manufacture secretions called hormones. These are released into the blood, which carries them to various tissues. Some endocrine glands, such as the adrenal and thyroid glands, are discrete. Others consist of clusters of cells distributed through larger organs. For example, endocrine cells in the placenta of mammals produce several hormones during pregnancy.

endoplasmic reticulum (*cells*) Extending throughout the fluid portion of the cytoplasm of most cells is the endoplasmic reticulum (ER), an elaborate network of branching tubes and flattened sacs. There are two types of ER: rough, covered with ribosomes; and smooth, without ribosomes. ER modifies proteins made in ribosomes and transports the proteins to other parts of the cell. ER also synthesizes lipids, particularly lipids found in cell membranes and subcellular membranes surrounding organelles, and breaks down drugs such as alcohol and barbiturates. The amounts of ER, and the relative proportions of smooth and rough ER, vary depending on the type of cell. For example, pancreatic cells that synthesize digestive enzymes have more rough ER, whereas brain cells that specialize in lipid metabolism have more smooth ER.

endorphin (*biochemistry*) In the 1970s, after discovering that morphine and other opium-derived drugs (opiates) work by binding to specific receptor molecules in the brain, scientists found that vertebrates produce neurotransmitters that function much like opiates. Named endorphins ("the morphine within"), these proteins relieve pain and are implicated in a variety of other behaviors, including the sense of euphoria known as "runner's high."

endosperm (*botany*) Endosperm is a tissue that surrounds the embryo in most seeds. It stores food that is digested and absorbed by the embryo either before or during germination. In some plants—particularly corn, wheat, and other grasses—numerous grains containing protein reserves are present in the endosperm; the protein is used chiefly to form new cells in the germinating embryo.

endospore (*monera*) During unfavorable environmental conditions, many species of bacteria form small structures called endospores in their cytoplasm. After a bacterium forms an endospore, the bacterium breaks apart and the endospore is released. Endospores are dormant; that is, they do not carry out life processes. They have a thick outer coat and are extremely durable, and can survive drought, temperature extremes, exposure to harmful chemicals, and other adverse conditions. When once again in a favorable environment, endospores develop into active, growing bacterial cells.

energy (*thermodynamics*) Energy in physics is defined as the ability to do work. It is measured in the same units as work—foot-pounds; or joules (newton-meters) in SI measurement; or, especially when the energy is in the form of heat, British thermal units (BTUs) or calories. Energy is broadly classed as kinetic energy or potential energy, but more specifically it is mechanical energy, energy of position, chemical energy, electricity, radioactivity, nuclear energy, and heat. Matter and energy are 2 aspects of the same entity, but when energy is locked up in matter it cannot perform work.

entomology (*sciences*) Entomology is the branch of zoology that deals with the study of insects. The branch is subdivided into often overlapping fields. Some entomologists specialize in fields such as insect taxonomy or physiology. Others study specific types of insects, such as flies (dipterology) or butterflies and moths (lepidopterology). Still others devote their efforts to applied entomology, focusing on medical, agricultural, or economic issues.

entropy (*thermodynamics*) The measure of how concentrated or diffuse energy is in a given system is called entropy—the higher the entropy, the more diffuse the energy and therefore the less usable. A system suffused with uniform heat has the highest entropy. Mathematically the change in entropy of a system is the amount of heat added to the system divided by the system's temperature. In a closed system entropy tends to increase (the second law of thermodynamics), resulting in less energy available to do work.

The concept of entropy applies far beyond thermodynamics. A crystal has lower entropy than the more disordered liquid, which has lower entropy than a gas. All ordered systems tend to increase in entropy; that is, to become less ordered. Solids crumble, liquids and gases spread. Order can be equated with information: the less information that is available, the higher the entropy, and the second law becomes "information tends to be lost during transmission."

envelope (*geometry*) A large number of intersecting straight lines can be arranged to mark off a curve. If all the lines are tangent to the curve, the curve is the envelope of that family of lines. This idea can be generalized to describe

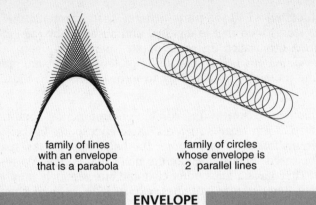

family of lines
with an envelope
that is a parabola

family of circles
whose envelope is
2 parallel lines

ENVELOPE

an envelope of a family of curves where the envelope and curves are both tangent to the same line at the same point.

environment (*ecology*)   The sum of all the conditions in which an organism lives comprises its environment. Environment includes nonliving factors such as water, temperature, pH, and light, plus all the living things with which the organism interacts. The organism has various adaptations that enable it to live in its environment. If the environment changes, the organism must either adjust to the changed conditions or move to a suitable environment. Otherwise it will not survive.

Enzymes

CLASS OF ENZYME	FUNCTION	EXAMPLES
hydrolases	catalyze hydrolysis reactions, in which large molecules are broken into smaller ones by reacting with water	lipases, which convert fats into fatty acids and glycerol; peptidases, which break peptides into amino acids
isomerases	catalyze the conversion of molecule to one of its isomers	racemases, which act on amino acids; epimerases, which act on carbohydrates
ligases	catalyze joining together of 2 molecules, coupled with cleavage of ATP	tRNA-ligases that form carbon-oxygen bonds; cyclo-ligases that form carbon-nitrogen bonds
lyases	catalyze removal of groups from molecules to form double bonds	decarboxylases, which remove carbon dioxide from molecules
oxidoreductases	catalyze oxidation-reduction reactions, in which electrons are transferred between molecules	dehydrogenases, which remove hydrogen from molecules; oxidases, which add oxygen to a substance
transferase	catalyze reactions in which a subunit, or functional group, is transferred from one molecule to another	kinases, which transfer phosphate groups; acyltransferases, which transfer acyl groups; transaminases, which transfer amino groups

enzyme (*biochemistry*) Proteins that act as **catalysts**, or promoters of biological reactions, in living organisms are called enzymes. Although grouped in 6 classes according to function, there are thousands of different enzymes, each highly specific; many catalyze only a single reaction while others are able to catalyze a few closely related reactions. Each enzyme molecule has an active site, which has a shape that roughly mirrors the shape of part of the molecules on which it acts. The substances that an enzyme causes to react are called substrates. Some enzymes break compounds into smaller products. An enzyme and compound bind at the active site, forming an unstable enzyme-substrate complex. Some enzymes cause the substrate to break in two. Other enzymes act to combine small substrate molecules into a larger molecule. In either situation, when the substrate products are released, the enzyme can be reused; it is not destroyed or permanently changed by the process.

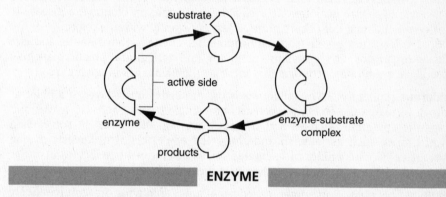

ENZYME

Eohippus (*paleontology*) The earliest known ancestor of today's horse is familiarly called Eohippus, or "dawn horse." Scientists classify the several known species, which ranged in size from a house cat to a fox, in the genus *Hyracotherium*.

Eohippus lived about 60 million years ago in forests and woodlands of Europe and North America, where it browsed on trees and shrubs. Its back legs were longer than its front; its feet were padded; and there were 4 toes on each front foot and 3 toes on each hind foot. In gradual stages, broad molar teeth suitable for grazing, a single toe surrounded by a hoof, and other anatomical changes evolved, resulting in the modern horse, *Equus*.

ephemeral (*botany*) A plant that completes its life cycle—from germination to flowering to production of seeds—in a short period of time is an ephemeral. A number of desert species are ephemerals, completing their life cycles in the few days or weeks following a spring rain. Their seeds lie dormant through the long dry season, until the next good rain.

epicenter (*geology*) The source of an **earthquake**, which is called its focus, can range from close to the surface to deep earthquakes in **subduction** zones some 400 mi (650 km) below. The point on the surface directly above the focus is called the epicenter. Typically the greatest damage caused by an earthquake is at or near the epicenter since that is the closest point to the break. Seismic waves gradually weaken as they travel away from the focus.

epicycle (*solar system*) Sometimes it appears from Earth that a planet reverses direction for a time and then returns to its original path. This occurs in a pronounced fashion with the inner planets. Small circular orbits called epicycles were imposed atop circular orbits around Earth by ancient Greek astronomers to explain this apparent backward motion of the planets. Today the explanation for the paths of Mercury and Venus is that they orbit the Sun inside Earth's orbit and so appear to reverse direction when their orbits are near the extremities as viewed from Earth; however, planets outside Earth's orbit appear to move backward at times because Earth travels faster in its orbit than they do in theirs.

epidemic (*ecology*) An outbreak of a disease with a large number of cases within a relatively small area is called an epidemic. If the outbreak extends over a wide geographical area, it is called a pandemic. In contrast, a disease is considered endemic if it occurs more or less constantly within a region.

Epidemics often result from environmental changes that benefit disease-causing organisms. Overcrowding, pollution, and agricultural practices such as monoculture are human activities that increase the likelihood of epidemics.

epidermis (*anatomy*) The epidermis is the protective outer layer of a plant or animal. In plants the epidermis consists of a single layer of cells; in leaves and stems the epidermis secretes a thin waxy covering, the cuticle. In soft-bodied invertebrates, such as jellyfish and slugs, the epidermis also consists of one layer of cells. The epidermis secretes an exoskeleton in insects and other arthropods. In vertebrates the epidermis is the outer layer of skin.

epiphyte (*botany*) Sometimes called air plants, epiphytes are comparatively small, nonwoody plants that grow on other plants, often high above the surface of the ground; they have no roots in the soil. Mosses, bromeliads, and many orchid and fern species are examples of epiphytes. Epiphytes are not parasitic; they make their own food, obtaining water and minerals from the air and from organic debris that becomes entangled in their aerial roots. Many epiphytes have adaptations to conserve water, such as thick waxy cuticles on leaves. The flowerpot plant (*Dischidia rafflesiana*) has leaves modified as water collecting and storage organs and roots that grow into these "pots" to absorb the water.

equation (*algebra*) Any statement that 2 entities are equal is an equation. In algebra equations are statements of equality for expressions formed from numbers and 1 or more variables, such as $4x^2 - 2x^2 - 5 = 0$ or $2x^2y + 3xy^2 = 4xy + 5$. Some equations are always true, such as $x + 1 = 1 + x$. Others are always false, such as $x = x + 1$. But most equations of interest, such as $x + 1 = 2$, are true for some values (called solutions) and false for others.

equation (*chemistry*) A chemical equation is a shorthand description of ingredients, steps, and results of a chemical reaction. Chemical formulas show ingredients to be combined, or that result from reactions, connected with plus signs (+). A horizontal arrow, meaning "yields," represents the reaction. Arrows pointing up or down indicate gases escaping ($\uparrow$) or solids leaving solution

($\downarrow$). Thus Fe + S → FeS means that mixing iron (Fe) with sulfur (S) (and heating, sometimes shown with a delta, $\triangle$) yields ferrous sulfide (FeS).

Equations are balanced when numbers of atoms involved in the reaction are indicated. For example, heating mercuric oxide (HgO) produces mercury and oxygen, but oxygen gas consists of molecules of 2 atoms. The balanced equation shows 2HgO → 2Hg + O_2 ↑ (2 molecules of mercuric oxide yield 2 atoms of hydrogen and 1 molecule of oxygen gas).

equator (*geography*)   The largest circle (parallel) of latitude, situated at 0°, is called the equator. If Earth were a perfect sphere the equator would be exactly halfway between the north and south poles as measured along the surface. The diameter of the equator, 7927 mi (12,756 km) through the center of Earth, is used as Earth's diameter.

equilateral (*geometry*)   Any polygon whose sides are all equal in length is equilateral. Most often encountered is the equilateral triangle, which also has all its angles equal—all are 60° or $\pi/3$ radians—and also is called equiangular.

equilibrium (*mechanics*)   A stationary body treated as a particle is in equilibrium when all forces on the body, taking direction of the forces into account, add up to 0. For a system of particles, such as a rigid body, the sum of the torques about any line must also be 0.

equinox (*solar system*)   As its name suggests, an equinox occurs when night is equal, specifically when the time of darkness is equal to the time of sunlight. Because Earth is tilted with respect to the ecliptic and maintains the same angle throughout its orbit, there are two times each year when the tilt is neither toward nor away from the Sun. At these times, currently March 21 and September 23, an equinox occurs.

equivalence principle (*physics*)   The equivalence principle, which recognizes that gravitational force and inertia have indistinguishable effects, is the foundation of Einstein's general relativity. There is no difference between the force caused by uniform acceleration and the acceleration due to gravity. People experience this in an accelerating elevator when they feel their body grow heavier if the elevator is rising or lighter if falling. Weightlessness in a satellite, often called free fall, occurs because the acceleration caused by a curved orbit exactly balances the gravitational force needed to maintain the orbit.

Eratosthenes of Cyrene (*astronomer/mathematician*)   [Hellenic: c276–c195 BC] The best-known accomplishment of Eratosthenes is his measurement of the size of Earth. He used the difference in the 90° angle of the Sun at its highest point in Cyrene (now Aswan) on a midsummer day and its lower angle in Alexandria, Egypt, on the same day. Knowing the distance between the 2 places, he calculated that the circumference of Earth is 250,000 stades. The most commonly accepted value for the stade would make this determination accurate within about 50 mi (80 km).

In mathematics Eratosthenes is known for the sieve of Eratosthenes, a procedure for finding all the prime numbers. From a list of all the counting numbers greater than 1, strike out all the multiples of each number except the first; this leaves only primes.

erbium (*elements*) Erbium is a rare earth, atomic number 68, recognized as an element in 1842 by Carl Gustav Mosander [Swedish: 1797–1858]. It was named for Ytterby, a village in Sweden where erbium ore was combined with lutetium, terbium, ytterbium, and yttrium. The pure metal was not isolated until 1934. Erbium has limited use in alloys and in production of pink glass. Chemical symbol: Er.

ergot (*fungi*) Sac fungi of the genus *Claviceps*, particularly *C. purpurea*, cause a disease of rye, wheat, and other grasses called ergot. The fungi attack the flowers of the grasses, particularly the ovaries, transforming them into large, hard purplish bodies called sclerotia, or ergots. A sclerotium contains toxins that can be deadly to animals, including humans, but that in small quantities have medicinal value.

ermine (*animals*) A member of the weasel family, the ermine or stoat (*Mustela erminea*) lives in woods and brush areas in North America, Europe, and Asia. It has a long, slim body and is about 1 ft (30 cm) in length, including its tail. The ermine is quick, agile, and ferocious, preying mainly on rodents. In warmer climates the ermine's coat may remain brown with a white belly year-round. In cold climates it is brown during the summer and white in winter—except for the black tip of its tail.

erosion (*geology*) Erosion combines breaking up rock or soil by natural means, called weathering, with removal of material to some other location. Removal may take place simultaneously with weathering or at some later time. Causes of erosion are called agents; agents include wind or wind carrying sand; moving water, including streams, waves, currents, and glaciers; and gravity, including landslides, mudflows, and slow movement of soil down a slope, called creep.

erratic (*geology*) A large rock or boulder that is not related to the other rocks found in a given vicinity is termed an erratic. Usually an erratic is a glacial erratic, carried in the past by a river of ice from a location nearer the poles or higher on a mountain to its current position.

error (*measurement*) The error of a measurement is half the unit used in measuring, since it is assumed that the observer can determine whether a measurement is closer to one unit or another. Using a ruler marked in sixteenths of an inch, the error is normally $1/32$ in., but the person measuring may choose the measure to the nearest $1/8$ in., using the marks at the sixteenths as a guide or, when precision is not needed, to the nearest $1/4$ in. or $1/2$ in.

eruption (*geology*) A volcanic eruption is the flow of large amounts of gas, rock fragments, or lava from a volcano. Eruptions may include a devastating burst of very hot gases and rocks that moves down the side of the mountain (a pyroclastic flow); only gas (a phreatic eruption); both gas and ash rising high above the crater; or flowing lava rivers that ignite forests or cover buildings for miles around. A heavy ash fall from an eruption can bury whole towns. Sometimes the mountain seems to explode, but in most cases a side of the mountain bursts from pressure within or the mountain collapses because the magma reservoir has emptied.

escape velocity (*astronomy*) The speed needed for an object to be propelled from the surface of a planet or star and not fall back is called its escape velocity. The escape velocity for Earth, for example, is 6.96 mi (11.2 km) per second. Since gravitational force between 2 objects is determined by the product of the masses divided by the square of the distances between their centers, escape velocity depends on the mass of the planet or star and also on its radius—it is assumed that the mass of the object escaping is insignificant compared with the mass of the planet or star. On some very large stars, despite enormous mass, the distance of the surface from the center is so great that the escape velocity is 0. A black hole combines a radius of 0 with an infinite escape velocity.

Escherichia (*monera*) Members of the genus *Escherichia* are gram-negative rod-shaped bacteria widely distributed in nature. The most important species is *E. coli*, of which there are many strains. Some strains are normal, harmless inhabitants of the intestines of humans and other warm-blooded vertebrates. Toxic strains can cause serious infections of the gastrointestinal and urinary tracts. *E. coli* has been much studied and is used widely in genetics experiments. By inserting human genes into *E. coli*, scientists have turned the bacterium into a factory for the production of insulin and other valuable compounds.

esker (*geology*) Pressure from above and friction from below melts water at the base of a glacier. This water forms rivers underneath the glacier, rivers that carry rocks and rock dust. When the glacier melts, the path of the river is marked with long, winding piles of drift called eskers. Where the subglacial rivers exit at the front of the glacier, they carry drift into long, flat deposits called outwash plains.

ester (*compounds*) An ester is an organic compound produced by action of an acid on an alcohol. For example, the ester ethyl acetate (nail polish remover) begins as ethyl alcohol and acetic acid. Esters include chemicals with familiar odors or flavors (wintergreen is methyl salicylate), common solvents, and fats. Fats and oils used in cooking are generally combinations of several chemical fats, but each can be formed with the alcohol glycerol and a fatty acid, such as stearic acid.

estimation (*arithmetic*) There are several kinds of estimation in arithmetic, including guessing the number of members in a large set. A different type of estimation begins with exact numbers that are large or that have many decimal places. The answer to an operation with such numbers can be estimated by replacing the numbers with approximations. For example, 3593×236 can be estimated as $3600 \times 200 = 720{,}000$; or $1.3498 + 4.5805 + 2.9836$ can be estimated as $1.3 + 4.6 + 3.0 = 8.9$.

estivation (*physiology*) During hot, dry seasons, when food and water are scarce, some animals enter a state of dormancy called estivation. The African lungfish (genus *Protopterus*) burrows into mud and secretes a coat of mucus. As the mixture of mud and mucus hardens, the fish forms a breathing hole to the surface, enabling it to obtain the oxygen it needs for minimal life functions. When rains return, the mud softens and the fish emerges to live in the water. Other animals that estivate include certain species of snails, frogs, and turtles.

estrous cycle (*physiology*)   Unless pregnant, the adult females of many mammals pass through a reproductive cycle that usually lasts 5 to 60 days, depending on the species. In the first phase of the cycle—the period of heat, or estrus—1 or more eggs are released from an ovary and the female is eager to copulate. She will not accept a male at other times. Hormones produced by the pituitary and other glands control the estrous cycle.

estuary (*oceanography*)   An estuary is that part of a river that meets the ocean and experiences tides. The Hudson River in New York, for example, is an estuary for almost 200 mi (320 km) from the mouth, although most rivers have much shorter estuarine sections. The tides carry salt water from the ocean into the freshwater of the river. The mixture of salt and freshwater, called brackish water, is a nutrient-rich region where many species breed.

ether (*waves*)   When James Clerk Maxwell developed the laws of electromagnetic waves, he assumed that waves travel though some otherwise undetectable medium, which he called ether. The Michelson-Morley experiment cast doubt on the existence of ether and the special relativity theory of Einstein showed that it is not required. Electromagnetic waves create their own field as they move through empty space.

ethology (*sciences*)   Ethology is the study of how animals behave in their natural habitats. Ethologists are particularly interested in the origins and evolution of fixed-action patterns of behavior, which they believe to be as uniquely characteristic of a species as the species' anatomy. Fixed-action patterns are those that are inherited—that is, genetically based—rather than learned; they are performed automatically the first time an animal encounters the appropriate stimuli. Courtship rituals, territorial defense, and grooming behaviors are examples.

eucalyptus (*plants*)   Almost all of the 500-plus species that constitute the genus *Eucalyptus* are native to warm, dry areas of Australia. Some, often called gums, are among the world's tallest trees, reaching heights of more than 300 ft (91 m). Others, commonly called mallees and marlocks, may reach only 6 ft (1.8 m). Eucalyptuses are highly aromatic due to a rich supply of volatile oils. Characteristics of the bark and leaves vary with the species and with the age of a plant. The unique flower bud has a lid formed of fused petals, sepals, or bracts; the lid falls off as the bud opens, exposing the reproductive organs. The fruit is a capsule containing numerous tiny seeds.

Euclid (*mathematician*)   [Greek: c300 BC] Euclid organized what was known about mathematics in his time into a complete axiomatic system in a work called *Elements*. His selection of axioms and postulates, although criticized for a few unstated assumptions, shows great sophistication. Many theorems in *Elements* are thought to be original with Euclid, especially those in number theory.

Euclidean space (*geometry*)   A Euclidean space of n dimensions is a set of points, each point represented by a coordinate of n components, for which distance has its ordinary meaning. Distance between 2 points is positive unless the points are identical; the distance from point a to b is the same as that from b to a; the distance between them does not change if the points are moved by translation; and the Pythagorean theorem holds for 3 points that are the vertices of a right triangle. The ordinary spaces of 2 and 3 dimensions used in analytic geometry or in nonrelativistic physics are Euclidean.

Eudoxus of Cnidus (*mathematician/astronomer*)   [Greek: c400–c347 BC] Eudoxus was a student of Plato [Greek: c427–c347 BC] and may have been in charge of the Academy during Plato's absence. Unlike Plato, Eudoxus made many major contributions to geometry, including many proofs that we know from Euclid's *Elements*. He resolved the difficult problem of reconciling incommensurable quantities, such as the diagonal and side of a square, creating a logically sound treatment of what we would today call irrational numbers. In astronomy he was the first to use circles within circles to explain observed planetary motions and to map the sky using degrees of latitude an longitude.

Euglena (*protists*)   One-celled species of the genus *Euglena* have an oval body with a rounded anterior and a pointed posterior. A long flagellum is attached to a gulletlike pocket at the anterior end. Like other members of Phylum Euglenophyta, *Euglena* typically contain chlorophyll and carry out photosynthesis, but they can also absorb food through their cell membrane. A light-sensitive eyespot provides information about light intensity and helps the organisms orient themselves. *Euglena* are common in ponds and other freshwater habitats.

eukaryote (*biology*)   Except for archaea and bacteria and blue-green algae of Kingdom Monera, organisms are eukaryotes, made up of eukaryotic cells in which the genetic material is located in a defined nucleus. The cells also contain a variety of other membrane-bound organelles (mitochondria, vacuoles, plastids, etc.). The cells usually divide by either mitosis or meiosis; in prokaryotes, cells divide by fission.

Euler, Leonhard (*mathematician*)   [Swiss: 1707–83] Euler created much of the calculus that is taught in colleges today, including many specific sequences and series, advanced trigonometry, and much of standard notation, such as $f(x)$ for a function. More than a dozen well-known results bear his name and e is called Euler's number. In addition to calculus Euler contributed to all branches of mathematics, notably to number theory and geometry, averaging about 800 pages of new mathematics each year from the time he was 16 until the day of his death.

Eurasian Plate (*geology*)   The largest of the continents, Asia, which combines with Europe to form Eurasia, covers most of the Eurasian Plate. This giant plate, second in size to the Pacific Plate, extends from the eastern side of the Japanese islands to a western boundary that passes through Iceland and the middle of the North Atlantic. There are, however, possible rifts in the Eurasian Plate, one along the northern boundary of China and another through the western edge of Siberia. All edges but the northern boundary of the plate are tectonically active, with many volcanoes and frequent earthquakes.

Europa (*solar system*)   Europa, which orbits Jupiter inside Ganymede, is about the size of Earth's Moon—1945 mi (3130 km) in diameter—and has a surface of water ice marked by networks of cracks. It is the most intriguing of

Jupiter's 16 moons because its ice-covered liquid water oceans are among the few places in which life might exist in the solar system.

europium (*elements*) Europium, atomic number 63, is a rare earth that was analyzed by spectroscopy in 1896 by Eugène-Anatole Demarçay [French: 1852–1904], who obtained a relatively pure sample in 1901. The metal is named for Europe and is represented by the symbol Eu.

eustachian tube (*anatomy*) Named for Bartolomeo Eustachio [Italian: 1520–74], the eustachian tube is a canal lined with mucous membrane that connects the middle ear to the throat. It equalizes air pressure on both sides of the eardrum.

eutrophication (*environment*) When excessive amounts of nutrients enter a lake or other body of water, growth of algae and plants explodes. This reduces the water's oxygen supply. As algae and plants die and accumulate in a thick layer on the bottom, decomposers use additional oxygen. Fish and other aquatic animals, unable to obtain the oxygen they need, die of suffocation. The process, called eutrophication, often results from sewage effluents and runoff from fertilized agricultural land; these dump organic and mineral nutrients, particularly nitrogen and phosphorus, into the water.

evaporation (*physics*) Molecules of a liquid can move in any direction, but at the surface of a liquid and a gas or a vacuum, the force attracting molecules of the liquid to each other is great enough to stop most of them and most fall back on the surface. However, a few move fast enough to escape and pass into the gas or vacuum, a process called evaporation. Since only faster moving molecules leave the liquid, they persist as a gas. Evaporation increases with temperature, since the average speed of all molecules is greater at higher temperatures.

Eve, African (*paleoanthropology*) DNA research suggests that every human alive today is descended from one woman who lived in Africa about 200,000 years ago, the "African Eve." Studies in modern humans of a form of DNA transmitted only from mother to child provided the original evidence for this theory.

Many anthropologists believe on the basis of other evidence that all humans today are descendants of a group that evolved in Africa some 300,000 to 100,000 years ago. In the "Out of Africa" hypothesis, this group then spread throughout the world, displacing all other hominids except the great apes. Alternative theories propose that modern humans evolved several times from an earlier population that was widespread in the Old World; or that modern humans developed first in Asia.

evening star (*solar system*) Planets that orbit the Sun inside Earth's orbit must always be in about the same direction from Earth as the Sun. Mercury and Venus are the two planets closer to the Sun than Earth, so they can be seen only when the Sun is a little below the horizon. After sunset, they appear as "evening stars." (We see planets at night as bright spots of light; they usually twinkle less, but otherwise appear as stars.) Venus is so much brighter than Mercury or than any real star, that it is known as *the* evening star. An evening star sets soon after sundown.

event (*probability*)   Any result of an experiment or observation treated in probability theory is an event, whether it is heads for 1 coin toss or a string of 9 tails in 100 tosses or something less obvious, such as a temperature above 90°. Events that cannot be simplified further are called simple; those that can be decomposed into simple events are compound. For example, heads and a temperature of 83° are simple events, but a string of 9 tails and temperature above 90° are both compound.

evergreen (*botany*)   Woody plants that retain their leaves throughout the year are called evergreens. Leaves formed during one growing season remain on the branches at least until new leaves form in the following growing season. Most conifers (pines, etc.) are evergreens; their leaves (needles) are compact and covered by a thick cuticle; therefore, unlike deciduous plants, evergreens lose very little water by transpiration. Most broad-leaved flowering trees of the moist tropics are evergreens, as are a few such plants of temperate regions, such as holly and live oak.

evolution (*evolution*)   A vast body of scientific evidence indicates that the millions of different kinds of organisms on Earth today did not come into existence in their present form but, rather, descended from earlier forms in a process of genetic change called evolution. Evidence of evolution comes from the fossil record and the study of similarities and differences among organisms, including comparative morphology, embryology, physiology, and biochemistry. Further evidence is provided by laboratory experiments and direct observations of changes that have occurred in populations in recent history, such as the emergence of DDT-resistant flies.

The foundation for our understanding of evolution was laid by Darwin. He recognized that there are inheritable variations among members of a species. Because most species produce more offspring than the environment can support, there is competition for available resources among the offspring of each generation. Those offspring with favorable variations are more likely to survive and reproduce, passing on the favorable variations to the next generation. Over time, these variations tend to become more and more common within the population. Darwin called this process natural selection.

excretion (*physiology*)   Metabolism produces various waste products that must be removed from an organism, a process called excretion. In animals the main wastes are water, carbon dioxide, nitrogen compounds, and salts. Wastes may be removed by diffusion; in this manner carbon dioxide is excreted from fish through gills and from mammals through lungs. Or wastes may be removed by specialized organs; for example, water, nitrogen compounds, and salts are removed from fish and mammal blood via kidneys and the rest of the excretory system.

excretory system (*anatomy*)   The function of an animal's excretory system is to remove, or excrete, wastes from the body. Other organs also have excretory functions; in humans, for example, lungs excrete carbon dioxide and sweat glands excrete salts. A variety of excretory systems are found among invertebrates. In earthworms almost every body segment has a pair of coiled nephridia that excrete wastes. Insects and spiders have threadlike Malpighian tubules that empty wastes into the intestine. The major organs of the vertebrate excretory system are 2 kidneys.

existence theorem (*logic*)   A proof that some entity must exist that does not describe how to locate the entity is an existence theorem. For example, one can prove that if a polynomial is positive for some value of the variable and negative for another, there exists a value of the variable for which the polynomial equals 0. This statement, however, does not locate the value of the variable; a different proof, in fact, shows that there is no general algebraic method for locating such a value for all polynomials.

Some mathematicians have rejected existence theorems as worthless, calling for restricting mathematics to proofs in which all entities can be constructed. However, most mathematicians accept existence theorems while preferring theorems that provide specific methods.

exobiology (*sciences*)   Earth is the only entity in the universe known to harbor life, but people have long been fascinated by the possibility of life on other worlds. The study of extraterrestrial biology—that is, life elsewhere in the universe—has 2 main focuses at present. One is the study of how life originated and evolved on Earth, for this may provide clues to how life could have developed elsewhere. The second is on detecting life and environments suitable for life using such tools as spacecraft and radio astronomy.

exocrine gland (*anatomy*)   Unlike endocrine glands, which are ductless, exocrine glands have ducts (tubes) to carry secretions directly to target areas. Examples include the sweat glands and mammary glands, with ducts to the surface of the skin, and the salivary glands, with ducts into the mouth. Some glands, including the pancreas, are considered to be endo-exocrine glands because they have both endocrine and exocrine components.

exon and intron (*genetics*)   A gene consists of a long sequence of base units. Some segments of this sequence, called exons, are transcribed to messenger RNA (mRNA) and translated into a protein. Many genes also contain intervening segments called introns that do not code for protein synthesis. The introns are transcribed to the mRNA but are removed before the mRNA leaves the nucleus of the cell. The exons are then spliced together without any rearrangement of their order.

exoskeleton (*anatomy*)   Several types of invertebrates have external skeletons, or exoskeletons, that provide protection and support. In insects and other arthropods, the exoskeleton is outside the skin. Starfish and other echinoderms have an exoskeleton embedded in the skin. Mollusk exoskeletons are called shells.

Certain vertebrates have an exoskeleton in addition to an internal skeleton. An armadillo, for example, has a shell composed of bony plates and small scales that develop from the dermis layer of the skin.

exosphere (*atmospherics*)   The exosphere is the atmosphere from about 300 mi (500 km) up to the region several thousand miles above Earth. It is dominated by high-energy particles and magnetic fields. Widely spaced molecules of light gases such as hydrogen leak into space from the exosphere. Since the exosphere is above the region where many artificial satellites, including space stations, orbit Earth, people usually think of it as "space." The very thin atmosphere would be considered a vacuum on Earth.

exotic species (*ecology*)   Any nonindigenous species living in a habitat or region where it did not previously exist is referred to as an exotic, or alien. Such species typically have no natural predators in their new homes, and as a result multiply rapidly. They compete against native species for food and other resources, and usually win, leading to a loss of species and altered environments. For example, the brown tree snake (*Boiga irregularis*) of New Guinea found its way to Guam, probably as a stowaway on an airplane. It has already exterminated 8 of Guam's 11 species of native birds. Most often, humans are responsible for introducing aliens. Some dispersals are inadvertent—zebra mussels (*Dreissena polymorpha*) accidentally carried from Europe to North America's Great Lakes in ships' bilge water. But in many cases, introductions have been deliberate, such as gypsy moths (*Lymantria dispar*) introduced to the United States by a French biologist who hoped to establish a silk-producing industry.

expansion of universe (*cosmology*)   In the 1920s Edwin Hubble measured the motions of galaxies for the first time and discovered that the universe is expanding, with everything moving away from (recessing from) everything else. For distant objects, such as galaxies and quasars, expansion allows use of the Doppler effect to measure distance, so astronomers commonly use the red shift of light from faraway objects as the measure of their distance. The present evidence is that expansion will continue and perhaps accelerate, but many theorists believe that expansion will gradually slow toward a rate of zero. A few theorists support the possibility that the expansion will halt and then reverse, leading to a "big crunch."

expectation, mathematical (*probability*)   When an outcome can be assigned a numerical value, the mathematical expectation is the sum of the products of each outcome with its probability. For example, in a board game the number of moves forward may be controlled by an ordinary die, which has a probability of 1/6 for each face and from 1 to 6 dots on each face. The mathematical expectation is the sum of each number from 1 through 6 multiplied by 1/6, or 21/6 (3.5). This corresponds to the average (mean) length of a move for each toss of the die.

experiment (*philosophy of science*)   An experiment may be the physical test of a theoretical hypothesis, such as dropping 2 bodies with different masses to see if they fall at the same rate. But more often experiments are conducted to learn something about which there is no specific knowledge, as in deleting a gene from an organism to determine the effect on the organism. Even in branches of science where experiments are rare, such as cosmology, measurements of specific quantities, such as the wavelength of cosmic background radiation, are considered experiments.

exponent (*algebra*)   An exponent is a superscript written to the right of a real number, or algebraic expression representing a real number, and used as a particular function. For example, the exponent 2 in x^2 means the squared function found by multiplying x by itself. Whenever the exponent is a natural number 2 or greater, it indicates repeated multiplication. The base is used as a factor as many times as the number named by the exponent. Hence 2^5 means $2 \times 2 \times 2 \times 2 \times 2 = 32$.

Simple rules extend exponents to all the integers: $x^1 = x$, $x^0 = 1$ (unless $x = 0$, which is undefined), and $x^{-n} = 1/x^n$ (for example, $2^{-5} = 1/32$). Rational exponents have a different meaning but one compatible with previously stated rules. First of all, $x^{1/n}$ is the nth root of the base x (for example, $32^{1/5} = 2$). Combining the definition of $x^{1/n}$ with a property of exponents, $(x^n)^m = x^{nm}$, establishes the meaning of any rational-number exponent, $x^{(n/m)} = (x^{1/m})^n = (x^n)^{1/m}$. For example, $32^{2/5} = (32^{1/5})^2 = 2^2 = 4$. Real-number exponents that are irrational numbers can be found as the limit of a sequence of natural exponents. Exponents and bases may also be complex *numbers.*

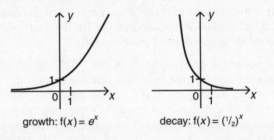

growth: $f(x) = e^x$ decay: $f(x) = (1/2)^x$

EXPONENTIAL FUNCTION

exponential function (*mathematics*) The relationship between values of a variable and the numbers formed by raising some positive number to the power of those values is called an exponential function. In function notation, $f(x) = a^x$, where a is a positive number. When a is greater than 1, the function is a growth function, which doubles in each fixed amount of time. If a is less than 1, the function describes decay, in which there is half as much after a fixed amount of time.

exterior and interior angle (*geometry*) For any polygon, including a triangle, the angles that enclose other sides of the polygon are the interior angles; these are the angles one normally means. An exterior angle is not part of the original polygon, but is produced by extending one side past a vertex. An exterior angle is the angle between that ray and another side that meets that same vertex.

extracellular fluid (*cells*) The cells of a multicellular organism are bathed by fluid in much the same way that pond water bathes 1-celled organisms. This extracellular ("outside the cell") fluid contains nutrients, oxygen, and other materials needed by the cells. The extracellular fluid is continually renewed—in mammals, for example, by blood plasma that diffuses out of blood capillaries at the same time that a nearly equal amount of extracellular fluid moves into the capillaries. Similarly, an exchange of materials occurs across cell membranes between the extracellular fluid and the content of the cells.

extrachromosomal gene (*genetics*)   Although most genes are located on chromosomes, additional genes may be present in a cell's cytoplasm. In prokaryotes extrachromosomal genes are carried on plasmids. In eukaryotes such genes are found in plant chloroplasts and animal mitochondria. An organism receives half its chromosomal genes from its mother and half from its father. But the organism generally receives the bulk of its extrachromosomal genes from its mother because the mother's egg contributes much more cytoplasm to the zygote (fertilized egg) than does the father's sperm. Certain characteristics are determined by this so-called maternal inheritance. For example, the makeup of the mother's mitochondrial DNA determines whether the shell of a wandering snail (*Limnaea pereger*) coils to the left or the right.

extrasolar planet (*astronomy*)   A planet outside the solar system is extrasolar. In 1983 a satellite reported dust clouds around several stars that seemed to relate to planet formation. In 1994 a wobble in a pulsar was recognized as being caused by planets in orbit about it. In 1995 similar evidence revealed a planet about the ordinary star 51 Pegasi. Then in 1999 the star Upsilon Andromedae was discovered to have 3 planets in orbit, the first known planetary system around an ordinary star. Nearly all known extrasolar planets are as large as Saturn or several times larger. Small planets are still beyond detection.

eye (*anatomy*)   An eye is an animal sense organ that contains photoreceptors, light-sensitive cells that have pigments that absorb light energy. Associated with the photoreceptors are nerve cells. Eyes range in complexity from simple organs in the skin of certain worms to vertebrate eyes that possess millions of photoreceptors.

A transparent cornea forms the front, exposed portion of a vertebrate eye. Behind is a doughnut-shaped iris composed of muscle fibers arranged around a central opening, the pupil; contraction and relaxation of the muscle fibers controls the amount of light passing through the pupil. The light is focused by a transparent lens onto the back inner layer, the retina. Nerve impulses originating in photoreceptors of the retina are passed to the optic nerve and then to the visual sensory areas of the cerebral cortex of the brain.

eyespot (*anatomy*)   The term eyespot describes several different structures. *Euglena* and some other chlorophyll-containing protists have a red eyespot near a flask-shaped cavity, or reservoir; working together, the eyespot and the reservoir enable the protist to orient toward light for food-making. Jellyfish and many other invertebrates have simple light-sensitive organs called eyespots (or eyes) that sense light intensity and stimulate the nervous system.

The wings of some insects, especially butterflies and moths, have eyelike patterns called eyespots. Certain fish have similar eyespots near the base of the tail. Such markings are probably meant to fool predators.

Fabricius, Hieronymus (*anatomist*) [Italian: 1537–1619] A founder of the science of embryology and a renowned teacher, Fabricius made major contributions in the fields of development and comparative anatomy of animal fetuses. He was the first person to describe the placenta in detail; he also gave a detailed description of the semilunar valves in veins. William Harvey was among the students who traveled to Italy to study with him.

fact (*arithmetic*) An operation on 2 whole numbers and the correct answer form a fact. If 2 of the 3 numbers involved are single digit, such as 13 – 7 = 6, the fact is called basic. There are 100 each of the basic addition, subtraction, and multiplication facts, but only 90 basic division facts because division by 0 must be eliminated.

factor (*numbers*) As a noun, a factor is one of the numbers used to form a particular product; as a verb, to factor is to locate all factors of the number. For example, the factors of 12 are 1, 2, 3, 4, 6, and 12, found by factoring 12. In most cases only prime number factors are found and they usually are shown as an indicated product, so the [prime] factorization of 12 is $2 \times 2 \times 3$, or $2^2 \cdot 3$. Finding prime factors of small numbers is easy using division by successive primes, but factoring numbers of 100 or more digits is so difficult that it is used as the basis of commercial codes.

factorial (*algebra*) A factorial of a number is the product of consecutive natural numbers starting with 1 and continuing through the number. For example, the factorial of 5, called 5 factorial and written 5!, is $1 \times 2 \times 3 \times 4 \times 5$, or 120. By definition 0! = 1 and 1! = 1. Numbers that are factorials begin 1, 2, 6, 24, 120, 720, 5040, . . . and rise rapidly, exceeding 1,000,000 after 10!, which is 3,628,800. Factorials often occur in counting theory and probability; for example, the number of permutations of n objects is $n!$.

fairy ring (*fungi*) The vegetative body, or mycelium, of mushrooms tends to grow outward in all directions, dying out at the center as the food supply is depleted. In grassy places, circular fairy rings may result—so named because at one time many people believed that fairies or other supernatural phenomena produced them. The circles, marked by successive crops of fruiting bodies (the "mushrooms") enlarge over the years, until some environmental barrier breaks them up or changes their shape.

falcon (*animals*) Birds of prey that hunt during the day, falcons have superb eyesight, a strongly hooked bill, long, pointed wings, and feet armed with curved claws called talons. They are swift fliers—during a dive in pursuit of prey, a peregrine falcon (*Falco peregrinus*) reaches speeds well over 100 mph (160km/hr). The falcon family, Falconidae, also includes the caracaras of Central and South America.

fallacy (*logic*)   An apparently logical argument that contains a hidden error is a logical or mathematical fallacy. The conclusion of such an argument is also called a fallacy. A common fallacy in logic results from incorrectly thinking that if p implies q is true, then q implies p must also be true. In algebra hidden division by 0 can lead to a fallacious conclusion.

Faraday, Michael (*chemist/physicist*)   [English: 1791–1867] Although Faraday's work with electricity is better known, he began his scientific career as a chemist, synthesizing chlorocarbons and, in 1825, discovering benzene, the chemical that led to understanding all the aromatics. His studies of electricity began in 1830. Faraday then developed electric motors and generators and found the basic laws of electrolysis and induction. Faraday is renowned as an experimentalist, but his concept of a field based on lines of force has become one of the main underpinnings of theoretical physics.

fat (*biochemistry*)   Animal bodies store some excess food as fats, or triglycerides; these contain much more chemical energy than other nutrients. When oxidized, fat yields an average of about 9.3 kilocalories per gram (kcal/g), compared with 3.8 kcal/g of carbohydrate and 3.1 kcal/g of protein.

A fat molecule is composed of glycerol and fatty acids. Saturated fatty acids, such as palmitic acid, have no double bonds between carbon atoms; each carbon atom is bonded to 4 other atoms. Unsaturated fatty acids, such as oleic acid, have 2 carbon atoms joined by a double bond, which means these carbon atoms have the potential to form additional bonds with other atoms. Polyunsaturated fatty acids, such as linoleic acid, have 2 or more double bonds.

Triglycerides are solid at room temperature. Fats that are liquid at room temperature are called oils. Many plants store reserve food as oils in seeds (castor bean, peanut) and fruits (avocado, olive).

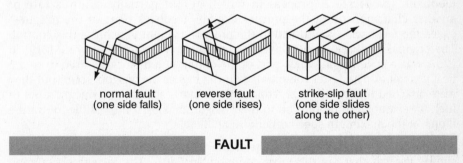

| normal fault | reverse fault | strike-slip fault |
| (one side falls) | (one side rises) | (one side slides along the other) |

FAULT

fault (*geology*)   Any rock formation caused by one region moving with respect to another is a fault. The movement breaks bonds that held rock on one side of the fault to the other, producing an earthquake. A fault once formed becomes the center for additional earthquakes, since rock along an existing fault is more likely to slip in response to stress.

Faults are classed by how the 2 sides of the rock formation move with respect to Earth itself. If 1 side falls because the rock is being pulled apart, the fault is normal, but if a side rises as rock is pushed toward the fault, the fault is a reverse, or thrust, fault. Large-scale or repeated reverse faults produce a characteristic type of mountain range called fault-block mountains. When the rock movement is largely horizontal, the fault is called a strike-slip or lateral slip fault.

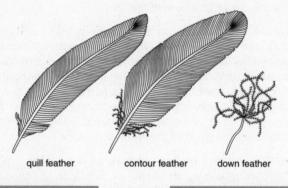

quill feather   contour feather   down feather

FEATHER

feather (*anatomy*)   Feathers grow from a bird's skin and are composed mainly of proteins called keratins. Common kinds include contour, quill, and down. Contour feathers make up most of the plumage; they cover the body of the bird, creating its streamlined shape and providing coloration. Quills are large, stiff contour feathers on the wings and tail that are important in flight. Down feathers are small, soft, and fluffy; in adult birds their primary function is insulation. Birds are the only living animals with feathers; however, some dinosaurs, including species of coelurosaurs, are known to have had feathers.

feedback (*physics*)   A process in which output partially controls rate or another characteristic of the process exhibits feedback. Part of the output—called the feedback—is returned to the process as input to achieve this control. The original feedback device, invented by James Watt [Scottish: 1736–1819] in 1787, was a governor in which part of the output of a motor was used to spin 2 heavy weights. The faster the motor ran, the higher the weights spun; but they were attached to a device that reduced the motor's speed (by reducing input of fuel, for example) in proportion to their height, making it impossible to run the motor at the maximum speed otherwise attainable.

feldspar (*rocks and minerals*)   The name feldspar refers to a large group of similar minerals that are the most common on Earth. The feldspars' common chemistry is a combination of aluminum, silicon, and oxygen together with an alkali metal or an alkaline earth metal. The different varieties are often named for their crystal shapes, such as orthoclase feldspar (monoclinic) and plagioclase (triclinic). Feldspar is often the colored (pink, green, or blue) component of granite but it also occurs in basalt and gabbro. After weathering into small particles, feldspars form clays.

Fermat, Pierre de (*mathematician*) [French: 1601–65] Fermat was a lawyer whose thoughts on mathematics were transmitted mainly by letter. He discovered analytic geometry a year before Descartes and some of the main ideas of calculus before either Newton or Leibniz were born. His main fame today comes from his work in number theory, especially Fermat's Last Theorem: for natural numbers x, y, and z there is no natural number n greater than 2 for which $x^n + y^n = z^n$ is true. Fermat claimed a proof of this theorem, but did not reveal it. It was finally proved in 1995 by Andrew Wiles [American: 1953–].

fermentation (*physiology*) Sometimes called anaerobic respiration, fermentation is a complex multistep process that breaks down glucose and other carbohydrates, releasing energy, in the absence of oxygen. The end products of fermentation are alcohols or organic acids plus carbon dioxide. The amount of energy released is much less than in aerobic respiration, which requires oxygen.

Some 1-celled organisms are anaerobes that obtain all their energy by fermentation. In multicellular organisms, cells may use fermentation when oxygen is in short supply. For example, vertebrate muscle cells usually oxidize the sugar glucose, producing carbon dioxide, water, and energy. But during strenuous exercise, when energy needs increase dramatically, the muscle cells may convert glucose to lactic acid ($C_3H_6O_3$) plus carbon dioxide and energy. Later, when sufficient oxygen is available, the lactic acid can be broken down to carbon dioxide, water, and energy. Or the lactic acid may be carried by the blood to the liver, where it is converted back to glucose.

Fermi, Enrico (*physicist*) [Italian-American: 1901–54] Fermi was the only physicist of the 20th century whose experimental and theoretical work were equally valuable. After some important clarifications of Einstein's general theory of relativity, Fermi in the 1920s was the first to develop a mathematical treatment of how particles such as electrons—those that obey the Pauli exclusion principle, and now called fermions—interact physically. The topic was investigated independently by Paul Adrien Maurice Dirac and consequently called Fermi-Dirac statistics. Electrons in metals obey Fermi-Dirac statistics and Fermi's work is the basis of modern condensed matter physics. Fermi also studied the atomic nucleus, contributing to the theory of beta decay and showing experimentally that slow-moving neutrons can change one element or isotope into another (for which he received the 1938 Nobel Prize in physics). In 1942 Fermi was the chief designer of the first working nuclear reactor; it was based on several techniques that he developed to control nuclear fission in uranium.

fermion (*particles*) The particles that make up matter are called fermions (as opposed to bosons, which create forces). Fermions are named after Enrico Fermi, who developed the mathematics of their interactions. All obey the Pauli exclusion principle, which means that fermions occupy a definite space. Unlike bosons, 2 fermions cannot be in the same place at the same time. Fermions can be recognized as particles with an intrinsic spin of $1/2$. Subatomic particles that are fermions include such elementary particles as leptons and quarks as well as baryons, which are made from quarks.

fermium (*elements*) Fermium is element number 100, a synthetic radioactive actinide metal similar to uranium and plutonium. It was discovered by an American team led by Albert Ghiorso [American: 1915–] in 1952 in the debris from an H-bomb explosion, and classified as a military secret. The existence of element 100 was unclassified soon after physicist Enrico Fermi died and the name was given in his honor. Chemical symbol: Fm.

fern (*plants*) The more than 10,000 species of living ferns range from minute water ferns to giant tree ferns that attain heights up to 60 ft (18 m). Ferns are most abundant in tropical rain forests but also are found in many other moist habitats. Like flowering plants and conifers, ferns have true roots, stems, and leaves, with vascular tissue that conducts water and nutrients and that provides support. Unlike flowering plants and conifers, ferns do not produce seeds; rather, they reproduce by spores and have an obvious alternation of generations.
  Vegetative propagation also is common among ferns: their stems, called rhizomes, grow horizontally at or just below the soil's surface, sprouting roots and leaves at intervals. The young leaves, known as fiddleheads, are tightly coiled. Mature leaves, called fronds, vary greatly in size and shape, depending on the species; some species even have distinct sterile and fertile (spore-bearing) leaves.

ferromagnetism (*materials science*) Ferromagnetic materials—principally iron, cobalt, nickel, and various alloys—have magnetic fields surrounding each atom; these tend to line up naturally and align even more in response to an external magnetic field. Ferromagnets can become saturated with magnetism when all internal fields are aligned. In some cases, applying or producing a magnetic field with ferromagnetic materials releases energy as heat (these magnets exhibit hysteresis, failure to return to their original state when the magnetic field is removed). Ferromagnets called permanent magnets can continue to produce a magnetic field even when external magnetism is no longer applied.

fertilization (*biology*) An essential part of sexual reproduction is fertilization: the union of a sperm (male sex cell) and an egg (female sex cell) to form a single cell (zygote, or fertilized egg). Fertilization is either internal or external. In internal fertilization eggs are retained within the female's reproductive organs; they are fertilized by sperm that travel to the eggs through a watery fluid. Reptiles, birds, and mammals practice internal fertilization. In external fertilization both parents release their sex cells into water; the sperm swim through the water to the eggs. This is much less efficient than internal fertilization; thus, organisms that practice external fertilization produce many more eggs and sperm than organisms with internal fertilization. A female *Crassostrea virginica* oyster, for example, may release as many as 50,000,000 eggs at a time!

fescue (*plants*) The 360 or more species of fescues (*Festuca*) are annual and perennial grasses widespread in temperate and cool regions. Some species are cultivated for fodder or lawns; other species are weeds. They grow from 0.8 in. to 6.5 ft (2 cm to 2 m) high. The small bisexual flowers are borne in narrow clusters called panicles.

Feynman, Richard (*physicist*) [American: 1918–88] In 1947 Feynman was among several physicists whose combined work created quantum electrodynamics. One of his contributions, first published in 1948, is the Feynman diagram, a graphic way to calculate the results of particle interactions. Later Feynman developed a theory that protons and neutrons are made from smaller particles, although the quark theory of such particles is owed to Murray Gell-Mann. As a participant in the commission investigating why the space shuttle *Challenger* exploded, Feynman determined that cold weather had made certain parts brittle, leading to a disastrous fuel leak.

fiber (*biology*) In plants fibers are very elongated cells found in strengthening tissue called sclerenchyma. Flax fibers, which may be 3 ft (0.9 m) long, come from the stems of the flax plant (*Linum usitatissimum*); manila hemp fibers come from the leaf stalks of manila hemp plants (*Musa textilis*).

 In animals long, thin cells are typically called fibers—for example, nerve fibers are actually nerve cells (neurons). Also, certain molecules, such as those of collagen, occur as fibers.

Fibonacci sequence (*numbers*) The Fibonacci sequence is formed by adding the 2 preceding numbers to find the next, starting with an initial pair of 1s: hence, 1, 1, 2, 3, 5, 8, 13, 21, 34, 55, 89, 144, 233, ... Members of the sequence are Fibonacci numbers. The sequence originated in a problem proposed by the most influential mathematician of the Middle Ages, Leonardo of Pisa [Italian: c1170–c1250], also known as Fibonacci ("son of Bonaccio"): How many pairs of rabbits will be produced in a year, beginning with a single pair, if every month each pair bears a new pair that becomes productive from the second month on?

fictitious force (*physics*) An apparent force determined by something other than acceleration of a mass is called fictitious. Examples include centrifugal and centripetal forces, the Coriolis effect, and the acceleration of a falling object due to gravity. Centrifugal force, for example, is the result of an object in motion being restrained from its normal straight-line path. Centrifugal force depends on the radius of the path of the object and the speed of rotation. Location, direction, and speed determine the Coriolis effect. Gravity speeds up different masses equally as they fall to Earth.

field (*algebra*) Any algebraic structure that shares the common rules for operations of the rational numbers, real numbers, and complex numbers is a field. A field has 2 operations, such as addition and multiplication, each of which must exhibit closure for the set of elements in the field. Both operations must obey the commutative property and associative property. The set must possess an identity element for each operation and inverse elements with respect to every element in the set with 1 exception—the identity element for addition need not have an inverse element for multiplication. Finally, the 2 operations must be connected by a distributive property of multiplication over addition.

field (*physics*) Michael Faraday originated the idea of a field in physics with the concept of invisible lines of force surrounding a magnet—the magnetic field. An electric current produces a similar field, a power that affects objects and diminishes with distance from the source. A more formal definition of a field

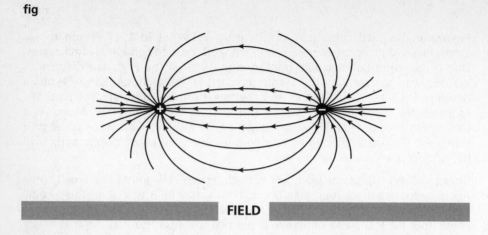

FIELD

assigns to every point in space a vector that contains an amount and a direction; for example, the gravitational field around Earth consists of the force of gravity at any point directed toward Earth's center. One interpretation of the physical world is that fields are the fundamental entities and particles or forces derive from them.

fig (*plants*) Members of the mulberry family, figs (*Ficus*) are a diverse group of subtropical and tropical vines, shrubs, and trees. Many figs begin life as epiphytes on branches of host plants. They send roots down to the ground and eventually become independent—often killing the host in the process. Figs have minute flowers that form inside a hollow structure, the syconium, at the top of the flower stem. Tiny fig wasps enter the single small opening in the syconium and pollinate the flowers. There are some 900 species of figs, each dependent on its own species of fig wasp for pollination and, therefore, for production of seeds and survival of the species. The common fig (*F. carica*) is cultivated for its "fruit"—actually, the edible syconium. Other well-known species include the banyan, bo tree (*F. religiosa*), and Indian rubber plant (*F. elastica*).

figurate number (*numbers*) Early Greek mathematicians represented natural numbers with patterns of dots. Those patterns that suggest geometric shapes are called figurate numbers. The best known figurate numbers are the square numbers (1, 4, 9, 16, . . .), for which the dots can be arranged as squares. Another well-known pattern is formed by the triangular numbers. The triangle representations have rows of dots that are successive natural numbers; for example, 10 is a triangular number since it can be shown as rows of 1, 2, 3, and 4 dots. For any natural number n the number $n(n + 1)/2$ is a triangular number.

filovirus (*viruses*) The virus family Filoviridae includes the viruses that cause 2 recently recognized diseases: Marburg and Ebola, both of which are characterized by severe fever and bleeding. The Latin *filo* means "threadlike," but while some filoviruses are filamentous, others are spherelike. The genetic material consists of linear, single-strand RNA.

filter feeder (*zoology*) Animals that feed on small organisms they strain from the water are filter feeders. Some animals are filter feeders during only part of their life cycle; for example, lamprey larvae are filter feeders, then mature into parasitic adults that feed on blood. The blue whale—the largest, heaviest animal on Earth—is a filter feeder whose diet consists mainly of tiny crustaceans called krill.

finch (*animals*) The finch family, Fringillidae, comprises more than 500 species of perching birds ranging in length from 4 to 10 in. (10 to 25 cm). The family includes buntings, cardinals, finches, grosbeaks, juncos, redpolls, siskins, towhees, and most sparrows. All have 9 flight feathers per wing and 12 tail feathers. Many are known for their attractive songs. Most are primarily seedeaters, with strong cone-shaped bills. Crossbills (*Loxia*) have pointed, crossed mandibles designed for removing seeds from pinecones.

fiord (*geology*) When a glacier enters the sea, the trough eroded by the glacier may be hundreds of feet below sea level. After the glacier melts, ocean water enters this trough and forms a long arm of the sea called a fiord. Rivers flowing through glacial troughs extend the fiords farther inland. Although the best-known fiords are in Norway, the lower Hudson River in New York is also a fiord.

fir (*plants*) Members of the pine family, firs (*Abies*) are tall evergreen conifers native to temperate and subalpine areas of the Northern Hemisphere. They have a symmetrical, pyramidlike shape, with a straight trunk and dense branches. The flat linear leaves usually are shiny dark green on the upper surface and whitish underneath. The seed-forming cones grow upright. The tallest fir probably is the noble fir (*A. nobilis*) of western North America, which can attain heights approaching 300 ft (90 m).

Douglas fir (*Pseudotsuga taxifolia*), also in the pine family, is native to western North America. It can grow more than 300 ft (90 m) tall. Its flat, blunt-tipped needles grow in a spiral around the twig. The seed-forming cones hang from branches and have 3-pointed bracts projecting from between the scales.

fire (*chemistry*) Fire, or combustion, is a chemical reaction that produces gases, heat, and incandescence. Most fires burn carbon or carbon-hydrogen compounds—such as parts of wood, wax, or gasoline—in oxygen to produce carbon dioxide and water. Fire in a solid or liquid starts when heating releases hot gases, allowing the gases to react with oxygen. The gases also lift small solid or liquid particles that burn in air. Part of the light (yellow or red) and heat comes from the burning particles. The gas reactions produce additional light (blue) and heat. Not all fire is carbon-based (powdered magnesium burns in air); fire can also occur without oxygen, in fluorine, for example.

fireball (*solar system*) Fireballs are bright meteors with an apparent magnitude greater than –4 (approximately brighter than Venus, generally the brightest light in the night sky other than the Moon). Often they appear as bright disks, not points of light. Fireballs are caused by meteoroids heating the air around them as they pass through the upper atmosphere. They can often be observed for several seconds and may leave trails that last for minutes. Most fireballs, however, burn completely and do not fall to Earth as meteorites.

firefly (*animals*) Members of the beetle family Lampyridae, fireflies—also called lightning bugs—are relatively drab-looking insects up to 1 in. (2.5 cm) long. They are famed for their **bioluminescence**, or ability to produce light. On spring and summer evenings they emit light signals to attract mates or prey. Each species has its own unique rhythm, or pattern of flashes. The larvae of some species also are light producers; they are called glowworms, a term also used for luminescent larvae of the closely related family Phengodidae.

fish (*animals*) Fish are aquatic vertebrates with fins and, generally, a body covering of overlapping scales. They have a 2-chambered heart and remove oxygen from the water by means of gills. Most fish have a streamlined torpedo shape that allows movement through water more easily than other shapes. But some fish are round, flattened from top to bottom or from side to side, elongated into long ribbons, or many other shapes. Fish thrive in every aquatic habitat, from shallow ponds and rushing rivers to ocean depths and underground streams.

The approximately 25,000 known species of fish account for about half of all known living vertebrates. They are divided into 3 classes: jawless fish, cartilaginous fish, and bony fish. The smallest species is the Philippine pygmy goby (*Pandaka pygmaea*), less than 0.5 in. (1.3 cm) long. The largest is the whale shark (*Rhincodon typus*), which may attain lengths of more than 50 ft (15 m).

fish kill (*environment*) The death of many fish in an area over a short period of time is called a fish kill. For example, in mid-1998 more than 1,000,000 fish died in a 3-mi (5-km) stretch at the north end of California's Salton Sea. Fish kills may be caused by disease-causing microorganisms, such as the **dinoflagellates** *Amyloodinium ocellatum* (in the Salton Sea case) and *Pfiesteria piscicida*. Other causes include sudden changes in water temperature or salinity, insufficient dissolved oxygen in the water, and sewage and chemical spills.

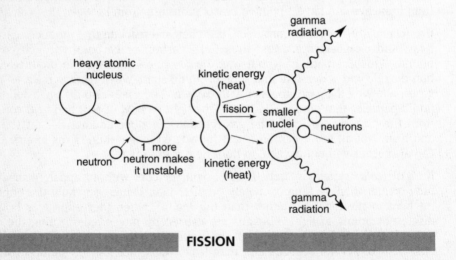

FISSION

fission (*particles*) Nuclear fission occurs when an atomic **nucleus** breaks into 2 or more pieces, releasing energy. Lise Meitner, who named the process in 1939, was the first scientist to understand fission; her work was based upon experiments earlier that year by Otto Hahn [German: 1879–1968]. Fission occurs because the strong force that holds a nucleus together acts at short distances only, but charge is effective at greater distances. If protons in the nucleus become too far apart, their positive charges overcome the nuclear force and the atom splits into 2 roughly equal-size pieces. The new atoms that form have less mass than the original atom. Some lost mass becomes energy of motion, pushing the new atoms apart, producing heat, and often ejecting fast-moving neutrons, while the rest turns into electromagnetic radiation.

fissure (*geology*)   When lava erupts from a long crack in Earth's surface instead of from a central location, the rift is called a fissure. Active fissures are common along mid-oceanic ridges , but are currently inactive on land other than in Iceland. In the past fissure eruptions of very fluid lava, called flood basalt, covered large regions of North America (Columbia Plateau), South America (Paraná of Brazil and Paraguay), and Asia (Deccan Plateau of India).

flagellate (*protists*)   Members of the protist phylum Mastigophora are commonly called flagellates because all have long, whiplike flagella that aid in locomotion. Many species are free-living organisms but some are parasites. Serious human illnesses caused by flagellates that infect the bloodstream are trypanosomiasis (African sleeping sickness), caused by members of the genus *Trypanosoma*, and kala-azar, caused by members of the genus *Leishmania*.
     Flagellated protists such as *Euglena*, which contain chlorophyll and are considered to be algae, may be classified separately, in Phylum Euglenophyta.

flagellum (*cells*)   Flagella (from a Latin word meaning "little whips") are long whiplike organelles present in a variety of organisms. In 1-celled protists such as *Euglena*, the flagellum beats back and forth to propel the organism through water. Some bacteria move by rotating flagella that extend singly or in tufts from one or both ends of the cell. In sponges, flagellated cells line the internal cavity, producing a flow of water important to food gathering.

flamingo (*animals*)   Tall birds with pinkish plumage, flamingos (*Phoenicopterus*) are among the most striking of birds. They have a large, downward-bending bill; a long flexible neck; and very long legs. Flamingos live mainly in tropical regions, around salty and brackish water, often forming colonies of thousands of individuals. To feed, they dip the head upside down underwater and rake the bill back and forth, straining small mollusks, worms, fish, and other food out of the water.

flare, solar (*solar system*)   Storms on the Sun are caused by its intense magnetic fields, not by temperature differences (which cause storms on Earth). When magnetic storms occur, they can break lines of force, propelling great clouds of ionized hydrogen and other gases millions of miles out into space. Such events are called solar flares. Large solar flares sometimes reach the vicinity of Earth, where they compress Earth's magnetic field and disrupt electrical communications all over the planet. Satellites, especially those far from Earth, are especially vulnerable to disruption by flares. Flares can last for hours, although 20 minutes is more typical.

flatfish (*animals*) Marine fish constituting Order Pleuronectiformes, including flounders, soles, halibuts, tonguefish, and turbots, have bodies strongly flattened from side to side. During the early larval stage, spent in surface waters, the eye on one side of the body—either right or left, depending on the species—moves to the opposite side of the head, so that both eyes are on the same side. Following this and other changes, a flatfish settles on the ocean floor, with the eyeless side resting on the bottom. Most markings develop on the upper side, and often match the surroundings. Most flatfish are carnivores, and often bury themselves in sand as they wait for prey. Species range in length from 2-in. (5-cm) tonguefish to the Atlantic halibut (*Hippoglossus hippoglossus*), which can reach 8 ft (2.4 m) and a weight of 700 lb (320 kg).

flatworm (*animals*) Invertebrates of Phylum Platyhelminthes ("flat worms") are the simplest animals with heads and bilateral symmetry (having right and left sides). Their soft bodies are flattened and usually elongated, with distinct front and rear ends. Acoels and planarians are free-living flatworms; flukes and tapeworms are parasitic. Almost all species are hermaphrodites and, in addition to reproducing sexually, can reproduce asexually by splitting in 2 to form 2 daughter organisms. Flatworms also have great powers of regeneration.

flax (*plants*) Native to temperate and subtropical regions, flaxes (*Linum*) are erect herbs and small shrubs with narrow leaves and brightly colored flowers. The fruit is a capsule filled with oily brown seeds. The species *L. usitatissimum*, which grows to 4 ft (1.2 m), has been cultivated since prehistoric times. Its fibers are used to make linen and its seeds are the source of linseed oil, used in paints and inks.

flea (*animals*) The more than 2000 species of fleas are wingless insects that live on the skin of birds and mammals and feed on their blood. They are tiny—generally less than 0.16 in. (0.4 cm) long—with piercing mouthparts, a tough yellow-brown exoskeleton, and numerous hairs and bristles on the body. The large, strong back legs are designed for jumping; when trying to leap from the ground onto a potential host, a flea moves at 140 times the force of gravity!

Fleming, Sir Alexander (*bacteriologist*) [English: 1881–1955] In 1921 Fleming discovered an enzyme he called lysozyme. Found in tears, saliva, and other natural substances, lysozyme breaks bonds in the cell walls of bacteria, causing the bacteria to lyse, or break apart.

Fleming's most famous discovery came in 1928, when a *Penicillium* mold contaminated a dish in which he was growing the bacterium *Staphylococcus aureus*. Fleming noticed that no *Staphylococcus* grew near the mold, and discovered that the mold produced a powerful antibacterial chemical, which he named penicillin. Penicillin was first purified and produced in sufficient quantities to be used as an antibiotic to treat infectious diseases in the 1940s by Howard Florey [Engish: 1898–1968] and Ernst Chain [Engish: 1906–79]. They and Fleming shared the 1945 Nobel Prize in physiology or medicine.

flight (*physiology*) Bats, birds, and insects are the only animals that have wings and are capable of true flight. In addition, certain mammals and birds can glide, an activity that does not require the enormous expenditure of energy needed for flight. There are various types of flight: for example, the up-and-down flapping flight of a flock of Canada geese, the hovering of hummingbirds, the soaring of eagles, and the complex figure 8 pattern used by most insects. Among the swiftest fliers is the peregrine falcon (*Falco peregrinus*), which attains speeds of well over 100 mi (160 km) per hour during dives in pursuit of pigeons and other prey.

flint (*rocks and minerals*) Flint is a hard silica rock, often chalcedony, that occurs in separate lumps, called nodules, in beds of limestone. Essentially the same material is also called chert, especially if it occurs in a bed of its own instead of as nodules. The exact origin of flint and chert is unknown, but fossils are sometimes found in the nodules and perhaps represent where the skeleton of a sponge or echinoderm introduced a variation in calciferous sediments. Flint, hard yet easily worked, was the basis of the best prehistoric stone tools and probably was the first mineral to be mined.

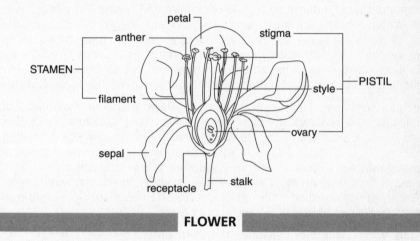

FLOWER

flower (*botany*) The reproductive unit of angiosperm plants, a flower exists for only a brief period of time before parts of it develop into a fruit. Most flowers have 4 types of parts. Outermost are the sepals, which usually resemble green leaves; they enclose and protect the flower bud. The petals are often colorful, and some have glands that secrete nectar; they attract insects and other animals that aid pollination. The stamen is the male organ; it produces pollen, which gives rise to sperm. The pistil is the female organ; it produces eggs. In flowers such as roses, all the organs are separate and distinct; in others, such as morning glories, some parts are fused together. Some flowers, such as those of corn, are either male or female, having stamens or a pistil but not both. Sepals and petals also are missing in some species. The many differences among flowers, including how they are clustered into inflorescences, are useful for plant identification.

fluid (*materials science*) A fluid ("flowing") is a substance that, when confined to a container by walls or forces (such as gravity), takes the shape of the container. Liquids are fluids, but not all fluids are liquid. A solid divided into many particles that are small in comparison to the container behaves as a fluid—sand or dust can be fluid in ordinary containers, while in a landslide boulders can fill valleys. Gas is always fluid. Glass and hot rock act as fluids over longer periods of time, gradually fitting a container or flowing downward.

fluke (*animals*)   Parasitic flatworms constituting Class Trematoda, flukes are somewhat leaf-shaped, with 1 or more suckers for attaching themselves to a host. They range in length from less than 0.1 in. (0.25 cm) to 6.5 in. (16.5 cm). Some flukes are external parasites on the skin or gills of fish. Others are internal parasites, mainly of fish but also of other vertebrates. These internal parasites usually have a complex life cycle involving several hosts. The most serious human fluke parasites are the blood flukes (genus *Schistosoma*), which cause schistosomiasis.

fluorescence (*materials science*)   Fluorescence is used to describe visible phosphorescence that is induced by high-energy input and released at a lower energy level without producing much heat. The incident energy, such as ultraviolet light or electrons, raises the energy level of atomic electrons; when the electrons return to a lower state, they emit the energy as photons of visible light. This mechanism is used in cathode-ray tubes as well as fluorescent lighting. Glowing nebulas in stars also produce light this way. (*See also* **incandescence**.)

fluorine (*elements*)   Fluorine is a yellowish halogen gas, atomic number 9, first isolated in 1886 by Henri Moissan [French: 1852–1907]. Smelters used minerals called fluors, from Latin *fluere*, "to flow," to help keep metals liquid, leading to names such as fluorite and fluorospar, from which elemental fluorine was finally produced. Chemical symbol F, fluorine is the most reactive of all elements; its tight hold on other elements has produced inert compounds such as Teflon and fluorides that aid in making teeth and bones strong. Fluorine is so reactive it even combines with several noble gases.

fly (*animals*)   True flies constitute the insect order Diptera ("2 wings"). They have 1 pair of functional wings; the back wings are modified into knobbed organs called halteres, which act as stabilizers during flight. The mouthparts typically are designed for piercing and sucking. The legs usually have spurs or stout spines plus hairy pads on the bottom of the feet; many species have taste organs on the legs. The larvae, called maggots, are worm-shaped, eyeless, wingless, and legless.

Many dipterans are bloodsuckers—including black flies, gnats, punkies, and mosquitoes—and as such transmit disease. Others, such as houseflies and blowflies, feed on animal feces and human food; they, too, are agents of disease. Midges and Hessian flies feed on plants. Soldier flies and stiletto flies are predators of insects. (In addition to dipterans, many other insects are called flies, including the dragonfly and firefly.)

flycatcher (*animals*)   Two groups of perching birds are commonly called flycatchers. They typically sit on a branch at the edge of the forest or some other vantage point watching for flying insects. On spotting a potential meal, a flycatcher darts off and grabs the insect with its flattened bill. Flycatchers of the family Muscicapidae are native to the Old World. Tyrant flycatchers of the family Tyrannidae, including kingbirds, pewees, and phoebes, are native to the Americas.

fog (*meteorology*) Fog consists of tiny water droplets suspended in air for several feet or more above land or water—ground-level clouds, in effect. Some fogs are formed by advection as warm, moist air blows over cooler surfaces. Others, called radiation fogs or ground fogs, result from conditions similar to those producing dew, but the air below the dew point rises from the cool ground or water and forms a layer of cool, foggy air near the surface. Radiation fog often forms in valleys, so that a traveler passes in and out of fog as elevation changes.

food chain (*ecology*) In an ecological community, energy is passed from one organism to another in feeding patterns called food chains. A typical grassland food chain is grass → grasshopper → frog → snake → hawk. All food chains begin with a producer, a green plant or other organism that converts external energy (solar energy, for example) into chemical bonds in food. The producer is eaten by a primary consumer, which is eaten by a secondary consumer, and so on. At each step in the food chain, some energy is dissipated during respiration.

A food chain does not really end with the final consumer. When a hawk dies (assuming a vulture or other animal does not eat it), bacteria and other decay organisms use its tissues as food and in the process return materials to the soil for use in future food chains. (*See also* **food web**.)

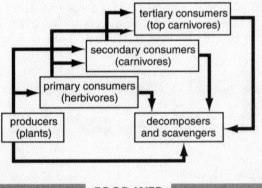

FOOD WEB

food web (*ecology*) To understand a community it is important to look at the flow of energy through its many overlapping food chains. A community has numerous different producers, consumers, scavengers, and decomposers. For example, producers in a pond community may include various species of algae, grasses, and water lilies. The producers are eaten, directly or indirectly, by consumers ranging from 1-celled protozoa to insect larva to fish and frogs—plus visiting birds and mammals. When consumers and producers die, bacteria, fungi, and other scavengers and decomposers feed on them.

Most consumers have a varied diet. A barn owl (*Tyto alba*) eats mostly small mammals such as mice, moles, rats, gophers, and rabbits, but it also eats pigeons, jays, sparrows, starlings, and other birds. Thus the barn owl is part of numerous food chains. All the food chains in a community form a food web. Generally, the greater the variety of species in a food web, the more stable the community.

foraminiferan (*protists*)   Most foraminifera are one-celled amoeboid protists with shells made of calcium carbonate (limestone). Slender pseudopods extend through holes in the shells and twist around one another to create a net for trapping prey.

Foraminifera live in marine habitats throughout the world. Some species live on the bottom of the sea or attached to plants while others are free-floating. The accumulated shells of dead foraminifera have formed dense limestone deposits on ocean floors. The White Cliffs of Dover, England, are such deposits that have been uplifted by geological forces. And the Great Pyramids of Egypt were carved from limestone made by foraminifera that lived 50,000,000 to 60,000,000 years ago.

Naked foraminifera—without shells—were identified in 1999 in freshwater environments. Before DNA tests, the organisms had been classed as amoebas.

force (*mechanics*)   At its simplest, a force is any push or pull. Both the amount, measured in poundals or newtons (kilogram-meters per second per second), and direction are parts of the description of a force. A force acting on a mass produces acceleration unless balanced by a force in the opposite direction. For a given mass, the acceleration is proportional to the force. Acceleration includes any change in direction, so a force can also be defined as that which causes a change in movement or direction. The extension of a spring or elastic substance is also proportional to the force exerted on it.

Fundamental Forces between Particles

FORCE	OCCURS BETWEEN	PARTICLES EXCHANGED	REST MASS OF EXCHANGED PARTICLES	RANGE	RELATIVE STRENGTH
color	quarks	gluons	0	c 10^{-17} centimeters	increases with separation of quarks
strong	nucleons (protons and neutrons)	pions	135 to 189.6	c 1.4×10^{-13} centimeters	1
electromagnetic	charged particles	photons	0	infinite (diminishes as inverse of the square of separation)	1/100 of strong force
weak	leptons and quarks	W and Z	80,410 to 91,160	c 2.5×10^{-16} centimeters	$1/10^{15}$ of strong force
gravitational	particles with mass > 0	gravitons (not observed so far, but expected)	0	infinite (diminishes as inverse of the square of separation)	$1/10^{40}$ of strong force, or $1/10^{25}$ of weak force

force, fundamental (*particles*)   Forces arise between particles when two fermions exchange bosons and the range—effective distance the force travels—depends on the mass of the boson that carries the force. Physicists generally recognize 3 fundamental forces, sometimes called *interactions*. Other forces can ultimately be traced to these 3, each of which operates by the exchange of one kind of particle between particles of another kind. The forces are gravity, the electroweak interaction (which at ordinary energies is either electromagnetism or the weak interaction), and the color force, which also accounts for the strong force between nucleons.

Physicists have long tried to reduce the number of fundamental forces to one in a Theory of Everything. Some mathematical approaches based on group theory unify electroweak and color forces and supersymmetry brings in gravity, but neither of these unifications have been established by experiment or observation.

forensic science (*sciences*)   Forensic science is the application of scientific principles and techniques to the analysis of evidence in criminal cases. The discipline had its origins in the 19th century, with people such as Mathieu Orfila [Spanish: 1787–1853], who published the first scientific treatise on the detection of poisons, thereby founding forensic toxicology. The first scientific system of personal indentification, using a series of precise body measurements, was developed by Alphonse Bertillon [French: 1853–1914]. In the early 1900s Bertillon's system was replaced by fingerprint identification; since the late 1900s forensic scientists also use DNA fingerprints—profiles of DNA fragments that contain different numbers of sequences in different people. Forensic dentistry is the study of teeth and bite marks to help identify human remains and forensic anthropology is concerned with the identification of skeletal remains. Other specialties analyze physical evidence such as fibers, firearms, and documents.

forest (*ecology*)   A forest is a biome in which the dominant food producers are trees. The forest consists of horizontal layers that typically include a canopy formed by the tops of the tallest trees; an understory consisting of the tops of shorter trees; possibly a shrub layer; a ground layer of mosses and herbs; and the forest floor, covered with litter from the upper layers. At each level exists a diversity of animal as well as plant life, though populations of large herbivorous animals are low, enabling young trees to survive and grow.

Trees have comparatively high water needs, and forests form and survive only where moisture is available throughout the year. Major types of forests include rain forests, temperate deciduous forests, and northern evergreen forests called taiga.

formaldehyde (*compounds*)   Formaldehyde is the simplest of the aldehydes, chemicals formed by using oxygen to remove 2 hydrogen atoms from an alcohol to form H_2O, water. The name aldehyde is short for alcohol dehydrogenation. Oxygen removes 2 hydrogens from methanol (CH_4O). The result is formaldehyde, usually written HCHO since the radical CHO, called the formyl group, always appears in an aldehyde. Formaldehyde, used for preserving biological specimens, is also important in the manufacture of plastics and building materials.

formula (*chemistry*)   A chemical formula is a shorthand method for describing the components of a molecule or ionic compound. Each element is represented by 1 or 2 letters, called the symbol. Subscripts are used to indicate how many atoms of each element are present. Thus water with 1 atom of oxygen (symbol O) to 2 of hydrogen (H) has the formula H_2O and glass with 2 oxygen atoms to 1 of silicon (Si) is SiO_2. The arrangement of atoms is important for organic molecules, so their structural formulas show each atom or radical separately in symbols and often indicate bonds between specific atoms.

formula (*mathematics*)   The statement of a rule in terms of variables only is a formula, such as the formula $d = rt$ for finding distance (d) as the product of a rate (r) and time (t). Similar formulas are used in physics. In logic any statement consisting solely of variables and such connectives as and, not, or if . . . then is called a formula.

fossil (*paleontology*)   The remains or traces of organisms that lived in the past are called fossils (from the Latin word meaning "dug up"). They include imprints, tracks, molds, and casts, as well as evidence of petrification and actual parts such as bones and teeth. Most fossils were formed in sedimentary rock, but organisms also have been preserved in glaciers, bogs, tar pits, volcanic ash, and amber. Though thousands of different fossil organisms have been identified, they are believed to represent only a small fraction of the many types of organisms that lived in the past. Most organisms died without leaving any traces. Nonetheless, fossils provide at least a partial history of the development of life on Earth as well as evidence of geological and climatic changes. For example, fossils of ancient corals found in the Arctic suggest that the Arctic was once much warmer than it is today.

fossil fuel (*environment*)   Fuels formed in Earth's crust by the decay and chemical change of dead organisms that lived long ago are called fossil fuels. Petroleum, natural gas (methane, propane, etc.), and coal are the main fossil fuels. Because they take millions of years to form and can be used only once, they are nonrenewable natural resources.

Fossil fuels are hydrocarbons, composed of hydrogen and carbon. When burned in the presence of oxygen, they are broken down into carbon dioxide and water; in the process they give off light and heat energy. Since the start of the Industrial Revolution in the late 18th century, fossil fuels have been used in increasing quantities. The carbon dioxide released as a result has been a major contributor to global warming.

Foucault, Léon (*physicist*)   [French: 1819–68] From 1849 through 1862 Foucault studied the speed of light, first collaborating with Armand Fizeau [French: 1819–96] in determining an improved value and then in 1850 finding the speed of light in water. Because the speed in water is slower than in air, Foucault's work confirmed that light behaves as a wave (today it is recognized that light also behaves as particle). In 1851 Foucault constructed the first of several large pendulums with which he demonstrated experimentally that Earth rotates on its axis. As an outgrowth of his work with pendulums, he invented the gyroscope (in 1852).

Fourier series (*trigonometry*) Joseph Fourier [French: 1768–1830] analyzed heat transmission, inventing much of the necessary mathematics to do so. His most enduring mathematical innovation is the Fourier series, an infinite sum of sines and cosines of a variable x adjusted so that the value of the sum for each x is the same as the corresponding value of almost any useful function of x. Trigonometric functions respond well to the tools of calculus, making it often easier to work with Fourier series than with original functions. Operations called Fourier transforms translate between representations. Among the many applications of this idea is tomography, the basis of the CT scan.

fowl (*animals*) The name fowl generally refers to birds of the order Galliformes, including the domestic fowl, or chicken, which is a descendant of jungle fowl native to Asia. Other members of the order include megapodes, curassows, grouse, peafowl, pheasants, and turkeys. Fowl generally live on or near the ground. They have short wings and often elaborate tails. All have strong legs adapted for walking and running. Each foot has 3 toes in front and 1 behind. The toes have hard nails and are designed for scratching in the ground for seeds and other food.

fox (*animals*) Widespread throughout a variety of habitats in the Americas, Eurasia, and Africa, foxes are the smallest members of the dog family. They have an elongated muzzle, erect ears, short legs, and a long bushy tail. They are solitary, preying on mice and other small animals and also eating eggs, fruits, and carrion. The most common species, the red fox (*Vulpes vulpes*), is about the size of a small dog.

foxglove (*plants*) Foxgloves (*Digitalis*) are biennial and perennial herbs native to Europe, Asia, and northern Africa. The common foxglove (*D. purpurea*) is a popular ornamental, thanks to its 4-ft (1.2-m) spikes of hanging tubular flowers. Both this species and *D. lanata* are also grown for their leaves, from which digitoxin and other compounds called glycosides are extracted for medicinal uses. All foxgloves are poisonous if eaten.

   Foxgloves are classified in the figwort family, Scrophulariaceae, which comprises more than 4000 species of mostly temperate herbs. All species have flowers with 5 partly or wholly united petals; the petals form a structure that usually has 2 distinct lips, with the upper lip divided into 2 lobes and the lower lip into 3 lobes. The family includes figworts (*Scrophularia*), mulleins (*Verbascum*), snapdragons (*Antirrhinum*), speedwells (*Veronica*), and Indian paintbrushes (*Castilleja*).

fractal (*mathematics*) The idea of fractal ("broken") is bound up with the concept of a figure that, for a given ratio, has a small part that is a similar figure to a corresponding larger part no matter how small one goes (called self-similarity). Any coastline has a rough sort of self-similarity, since the whole coast with peninsulas and bays resembles a small portion of itself, which in turn is like an even smaller portion of the shore. The technical definition of a fractal is based on a numerical way to identify such figures in terms of dimension. Fractal geometry, introduced by Benoit Mandelbrot [Polish-American: 1924–] in 1975, has proven to be important in understanding the behavior of complex systems.

fraction (*numeration*)   Originally fraction meant part of a unit ("broken"). Today a fraction is a type of numeral. A rational number between 0 and 1 can be written as a proper fraction by representing division of 2 numbers with one called the numerator above the line and another called the denominator below the line, as in $5/6$. If the numerator is larger than the denominator, as in $23/7$, the fraction is called improper. Fractions written as indicated division are called common fractions. Different fractions can show the same number, for example $2/3$ and $8/12$. The same numerator-denominator method is used in algebra, where it is the most frequent way to show division. (*See also* **decimal fraction.**)

fragmentation (*biology*)   Certain multicellular organisms can reproduce asexually by fragmentation. In this process pieces of the parent break away and develop into new individuals. Fragmentation is common in algae that consist of threadlike filaments, such as *Spirogyra* and *Ulothrix*.

frame of reference (*physics*)   A coordinate system used as the basis of measurements is a frame of reference. For example, descriptions of motion of planets in the solar system usually use the Sun as the center of the frame, while experiments on Earth's surface normally use coordinates that rotate along with Earth. Different reference frames could be employed, but motions would become more difficult to describe.

If movement of a reference frame is constant, the frame is called inertial. A vehicle that moves without acceleration provides an inertial frame. In such a vehicle, Newton's laws of motion hold and a passenger who cannot see outside cannot detect the motion. The theory of how an observer in one inertial frame relates to an observer in a different inertial frame is Einstein's special theory of relativity. Investigation of the effects of acceleration on reference frames led Einstein to the general theory of relativity.

francium (*elements*)   Francium is the heaviest alkali metal, atomic number 87. All isotopes are radioactive with a short half-life. The most stable occurs naturally only as part of the disintegration of uranium—at any given time there is probably about 15 oz (400 g) on Earth. Francium was synthesized in 1939 by Marguerite Perey [French: 1909–75] and named for France (chemical symbol: Fr). Its properties are inferred from the periodic table to be similar to those of cesium.

frankincense (*plants*)   Several evergreen shrubs and small trees of the genus *Boswellia*, native to eastern Africa and the Arabian Peninsula, exude an aromatic resin called frankincense, used in incense and perfumes. Other members of the same family, Burseraceae, also produce aromatic resins, including species of the genus *Commiphora*, also of eastern Africa and Arabia, which secrete a resin called myrrh.

Franklin, Benjamin (*physicist*)   [American: 1706–90] Franklin, of course, was more than a physicist, but he made major contributions to the theory of electricity, including the first recognition of conservation of charge, a theory of static electricity, and proof that lightning is an electrical phenomenon. He also recognized the importance of ocean currents, and had the Gulf Stream mapped for the first time.

Fraunhofer lines (*optics*)   In 1814 optical instrument maker Joseph von Fraunhofer [German: 1787–1826] noticed that the Sun's spectrum produced by a high-quality prism is crossed by hundreds of dark lines. He named the most prominent with letters and carefully measured their positions. The dark lines, now called Fraunhofer lines, always appear the same in the solar spectrum, but when Fraunhofer made similar spectrums of other stars, he found that the pattern of lines varied slightly.

In 1859 Gustav Robert Kirchhoff explained the lines after he recognized that Fraunhofer's D line is the same as a bright line produced by incandescent sodium. Dark Fraunhofer lines occur when light passes through a gas, which absorbs the same wavelengths that the heated gas emits. Thus, Fraunhofer lines are used to determine which elements are in stars from their spectrums. In 1913 Niels Bohr showed that the lines reflect changes in the orbits of atomic electrons. (*See also* **red shift**; **Zeeman effect**.)

free radical (*biochemistry*)   A molecule with at least 1 unpaired electron—a condition that gives the molecule an electrical charge—is a free radical. Such a molecule constantly tries to neutralize its electric charge, either by giving up the unpaired electron to another molecule or by stealing an electron from another molecule.

Produced as by-products of normal metabolism, free radicals play valuable roles in activities such as fighting infection. But they also have harmful biological effects, including damaging cell membranes and causing mutations in DNA.

frequency (*waves*)   Frequency is the number of waves passing per second, which is measured as hertz (Hz)—10 Hz is 10 cycles per second. Frequency is determined by wavelength and the speed a wave travels: speed = frequency × wavelength. Light in a vacuum travels at a constant speed so frequency and wavelength describe the same aspect of light in 2 ways. The period of a wave is the reciprocal of the frequency: period = 1/frequency. A radio station broadcasting at 880 KHz (880,000 Hz) uses waves with a period of 1/880,000, so 880,000 waves per second are generated.

friction (*mechanics*)   Friction is a force caused by sliding one solid over another or by motion of solids, liquids, or gases with respect to each other. It always opposes the direction of motion.

The cause of friction in solids includes tiny bumps or ridges and electrostatic attraction between surfaces. There must be a force perpendicular to the direction of motion for friction to occur. Friction is proportional to this force. The ratio is a parameter called the coefficient of sliding friction. Friction does not depend on either the speed or the area of contact.

If both solids are stationary, there is no net force (Newton's third law of motion). But stationary solids cling tighter, so the coefficient of static friction, based on the amount of force needed to induce motion, is higher than for sliding friction. When one or both of the materials is a fluid, friction is determined by viscosity.

Fries, Elias (*botanist*)   [Swedish: 1794–1878] Fascinated by fungi from the age of 12, Fries was the first to recognize rusts and smuts as a related group of organisms. He developed the first modern system of classifying fungi based on developmental and structural similarities and differences.

frigatebird (*animals*)   Masters of the sky, frigatebirds (genus *Fregata*) can soar for hours, perform aerial maneuvers, steal food in midair from other birds, and plunge straight down to grab a flying fish at the water's surface. But these seabirds of tropical and subtropical oceans can neither walk nor swim. Ashore, they roost in trees or on rocks. During mating season the male attracts a female by inflating a neck pouch that turns bright red. The largest species is the magnificent frigatebird (*F. magnificens*), which may be 45 in. (114 cm) long with a wingspan of 96 in. (244 cm).

frog and toad (*animals*)   Amphibians of the order Anura, commonly called frogs or toads, have a large head, a fleshy tongue attached to the front of a wide mouth, a short, tailless body, and powerful hind jumping legs with long feet and toes. Their backbones are the shortest of all vertebrates, composed of only 5 to 9 vertebrae, and their feet typically are webbed, to aid in swimming. Thin skin contains numerous glands, some secreting mucus, others secreting toxins meant to discourage predators. Anurans hunt by sight, feeding mainly on insects. Toads generally have tougher, drier skin than frogs, often marked with scattered bumps and rough spots. They tend to be broader and flatter and spend most of their life on land, even thriving in comparatively dry habitats such as prairies. Both frogs and toads breed in water and their eggs hatch into tadpoles that breathe through gills before undergoing metamorphosis and becoming air-breathing adults.

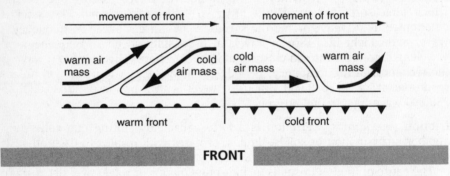

FRONT

front (*meteorology*)   A front is the boundary surface between 2 air masses that have differing temperatures and air pressures. Winds and precipitation often occur at a front. The air mass with the higher pressure and cooler temperature pushes under the one with lower pressure, causing warm air to rise; the air then cools and releases moisture. The overall movement, if any, of the boundary line is used to name the type of front: warm air advancing is a warm front; cold air advancing is a cold front.

fruit (*botany*)   The ripened ovary of a flower, together with any nearby structures that ripen and fuse with it, is called a fruit. Typically, a fruit forms following fertilization of 1 or more eggs within the ovary. Its development is a complex phenomenon in which plant hormones (**auxins**) play critical roles. The plant moves food into the ovary tissues; for example, an accumulation of sugar caused this way is responsible for the sweetness of grapes, while an accumulation of water gives the grapes a pulpy texture. As the fruit ripens, pigmentation changes may also occur. For example, chlorophyll may disappear from grapes and be replaced by pigments with a reddish or purplish color.

Fruits exhibit an almost infinite variety of structure, shape, and texture. A simple classification of the main types is presented in the accompanying table. Note that botanical definitions do not necessarily match common usage. A walnut is not a nut, and raspberries are not berries. Also, in botanical terms there is no such thing as a vegetable. Many foods commonly called vegetables—corn grains, tomatoes, string beans, for example—are fruits. Other vegetables come from other plant parts; carrots are roots, lettuces are leaves, potatoes are underground stems, etc.

TYPE OF FRUIT	DESCRIPTION	EXAMPLES
Simple: develops from flower with single ovary		
Fleshy		
berry	entire ovary fleshy and often juicy; usually contains many seeds	blueberry, cucumber, grape, lemon, orange, tomato, watermelon
drupe	ovary forms 2 layers: an outer fleshy layer and a hard inner stone, or pit, that usually encloses a single seed	almond, apricot, cherry, coconut, olive, peach, pecan, plum, walnut
Dry, splits open when ripe to release seeds		
capsule	multichambered ovary with many seeds; splits in various ways, such as along many seams or by forming pores near top of fruit	azalea, cotton, iris, lily, plantain, poppy, snapdragon, tulip, violet
follicle	podlike, with a single chamber containing many seeds; splits along single seam	columbine, larkspur, milkweed, peony, spiraea
legume	podlike, with a single chamber containing many seeds; splits along 2 seams	bean, clover, locust, pea, peanut, soybean, vetch
Dry, does not split open to release seeds		
achene	single seed is attached to ovary wall only at the base	buttercup, clematis, dandelion, sunflower
grain	wall of single seed is completely fused to thin ovary wall	corn, oat, rice, rye, wheat
nut	ovary forms a hard, thick wall that usually encloses a single seed	chestnut, hickory, oak (acorn)
samara	similar to achene, but ovary wall has a prominent winglike outgrowth	ash, elm, maple, sycamore
Aggregate	compound fruit formed from many ovaries within a single flower	blackberry, raspberry

(continued)

TYPE OF FRUIT	DESCRIPTION	EXAMPLES
Multiple	compound fruit formed when ovaries of many flowers on a common stalk fuse together	fig, mulberry, pineapple
Accessory	fleshy part forms from receptacle (top of flower stalk), which grows to surround or adhere to ovaries	apple, pear, quince, rose, strawberry

frustum (*geometry*)   The frustum of a solid is the part between 2 parallel planes, but in applications 1 of the planes is the base of the solid and the solid is a pyramid or cone. If the parts of a pyramid or cone intercepted by the planes are B and b and the altitude is h, the volume of the frustum is $1/3\ h(B + b + \sqrt{Bb})$. Determining the formula for the volume of the frustum of a pyramid was the height of Egyptian achievement in geometry about 1890 BC.

fuel cell (*chemistry*)   In 1839 William Groves [English: 1811–96] showed that hydrogen and oxygen can combine in a device called a fuel cell to produce both water and electric current. This reverses electrolysis of water. In one type of fuel cell the anode and cathode are immersed in potassium or sodium hydroxide. At the anode, hydrogen combines with hydroxide ions to produce water and free electrons. The electric current releases free electrons at the cathode, where they combine with water and oxygen to produce hydroxide ions.

fulmar (*animals*)   The 2 species of fulmars (genus *Fulmarus*)—one of the northern Atlantic and Pacific, the other of southern oceans—are large, stocky seabirds with long, powerful wings and webbed feet. They spend almost their entire lives at sea, coming ashore only to breed and raise their young. Fulmars feed on fish, invertebrates, and garbage from ships. When attacked, they discharge a foul-smelling oil from the stomach.

function (*analytic geometry*)   A function is any rule that unambiguously determines a value from 1 set when applied to a member of another set. Some typical functions on numbers are the square (the product of the number with itself); the number of prime number factors, and the cube root. The square root of a real number is not a function because there are 2 possible values (the square root of 4 can be either $+2$ or -2), so it is ambiguous. All functions can be treated as sets of ordered pairs where the first member of the pair is connected to the second by the rule. For example, the square function for natural numbers corresponds to the set (1, 1), (2, 4), (3, 9), and so forth. Such sets are functions when they have no 2 pairs with the same first member. The graph of all the points in such a function and the function itself are the basic entities of analytic geometry, while calculus consists of operations on functions.

fungus (*fungi*)   A diverse group of single-celled and multicellular organisms, fungi are found almost everywhere on Earth. Most fungi, such as mushrooms and puffballs, are saprophytes that feed on decaying leaves and other dead organic matter. Some fungi—rusts and smuts, for example—are parasites, obtaining nutrients from other living organisms. Because fungi cannot take in solid food, they secrete digestive enzymes onto the food source. The enzymes break down food into simple organic substances that the fungi can absorb. The kinds of enzymes produced by a species determine what substances it can use as food.

Multicellular fungi usually consist of a vegetative body and a fruiting body. The vegetable body, or mycelium, obtains food. It is a tangled mass of branching threadlike filaments called hyphae, and is usually buried in soil, tree bark, or some other substrate. The conspicuous fruiting body—for example, the part of a mushroom that appears above the ground—produces spores.

The scientific study of fungi is called mycology (from the Greek *mykes*, meaning "mushroom"). The approximately 100,000 known species are generally classified in their own kingdom, Fungi, though in the past they were considered part of the plant kingdom. Classification within the kingdom is based primarily on the structure of the fruiting body.

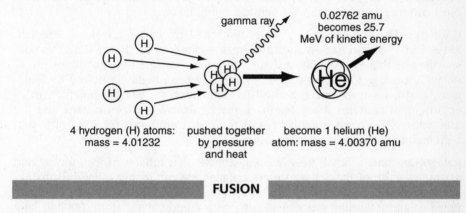

4 hydrogen (H) atoms: mass = 4.01232 — pushed together by pressure and heat — become 1 helium (He) atom: mass = 4.00370 amu

gamma ray — 0.02762 amu becomes 25.7 MeV of kinetic energy

FUSION

fusion (*particles*)   When atomic nuclei are forced to overcome repelling forces caused by charge and become very close to each other, the strong force takes over and causes the nuclei to merge in a process called nuclear fusion. Fusion can be induced by a combination of heat and pressure or, in the centers of stars, by great pressure alone. Lighter elements fuse more easily than heavy ones for the most part, so the most common fusion reaction is for 4 hydrogen nuclei to fuse into a nucleus of helium, releasing energy in the process. This process is also the basis of thermonuclear devices (hydrogen bombs). After most hydrogen in a star has fused, new reactions at higher temperatures produce carbon and eventually other elements up to iron. Fusion of elements heavier than iron requires the energy of a supernova.

G

gadolinium (*elements*)   The rare earth with atomic number 64, gadolinium was discovered in 1880 by Jean-Charles Marignac [Swiss: 1817–94], who produced the oxide from gadolinite, a mineral named for chemist Johan Gadolin [Finnish: 1760–1852]. Although gadolinium metal has many useful magnetic properties, it has been very difficult to separate and therefore too expensive for most commercial applications. Chemical symbol: Gd.

Gaia hypothesis (*ecology*)   First propounded by James Lovelock [English: 1919–] in 1969 and named after the Greek goddess of Earth, the Gaia hypothesis suggests that Earth is like a living organism, sustained by a complex physiology. According to the hypothesis, which has been highly criticized by biologists, Earth, its atmosphere, and all its organisms are interdependent. Earth's environment contains many feedback mechanisms—between oceans and the atmosphere, between organisms and their surroundings, for example—that help maintain a stability, or homeostasis.

galago (*animals*)   Small tree-dwelling primates that inhabit African forests and savannas south of the Sahara Desert, galagos are commonly called bushbabies because their cry sounds like that of a human infant. The largest is the thick-tailed galago (*Otolemur crassicaudatus*), with a body about 15 in. (38 cm) long and an 18-in. (46-cm) tail. Galagos have large ears and eyes, long back legs, and feet adapted for grasping. Except for the second hind toe, which has a long claw used for grooming, the digits have flat nails. Galagos move quickly through trees, leaping nimbly from one branch to another. They are primarily nocturnal, with species varying in food habits. Galagos are classified either with the lorises or as a separate family, Galagonidae.

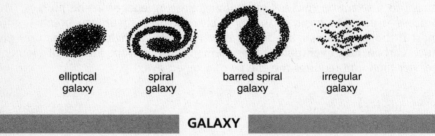

| elliptical galaxy | spiral galaxy | barred spiral galaxy | irregular galaxy |

GALAXY

galaxy (*galaxies*)   Galaxies, sometimes called island universes, are systems of very many stars separated from one another by what appears to be empty space. In the 18th century **William Herschel** concluded that many cloudy patches of light seen among the stars are giant systems of billions of stars so far away from Earth as to look like clouds. Better telescopes proved him right in the early 20th century, and these far-off, great masses of stars became known as galaxies, after our own Milky Way, the galaxy that includes the Sun. Observations with large telescopes in the 20th century have revealed 2 main types of galaxies—spiral and elliptical—although some galaxies are neither (irregular). Spiral galaxies are large, bright, and well-formed, so they dominate the scene; but there are more elliptical galaxies in all. Just as stars are not randomly distributed through space, but collected in galaxies, the galaxies themselves form into clusters and the clusters into superclusters.

Galen (*anatomist/physician*)   [Greek: 129–c199] The last outstanding biologist of antiquity, Galen dissected many different animals (dissection of human bodies was not allowed). He described the structural differences between arteries and veins, and demonstrated that blood vessels contain blood—not air, as had been believed. He differentiated sensory and motor nerves, identified 7 pairs of cranial nerves, and showed that different parts of the spinal cord control different muscles. Galen made many other valid observations, but he also had erroneous notions, such as believing that food is converted into blood in the liver, which he said is the central organ of the blood system. After Galen's death, anatomical research ceased and for almost 1400 years his writings were considered infallible.

Galileo Galilei (*astronomer/physicist*)   [Italian: 1564–1642] Most historians consider Galileo (usually known by first name only) as the first scientist of the Scientific Revolution. His greatest fame is for discoveries in astronomy (moons of Jupiter, phases of Venus, and much more), but his influence on physics is pervasive; the observation that all bodies fall at the same speed in a vacuum is just one of Galileo's ideas that led to the laws of motion and eventually to relativity theory. He also contributed to the study of mathematical infinity. Galileo's influence comes not only through his persuasive and popular books about the solar system, kinematics, and materials, but also as a result of his inventions (the astronomical telescope and the thermometer), his correspondence (notably with **Kepler**), and his pupils and assistants (especially **Evangelista Torricelli**). Galileo's observations led **Huygens** to develop the pendulum clock and Torricelli to invent the barometer.

Galileo promoted Copernican views as early as 1604 and did not stop when in 1616 the church declared such ideas to be heresy. Galileo's *Dialogue on the Two Chief World Systems* of 1632 defended the heliocentric view of the solar system. As a result, he was put before the Inquisition and informed that he must recant or be tortured. He recanted, but spent the last years of his life under house arrest, during which time he wrote and published (in 1638) his most influential work on physics, *Dialogue on Two New Sciences*.

gall (*botany*)   An abnormal, localized growth of plant tissue caused by bacteria, fungi, insects, or other parasites is termed a gall. The growth may consist of enlarged cells, an increased number of cells, or both. Galls come in many sizes and shapes, but often it is possible to identify the parasitic species by the structure of the gall. For example, the oak hedgehog gall on leaves of white oak trees is produced in response to the deposition of eggs by the gall wasp *Acraspis erinacei*. This gall is round or oblong, 0.4 to 0.6 in. (10 to 15 mm) long and covered with spines. Internally it has 2 to 8 cells from which adult wasps emerge in late autumn.

gallbladder (*anatomy*)   Present in most vertebrates, the gallbladder is a saclike organ located between the lobes of the liver. It concentrates and stores bile produced by the liver. The gallbladder releases bile into the bile duct, which connects to the small intestine, in response to the hormone cholecystokinin, which is secreted as fats enter the small intestine.

In horses, elephants, and other animals that lack gallbladders, the liver secretes a comparatively large amount of bile.

gallium (*elements*)   The metal gallium, atomic number 31, is a liquid at a few degrees above room temperature. It was the first element to be predicted—by Mendeleyev on the basis of his periodic table—before discovery. The name was explained by the discoverer, Paul-Emile Lecoq de Boisbaudran [French: 1838–1912], as based on an ancient Latin name for France, but *le coq,* or *gallus* in Latin (meaning "rooster"), the discoverer's name, is a more probable derivation.

Gallium has useful electronic and optical properties. The arsenide of gallium is the basis of fast computer chips and some optical computing systems. Some neutrino telescopes consist of tons of gallium in underground tanks. Chemical symbol: Ga.

Galois, Evariste (*mathematician*)   [French: 1811–32] Galois died young with almost none of his work published or even read during his lifetime. In 1846 his papers and a letter describing his main discovery were edited and published. His main discovery was that equations of degree 5 or higher cannot be solved using the ordinary tools of algebra; the method he used to reach the conclusion was more influential than the theorem. The method, group theory, has been at the core of mathematical thought for the past century and a half.

Galvani, Luigi (*anatomist/physiologist*)   [Italian: 1737–98] While studying movement in frogs, Galvani discovered that muscles of a frog's legs contracted when they were in contact with 2 dissimilar metals, and electric current was produced. Galvani incorrectly believed that the frog's tissues generated the electricity and the metals discharged it. In 1800 Volta showed that electricity is produced from the contact of 2 different metals in a moist environment. By analogy, a person is said to be *galvanized* by any stimulus that causes a strong reaction.

gamete (*cells*)   Sperm and eggs—the reproductive cells whose nuclei fuse during the process of fertilization—are called gametes. They form by a special kind of cell division called meiosis. Gametes or the cells that give rise to them are sometimes called germ cells.

game theory (*mathematics*)   The theory of games describes strategies for winning rewards as a result of 2 or more people competing according to a set of rules that describe permissible actions, probabilities of outcomes that result from such actions, and the circumstances that will conclude the competition. The rules, probabilities, and stopping place constitute a game. Often games are described in terms of the number of players (a 2-person game or an *n*-person game). The games for which the mathematical descriptions are most complete are usually *zero-sum* games in which the amounts lost by some of the players are won by the other players (poker is a familiar example).

gamma radiation (*waves*)   Gamma radiation is the part of the electromagnetic spectrum with a frequency above about 100,000,000,000,000,000 Hz. Radiation from uranium that is more penetrating than X rays was discovered in 1900 by Paul Ulrich Villard [French: 1860–1934]. In 1903 Ernest Rutherford labeled the 3 types of radioactivity as alpha (helium nuclei), beta (electrons), and gamma (high-energy electromagnetic waves). Gamma radiation is produced by energetic motions or experiences of electrons, so sources include explosions of stars and any process where an electron combines with its antiparticle the positron. Extremely energetic gamma-ray bursts are thought to be caused by collisions of neutron stars or stars falling into black holes.

gamma-ray burst (*stars*)   Gamma-ray bursts seem to come at random from all parts of the universe; they are extraordinarily powerful. One observed in 1998 is thought to be the largest known energy release since the big bang. Bursts were first observed in 1968 by Cold War-inspired military satellites seeking secret nuclear blasts on Earth. Most bursts last only a few seconds, although a few continue for minutes. The evidence suggests that the triggering event occurs far outside our galaxy. Among the popular theories as to cause are that the bursts result from the collisions of neutron stars or from the collapse of supergiant stars into black holes.

ganglion (*anatomy*)   A cluster of nerve cell bodies is a ganglion (from the Greek word for "swelling"). In vertebrates, ganglia are located outside the central nervous system. In invertebrates, ganglia may be part of the central nervous system. The earthworm, for example, has a cerebral ganglion in its head that functions as a primitive brain.

Ganymede (*solar system*)   Ganymede, with a diameter of 3273 mi (5268 km), is the largest moon in the solar system, larger than the planets Pluto and Mercury. It is a huge, cratered ball of water ice, perhaps with a rock core, that orbits Jupiter at a distance that reaches 665,000 mi (1,070,000 km).

gar (*animals*)   Freshwater fish native to North and Central America, gars—also called gar pikes—constitute the Family Lepisosteidae. Gars have a long slender body covered with hard diamond-shaped scales, forming an armor that protects against attacks by other fish. Their long jaw is lined with sharp teeth. Gars are predators, feeding on all kinds of fish. The largest species, the alligator gar (*Atractosteus spatula*), grows to almost 10 ft (3 m) long.

gas (*materials science*)   A gas is any material for which molecules move in straight lines until they encounter other molecules. In a gas molecules are farther apart than in a liquid. Gases are fluids that take the shape of a container if they are completely enclosed, but expand into a vacuum or mix with other gases in air if there are openings. One difference between a gas and a liquid is that molecules of a gas are not attracted to each other enough to create surface tension.

gas giant (*solar system*)   The planets Jupiter, Saturn, Uranus, and Neptune are deemed gas giants because they are largely hydrogen gas (or hydrogen lique-fied by pressure at the bottom of a largely hydrogen atmosphere) and because they are huge. The smallest two, Uranus and Neptune, each have about 4 times the diameter and 16 times the mass of Earth.

gas laws (*chemistry*)   Chemistry became an effective science when chemists began to investigate gases instead of concentrating on precious metals and acids. Gases obey simple mathematical relationships that suggest basic princi-ples for all materials—for example, Dalton's atomic theory originated in his study of numerical gas laws.

In 1662 Robert Boyle discovered Boyle's law (Mariotte's law in France, after Edmé Mariotte [French: 1620–1684] who stated it in 1676): at a constant tem-perature, the volume of a confined gas shrinks in proportion to applied pres-sure. In 1801 Jacques-Alexandre Charles added Charles's law: all gases expand at a rate of 1/459 for each degree Fahrenheit (1/273 for each degree Celsius) if pres-sure is kept constant. In 1802 Joseph-Louis Gay-Lussac [French: 1778–1850] published a version of the same law for constant volume instead of constant pressure. The gas laws can be combined into a single formula that states that the new volume of a gas is equal to the old volume times the ratio of the origi-nal pressure to the new one times the ratio of the new temperature to the origi-nal one (the combined gas law). Real gases obey these rules only approxi-mately, so they are often stated for an ideal gas that would obey them exactly.

gastrointestinal tract (*anatomy*)   The digestive system of vertebrates consists mainly of a long tube, the gastrointestinal tract or alimentary canal, that extends from the mouth, where food is ingested, to the anus, where wastes are excreted. The tract includes the pharynx, esophagus, stomach, and small and large intestines. Within this standard framework, there is much variation. In birds, for example, the lower part of the esophagus is expanded into a storage organ, the crop. The stomach in birds has 2 sections—the proventriculus, where food mixes with gastric juice, and the muscular gizzard, where food is ground up, aided in many cases by small stones the bird has swallowed.

gastropod (*animals*)   Snails, slugs, and nudibranchs are members of the mollusk class Gastropoda (from Greek words meaning "stomach foot"). Most gas-tropods have a 1-piece, spirally coiled shell that comes in a great variety of sizes, shapes, and colors, often ornamented with ribs, spines, knobs, or bris-tles. Slugs and nudibranchs lack shells. Gastropods have a well-developed head region, usually with 2 eyes, 2 tentacles, and a mouth containing a rasplike radula covered with rows of tiny teeth. Gastropods are common worldwide in marine and freshwater habitats, and are the only mollusks plentiful on land.

Gauss, Karl Friedrich (*mathematician/physicist/astronomer*)   [German: 1777–1855] Gauss is ranked with Archimedes and Newton as the greatest of mathematicians. At age 19 he constructed a regular polygon with 17 sides, the first major new construction since Greek times. In 1799 he provided the first proof that all equations formed by setting a polynomial equal to 0 have a solution, the fundamental theorem of algebra. In 1801 he completely restructured number theory. An interest in surveying led to his 1827 theory that mathematically describes properties of surfaces.

The asteroid Ceres was discovered in 1801, but astronomers soon lost the location of the dim body. Gauss invented a way to calculate the path of a planet from a few observations. When Ceres was found exactly where he predicted, his fame was assured.

Gauss was the first to establish a non-Euclidean geometry but he did not publish it. Other mathematicians soon rediscovered it. Gauss similarly discovered much of advanced calculus but failed to publish his results. It is said that his discoveries would have advanced mathematics by 50 years if announced when made.

Gauss also studied Earth's magnetic field. He and a collaborator built a working telegraph more than a decade earlier than the 1843 system of Samuel F.B. Morse [American: 1791–1872].

gazelle (*animals*)   Gazelles are graceful, long-legged antelopes native to grasslands and deserts of Asia and Africa. Their horns, generally present on both males and females, curve forward and usually are ringed. Among the largest gazelles is the dama (*Gazella dama*) of the Sudan, which can reach a shoulder height of 4 ft (1.2 m).

gecko (*animals*)   Lizards of Family Gekkonidae get their name from the loud calls made by the Tokay gecko (*Gekko gekko*) of Southeast Asia. Geckos typically are about 6 to 8 in. (15 to 20 cm) long. They are excellent climbers, thanks to sharp claws and numerous microscopic structures on the underside of the toes that cling to surfaces by friction. Geckos have large eyes and usually are active at night, hunting insects and spiders. Their tail, easily broken, can be regenerated. Most common in warm areas, geckos live in a great variety of habitats, from deserts to forests—on the ground, in trees, and in people's homes.

Geiger counter (*particles*)   Physicist Hans Geiger [German: 1882–1945] invented this device for detection and rough measurement of radioactivity around 1910. It is a gas-filled tube that conducts electricity only when ionized by a charged subatomic particle. A version developed in 1928 with Wilhelm Müller, the Geiger-Müller counter, uses an amplifier to produce an audible click when a particle passes through, so that one can gauge the degree of radioactivity by the loudness and frequency of the clicks.

Gell-Mann, Murray (*physicist*)   [American: 1929–] Gell-Mann's main work has been a theoretical analysis of heavy subatomic particles using the mathematical idea of group, which since has become a basic underpinning of particle physics. In 1963 he (and independently Yuval Ne'eman [Israeli: 1925–]) showed that short-lived heavy particles do not decay as fast as previously predicted because of a quality Gell-Mann named strangeness. Arranging these particles into a group allowed Gell-Mann to predict the properties of a previously unknown particle, the omega-minus particle, discovered in 1964. That year Gell-Mann (and independently George Zweig [American: 1937–]) proposed that mesons, protons, and other heavy particles are constructed from smaller particles called quarks, an idea now generally accepted as proven by experiment.

gene (*genetics*)   The basic unit of inheritance is the gene, a term coined in 1909 by Wilhelm Johannsen [Danish: 1857–1927]. A gene consists of a length of nucleic acid (RNA in some viruses, DNA in all other organisms). A gene is often referred to as a blueprint; it controls a characteristic of an organism by regulating the synthesis of a specific polypeptide chain, a building block of proteins. The gene may have 2 or more forms, or alleles, each specifying a variation of the characteristic—for example, alleles for brown and blue eyes in humans.

gene pool (*genetics/evolution*)   All the genes of all the individuals in a population make up that population's gene pool. For example, all of the alleles (forms of a gene) of all the genes present in yellow perch living in a lake is the gene pool for that yellow perch population.

At any point in time, each allele occurs in a gene pool with a certain frequency. But with new generations, allele frequencies may change. Over time such changes in the gene pool result in the evolution of the population. Gene pools can change dramatically or remain extremely stable. Today's horses have a very different gene pool than that of the first horse, Eohippus. In contrast, today's horseshoe crabs are very similar to horseshoe crabs that lived 400,000,000 years ago.

gene splicing (*genetics*)   Splicing is the method by which researchers use a variety of enzymes to cut a molecule of DNA, insert a foreign gene, and then join together the pieces of the molecule. This recombination of genetic material is a basic technique of genetic engineering.

genetic code (*genetics*)   The sequence of nucleotides in a gene or in messenger RNA (mRNA) is a code that spells out the sequence of amino acids in a protein. For instance, UUU is a codon for the amino acid phenylalanine, and CAC is a codon for histidine. If an mRNA molecule includes the sequence UUUCAC-CAC it is translated into a protein with a phenylalanine-histidine-histidine sequence. This genetic code is basically the same in all organisms, from bacteria to mammals.

genetic drift (*evolution*)   In small, semi-isolated populations, the proportion of different alleles (gene forms), can change, or drift, due to variations in births and deaths. Consider, for example, a population of dogs with 2 alleles for ears: one allele for erect ears, the other for floppy ears. If by chance dogs possessing floppy-ear alleles die without leaving offspring, the allele will no longer exist in the population. The smaller the population, the greater the potential role of genetic drift in evolution.

genetic engineering (*genetics*)   Genetic engineering is the deliberate alteration of an organism's genetic material. The term refers primarily to recombination, which involves removing a segment of DNA from one organism and inserting it into the DNA of another organism. An organism that receives a foreign gene in this manner is said to be transgenic.

genetics (*sciences*)   Genetics is the study of genes and their variations. Geneticists are concerned with the composition of genes, how they work, how they change, and how they are passed from parent to offspring. Some geneticists specialize—focusing, for example, on the chemical structure of genes (molecular genetics) or how genes are distributed in populations of a species (population genetics). Other geneticists devote their efforts to applied sciences such as agricultural genetics and medical genetics.

genome (*genetics*)   The complete complement of DNA of an organism or a species is its genome. The genome of the bacterium that causes tuberculosis (*Mycobacterium tuberculosis*) consists of 4,411,529 units, or bases, of DNA. In contrast, the human genome contains an estimated 3,000,000,000 units. Only 10 to 15% of these units are believed to make up genes. Some of the remaining units have important functions, such as turning genes "on" or "off." But a large portion of DNA—much of it consisting of short sequences repeated over and over again—has no known purpose. Efforts to decode the human genome are under way.

genotype and phenotype (*genetics*)   The genetic makeup of an organism or a specific trait of that organism is its genotype. The physical appearance of the organism or trait is its phenotype. For example, a person's genotype for eye color may consist of 2 brown genes (BB), 2 blue genes (bb), or 1 brown and 1 blue gene (Bb). Because brown eye color has dominance, BB and Bb people have the same phenotype (brown eyes) even though they have different genotypes. Environmental conditions also can influence phenotypes. For example, poor nutrition stunts growth.

gentian (*plants*)   Herbaceous, mostly perennial plants found in cool, moist habitats, gentians (*Gentiana*) grow to about 2 ft (0.6 m), though many species are much shorter. They often are cultivated for their attractive funnel-shaped flowers, which are usually bright blue.

geological time scale (*geology*)   Geologists in the 1700s recognized that Earth's history extends far back in time, thousands of millions of years. Such spans, vastly longer than human life or history, came to be called geological time. Using rock strata and embedded fossils as guides, geologists described a hierarchy of nested intervals that form the geological time scale. From shortest to longest stage, these are called epoch, period, era, and eon.

Geological Time Scale

YEARS BEFORE PRESENT	GEOLOGICAL INTERVAL NAMES	SIGNIFICANT EVENTS
4,600,000,000	**ARCHEAN EON**	Earth cools to solid; monerans develop within a few million years
2,500,000,000	**PROTEROZOIC EON**	archaea, protists, and first animals evolve; first mass extinction (mainly of algae) occurs c700,000,000 years ago

(*continued*)

geological time scale

Geological Time Scale (continued)

YEARS BEFORE PRESENT	GEOLOGICAL INTERVAL NAMES	SIGNIFICANT EVENTS
545,000,000	**PHANEROZOIC EON** **Paleozoic Era** Cambrian Period	explosion of animals into all major phyla of today; Late Cambrian extinction (c505,000,000 years ago) affects mainly trilobites
500,000,000	Ordovician Period	conodonts, possible ancestors of vertebrates appear; final Ordovician extinction (c435,000,000 years ago) of marine species affects nearly half of all genera
425,000,000	Silurian Period	plants and animals invade land
395,000,000	Devonian Period	true fish and amphibians evolve; Late Devonian extinction (c367,000,000 years ago), linked to impact craters, affects about half marine genera—corals, brachiopods, armored fish, and trilobites
350,000,000	Carboniferous Period	giant club mosses and ferns lay down major coal beds; early reptiles appear
290,000,000	Permian Period	early conifers and mammal–like reptiles appear; Final Permian (*P/T⁻) extinction of c250,000,000 years ago brings Paleozoic Era to a close with loss of perhaps as many as 90% of marine species, including last of trilobites, and many land animals
249,000,000	**Mesozoic Era** Triassic Period	early mammals and early dinosaurs appear; Late Triassic mass extinction of 198,000,000 years ago removes quarter of land families and half of marine genera
190,000,000	Jurassic Period	flowering plants and classic dinosaurs appear; Final Jurassic extinction of c144,000,000 years ago closes period
130,000,000	Cretaceous Period	height of age of reptiles, including dinosaurs, flying reptiles and birds, and marine reptiles; era ends with Final Cretaceous (K/T) extinction of c65,000,000 years ago when asteroid strikes Earth near Yucatán; dinosaurs, flying reptiles, and marine reptiles disappear
66,500,000	**Cenozoic Era** Tertiary Period Paleocene epoch	early primates and horses, rodents, and sycamores evolve
62,400,000	Eocene epoch	whales, penguins, roses, bats, camels, early elephants, dogs, cats, and weasels appear
36,600,000	Oligocene epoch	deer, pigs, saber–toothed cats, and monkeys evolve
23,500,000	Miocene epoch	seals, dolphins, giraffes, bears, hyenas, early apes, grasses, and composites occur

YEARS BEFORE PRESENT	GEOLOGICAL INTERVAL NAMES	SIGNIFICANT EVENTS
5,200,000	Pliocene epoch	apes, early hominids, including first humans, mammoths, giant sloths, and armadillos evolve
1,600,000	Quaternary Period Pleistocene epoch	*Homo erectus,* Neandertal and modern humans arise; most recent Ice Age ends c12,000 years ago, marked by extinction of many large vertebrate species
11,000	Holocene epoch	flora and fauna of today

geology (*sciences*)   Although geology is defined as the study of the *geos* or whole Earth, in practice geology is mainly concerned with the solid and nearly solid rocks, soils, and magma, leaving the air to meteorology, the seas to oceanography, and Earth's place in space to astronomy. Because fossils are used to identify rock strata, paleontology is closely aligned to geology. Branches include mineralogy, petrology (study of rocks), volcanology, and plate tectonics, which uses the results of seismology (study of earthquake waves).

geometric sequence and series (*calculus*)   A geometric sequence (or geometric progression) consists of numbers for which the ratio of each to the one before is a constant. If a is the first term and r the constant ratio, the sequence is a, ar, ar^2, ar^3 . . . . For example, if the first term is 4 and the ratio $\frac{1}{2}$, the sequence begins 4, 2, 1, $\frac{1}{2}$, $\frac{1}{4}$, $\frac{1}{8}$, ...

A geometric series is the indicated sum of a geometric sequence, or $a + ar + ar^2 + ar^3 + \ldots$. The sum of n terms of a geometric series is $S = a(1 - r^n)/(1 - r)$. For example, the sum of $4 + 2 + 1 + \frac{1}{2} + \frac{1}{4} + \frac{1}{8}$, which has 6 terms, is $4[1 - (\frac{1}{2})^6]/(1 - \frac{1}{2})$, or $7\frac{7}{8}$. If r is less than 1, a geometric series that increases to infinity has the sum $a/(1 - r)$. The series $4 + 2 + 1 + \frac{1}{2} + \frac{1}{4} + \frac{1}{8} + \ldots$ has the sum $4/(1 - \frac{1}{2})$, or 8.

geometry (*mathematics*)   Geometry means "earth measuring." Greeks thought geometry originated in methods for laying out boundaries of fields after the annual flood of the Nile in Egypt. Although Egyptian and Mesopotamian mathematicians found many practical rules for lengths, areas, and volumes, geometry as reasoning about properties of figures formed from 0-dimensional points began about 600 BC with the Ionian Greeks on the southern coast of what is now Turkey. Euclid arranged the various proofs concerning figures formed from curves with no width and surfaces with no depth into a powerful axiomatic system that was the basis of geometry for the next 2000 years.

In the 19th century non-Euclidean geometry expanded the field considerably. Other forms of geometry, such as projective geometry, employed simplified assumptions that generalized results. In 1872 Felix Klein gave the definition that is the basis of present-day understanding: Geometry is the study of those properties of figures that remain invariant under a particular group of transformations.

geophysics (*sciences*)   The use of techniques and theories from physics to study Earth is termed geophysics. Thus, seismology (study of earthquake waves) and the study of Earth's interior mantle and core, including the physics of volcanoes, represent the largest part of geophysics. Other branches involve studying Earth's magnetic field and the exact shape of the planet (geodesy). Parts of oceanography (wave movements, for example) and meteorology are often treated as geophysics as well.

geothermal energy (*environment*)   Natural radioactive decay deep within Earth heats subterranean rocks. When water comes in contact with the rocks, it turns to steam. Where cracks open to the surface, some of the steam is released in hot springs and geysers. This geothermal energy can be used for heating and electric power generation. People have tapped natural sources of geothermal energy and have created artificial sources by drilling deep wells, pumping water down the wells and collecting the steam and hot water that result. Geothermal energy is environmentally attractive because resources are practically unlimited and pollution is much less than that created by burning fossil fuels.

geranium (*plants*)   Plants of the genus *Geranium*, called wild geraniums or cranebills, mostly are perennial herbs of temperate regions. The closely related genus *Pelargonium*, native to southern Africa, includes the familiar garden and house geraniums. A long capsular fruit that projects from the flower stalk like the bill of a stork characterizes members of both genera.

gerbil (*animals*)   Native to hot, arid regions of Africa and Asia, gerbils are rodents of Family Muridae. They are 2 to 8 in. (5 to 20 cm) long, with an equally long tail. Most gerbils live in colonies in underground tunnels and burrows. The Mongolian gerbil (*Meriones unguiculatus*) is a popular pet.

germanium (*elements*)   A metal, atomic number 32, germanium was discovered in 1886 by Clemens Winkler [German: 1838–1904], who named it for Germany. It had been predicted by Mendeleyev in 1871 as "eka-silicon," since the metal occupies the place between silicon and tin in the carbon family. Like silicon, it is an important semiconductor that can be used in electron devices; it also is employed in optical switches. Chemical symbol: Ge.

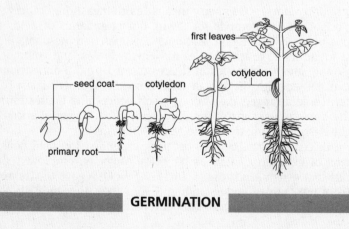

GERMINATION

germination (*physiology*)   The process in which a seed or spore begins to grow, called germination, involves a series of steps, as exemplified by a sprouting bean seed. First the seed absorbs water. This activates enzymes that break down food stored in the cotyledons, releasing energy. The seed swells and the protective seed coat splits. The embryo root grows out and downward, eventually giving rise to the entire root system. The embryo stem lengthens, lifting the cotyledons out of the soil. Two small green leaves appear between the cotyledons. A bud that develops into the bean plant's main stem forms between the leaves. As soon as the leaves are exposed to light, the plant shifts to adult metabolism and begins to carry out photosynthesis.

germ theory (*biology*)   Until fairly recently, people did not know what caused diseases such as plague, swine fever, and tomato wilt. Theories blamed everything from alien spirits to a deficiency of nervous energy to spontaneous generation from rotting material. In the 1870s Pasteur and Koch independently established the germ theory of disease, which holds that microbes ("germs") cause infectious diseases. This soon led to the isolation of the causative agents of diphtheria, tuberculosis, anthrax, and other age-old scourges.

gestation period (*physiology*)   The length of time that a developing embryo is carried in the uterus of a mammal is called the gestation period or pregnancy. The gestation period usually varies directly with size of the animal, whereas the number of young borne at one time usually is inversely proportional to size: mice and other rodents typically average 6 or more young per litter while elephants and other large mammals usually produce 1 young per pregnancy.

MAMMAL	GESTATION PERIOD (AVERAGE, IN DAYS)
mouse, house	21
rabbit, cottontail	30
squirrel	40
dog, domestic	63
cat, domestic	63
raccoon	63
jaguar	100
lion	110
pig	115
goat	150
monkey, rhesus	164
hippopotamus	240
cattle	280
human	280
horse	336
elephant	640

geyser (*geology*) A recurrent fountain of hot water caused by a hot spot in Earth's crust is called a geyser. Magma near the surface heats groundwater to temperatures well above ordinary boiling because high pressure raises the boiling point. When some of the superheated water reaches regions of lower pressure, a large amount turns to vapor and the hot vapor propels fluid out of the geyser's vent, emptying the geyser. The pattern repeats as more water fills the geyser's lower chambers.

gibberellin (*biochemistry*) One of the main classes of plant growth regulators, gibberellins induce the cell elongation that promotes stem growth. They also stimulate production of vascular tissue, increase the rate of fruit growth, and promote seed germination. Other gibberellins are found in the fungus *Gibberella fujikuroi*, from which they get their name, and in certain bacteria.

gibbon (*animals*) Closely related to hominids, gibbons (Family Hylobatidae) live in tropical forests of Southeast Asia. They are active during the day, feeding mainly on leaves and other plant matter. Gibbons have extremely long arms, which they use to swing from tree to tree. To maintain balance when running, they hold the arms over their head. The largest species, the siamang (*Hylobates syndactylus*), grows to 35 in. (90 cm) tall, with an arm span up to 5 ft (1.5 m). The siamang has a hairless throat sac that inflates and serves as a resonating chamber when it calls, creating a hoot that can be heard 2.5 mi (4 km) away.

Gibbs, Josiah Willard (*chemist*) [American: 1839–1903] Although Gibbs was the founder of chemical thermodynamics, his highly theoretical work seems to belong more to the world of mathematical physics, of which he was a professor (at Yale) than to chemistry. He developed mathematical techniques for predicting the outcome of chemical reactions before making a physical trial. He also derived the laws of thermodynamics from the basic concept of heat as movement of particles.

Gilbert, William (*physicist*) [English: 1544–1603] Gilbert's book *De magnete* (in English called *On the Magnet, Magnetick Bodies Also, and on the Great Magnet the Earth*), published in 1600, is among the seminal works of the Scientific Revolution. Gilbert describes his experiments with lodestones and compasses, gives methods of making iron display magnetic effects, introduces the concept of magnetic poles, and recognizes the magnetic field of Earth.

gill (*anatomy*) Respiratory organs of aquatic animals, gills are usually made of many thin layers; these create a large surface area through which gases can diffuse. Gills are richly supplied with blood for transport of gases to and from other parts of the body. They occur in many mollusks and crustaceans, some annelid worms, fish, and some amphibians.

ginger (*plants*) A perennial herb native to tropical Asia, ginger (*Zingiber officinale*) has a thick, aromatic rhizome (underground stem) that has been used in cooking since ancient times. Upright stems bearing large elongated leaves grow to about 3.3 ft (1 m) tall. The small flowers are borne in dense spikes composed of overlapping bracts (leaflike structures), with each bract enclosing a single flower. The ginger family, Zingiberaceae, also includes turmeric (*Curcuma longa*) and cardamom (*Elettaria cardamomum*).

ginkgo (*plants*) The sole survivor of a group of gymnosperm plants dating from 200,000,000 years ago, the ginkgo or maidenhair tree (*Ginkgo biloba*) is characterized by its fan-shaped leaves, which turn yellow in autumn. A ginkgo tree is either male or female; the latter produce foul-smelling plumlike fruits if a male tree is nearby. Native to China, ginkgos grow to heights of 100 ft (30 m) or more.

ginseng (*plants*) Ginsengs (*Panax*) are perennial herbs that grow to heights of about 3 ft (0.9 m). They have a fleshy aromatic root, compound leaves, and clusters of small greenish yellow flowers that develop into red berries. The root of the Chinese species (*P. ginseng*) has been the source of a medicinal tonic for some 5000 years. Today the root of the American ginseng (*P. quinquefolius*) is used similarly.

giraffe (*animals*) The tallest mammal is the giraffe (*Giraffa camelopardalis*), which may stand more than 17 ft (5 m) tall and weigh 1800 lb (800 kg). One-third of the giraffe's height is neck; despite its length, however, the neck contains only 7 bones, the same number found in the neck of other mammals, including humans. Giraffes are native to dry savannas of Africa. They feed mainly on plants such as acacias and mimosas, using their long, flexible tongue and muscular upper lip to remove leaves from branches. They can go for weeks without drinking water; to drink—or to graze on the ground—a giraffe must spread its forelegs far apart.

The giraffe family, Giraffidae, also includes the okapi (*Okapia johnstoni*), which resembles the giraffe but has a shorter neck and zebralike stripes on its legs. It inhabits forests of equatorial Africa.

gizzard (*anatomy*) The gastrointestinal tract of many animals has a muscular region called the gizzard, where food is ground up before entering the stomach. The gizzard serves the same purpose as the teeth and jaws of mammals. In the gizzards of earthworms and many birds and reptiles, particles of soil, sand, or gravel ingested with food help grind up the food.

glacier (*geology*) If more snow falls winter after winter than melts each summer, a large mass of ice called a glacier develops. Lower layers become compacted into a solid mass called firn; further pressure over time changes the firn

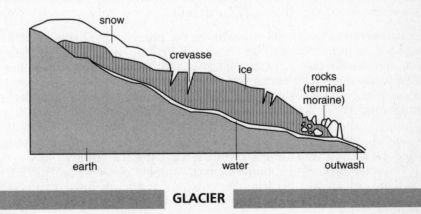

GLACIER

to ice. The glacier's weight eventually causes the bottom layer to melt, permitting the glacier to slide and float downhill. When the glacier reaches level ground or an uphill slope, the weight of the ice behind it pushes the front forward until it finally reaches a location—usually miles from its origin—where more snow melts each year than the glacier moves. Glacial movement is usually slow, a matter of a few feet a day, but sometimes glaciers spurt hundreds of feet in a day. As the glacier moves, it creates geological formations that persist long after warming has melted the entire glacier.

gland (*anatomy*)   Any organ that produces and secretes 1 or more substances is a gland. In animals there are 2 basic types of glands: endocrine glands secrete substances directly into the blood; exocrine glands secrete substances via ducts. Nectaries—which secrete nectar—are an example of plant glands. Some organisms have individual cells that act as glands; for example, *Hydra* has gland cells that discharge secretions containing digestive enzymes.

glass (*materials science*)   A liquid that cools quickly enough into a solid maintains unordered locations of molecules, ions, or atoms rather than forming a crystal. Such a solid is called a glass, after the similarly transparent, brittle manufactured glass. Glasses fail to have definite melting points, but when heated they gradually soften and begin flowing.

Manufactured glass is the oldest synthetic produced by combining materials. It was first made in bulk about 2000 BC but was present in ceramic glazes and in imitation lapiz lazuli thousands of years earlier. Silica, soda ash (sodium carbonate), and limestone (calcium carbonate) are mixed and heated, driving out the carbon as carbon dioxide and combining silicon, oxygen, and metals into the silicates that form ordinary glass. Other elements, such as lead or boron, are often added to produce specific properties.

glial cell (*cells*)   Surrounding most nerve cells, or neurons, in a vertebrate's nervous system are glial cells (also called glia, from the Latin word for glue). The only parts of nerve cells not covered by glial cells are the areas adjacent to synapses. Glial cells come in many different forms and perform various functions, including mechanical and nutritional support, electrical insulation, and aid in embryonic development. One type, Schwann cells, surrounds the axons of neurons outside the brain and spinal cord.

global warming (*environment*)   During the 20th century, Earth's average surface temperature increased by about 1°F (0.5°C), with the 1990s being the warmest decade on record. Scientists believe that this global warming is at least partly induced by human activities, particularly the burning of fossil fuels but also rice farming, cattle ranching, the use of fire to clear land, and other endeavors. These activities release carbon dioxide and other heat-trapping gases into the atmosphere, creating a greenhouse effect. Without significant action to reduce emissions, scientists predict that Earth's surface temperature will increase another 2° to 7°F (1° to 3.5°C) in the 21st century.

The consequences of continued global warming would be profound, including more extreme weather with more frequent and severe droughts and storms. Shrinkage of ice sheets plus expansion of water as it warms could cause sea levels to rise as much as 2 ft (0.6 m), inundating many low-lying land areas. Plant and animal species unable to adapt or migrate fast enough to keep pace with climatic changes would be threatened with extinction. There is growing evidence that such changes already are occurring.

globular cluster (*stars*)   Some 100,000 to 10,000,000 stars gather into a tight sphere, denser toward the center, known as a globular cluster. Since the stars in a globular cluster show all the signs of old age, one theory is that these formations, more than 100 of which are found near the center of the Milky Way galaxy along with similar ones in other galaxies, were the first to condense out of the galaxy-forming cloud. The age of the globular clusters is so great—perhaps 12,000,000,000 years, as to put a lower limit on the age of the universe.

glucose (*biochemistry*)   A simple sugar with the formula $C_6H_{12}O_6$, glucose is formed during photosynthesis in green plants and algae. It is the main fuel for nearly all organisms. During cell respiration glucose is broken down to carbon dioxide and water, with the release of energy. Glucose is also the substance from which other carbohydrates, such as starch and glycogen, are made.

gluon (*particles*)   There are 8 different bosons exchanged between quarks to produce the force that keeps them tightly bound. These bosons are called gluons. Trading gluons changes a quality that each quark has called its color (on the analogy of colors mixing to produce white, since combinations of quarks do not exhibit color themselves). The color force is also the basis of the strong force that holds the nucleus together, although the strong force also involves the exchange of pairs of quarks bound into virtual pions.

glycogen (*biochemistry*)   Often called animal starch, glycogen is the main storage carbohydrate in animal tissues. It is also found in certain bacteria, blue-green algae, and fungi, but not in plants. Glycogen is formed from the sugar glucose and stored mainly in liver and muscle cells. In the liver glycogen can be converted to glucose by hydrolysis. Glycogen in muscle is broken down to lactic acid by glycolysis.

glycoprotein (*biochemistry*)   A protein that has sugars as part of its molecule is called a glycoprotein. There are 2 major types. Intracellular glycoproteins are present on the surface of cell membranes. They play a role in the recognition of molecules (hormones, viruses, etc.) that interact with the cells. Secretory glycoproteins are secreted by various cells. For example, immunoglobulins are secreted by certain white blood cells; jellylike mucins are secreted by mucous membranes.

gneiss (*rocks and minerals*)   Gneiss (pronounced "nice") is a metamorphosed granite that preserves the original quartz, feldspar, and a third mineral (typically a dark mineral, such as a mica or hornblende). Heat and pressure cause these minerals to migrate into alternating light and dark bands.

gnetophyte (*plants*)   A small group of gymnosperms, sometimes classified in their own phylum, Gnetophyta, gnetophytes grow mainly in dry tropical and subtropical habitats. There are 3 genera, each sharing certain characteristics with angiosperms (flowering plants). This has led some scientists to theorize that gnetophytes share a common ancestry with angiosperms.

Ephedra species are primarily low shrubs, native to China and western North America. *Gnetum* species are vines, shrubs, or small trees, native to the tropics. The single *Welwitschia* species lives in the Namib Desert of southwest Africa. It has a large underground root and produces only 2 leaves during its entire life, though it may live hundreds of years. Each year the leathery leaves grow at their base, even as their tips wither or become shredded into strips by the harsh environment.

goat (*animals*)   Members of the bovid family of ruminant mammals, goats (*Capra*) are surefooted animals that generally live in small herds in mountainous regions. Males often have a beard and, during the breeding season, a strong odor. The best-known species is the ibex (*C. ibex*), native to Europe, Asia, and northern Africa. The domestic goat (*C. hircus*), now in numerous, varied breeds, was among the first hoofed mammals to be tamed, more than 8000 years ago.

The mountain goat (*Oreamnos americanus*) of western North America is not a true goat but, rather, a goat antelope. It has long white hair and slender black horns that curve backward. Its hooves, with a sharp rim surrounding a soft inner pad, enable it to climb nimbly over steep, icy terrain and even to scale cliff faces when danger threatens.

goby (*animals*)   Constituting the family Gobiidae, gobies are tropical fish that range from 0.5 to 19 in. (1.7 to 48 cm) long. Most species are marine; they are particularly abundant in coral reefs, where they inhabit holes and crevices. Species that live in freshwater spend their larval stage in the sea. A goby has 2 dorsal fins. The pelvic fins are connected by a membrane, forming a suckerlike disk that enables the goby to cling to the substrate.

Gödel's theorem (*logic*)   Although Kurt Gödel [Austrian-American: 1906–78] made many contributions to logic (including showing that every statement in the most basic form of logic can either be proved or disproved) and some to physics, his 1931 proof that any system that contains the arithmetic of natural numbers is either not complete or not consistent, known as Gödel's theorem, is his most famous. The proof is based on an axiomatic system of numbers taken to be equivalent to the natural numbers. Gödel used a complex system to assign a number to every statement derived from the axioms and established that a number (and therefore a statement) exists that cannot be derived from the axioms if the axioms are consistent. If the unprovable statement is added to the axiom list, however, it will either be inconsistent with the previous axioms or create yet another unprovable statement.

gold (*elements*)   Gold (or else copper) was the first element to be recognized as a separate substance. Each occurs in a relatively pure form called a native element. Gold, atomic number 79, has been used primarily for jewelry and decoration and in coins. It is the most malleable and ductile element. The chemical symbol, Au, is from the Latin name *aurum*.

golden algae (*protists*)   Members of the phylum Chrysophyta appear yellowish or golden because orange carotene and yellow xanthophyll pigments mask the green chlorophyll. Most members of the phylum, including the diatoms, are unicellular. The cell wall generally consists of two overlapping halves and frequently contains silica. Golden algae inhabit both fresh and salt water.

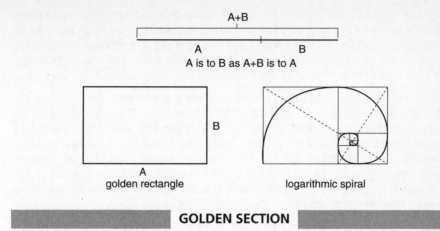

A+B

A   B
A is to B as A+B is to A

B

A
golden rectangle

logarithmic spiral

GOLDEN SECTION

golden section (*geometry*)   The golden section, or golden ratio, is a number used to divide a line segment in 2 so that the length of the whole segment is to its larger part in the same ratio as the larger part is to the smaller. The ratio is about 1.61803398, symbolized as φ. Many writers since the early 19th century have claimed that architects and painters base works on the golden section, perhaps unconsciously. The golden rectangle has sides with the ratio φ.

Golgi body (*cells*)   Named for their discoverer, Camillo Golgi [Italian: 1843–1926], Golgi bodies are **organelles** of almost all protist, fungal, plant, and animal cells. Located near the nucleus, they consist of stacks of flattened sacs enclosed by membranes. Their function is to store, modify, package, and distribute proteins made in the **endoplasmic reticulum**. Golgi bodies also synthesize long-chained sugars called polysaccharides, such as the cellulose used to make cell walls in plants. The number of Golgi bodies varies but averages 10 to 20 in animal cells and hundreds in plant cells.

gonad (*anatomy*)   The reproductive cells, or gametes, of animals develop in organs called gonads. A male animal has gonads called **testes**, which produce sperm. A female animal has gonads called **ovaries**, which produce eggs (ova). Some invertebrates, such as earthworms, are **hermaphrodites**, with both male and female gonads.

Gondwanaland (*geology*)   The continents of what is now the Southern Hemisphere were joined before 120,000,000 years ago in a large landmass called Gondwanaland. Surprisingly, the recognition of Gondwanaland precedes the theory of **continental drift**. In 1885 Eduard Suess [Austrian: 1831–1914] traced the occurrence of a fossil fern through Africa, South America, Australia, and India. Since the fern could not have spread over oceans, Suess postulated that the continents had been joined in some way. He named the supercontinent Gondwanaland after a region in India.

Goodall, Jane (*primatologist*)   [English: 1934–] Goodall is best known for her long-term observations of chimpanzees in their natural habitat, particularly at the Gombe Stream Research Center in Tanzania. She has stressed that chimpanzees—

humans' closest living relatives—resemble humans not only genetically and physiologically but also behaviorally. For example, chimpanzees use such gestures as kissing, tickling, embracing, holding hands, and patting one another on the back. Goodall discovered that chimpanzees use twigs and other tools to capture termites, and that they are not strict vegetarians, as had been believed.

gooseberry (*plants*) Generally classified in the genus *Ribes*, gooseberries are deciduous shrubs native to Eurasia and North America. They have long, often arching branches usually armed with large thorns. The small, inconspicuous flowers—pollinated mainly by bees—give rise to small clusters of edible berries. In the same genus are the currants—so named around 1550 because their berries appear to resemble dried currants (raisins made from small, seedless grapes). Currant shrubs usually lack thorns. The berries are smaller than those of gooseberry shrubs and borne in larger clusters.

gorilla (*animals*) The largest living ape, the gorilla (*Gorilla gorilla*) grows to a height of 6 ft (1.8 m) and a weight of more than 400 lb (180 kg). Native to forests of tropical Africa, gorillas spend most of their time on the ground. They are shy vegetarians that travel about in groups called troops. Gorillas have a black hairless face with a prominent brow ridge over the eyes, a wide mouth, and a heavy jaw. The powerful arms are longer than the legs.

gourami (*animals*) Freshwater fish native to southern Asia, gouramis belong to Family Anabantidae, which also includes climbing perches and bettas, or Siamese fighting fish; many members of the family are popular aquarium fish. These fish often inhabit oxygen-poor water and rise to the surface to take gulps of air. All have an auxiliary respiratory structure called a labyrinth. Located in the upper part of the gill cavity, the labyrinth is used to remove oxygen from air.

gourd (*plants*) The gourd family, Cucurbitaceae, comprises about 700 species native to tropical and subtropical habitats. Most species are rapidly growing vines with coiling tendrils. The flowers, usually yellow, tend to be showy but open only briefly. The fruit is a type of berry called a pepo; it has a hard rind surrounding fleshy pulp with many seeds. Many gourds, including pumpkins, melons, cucumbers, and squashes, are cultivated for their edible fruits and exist in numerous variants.

grain (*botany*) The fruit of a grass is called a grain, or caryopsis. It is a dry fruit containing a single seed; the seed wall is completely fused to the thin wall of the ripened ovary. A grain is rich in nutrients, making it an excellent food for animals. Of particular importance to humans are the grains of the so-called cereal grasses, including barley, corn, oats, rice, and wheat. The discovery that the grains of wild grasses would grow into new plants if placed in soil probably marked the start of the Agricultural Revolution some 11,000 years ago.

Gram staining (*monera*) Introduced in 1884 by Hans Gram [Danish: 1853–1938], Gram staining employs the compound crystal violet to differentiate bacteria. Gram-positive bacteria retain the purple-violet stain but gram-negative bacteria lose the color when subjected to alcohol. The gram-negative bacteria can then be stained with safranin, which gives them a red color. The different reactions may be due to structural differences. In addition to staining reactions, the 2 groups differ in other ways. For example, gram-positive bacteria generally are more susceptible to penicillin whereas gram-negative bacteria are more susceptible to antibiotics such as streptomycin. Thus distinguishing between them is important in treating certain diseases.

granite (*rocks and minerals*)   Granite is the most common igneous rock of the continents, forming a less dense mass that rises above the low basalts of the ocean floor. Granite is formed when large masses of lighter minerals, especially feldspar and quartz, and a dark mineral, such as a mica or hornblende, rise into the crust and cool underground. Slow cooling makes crystals large enough to be seen easily, especially since the 3 minerals crystallize separately. Consequently granite has a mottled appearance, often pink, white, and black. The same materials form rhyolite if the rock cools faster, often because it reaches the surface. Heat and pressure turn granite into gneiss.

grape (*plants*)   Members of the grape family, Vitaceae, are woody vines that climb by means of tendrils. The vines often grow to lengths of 60 ft (18 m) or more. The leaves may be simple or compound. The clusters of small, insignificant flowers give rise to fleshy berries that contain tiny seeds called pips. The first species to be domesticated probably was the grape *Vitis vinifera* of Eurasia, which has been cultivated for more than 4000 years.

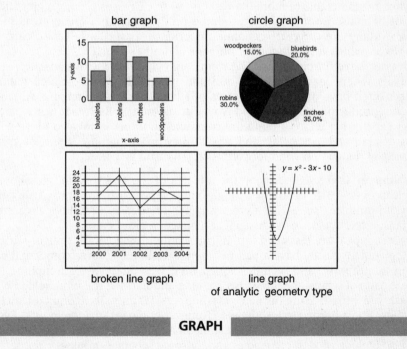

bar graph   circle graph

broken line graph   line graph
of analytic geometry type

GRAPH

graph (*analytic geometry/statistics*)   The noun *graph* has several meanings in mathematics. In elementary mathematics, students study ways to display data, including the bar graph, broken line graph (often called a line graph), and circle graph. In algebra and in mathematics based on analytic geometry, a 2-dimensional graph is a curve defined by 1 or more equations, while a 3-dimensional graph is a surface. The idea behind all graphs is use of a geometric figure to represent a set of numbers paired with classes or pairs (or triples) of numbers. In a bar graph the length of a rectangle indicates a number matched with a class, such as types of birds in a bird count. A broken line graph shows number pairs in which 1 number means time, such as number of birds per year. A circle graph shows pairs that are ratios or percents compared with a whole, such as bluebirds with all birds in a count. In analytic geometry, the numbers are abstract and usually run continuously. (*See also* **network**.)

graptolite (*paleontology*)   Named in 1725 by Linnaeus—who considered them to be minerals—graptolites are an extinct group of marine invertebrates that lived in large colonies. In a typical colony, the tiny animals were enclosed within branching tubes that easily fossilized. Because they were common worldwide from the Cambrian into the Carboniferous, the numerous graptolite species often are used to identify rock strata.

grass (*plants*)   The grass family, Gramineae or Poaceae, comprises some 9000 species of annual and perennial flowering plants. Most, including the familiar lawn grasses, are low-growing herbs. But corn grows to 15 ft (4.5 m) tall and bamboos reach heights of more than 100 ft (30 m). Grasses occupy a great diversity of habitats, from seashore to alpine tundra and from arctic areas to the tropics. Many are of great economic importance, particularly barley, oats, rice, and wheat, which furnish food for humans and other animals.

Grasses have fibrous roots. The stems are cylindrical, with nodes at regular intervals where leaves form. In addition to upright stems—called culms—grasses often have underground stems or stems that creep along the surface of the soil, producing new roots and leaves at the nodes. Each leaf has 2 parts: a lower sheath that wraps around the culm and an upper, elongated blade. Flowers are small and lack petals and sepals; they are borne in clusters, or inflorescences, and depend on wind for pollination. The fruit is a grain.

grasshopper (*animals*)   Among the athletes of the insect world, grasshoppers can leap over 3.3 ft (1 m) in a single jump. They also are noisy, using their wings to produce "songs" in a process called stridulation. Grasshoppers are vegetarians; a few species can cause great damage to grass crops.

Grasshoppers are classified in the insect order Orthoptera ("straight wings"). Orthopterans typically have a pair of antennae ("horns"), large compound eyes, biting mouthparts, and well-developed hind legs adapted for jumping. They have 2 pairs of wings; the forewings are long and narrow; the hind, folded when at rest, are broad. Long-horned grasshoppers (Family Tettigoniidae) include katydids. Short-horned grasshoppers (Family Acrididae) include locusts. Pygmy grasshoppers (Family Tetrigidae) are sometimes called pygmy locusts. Orthopterans also include crickets, cockroaches and, in some schemes, mantises.

grassland (*ecology*)   One of Earth's major biomes, grassland forms in continental interiors in temperate and tropical regions where precipitation is insufficient to support tall trees. Temperate grasslands, such as the prairies of North America, average 10 to 30 in. (25 to 75 cm) of precipitation annually. Summers are hot, winters cold, and prolonged droughts are not unusual. Tropical grasslands, such as the savannas of Africa, average 35 to 60 in. (89 to 150 cm) of rainfall annually. They are warm year-round but have distinct wet and dry seasons.

The grasses that constitute the dominant vegetation have extensive root systems to soak up water. Also, many of the plants' growing and reproductive structures are underground. This enables the plants to survive the fires that periodically sweep across grasslands and to grow back after feeding by large herds of grazing animals.

Gravettian (*paleoanthropology*)   The Gravettian culture comprised a widespread way of life centered in eastern Europe about 25,000 BC. It is characterized by Venus figurines, statues of women with exaggerated female characteristics; streamlined stone points, including microliths; baked clay figurines of animals, often ritually broken; use of bone for fuel; buildings often made from mammoth bones next to pits for meat storage; and trade with distant sources for seashells.

graviton (*particles*)   The graviton is a boson that produces gravitational force by its exchange with all kinds of particles, but so far it is known only in theory. No experiments have detected an actual graviton either as a particle or as a wave. Current experiments have some chance of detecting very strong gravity waves from astrophysical events, from which a graviton could be inferred.

gravity (*physics*)   Gravity once meant attraction between a body and Earth while gravitation was used for the universal force between any 2 bodies with mass; but modern usage, when any distinction is made at all, calls the concept of mutual attraction gravity and the force of attraction gravitation. Newton recognized in 1665 that the same force that causes an apple to fall also keeps the Moon in its orbit; his discovery is called universal gravitation. Newton showed that gravitational force between 2 objects is directly proportional to the product of masses (M and m) and inversely proportional to the square of distance d between their centers ($F = GMm/d^2$ where G is a universal constant, now calculated as 6.672×10^{-11} N-m^2/kg^2).

Einstein's general theory of relativity treats gravity as inertia acting in space curved by mass. Two masses together curve space so that "straight lines," in which inertia propels masses, curve toward each other, producing gravitational force. Einstein's theory calculates slight differences from Newton's. Experiments support relativity. Modern particle physicists explain gravity in terms of exchange of particles called gravitons, yet to be detected.

Gray, Asa (*botanist*)   [American: 1810–88] A noted plant collector, Harvard University professor, and author of a series of botany textbooks, Gray helped found the study of plant geography. He noted similarities in plants of eastern Asia and eastern North America, theorizing that the plants were descendants of a single species that had lived across the Northern Hemisphere prior to the Ice Age.

Great Attractor (*cosmology*)   The Local Group of galaxies and 2 nearby superclusters of galaxies all seem to be moving in the same direction with respect to cosmic background radiation, apparently pulled by the gravitational force of an unseen object nicknamed the Great Attractor. The Great Attractor produces the effect of an object of giant mass situated behind the Hydra-Centaurus super-cluster of galaxies at about twice the distance from that supercluster as the supercluster is from Earth. However, there may be some other unknown cause for galaxies' streaming in that direction.

Great Wall (*cosmology*)   In 1986 astronomers discovered that the galaxies form a great sheet about 500 million light-years long; they named this the Great Wall. A similar structure is found in the other direction—the Southern Wall, which is more than 300 million light-years long, is made from more than 1000 galaxies. In between the walls are empty regions—great voids. This combination has caused some to propose that the universe is a foam whose walls are constructed from galaxies.

grebe (*animals*)   Found worldwide except in Antarctica, grebes (Family Podicipedidae) are aquatic birds that look somewhat like ducks but have a slender, pointed bill; usually a long, slender neck; and feet that are only partly webbed. Although many species migrate with the season, grebes are rather weak fliers. They spend most of their lives on water, even sleeping and raising their young there, and often carrying the chicks around on their backs.

green algae (*protists*)   The more than 7000 species of grassy-green algae that make up Phylum Chlorophyta are a diverse group, ranging from one-celled types to filamentous forms to many-celled flat sheets. Some green algae, such as sea lettuce (*Ulva*), are marine, but most live in fresh water, either as part of the plankton or attached to submerged rocks and other substrates. Still other species grow in moist places on land, including soil, rock, tree bark, animal fur, and even snow and ice. The algal component of many lichens are green algae.

Green algae share important characteristics with plants; for example, they both have chlorophyll *a* and *b*, and food is usually stored as starch. It is possible that plants evolved from green algae.

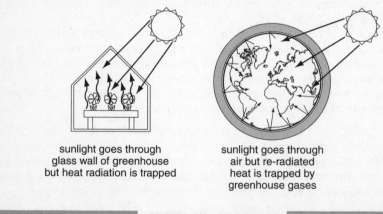

sunlight goes through
glass wall of greenhouse
but heat radiation is trapped

sunlight goes through
air but re-radiated
heat is trapped by
greenhouse gases

GREENHOUSE EFFECT

greenhouse effect (*waves*)   A greenhouse is a building made largely from glass or plastic used for keeping plants warm (which is why it is "green"). Sunlight can enter the greenhouse but glass or plastic is not transparent to heat radiation, so a greenhouse stays warm on a sunny day even when it is cold outside. The greenhouse effect employs the same process on a global scale. Air is largely transparent to sunlight. Energy from the sunlight heats land and water, which then radiate heat as infrared radiation. Certain gases, such as carbon dioxide and methane, trap infrared radiation, heating the air. A concern is that human activities are raising levels of the "greenhouse gases" and causing unwanted global warming.

grooming (*ethology*)   Members of certain mammal species, particularly chimpanzees, macaques, and other primates, spend much time grooming one another, a behavior similar to preening among birds. Grooming has much more significance than simply the removal of dirt, dandruff, and parasitic insects. It is a social gesture, strengthening peaceful relationships between individuals.

group (*algebra*)   An algebraic structure consisting of a set and 1 operation is a group if (1) the structure exhibits closure; (2) obeys the associative property for that operation; (3) contains an identity element; and (4) for every element possesses an inverse with respect to the identity element. Familiar groups include the integers and addition, which form a group whose identity element is 0. The rational numbers, excluding 0, form a group whose identity element is 1. Both groups such as these are abelian (meaning that the operations obey the commutative property). Also, each has an infinity of elements.

Many groups are finite and do not involve numbers. One example is based on clockwise rotations of an equilateral triangle *ABC* about its center. Group elements are *ABC, CBA,* and *BAC,* where the first letter is the top vertex after rotation (*CBA* is a rotation of 120° from *ABC*), and an operation • means "followed by." *ABC* • *BAC* = *BAC* means that starting with *ABC* a rotation of 240° brings the triangle to *BAC,* while *CBA* • *BAC* = *ABC* means that starting with *CBA* a rotation of 240° returns to *ABC.* Closure is obvious and *ABC* is easily recognized as the identity element. Every operation has an inverse, since the triangle can always be rotated into the *ABC* position. To prove that this structure is a group, however, list all combinations of 3 elements and show that the operation is associative.

grouse (*animals*)   The grouse family, Tetraonidae, includes grouse, heath hens, prairie chickens, and ptarmigans. These chickenlike birds differ from other fowl in having densely feathered feet and nostrils hidden by feathers. They live in cold regions of Europe, Asia, and North America, mostly in forests, though some species—prairie chickens, for example—inhabit grasslands. Many are noted for their elaborate courtship displays.

growth and decay (*analytic geometry*)   When numerical growth depends on a constant ratio, such as population growing 4% a year or savings growing with compound interest, the amount of change over time can be defined by an exponential function of the form $y = Ae^{kt}$ where A is the amount at the beginning, k is a positive number related to the rate of growth, the constant number e is about

2.71828 . . . , and t is time. Decay, such as reduction of the amount of an element through radioactivity, has the same formula, but k is a negative number.

growth rate (*ecology*)   The growth rate, or rate of change in the size of a population, is defined as birthrate minus death rate within a specified time period. A population grows when natality (birthrate) is higher than mortality (death rate). In some species, a population's growth rate can also be affected by migration.

Under optimal conditions, growth is exponential, dependent only on the number of individuals and their rate of reproduction. In reality exponential growth occurs only briefly, for growth is quickly restricted by dwindling food supplies and other limiting factors. In stable populations, the number of individuals represents the carrying capacity of the area and the growth rate is around zero.

grub (*animals*)   The larvae of beetles are generally called grubs. Like the adults, they exhibit much variety. Carnivorous ground beetle larvae are elongated and agile. Snout beetle larvae often are legless. Aquatic larvae of whirligig beetles have feathery tracheal gills for respiration.

guano (*zoology*)   Deposits of the dried excrement of seabirds and bats are called guano (from *huanu*, a word for dung in the Quechuan language of Peru). Accumulations, particularly on islands where large colonies of seabirds nest and in caves that are home to large numbers of bats, can reach depths of 100 ft (30 m) or more. Guano is rich in nitrogen and phosphorus and has been used by humans for centuries as fertilizer.

guava (*plants*)   Native to tropical South America, guava (*Psidium guajava*) is a shrub or tree that grows about 30 ft (9 m) tall. Its fragrant white flowers develop into fleshy, rather acidic fruits up to 4 in. (10 cm) long that contain numerous small seeds. The strawberry guava (*P. cattleianum*), also native to South America, bears smaller, sweeter fruits.

gull (*animals*)   Gulls are medium to large seabirds with long wings and web feet. Most are gray and white, usually with black wing tips. Most inhabit shallow seas and bays but it is not uncommon for gulls to nest and feed inland, around freshwater lakes and rivers. Gulls eat almost anything, including dead fish, garbage, earthworms, and the eggs of other birds. The gull family, Laridae, also contains the terns.

gymnosperm (*plants*)   Seed-producing plants that bear their seeds in reproductive structures called cones are termed gymnosperms ("naked seeds"). They include cycads, ginkgos, conifers, and gnetophytes. Except for some gnetophyte vines, they are shrubs or trees and include the tallest living things on Earth: the coastal redwoods of California, one of which measured 385 ft (117 m). Gymnosperms are found worldwide, but most species live in temperate and subarctic habitats. There are 600 to 700 known species.

gypsum (*rocks and minerals*)   Gypsum is a hydrate of calcium sulfate. The unhydrated form is the mineral anhydrate. Both are deposited when ocean water evaporates. Since calcium sulfate is less soluble in water than sodium chloride, gypsum forms a bed below halite (rock salt), often with anhydrate between. Gypsum is widely used as plaster of Paris, in plasterboard, as an ingredient in cement, and as a soil conditioner.

H

habitat (*ecology*)   The particular area where an organism normally lives is the organism's habitat. It consists of both living and nonliving components; competitors, predators, soil, climate, and topography are all parts of an organism's habitat. Some organisms have habitats as large as an ocean. The habitats of others are the size of a meadow or tide pool. Very small habitats, such as a log or the area beneath a rock, are called microhabitats. Within its habitat, an organism occupies a unique position, or niche.

habitat loss (*environment*)   When people cut down forests, build dams, dig mines, construct roads, and replace grasslands with subdivisions, they permanently alter or destroy the habitats of plants and animals. As a result of such habitat loss, species may suffer dramatic population declines. For example, destruction of bamboo forests in China have brought the giant panda (*Ailuropoda melanoleuca*), whose diet consists primarily of bamboo, to the brink of extinction.

hadron (*particles*)   The only particles to respond to the strong force are baryons and mesons, collectively called hadrons (*hadro-* can mean "strong" or "bulky"). With the development of the quark theory in 1964 it became clear that hadrons are all particles formed from quarks, so an alternate definition would be any particles formed from 2 or 3 quarks.

hadrosaur (*paleontology*)   The duck-billed dinosaurs, or hadrosaurs, were plant eaters with wide, flat snouts. Their front limbs were short but had hooflike claws; this suggests that while hadrosaurs usually walked on their back legs, they may have moved on all fours while feeding. Hadrosaurs apparently traveled in herds, and may have spent much of their time in swamps. They lived during the middle and late Cretaceous. Species, some of which had prominent head crests, ranged in length from about 20 ft (6 m) to 40 ft (12 m).

hafnium (*elements*)   Hafnium is a metal, atomic number 72, discovered in 1923 by Dirk Coster [Dutch: 1889–1950] and György Hevesy [Hungarian: 1885–1966] and named for Hafnia, ancient name of Copenhagen, where the metal was first identified. It is chemically similar to and hard to separate from zirconium, the much lighter element above hafnium in the periodic table. Hafnium resists corrosion well and several compounds withstand high temperatures, useful properties that fill special needs. Chemical symbol: Hf.

hagfish (*animals*)   Hagfish (Family Myxinidae) are jawless fish with a long cylindrical body. They lack true teeth and have a cartilage skeleton. The eyes, covered by skin, are vestigial. Hagfish do not have scales; slime glands in the skin produce enormous amounts of slime, particularly during an attack. Hagfish live near the ocean bottom. They are both predators and scavengers, using their rasplike tongue to strip flesh from living or dead fish.

hail (*meteorology*)   Hail is ice pellets that fall from the sky—from pea size to "stones" as large as 5.5 in. (15 cm) across. Hail forms in thunderstorms, usually during summer. Small ice pellets fall for a time in a cumulonimbus cloud, but then are carried upward by vertical winds called updrafts. The trip down and ride up may be repeated several times, with a new coat of ice forming on the pellet each time. When the updraft is finally too weak to lift the enlarged hailstones, they fall to the ground, often damaging crops and occasionally causing human fatalities.

hair (*biology*)   Both plants and animals have filamentous structures called hairs. Plant hairs are outgrowths of epidermal cells. Root hairs are designed for absorption. Some leaf hairs, as in geraniums (*Geranium*), are glandular and secrete sticky materials. Other leaf hairs, as in mullein (*Verbascum*), occur as a dense covering that slows evaporation of water.

Among animals hair is unique to mammals. Outgrowths of the epidermis of the skin, hairs develop in follicles. Most mammals have a thick coat of hair—often called fur or wool—that insulates the body and helps keep body temperature constant. In contrast, humans have thick hair only in patches and whales are hairless except for a few bristles around the mouth. Specialized hairs serve a variety of purposes: eyelashes keep grit out of eyes; nose hairs trap dust, keeping it out of lungs; quills protect porcupines against predators; and vibrissae (whiskers) help cats find their way at night.

Hale, George Ellery (*astronomer*)   [American: 1868–1938] Hale's name is most often associated with the large Hale Telescope at Mount Palomar Observatory in California, and perhaps Hale's greatest contribution to astronomy was his tireless work in obtaining funding for the largest and best telescopes of his day. But he also invented the basic tool for studying the spectrum of the Sun and with it discovered that sunspots are caused by strong magnetic fields and that the Sun's magnetic field regularly reverses in a 22-year cycle.

half-life (*particles*)   The half-life of a radioactive element is the time that any amount will decay to a mixture of half the original element and half other elements. Radioactive elements decay atom by atom in random events. Which individual atom decays is inherently unpredictable, but statistically the change follows a specific decay function for each isotope of an element. A characteristic of all such exponential functions is that the time for the function's value to decrease by half is constant.

Sometimes half-life refers to total radioactivity of a substance, not changes in a specific element.

Halley, Edmond (*astronomer/earth scientist*)   [English: 1656–1742] Although best known today for his recognition that the same great comet repeatedly visits the vicinity of Earth (hence, comet Halley), Edmond Halley's greatest contribution to science may have been his persuading Newton to write the *Principia* and even to pay for publication of it. Here is a partial list of Halley's achievements in astronomy: first catalog of the skies of the Southern Hemisphere and first based on telescopic observation (1679); recognition of proper motion of stars (1718); and measurement of parallax of the Sun (1677).

As an earth scientist, Halley was an indefatigable mapmaker, charting Earth's winds, its magnetic field and tides, and coastlines of the English Channel. Halley developed a method for obtaining the longitude at sea from the position of the Moon, which he also mapped over a 19-year period.

halogen (*elements*)   The 5 halogens are elements that each have exactly 7 electrons in the outer shell and thus easily combine with alkali metals or any other substance that can provide a spare electron to fill the shell with a stable 8 electrons. Fluorine, the lightest halogen, is the most reactive of all elements, while chlorine, bromine, and iodine are sufficiently reactive to be toxic to most life forms. The final halogen, astatine, is less reactive but very radioactive—its longest-lived isotope has a half-life of slightly more than 8 hours.

halophyte (*botany*)   Plants adapted to growing in soils with high concentrations of salt are halophytes, a term derived from Greek words meaning "salt plants." Such plants are common in salt marshes and other habitats close to the sea; cordgrasses (*Spartina*) are an example. Halophytes also live in alkaline deserts and salt flats; saltbushes (*Atriplex*) are an example.

hamster (*animals*)   Native to Europe and Asia, hamsters are nocturnal, burrowing rodents classified in Family Muridae. Hamsters have large internal cheek pouches used to store food while foraging. The largest species, the European hamster (*Cricetus cricetus*), grows to 12 in. (30 cm) long. The much smaller golden hamster (*Mesocricetus auratus*) has been domesticated as a pet and laboratory animal.

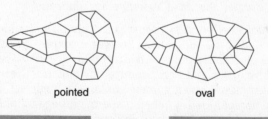

pointed   oval

HAND AX

hand ax (*paleoanthropology*)   The most common tool recovered from sites dating from about 1,000,000 to 100,000 years ago is a heavy stone point shaped to have 2 working faces, typically longer than 3 in. (10 cm) and weighing up to 2 lb (1 kg). The traditional name for this tool is hand ax, which suggests chopping wood or bone, but no one knows for sure how early humans used hand axes. One theory is that they were thrown to kill small game, with today's recovered hand axes being misses that were lost.

handedness (*chemistry*)   Many complex molecules, especially organic ones, come in mirror-image forms, showing handedness or chirality (from the Greek *cheir*, "hand"). Each form contains the same atoms linked in the same ways but the shapes differ as human hands do—just as a right-handed glove does not fit the left hand, a right-handed molecule of tartaric acid cannot be superimposed on a left-handed one.

The 2 isomers are called enantiomers or enantiomorphs and labeled with a D (dextrorotatory) for right-handed or an L (levorotatory) for left, based on how a beam of polarized light that passes through a pure D or L solution rotates. Although the D and L molecules behave the same in most chemical reactions, organisms treat them differently. The sugar produced by cane or beets is a D-form, for example, and its enantiomer, L-sucrose, is not metabolized by animals. Most amino acids are L-forms, which connect to form right-handed structures in proteins. Similarly, nucleotides are L-forms, so DNA and RNA are right-handed helices.

hand stencil (*paleoanthropology*)　From early Cro Magnon days of about 35,000 years ago to recent times in Australia and North America, a common form of rock painting has been the hand stencil. The stencil is formed by placing 1 hand on a flat rock wall and using a tube to blow pigment over it. When the hand is removed, the outline remains, surrounded by pigment. Many hand stencils have a finger missing, either because it was missing from the hand or because it was bent out of the way, possibly as a form of code.

hanging trough (*geology*)　A glacier moving down a mountain slope causes a U-shaped valley, or trough. Often a large glacier has several smaller tributary glaciers, each with its own trough, rather like river valleys coming together. The main glacier erodes the mountain so much deeper than the tributaries that the tributary troughs have floors hundreds of feet above the base of the main trough. After the glaciers melt, the tributary troughs are called hanging troughs or hanging valleys. Frequently rivers run through the troughs, producing a series of dramatic waterfalls, as at Yosemite in California.

haploid and diploid cells (*cells*)　Most cells contain a double set of chromosomes, and are said to be diploid, or 2n. For example, all human body cells contain 23 pairs of chromosomes. An exception is the sex cells, or gametes: sperm produced by males and eggs produced by females. Created in a cell division process called meiosis, each gamete contains a single set of chromosomes, and is said to be haploid, or 1n. When a sperm and egg unite in the process of fertilization, they form a single diploid cell (the zygote or fertilized egg).

Asexual spores and the organisms that develop from such spores also are haploid. For example, plants such as ferns, which have an alternation of generations, have both a haploid and diploid stage in their life cycles.

harbor wave (*oceanography*)　Often confused with a tsunami, a harbor wave is a giant wave (or waves) caused when a landslide falls into a bay or strait, often as the result of an earthquake. Because the water displaced by the rock is confined, the resulting wave can be even higher than a tsunami. In 1958 a harbor wave in Lituya Bay, Alaska, reached a height of 1740 ft (530 m), by far the highest wave ever recorded.

Hardy-Weinberg law (*genetics/evolution*)　The foundations for the science of population genetics were laid in 1908, when Godfrey Hardy [English: 1877–1947] and Wilhelm Weinberg [German: 1862–1937] independently explained how the proportion of different alleles of a gene in populations remains constant from generation to generation. Allele frequencies change only under certain conditions: (1) the population is small; (2) mating is not random; (3) mutations occur; or (4) new genes enter the gene pool, perhaps due to migration of individuals between populations. Using mathematical equations to study allele frequencies, scientists can identify populations that are changing genetically, or evolving.

Harvey, William (*physician/anatomist*)   [English: 1578–1657] Harvey was the first to propose that the heart is a pumping organ that propels blood on a circular course through the body, leaving through arteries and returning to the heart through veins. He noted that blood spurts from a cut artery in conformity with muscular contractions of the heart, and observed that clamping a vein causes it to swell with blood on the side away from the heart. Harvey's theory was confirmed when Marcello Malpighi discovered the capillaries that complete the path between arteries and veins.

hassium (*elements*)   A radioactive metal, atomic number 108, synthesized in 1984 by the Society for Heavy Ion Research in Darmstadt, Germany, hassium is named for the German province of Hesse, where Darmstadt is located. Very small amounts have been made, a few molecules that quickly transmute into lighter elements. Chemical symbol: Hs.

hawk (*animals*)   Birds of prey that hunt during the day, hawks have strongly hooked beaks, broad wings, and strong feet with sharp, curved claws called talons. Their vision is perhaps the best of all animals, particularly for detecting movement. Hawks of the genus *Buteo* (sometimes called buzzards) soar high in the sky, watching for mice and other prey on the ground. *Accipiter* hawks have shorter wings and longer tails; they seek prey in wooded areas, darting between trees and snatching birds out of the air. The hawk family, Accipitridae, also includes eagles, kites, and Old World vultures.

Hawking, Stephen William (*cosmologist*)   [English: 1942– , knighted 1972] In 1974 Hawking calculated that black holes emit a form of energy now known as Hawking radiation. Since then he has analyzed the shape and fate of the universe. Although Hawking has been afflicted with severe progressive muscle degeneration since the early 1960s, he manages with the use of a speech synthesizer to continue as Lucasian professor of mathematics at Cambridge University, the chair once held by Isaac Newton.

hawthorn (*plants*)   Members of the rose family native to the Northern Hemisphere, hawthorns (*Crataegus*) are deciduous shrubs and trees generally less than 33 ft (10 m) tall. Most have branches armed with thorns. The small, usually white flowers are borne in clusters and develop into fruits called haws, which look like small apples.

haze (*meteorology*)   Visibility is reduced when air contains small particles of liquids or dust or even traces of visible gases, an atmospheric condition called haze. Haze results from many different phenomena. Chemicals released by conifers that produce hazes have given the Blue Ridge and Great Smoky Mountains their name.

Photochemical smog is a haze produced by the action of sunlight on automobile exhaust fumes. Fine dust is kicked up by light winds over dry land. Ash particles or sulfuric acid droplets from volcano eruptions can linger in the sky for months.

hearing (*physiology*)   The sense by which animals recognize and distinguish among different sounds is called hearing. Specialized receptor cells detect changes in vibrations transmitted from a sound-producing object via air or another medium; this information is then sent to the brain, where it is interpreted. Hearing may be used to recognize other animals, detect danger, locate prey, or find a mate. Audible frequencies vary with the species: the human range is 20 to 20,000 hertz while the domestic cat's range is 60 to 65,000 hertz and a bat's range is 1000 to 120,000 hertz.

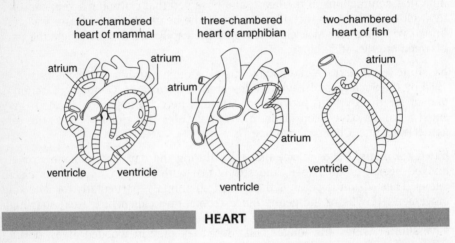

HEART

heart (*anatomy*)   The heart is a muscular organ that pumps blood through an animal's circulatory system. In insects, mollusks, and most other invertebrates that have a heart, this organ lies near the back, or dorsal, surface. Vertebrate hearts, composed of cardiac muscle, are near the lower (ventral) surface. They have chambers that receive blood from the veins, called atria, and chambers that pump blood through the arteries, called ventricles. A fish heart is 2-chambered; deoxygenated blood is pumped forward to the gills and from there oxygenated blood flows to the rest of the body. In vertebrates with lungs, both oxygenated blood and deoxygenated blood flow through the heart; it is necessary to keep these separate—that is, to have double circulation. Amphibians and most reptiles have 2 atria but only 1 ventricle. Birds and mammals have a 4-chambered heart: the right side (seen at left above) receives deoxygenated blood from the body and pumps it to the lungs; the left side receives oxygenated blood from the lungs and pumps it to the body.

heat (*thermodynamics*)   Heat is a form of kinetic energy that arises from the sum of the individual kinetic energies of the particles in a body. However, in thermodynamics heat is defined as the transfer of energy that results from temperature differences. Heat can only be measured (in energy units, such as joules, British thermal units, or calories) as it is being transferred. For example, the British thermal unit is the heat needed to change the temperature of 1 lb of water 1°F, while the physicist's calorie is the amount needed to raise 1 g of water 1°C or 1K (1 calorie = 1000 dietician's calories = 4.187 joules). The amount of energy stored in a body as heat is called its internal energy.

heath (*plants*)   Native to the Mediterranean region and southern Africa, heaths (*Erica*) are evergreen shrubs with small narrow leaves and small white, yellow, or rose flowers. Like other members of their family, Ericaceae, they prefer acidic habitats, their leaves are adapted for minimizing water loss, and they have tubular or bell-shaped flowers.

Some of the approximately 3500 species of Ericaceae, including rhododendrons, are popular ornamentals. Others, including blueberries, are raised for their edible fruits. Heather, Indian pipe, mountain laurel (*Kalmia latifolia*), and wintergreen (*Gaultheria*) also are members of this family.

heather (*plants*)   Native to Europe and Asia Minor, heather (*Calluna vulgaris*) is a low-growing evergreen shrub in the heath family. It has overlapping scale-like leaves and plumelike flower clusters called racemes. The flowers are small and usually purplish; however, horticulturists have created numerous varieties, or cultivars, in other colors.

hedgehog (*animals*)   Small insectivore mammals of Europe, Asia, Africa, and New Zealand, hedgehogs weigh up to about 2 lb (0.9 kg). They have a dense coat of stiff spines on the back and sides. When threatened, hedgehogs roll up into spiny-covered balls that deter most enemies. They rely on their keen smell and hearing to identify mates, babies, prey, and enemies. Nocturnal, they feed on insects and other small animals.

Heisenberg, Werner (*physicist*)   [German: 1901–76] Heisenberg, with mathematical help from Max Born, developed in 1925 the first version of quantum mechanics, a matrix method of calculating the behavior of electrons and other subatomic particles. The method was superseded as a practical tool later that year by the more intuitive wave equation of Erwin Schrödinger, but it remains a great intellectual accomplishment. Heisenberg's most lasting contribution was his discovery in 1927 of the uncertainty principle, a foundation of quantum theory. A few years later he introduced a new quantum number called isotopic spin. Heisenberg continued to contribute to particle physics, introducing useful computational techniques in the 1950s.

helium (*elements*)   Helium is a noble gas, atomic number 2, first observed in the Sun's atmosphere in 1868 by Pierre-Jules-César Janssen [French: 1824–1907] and Sir Norman Lockyer [English: 1836–1920] and so named from the Greek *helios*, "Sun." Helium, along with hydrogen, was formed in the big bang; since alpha particles are helium nuclei, it is also produced via radioactive decay. Relatively uncommon on Earth, helium is the second most abundant element in both the solar system and the universe. Helium remains liquid to absolute zero at ordinary pressures and is the principal tool used to cool superconductors. It is also used in other low-temperature applications. Chemical symbol: He.

helix (*geometry*)   The thread of a screw or bolt forms a curve in 3 dimensions called a helix (Latin for "spiral"). More formally, the curve of a helix lies on the surface of a cone or cylinder and makes a constant angle with each of the lines that lie along the curved surface. Helices can have either right or left handedness. The famous double helix of DNA is right-handed.

Helmholtz, Hermann von (*physicist/physician*)   [German: 1821–94] Helmholtz based his essentially correct theories of color vision and hearing on principles of physics. In 1847 he developed the clearest statement to that date of the conservation law for energy as a result of his studies on muscle action. Having measured for the first time the speed of transmission of a nerve impulse, Helmholtz became interested in James Clerk Maxwell's theory of electromagnetism and induced his pupil Heinrich Hertz to search successfully for radio waves.

hemlock (*plants*)   Members of the pine family, hemlocks (*Tsuga*) are evergreen trees native to Asia and North America. They have short, flat needles that grow in 2 opposite rows on the branches; the needles are dark green on the upper surface, silvery below. The seed-forming cones are typically less than 1 in. (2.5 cm) long. The tallest species is the western hemlock (*T. heterophylla*) of coastal regions of western North America, which can exceed 250 ft (75 m) in height.

hemoglobin (*biochemistry*)   The characteristic red color of vertebrate blood is owed to millions of molecules of hemoglobin in each red blood cell. These molecules get their red color from atoms of iron. It has long been known that hemoglobin is responsible for carrying oxygen from the lungs to the body tissues, and carbon dioxide from the tissues to the lungs. In 1996 scientists detected an additional task performed by another part of the hemoglobin molecule: the transport of nitric oxide, a gas that promotes muscle relaxation and the dilation of blood arteries. Hemoglobin also is found in the blood of some invertebrates, but usually in solution rather than as part of red blood cells.

hemp (*plants*)   Native to Asia, hemp (*Cannabis sativa*) is an erect annual that grows up to 20 ft (6 m) tall. Stems contain strong fibers up to 8 ft (2.5 m) long that are used to make rope. The leaves, stems, and female flowers are the source of the drugs marijuana and hashish.

The hemp family, Cannabaceae, also includes hop. Various plants of other families that are cultivated for their tough fibers also are called hemps. These include Manila hemp (*Musa textilis*) of the banana family and sisal hemp (*Agave sisalana*) of the agave family.

Henry, Joseph (*physicist*)   [American: 1797–1878] Although Samuel F.B. Morse [American: 1791–1872] successfully built a telegraph and started the business of telegraphy, his device leaned heavily on the ideas and inventions of Joseph Henry, the first great scientist of the United States after Benjamin Franklin. Like Franklin, Henry studied electricity. He improved the electromagnet in 1829 by insulating the wire coil. With electromagnets he discovered electromagnetic induction and a phenomenon called self-induction (a coil carrying a current also induces a current in itself). He made these discoveries both independently from and earlier than Michael Faraday, who was studying electricity in England. Henry built the first electric motor in 1831 and developed the electric relay that makes long-distance telegraphy possible in 1835. Henry also was the first to study the temperature of sunspots, finding them cooler than the surrounding surface of the Sun.

herb (*botany*)   In popular usage, a herb is a plant—or its products—used in cooking or medicine. Plants such as basil (*Ocimum*), tarragon (*Artemisia*), and thyme (*Thymus*) are examples. Botanically, a herb is a seed-bearing plant that has a herbaceous (nonwoody) stem. Its aboveground parts die at the end of each growing season, though roots and other underground parts may persist for many years, sending up new shoots at the start of each growing season.

herbivore (*ecology*)   Organisms that feed only on green plants and algae are called herbivores. Grasshoppers, sea urchins, parrots, and elephants are examples. Each herbivore has special adaptations for its chosen diet. For example, by beating its wings up to 50 times a second, a hummingbird can hover almost motionless over a flower as its long tongue, which is equipped with a brushlike tip, probes for nectar.

hermaphrodite (*biology*)   An organism that has both male and female reproductive organs is a hermaphrodite. Hermaphroditism is common among protists, fungi, plants, and invertebrates, and also exists in some vertebrate species. Self-fertilization may occur, but most species have mechanisms to avoid this. For example, a hermaphroditic sponge produces sperm and eggs at different times.

In some species an individual is either male or female when young, but later changes its sex at least once. The tropical fish *Labroides dimidiatus* lives in groups consisting of a male and a harem of females. If the male dies, changes occur in the body of one of the females; she becomes a male and takes control of the group. This is termed sequential hermaphroditism.

hermit crab (*animals*)   Hermit crabs are marine crustaceans that resemble true crabs but have a long soft abdomen with atrophied appendages. A hermit crab protects its abdomen by inserting it into an empty snail shell. The front part of the body protrudes as the animal walks about, but it can be retracted into the shell at any sign of danger. A hermit crab periodically leaves one shell for another, either for a better fit or because it has outgrown its former home.

heron (*animals*)   The heron family, Ardeidae, includes herons, bitterns, and egrets. All are wading birds that live around watery habitats. The birds have a long slim bill, a long neck, and long legs with 4 long unwebbed toes on each foot. They feed mainly on fish, frogs, snakes, and other small vertebrates, grabbing the prey with the bill and swallowing it whole. Among the largest is the great blue heron (*Ardea herodias*) of North America, which stands about 4 ft (1.2 m) tall.

herpesvirus (*viruses*)   The virus family Herpesviridae gets its name from the Greek *herpein*, which means "to creep" and refers to the chronic and recurrent aspects of infections caused by some members of the family. The genetic material of the virus particle consists of double-stranded DNA. The particle is a 20-sided polyhedron, with a lipid and glycoprotein envelope, and spikelike projections over its entire surface. Herpesviruses cause a variety of diseases in animals, including the human diseases chicken pox, shingles, herpes, and infectious mononucleosis.

herring (*animals*)   Mostly marine, members of the herring family, Clupeidae, are streamlined silvery fish that travel in schools. They include herrings, sardines, shads, and alewives. Typically, the fish are tapered at both the head and tail ends and strongly compressed from side to side. The fins lack spines. Most species have a sawlike band of modified scales called scutes along the belly. The commercially important Atlantic herring (*Clupea harengus*) is native to arctic and cold temperate waters; often sold as a "sardine," it grows to 18 in. (45 cm) long.

Herschel, William (*astronomer*)   [English: 1738–1822] Herschel was the greatest astronomer of his day. Furthermore, both his sister Caroline and his son John were astronomers of great accomplishment. Herschel gained lasting fame as the first person to discover a new planet in the solar system, Uranus, in 1781. Six years later he also discovered and named Titania and Oberon, satellites of Uranus. Additionally, he discovered 2 of Saturn's moons.

Herschel's primary interest was in stellar astronomy and nebulas. In 1785, after studying thousands of previously undiscovered nebulas, he wrote that the Milky Way was a great collection of stars and that the nebulas were similar vast collections seen from a great distance. He retreated somewhat from this correct claim after studying planetary nebulas, which are true clouds of gas and not galaxies of stars. Nevertheless, he was the first to recognize the nature and shape of the Milky Way.

Herschel made one major discovery outside of astronomy. While studying the temperature of sunlight he found infrared radiation, the first known invisible electromagnetic radiation.

Hertz, Heinrich (*physicist*)   [German: 1857–94] The equations for electromagnetism that James Clerk Maxwell developed from 1856 to 1873 showed that an electromagnetic wave is caused by the oscillation of an electric charge and propagates at the speed of light. Maxwell speculated (correctly) that light is an electromagnetic wave and that other forms of radiation have shorter or longer wavelengths. Hertz set out to create an electromagnetic wave with an apparatus that developed an oscillating spark. In 1888 he produced, detected, and identified what we now call radio waves. Hertz also observed that his device made better sparks when exposed to ultraviolet radiation, a result of the photoelectric phenomenon caused by energetic photons knocking electrons off a metal surface.

Hertzsprung-Russell diagram (*stars*)   The most famous graph in stellar astronomy is the Hertzsprung-Russell (HR) diagram. A point for each star is placed a distance corresponding to brightness above a horizontal axis and a distance corresponding to temperature to the right of a vertical axis. When Ejnar Hertzsprung [Danish: 1873–1967] and Henry Norris Russell [American: 1877–1957] independently drew such graphs, they were surprised to find that the stars are not distributed randomly; most lie on a curve that is nearly a straight line. If plotted with the hottest stars on the left and the brightest stars at the top, this line—now known as the main sequence—goes from upper left to lower right.

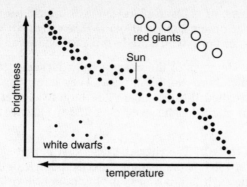

brightness

temperature

red giants

Sun

white dwarfs

HERTZSPRUNG-RUSSELL DIAGRAM

Hesperornis (*paleontology*)   A flightless seabird that lived about 70 million years ago in a large inland sea that covered what is today the central United States, *Hesperornis* had powerful legs and feet adapted for paddling; its wings were small and nonfunctional. Like its reptilian ancestors, it had teeth in its elongated jaws. *Hesperornis* preyed on fish, which it caught underwater. The largest species grew to lengths of about 6 ft (1.8 m).

heterotroph (*biology*)   Heterotrophs are organisms that derive the energy they need to live from the metabolism of organic substances. They depend, directly or indirectly, on green plants and other autotrophs. Heterotrophs include herbivores, carnivores, omnivores, and detritus feeders. All animals and fungi are heterotrophs; so too are many bacteria and protists.

heterozygous and homozygous organisms (*genetics*)   An organism that has 2 different alleles (forms of a gene) for a trait is said to be heterozygous for that trait. A homozygous organism, by contrast, has 2 identical alleles for a given trait. For example, a pea plant with 1 allele for wrinkled seeds and 1 allele for round seeds is heterozygous for seed shape; a pea plant with 2 alleles for round seeds is homozygous.

hibernation (*physiology*)   Some mammals escape from winter—when temperatures drop and food supplies vanish—by entering a state of dormancy called hibernation. During hibernation metabolism may slow drastically. For instance, the heartbeat of a common dormouse (*Glis glis*) falls from 450 to 25 beats a minute and body temperature drops to within 1.8°F (1°C) of the surrounding air. The dormouse remains curled up in a tight ball and in a deep sleep until spring arrives. Mammals such as bears and skunks also go into a long winter sleep, but their metabolism slows only a little. They can awaken comparatively easily and will leave their dens to look for food on warm winter days.

hibiscus (*plants*)   The more than 200 species of the genus *Hibiscus* are mostly tropical and subtropical shrubs with large, showy flowers that last only 1 day. Extensive hybridization of the Chinese hibiscus (*H. rosa-sinensis*) has created hundreds of varieties, or cultivars. The genus also includes rose of Sharon (*H. syriacus*), a popular ornamental, and okra (*H. esculentus*); the latter is raised for its edible fruits, or pods, which are best picked while still immature.

hickory (*plants*) Members of the walnut family, hickories (*Carya*) are deciduous trees native mainly to North America. They have long compound leaves and meaty seeds called hickory nuts. The largest species is the pecan (*C. pecan*), which can attain heights well over 100 ft (30 m); its edible nuts are highly prized.

Higgs particle (*particles*) The Higgs particle, first proposed by Peter W. Higgs [English: 1929–] in 1964, is the main undetected particle of the standard model of subatomic particles. It is a boson that physicists believe confers mass to all other particles. The reason that it has not been detected so far is that its mass is thought to be very great, perhaps even greater than that of the top quark.

high-temperature superconductor (*materials science*) Superconductivity was originally discovered in 1911 at temperatures near –454°F (–270°C), so when in 1986 Karl Alexander Müller [Swiss: 1927–] and Georg Bednorz [German: 1947–] created a ceramic that becomes superconducting at –405°F (–243°C), the new material came to be called a high-temperature superconductor. Since then, high-temperature superconducting ceramics have been discovered with superconducting properties at temperatures of about –292°F (–180°C). These materials appear to have great technological promise. The cause of superconductivity in such ceramics is poorly understood.

Hilbert, David (*mathematician*) [German: 1862–1943] Hilbert led an ultimately unsuccessful attempt to put all mathematics on a provably consistent and finite logical basis. In 1898 Hilbert himself developed the first completely logical version of Euclidean geometry. In 1900 Hilbert proposed a famous list of 23 problems that needed solution; resolving these became one of the main threads of 20th-century mathematics. Later he contributed greatly to applied mathematical physics; the Hilbert space of vectors with infinitely many components is perhaps his most useful advance.

Hipparchus of Rhodes (*astronomer/mathematician*) [Greek: c170–c125 BC] The Greek model of the solar system known as the Ptolemaic system was originally developed, according to Ptolemy [Hellenic: c100–c170], by Hipparchus, although little original work survives. Hipparchus created the basis of trigonometry and used this mathematical technique and parallax to determine correctly the distance to the Moon. He also compiled the first good star map, inventing the system of magnitudes for stars. His work on the map caused Hipparchus to recognize that the sky appears to be shifting in a circle over a period of about 26,000 years, an effect of the precession of Earth as it spins.

hippopotamus (*animals*) A massive, short-legged hoofed mammal of sub-Saharan Africa, the hippopotamus (*Hippopotamus amphibius*) derives its name from 2 Greek words meaning "river horse." It is true that the hippo, an excellent swimmer, spends much of its time in water, but it is more closely related to the pig than the horse and recent DNA analysis suggests that its closest relative is the whale. The hippo grows to a shoulder height of 5 ft (1.5 m) and a weight of 7700 lb (3500 kg). It has thick, virtually hairless skin, a large head, and a short tail. A smaller species, the pygmy hippopotamus (*Choeropsis liberiensis*), lives in dense rain forests of western Africa.

histamine (*biochemistry*)   A chemical released by white blood cells and other cells in response to foreign substances (**antigens**), histamine stimulates secretions from salivary and gastric glands, causes smooth muscles in the lungs and intestines to constrict, and dilates and increases permeability of blood capillaries. It is responsible for the itching, sneezing, and other symptoms of an allergy.

hoatzin (*animals*)   Living near rivers and swamps in northern South America, the hoatzin (*Opisthocomus hoazin*) is a plump bird up to 2 ft (60 cm) long and weighing about 2 lb (0.9 kg). It has a blue face, a crest of feathers on its head, and a long tail. As a chick it has functional claws on its wings. The hoatzin is one of the few known birds that feed almost exclusively on green leaves; like a cow it has bacteria in its gut to help digest cellulose.

Hodgkin, Dorothy (*chemist/crystallographer*)   [English: 1910–94] Hodgkin demonstrated the value of X-ray crystallography in determining the structure of complex organic molecules. In 1933 she helped make the first X-ray diffraction photograph of a protein crystal, pepsin. Later, she used X-ray analysis to discover the molecular structure of several important compounds, including penicillin, vitamin B_{12}, and insulin. Hodgkin was awarded the Nobel Prize in chemistry in 1964.

holly (*plants*)   Hollies (*Ilex*) are shrubs or trees with thick, leathery leaves often edged with spiny teeth. The flowers usually are unisexual, with individual plants bearing either male or female flowers. The best-known species is the English holly (*I. aquifolium*), a native of Europe and Asia that can grow to 80 ft (24 m) tall. The female plants bear dense clusters of bright red berries that remain through the winter.

holmium (*elements*)   Holmium is a rare earth, atomic number 67. Its discovery in 1879 is usually credited to Per Teodor Cleve [Swedish: 1840–1905], who named it for *Holmia* (Stockholm) his native city. The element had been recognized spectroscopically a year earlier. Despite interesting magnetic properties, the metal has been too expensive for most applications. Chemical symbol: Ho.

homeobox (*genetics*)   The homeobox is a small segment of DNA found in certain genes called **homeotic genes**. This DNA consists of a specific sequence of 180 nucleotides (building blocks of DNA); this means that it codes for about 60 amino acids (building blocks of proteins). Discovered in *Drosophila* fruit flies in 1983, the homeobox is believed to have arisen early in the evolution of multicellular organisms. It has since been identified in many species; probably all animals, plants, and other multicellular eukaryotes that have a segmented pattern during some stage of their development have the homeobox.

homeostasis (*physiology*)   The ability of a cell or an organism to maintain a relatively stable internal environment regardless of external changes is termed homeostasis (from Greek words meaning "constant state"). For example, if an animal cannot find food, it does not immediately stop functioning; rather, it depends on stored food reserves to continue normal functions. Homeostasis is vital for survival; if a cell or organism's internal environment changes significantly, it dies. In mammals and other complex organisms, many internal processes regulated by the endocrine glands and nervous system must work together to maintain homeostasis.

homeotic gene (*genetics*)   In 1978 Edward B. Lewis [American: 1918–] reported the discovery of genes in the fruit fly *Drosophila* that control the expression of other genes involved in the development of a part of a plant or animal body. Lewis also noted that these control genes, named homeotic genes, are arranged on the fly's chromosome in the same order as the body segments they control. The first genes control head development, the middle genes control thorax development, and the last genes control abdominal development—a phenomenon known as the colinearity principle.

Any mutation in a homeotic gene causes severe malformations, such as a fly developing 2 sets of wings or legs in the position where antennae normally form. In 1983 the homeobox was first identified within homeotic genes of *Drosophila*.

hominid (*paleoanthropology*)   The hominids (Family Hominidae) comprise 3 extinct genera: *Ardipithecus, Australopithecus,* and *Paranthropus*; and 4 surviving genera: *Homo, Gorilla, Pan* (chimpanzees), and *Pongo* (the orangutan— sometimes treated as a separate family). In other words, the hominids consist of extinct australopithecines, humans, and great apes. The family is part of the superfamily of hominoids in the primate order of mammals. It is characterized by large body size as compared with other primates; sexual dimorphism; bipedal walking or, in *Gorilla* and *Pan*, knuckle walking; absence of tails; flat faces; and large ratios of brain size to body size.

An important subfamily is the Hominidae, which includes *Ardipithecus, Australopithecus, Homo, Gorilla,* and *Pan*. Recognition that *Homo, Gorilla,* and *Pan* should be grouped together because of bone structure and other physical evidence is confirmed by DNA analysis.

hominoid (*paleoanthropology*)   The hominoids are a superfamily that includes the gibbons (Hylobatidae), hominids (Hominidae), and, if the orangutan is not included with the hominids, that species (as Family Pongidae). Numerous fossil apes, including *Proconsul, Afropithecus, Kenyapithecus, Dryopithecus,* and *Ramapithecus*, were also hominoids.

Homo antecessor (*paleoanthropology*)   Hominid fossils from 500,000 years ago found in the Atapuerca hills of Spain, christened *Homo antecessor* ("pioneer" or "explorer man"), represent the earliest humans to inhabit Europe. It is thought that *H. antecessor* arose first in Africa and then spread to Europe, where its descendants became the Neandertals.

Homo erectus (*paleoanthropology*)   Until recently most paleoanthropologists identified nearly all fossil prehumans from 2,000,000 to 200,000 years ago as the same species, *Homo erectus*. This species had become famous in the 1890s as the Java Ape Man and in the 1930s as Peking Man. Today, however, most African fossils from this period have been reclassified as *Homo ergaster*. Newer dates suggest that *H. erectus* may have lived in Java as recently as between 57,000 and 21,000 years ago. Until recently it was thought that East Asian fossils, still identified as *H. erectus,* did not manufacture the hand axes that are prominent in Africa and Europe; recognized stone tools connected to *H. erectus* are all roughly made.

Homo ergaster (*paleoanthropology*)   Human ancestors living in Africa in the period from about 1,500,000 years ago to about 200,000 years ago are generally put into the species *Homo ergaster* ("working man"), although earlier the fossils were classed as *Homo erectus* (now thought to have been confined to Asia). *H. ergaster* has longer legs and a larger body and brain-to-body ratio than *Homo habilis.* H. ergaster is the immediate ancestor of *H. sapiens* in the African Eve scenario of human evolution, although some paleoanthropologists believe that *H. sapiens* evolved several times, including, for example, evolution in Asia from *H. erectus.*

Homo habilis (*paleoanthropology*)   The earliest known member of the genus *Homo* is *Homo habilis* ("handy", or "clever man," because fossils are often associated with stone tools). *H. habilis* lived in southern and eastern Africa from about 2,500,000 years ago until about 1,500,000 years ago. The principal difference between *H. habilis* and the australopithecines that flourished in Africa at about the same time is a much larger brain size in *H. habilis* along with more humanlike teeth. Evidence suggests that *H. habilis* obtained food by scavenging and left caches of crude stone tools of the Oldowan type by sites near water where scavenging opportunities were most likely.

Homo heidelbergensis (*paleoanthropology*)   In northern Africa and Europe, possibly from as early as 900,000 years ago until perhaps as recently as 100,000 years ago, the species *Homo heidelbergensis* ("Heidelberg man") flourished, sometimes along with its probable descendants, archaic *Homo sapiens* and the Neandertals. Members of this little-known species (often previously classed with *H. erectus*) were more advanced than *H. ergaster* but not equal to anatomically modern humans.

homologous chromosomes (*genetics*)   Except in germ (sex) cells, chromosomes typically occur in pairs. And except in some cases involving sex chromosomes, the 2 chromosomes of a pair are homologous. That is, they are identical in size and appearance with the same order and position of genes (though perhaps different alleles, or forms, of the genes).

homology and analogy (*evolution*)   Different organisms often have features that are similar in structure and position. Similarities that result from common ancestry are called homologous structures. The front limbs of reptiles, birds, and mammals are examples; though quite different in function and appearance, they share the same basic structure and embryonic origin. Similarities that reflect adaptation to similar environments but are completely different in evolutionary background are called analogous structures. The wings of insects,

Homo sapiens

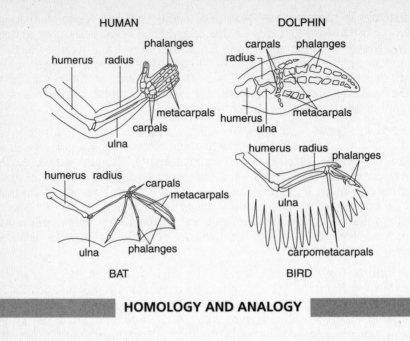

HUMAN

DOLPHIN

BAT

BIRD

| HOMOLOGY AND ANALOGY |

which develop as extensions of the body wall, are analogous to bird wings. Similarly, rose thorns (modified stems) and cactus spines (modified leaves) are analogous structures.

Homo sapiens (*anthropology*)   All living humans belong to the same species, named *Homo sapiens* (wise man) by Linnaeus. Fossil evidence suggests that anatomically modern humans, known as archaic *Homo sapiens,* evolved 200,000 years ago or more, although it is not clear that they possessed language or other forms of human culture. Another group, the Neandertals, are often thought to be members of our species as well. About 35,000 years ago, clearly modern humans with much better tools replaced the Neandertal population. Modern *Homo sapiens* quickly occupied all the major continents, certainly coming to the Americas before 12,000 years ago and perhaps as early as 30,000 years ago. All habitable islands, even in the remotest parts of the oceans, were populated by about 1000 years ago.

Hooke, Robert (*physicist*)   [English: 1635–1703] Hooke, like most other scientists of the 17th century, investigated what are now regarded as several different branches of science. Hooke's law is that stretching is directly proportional to force, a discovery in materials science. Hooke introduced the word cell to describe tiny compartments in cork. His invention of a compound microscope with a separate light source enabled him to observe not only cells but also the tiny parts of insects and to confirm the existence of microbes. Hooke's theory of combustion preceded the phlogiston idea by 35 years or so and was closer to what we now know to be true. Hooke's correct guess that gravity obeys an inverse square law contributed to his famous feud with **Newton**, who had actually proved this result. Hooke was one of the first to recognize a double star; he also proposed a wave theory of light.

Hooker, Joseph (*botanist*)   [English: 1817–1911] Hooker's collections of specimens and his extensive notes and sketches of the plants he saw during travels to the Southern Hemisphere and the Indian subcontinent contributed much to the fields of plant taxonomy and plant geography. From 1865 to 1885 he was director of the Royal Botanic Gardens at Kew, which had been created by his father, Sir William Hooker [English: 1785–1865]; he turned the gardens into an international center for botanical research. Hooker was a close friend of **Darwin**, and together with **Charles Lyell** persuaded Darwin to present his views on evolution at the 1858 meeting of the Linnaean Society.

hookworm (*animals*)   Tiny parasitic **roundworms** known as hookworms are named for the toothlike plates in the mouth. These are used to hold onto the inner wall of a host's small intestine while the hookworms feed on blood, lymph, and other host tissues. The hookworms mate in this environment, with eggs passing out in the feces. The larvae develop in the soil and invade new hosts by burrowing through the skin. Hookworms that infect humans are found in moist tropical and subtropical regions around the world.

hoopoe (*animals*)   Found in warm, dry areas of Europe, Africa, and Asia, the hoopoe (*Upupa epops*) is an elegant bird about the size of a jay. It has a crest of large pinkish brown feathers tipped with black, a pinkish brown body, and wings and tail of alternating black and white stripes. The long bill, which curves downward, is used to probe for worms, insects, and other small invertebrates. The closely related wood hoopoes, native to Africa, lack a crest and generally have rather uniform dark blue or green plumage.

hop (*plants*)   Hop (*Humulus lupulus*) is a twining vine cultivated for its female flower clusters, or catkins, which are dried and used to impart a bitter flavor to beer. The plant has a large perennial root from which new vines are produced annually. The vines may grow to 20 ft (6 m) long.

hopper (*animals*)   Certain small insects of the **cicada** order, Homoptera, are called hoppers in recognition of their jumping skills. All feed on plant juices. Many act as serious pests, either by stunting plant growth or as vectors of disease-causing microorganisms. Treehoppers, particularly tropical species, are characterized by weird shapes resulting from extensions of the part of the thorax just behind the head.

horizon (*archaeology*)   Archaeologists, especially those working in the Americas, call a widespread culture represented by characteristic artifacts a horizon. This probably derives from the use of "horizon" in geology to denote a specific layer of sediment. A horizon occurs at one time but over a wide region; it can be contrasted with a tradition, which occurs in one locality for a long period of time.

Major Vertebrate Hormones

GLAND	HORMONE(S)	MAIN EFFECTS
hypothalamus	oxytocin and vasopressin	stored in posterior pituitary (see below)
	releasing and inhibiting factors	stimulate or inhibit release of certain hormones by anterior pituitary
Pituitary: anterior lobe	ACTH (adrenocorticotropic hormone)	stimulates adrenal cortex to secrete hormones
	FSH (follicle–stimulating hormone)	stimulates production of eggs and sperm; stimulates secretion of estrogen in females
	GH (growth hormone)	stimulates muscle and skeletal growth
	LH (luteinizing hormone)	stimulates development of corpus luteum in females; stimulates secretion of sex hormones by ovaries and testes
	melanophore-stimulatinghormone	stimulates skin cells called melanocytes, causing darkening of the skin
	prolactin	stimulates milk production; controls certain types of parental behavior, such as nest-building in birds
	TSH (thyroid–stimulating hormone)	stimulates thyroid to produce hormones
Pituitary: posterior lobe	oxytocin	stimulates contraction of uterus and release of milk by mammary glands
	vasopressin (antidiuretic hormone, or ADH)	stimulates retention of water
pineal body	melatonin	regulates biological rhythms and other functions related to light
thyroid	calcitonin	decreases concentration of calcium in blood
	thyroxin	stimulates and maintains metabolism
parathyroid	parathyroid hormone	increases concentration of calcium in blood
thymus	thymosin	stimulates development of T cells
pancreas (islets of Langerhans)	glucagon	increases blood sugar levels by stimulating conversion of glycogen to glucose
	insulin	lowers blood sugar levels by facilitating absorption of glucose by cells and by promoting formation and storage of glycogen; stimulates protein synthesis and fat storage
adrenal: medulla	epinephrine	released in times of stress, it stimulates "fight-or-flight" reactions (speeding conversion of glycogen to glucose; increasing blood flow to brain, heart, and skeletal muscles while decreasing flow to skin, kidneys, and digestive organs)
	norepinephrine	sustains blood pressure
adrenal: cortex	glucocorticoids (cortisone, corticosterone, hydrocortisone)	raises sugar levels in blood; stimulate glucose production by cells
	mineralcorticoids (aldosterone, deoxycorticosterone)	regulates water and salt balance
kidneys	erythropoietin	stimulates red blood cell production
stomach (mucosa)	gastrin	stimulates secretion of gastric juice
duodenum (mucosa)	CCK (cholecystokinin)	stimulates secretion of pancreatic enzyme and contraction of gallbladder
	secretin	stimulates secretion of pancreatic juice
ovaries	estrogen	stimulates development of secondary female sex characteristics
	progesterone	maintains uterine lining for embryo implantation; aids in placental function; prevents miscarriage
testes	testosterone	stimulates development of secondary male sex characteristics

hormone (*biochemistry*)   Hormones are organic compounds produced in minute amounts by endocrine glands. They are sometimes called chemical messengers, since they are carried to other tissues or organs ("targets"), where they regulate a broad range of activities. The first hormone was identified in 1902 by Ernest Starling and William Bayliss [English: 1860–1924]. Since then, approximately 50 vertebrate hormones plus numerous invertebrate hormones have been identified.

In terms of chemical composition, there are 2 main types of hormones: nitrogen-containing compounds such as proteins, peptides, and amino acid derivatives; and steroids, which are fatty compounds synthesized from cholesterol. Each hormone is designed to bind only with specific receptor molecules on or in target cells.

hormone disruptor (*environment*)   Exposure to certain chemicals in the environment has been linked to the disruption of the endocrine systems of animals. These chemicals, called hormone disruptors, appear to mimic estrogen and other hormones and even in small amounts interfere with fertility, embryo development, and other life processes. The chemicals include various pesticides, among them DDT and lindane, and industrial pollutants such as PCBs and dioxins. The effects of these chemicals on humans is a matter of controversy.

hornbill (*animals*)   Native to tropical Africa and Asia, hornbills (Family Bucerotidae) are large birds, 15 to 60 in. (38 to 150 cm) long. They have a huge colorful beak, which in some species is further adorned with a horny "helmet" or casque. Following mating, the female builds a nest in a tree cavity high above the ground. She and her mate seal the opening to the nest with a wall of mud and excrement, leaving only a narrow slit. The female, a prisoner in the nest, lays eggs and incubates them for several weeks. At frequent intervals during each day the male brings her food, passing it through the opening. He continues to bring food for weeks after the young are born. When the young are ready to leave the nest, the female breaks down the wall.

hornblende (*rocks and minerals*)   Hornblende is similar chemically to mica, but contains calcium instead of potassium and always contains iron and magnesium. It is a dark mineral, green to black, with long, needle-shaped hexagonal crystals. It is common in igneous rock and also occurs in metamorphic rock.

hornwort (*plants*)   Small bryophyte plants seldom more than 1 in. (2.5 cm) long, hornworts are named for their spore-producing generation, which looks like green horns sprouting from the flattened body of the dominant gamete-producing generation. Hornworts grow mainly in moist shady soil. Colonies of the blue-green algae *Nostoc* often are associated with hornworts in what may be a mutually beneficial relationship.

horse (*animals*)   Members of Family Equidae, horses are large hoofed mammals with long limbs specialized for running. The upper part of the leg has powerful muscles; the lower part is slender; and the foot has 1 functional toe that ends in a hoof. Horses feed on grasses.

The earliest horse was Eohippus, which lived about 60,000,000 years ago. The only true wild horse of today is Przewalski's horse (*Equus przewalskii*), which has been reintroduced to the wild in Mongolia through the efforts of captive breeding programs. It is believed by some authorities to be the ancestor of all domesticated horses (*Equus caballus*).

horse latitudes (*geography*)   According to sailing lore, the North Atlantic along the latitude of the Sahara and American desert regions is called the horse latitudes because becalmed and thirsty crews threw horses overboard to save food and water. Hot air from near the equator that has cooled in the upper atmosphere descends here. A similar pattern of descending air starts where cold polar air pushes warmer air above it. Thus the horse latitudes, now recognized at about latitude 30° to 35° in each hemisphere, experience only descending air currents, making for weak winds, clear skies, and warm weather.

horseradish (*plants*)   Known for its white fleshy taproot, which is grated and used as a pungent condiment, horseradish (*Armoracia rusticana*) is a perennial herb native to southeastern Europe. A member of the mustard family, it grows to 4 ft (1.2 m) tall and has large leaves and tiny white flowers produced in inflorescences called racemes.

horseshoe crab (*animals*)   Considered "living fossils" because they are much like their ancestors of 500,000,000 years ago, horseshoe crabs (Class Merostomata) are marine arthropods found in shallow waters of the eastern North Atlantic and the Indo-Pacific. However, their name is a misnomer, for they are more closely related to spiders than to crabs. Growing to about 24 in. (60 cm), horseshoe crabs have a fused head and thorax covered with a hard horseshoe-shaped shell; the small abdomen is covered by a narrow shell. Horseshoe crabs feed at night, eating small invertebrates that they dig up from the ocean floor.

horsetail (*plants*)   Easily recognized by their green jointed stems, with whorls of tiny scalelike leaves radiating from each joint, horsetails (*Equisetum*) typically grow in moist locations. They seldom are more than 3.3 ft (1 m) tall. Some species have thin branches arising from the upper stem, creating a resemblance to horse tails. Ancient relatives, *Calamites*, looked identical except in size, reaching heights of 60 ft (18 m); they dominated swamp forests of the Carboniferous.

host (*biology*)   The organism on or in which a parasite lives is called a host. The host is harmed by the relationship, while the parasite obtains food or some other benefit. For example, elm trees are the hosts of the fungus that causes Dutch elm disease.

hot spot (*geology*)   Hot spot is the name geologists give to a location where magma from the mantle rises in a column called a plume that pushes its way to Earth's surface. Plate tectonics, or the motion of sections of the crust above the mantle, makes the hot spot seem to move, although the actual motion is of the plate. One famous hot spot in the Pacific continues to produce new volcanic islands in a chain that includes the Hawaiian Islands. Another is currently beneath Yellowstone National Park in the United States.

hot spring (*geology*)   Where groundwater is heated by a reservoir of magma, the water may emerge boiling hot and filled with dissolved minerals, forming hot springs that often bubble up through colored "mud" caused by the minerals coming out of the solution. But other hot springs may simply originate deep in the crust where rock is also hot. These hot springs or warm springs also tend to contain dissolved minerals, but are much less hot than those heated by magma.

Hubble, Edwin (*astronomer*)   [American: 1889–1953] Hubble used Henrietta Leavitt's discovery of how to find distances with Cepheid variables to obtain in 1923 the distance to the Andromeda galaxy (M31), showing that it is far outside the Milky Way. By 1929 he had found approximate distances to 18 galaxies and determined that the red shift of light from galaxies increases in a direct ratio to how far away they are. This effect is most easily explained as coming from the expansion of the universe, a concept that was entirely theoretical before Hubble's work. Hubble also developed the method still used for classifying galaxies.

Hubble constant (*cosmology*)   The expansion rate of the universe is given as a ratio named the Hubble constant (H) after Edwin Hubble, who in 1929 discovered the expansion. The Hubble constant is the ratio of the expansion velocity of any faraway object to its distance. (Technically H is not a constant, since the expansion rate changes, so some astronomers call H the Hubble parameter.) The Hubble constant reveals how much time has passed since the big bang at the start of the universe. With a careful choice of units, the age of the universe is $1/H$ times 100,000,000,000 years. The exact number for H is in dispute, but common estimates are 50 to 80, which gives an age that ranges from 12,000,000,000 to 20,000,000,000 years.

humidity (*atmospherics*)   Specific humidity is the ratio of the mass of water vapor in air to the total mass of air, usually expressed as a percent. But the amount of water vapor that can be dissolved in air before it begins to precipitate—called condensation for water—is highly dependent on temperature. A more common measure of humidity is the ratio, expressed as a percent, of water vapor to the total amount of water vapor air can hold at a given temperature; this is called relative humidity. When relative humidity is over 50% and the temperature is warm, humans become uncomfortable because sweat evaporates too slowly to cool the body efficiently.

hummingbird (*animals*)   Deft, highly maneuverable fliers, hummingbirds are the only birds that can hover. They are named for the humming sound made by their rapidly beating wings—as much as 80 beats per second in some species. Enormous amounts of food are needed to maintain their metabolism; a hummingbird must take in about half its weight—mainly sugary nectar—every day. Hummingbirds are small, with a slender, pointed bill, long wings, short legs, and colorful plumage. The approximately 320 species constitute Family Trochilidae. They are found only in the Americas, mainly in the tropics. Some species migrate with the seasons.

humus (*soils*)   The part of soil that consists of small particles formed when organisms or their wastes decay is called humus. Humus is a slightly acidic dark material that tends to form crumbs, holding small particles of clay and grains of sand together. These crumbs soak up water and retain it. Humus also contains available forms of nitrogen, phosphorus, and potassium that plants require, so better agricultural soils contain lots of humus.

hunter-gatherer (*anthropology*)   A hunter-gatherer society is one that subsists by hunting large or speedy animals and gathering plant materials, fungi, or easily captured animals. Hunting may be conducted with weapons or traps or by special techniques such as driving herds of horses or bison over cliffs. Gathering includes not only picking berries, nuts, mushrooms, fruits, and shoots, but also digging up roots, harvesting shellfish, collecting eggs, and catching (or trapping) slow game, such as turtles. Fishing is also included.

   Most hunter-gatherers do not depend on farming or herding animals, although these may also contribute to the diet. In a typical hunter-gatherer society, men are the hunters and women do most of the gathering. Most of the day-to-day food supply comes from gathering, but foods eaten at festivals are usually obtained by hunting.

hurricane (*meteorology*)   Europeans sailing in the West Indies learned of a type of massive storm called *huracan* by Caribs. These Atlantic storms, which strike from the northeast coast of South America to New England, are called hurricanes in English. Similar storms that strike the west coasts of Mexico, Central America, and northwestern South America are now called hurricanes as well. An Atlantic hurricane is caused by a low-pressure cell that starts as a hot, damp air mass in the eastern Atlantic Ocean; this air mass creates an anticyclone that travels west until it passes over land or cold water. Meteorologists class hurricane winds as those exceeding 75 mph (120 km/hr) and gauge the intensity using a scale that runs from 1 to 5, with 5 the most destructive.

hutia (*animals*)   The only living rodents native to the West Indies, hutias (Family Capromyidae) are 8 to 24 in. (20 to 60 cm) long, with a large head, stocky body, and short legs. They typically live in pairs and are active during daylight. Hutias do not dig burrows but usually live in hollow trees or rock crevices.

Hutton, James (*geologist*)   [Scottish: 1726–97] After a successful career as a businessman, Hutton spent nearly 20 years as an amateur geologist before, in 1785, summarizing his views in *Theory of the Earth*: Earth is very old and present-day geological structures formed slowly by processes observable today, such as erosion and deposition. His phrase describing Earth as having "no vestige of a beginning, no prospect of an end" and related ideas came to be known as uniformitarianism. His idea that mountain building is caused by terrestrial heat is the Plutonic theory.

Huygens, Christiaan (*physicist/astronomer/mathematician*)   [Dutch: 1629–95] Huygens's first successes came using telescopes he made that had better lenses than those of Galileo and Kepler. In 1656 Huygens became the first to recognize that Saturn is surrounded by a ring and has a moon; he also observed the Orion nebula. Huygens recognized that he needed a better clock to time astronomical phenomena, leading him to work out the mathematical theory for and build the first accurate pendulum clocks. Some of his clocks used a cycloid, a curve that he proved would make a pendulum swing exactly the same amount no matter the size of the swing. Huygens also was the first to understand the catenary as well as other interesting curves. Huygens's theory of light was that an ether consisting of tiny elastic balls pervades space and that longitudinal waves propagate through it. Although he was wrong on both counts, his wave theory of light was the best alternative to the particle theory of Newton.

hyacinth (*plants*)   Members of the lily family, hyacinths (*Hyacinthus*) are perennial herbs that grow from underground bulbs. Many varieties, or cultivars, exist, mainly of *H. orientalis*, native to the eastern Mediterranean region. These hyacinths have a basal cluster of slender leaves and 1 or more erect flower stalks. Each flower stalk bears a dense cluster of small, highly fragrant flowers.

Water hyacinths (*Eichhornia crassipes*), native to the American tropics, are classified in the pickerelweed family, Pontederiaceae. Their leaves float on the water's surface and their roots hang downward; if the water is shallow, the roots may penetrate the mud. A flower stalk rises above the leaves and bears a cluster of lavender flowers. Water hyacinths grow and reproduce rapidly; in many waterways they interfere with navigation.

hybrid (*genetics*)   A hybrid is the offspring of 2 genetically dissimilar parents. The term usually refers to the offspring of parents who are markedly different from one another—that is, of different breeds, varieties, or species. If the parents are from different species, the hybrid usually is infertile. For example, mating a horse and a donkey produces a sterile hybrid called a mule. Among some hybrid plants, fertility can be restored by doubling the chromosome number and producing polyploidy.

Hybridization—the breeding of 2 different plants or animals—occurs in nature and also is used extensively by humans to develop more productive dairy cattle, disease-resistant crops, showier garden flowers, and so on. Hybrids often are fitter than either parent, a condition called hybrid vigor. For example, mules have more endurance than horses and are stronger than donkeys.

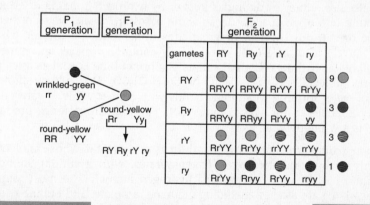

HYBRID AND DIHYBRID CROSSES

hybrid and dihybrid crosses (*genetics*)   The mating of 2 organisms with different traits is called a cross. A convenient way to determine the various combinations of genes that can result for a particular cross is by using the Punnett square, named after its inventor, Reginald C. Punnett [English: 1875–1967]. A

Punnett square for crossing a plant that is pure dominant for round peas (RR) with one that is pure recessive for wrinkled peas (rr) would have 4 boxes, showing that all possible combinations of gametes result in the same hybrid (Rr). A dihybrid cross follows what happens when 2 characteristics of the parental (P_1) generation are considered, such as round-yellow peas (RRYY) and wrinkled-green peas (rryy). All the offspring (F_1) are hybrid for both traits (RrYy). But 4 different gametes are produced by the F_1 generation: RY, Ry, rY, and ry. When these are crossed, the F_2 generation shows a 9:3:3:1 ratio of possible phenotypes. This was originally determined by Mendel.

hydra (*animals*)　A freshwater cnidarian generally less than 1 in. (2.5 cm) long, the hydra (*Hydra*) has a thin bag-shaped body with a mouth at the top, surrounded by tentacles armed with nematocysts. Unlike other cnidarians, the hydra has only a polyp stage. It lives attached to rocks and other substrates. The hydra can move by gliding on its base and by somersaulting: bending over, attaching its tentacles to the substrate, loosening its base, flipping the base up and over, then reattaching the base to the bottom and loosening the tentacles.

hydrangea (*plants*)　Native to Asia and the Americas, hydrangeas (*Hydrangea*) are deciduous shrubs known for their dense, showy clusters of white, pink, or blue flowers. Flowers that lack the red pigment anthocyanin are white. Flowers with anthocyanin are pink on plants grown in neutral and alkaline soils but blue on plants grown in acidic soils. (In acidic environments the plants absorb aluminum, which reacts with anthocyanin to form a blue compound.) Once classified in the saxifrage family, today hydrangeas usually are placed in the family Hydrangeaceae, which also includes mock oranges (*Philadelphus*).

hydrate (*chemistry*)　Hydrates are formed when water molecules bond weakly to other molecules, either by acting as ions or by hydrogen bonding or even by simply tucking into open spaces. Because the bonds are weak, water of hydration can be driven out by heat or great pressure, producing the anhydrous ("without water") form of a compound. Many common crystals and minerals are encountered as hydrates. Copper sulfate is a bright blue crystal as a hydrate with 5 water molecules ($CuSO_4 \cdot 5H_2O$ or $Cu(H_2O)_4SO_4 \cdot H_2O$, which shows 4 molecules of water forming weak ionic bonds with copper while 1 water molecule loosely attaches to the whole molecule). The anhydrous form, produced by heating, is a white powder.

hydrocarbon (*compounds*)　A hydrocarbon is a chemical whose molecules contain only carbon and hydrogen (carbohydrates, with a similar sounding name, contain half as many oxygen as hydrogen atoms). The most familiar pure hydrocarbons are alkanes, including methane, propane, and butane with 1, 3, and 4 carbon atoms combined with 4, 8, and 10 hydrogen atoms, respectively. More common even than these, however, are complex mixtures of different hydrocarbons, such as natural gas, petroleum, and gasoline.

hydrogen (*elements*)　The simplest element is the very lightweight gas hydrogen, atomic number 1. It was recognized as a distinct element in 1766 by Henry Cavendish and named from the Greek *hydor*, "water," plus *gen*, "forming"— Cavendish observed that hydrogen burned in air forms water. Hydrogen was produced by the big bang and remains the most abundant as well as the oldest element in the universe—9 out of every 10 atoms are hydrogen. Hydrogen-2, or deuterium, and radioactive hydrogen-3, or tritium, also occur naturally. Chemical symbol: H.

hydrogen peroxide (*compounds*)   Hydrogen peroxide (H_2O_2) is water made zippy by a detachable oxygen atom. Peroxides are characterized by 2 oxygen atoms bonded to each other, a combination acting as a radical with a valence of −2. Atomic oxygen, formed when hydrogen peroxide degrades or when oxygen is removed in reactions, makes hydrogen peroxide a useful mild antiseptic and bleach.

hydrolysis (*chemistry*)   Hydrolysis is the use of water to break apart molecules (the name, however, means unbinding water). In certain salts, hydrolysis occurs when part of a compound dissolved in water combines with either positive hydronium ions (H_3O^+) or negative hydroxide ions (OH^-), unbalancing water's normal neutrality. Excess hydronium ions produced when all hydroxide ions have combined create a weak acid, while the reverse situation results in a base. Some compounds break and reform as new compounds when dissolved in water. Organisms use hydrolysis combined with enzyme action to break large molecules, such as fats or carbohydrates, into smaller ones.

hydronium ion (*chemistry*)   Pure liquid water does not ordinarily exist solely as the familiar molecule of 2 hydrogen atoms bonded to 1 oxygen (H_2O). Some water molecules are pulled apart by weak charge interactions, and 1 hydrogen atom attaches to a complete water molecule, forming a hydronium ion (H_3O^+). The water remains neutral, however, as there are equal numbers of negatively charged hydroxide ions (OH^-).

hydrosphere (*geology*)   Most of the hydrosphere ("sphere of water") is the 70% of Earth's surface covered with ocean; oceans account for almost 98% of all water at or near Earth's surface. Regions around both poles and atop high mountains also contain water, in the form of snow and ice; such solid water is most of the rest. Lakes, rivers, swamps, and water in Earth's crust together amount to less than 0.5% of the hydrosphere. Water vapor in air counts as part of the atmosphere.

hydrothermal vent (*geology*)   Where tectonic plates are pulled apart at rift valleys beneath the sea or rift lakes such as Lake Baikal in Asia, very hot springs laden with dissolved minerals, such as sulfates and sulfides, pour out into the surrounding cold water, causing the minerals to precipitate. Such regions are called hydrothermal vents. Similar vents in the past may have produced today's ores. The region around a hydrothermal vent teems with organisms whose food chain is based on bacteria that metabolize sulfur. Animals include some types of giant clams, tube worms, and crabs that are found only at hydrothermal vents.

hydroxide ion (*chemistry*)   Pure liquid water contains hydroxide ions (OH^-, and equal numbers of hydronium ions (H_3O^+), keeping the liquid electrically neutral. When the hydroxide ion becomes part of an organic compound, such as alcohol, it is known as a hydroxyl group.

hyena (*animals*)   Inhabiting grasslands and open woodlands of Africa and Asia, hyenas (Family Hyaenidae) are doglike in appearance. The largest of the 3 species is the spotted hyena (*Crocuta crocuta*), also known as the laughing hyena because of its call, similar to a human laugh; the animal is about 6 ft (1.8 m) long and weighs up to 185 lb (84 kg). Hyenas sometimes make their own kills, but more often they scavenge, feeding on dead animals and carcasses abandoned by lions and other predators. Their powerful jaws enable hyenas to crack large bones, so that not even a skeleton is left behind after hyenas have finished eating.

hyperbola (*geometry*)   A hyperbola is a conic section rather like an ellipse that has been turned inside out so that the 4 ends of the open curve extend indefinitely into space. Thus it is considered a single curve with 2 branches. The branches have a pair of intersecting lines as asymptotes, and the point of intersection is the center of the hyperbola. The 2 points closest to the center are vertices. The hyperbola's equation is a second-degree equation in 2 variables set equal to 0 for which the coefficients of the 2 squared terms have opposite signs.

hyperbolic function (*trigonometry*)   Parallel to the 6 trigonometric functions defined on the basis of a circle, there are 6 hyperbolic functions related to the hyperbola and indicated by adding an h to the end of the abbreviation of a trigonometric function. The functions sinh z and cosh z are defined as half the difference $e^z - e^{-z}$ and half the sum $e^z + e^{-z}$, respectively, while tanh z = sinh z/cosh z and the other 3 are reciprocals of sinh z, cosh z, and tanh z.

hyperon (*particles*)   Starting in 1950 experimenters discovered a number of previously undetected particles that did not behave as particles were expected to. Because these particles have masses greater than that of the proton and neutron, they were called hyperons. A classification scheme for these particles developed in 1961 helped physicists understand them better, but their essential difference was already labeled "strangeness." Quark theory ascribes this quality to the presence of the strange quark in the hyperons, which are the lambda, sigma, xi, and omega particles.

hypothalamus (*anatomy*)   The part of the vertebrate brain just below the thalamus, the hypothalamus produces a number of hormones that directly or indirectly control many body activities. The hypothalamus also contains neurons (nerve cells) that control activities associated with sleep, appetite, thirst, body temperature, reproduction, and certain emotions, including pleasure and fear.

hypothesis (*philosophy of science*)   An informed guess or even speculation about the effect of a particular experiment is called a hypothesis. A scientific hypothesis must, however, be something that can be found true or false by a well-designed experiment.

hyrax (*animals*)   Also called dassies (and, in the Bible, coneys), hyraxes are small mammals of Africa and the Middle East, constituting Order Hyracoidea. Ground-dwelling species live in savannas and rocky areas and feed in daytime; tree-dwelling species live in forests and are nocturnal. All are herbivores. Hyraxes look like guinea pigs but they actually are ungulates, with most toes ending in a tiny hoof.

ibis and spoonbill (*animals*)   Family Threskiornithidae ("sacred bird") comprises ibises and spoonbills. These are wading birds 19 to 42 in. (48 to 107 cm) long. They have a long neck, long legs, partly webbed toes, and a short tail. Ibises have a long thin bill that curves downward. Spoonbills have a long broad bill flattened at the tip. Found near water in tropical and warm temperate regions, ibises and spoonbills are gregarious birds, nesting in colonies and traveling in flocks. They eat a variety of plant and animal materials. The sacred ibis (*Threskiornis aethiopica*) was worshiped in ancient Egypt some 5000 years ago.

ice (*physics*)   Ice is solidified water. Molecules in ice are farther apart than they are in liquid water at its most dense, which occurs near 40°F (4°C). Ice molecules form a 6-sided ring with a large open space in the middle; such rings are cross-linked to form even more open spaces. Thus ice floats in liquid water—its density is about 0.92 that of the 40°F water. When pressure on ice pushes molecules together, the freezing point lowers—decreasing approximately 1°F (0.5°C) for each increase of 80 bar.

Ice that freezes in motionless water has all its crystals oriented in the same way and is transparent, but when this structure is broken or arranged into complex snow crystals, ice scatters light and appears white. Several different crystalline forms of ice have been discovered at extremely low temperatures or high pressures or both, but only one form, known as ice I, is found in nature.

ice age (*geology*)   Periods in Earth's climate history so cool that glaciers advanced into temperate or low-lying regions are termed ice ages or glaciations. An ice age usually is said to continue until glaciers' final retreat to high mountains and regions around the poles.

The most recent ice age occupied the Pleistocene epoch, possibly ending some 12,000 years ago. In the 18th and 19th centuries, geologists recognized that typical glacial features, such as **drift** and **erratics**, covered much of the Northern Hemisphere. The material came from the Pleistocene glaciation, now known as the Ice Age (capitalized). At the height of the Ice Age, roughly 1,000,000 years ago, glaciers covered North America as far south as southern Illinois, and ice was similarly spread over much of both hemispheres. Since then there have been 3 major advances sandwiched around warm periods called interglacials. It is not known whether the present warm spell is another interglacial or if the Ice Age has concluded.

Discovery of other ice ages before the Pleistocene followed. Today about 20 ice ages scattered throughout Earth's history are known. Many coincide with periods when Earth's distance from and orientation to the Sun reduced the amount of solar energy reaching Earth. Other factors may be important triggers for the start of an ice age, including changes in ocean basins and mountain ranges caused by plate tectonics.

iceberg (*geology*)   When glaciers, ice caps, or ice sheets reach the sea, the leading edges are buoyed up, breaking off large chunks called icebergs ("ice mountains") that float out to sea. These float because glacial ice is fresh water, which is less dense than salt water, and because water expands slightly when it freezes. The density differences are not great, however, so about 4/5 of the iceberg lies underwater. Giant icebergs that break off the Antarctic ice sheet may be the size of a large island; hence these are sometimes called ice islands. Currents often carry icebergs or ice islands into temperate waters, where they eventually melt.

ice cap (*geology*)   Relatively small ice sheets termed ice caps cover parts of several large islands that are near the north pole, including Iceland, Baffin Island, and Spitzbergen. No islands large enough to have an ice cap are near the south pole except for those covered by the Antarctic ice sheet.

ice sheet (*geology*)   Like a glacier, an ice sheet begins in regions where more snow falls each year than melts. But an ice sheet covers a large region and does not advance in one direction. Since the central portion of the ice sheet remains in a cold, snowy region, ice can grow thousands of feet thick. While the sheet as a whole does not move in one direction, its edges spread, often stopping only when the sheet reaches the sea. The main ice sheets on Earth are the continental glaciers that cover most of Antarctica and Greenland.

Ichthyornis (*paleontology*)   A ternlike seabird that lived in North America during the late Cretaceous, *Ichthyornis* ("fish bird") stood about 8 in. (20 cm) tall. It had powerful wings and—like its reptilian ancestors—small, sharply pointed teeth that were probably used to catch and eat fish.

ichthyosaur (*paleontology*)   Ichthyosaurs ("fish lizards") were marine reptiles with streamlined bodies that arose in the Triassic, reached their peak in the Jurassic, and declined and became extinct in the Cretaceous. Ranging in length from 15 to 30 ft (4.5 to 9 m), they looked somewhat like dolphins, with long jaws with numerous teeth for grabbing prey, a stabilizing dorsal fin, limbs modified into flippers, and a strong tail that propelled them rapidly through the sea. They even gave birth in the sea, bearing live young instead of laying eggs.

icosahedron (*geometry*)   Although any polyhedron with 20 faces is an icosahedron ("12 bases"), most of the time the icosahedron is a regular figure, the convex Platonic solid with the most faces. For that icosahedron, all 20 faces are equilateral triangles. Among the interesting properties of the icosahedron is one discovered by Luca Pacioli [Italian: c1445–1509]: the 12 vertices of the regular icosahedron are the same as the vertices of 3 rectangles, orthogonal to each other at their centers, whose sides have the ratio of the golden section (known as golden rectangles).

ideal gas (*physics*)   The theoretical gas that exactly obeys the gas laws is called an ideal gas since no physical gas can perfectly do this. A true ideal gas would consist of molecules that are points and would thus occupy zero volume at absolute zero. Furthermore, the molecules would interact only by colliding with complete elasticity.

identity (*algebra*) An equation that is true for every value of the variable, such as $x + 1 = 1 + x$, is an identity. Equations true for some values but not others are called conditional. Since either part of an identity can be used in a mathematical situation, identities are useful in simplifying problems, especially identities for trigonometry.

identity element (*algebra*) An element of a set that does not change the value of any other element for a particular operation is called the identity element for the operation. The number 0 is the identity element for addition of numbers and 1 is the identity element for multiplication. In set theory the empty set is the identity element for union of sets while the universal set is the identity for intersection of sets.

if . . . then (*logic*) The combination if . . . then, symbolized by →, is the logical connective known as a conditional. A sentence such as "If the sky is blue, then the Sun is shining" could be shown as "The sky is blue" → "The Sun is shining." The part of the conditional following "if" or preceding → is called the antecedent and the part following "then" or → is the consequent or conclusion.

Such a combined sentence is always true in logic except when the second sentence is false, so the sentence "If the sky is not blue, then the Sun is shining," which is incorrect in ordinary speech, is nevertheless true in mathematical logic when reporting on a sunny day.

igneous rock (*rocks and minerals*) Rock that originates as a molten mass and forms by cooling is called igneous. Igneous rocks are classified in part by the size of their crystals, with larger crystals reflecting slower cooling. It is believed that all igneous rock originates as magma from Earth's mantle, but differences in density cause lighter igneous rocks, such as granite, to separate from heavier ones, such as basalt. Different minerals may also separate because of chemical affinities.

iguana (*animals*) Lizards of Family Iguanidae are native mainly to the American tropics and subtropics. They are comparatively large lizards: the common, or green, iguana (*Iguana iguana*) attains lengths of more than 4 ft (1.2 m), of which almost ⅔ is tail. A dewlap (fold of skin) hangs from the throat and a crest runs along the middle of the back. Iguanas eat both plant and animal matter.

imaginary number (*numbers*) All nonzero real-number multiples of the square root of −1, known as i, comprise the imaginary numbers. Examples include $2i$, $-i$, $i/3$, and $4i\sqrt{2}$. The existence and utility of such numbers is beyond question, but the name (to contrast with real numbers) was coined when mathematicians doubted that such numbers make sense. On the plane of complex numbers, the imaginaries occupy the vertical, or y, axis, while the reals are on the horizontal, or x, axis.

immune response (*physiology*) When bacteria or other foreign organisms or chemicals invade a vertebrate, they trigger a highly specific counterattack, or immune response, that has 2 parts. The primary response includes recognizing an invader as foreign, synthesizing antibody proteins that combine with antigen molecules on the surface of the invader, and destroying the invader. The secondary response, called cell-mediated immunity, involves the formation of long-lived "memory cells." If the host is exposed to the same antigen again at a later time, the memory cells immediately go into action, releasing antibodies. This is the basis for immunity to certain infectious diseases—once you have had measles you are not likely to have the disease again—and the basis for immunization programs that vaccinate against diseases.

immune system (*anatomy*)   One of a vertebrate's basic lines of defense against bacteria and other foreign substances is the immune system, which includes various white blood cells and tissues of the lymphatic system. The immune system recognizes foreign substances, tries to destroy or neutralize them, and remembers them in case they invade again in the future.

impact crater (*solar system*)   The surfaces of most solid bodies in the solar system contain impact craters, large, circular regions surrounded by walls or by rays of debris. Although craters on the Moon were once thought volcanic in origin, there is strong evidence that most craters there and elsewhere were caused when large bodies—most likely asteroids—struck planets, satellites, or other, larger asteroids. Many impact craters date from early in the history of the solar system. As the number of impacts mounted, the number of bodies available for collision decreased. Some impacts are much more recent, however, and large impacts probably still occur from time to time.

impala (*animals*)   A graceful antelope native to grasslands and open woodlands of sub-Saharan Africa, the impala (*Aepyceros melampus*) stands 30 to 40 in. (77 to 100 cm) tall at the shoulder and weighs up to 175 lb (80 kg). Males have long ridged, lyre-shaped horns; females, which are smaller, lack horns. Impalas are prodigious jumpers, able to cover more than 30 ft (9 m) in a single leap.

implantation (*physiology*)   The attachment of the mammalian embryo to the lining of the uterus (endometrium) is called implantation. Typically implantation occurs within days of fertilization; in humans, for example, it occurs about a week after fertilization. Some mammals, however, have a period of delayed implantation—an adaptation to availability of food or other environmental conditions. For example, long-tailed weasels (*Mustela frenata*) mate in the summer, but their fertilized eggs do not become attached to the uterus until the following spring. Then, in less than 4 weeks, they develop into full-term embryos and the mother gives birth—after a gestation period of approximately 279 days.

impossible construction (*mathematics*)   Plato [Greek: c427–c347 BC] prescribed using just a straightedge and compass for geometric constructions. With this restriction, there are impossible constructions. These include trisection (division into 3 equal parts) of an angle, duplication of a cube (constructing a cube with a volume exactly double the volume of a given cube), and squaring the circle (constructing a square with the same area as a given circle). Impossibility for all 3 was proved during the 19th century.

imprinting (*ethology*)   Imprinting is a type of learning that occurs during a brief period early in the life of certain animals. Young geese, for example, learn to follow only the first relatively large moving object they see shortly after hatching. Normally this object is their mother, who will protect and feed them. But as Konrad Lorenz demonstrated, newly hatched geese that see only a human will imprint on the human and follow him about even when they later see their mother.

impulse (*mechanics*)   Impulse is the action of a force over a time interval—that is, the product of a force with the difference between the force's starting and ending times. A second definition of impulse is change in momentum, since for a body of mass m and a velocity of v_0 at time t_0 acting under a force F until time t_1 (when the velocity is v_1), the impulse $F(t_1 - t_0)$ is equal to the change in momentum, $mv_1 - mv_0$. Impulse is useful in discussing collisions between bodies.

inbreeding (*genetics*)   The mating of closely related individuals is called inbreeding. In organisms requiring cross-fertilization, matings such as mother-son, father-daughter, and brother-sister would be extreme examples of inbreeding. In plants, self-pollination is the ultimate example. Inbreeding decreases genetic variation and increases the number of homozygous genes, often resulting in the expression of harmful recessive genes. In contrast, out-breeding (matings that do not involve close relatives) increases genetic variation and hybrid vigor.

incandescence (*materials science*)   Light produced by heating a material is called incandescence in contrast to fluorescence and phosphorescence. As the material is heated, the speed of the particles in it increases. Collisions become more frequent and more intense, shaking electrons into temporary higher states. As the electrons fall back to lower energy states, they emit photons. The higher the temperature, the more energetic the majority of emitted photons become. Thus, as a material is heated it first produces infrared radiation, then red light, then white, and finally higher energies. Photons of all energies are emitted, but the proportion at higher energy levels increases with temperature. Incandescence of metal wires produces the light of an incandescent bulb. Incandescence is also the main cause of light from stars.

inchworm (*animals*)   The long, slender caterpillars of moths of the family Geometridae are called inchworms or loopers. They have 3 pairs of legs on the thorax and 2 or 3 pairs of appendages called prolegs on the rear of the abdomen. To move, an inchworm anchors its legs to a leaf or other surface, draws the abdomen forward in an arching loop, anchors the prolegs, stretches the body forward . . . and repeats the process.

incubation (*ethology*)   The eggs of some animals must be warmed, or incubated, until they hatch. For example, a female eastern bluebird (*Sialia sialis*) sits on her eggs for 13 to 16 days; if the eggs are not kept warm and turned regularly, they will not hatch. Turtles depend on the Sun's heat to incubate their eggs. The female digs a hole in soil or damp sand, lays her eggs, and leaves them. Among seahorses, the young are incubated in a pouch under the father's tail.

indefinite integral (*calculus*)   The indefinite integral is the same as the anti-derivative but interpreted differently. The indefinite integral is a function that describes the change in area under a function's curve from some unspecified point to another arbitrary point. Since the first point is unspecified, there is an arbitrary constant, C, that is always part of the indefinite integral. For a given function $f(x)$ the indefinite integral is symbolized as $\int f(x)\,dx$ where dx is called the differential with respect to x. The theorem that the antiderivative is the same as the indefinite integral is the fundamental theorem of calculus.

index of refraction (*optics*)   The index of refraction is the ratio of the speed of light in a vacuum to speed in a particular material. The index depends mainly on the material and somewhat on the wavelength and temperature. Air has an index slightly higher than 1, and glass about 1.5, while diamond glitters in part because of its index of about 2.4.

Indian pipe (*plants*)   Among the most unusual looking flowering plants, Indian pipe (*Monotropa uniflora*) has a whitish, almost leafless stem up to 8 in. (20 cm) high bearing a single white flower that faces the ground. The plant has no chlorophyll and therefore cannot make its own food. Rather, Indian pipe depends on mycorrhizal fungi on its roots for food as well as for water and minerals. Indian pipes grow in woodlands in North America and eastern Asia.

Indian Plate (*geology*)   The section of Earth's lithosphere that contains India and much of the Indian Ocean, known as the Indian Plate, is pushing under the Eurasian Plate, raising the Tibetan Plateau and the Himalayas, Earth's tallest mountains. The boundary between the Indian Plate and the Australian Plate is not a complete break and the two are often considered as a single plate.

indicator (*chemistry*)   A chemical such as litmus that changes color to show the degree of acidity (positive hydrogen ions) is an indicator. Some indicators are used for strong acids (such as thymol blue) or strong bases (such as indigo carmine). Chemicals whose color varies with concentrations of other ions are also called indicators.

indigenous and naturalized species (*ecology*)   An organism that is native to a particular region or environment is said to be indigenous. A species that has been introduced to a place by humans and establishes itself so that it produces generation after generation is considered to be naturalized. For example, the gray squirrel (*Sciurus carolinensis*) is indigenous to North America. It was introduced to England in 1876, where it became naturalized and now greatly outnumbers England's native red squirrels (*Sciurus vulgaris*). (*See also* **exotic species**.)

indirect proof (*logic*)   Indirect proof is a procedure for proving a proposition p in which you begin by accepting as true (for the sake of argument) a proposition, or premise, that is the opposite of what you are trying to prove. You must show that direct proof based on the premise not p leads to a conclusion not consistent with that premise or else leads to an inconsistent set of propositions. Then the premise (not p) cannot be true, which implies that the negation of it (p) must be true.

indium (*elements*)   Indium is a metal, atomic number 49, discovered in 1863 by Ferdinand Reich [German: 1799–1882] and Hieronymus Theodor Richter [German: 1824–98]. The discovery came when Richter observed a previously unknown indigo (Latin *indicum*) line in an ore's spectrum. Among many applications, indium, as shiny as silver but resistant to tarnish, is used in mirrors. Chemical symbol: In.

Indricotherium (*paleontology*)   The largest land mammal that ever existed, *Indricotherium*—formerly known as *Baluchitherium*—was a long-necked, hornless relative of modern rhinoceroses that roamed across Asia some 35 million years ago. A plant eater, it was about 23 ft (7 m) long, with a shoulder height of 18 ft (5.4 m), and weighed more than 15 tons.

induction, electromagnetic (*electricity*)   Electromagnetic induction is the creation of an electric current by a magnet moving with respect to a conductor so that the lines of force of the magnetic field cut across the conductor. The phenomenon was discovered simultaneously in 1831 by Michael Faraday and Joseph Henry. The direction of the current is determined by the direction of motion. When the conductor is a coil of wire, the lines of force can cut across the wires many times, greatly increasing the potential difference of the current. Electromagnetic induction is the basis of the generators that produce electric current in power plants.

induction, mathematical (*logic*)   Mathematical induction is a form of proof that slightly resembles ordinary induction (establishing a principle on the basis of observation of many examples). To prove a statement true by mathematical induction, first establish that a statement about natural numbers is true for 1 example and then show that if true for any natural number, the statement must also be true for the next higher number. In consequence, the statement is true for all natural numbers greater than the one used in the first example.

industrial melanism (*evolution*)   In places with heavy industrial air pollution, where trees become blackened with soot, the proportion of dark, or melanistic, forms of certain insects increases. The classic example concerns the peppered moth, *Biston betularia*. As areas in England became industrialized and polluted, the light-colored peppered form was replaced by a black-winged form. But after clean air laws were passed, populations gradually reverted to the peppered form. It was originally hypothesized that this natural selection resulted from the ease with which hungry birds could spot the moths. However, it now appears that the causes of industrial melanism are more complex.

inequality (*algebra*)   A statement that 1 expression is greater than ($>$) or less than ($<$)—or greater than or equal to ($\geq$) or less than or equal to ($\leq$)—another is an inequality. (A statement that 2 expressions are not equal ($\neq$) is also an inequality, but such a statement is not usually solved in algebra.) For inequalities such as $2x + 3 > 5$ or $7 - 3x \leq 13$, the solution can be expressed by a simple inequality for x—the solution of $2x + 3 > 5$ is $x > 1$ while the solution of $7 - 3x \leq 13$ is $x \geq -2$. The relation indicated by the inequality symbol, called its sense, is reversed when multiplying or dividing by a negative, so $-3x \leq 6$, with the opposite sense from $x \geq -2$, has the same solution.

inertia (*mechanics*)   Inertia is the quality that makes objects at rest stay at rest and objects in motion continue in motion in a straight line. The concept was introduced by Galileo and refined by Newton as his first law of motion.

Inertia arises because any rigid body can be a frame of reference. The frame, called an inertial frame, can be considered as stationary, with every other frame moving or still with respect to it. Turning from straight-line motion, slowing down, or speeding up is acceleration, requiring a force and making the frame no longer inertial. Since resistance to acceleration depends on mass and inertia also resists acceleration, mass is proportional to inertia. Moment of inertia concerns the resistance of a rotating body to acceleration, but is otherwise quite different from simple inertia.

infinitesimal (*calculus*)   An infinitesimal usually is a variable that has as its limit 0, although historically it was an "infinitely small quantity." The derivative was originally based on the ratio of 2 infinitesimals (called differentials in this context), while in integral calculus the basic operation was adding an infinity of regions with infinitesimal areas. Today's textbooks warn that calculus has nothing to do with the infinitely small quantities. Despite this, since 1966 a branch of mathematics called nonstandard analysis has offered an interpretation of calculus based on infinitesimals that are infinitely small numbers.

infinity (*sets*)   The infinity commonly encountered in mathematics is potential infinity. A potential infinity of natural numbers exists because you can always name a number greater than any given number. A line extends to infinity because any given line segment can be extended. The potential infinity is symbolized with ∞.

A completed infinity refers to the size of an infinite set, such as all natural numbers or all points on a line. Around the end of the 19th century, Georg Cantor defined infinite sets of different sizes. Natural numbers are a countable infinity, and all sets in one-to-one correspondence with the natural numbers are also countable. An infinity of this size is symbolized by $\aleph_0$. Points on a line form a larger infinite set, whose size is c for "continuum." The sets of all real numbers and all complex numbers have c elements.

A defining property of an infinite set is that it can be put into one-to-one correspondence with a proper subset of itself. For example, the natural numbers can be put into one-to-one correspondence with their proper subset the even numbers.

inflation (*cosmology*)   According to a theory first proposed in 1980 by Alan H. Guth [American: 1947–], the big bang was preceded by a period of inflation, when the universe increased in size exponentially from a point with negative energy to a size about that of a grapefruit, more or less. During this expansion period, or inflationary universe, forces such as charge and gravity appeared, slowing the expansion to the nearly constant rate observed today.

inflorescence (*botany*)   Many types of plants bear flowers in clusters, or inflorescences. There is an almost infinite variety of shapes and structures among inflorescences, but most fall into 2 groups: indeterminate and determinate.

Simple indeterminate inflorescences have 1 main stalk; flowers low on the stalk bloom before flowers higher up. Racemes (snapdragons), corymbs (cherries), spikes (cattails), and umbels (ginseng) are examples. Compound indeterminate inflorescences have several main stalks; compound umbels (carrots) are

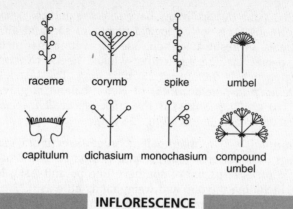

raceme corymb spike umbel

capitulum dichasium monochasium compound umbel

INFLORESCENCE

an example. In a capitulum (dandelion), all the flowers are on the same level, with the growing point in the center.

In determinate inflorescences, flowers at the apex develop first; dichasium (hawthorn) and monochasium (iris) are examples.

infrared radiation (*waves*)   Infrared radiation is that part of the electromagnetic spectrum below visible light, ranging from about 100,000,000,000 to 100,000,000,000,000 Hz. It has an energy level that is easily absorbed by molecules, neither knocking them apart nor passing by with little interaction. Since the absorbed energy causes molecules to move faster, infrared radiation is associated with radiation of heat. Our skin perceives infrared radiation as heat, also. Infrared radiation is produced by acceleration of charges in atoms. Since atoms always move, even close to absolute zero, every body emits some infrared radiation.

Ingenhousz, Jan (*plant physiologist*)   [Dutch: 1730–99] Building on experiments of Joseph Priestley that showed that plants give off oxygen, Ingenhousz laid the groundwork for understanding photosynthesis when he demonstrated that green leaves absorb carbon dioxide and release oxygen only in sunlight. In darkness, he showed, leaves—like animals and nongreen plant parts at all times—absorb oxygen and give off carbon dioxide.

Also a noted physician, Ingenhousz was among the first to inoculate against smallpox; unlike the safer method later developed by Edward Jenner, however, Ingenhousz used live smallpox viruses.

inorganic compound (*chemistry*)   Any compound that does not contain carbon is called inorganic; simple carbon compounds, such as carbon dioxide, may also be treated as inorganic. Because carbon compounds tend to be complex and have properties dependent on the shapes of molecules, the study of inorganic compounds is treated as a separate branch of chemistry.

insect (*animals*)   The arthropod class Insecta comprises some 1,000,000 described species—and probably many millions more yet to be documented. Insects have a body divided into 3 parts (head, thorax, and abdomen), a hard exoskeleton, and 3 pairs of legs. Many have **antennae** and **compound eyes**. Insects occupy an enormous range of ecological niches in soil and air, on bodies of plants and animals, in freshwater and crude oil, even in marine coastal waters. The smallest species include the feather-winged beetle (*Nanosella fungi*) of North America, which is 0.01 in. (0.25 mm) long. The heaviest may be the giant weta (*Deinacrida heteracantha*) of New Zealand; a gravid female can weigh 2.3 oz (70 g). The longest is the stick insect *Pharnacia serratipes* of Malaysia, which can reach almost 22 in. (56 cm).

insectivore (*animals*)   The mammalian order Insectivora includes shrews, moles, hedgehogs, solenodons, tenrecs, and desmans. The smallest is Savi's pygmy shrew (*Suncus etruscus*) of southern Europe and Asia, less than 3 in. (7.5 cm) long, including the tail, and weighing as little as 0.32 oz (10 g) when fully grown. The largest is the Cuban solenodon (*Solenodon cubanus*), which can grow to 2 ft (0.6 m) long and weigh 2.2 lb (1 kg). Most insectivores have whiskered faces and a keen sense of smell; their sight is poorly developed. They are primarily nocturnal, feeding not only on insects but also on plants and other small animals.

instinct (*ethology*)   Behavior that is inborn rather than learned during an animal's life is termed instinctive. It is passed in genes from one generation to the next and is characteristic of the species. It appears the first time the animal encounters the appropriate environmental stimulus. For example, in mating season a male stickleback fish (genus *Gasterosteus*) develops a bright red belly. If he sees a male stickleback without a red belly he will not attack. But he instinctively attacks anything with a red belly, even unrealistic looking models of sticklebacks.

integer (*numbers*)   The natural numbers, including 0, and their additive inverses comprise the integers, sometimes known as the positive and negative whole numbers. They extend infinitely in both directions, so they can be shown as . . . $-3, -2, -1, 0, +1, +2, +3, ...$

integrated pest management (*environment*)   To reduce use of synthetic pesticides and thus limit the harmful environmental effects of these chemicals, many farmers use an approach called integrated pest management (IPM) to control pests. IPM incorporates a combination of physical, chemical, and biological controls. For example, crops are rotated to limit pests that require a particular plant species; mulch or hoeing is used to control weeds; weather conditions are tracked to forecast certain disease outbreaks; synthetic pesticides are applied as needed rather than routinely.

intein (*genetics*)   Most genes contain sequences of DNA called introns. These are not part of the proteins for which the genes code; the introns are transcribed to messenger RNA (mRNA) but removed before the mRNA leaves the cell nucleus. Since 1990 researchers have discovered dozens of genes with intervening sequences that do get translated by mRNA. These sequences, named inteins, are between precursor protein sequences, named exteins. Following translation, an intein forms a loop, bringing together the neighboring exteins. The intein then catalyzes the formation of a peptide bond between the exteins, creating a mature protein.

intensity (*waves*)   The intensity of light is measured as one of the 7 basic SI measurement units, the candela, defined as 1/60 of the radiation per cm² of a black body that has the temperature of melting platinum, taken to be 3715°F (2046°C). This is about the same as the light from a candle, as in the customary measure of light intensity as a foot-candle. A higher intensity means more radiation.

intercept (*analytic geometry*)   An intercept is the distance of a point on a coordinate system axis from the origin to where a curve or surface intersects the axis; sometimes the intercept is taken to be the point instead of its distance. Intercepts are named after the axis; that is, the x intercept is on the x axis, and so forth. Taken as points, the x intercept in a plane has the form $(a, 0)$ with distance a from the origin, while the y intercept is $(0, b)$ at distance b from the origin.

interference (*waves*)   If 2 waves of any kind travel through the same medium at the same location, crests and troughs combine additively, a phenomenon called interference. If 2 crests match, the combined wave has a crest the height of the sum of both heights, but crests paired with troughs flatten the wave. A single wave can also be made to interfere with itself, either by combining parts of the wave observed from different locations or by using mirrors to reflect part of a wave back on itself.

interferometry (*measurement*)   Wave interference is easy to observe and enables measurements as precise as half a wavelength in a process named interferometry. Astronomers sometimes use 2 optical or radio telescopes to observe the same object from different locations. Combining the 2 observed waves with interferometry provides resolution equivalent to a single telescope with a lens or antenna whose diameter is equal to the distance between the telescopes.

interferon (*biochemistry*)   Discovered in 1957, interferons are proteins produced by animal cells in response to the presence of foreign agents, mainly viruses. Various species have their own unique interferons, and different cells produce different interferons. Unlike antibodies, interferons do not attack viruses directly. Rather, they lead to changes in cells that interfere with growth and replication of the viruses. Interferons also have some antitumor properties.

interleukin (*biochemistry*)   A diverse group of proteins produced mainly by white blood cells called lymphocytes, interleukins regulate interactions between cells involved in immune responses. For example, interleukin-1 triggers fever and other aspects of inflammation; interleukin-3 promotes the formation of some white blood cells.

international date line (*geography*)   A traveler heading east or west passes into a different time zone every 15° or so of longitude. Moving west the traveler gains 1 hour for each of the 24 zones; the traveler loses the same amount heading east. If not for an agreed change of date along a line called the international date line, which lies in the sea on or near 180° longitude, the traveler's date would become incorrect. Crossing the line west, the date moves back 1 day; crossing east moves it forward 1 day. The date line is in a single time zone, so only the calendar and not the clock needs changing.

interpolation (*analytic geometry*)   Interpolation is finding an approximation for a value of a function when an unknown value is between 2 known numbers. The most common type, called linear interpolation, treats numbers between the known ones as lying on a line so the value, or y, is proportional to the ratio of the distance of x from one number to the distance between the x values of the known numbers. For example, to find 2.3^2, assume that since 2.3 is 0.3 of the way between 2 and 3, 2.3^2 is about 0.3 of the way between $2^2 = 4$ and $3^2 = 9$, or $0.3 \times 5 = 1.5$ more than 4, giving an interpolated value of $4 + 1.5 = 5.5$ (actual value, 5.29). Linear interpolation is often used for logarithms, where interpolation is much easier than computing actual values.

intersection of sets (*sets*)   The set consisting of all elements of set A that are also in set B is called the intersection; on a Venn diagram this set consists of the points in the intersection of 2 disks. Creating an intersection is a binary operation on sets, symbolized $\cap$. If A is the set of prime numbers and B is the set of even numbers, then $A \cap B = \{2\}$ since 2 is the only even prime number. If the intersection of 2 sets is empty, the sets are called disjoint. Intersection corresponds to the logical operator and.

intertidal zone (*ecology*)   The area of an ocean between the high-tide line and the low-tide line is the intertidal zone. Twice each day, organisms in this zone are covered with water, then exposed to air. Temperatures can vary greatly—as water depth changes, as water is replaced by air, and as air temperature changes with the seasons. Intertidal organisms have evolved numerous adaptations for this rugged environment. Barnacles attach themselves to rocks and crabs burrow into sand to avoid being washed out to sea. Coral polyps retract into their skeletons. Fish and sea anemones live in tide pools (small bodies of water that do not empty as the tide goes out).

invariance (*mathematics*)   When a geometric figure, an equation, or function is changed by a transformation, the properties that always remain the same for that particular transformation are called invariant. Thus invariance is a characteristic of the transformation. For example, in geometry congruence is invariant (or preserved) by such transformations as translation and rotation, but not by dilatation.

inverse (*algebra*)   In algebra an inverse has the underlying meaning that the inverse "undoes" a related element or operation.
     Most algebraic systems (groups, rings, fields, and so forth) require inverse *elements* defined in terms of the identity element for an operation. For example, the additive inverse of x is $-x$ because $x + (-x) = 0$, where 0 is the additive identity element. The multiplicative inverse of x ($x \neq 0$) is $1/x$ since $x(1/x) = 1$. If $f(x)$ is a function, its inverse—if one exists—is $f^{-1}(x)$. The function and its inverse "undo" each other because $f(f^{-1}(x)) = x$, and $f(x) = x$ is the identity function.
     Inverse *operations* undo each other more directly. Subtraction is the inverse of addition because $a + b - b = a$ is true, while division is the inverse operation to multiplication.

inverse square law (*waves*)   Waves and many field phenomena tend to diminish in intensity proportional to the square of distance from the source. As distance increases, intensity diminishes (mathematically inverse variation), so light, electric fields, magnetic fields, and gravity obey inverse square laws. Such laws result because the area of a wave increases as the square of distance, decreasing effects per unit of area.

inverse variation (*analytic geometry*) A relation between 2 variables such that 1 of them equals a constant divided by the other or, equivalently, such that the ratio of 1 to the reciprocal of the other is a constant, is inverse variation; for example, $y = c/x$ where c is a number. (*See also* **direct variation**.)

inversion (*meteorology*) Any situation in the troposphere in which a layer of warm air is above a layer of cooler air is an inversion, short for thermal inversion. Technically the warmer stratosphere above the cool top layers of the troposphere is an inversion, but usually the term refers to a weather phenomenon.

An inversion prevents air from rising; thus it traps smog and other air pollution near the surface. Typically inversions form in large valleys surrounded by higher elevations. In large cities located in such bowls, such as in Mexico City or Los Angeles, air pollution trapped by inversions often reaches unhealthy levels.

invertebrate (*zoology*) Any animal that does not have vertebrae—bones that form an internal backbone—is an invertebrate. Invertebrates make up more than 95% of the known species of animals. The group comprises a wide variety of organisms in every habitat on Earth, including sponges, cnidarians, worms, mollusks, echinoderms, and arthropods.

Io (*solar system*) Jupiter's moon Io is famous for its volcanoes, powered by tidal forces caused by its proximity—less than 262,000 mi (421,600 km), about the same distance as Earth's Moon—to the giant planet. Orange-red patches on Io's mottled surface are apparently molten sulfur beds connected to the spectacular volcanoes, but other parts of Io's surface are very cold (about -229°F, or –145°C).

iodine (*elements*) Iodine is a solid halogen nonmetal, atomic number 53, discovered accidentally in seaweed in 1811 by Bernard Courtois [French: 1777–1838]. Humphry Davy named it for the color of its vapor, using the Greek *iodes*, "violet." Its principal source is the sea; since it is essential to thyroid hormones, people living in the interiors of continents can develop enlarged thyroids from too little dietary iodine (prevented today with iodized salt). One radioactive isotope is a serious hazard in fallout from nuclear explosions. Chemical symbol: I.

ion (*particles*) An atom that loses or gains 1 or more electrons becomes an ion. While an atom has the same number of positive and negative charges, which cancel to give 0 charge, an ion is positively or negatively charged. Free ions soon take or release electrons to return to neutral, but many compounds, including sodium chloride, consist of stable positive and negative ions arranged into a crystal.

ionizing radiation (*particles*) A flow of subatomic particles powerful enough to knock electrons out of atoms is called ionizing radiation; this includes X rays and gamma rays that are usually thought of as waves, but which are also energetic photons. Other common forms of ionizing radiation are alpha particles (helium nuclei), beta particles (electrons and positrons), and neutrons from radioactivity. Normally the phrase excludes ultraviolet radiation and is used only to mean other radiation dangerous to humans.

ionosphere (*atmospherics*)   The ionosphere consists of layers at the base of the thermosphere where ultraviolet radiation from sunlight is great enough to separate electrons from oxygen and nitrogen molecules, and where air pressure is sufficient for noticeable effects (about 0.00001 bar). This region reflects long-wavelength electromagnetic radiation, enabling AM radio waves to travel beyond Earth's horizon. At night the lowest level, the D layer, has too few electrons to reflect radio waves, which then reflect from the higher E or F layers, causing radio signals to travel farther.

iridium (*elements*)   Iridium is a platinumlike metal, atomic number 77, discovered in 1803 by Smithson Tennant [English: 1761–1815]. Its name is based on the Greek *iris* ("rainbow"), given for the bright colors of its salts. The standard kilogram of SI measurement is platinum alloyed with 10% iridium to make it corrosion-resistant. A thin layer of iridium found worldwide is the sign of an asteroid impact 65,000,000 years ago known as the K/T event. Chemical symbol: Ir.

iris (*plants*)   Perennial herbs found mainly in northern temperate regions, irises or flags (*Iris*) are widely grown for their showy flowers, which come in a wide range of colors. The flowers—the origin of the fleur-de-lis—have 3 erect petals and 3 sepals that usually hang downward. Irises have underground bulbs or rhizomes from which the sword-shaped leaves and flower stalk arise. The iris family, Iridaceae, also includes crocus (*Crocus*), freesia (*Freesia*), and gladiolus (*Gladiolus*).

Irish elk (*paleontology*)   The misnamed Irish elk (*Megaceros*) was the largest deer that ever lived. It weighed 700 to 1200 lb (318 to 544 kg). Its spectacular antlers—grown anew each spring—spanned 10 to 12 ft (3 to 3.6 m) and weighed more than 100 lb (45 kg). It lived across northern Europe and entered into Asia before becoming extinct about 10,000 years ago.

iron (*elements*)   Iron, atomic number 26, is the most abundant metal on Earth, second in the crust and the main element in Earth's core. It is the heaviest element created by fusion in stars that have not become supernovas—above iron, more energy is required for fusion than is released. Iron is not found as a pure metal in the crust, but metallic iron alloys land on Earth regularly in meteorites. Humans learned to extract iron from ores some 5500 years ago, but regular use dates from about 1200 BC in the Near East. Alloys of iron including cast iron and steel have been the dominant materials in terms of strength and durability since that time. Iron is also essential to vertebrate life as the main element in hemoglobin. The chemical symbol, Fe, is from the Latin *ferrum*.

ironwood (*plants*)   Various shrubs and trees with wood so hard and heavy that it does not float in water are commonly called ironwoods. These plants include the American hop hornbeam (*Ostrya virginiana*) of eastern North America, the black ironwood (*Krugiodendron ferreum*) of the Caribbean, and certain casuarinas of Australia.

irrational number (*numbers*)   Any real number that cannot be represented as the ratio of 2 integers is an irrational number. That such entities exist was discovered for lengths by the Greeks in the 400s BC, including a proof that if the side of a square is used as a unit of measure (or 1), the diagonal (or √2) cannot be measured exactly with that unit. When real numbers are expressed as infinite decimals, any decimal that fails to repeat the same finite pattern of digits over and over represents an irrational number.

irritability (*biology*)   Response to changes in the environment, called irritability, is a basic property of all living things. The type of response depends on the nature of the organism and the strength of the stimulus. The response may be slow, as in the opening of a rosebud, or rapid, as in the darting movement of a fly escaping a flyswatter.

island (*geography*)   An island is a body of land, smaller than a continent and larger than a reef, surrounded by water. Several geological formations meet this definition. Islands in rivers or lakes or near continents are nearly always a part of the continent that happens to be higher than the water. Islands in the deep ocean may be small versions of continents (Greenland or New Guinea, for example); volcanoes that rise above the sea; or atolls produced by coral. Volcanoes occur in island arcs, over or behind hot spots, along mid-oceanic ridges, or as isolated mountains that rise from the thin ocean floor.

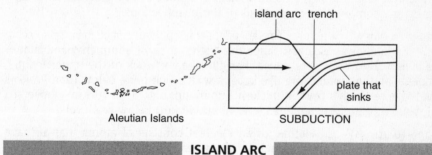

Aleutian Islands   SUBDUCTION

ISLAND ARC

island arc (*geography*)   Many ocean islands occur in long curved chains called arcs. Examples are the Aleutians and the islands forming Japan. Such island arcs are volcanic, produced by magma rising where an ocean plate is undergoing subduction. As one plate bends and sinks below a second ocean plate, a deep trench is formed. The front edge of the lower plate heats up and partly melts as it descends and expands, raising the islands on the far side of the trench.

isomer (*chemistry*)   The same set of atoms often can bond into a molecule in more than one spatial arrangement. Such different arrangements are called isomers. Often isomeric compounds have different physical properties, such as color or density. The 2 basic types are stereoisomers and structural isomers.

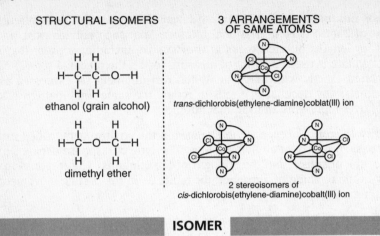

STRUCTURAL ISOMERS

ethanol (grain alcohol)

dimethyl ether

3 ARRANGEMENTS
OF SAME ATOMS

trans-dichlorobis(ethylene-diamine)coblat(III) ion

2 stereoisomers of
cis-dichlorobis(ethylene-diamine)cobalt(III) ion

ISOMER

Stereoisomer molecules are exact mirror images and many can be identified by the direction in which the compounds twist polarized light, with one isomer twisting it left and the other right (optical isomers). Most biological molecules can exist in both right-handed and left-handed versions, but typically only 1 form is biologically active.

Structural isomers of the same molecule cannot be transformed into each other by any combination of rotations or mirror reflections.

isostasy (*geology*)   When Archimedes' principle is applied to mountains or to continents, it is known as isostasy. Since the "light" rock of continents and mountains is only 10% or so less dense than the "heavy" rock of the upper mantle, continents and mountains are like icebergs with much more below than projects above. Where Earth's crust is thickest, mountains rise 5 mi (8 km) above sea level while their roots extend as far as 30 mi (50 km) below sea level.

isotope (*particles*)   An isotope of an element consists of atoms that all have the same number of protons and neutrons per atom. The number of protons determines the name of the element and most of the properties. Atomic mass, the sum of the number of protons and neutrons per atom, is used to identify a specific isotope. For example, the common isotopes of carbon are carbon-12 with 6 protons and 6 neutrons; carbon-13 with 6 protons, 7 neutrons; and carbon-14, with 6 protons, 8 neutrons.

isotopic dating (*geology*)   Methods for finding the age of rock are based on ratios of radioactive elements with long half-lives to stable decay products. Since each radioactive isotope has a specific half-life and decay sequence, this technique is called isotopic dating. (Radiocarbon dating uses 2 different isotopes of carbon instead of radioactive carbon and its decay product.) Isotopic dating was invented in 1906 by Benjamin Boltwood [American: 1870–1927], who used uranium-lead dating to find the first accurate ages for rocks. Different combinations of isotopes are useful for particular time spans or situations. Potassium-argon dating of lava or volcanic ash is important to paleoanthropologists because the method can be used for ages ranging over hundreds of thousands of years past.

iteration (*algebra*)   Iteration means "repeating," but mathematical iteration invariably involves feedback as well. Iteration is a procedure in which the result of each step is fed to the next step, and then the procedure is repeated. Calculating a square root by iteration is an example. To find the square root of 50, guess that it is slightly more than 7 ($7^2 = 49$). Divide 50 by 7, obtaining 7.14. Average 7 and 7.14 to get 7.07. Iterate. 50 ÷ 7.07 is 7.072. Another iteration of averaging shows the square root is 7.071 to the nearest thousandth.

ivory (*biology*)   Animal ivory is a hard, white substance that is a modified form of dentine (the main component of teeth). Ivory makes up the tusks of elephants, walruses, narwhals, and several other large mammals. People have carved ivory for more than 20,000 years.

Vegetable ivory comes from the nut of the ivory nut palm (*Phytelephas macrocarpa*), a native of northern South America. The nut (also known as tagua) becomes extremely hard on drying and can be carved like animal ivory.

ivy (*plants*)   Various creeping and climbing vines are commonly called ivy. English ivy (*Hedera helix*), a member of the aralia family, is a woody evergreen with glossy leaves native to Europe, Africa, and Asia. Boston ivy (*Parthenocissus tricuspidata*), native to Asia, and Virginia creeper (*P. quinquefolia*), native to North America, are woody deciduous vines of the grape family. Poison ivy (*Toxicodendron radicans*), in the cashew family, is a woody deciduous vine or shrub with leaves divided into 3 leaflets, native to North America.

jacana (*animals*) Inhabiting shallow lakes in tropical regions worldwide, jacanas (Family Jacanidae) are slender birds up to 14 in. (36 cm) long. They have long legs and very long toes that enable them to hop nimbly on water lily leaves and other floating vegetation—hence their other common name, lily-trotters. Jacanas are good swimmers and divers, and often hide by submerging themselves in the water. They eat seeds and small animals.

jackal (*animals*) Members of the genus *Canis* in the dog family, jackals are native mainly to deserts and grasslands of Africa. They generally are small and look rather like a cross between a fox and a wolf. They hunt prey but also eat eggs, leaves, fruits, and carrion, locating the latter by watching vultures as they stop circling in the air and swoop to the ground.

jaguar (*animals*) Growing to a length of 6 ft (1.8 m), not including its tail, the jaguar (*Panthera onca*) is the biggest cat native to the Americas. It is found in forested and brushy habitats from the southwestern United States through Central America and most of South America. The jaguar is an excellent climber and swimmer, and preys on birds, turtles, and alligators as well as deer and other ground-dwellers.

jawless fish (*animals*) The earliest known fish were ostracoderms ("shell skins"), small jawless creatures covered with scales and bony plates. They lived on ocean bottoms 510,000,000 to 350,000,000 years ago, sifting the mud for food particles. Today this once large group is represented only by some 70 species of lampreys and hagfish. Unlike their ancestors, lampreys and hagfish have a skeleton made of cartilage.

jellyfish (*animals*) Graceful in appearance but deadly in behavior, jellyfish are cnidarians found in both marine and fresh water. They are characterized by a bowl-shaped, transparent body that consists of 2 cell layers separated by a layer of jellylike material. The mouth is on the under side, surrounded by tentacles. Some jellyfish have only the free-swimming medusa form in their life cycle; others have both a medusa and a bottom-dwelling polyp stage. Species such as the common *Aurelia aurita* often drift at the ocean's surface in shoals of hundreds of thousands of individuals.

Jenner, Edward (*physician*) [English: 1749–1823] In the late 1700s it was common knowledge in the English countryside that dairymaids who had had cowpox did not catch the much deadlier smallpox. Jenner believed that people could be protected against smallpox by inoculating them with cowpox, a technique that came to be called vaccination. Jenner inoculated several people with material from dairymaids' cowpox lesions and detailed cases of people who had cowpox and later resisted smallpox. Publication of his findings in 1798 created much controversy, but vaccination against smallpox soon became widespread.

jerboa (*animals*)   Rodents native to semidesert areas of Africa and southwestern Asia, jerboas (Family Dipodidae) are 1.5 to 6 in. (4 to 15 cm) long. They have large ears and eyes, a long tufted tail, and powerful hind legs 4 times longer than the front legs. A jerboa can jump more than 10 ft (3 m) in a single bound. Jerboas are nocturnal, spending the day in underground burrows.

jet (*astronomy*)   Any high-velocity narrow stream of a fluid is a jet. In astronomy several kinds of jets are observed, but the word jet is most often used for giant glowing streams, thousands of light-years long, that are emitted from the centers of active galaxies, such as Seyfert galaxies. Such cosmic jets form when material being sucked into a black hole develops a ring around the hole; some largely disassembled material is shot in a narrow band perpendicular to the ring, forming 2 jets. Often only 1 jet is observed because the other is on the far side of the black hole and moving away at nearly the speed of light. Because of an optical illusion, some jets that point toward us contain particles that appear to be moving faster than the speed of light. Our own Milky Way galaxy has a jet from its center that consists of positrons instead of the more typical gases and electrons.

jet stream (*meteorology*)   In 1944 high-flying airplanes encountered narrow, rapid currents in the upper atmosphere. Such currents, which range from 60 mph (100 km/hr) to nearly 300 mph (480 km/hr), are called jet streams. At times as many as 6 west-to-east jet streams, each from 180 mi (290 km) to 300 mi (480 km) wide and from 30,000 ft (9000 km) to 45,000 ft (14,000 km) high, meander around Earth. The Indian Ocean monsoon, when it flows from east to west, is also considered a form of jet stream.

The main jet streams are the boundaries between polar and temperate air. The north-south location of the jet streams between cool and warm air masses is often blamed for bringing cold weather when they dip toward the equator; however, it is more accurate to say that the position of the jet streams is caused by the location of the air masses.

Johanson, Donald (*paleoanthropologist*)   [American: 1943–] Donald Johanson was a coleader of the team that in 1974 discovered in Ethiopia the first specimens of *Australopithecus afarensis*, including the 1974 find of Lucy, a 40%-complete fossil from 3,300,000 years ago. The year before Johanson had found the knee joint of an australopithecine, which provided the first hard evidence of upright stance.

joint (*anatomy*)   In vertebrates, a joint is an area where 2 bones meet. Most joints are movable but some, as in the skull, are fused and immovable. Ball-and-socket joints, as in the shoulder and hip, permit rotational movement. Pivot joints, such as the one between the skull and the vertebrae, allow another type of rotational movement. Hinge joints, found in the elbow and knee, allow movement in 1 direction only. Saddle joints in the thumb allow movement in 2 directions. Bands of tough, flexible tissue called ligaments hold bones together at movable joints and a lubricating fluid, sinovial fluid, usually is present inside the joints. In arthropods ("joint-footed animals") there are joints between parts of the exoskeleton. For example, the walking legs of a crayfish have several hinge joints.

Joliot-Curie, Irène (*physicist*) [French: 1897–1956] Together with her husband Frédéric Joliot [French: 1900–58], Joliot-Curie found a way to create radioactive isotopes of elements whose common isotopes are stable, beginning with radioactive phosphorus in 1933. These artificial isotopes have become the main sources of radioactivity for medicine and industry.

Josephson effect (*particles*) In 1962 Brian Josephson [English: 1940–] calculated that when 2 superconductors are separated by a very thin insulator, they will develop a flow of current from one to the other by electrons tunneling through the insulator, an effect since observed. The current can be controlled by weak magnetic fields, which has been exploited in sensitive measuring devices and very fast switches for computers.

Joule, James Prescott (*physicist*) [English: 1818–89] From an early age Joule was a careful experimenter and his precise measurements of heat clarified concepts that were vague in the formulations of other physicists. Notably during the decade 1837 to 1847 he established that heat is a form of energy and proved experimentally the conservation law for energy. For example, in 1843 he calculated precisely how much heat is produced when a falling weight turns a waterwheel. Joule's other important experimental discoveries include Joule's law: the heat of an electric current is proportional to the product of the resistance and the square of amperage; the Joule-Thomson effect: temperature decreases when a gas expands; and magnetostriction: a tiny increase in the length of a magnetized ferromagnet.

J/psi particle. (*particles*) Like the hyperon, J/psi is a heavy particle that appears at high energies and is produced by a different kind of quark, charm. The name J/Psi results from the particle's independent discovery by 2 investigators in 1974, one who called it J and the other who named it psi.

Juan de Fuca Plate (*geology*) One of the smallest recognized divisions of Earth's lithosphere, the Juan de Fuca Plate off Oregon, Washington, British Columbia, and Alaska has raised the volcanic Cascade Mountains and other mountains along the northwest coast. This region is marked by several pieces of crust called terrains that have moved long distances to plaster themselves along the west coast of North America.

jungle (*ecology*) A jungle is a very dense, tangled mass of vegetation found in certain tropical areas. It is most common on the edges of rain forests and along riverbanks and other clearings in forests. In these places sunlight reaches the ground, enabling rapid growth of vines and other plants whose growth is limited by a dark forest canopy.

juniper (*plants*) Members of the cypress family, junipers (*Juniperus*) are evergreen shrubs and trees widespread in northern temperate regions. The leaves are needles or scales. The male cones form catkins; the female seed-bearing cones resemble fleshy berries. The common juniper (*J. communis*) grows to 100 ft (30 m) tall; its aromatic cones are used to flavor gin.

Jupiter (*solar system*) The largest planet in the solar system, Jupiter has 2.5 times the mass of all the other planets together and is 11 times as wide as Earth. Jupiter's thick atmosphere is mainly hydrogen, with some helium. Cloud tops are cold (about -202°F, or -130°C), but temperatures increase deeper inside the atmosphere. About 600 mi (1000 km) down, great oceans of liquid hydrogen form Jupiter's surface. The oceans may be some 12,000 mi (20,000 km) deep. Beneath them lies a rock and iron ball about the size of Earth. Jupiter spins so fast that its clouds form bands, giving the planet a striped appearance. Jupiter's Great Red Spot is a massive hurricane first observed some 300 years ago that continues unabated. Jupiter has 16 moons; the 4 largest, called Galilean moons, were first observed by Galileo in 1610. Jupiter also has a faint ring system that extends almost 186,000 mi (300,000 km) from the planet's surface.

juvenile hormone (*biochemistry*) Molting and metamorphosis in immature insects are controlled by hormones. Under the direction of a hormone produced by certain brain cells, glands in the thorax produce molting hormone, or ecdysone, which stimulates growth and molting. Juvenile hormone, produced in paired endocrine glands in the head, suppresses metamorphosis. This enables the immature insect to grow in size as it molts while remaining in a larval or nymph state. When secretions of juvenile hormone decline, the larval insect metamorphoses into a pupa. Secretions of juvenile hormone then cease, allowing adult characteristics to develop. When ecdysone induces the next molt, an adult insect emerges.

kame (*geology*)   Small cone-shaped hills of drift from a glacier are called kames. Water near the top of the glacier melts in warmer weather and forms rivers that can flow along the glacier much of the year. The rivers carry rock and rock dust off the edge of the glacier, where they leave kames made of layers of drift.

Kamerlingh Onnes, Heike (*physicist*)   [Dutch: 1853–1926] Kamerlingh Onnes continued the work of James Dewar [Scottish: 1842–1923] in liquefying gases, achieving liquid helium (the coldest liquid gas) in 1908. Three years later he recognized that certain metals, such as mercury, tin, or lead, when cooled to the temperature of liquid helium—below −454°F (−268.9°C)—conduct electricity with 0 resistance, the phenomenon known as superconductivity.

kangaroo (*animals*)   Marsupials known as macropods ("long feet"), kangaroos are native to Australia and nearby islands. They typically have short arms, large powerful legs, and a large muscular tail that acts as a balance as the animal moves along in a series of hops or leaps. The smallest are rat kangaroos, less than 18 in. (45 cm) long. Among the medium-sized species are wallabies. The largest are red and gray kangaroos, which may be nearly 7 ft (2 m) tall and weigh up to 200 lb (90 kg). Kangaroos are herbivores. They live in a variety of habitats: red kangaroos on open plains, wallaroos in rocky mountainous areas, tree kangaroos in rain forests, and so on.

kangaroo rat (*animals*)   North American rodents, kangaroo rats (genus *Dipodomys*) are exquisitely adapted for life in arid, desertlike environments. They emerge from their burrows only at night to feed on seeds, leaves, and fruits, some of which they stuff into cheek pouches for later use. They have large hind feet with flat, hairy soles that prevent them from sinking into soft sand as they hop about, and a tail often longer than their body to provide balance. Kangaroo rats do not drink water, but obtain it from their food. To keep their fur clean, they take dust baths.

kaon (*particles*)   Mesons that combine a light up or down quark with the heavier strange quark are called kaons (originally kappa mesons, shortened to K mesons, then kaons). There are 2 different neutral kaons, often called K-long and K-short because of the different lengths of time they exist before changing into other particles. Studies of the neutral kaons have been important in modern theories of particle interaction.

Karst topography (*geology*)   Regions around the world that have the same landscape features as the Kars Plateau in eastern Europe are said to have Karst topography. In a region with Karst topography, the land is pitted with sinkholes, while streams often flow underground, sometimes reaching the surface in the form of large springs. In some cases, as in Yucatán, there is almost no surface water except in sinkholes. Caves and caverns abound, as well as vertical shafts. Karst topography results from water acting on limestone, especially where there is little soil above bedrock.

karyotype (*genetics*)   The full complement of chromosomes of a eukaryotic cell or organism is its karyotype. The karyotype of the mosquito *Culex pipiens* consists of 3 pairs of chromosomes; rice (*Oryza sativa*), 12 pairs; humans, 23 pairs; and dogs (*Canis familiaris*), 39 pairs. A photographic representation of all the chromosomes of a cell—showing their number, size, and shape—is also called a karyotype (or karyogram). A standard format is used, with the pairs arranged in order of decreasing size.

Kekulé, Friedrich A. (*chemist*)   [German: 1829–96] The field of organic chemistry was first put on a sound basis by Kekulé in 1858 with his determination that carbon has a valence of 4. Kekulé also found that of the 4 bonds for each carbon atom, 1 bond can link to another carbon atom, permitting molecules to form as chains. Kekulé's structures provided an explanation for isomers and other phenomena. His half-dreamed vision in 1861 of the benzene ring (in his dream, a snake eating its tail) became the basis for the chemistry of aromatic compounds.

kelp (*protists*)   Members of the brown algae (Phylum Phaeophyta), kelp are the largest species of protists; some kelp attain lengths of 200 ft (60 m) or more—and grow at a rate of 2 ft (0.6 m) a day. Kelp are marine organisms that mainly inhabit cold water.

The body of most kelp is differentiated into several structures, including a holdfast that anchors the kelp to the ocean floor, a stemlike stipe, and leaflike blades. Air-filled bladders help keep the blades near the water's surface. Kelp contain chlorophyll, but this green pigment is masked by a brown pigment, fucoxanthin.

Kelvin, William Thomson, Baron (*physicist*)   [Scottish: 1824–1907] Thomson did most of his scientific work before becoming a baron. However, he is best known by the name Kelvin. That name is used for the measure of absolute temperature, for a half dozen devices he invented, and for Kelvin's law describing the size of an economical electrical conductor. Additionally, there are 2 other well-known English physicists named Thomson.

Kelvin's main accomplishments were in thermodynamics, where he developed the idea of absolute zero and independently recognized its first and second laws. With James Prescott Joule Kelvin discovered that gases cool when allowed to expand, the Joule-Thomson effect. Among his best-known work is attributing a far too young age for Earth based on how long the planet would take to cool to its present temperature; Kelvin's error occurred because he calculated this in 1846, a half century before radioactivity, which keeps Earth hot, was discovered.

Kepler, Johannes (*astronomer/mathematician/physicist*)   [German: 1571–1630] Kepler correctly recognized that planetary orbits are ellipses with the Sun at one focus; calculated that a line from any planet to the Sun sweeps out equal areas in equal times; and determined that the squares of revolution times are proportional to the cubes of average distances from the Sun. Kepler believed that mathematical relationships explain the universe, but built his laws on observation, using Tycho Brahe's improved tables. In mathematics, Kepler developed ideas that led directly to calculus. Sent one of the first astronomical telescopes by Galileo, Kepler improved the design, worked out the theory of lenses, invented the compound microscope, and described how a parabolic mirror focuses light at a point.

keratin (*biochemistry*) Hair, feathers, nails, claws, hooves, and horns are formed from surface skin cells that become filled with keratins. These fibrous, highly insoluble proteins are resistant to wear and chemical disintegration. Keratins also are the main structural component of other rugged animal tissues, including the scales of reptiles, the coverings of fish eggs, and the skeletons of sponges. Organisms that grow on hair and other keratinous material, such as certain fungi, are called keratinophilic.

ketone (*biochemistry*) Ketones are a class of organic compounds characterized by a fragrant odor and the presence of a carbonyl group ($C=O$) within the molecule. Many ketones are formed as intermediaries in the synthesis of organic compounds. Ketones also are products of metabolism; for example, the simplest ketone, acetone (CH_3COCH), is a product of protein metabolism. Sugars that contain a ketone group are called ketoses; fructose (fruit sugar) is an example.

kettle (*geology*) Sometimes glacial movement pushes a large block of ice from the glacier into the ground or into a moraine. After the glacier melts, it leaves behind a deep hole in the ground. The roughly circular depression is called a kettle or, if filled with water, a kettle lake.

keystone species (*ecology*) The interdependence of living things within a natural community is a fundamental principle of ecology. But members of a community are not equally influential. Some members may play disproportionately large roles. If they disappear, the community undergoes rapid, dramatic changes. For example, fig trees are pollinated by tiny fig wasps. If the fig wasps disappear, the fig trees will not produce flowers or bear fruit. Without the fruit, a variety of birds, monkeys, and bats will face a major shortage of food. The fig wasps, then, are keystone species. Their demise could lead to the extinction of other species.

Khorana, Har Gobind (*genetics*) [Indian-American: 1922–] In the 1960s Khorana was among the first to analyze the genetic code, synthesizing each of the 64 possible combinations of 3 nucleotides that form its "letters." In 1970 he created an artificial version of a 126-base gene found in a bacterium. Today such strings of nucleotides, called oligonucleotides, are easily produced by machines based on Khorana's approach.

al-Khwarizmi, Muhammad ibn-Musa (*mathematician*) [Persian: c800–c850] Two books by al-Khwarizmi had a major impact on mathematics. One, whose Arabic manuscript was lost, "Concerning the Hindu Art of Reckoning" (published in Latin in 1240), introduced Europe to the numeration system. This system, which most of the world uses today, is called the Hindu-Arabic system, although al-Khwarizmi gave all credit to the Indians. His other book, *Al-jabr wa'l muqabalah* ("Restoration and Balancing"), provides the origin of the word "algebra," since it contains the solutions to quadratic equations and methods for multiplying binomial expressions.

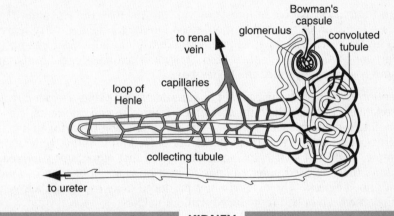

KIDNEY

kidney (*anatomy*)   The main organs of the vertebrate excretory system are 2 kidneys. In humans they are bean-shaped organs about 4 in. (10 cm) long and 2 in. (5 cm) wide that lie against the back in the upper abdomen. The kidneys remove wastes from the blood, producing a fluid urine (in most mammals and amphibians) or crystalline uric acid (in birds and snakes). The kidneys also regulate water balance and concentrations of salts and other chemicals.

The structural and functional units of a kidney are microscopic nephrons. Water, salts, and other substances diffuse from the glomerulus (a cluster of blood capillaries) into the nephron's Bowman's capsule. As these substances pass through tubules and the loop of Henle, some—dissolved foods, for example—are reabsorbed into capillaries. The remaining substances empty into a collecting tubule and out of the kidney.

kinesthesia (*physiology*)   The sense of perception of movement, weight, and position is termed kinesthesia. Specialized receptor cells, proprioceptors, are located in muscles, tendons, joints, the wall of the gastrointestinal tract, and other internal organs. These cells detect changes and transmit the information to the central nervous system, where it is interpreted and acted on. Kinesthesia is vital for maintaining balance and posture and for coordinating body movements.

kinetic energy (*mechanics*)   Energy that derives from an object's motion is called kinetic. The amount of energy depends on the mass of the object (m) and the square of the speed (v), and is measured as $1/2 \times mv^2$. Another way to view kinetic energy is that the amount of work done in moving an object from rest to a particular speed is the same as the kinetic energy of the object.

kingfisher (*animals*)   Found mainly in the tropics, kingfishers (Family Alcedinidae) have a long, straight, pointed beak, a short neck, short wings, and short legs. Their plumage is often very colorful and some species have long tails. Many species feed primarily on fish, which they grab after diving head-first into the water. But some species are not fishers, preying instead on other small animals. Among the largest is the giant kingfisher (*Ceryle maxima*) of Africa, almost 20 in. (50 cm) in length.

kinkajou (*animals*)   Native to forests of tropical America, the kinkajou (*Potos flavus*) is a carnivorous mammal in the raccoon family. Its slender, furry body is 15 to 30 in. (38 to 76 cm) long; its prehensile (grasping) tail is 15 to 22 in. (39 to 57 cm) long. Active at night, it feeds on small animals, eggs, fruits, and honey, and is sometimes called the honey bear.

kin selection (*ethology*)   The concept of kin selection explains altruism directed toward relatives. Because they are descended from common ancestors, closely related animals share the same genes. Thus self-sacrificing behavior that bene-fits relatives helps ensure that those genes are passed on to future generations. For example, studies of jackals in Tanzania showed that more pups survived in family units that included older siblings from a previous litter than in family units with no "helpers." The older siblings produced more copies of their genes than if they themselves had reproduced without helpers.

Kirchhoff, Gustav Robert (*physicist*)   [German: 1824–87] In 1845 Kirchhoff developed the laws governing current and resistance in electrical networks. In 1859 Kirchhoff and Robert Bunsen invented the spectroscope, obtaining spectrums from glowing samples. Kirchhoff recognized that elements can be identified this way and that dark lines in a star's spectrum represent elements in gas sur-rounding the star. He calculated the type of radiation to be emitted by a heated black body. Resolving contradictions between Kirchhoff's black-body theory and experiment led to the development of quantum theory.

Kitasato, Shibasaburo (*bacteriologist*)   [Japanese: 1852–1931] In 1889, while working with Koch in Berlin, Germany, Kitasato became the first person to grow a pure culture of *Clostridium tetani,* the bacillus that causes tetanus. The fol-lowing year he and Emil von Behring showed that animals injected with dilute tetanus toxin develop antitoxins that protect them against the disease. In 1894, after returning to Japan, Kitasato discovered *Pasteurella pestis*, the bacillus that causes bubonic plague.

kite (*animals*)   Members of the hawk family, kites are graceful fliers with long wings and a long tail. Their legs are shorter and their feet weaker than those of other members of the family. As a result, they hunt smaller prey. The red kite (*Milvus milvus*) of Eurasia is a pirate that often steals food from other birds. The Everglade kite (*Rostrhamus sociabilis*) of the Americas feeds almost exclu-sively on freshwater snails of genus *Pomacea*.

kite (*geometry*)   A kite is a quadrilateral for which adjacent sides are equal in length while opposite sides are not. Line segments connecting the opposite corners of a kite are diagonals and the area, A, of a kite is half the product of the diagonals, d and D, or $A = 1/2\, dD$.

kiwi (*animals*) Flightless birds with tiny vestigial wings hidden by hairlike feathers, kiwis (genus *Apteryx*) are native to New Zealand forests. About the size of chickens, they can run swiftly on sturdy, muscular legs. Their nostrils are at the tip of a long, flexible bill. Kiwis have an acute sense of smell and hearing, used to detect food (mainly worms and insects) and potential danger.

kiwi (*plants*) Various species of the genus *Actinidia* in the family Actinidiaceae produce edible fruit. The best known is *A. chinensis*, a large deciduous vine native to China that has deep green leathery leaves and fragrant whitish flowers. The first commercial varieties were developed in New Zealand following the plant's introduction there early in the 20th century. The oval fruit has brownish skin densely covered with short, stiff hairs—rather like the hairy plumage of the kiwi, the bird after which the fruit is named. (The fruit also is called Chinese gooseberry and, in China, yang tao.) The flesh of the fruit, usually bright green, contains many minute seeds at the center.

Klein, (Christian) Felix (*mathematician*) [German: 1849–1925] Klein worked with transformations while a student and gained fame by unifying the different types of geometry (Euclidean, non-Euclidean, projective, and topological) in terms of groups of transformations. Euclidean geometry, for example, can be analyzed as based in the group of reflections. Klein also invented the Klein bottle and developed the theory of the gyroscope.

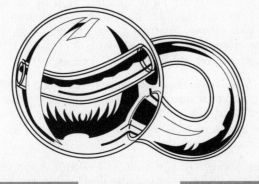

KLEIN BOTTLE

Klein bottle (*geometry*) A Klein bottle is a surface with only 1 side; it was invented by Felix Klein in 1882. A tapered tube is bent so that the small end passes the tube's side without intersecting it and joins smoothly to the larger end from inside. The result is 2 mirror-image Möbius strips joined at their edges. A Klein bottle cannot be created in 3-dimensional space but is possible in 4 dimensions.

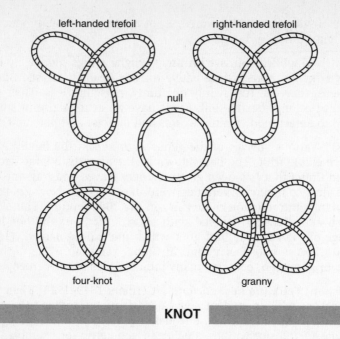

left-handed trefoil   right-handed trefoil

null

four-knot   granny

KNOT

knot (*geometry*) A mathematical knot is a closed 1-dimensional curve in space, so the knot can never be untied. A circle is an example of the simplest knot, called the null knot. Any more complex knot has at least 1 part going through the hole. Knots are classified by flattening them and counting the number of times one part crosses another. There are only 2 knots, called trefoils, that have a single crossing, but 15 exist with 7 crossings. The number rises rapidly: there are over 1,700,000 knots with 16 crossings.

koala (*animals*) The koala (*Phascolarctos cinereus*) is a tree-dwelling marsupial found mainly in eucalyptus forests of eastern Australia. It has dense light gray fur, large ears, and short legs with powerful feet adapted for climbing. It is up to 32 in. (82 cm) long and weighs up to 33 lb (15 kg). The koala forages at night, feeding on eucalyptus leaves.

Koch, Robert (*bacteriologist*) [German: 1843–1910] A founder of the science of bacteriology, Koch proved that the cattle disease anthrax is caused by a bacterium (*Bacillus anthracis*) that forms hardy spores able to survive in soil for many years. Koch also discovered the bacteria that cause tuberculosis (*Mycobacterium tuberculosis*) and cholera (*Vibrio cholerae*). He developed improved methods for staining bacteria and introduced the use of gelatin and, later, agar as growing media for bacterial colonies. His laboratory in Berlin, Germany, attracted many brilliant young researchers, including Emil von Behring, Paul Ehrlich, and Shibasaburo Kitasato.

Koch's postulates (*monera*) In 1890 Koch laid down 3 conditions—Koch's postulates—that must be met to prove that a disease is caused by a particular microorganism. One: the microorganism must always be found in diseased individuals and must be grown in pure culture in the laboratory. Two: a sample of the culture, when injected into a healthy individual, must produce the disease. Three: the microorganism must be recovered from the newly diseased individual, grown as a pure culture, and shown to be identical to the original microorganism.

Kornberg, Arthur (*biochemist*)   [American: 1918–] Kornberg discovered the enzyme DNA polymerase and determined that it plays a role in the replication of DNA. In 1957 Kornberg produced the first artificial DNA; the giant molecule was chemically exact but biologically inert (it could not replicate itself). In 1967, Kornberg and 2 associates succeeded in synthesizing DNA that was biologically active. Kornberg shared the 1959 Nobel Prize in physiology or medicine with Severo Ochoa [American: 1905–93], who synthesized RNA in 1956.

Kovalevski, Sonya (*mathematician*)   [Russian: 1850–91] Kovalevski's mathematical work dealt with sophisticated generalizations of parts of calculus and with applied topics, including the equations describing rotations of a body about a fixed point, the propagation of electromagnetic waves, and the structure of the rings of Saturn.

Krebs cycle (*biochemistry*)   During the first stage of cellular respiration, living cells use a series of reactions called the Krebs cycle, or citric acid cycle, to obtain energy from food. The cycle was discovered by Hans Krebs [German-English: 1900–81] in the 1930s. In the cycle, which involves changes in several compounds, carbon-carbon bonds in food molecules are broken. The carbon is expelled as carbon dioxide. The energy that held the carbon atoms together is released and stored in adenosine triphosphate (ATP).

krill (*animals*)   Comprising about 90 species in the order Euphausiacea, krill are small shrimplike crustaceans 0.3 to 3 in. (0.8 to 8 cm) long. They live in open seas, typically in dense swarms, and are bioluminescent, producing a blue-green light that is thought to be used for communication during swarming and reproduction. Krill feed on microscopic plankton and are extremely important in marine food chains, serving as the main food for baleen whales and many fish and birds. A whale may consume 5000 lb (2268 kg) of krill at one feeding.

krummholz (*ecology*)   The transition zone between forest and alpine tundra on many mountains is marked by clusters of low, stunted trees called krummholz (from German words meaning "crooked wood"). In this zone, environmental conditions—mainly icy winds—prevent normal growth in the height of trees, which generally are species of spruce, pine, and fir. The krummholz effect also is common along coastlines, with windbent trees growing away from the ocean. Salt from the ocean also retards growth of branches on the side of the trees facing the water.

krypton (*elements*)   Krypton is a noble gas, atomic number 36, discovered in 1898 by Alexander Ramsay [Scottish: 1852–1916] and Morris William Travers [English: 1872–1961]. It was named from the Greek *kryptos*, "hidden." Chemical symbol: Kr.

K/T event (*paleontology*)   The end of the Cretaceous, 65 million years ago, was marked by a mass extinction known as the Cretaceous-Tertiary or K/T event (the K comes from the Greek for chalk, *kreta*). In the sea ammonites, many mollusks, and most large marine reptiles became extinct. On land dinosaurs, flying reptiles, and 2 out of 3 marsupial mammal families disappeared. But bony fish, sharks, frogs, lizards, snakes, crocodilians, turtles, birds, and placental mammals survived and went on to prosper.

Numerous theories have been proposed to explain the event. At present, the most widely held theory is that the collision of a comet or asteroid with Earth sent billions of tons of matter into the atmosphere, blotting out the Sun and causing temperatures around the globe to plunge.

kudzu (*plants*)   Known as "the vine that ate the South," kudzu (*Pueraria*) is a legume native to Asia. It was introduced into the United States in 1876 and became widespread in the American South beginning in the 1930s. It has compound leaves, hanging clusters of fragrant purple flowers, and brown flattened seedpods. Kudzu vines grow as much as 1 ft (30 cm) per day, and as many as 30 vines—each up to 100 ft (30 m) long—can grow from a single massive root that weighs as much as 400 lb (180 kg). In Asia kudzu has natural enemies that limit its growth, but in the United States it is a serious pest.

Kuiper, Gerard Peter (*astronomer*)   [Dutch-American: 1905–73] Although he contributed to astrophysics, established the atmosphere of Saturn's moon Titan, and discovered Saturn's moon Miranda and Neptune's Nereid, Gerard Kuiper's prediction of thousands of comets traveling near the orbit of Neptune, now called the Kuiper belt, has become his best-known achievement.

Kuiper belt (*solar system*)   The Kuiper belt is home to billions of comets, some very large—the planet Pluto may be the largest—located mostly in the region from the orbit of Uranus to the orbit of Neptune. In 1951 the belt was proposed by Gerard Peter Kuiper as the source of short-period comets (those lasting less than 200 years). In 1977 Charles T. Kowal [American: 1940–] discovered the large body Chiron ranging between Saturn and Neptune; it was eventually recognized as a giant comet from the Kuiper belt. Several other large and small Kuiper belt objects have since been discovered. (*See also* **Oort cloud**.)

L

lacewing (*animals*)   Members of the insect family Chrysopidae have 4 similar delicate wings laced with veins; at rest the wings are held tentlike above the body. Lacewings have metallic eyes and greenish bodies. Many species have glands that produce an unpleasant-smelling secretion, giving them the name "stinkflies." Lacewings are predators; their larvae feed chiefly on aphids.

lachrymal gland (*anatomy*)   The tear, or lachrymal, glands are located below the upper eyelids of mammals. They secrete tears, which keep the surface of the eye moist and wash away dust and other particles. Like saliva and other body fluids, tears contain lysozyme, an enzyme that destroys bacteria.

lactose (*biochemistry*)   Lactose, or milk sugar, is a double sugar with the formula $C_{12}H_{22}O_{11}$. It is found only in the milk of mammals and is produced from 2 simple sugars, glucose and galactose. During digestion, lactose is split into glucose and galactose by the enzyme lactase. The souring of milk results from the activity of lactic acid bacteria, such as species of *Lactobacillus*, which change lactose to lactic acid.

lagoon (*geography*)   A still body of ocean water surrounded on all sides, or nearly so, by land is a lagoon, especially if the land is a coral reef or sandbar. Lagoons form in coral atolls when a central mountain sinks. Wave action on a beach can create a sandbar offshore with a lagoon between the bar and the beach.

Lagrange, Joseph-Louis (*mathematician*)   [French: 1736–1813] Lagrange developed many general ideas for handling mathematical problems, such as the use of higher-dimensional spaces to describe the motion of a body. He put the subject known as calculus of variations into its present form, a powerful tool for finding minimum or maximum functions. Although his famous book without pictures is called *Méchanique analytique* ("Analytical mechanics"), he was very much a pure mathematician, contributing, for example, to number theory. His main influence at a more familiar level comes from the metric system, whose foundation was in large part according to his plan.

lahar (*geology*)   A mud avalanche pouring down the side of a volcano, called a lahar, is a common and very destructive event. Lahars may be caused when a crater lake overflows or breaks through the side of the crater, when an eruption melts a glacier atop a tall volcano, or when a pyroclastic flow of hot gas and rock encounters a river or lake. Mud from a lahar flows easily because the wall of a steep cinder cone or composite volcano is loosely packed; sometimes villages miles from the base of a volcano are completely covered with lahar mud.

Lamarck, Jean-Baptiste (*naturalist*) [French: 1744–1829] Lamarck was the first person to propose a theory of evolution. He suggested that evolution of a species, which he called transformation, takes place as the result of "a new need that continues to make itself felt" and that acquired characteristics gained during an organism's life can be inherited by the organism's offspring—a concept later disproved. Similarly, Lamarck noted that Earth's surface is constantly transformed, resulting in dramatic changes over time.

Lamb shift (*particles*) In 1947 Willis Lamb, Jr. [American: 1913–] carefully measured the spectrum produced by "excited" hydrogen atoms when an energized electron dropped into a lower level and found a shift of about 10% from the location predicted by quantum theory. This Lamb shift is caused by new particles appearing out of the vacuum and almost instantly disappearing.

lamprey (*animals*) Lampreys are jawless fish with a long tubelike body. They lack true teeth and have a cartilage skeleton. The adults of nonparasitic species do not eat; they die soon after reproducing. The adults of parasitic species suck blood from other fish. The walls of their circular mouth are lined with horny toothlike structures and their tongue is equipped with sharp plates that scrape away the prey's flesh. Lampreys are found in both fresh and marine waters, but they all breed in freshwater.

lancelet (*animals*) Lancelets are primitive chordates that share important features with vertebrates, particularly the presence of a notochord and spinal cord, but lancelets have no backbone. Lancelets have a slender fishlike body about 2 in. (5 cm) long, ending in a pointed, lancelike tail. Long fins run the length of the body on the upper and lower surface. Lancelets live in shallow marine waters, burrowing in sand by day and swimming about at night.

Landau, Lev Davidovich (*physicist*) [Russian: 1908–68] Landau is best known for explaining superfluidity of helium-4 near absolute zero. A superfluid liquid has no internal friction, so a vortex continues to spin ceaselessly. The liquid flows uphill, over the sides of containers, as well as through tiny openings too small to permit passage of helium gas.

landslide (*geology*) Loose rock, or even bedrock broken off its foundations, on steep mountains or cliffs can pour down in a landslide, also called a rock slide. Wet soil, especially on the sides of volcanoes, forms a kind of landslide called a mudflow. Sometimes hot gases from a volcano break through the side of the mountain, carrying hot rock and volcanic ash in a devastating landslide called a pyroclastic flow or a *nuée ardente*.

Landsteiner, Karl (*immunologist*) [Austrian-American: 1868–1943] In the early 1900s Landsteiner discovered that 4 major blood groups exist among humans—A, B, AB, and O—based on the presence or absence of certain antigens and antibodies. This led to blood typing and the advent of safe blood transfusions; Landsteiner was awarded a Nobel Prize in 1930 for the discovery. In 1940 Landsteiner and a colleague discovered another antigen on the surface of red blood cells: the Rh factor, so-named because it was first found in the blood of rhesus monkeys.

Langmuir, Irving (*chemist*)   [American: 1881–1957] As a research chemist at General Electric Company for over 40 years, Langmuir investigated and introduced new ideas to such topics as valence, thermionic emission, surface films on liquids, reactions between gases and solids, and plasmas caused by electrical discharge. In addition to his theoretical work, Langmuir invented the modern incandescent lamp with a tungsten coil surrounded by an inert gas and also a high-temperature welding torch.

La Niña (*oceanography*)   A worldwide weather pattern results from a cooling of the middle part of the Pacific Ocean, an effect known as La Niña ("the girl," since it is the opposite of the warming that causes El Niño, "the boy"). Sea-surface temperatures drop as much as 7°F (4°C) in the Pacific, reducing evaporation there. Rain still falls in the western United States, but there is not enough moisture to reach the east; thus La Niña often causes drought in the eastern half of the United States.

lanthanum (*elements*)   Lanthanum gives its name to the lanthanide series of metals, also known as rare earths, because its atomic number, 57, makes it the first in the series. It was also the first to be discovered (in 1839) by Carl Gustav Mosander [Swedish: 1797–1858], who named it from the Greek *lanthanein*, "concealed," because it was hidden in a cerium ore. Its atomic structure, lacking the inner electrons found in other rare earths, makes it especially reactive, oxidizing rapidly when exposed to air. Its main use is in cores of bright carbon arc lights for the motion picture industry. Chemical symbol: La.

Laplace, Pierre-Simon (*mathematician*)   [French: 1749–1827] Laplace used mathematics to study the origin and stability of the solar system, completing Newton's work in his book *Méchanique céleste* ("Celestial mechanics"). In the process he invented the general concept of potential as it applies to any field, such as a gravitational or electric field. Laplace was also the main contributor to probability theory before the 20th century.

lapse rate (*atmospherics*)   The normal lapse rate is steady decline in the lower atmosphere of temperature with height—about 3.5°F (1.9°C) per 1000 ft (300 m). Sunlight passes through the troposphere without heating it, but it does heat Earth's surface. Heat in the form of infrared radiation, conduction, and convection is returned to the air from the surface. The radiation portion is trapped by the greenhouse effect, more effective at lower levels because of higher concentrations of greenhouse gases. Furthermore, as cooler air sinks, it pushes up warm air to levels where air pressure is lower, causing the rising air to expand. The gas laws tell us that expansion cools air further. This additional cooling, of about 5.5°F (3°C) per 1000 ft (300 m), is called the adiabatic lapse rate. However, if water vapor condenses because of cooling, it releases latent heat, so the moist adiabatic lapse rate is somewhat less than the dry rate.

larch (*plants*)   Members of the pine family, larches (*Larix*) are deciduous conifers native to cooler regions of the Northern Hemisphere. They have short needles that grow in clusters at the end of spurs on the branches. The small seed-forming cones grow erect. The tallest species is the western larch (*L. occidentalis*) of western North America, which can grow to 200 ft (60 m).

lark

lark (*animals*)   Sparrowlike perching birds about 7 in. (18 cm) in length, larks have long pointed wings and a long claw on each hind toe. Except for 1 American species, the horned lark (*Eremophila alpestris*), larks are native to Europe, Asia, and especially Africa. Many species, particularly the skylark (*Alauda arvensis*), are famed for their beautiful songs.

larva (*zoology*)   Many animals that hatch from eggs look radically different from how they will appear as adults. The young, called larvae, are independent—they move about, find their own food, and so on—but usually are sexually immature. They will eventually undergo metamorphosis, either directly to the adult stage or to an intermediate stage. Shrimp, for example, have several distinctively different larval stages. Among vertebrates the best-known larvae are frog tadpoles. Among invertebrates the caterpillars of butterflies and moths are familiar to all.

larynx (*anatomy*)   In amphibians, reptiles, birds, and mammals, the upper part of the trachea (windpipe) is enlarged, forming a larynx. In many of these animals, including most mammals, the larynx contains folds of fibrous tissue called vocal cords. The vocal cords produce sound when set into vibration by the flow of air from the lungs.

The human larynx is referred to as the Adam's apple. Due to the action of the sex hormone testosterone, it is larger in men than women, which accounts for the lower pitch of men's voices.

Lascaux cave (*paleoanthropology*)   The Lascaux cave in the Dordogne region of France was discovered in 1940. Its walls contain large-scale paintings of horses, wild oxen, bison, other animals, and 1 human and combine features in imaginative ways. The paintings date from about 16,000 years ago at the height of the Magdalenian culture. The negative effects of human visitors to the cave caused it to be closed to tourists, who can visit instead a life-size reproduction built near the original cave in 1983.

laser (*optics*)   A laser is a device that produces a beam of light at a single wavelength—called monochromatic since wavelength determines color—with each wave having the same direction and phase. Such a beam is called coherent light because the beam "sticks together," or coheres, instead of spreading apart.

The word laser is an acronym that describes its operation: light amplification by stimulated emission of radiation. An electric current or other means is used to raise the energy levels of outer electrons in a tube or cylinder of atoms chosen so that the energized electrons decay slowly, emitting a photon after an appreciable time interval. Soon there are more excited atoms than lower-energy ones and they begin decaying and releasing photons. Einstein predicted in 1917 that these photons will induce other excited atoms of the same type to decay suddenly, also releasing photons of exactly the same energy level, producing a cascade of photons. Mirrors at the ends of the tube or cylinder direct the photons, bouncing most back and forth but allowing some to escape from one end. The escapees form the laser beam.

latent heat (*thermodynamics*)   Heat energy that is stored in molecular arrangements is called latent heat. The extra calories needed to melt ice or boil water are familiar examples. Water heated to 212°F (100°C) does not boil until more heat is added. When it does vaporize, the latent heat is stored in the water vapor. On condensation this heat is released. Efficient engines collect and reuse the latent heat from condensation.

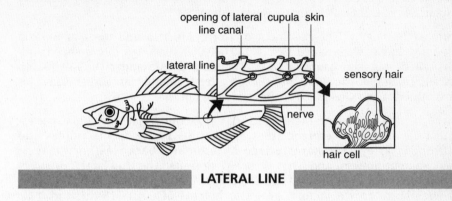

opening of lateral cupula skin
line canal

lateral line

sensory hair

nerve

hair cell

LATERAL LINE

lateral line (*anatomy*)   Fish and some amphibians have a lateral line system that detects slight changes in pressure in surrounding water, enabling the organisms to respond to currents, movements of nearby objects, etc. The system consists of canals, 1 along each side of the body, with extensions to the head. Cup-shaped organs (cupulas) along the canals contain highly sensitive hair cells covered by a gelatinous material. As water in the canals moves, the gelatin is deformed, causing hair cells to transmit impulses to a nerve.

The vestibular apparatus—the section of the vertebrate inner ear that detects the pull of gravity—is believed to have evolved from part of the lateral line system.

laterite (*soils*)   When iron-rich basalts weather under tropical conditions, soil particles called laterite form. Heavy precipitation in low areas leaches compounds from soil instead of washing soil away. Extensive leaching even removes iron, and aluminum-rich bauxite remains. Laterite soils depend on a constant supply of humus. When a tropical rain forest is cut, laterite particles bond together to form a stonelike surface, preventing regrowth.

latex (*biochemistry*)   Several hundred species of flowering plants have specialized cells or elongated ducts called laticifers; these contain a milky fluid called latex. Latex may be colorless, white, yellow, or reddish, and may contain salts, sugars, proteins, fats, tannins, and other dissolved or suspended substances. Latex of the Brazilian rubber tree (*Hevea brasiliensis)* and Indian rubber tree (*Ficus elastica*) contains high concentrations of rubber.

latitude and longitude (*geography*)   Ancient sailors in the Mediterranean navigated by distance along the width (or *latus*) and length (or *longus*) of the sea. By 300 BC Hellenic philosophers had mapped Earth on a sphere with latitude lines, circles cut off by parallel planes, indicating distances north and south. Latitude is measured in central angles from the largest such circle, the equator. The equator is at 0° and the north pole is 90°N latitude.

Longitude is based on equal-sized circles that pass through both poles. By the 16th century scholars recognized that longitude corresponds to time—1° of longitude equals 4 minutes. Longitude is measured from a half circle called a meridian that is 0° at Greenwich, England. The other half of that circle is 180°. In between longitude is identified as east or west—for example, New York City is about 74°W and St. Petersburg, Russia, about 30°E.

Laurasia (*geology*)   The continent of North America and most of Eurasia (excepting India) were joined before 120,000,000 years ago into a supercontinent called Laurasia. Laurasia was recognized by Alexander Du Toit [South African: 1878–1948] in 1937 and named after Laurentia (an earlier name for North America and Greenland) combined with Eurasia.

laurel (*plants*)   The laurel family, Lauraceae comprises at least 2000 species, almost all of them evergreen trees and shrubs native to the tropics and subtropics. They are characterized by aromatic bark, leathery leaves, small fragrant flowers, and berry or berrylike fruits. The true laurel, or sweet bay (*Laurus nobilis*), sacred to the Greek god Apollo and used by the Romans for wreaths to crown victors, grows to heights of 60 ft (18 m); today its stiff, oval leaves—called bay leaves—are used in cooking. The family also includes avocado, cinnamon, and sassafras (*Sassafras*). Unrelated plants also called laurels include the mountain laurel (*Kalmia latifolia*) of the eastern United States and the Alexandrian laurel (*Calophyllum ionophyllum*) of the Indo-Pacific.

lava (*geology*)   Any rock that when molten flows out of Earth's crust, either from a volcano or a fissure, is called lava, both while fluid and after solidifying. Lava's composition differs from magma because gases contained in the liquid rock are able to evaporate from lava but not from magma.

Lava that solidifies into smooth or ropy rock is pahoehoe (pah-HO-ay-HO-ay, a Hawaiian term). The ropiness comes from liquid lava flowing below a solid outer crust. Spiky or rough, jagged rock is aa (ah-ah, also Hawaiian). Lava that flows into water quickly forms a thin solid crust, but pressure from within causes the lava to break through that crust, creating rounded mounds of somewhat cooler rock. The crust that forms on these mounds is thick and the mounded forms, or pillows, freeze into rock formations known as pillow lava.

lavender (*plants*)   Members of the mint family, lavenders (*Lavandula*) are evergreen perennials and shrubs native from the Mediterranean region to India. They grow to about 3 ft (0.9 m) tall and have narrow leaves and dense clusters of small flowers that usually are blue or purple in color. Aromatic oil glands in the flowers yield an oil used in perfumes and medicinal products.

Lavoisier, Antoine-Laurent (*chemist*)   [French: 1743–94] Lavoisier used careful measurement and thoughtful experiments to turn chemistry into a science. He also explained the experiments of others, such as Joseph Priestley's discovery of oxygen and Henry Cavendish's production of water from hydrogen and oxygen. Lavoisier used the results of other chemists and his own experiments to create the modern theory of fire and to explain the role of air in combustion and respiration. He was the first to have a clear concept of a chemical element and the first to list the known elements. He also developed the idea of naming compounds from elements. In addition, Lavoisier was the first to use and to state clearly a conservation law for mass. For these reasons, he is known as the father of modern chemistry.

lawrencium (*elements*)   The last of the radioactive actinide series, with the highest atomic number, 103, lawrencium was synthesized in 1961 by Albert Ghiorso [American: 1915–] and coworkers. They named it for physicist Ernest O. Lawrence [American: 1901–58], who invented the cyclotron in 1931, the prototype for most particle accelerators. The half-life of lawrencium's most stable isotope is 35 seconds. Chemical symbol: Lr.

laws of motion (*physics*)   In 1687 Newton published 3 basic laws of motion in the *Principia*. These laws apply to objects moving through a frictionless vacuum. (1) An object at rest tends to stay at rest and a body in motion moves in a straight line at the same velocity unless acted upon by a force (law of inertia). (2) Acceleration (a) of an object is directly proportional to the force (F) acting on it and inversely proportional to the mass (m). This is $a = F/m$, more commonly expressed as $F = ma$. (3) For every action there is an equal and opposite reaction, also expressed as momentum does not change in a closed system.

leaching (*environment*)   The process by which water washes soluble minerals and organic compounds from leaf litter and soil is called leaching. Leaching usually carries materials downward, from upper to lower layers, bringing, for example, nutrients in newly fallen leaves down to plant roots. Water also leaches contaminants as it trickles through landfills, feedlots, and farmland. This can result in harmful substances affecting plants and soil organisms and entering groundwater and surface waters.

lead (*elements*)   Lead, atomic number 82, was known to the ancients (it is mentioned in the Bible) and was widely used in compounds (in cosmetics and flavorings) and as a metal in and of itself (in corrosion-resistant pipes and linings for containers). Although Hippocrates [Greek: 460–c370 BC] recognized about 400 BC that lead miners develop specific illnesses, lead's toxic nature did not become commonly known until the 1700s. Despite this, lead compounds were used in paint and gasoline until the 1980s. Recently the main use of lead has been in storage batteries. The chemical symbol, Pb, is from the Latin *plumbum*.

leaf (*botany*)   The main organ of photosynthesis (food making) for most plants is the leaf. In many plants, the leaf has 2 parts: a blade, flattened and broad to provide the photosynthetic cells maximum exposure to sunlight; and a slender stalk called the petiole. Leaves also can be shaped like needles (pines), scales (junipers), cylinders (onions), spines (cacti), or tendrils (peas).

Leaves arise from stems at nodes. Simple leaves have a single blade whereas compound leaves have a blade segmented into leaflets. Leaf shape, margins, arrangement of veins (venation), and arrangement on the stem are useful in species identification. Sizes range from duckweed leaves, which are less than

0.04 in. (1 mm) wide to leaves of the Amazon water lily, which may be 6.6 ft (2 m) in diameter, and the coconut palm, which are 20 ft (6 m) long.

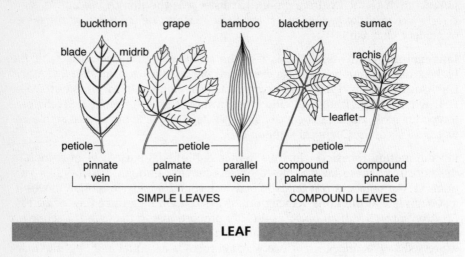

SIMPLE LEAVES   COMPOUND LEAVES

LEAF

Leakey family (*paleoanthropologists*)   Rarely, if ever, have so many members of the same family made important discoveries in one field as has the Leakey family in paleoanthropology. Louis Leakey [Kenyan: 1903–72] pioneered paleoanthropology in East Africa, discovering several important ape fossils and the Oldowan tool industry. In 1960 he described the first specimen of *Homo habilis* (the fossil cranium was found by his son Jonathan). His wife, Mary Nicol Leakey [English: 1913–96], found the first specimen of *Paranthropus* boisei in 1959 and led the expedition that found australopithecine footprints at Laetoli, Tanzania. Their son Richard Leakey [Kenyan: 1944–] found the first fossil of *Homo ergaster* in 1984. Richard's wife Maeve Leakey was in 1994 a codiscoverer of *Australopithecus anamensis*.

least common denominator (*arithmetic*)   The algorithm for addition or subtraction of fractions requires that fractions have the same denominator. Calculations are simplest if that denominator is the smallest natural number that will suffice, a number called the least common denominator.

least squares (*measurement*)   The method of least squares is used to find an equation whose graph fits closely to a set of data. The method assumes that the curve of best fit minimizes the sum of squares of deviations between data points and points on the curve. It was developed originally by Gauss and had its first success in Gauss's 1801 calculation, based on only a few observations, of the orbit of the asteroid Ceres.

Leavitt, Henrietta Swan (*astronomer*)   [American: 1868–1921] Despite deafness, Leavitt graduated from Radcliffe in 1892 and soon volunteered to work at Harvard Observatory. On staff by 1902, she began studying Cepheid variables. In 1912 she showed that their period of variation depends on magnitude, making it possible to use them as distance gauges. Leavitt established that the Magellanic clouds are beyond the Milky Way and opened the way to measuring distances to other galaxies.

Lederberg, Joshua (*geneticist*) [American: 1925–] In 1946 Lederberg and Edward L. Tatum [American: 1909–75] announced that they had discovered genetic recombination in bacteria. Several years later Lederberg discovered that viruses called bacteriophages could transfer genetic material from one bacterium to another, a phenomenon he called transduction. Lederberg shared the 1958 Nobel Prize in physiology or medicine with Tatum and George Wells Beadle.

leech (*animals*) Unlike other annelid worms, leeches lack bristles (setae). Also, they have a fixed number of body segments; they do not grow new segments throughout life. Leeches live in various aquatic and land environments in temperate and tropical regions. They feed on blood. Suckers on the head and rear ends of the body are used to attach a leech to its host. An anticoagulant, hirudin, in the leech's saliva prevents the host's blood from clotting while the leech feeds.

Leeuwenhoek, Anton van (*microscopist/biologist*) [Dutch: 1632–1723] Leeuwenhoek made single-lens microscopes with exquisite lenses that he ground as a hobby; with them he discovered an unknown world. Leeuwenhoek was the first person to see ciliated protists, which he called "animalcules" and "wretched beasties." He also discovered *Hydra*, rotifers, bacteria, and the sperm of dogs and other animals. His descriptions of blood capillaries and red blood cells extended the findings of Marcello Malpighi.

legume (*plants*) The approximately 18,000 species of the family Leguminosae (or Fabaceae) are commonly called legumes. They are widely distributed, particularly in temperate and subtropical regions, and are adapted to a great variety of habitats. Legumes range from miniature herbs to vines, shrubs, and trees. They include crop plants such as alfalfa, beans, clover, peas, and peanuts, plus ornamentals such as brooms, lupines (*Lupinus*), and wisteria (*Wisteria*).

Although extremely diverse, all legumes produce a pod fruit; when mature, the pod splits open along 2 seams to release its seeds. Legumes characteristically have compound leaves composed of numerous leaflets, creating a feathery appearance. The flowers, often in clusters, have an irregular shape, often referred to as butterflylike: the top petal curls upward and the 2 "wing" petals fit together, concealing 2 narrow, partly fused petals plus the reproductive organs. All legumes add nitrogen to the soil, thereby improving soil fertility in a process called nitrogen fixation.

Leibniz, Gottfried Wilhelm (*mathematician*) [German: 1646–1716] Leibniz is best known for having invented the calculus independently from Newton; much of the notation and vocabulary used today comes from Leibniz, who had a flair for both symbolism and language. He also took the first steps in symbolic logic. The calculating machine Leibniz invented was the first to multiply as well as add and subtract. In physics he contributed to developing the idea of kinetic energy.

lek (*ethology*) In some animal species courtship occurs in specific areas called leks or arenas, where males put on communal displays to attract females. For example, during breeding season several dozen or more male sage grouse (*Centrocercus urophasianus*) flock to a lek where they puff themselves up, spread their tail feathers, and strut around. The most successful of these male grouse are those selected as mates by a number of the visiting females.

lemming (*animals*)   Native to arctic regions of the Northern Hemisphere, lemmings are small volelike rodents in Family Muridae. They have a stocky body and a very short tail. The fur of some species turns white in winter. Lemmings are active both day and night. They generally live in colonies. When a lemming population becomes too large for the local food supply, the lemmings migrate; they can swim across small lakes and rivers but they die when attempting to cross wide expanses of water or ice.

lemur (*animals*)   According to a belief once widespread on Madagascar, spirits of dead people reside in lemurs, hence the animals' common name (from the Latin *lemures*, meaning "ghosts"). Found on Madagascar and the Comoro Islands, these primates are small to medium in size. They are excellent climbers and leapers, with opposable thumbs and big toes. The fingers and toes have nails, except for the second toe in some species, which has a long claw used for grooming. Most species spend the majority of their lives in trees. So-called true lemurs (Family Lemuridae) are active during the day. Indris and sifakas (Family Indriidae) also are diurnal. Dwarf and mouse lemurs (Family Cheirogaleidae) are nocturnal.

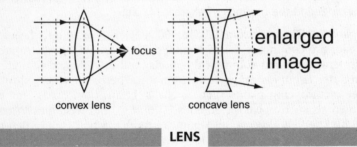

focus

enlarged image

convex lens   concave lens

LENS

lens (*optics*)   In antiquity people observed that water-filled glass globes concentrate light into a small area. Light changes speed as it travels through different transparent materials, causing waves that strike the interface at angles other than 90° to bend (refraction). A lens is a curved transparent device that uses refraction to concentrate light or to change the size of an image by bending light inward or outward. From the 13th century onward lenses have been used to correct impaired vision, and since the end of the 16th century lenses have been employed in telescopes and microscopes to observe distant or tiny objects.

lenticel (*botany*)   The bark of small branches, or stems, of a tree or shrub is dotted with tiny openings called lenticels. The lenticels allow the exchange of gases between the atmosphere and inner tissues of the stem. Lenticels generally appear as circular bumps or elongated slits, and often form characteristic patterns that can be used to identify the species.

Leonardo da Vinci (*engineer/artist*)   [Italian: 1452–1519] Leonardo's great reputation in science and invention is posthumous, based on the translation and publication of his coded notebooks in the late 19th and early 20th centuries. In his lifetime, in addition to his famous paintings, he was known for his engineering of canal locks, cathedrals, and engines of war. The notebooks reveal Leonardo's correct interpretations of anatomy, explanations of physical concepts such as inertia, and sketches for working parachutes and helicopters, all well in advance of those ideas entering the scientific record.

Leopold, Aldo (*conservationist*)   [American: 1886–1948] An early champion of the preservation of nature, Leopold believed that conservation is important for ethical and aesthetic reasons as well as economic reasons. He is credited with founding the field of wildlife management and aiding the U.S. Forest Service in setting aside the first officially designated wilderness reserve on federal land, in New Mexico in 1924. Leopold expressed his philosophy in numerous articles and essays, the most powerful of which were published in *A Sand County Almanac* (1949).

Lepidoptera (*animals*)   Butterflies and moths constitute the insect order Lepidoptera ("scaly wings"), with more than 110,000 described species. Among the largest is the atlas moth (*Attacus atlas*) of Southeast Asia, which has a wingspan that exceeds 10 in. (25 cm). Lepidopterans typically have 2 pairs of wings covered with tiny, overlapping scales. The distinctive wing designs are created by differently colored scales over transparent, membranous wings. Lepidopterans undergo complete metamorphosis. The larvae, or caterpillars, have chewing mouthparts and are voracious feeders that can cause much damage to plants. Most adults have a long, flexible proboscis (tube) used to suck plant nectar; they seldom are pests.

lepton (*particles*)   Light elementary particles that are fermions are called leptons (*leptos* is a Greek word for "small"). In the standard model of particle physics, leptons occur in 3 families. Ordinary matter is based on the first generation, the electron and an associated neutrino. At higher energies, the second and third families consist of the muon, the tauon, and their associated neutrinos. Thus, there are 6 basic leptons, 3 of which are neutrinos. In the standard model their antiparticles are also leptons, making a total of 12 leptons in all. In some theories neutrinos are their own antiparticles.

lettuce (*plants*)   A member of the composite family, cultivated lettuce (*Lactuca sativa*) is an annual long grown for its edible leaves. The genus also includes numerous species of wild lettuces. Lettuces produce a milky sap, or latex, and bear clusters of tiny yellow flowers. (Cultivated lettuce is picked before flower stalks develop.)

liana (*botany*)   Long-stemmed woody vines (climbing plants) of tropical forests are commonly called lianas. They typically germinate from seeds on the forest floor, then grow up tree trunks to sunlight at the top of the canopy. Some lianas reach lengths of more than 650 ft (200 m).

Libby, Willard (*physicist*)   [American: 1908–80] In 1947 Libby developed dating based on carbon-14, a radioactive isotope created by sunlight acting on atmospheric nitrogen. The ratio of carbon-14 to stable carbon-12 is nearly constant in air and maintained by living organisms—plants incorporate carbon during photosynthesis and animals feed on plants or other animals. After death, however, no more atmospheric carbon is consumed, so reduction in carbon-14 (half-life 5730 years) can measure the time since death.

libration (*astronomy*)   The Moon rotates steadily once each 27⅓ Earth days, the same time as its period of revolution. But its elliptical orbit requires faster and slower speeds of revolution along the way, so the face turned toward Earth appears at different angles, an apparent motion called libration. As a result, an earthbound observer can see 59% of the Moon's surface, although not all at once.

lichen (*biology*)   Composed of a specific fungus and a green or blue-green alga, a lichen is an association of symbiosis, or mutual benefit. The fungus forms the main part of the lichen's body, or thallus; it absorbs water and minerals, mainly from rain and other precipitation. The algal cells, enmeshed in fungal hyphae (filaments), use the water and nutrients to make food in the process of photosynthesis, feeding themselves and the fungus.

There are 3 main types of lichens. Crustose lichens form hard crusts on rocks, trees, or soil. Foliose lichens have flattened, leaflike bodies that are loosely attached to the substrate. Fruticose lichens, attached to the substrate at only 1 point, grow either upright or hanging from tree branches. Lichens are very tolerant of extreme conditions, surviving in places where few other organisms can exist. They grow on bare rocks, on high mountains, in the Arctic and Antarctic, in deserts, etc. They are sensitive to sulfur dioxide and thus are rare in cities and other polluted areas. Large species such as reindeer moss (*Cladonia rangiferina*) are the main food of reindeer.

life (*biology*)   Despite their great diversity, living creatures share certain characteristics that distinguish them from nonliving objects. All living things are composed of structures called cells. Every cell is a complex, highly organized system made up of many different kinds of chemical molecules. Living things acquire energy from the environment and use it to grow and develop; they respond to changes in the environment; and they reproduce, creating offspring of the same kind.

life cycle (*biology*)   An organism has a definite life cycle that includes 4 basic stages: birth, growth, maturity, and death. This fundamental characteristic distinguishes organisms from nonliving matter. Though the basic stages are the same, there is enormous variety among life cycles. Some bacteria complete their life cycle in hours, while some trees live for more than 1000 years. An amoeba remains a single-celled organism, but an elephant develops from a single cell into an organism with billions of cells. A daisy never leaves the spot where its seed germinates, but the European eel *Anguilla anguilla* travels thousands of miles to reproduce.

life span (*biology*)   The period of time between an organism's birth and death is its life span. Life span varies enormously: some rotifers live about 8 days, deep-sea clams may live 100 years, bristlecone pines live more than 4000 years. Maximum life span, which is genetically determined, indicates the greatest age that members of a species can live. Average life span is significantly lower, however, because predation, disease, drought, and other environmental factors take their toll. In recent years scientists have discovered that slight alterations in an animal's DNA can greatly affect life span. For example, a single mutation in a gene doubles the life span of the roundworm *Caenorhabditis elegans*, from 2 weeks to 4 weeks.

ligament (*anatomy*)   In vertebrates, ligaments are strong flexible bands of connective tissue that hold bones together at joints and help keep the bones properly positioned. They allow the joints to move and also protect against injury. Ligaments consist mainly of collagen but may also contain elastin. In clams and other bivalve mollusks, a hinge ligament joins the 2 valves of the shell.

light (*optics*)   Light refers to any electromagnetic radiation, but for nonscientists light means the small part of the electromagnetic spectrum humans directly perceive, also called visible light. Visible light ranges from about 400 nanometers (nm) to about 700 nm (1 nm = 0.00003937 in.). Regions just below or above the wavelengths of visible light are infrared, with wavelengths as long as 1000 nm (0.03937 in. or 1 mm), and ultraviolet, with wavelengths as short as 4 nm (0.00015148 in.). Physicists commonly refer to X rays and gamma rays (below 4 nm in length) as light also.

Light can be understood either as waves or as particles called photons. Short wavelengths correspond to high-energy photons. Photons carry the electromagnetic force between charged particles. Light is produced when a charged particle gains or loses energy in these interactions; some high-energy light is also produced when a charged particle meets its antiparticle and they destroy each other. Light also appears when a charged particle is moved back and forth, with the wavelength proportional to the speed of vibration.

light and dark reactions (*physiology*)   The food-making process, photosynthesis, has 2 distinct phases. Light-dependent reactions occur only in the presence of light. Chlorophyll traps light energy, which is used to synthesize the energy-carrying compound adenosine triphosphate (ATP). At the same time, water molecules are split, producing oxygen (released to the environment) and hydrogen atoms. The hydrogen is transferred to a coenzyme, nicotinamide adenine dinucleotide phosphate (NADP), producing NADPH.

In the dark, or light-independent, reactions sugar is made from carbon dioxide plus the hydrogen carried by NADPH. Energy for this process comes from the ATP and NADPH produced in the light-dependent reactions.

lightning (*meteorology*)   A visible emission of an electric current from one cloud to another or between the ground and a cloud is called lightning. In the giant cumulonimbus clouds of thunderstorms, negative charges are transported downward (probably on raindrops), leaving the cloud tops positively charged. The negative charge repels free electrons, creating a positive charge on the ground beneath the clouds. Periodically a thin extension of negative charge, called a leader, extends down from a cloud, reaching a similar (but shorter) leader from some high point on the ground. The 2 leaders provide a channel of conduction and a lightning bolt reaches up to the cloud, raising the temperature of the air to about 50,000°F (30,000°C) in a fraction of a second. The sudden expansion of air produces the pressure wave we hear as thunder. The electric potential of perhaps 100,000,000 volts from lightning can kill and the heat is a common cause of fires. (*See also* **ball lightning; blue jet; sprite.**)

light-year (*measurement*) Sometimes mistaken for a measure of time, a light-year is the greatest measure of length in common use (astronomers also use parsecs, which are longer). A light-year is the distance a photon of light moving in a straight line travels through a vacuum in a year. The nearest star is about 4 light-years away and the farthest known galaxies are about 12,000,000,000 light-years away.

lignin (*biochemistry*) Lignin is a complex organic material found in the walls of many plant cells, particularly the walls of the xylem cells that make up wood. Constituting up to 25% of wood's dry weight, lignin binds together cellulose fibers, providing stiffness and strength to the cell walls.

lilac (*plants*) Constituting the genus *Syringa*, lilacs are deciduous shrubs and small trees, growing to heights of about 23 ft (7 m). Native to Asia and southeastern Europe, they are widely cultivated for their large clusters of small, fragrant flowers, which typically are white, pink, or purple.

lily (*plants*) Native to northern temperate regions, lilies (*Lilium*) are perennial herbs generally 1 to 4 ft (30 to 120 cm) tall. They grow from underground stems called bulbs: clusters of leaves develop from a bulb's leaf scales and floral shoots develop from a bud or buds. The showy flowers, usually fragrant, occur in a wide range of colors. The lily family, Liliaceae, also includes other popular ornamentals, such as hyacinths, grape hyacinths (*Muscari*), hostas (*Hosta*), lilies of the valley (*Convallaria*), and tulips.

limestone (*rocks and minerals*) Limestone is a highly variable rock consisting mainly of calcite (calcium carbonate) that has either precipitated from seawater or formed from fossil shells and skeletons glued together with mud or silica or perhaps dissolved and recrystallized. The most finely textured limestone is a chemical precipitate, while some limestones are simply shelly fossils glued together. Although calcite is white, many different compounds can color limestone yellow, brown, red, or gray.

limit (*calculus*) A limit is a number that is approached as closely as desired by a sequence or a function. The limit may or may not be in the sequence or a value of the function. For example, the sequence 0.3, 0.33, 0.333, 0.3333, . . . does not contain ⅓, but a term of the sequence can be found in any neighborhood of ⅓, no matter how small. Similarly if L is the limit of a function $f(x)$ when x approaches 0, shown as

$$\lim_{x \to 0} f(x) = L,$$

then for every neighborhood of 0 there is a positive number such that the distance of L from $f(x)$ is less than that number.

limiting factor (*ecology*)   Any environmental condition that limits the numbers or activities of organisms in a particular place is considered a limiting factor. If a trout population's demand for oxygen is greater than its availability, oxygen is a limiting factor. On a rainy day, low light levels limit a plant's ability to make food (photosynthesize). Other common limiting factors include availability of food, living space, and nesting sites, as well as the presence of competitors, predators, and disease organisms.

limnology (*sciences*)   Limnology is the study of freshwater environments, such as lakes, rivers, and swamps. A limnologist considers the interplay among various biological, chemical, physical, and geologic factors in these habitats. For example, the rate at which water drains into a stream—and the chemicals carried by that water—are critical factors in determining what organisms live in the stream. In a large lake, chemical and physical conditions change with depth; as a result, organisms that float on the surface are structurally different from those that live in the bottom region.

limpet (*animals*)   Gastropod mollusks with a caplike shell and a large muscular foot, limpets are common along rocky coasts worldwide. A limpet forms a shallow depression in a rock, to which it clings as tides rush in and out. It may leave its "home" briefly, to creep about in search of the algae on which it feeds.

linden (*plants*)   Members of the genus *Tilia* are deciduous trees variously known as lindens, basswoods, and limes (not to be confused with the citrus tree that produces fruits called limes). The leaves are usually heart-shaped and toothed. The small, fragrant flowers secrete large quantities of nectar and are favorites of pollinating insects. The fruits look like gray peas. The tallest species include the American linden (*T. americana*), native to eastern North America, and the bigleaf linden (*T. platyphyllos*) of Europe, which can attain heights of 120 ft (37 m).

line (*geometry*)   When geometry is put into an axiomatic system, *line* is usually an undefined term. A definition could begin with 2 different points. The shortest curve that connects them is a line segment; a line segment is straight. This means that if part of the line segment is placed anywhere on the line segment, it will lie on the segment and not drift away from it. The curve formed by extending the segment in both directions, keeping it straight, is a line (all geometric lines are straight lines). It can be extended indefinitely. In analytic geometry a line is the curve described by an equation of the form $ax + by = c$, where a, b, and c can be any numbers (a and b cannot be 0 at the same time).

linear (*algebra*)   An equation or function of the form $ax + by + c = 0$ or $f(x) = mx + b$ is called linear because its graph is a line. This idea has been generalized to a concept called linear combination, which is the sum of 2 or more entities with each multiplied by some number (with not all the numbers being 0). Linear combinations of vectors, equations, and functions are commonly employed. For example, if x and y are vectors and a and b are numbers, $ax + by$ is a linear combination.

linear programming (*mathematics*)   Problems concerning the most desirable combination of several conditions have been investigated since the 18th century.

In the 20th century general methods for solving the problems were introduced, eventually forming a branch of mathematics called linear programming (or linear optimization). One linear programming technique uses the several conditions to create a polygon or polyhedron; maximal and minimal combinations of conditions must be at a vertex. Since there are a finite number of vertices, the combinations can be checked using a computer.

	Hi-Energy Paste	Lo-Cal Spam	Requirements
carbohydrates	8 grams	2 grams	16 grams/meal
fats	1 gram	1 gram	5 grams/meal
proteins	2 grams	7 grams	20 grams/meal
mass of package	100 grams	200 grams	

How many of each package on spacecraft for minimum mass?

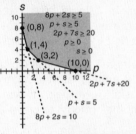

(0 x 100) + (8 x 200) = 1600
(1 x 100) + (4 x 200) = 900
(3 x 100) + (2 x 200) = 700
(10 x 100) + (0 x 200) = 1000
so (3, 2) or 3 pkg. High-Energy paste
and 2 pkg. Lo-Cal Spam is solution

LINEAR PROGRAMMING

Linnaeus, Carolus (*naturalist*)   [Swedish: 1707–78] The 1735 publication of *Systema Naturae* established the fame of Linnaeus with an organized classification structure for all living things. The binomial system of nomenclature, now the basis for naming and classifying all organisms, was first introduced by Linnaeus in 1749. In his *Species Plantarum* in 1753 Linnaeus attempted to name and describe all known plants, calling each kind a species and assigning to each a 2-part Greek or Latin name consisting of the genus (group) name followed by the species name. Many of his names of flowering plants survive with little if any change—for example, *Quercus alba* for white oak. The 1758 edition of *Systema Naturae* extended binomial classification to animals.

lion (*animals*)   The lion (*Panthera leo*) is a large, powerful member of the cat family. It is the only cat with strong sexual dimorphism. Males, weighing up to 550 lb (250 kg) and standing about 48 in. (122 cm) at the shoulder, are significantly larger than females; they also have a pronounced mane, which females lack. Also unlike other cats, lions are social animals; they form groups, called prides, consisting of several related females and their young plus temporary adult male members. The females hunt cooperatively for the pride, while the males defend the group's territory. Most lions live in savannas and other open country in sub-Saharan Africa; a small population lives in India's Gir Forest.

lipid (*biochemistry*)   Fats, oils, and waxes are types of lipids—one of the major groups of organic compounds found in living cells (the others being carbohydrates, proteins, and nucleic acids). Lipid molecules contain carbon, hydrogen, and oxygen in a different ratio than that found in carbohydrates; in some lipids elements such as nitrogen and phosphorus also are present. Lipids are insoluble in water but soluble in organic solvents such as ether and benzene. Lipids serve as energy-storage molecules and are important structural components of cells.

lipoprotein (*biochemistry*)   Lipoproteins are water-soluble proteins that contain 1 or more lipid groups in their molecules. They transport otherwise insoluble lipids in the blood. Chylomicrons carry dietary lipids, mainly triglycerides, absorbed from the intestines. Very-low-density lipoproteins (VLDL) mainly carry triglycerides produced in the body. Low-density lipoproteins (LDL) carry cholesterol. High-density lipoproteins (HDL) carry phospholipids and some cholesterol.

liquefaction (*geology*)   Although liquefaction means "becoming liquid," in geology the term generally refers to the liquidlike behavior of soil or sand during an earthquake. Shaking caused by seismic waves moves particles past each other easily. The process can result in collapse of structures built on sediments.

liquid (*materials science*)   A liquid is a material for which molecules move easily past one another but are attracted enough to produce surface tension. Thus liquids are fluids that can be kept in an open container in a gravitational field, where they form a nearly flat surface at the top. Molecules are closer together in a liquid than in a gas. Glasses are sometimes viewed as liquids with an extremely slow flow rate.

liquid crystal (*materials science*)   A liquid crystal is an organic liquid for which the molecules are long, fairly rigid, and different at each end. Such molecules form temporary structures that have forms similar to crystals, even though the bulk material remains liquid instead of solid. The difference at the ends of the molecule is usually in electric charge, making liquid crystals easy to manipulate with electricity. Their main use has been in display devices, such as certain kinds of computer monitors.

lithium (*elements*)   Lithium is the lightest known metal, atomic number 3. It is an alkali metal discovered in 1817 by Johan August Arfvedson [Swedish: 1792–1841]. Its name comes from the Greek *lithos*, "stony" (chemical symbol: Li), since it was first prepared from the mineral petalite. Lithium is best known today from the use of its salts to treat the mental illness bipolar disorder (manic depression).

lithosphere (*geology*)   The rigid outer layer of Earth is called the lithosphere ("sphere of rock"). It includes the part of the mantle above the asthenosphere as well as the crust. It is the lithosphere that is broken into about 20 tectonic plates. The mantle below the asthenosphere is chemically similar to rock, but so modified by heat and pressure that it is not counted as part of the lithosphere.

Little Ice Age (*meteorology*)   The Little Ice Age was a period of cooler temperatures in Europe (and perhaps elsewhere) starting sometime between 1250 and 1550 and ending between 1700 and 1850—a cold climate between 1550 and 1650 in Europe is well documented. Glaciers moved down mountains and obliterated villages, and the cold drove the Vikings out of North America and Greenland. The cause of the Little Ice Age is subject to considerable speculation. Some connect it to the Maunder minimum, a period of no sunspots between 1645 and 1715 first recognized by E. Walter Maunder [English: 1851–1928].

littoral zone (*ecology*) The part of a lake or sea that edges the shore forms the littoral zone. It typically is the most productive and richly inhabited zone in a body of water. In a lake the littoral zone extends to a depth of 20 to 33 ft (6 to 10 m). Cattails, rushes, and other wetland plants are rooted to the bottom, and water lilies and duckweed float on the water's surface. The tangle of plants harbors numerous animals, including snails, insect larvae, frogs, turtles, and water snakes.

An ocean's littoral zone extends from the high tide mark to a depth of about 600 ft (200 m)—approximately the maximum depth to which light penetrates. Seaweeds and grasses such as eel grass are the most conspicuous food producers. These organisms also provide shelter for many animals, including the young of fish that spend their adult lives in the open sea. The hardiest of all ocean organisms are those that live in the part of the littoral zone called the intertidal zone.

liver (*anatomy*) Found in many animals, the liver is a large organ with a number of vital functions. In vertebrates the liver manufactures bile, fats, cholesterol, and various proteins, such as fibrinogen, found in blood plasma. The liver converts excess sugar into glycogen and stores the glycogen until needed, when it converts the glycogen back to sugar. It also stores vitamins A, D, E, and K and removes toxic substances from the blood and changes them into inactive or less toxic compounds. The liver breaks down amino acids, forming urea, and also breaks down worn-out red blood cells.

liverwort (*plants*) Small bryophyte plants commonly called liverworts live on wet rocks, rotting logs, moist soil, and other damp places. Liverworts are found worldwide but they are most abundant in the tropics. Many species have flat, lobed bodies shaped somewhat like livers, hence the group's name. Other species resemble mosses, with stalks bearing tiny leaflike structures.

lizard (*animals*) The largest group of living reptiles, comprising some 3000 species of great variety, lizards typically have 4 legs that are held out to the sides. Their toes are clawed and may have adaptations for climbing, grasping, running, or digging. Their tail is another useful organ, used for fat storage, balance, and grasping branches. Many species have a fracture point where the tail breaks off easily—at which time muscles cause the fragment to coil and wiggle. Thus, if a predator grabs a lizard by its tail, this movement holds its attention while the rest of the lizard escapes. Within a few months, the lizard grows a new tail. Well-known lizards include chameleons, geckos, iguanas, skinks, monitors, and slowworms.

llama (*animals*) A ruminant mammal of the camel family, the llama (generally classified as *Lama glama*) exists only in the domesticated state. It stands about 4 ft (1.2 m) tall at the shoulder and has a small head, long neck, and long thin legs. Closely related to the llama are the domesticated alpaca (*L. pacos*) and 2 wild relatives, the guanaco (*L. guanicoe*) and the vicuña (*Vicugna vicugna*). All live in South America, in a variety of habitats from sea level to high altitudes.

lobster (*animals*) Marine crustaceans that include both coastal and deep-sea species, lobsters are mainly scavengers, though they also prey on fish and mollusks. The American lobster (*Homarus americanus*) is probably the largest, growing to 3 ft (90 cm) long and a weight of 44 lb (20 kg). Lobsters and their close relatives, crayfish, have an exoskeleton composed of a rigid, unjointed cephalothorax (fused head and thorax) and a flexible, jointed abdomen. A rostrum, a pointed extension of the cephalothorax, projects forward between the 2 eyes. There are 5 pairs of legs on the thorax—the first pair greatly enlarged for grasping, the others for walking—and 5 pairs of swimming legs, or swimmerets, on the abdomen. Appendages on the last segments of the abdomen form a fanlike tail used for swimming backward.

Local Group (*galaxies*)   Galaxies bind into clusters held together by mutual gravitational attraction despite great distances separating them. One such cluster, the Local Group, has as its major members our galaxy (the Milky Way) and 2 other large galaxies (Andromeda, or M31, and M33). Each of these large galaxies is surrounded by a swarm of dwarf galaxies, most of which orbit the large galaxy. The entire cluster contains between 20 and 32 galaxies. In addition, at least 3 dwarf galaxies appear to be passing through the Local Group at speeds too fast to join it.

locus (*genetics*)   The specific position or location of a gene on a chromosome is called its locus. Each locus can be occupied by only 1 gene. If a gene can have several forms (alleles), then only 1 of these is present at the locus.

locus (*geometry*)   A locus is a set of points that meet a stated condition. A condition such as being in a plane equidistant from a point describes the locus of a circle; the condition of satisfying the equation $y = ax^2 + bx + c$ $(a \neq 0)$ describes the locus of a parabola.

locust (*animals*)   Members of the short-horned grasshopper family, locusts exhibit polymorphism. They live solitary lives when populations are small. As populations increase and factors such as food availability become significant, secretion of certain hormones causes changes in structure and habits; the locusts enter a very active and gregarious phase. They may form migratory swarms so huge that they darken the sky and strip the land of every plant, as described in Exodus. A swarm of desert locusts (*Schistocerca gregaria*) may contain 28,000,000,000 individuals, each weighing a mere 0.08 oz (2.5 g) but in total equaling some 70,000 tons! The swarm phase may last for several generations before dying out and leaving behind only solitary phase locusts.

locust (*plants*)   Members of the legume family native to the Americas, locusts (*Robinia*) are deciduous shrubs and trees with feathery compound leaves that fold up at night. Clusters of fragrant pink or white flowers are followed by small, flat pod fruits. The most common species, the black locust (*R. pseudoacacia*), grows to 100 ft (30 m) tall.

Honey locusts (*Gleditsia*), native to the Americas, Africa, and Asia, are deciduous trees generally classified in the family Caesalpiniaceae. Their trunks and branches often are armed with large branching spines. They have compound leaves, clusters of tiny greenish flowers, and large pods. The honey locust of North America (*G. triacanthos*) grows to 150 ft (45 m) tall.

loess (*soils*)   Agriculturally productive soil formed from wind-borne particles is called loess. Loess has formed deep deposits in China, where it has blown from central Asian deserts, and thinner layers in U.S. Midwestern states, where it originated in Ice Age glaciers. Dust storms, including those produced in part by a combination of agriculture and drought (as in the Dust Bowl of the 1930s), produce loess when the wind dies down.

logarithm (*mathematics*)   A logarithm is the number that an exponent represents. The statements $y = x^r$ and $\log_x y = r$ ("logarithm to the base x of y equals r") mean the same; thus $\log_2 16 = 4$ because $2^4 = 16$. The logarithmic function $y = \log_a x$ is the inverse of the exponential function $y = a^x$. Logarithms with a base of 10 (usually not written) are common logarithms. Natural logarithms have a base of e and are often abbreviated as ln. Logarithms were invented in 1614 by John Napier [Scottish: 1550–1617] as an aid to computation, since adding or subtracting logarithms can be used to find products or quotients and multiplying logarithms can be used to find powers.

logic (*logic*)   Logic constitutes a branch of knowledge or inquiry that is separate from both mathematics and the sciences, although it closely resembles mathematics and is sometimes used as a basis for it. Modern logic, like modern mathematics, is often cast into axiomatic systems. Earlier systems, which date from Aristotle and before, consisted mainly of rules for syllogisms. Leibniz was among the first to use symbolic logic. The logical system based simply on truth values of not, and, or, and if . . . then is sometimes called the propositional calculus. It is an example of a complete system for which there is a method (the truth table) that can be used to determine the truth value of every statement. When statements that include "for all . . ." or "there exists . . ." are added to the propositional calculus, the resulting logic, sometimes called the predicate calculus, no longer possesses a method that can always be used for determining truth values.

longitudinal wave (*waves*)   The characteristic of a longitudinal wave is that entities forming the wave move forward and backward in the same direction as the wave itself moves. A longitudinal wave in 1 dimension could be a pattern of higher and lower density of a gas traveling in a pipe, for example. No molecule of gas moves very far, but the pulse of pressure transmits information, which we recognize as sound. Take away the pipe and the sound waves are expanding spheres of high and low pressure traveling through 3-dimensional space. Longitudinal waves traveling through rock form the **P**-waves of earthquakes.

loon (*animals*)   Called divers in Great Britain, loons are birds native to northern waters. They are clumsy on land. To fly, they must take off from water. Loons are excellent swimmers and superb divers, going as far as 240 ft (73 m) below the water's surface. They are up to 3 ft (0.9 m) long, with a streamlined body, long pointed bill, and webbed feet. Loons feed mainly on fish.

loosestrife (*plants*)   Loosestrifes (*Lythrum*) are herbs and small shrubs with 4-sided stems, white to purple flowers, and capsule fruits. Purple loosestrife (*L. salicaria*), an attractive wetlands plant, bears spikes up to 4 ft (1.2 m) tall that are covered with hundreds of magenta flowers in late summer. Native to Europe, Asia, and northern Africa, it is a serious pest in North America, where it crowds out native species.

lophophore (*anatomy*) The lophophore is a specialized food-catching organ that encircles the mouth of several types of aquatic invertebrates, including brachiopods and bryozoans. The lophophore bears long, slender, ciliated tentacles that sweep water toward the mouth. Microscopic food organisms are then sifted out of the water and swallowed. The lophophore also aids in respiration.

Lorentz, Hendrik Antoon (*physicist*) [Dutch: 1853–1928] In the 1890s Lorentz completed the theories of James Clerk Maxwell on electromagnetism, showing how electrons (which Lorentz named before electrons had been observed) interact to produce light and forces within materials. His theory suggested that a magnetic field would affect light; his student Pieter Zeeman [Dutch: 1865–1943] observed this experimentally—a splitting of spectral lines now called the Zeeman effect. Independently from George Francis FitzGerald [Irish: 1851–1901], Lorentz explained the negative result of the Michelson-Morley experiment as a shortening of matter in the direction of motion. Now called the FitzGerald-Lorentz effect, this was established as part of Einstein's special relativity theory.

Lorenz, Konrad (*ethologist*) [Austrian: 1903–89] Often considered the founder of ethology, Lorenz developed the concept of imprinting and found that animals have instinctive behavior patterns that are genetically determined. His comparisons of animal and human behavior, particularly in the area of aggression, were controversial. Lorenz shared the 1973 Nobel Prize for physiology or medicine with Nikolaas Tinbergen and Karl von Frisch [Austrian: 1886–1982].

loris (*animals*) Lorises, slow lorises, pottos, and angwantibos constitute the primate family Lorisidae, which in some classification schemes includes the galagos. Lorises are small arboreal animals native to Africa and Asia. They have very short tails and hands and feet designed for grasping, with a strongly opposed thumb and big toe. Lorises move slowly and silently, using hand-over-hand movements until they are close enough to prey to pounce.

lotus (*plants*) The lotus family, Nelumbonaceae, has a single genus of 2 species of aquatic herbs. They have thick tubers (potatolike stems) that are rooted in the mud. Bluish green leaves and large, showy flowers—all with long stalks—rise above the water. The sacred lotus (*Nelumbo nucifera*), native from southern Asia to Australia, has very fragrant white to pink flowers. The American lotus (*N. lutea*), native to eastern North America, has fragrant yellow flowers.

louse (*animals*) Most commonly the name louse refers to certain wingless insects that are external parasites of birds and mammals. All are small, with flattened bodies, reduced eyes, and short legs adapted for clinging to a host. Biting lice (Order Mallophaga) mainly parasitize birds, chewing on feathers. Sucking lice (Order Anoplura) feed on the blood of mammals and include the crab louse (*Phthirus pubis*) and other species that attack humans.

Booklice and barklice (Order Psocoptera) are tiny insects—some winged, some wingless—that feed on bookbinding, algae, dead insects, and other organic matter. Plant lice are better known as aphids. Fish lice and whale lice are crustacean parasites of, respectively, fish and baleen whales.

luciferase (*biochemistry*)   Fireflies (*Photinus*) and many other organisms produce light, or bioluminescence, by oxidizing a material called luciferin. The enzyme luciferase acts as a catalyst for this reaction. The chemical composition of luciferin and luciferase vary greatly depending on the kind of organism.

luminescence (*optics*)   Any production of light by a material from causes other than temperature is called luminescence. This may occur as bioluminescence, fluorescence, or phosphorescence.

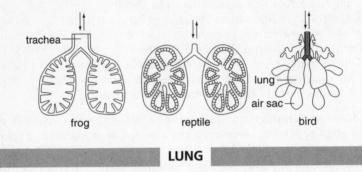

trachea   lung   air sac

frog   reptile   bird

LUNG

lung (*anatomy*)   All air-breathing vertebrates have 2 saclike organs of respiration called lungs. Air moves between the outside environment and the lungs via a series of tubes. Within the lungs these tubes end in alveoli, where gas exchange between the air and blood takes place. In birds the lungs are relatively small; air passes through them into air sacs, which are important in regulating body temperature. Lungs do not contain muscles. Air is drawn into and expelled from lungs through the action of rib muscles and, in mammals, the diaphragm.

lungfish (*animals*)   The 3 genera of lungfish are primitive air-breathing fish that have lungs as well as gills, enabling them to rise to the water's surface for a breath of air. Lungfish live in freshwater in parts of South America, Africa, and Australia. They are 2 to 6 ft (0.6 to 2 m) long. DNA studies suggest that lungfish, which were widespread during the Paleozoic, were ancestors of all 4-legged vertebrates.

Luria, Salvador (*molecular biologist*)   [American: 1912–91] A pioneer in the field of molecular biology, Luria conducted a variety of experiments using bacteriophages (viruses that infect bacteria). He believed that bacterial resistance to bacteriophages results from mutations, rather than adaptations on the part of individual bacteria (the leading theory in the early 1940s). With Max Delbrück, Luria developed the fluctuation test to confirm his theory. The test became a standard tool for determining mutation rates in bacteria and viruses. Luria shared the 1969 Nobel Prize in physiology or medicine with Delbrück and Alfred Hershey [American: 1908–97].

lutetium (*elements*)   Lutetium is a rare earth (atomic number 71) that is actually rare, since it is very difficult to isolate. Credit for its discovery in 1907 usually goes to Georges Urbain [French: 1872–1938], who identified the element in its combination with ytterbium, which had earlier been separated from erbium, terbium, and yttrium. Several scientists recognized lutetium as an element at about the same time, so its present name (from *Lutetia*, ancient name for Paris) was not settled on until 1949. Chemical symbol: Lu.

Lyell, Sir Charles (*geologist*)   [English: 1797–1875] Lyell is known as the principal advocate of uniformitarianism, the doctrine first propounded by James Hutton a dozen years before Lyell's birth. Lyell's *Principles of Geology*, first issued starting in 1830 and revised a dozen times, changed geology; it was also an important influence on Darwin as was Lyell himself, a good friend to Darwin throughout Darwin's adult life. Lyell's other work, *The Antiquity of Man*, issued in 1863, explored the implications of Darwinian evolution for humans.

lymphatic system (*anatomy*)   In a vertebrate some of the plasma (liquid portion) of blood passes through capillary walls into spaces between cells. This clear, colorless fluid, called lymph, bathes the cells, contains dissolved oxygen and food, removes wastes, and carries white blood cells that help fight disease. Lymph does not return to the circulatory system through capillary walls. Rather, it flows into a network of lymph vessels that eventually empties into large veins near the heart. Located along the lymph vessels in birds and mammals are masses of spongy tissue called lymph nodes. These nodes—which may become swollen during infection—filter out bacteria, dead cells, and other debris before the lymph reaches the bloodstream.

lymphocyte (*cells*)   Lymphocytes are a group of vertebrate white blood cells that play critical roles in immunity. There are two classes: B cells and T cells. B cells produce antibodies that combine with foreign substances called antigens. They are active mainly against bacteria, viruses, and toxins produced by these organisms. T cells react mainly with other cells of the body. Helper T cells and suppressor T cells regulate the immune response, including the activities of B cells. Killer T cells kill cells they attack, such as body cells infected with viruses.

lyrebird (*animals*)   Native to Australia, lyrebirds (*Menura*) are perching birds named for the adult male's extravagant tail, which may be 2.5 ft (0.75 m) long; when spread during courtship the tail resembles the stringed instrument called a lyre. Lyrebirds average about 3 ft (0.9 m) in length, not counting the tail, and look rather like brown chickens, with a small head and small rounded wings. They eat insects and other invertebrates. Lyrebirds are loud singers and good mimics of the songs of other birds.

lysis (*physiology*)   Lysis is the disintegration of cells with release of their contents. For example, certain proteins in vertebrate blood attach to and cause lysis of foreign invaders such as bacteria. Hemolysis is the rupturing of red blood cells with the release of hemoglobin.

Lysis is a critical process in the spread of bacteriophages (viruses that infect bacteria). A lytic cycle begins when a bacteriophage injects its DNA into a host bacterium. The phage DNA takes over the bacterium's metabolism, getting it to produce new phages. Enzymes then lyse the bacterium and the phages escape to infect other bacteria.

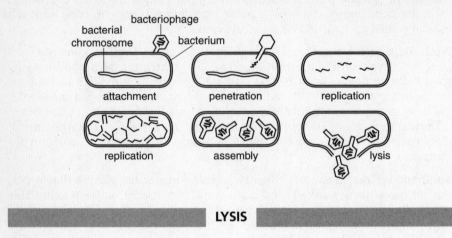

LYSIS

lysogenic bacteria (*monera*)   When certain types of **bacteriophages** infect bacteria, their DNA may become integrated into the bacterial chromosome. The integrated bacteriophages are called prophages. They are replicated with the rest of the bacterial chromosome and passed on to future generations. Bacteria infected with such viruses are said to be lysogenic. Every so often, the prophages break away from the bacterial chromosome. This initiates a cycle in which the bacteria rupture—in a process called lysis—releasing the viruses, which can then infect other bacteria.

lysosome (*cells*)   Small, spherical organelles in the cytoplasm of most eukaryote cells, lysosomes contain powerful **enzymes**. They are the digestive "system" of cells. Lysosomes break down molecules into smaller molecules. Foods are converted into simpler substances that can be used by the cell. They also make smaller molecules of waste products that need to be removed from the cell, invading bacteria, and damaged parts of the cell. If a lysosome breaks open and releases its enzymes into the rest of the cell, the cell is destroyed.

lysozyme (*biochemistry*)   Discovered and named by **Alexander Fleming** in 1921, lysozymes are enzymes produced by both animals and plants. In vertebrates lysozymes are mainly found in secretions such as saliva and tears. They act as catalysts in the digestion of compounds in the cell walls of bacteria. This weakens the bacteria, making them susceptible to lysis. Lysozymes are especially abundant in the whites of bird eggs.

M

macaque (*animals*)   Macaques are stocky monkeys that live in habitats as varied as tropical jungles, snowcapped mountains, and crowded cities. They have a heavy brow ridge over each eye, a projecting muzzle, large cheek pouches, short limbs, and a prominent bare spot on each buttock. The genus includes the rhesus monkey (*Macaca mulatta*), widely used in laboratory research, and the Barbary ape (*Macaca sylvanus*) of Gibraltar and northwestern Africa, the only wild monkey in Europe. The 81 species of the macaque, or Old World monkey family (Cercopithecidae), also include baboons, mangabeys, guenons, langurs, and proboscis monkeys. All have digits with flat nails and most have opposable thumbs and big toes. Many exhibit pronounced sexual dimorphism.

Mach, Ernst (*physicist*)   [Austrian: 1838–1916] Mach's name, used to describe the ratio of the speed of an object to the speed of sound, is better known than his work, which largely concerned the philosophy of science. In 1929 the system of Mach numbers for speed was introduced, commemorating Mach's 1887 discovery that air flow changes dramatically above the speed of sound (Mach 1). Mach's philosophy, which heavily influenced both Einstein and the founders of quantum mechanics, was based on using directly observed phenomena as the only basis of science.

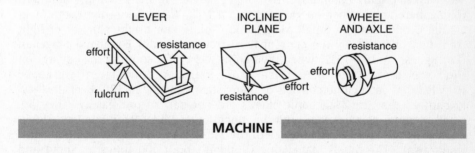

LEVER

INCLINED PLANE

WHEEL AND AXLE

effort

resistance

fulcrum

resistance

effort

resistance

effort

MACHINE

machine (*mechanics*)   Machines are devices that change the direction or magnitude of force. The force that operates a machine, called effort, is increased when the machine causes force to act over a smaller distance or decreased when the distance is lengthened. All machines are based on the 3 labeled simple: the lever, the inclined plane, and the wheel and axle. Other simple machines are variations on these, such as the wedge or screw—versions of inclined planes—and pulleys or gears—uses of the wheel.

mackerel (*animals*) Together with tunas and bonitos, mackerels make up Family Scombridae, a group of fast-moving, powerful schooling fish found worldwide in tropical and temperate marine waters. All have a streamlined body, a pointed snout, 2 dorsal fins, the front one spiny, and a stiff, deeply forked tail. The largest mackerel is the king mackerel (*Scomberomorus cavalla*), up to 5.5 ft (1.7 m) long and 100 lb (45 kg).

macrophage (*cells*) Among the vertebrate defenses against bacteria and other debris are large cells called macrophages ("big eaters"). They are found in lymph, blood, and other connective tissue. These phagocytes engulf and break down the unwanted particles. Some macrophages originate from white blood cells called monocytes.

madder (*plants*) Perennials native to warmer regions of Europe, Africa, and Asia, madders (*Rubia*) bear whorls of small prickly leaves and clusters of tiny flowers. The common madder (*R. tinctorum*) and several other species have fleshy roots that were the original source of the red dye alizarin.

The madder family, Rubiaceae, comprises some 6500 species of herbs, shrubs, and trees, most native to tropical areas. The family includes bedstraws (*Galium*), cinchonas (*Cinchona*), from which quinine is extracted, coffees, and gardenias (*Gardenia*).

Magdalenian (*paleoanthropology*) The last tool industry of the Paleolithic Age is termed Magdalenian (22,000 to 11,000 years ago). It is characterized by bone and horn needles and awls, harpoon points, and spear-throwers as well as the earliest microlith stone tools and well-made stone burins. Magdalenian culture featured jewelry and numerous engravings of animals, humans, and geometric designs on bone and antler as well as the most advanced cave art.

Magellanic clouds (*galaxies*) Two bright patches in the sky of the Southern Hemisphere are companion galaxies to our Milky Way galaxy—the Large Magellanic Cloud and the Small Magellanic Cloud. The bright patches of light were first observed by the crew of first circumnavigator Ferdinand Magellan [Portuguese: c1480–1521] when their ships passed far south of the equator on their way around South America. More than a century later, telescopes revealed that each cloud consists of many stars. Starting in 1912 Henrietta Leavitt showed that the Magellanic clouds are not part of our galaxy. They are small irregular galaxies that orbit the much larger Milky Way. The Large Magellanic Cloud, about 0.3% the size of the Milky Way, is roughly 160,000 light-years away. The Small Magellanic Cloud is about 100 times as small and somewhat farther away.

magic square (*numbers*) An arrangement of numbers in a square array such that the sum of any column, row, or diagonal is the same is called a magic square. The earliest known example is a Chinese invention known as the *lo shu*. Legend reports the *lo shu* was discovered by a King Yu on the back of a turtle in the 23rd century BC, but it probably dates from the 4th or 5th centuries BC.

4	9	2	=15
3	5	7	=15
8	1	6	=15

15″ 15″ 15″ 15″ 15

MAGIC SQUARE

magma (*geology*) Rock within Earth's crust or mantle that is fluid is magma; after solidifying, it becomes igneous rock. Magma rising through the mantle or crust is called a plume. After magma flows from a volcano or fissure, its gases evaporate and it becomes lava. Magma that is ejected into the air may cool quickly into glass, cool while frothy as pumice, or break into small pieces called ash.

magnesium (*elements*) One of the alkaline earth metals discovered in 1808 by Humphry Davy is magnesium, named after the mineral magnesia (magnesium oxide), most familiar combined with water to form milk of magnesia. Magnesium's atomic number is 12, making it low in density; its lightweight alloys are used in airplanes or in situations where weight must be conserved. Finely divided magnesium burns very brightly, a trait important to fireworks. Part of the chlorophyll molecule, magnesium is essential to most plants. Chemical symbol: Mg.

magnetar (*stars*) Certain neutron stars, according to a theory first proposed in 1992, spin fast enough to set up extremely large magnetic fields; the fields tear apart the upper levels of the stars, induce gases around the stars to release vast amounts of high-energy gamma rays, and finally slow the stars' rotation enough so that the magnetic fields revert to the normal, but still high, fields found in other neutron stars. Astronomers have detected only a few candidates for these bodies, called magnetars; but since this stage lasts only about 10,000 years, it is possible for most neutron stars to go though a magnetar phase.

magnetic pole (*geology*) Earth's magnetic field rises from currents of melted iron in its outer core, but the magnetic field does not exactly parallel Earth's rotation. The magnetic poles of this field are about 11° in latitude from the geographic poles and move slowly in a circle about the geographic poles at about 0.04° of longitude annually. Every 100,000 years or so, the field reverses, interchanging north and south magnetic poles.

A compass points to the magnetic pole, not the geographic pole. At any point there is a difference in compass direction, called declination, measured as a positive number of degrees east of true north (mariners call declination "variation"). A compass allowed to swing freely does not usually rest parallel to Earth's surface, but points up or down; this is called the inclination and is measured as + if down, – if up.

magnetism (*physics*) Magnetism is the name given to the effects of a magnetic field. Whenever a charge moves, the surrounding electric field also moves and produces a magnetic field. An object, such as an iron bar, acquires magnetism from the movements of the charge on its electrons, chiefly from spin. If spins are aligned to reinforce each other, the iron becomes a magnet. Spins can be aligned by a previously existing magnet or by an electric current. Only a few materials—notably iron, nickel, and cobalt and their alloys—can become strong, "permanent" magnets. Such magnets are not truly permanent, as heating destroys magnetic alignments.

Every magnet has 2 poles, called north and south in reference to Earth's magnetic field, which has magnetic poles near the geographic north and south poles. Unlike magnetic poles attract and like poles repel. The magnetic field consists of lines of force passing from pole to pole, with magnetic strength decreasing according to an inverse square law.

magnetite (*rocks and minerals*) Magnetite is the mineral Fe_3O_4, a combination of iron and oxygen that retains iron's ferromagnetism. Not all samples of magnetite are themselves magnetized, but those that produce a substantial magnetic field are called lodestones. Magnetite is a common mineral. Large deposits of it are used as iron ore.

magnetosphere (*astronomy/earth science*) The magnetosphere of a planet is not a sphere (the name parallels *atmosphere*) but the planet's giant elongated teardrop-shaped magnetic field, blunt end facing the solar wind. Five solar-system planets are surrounded by significant magnetospheres—Mercury, Earth, Jupiter, Saturn, and Uranus. Earth's lower magnetosphere is the outer portion of the atmosphere, also called the exosphere. Earth's magnetic field produces an electron and ion flow that protects Earth from the solar wind, turns aside charged cosmic rays, and causes the auroras. A magnetosphere turns with the rotation of the planet, so Jupiter's giant magnetosphere spins once every 10 hours. The magnetosphere of Uranus is oddly tipped, like the planet itself, making an angle of 60° with the axis of rotation.

magnitude (*stars*) The brightness of a star is called its magnitude, or apparent magnitude. A numerical rating for apparent magnitude was first developed by Hipparchus and Ptolemy [Hellenic: c100–c170]. They classed the brightest stars as the first magnitude and the dimmest seen from Earth with the naked eye as the sixth.

In 1856 George Phillips Bond [American: 1826–65] photographed stars, making magnitude directly measurable instead of subjective. Stars with a magnitude of 1.00 are exactly 100 times as bright as those with a magnitude of 5.00. Roughly, each whole number difference in magnitude corresponds to 2.5 times the brightness. The very brightest stars have negative apparent magnitudes. Since closer stars appear brighter, astronomers also use a system that measures brightness from the same distance. A star's brightness from this distance, 10 parsecs (32.6 light-years), is its absolute magnitude.

magnolia (*plants*) The magnolia family, Magnoliaceae, comprises some 225 species of evergreen or deciduous shrubs and trees native mainly to northern temperate regions. The plants generally have large showy flowers and large cone-shaped fruits. The family includes magnolias (*Magnolia*), tulip trees (*Liriodendron*), and champaks (*Michelia*).

mahogany (*plants*)   Famed for their hard, dark red wood, true mahoganies (*Swietenia*) are evergreen trees native to tropical America. They generally grow to about 60 ft (18 m) and have compound leaves, clusters of small greenish flowers, and fruits that split open into 5 parts. The related species African mahogany (*Khaya ivorensis*), native to tropical Africa, grows to 200 ft (60 m). The mahogany family, Meliaceae, comprising some 550 mostly tropical shrubs and trees, also includes Spanish cedar (*Cedrela odorata*) of tropical America, sometimes called cigar-box cedar because its aromatic wood is used to make cigar boxes, and neem (*Azadirachta indica*), probably native to southern Asia but widely grown in the tropics as an ornamental and for its wood and resin.

main sequence (*stars*)   When stars are graphed on a Hertzsprung-Russell diagram with luminosity on one axis and temperature on the other, nearly all appear on a line called the main sequence. There are more stars in the main sequence than elsewhere because it represents the longest phase of stellar evolution. Newly formed stars fall on the main sequence. As a star evolves, changing its method of nuclear fusion in response to depleting parts of its fuel supply, it moves away from the main sequence, usually becoming a red giant or a white dwarf. Giants move above the main sequence, while white dwarfs fall below it.

mallow (*plants*)   The mallow family, Malvaceae, comprises some 1000 species of herbs, shrubs, and trees, most native to tropical and subtropical lands. Many have fibrous stems and attractive, though short-lived, flowers. The family includes mallows (*Malva*); cottons; hibiscuses; marsh mallows (*Althaea*), including hollyhock (*A. rosea*); and Caesar weed (*Urena lobata*), a South American plant that yields a fiber used in making coffee sacks.

Malpighi, Marcello (*anatomist*)   [Italian: 1628–94] A pioneer in the use of the recently invented microscope, Malpighi discovered and described capillary circulation, thus confirming William Harvey's theory of blood circulation. Malpighi also identified sensory receptors (papillae) of the tongue and skin, and capsules of the kidney and spleen. He showed that bile is secreted by the liver and has a uniform yellow color; that is, there is no such thing as black bile, which since ancient times had been believed to be one of the 4 humors (fluids) of the human body, together with yellow bile, blood, and phlegm.

Malthus, Thomas (*economist*)   [English: 1766–1834] A pioneer in the study of population trends, Malthus is best-known for his 1798 *Essay on the Principle of Population*, in which he wrote that human populations increase geometrically until checked by limiting factors, such as insufficient food and other resources. The consequences of overpopulation, he believed, are starvation, disease, and war, which tend to restore the population to an optimal density. The essay influenced Darwin, who recognized that a struggle for existence exists within all populations, which favor certain variations or adaptations while eliminating others. Malthus's essay was extremely controversial, and continues to be debated to this day.

mammal (*animals*) Mammals are air-breathing vertebrates with fur or hair. They have the most highly developed nervous systems of all animals. Mammals are endothermic, maintaining a high body temperature by internal metabolic processes. Specialized mammary glands on the females provide milk for the young. Almost all mammals bear live young, but the monotremes lay eggs with shells. The marsupials give birth to extremely immature young, which typically continue their development in a special protective pouch on the mother's belly. Placental mammals give birth to well-developed young, which as embryos receive nourishment from the mother's body via the placenta.

The approximately 4500 living species are divided among some 19 orders (the number varies slightly depending on the classification scheme). The smallest mammals include species such as the pygmy shrew (*Sorex minutus*) of Eurasia, which is less than 2 in. (5 cm) long from the tip of its nose to the base of its tail. The biggest mammal is the blue whale (*Balaenoptera musculus*); the largest, heaviest animal that has ever lived, it may be 100 ft (30 m) long and weigh 260,000 lb (118,000 kg).

mammary gland (*anatomy*) Mammals are distinguished from all other animals by the presence on females of mammary glands, which secrete nourishing milk for the newborn. (The name comes from the Latin *mamma*, meaning "breast.") The production of milk—a process called lactation—begins late in pregnancy and is triggered by the hormone prolactin from the anterior pituitary. In primates and bats mammary glands are on the thorax; in other mammals they are on the abdomen.

mammoth (*paleontology*) During the Pleistocene Ice Ages close relatives of modern elephants called mammoths roamed over much of the Northern Hemisphere. Mammoths had long, shaggy outer hair, a dense inner coat of fine hair, and a thick layer of fat beneath the skin to insulate against cold. The long, inward-curving tusks were used to remove snow from the plants that made up their diet. Early cave paintings show people hunting mammoths, and it is possible that hunting caused the mammoths' extinction.

manatee (*animals*) Large aquatic mammals that live in tropical and subtropical coastal waters on both sides of the Atlantic Ocean, manatees (genus *Trichechus*) can exceed 13 ft (4 m) in length and 2200 lb (1000 kg) in weight. Manatees have a streamlined body, flippers with vestigial nails, and a large, single-lobed tail fin. The upper lip is deeply divided in the middle, with the 2 halves used to grasp plant food. The eyes are small and the numerous large whiskers probably are important sensory organs.

manganese (*elements*) Manganese is a metal, atomic number 25, that is part of many useful alloys and thought to be essential for bone and nerve formation. It was first isolated in 1774 by Johann Gottlieb Gahn [Swedish: 1745–1818]. Chemical processes precipitate lumps of manganese out of large bodies of water such as the Great Lakes and oceans. Its name is thought to be related to Latin *magnes*, magnet, because of confusion with magnetic iron ores. Ironically, although manganese by itself is not magnetic, its presence produces alloys with strong magnetic properties. Chemical symbol: Mn.

mango (*plants*)   A member of the cashew family, the mango (*Mangifera indica*) is an evergreen tree native to tropical Asia. It typically attains heights up to 80 ft (24 m) and can live to be more than 300 years old. It has long, leathery leaves. The fragrant flowers are borne in dense clusters called panicles; a single cluster may contain as many as 2000 tiny flowers. The large fruit has a tough skin surrounding a juicy, edible pulp and a single large, flat seed.

mangrove (*plants*)   Mangroves (*Rhizophora*) are evergreen shrubs and moderately tall trees native to the subtropics and tropics. They have thick leathery leaves, yellow flowers borne in clusters, and leathery fruits. They are adapted for life in salty and brackish coastal waters, a habitat unsuitable for the great majority of flowering plants. A mangrove sends down numerous aerial roots from its branches, forming a dense tangle of stilts that anchor the plant. These roots also trap sediment, gradually building up land. Another adaptation is seen in the fruit: its seed often begins forming roots while the fruit is still attached to the parent plant. When the fruit falls the juvenile root is ready to embed itself as soon as it touches mud. Mangroves are classified in the family Rhizophoraceae. Several species of other families are also commonly called mangroves; they live in similar habitats and have similar characteristics.

mantis (*animals*)   Large predatory insects, mantises or mantids have large compound eyes, chewing mouthparts, 2 pairs of wings, and extremely long front legs specialized for grasping prey. When hunting, a mantis seems to assume a praying position, folding its front legs up under its head. Camouflage coloration also aids in the hunt for insects and other small prey. Male mantises are smaller than females, and are often killed by females during or after mating. Mantises are classified either with grasshoppers and their kin in Order Orthoptera or in their own order, Mantodea.

mantle (*anatomy*)   The internal organs of a mollusk are surrounded by a thin fleshy tissue, the mantle. The mantle secretes the material that forms the mollusk's shell. The edges of the mantle often contain sense organs. In land-dwelling snails and slugs, the mantle forms a cavity that is richly supplied with blood and serves as a lung. Squid and other cephalopods have a thick, muscular mantle; as the animals contract the muscles, a jet of water is expelled forward—and the animals shoot backwards.

mantle (*geology*)   The largest part of Earth is a region called the mantle because it envelopes Earth's core. The mantle is about 2/3 of the total mass of Earth and 4/5 of the volume. A few samples of Earth's mantle reach the surface, but most information about it comes from seismic waves or from experiments using as much heat and pressure as can be produced with laboratory equipment. The bulk of the mantle is thought to consist of various oxides, especially silicon dioxide.

The mantle is neither homogeneous nor stationary. All proposed energy sources for plate tectonics require movement of at least part of the mantle. Not only is there a partly fluid layer near the top of the mantle (the asthenosphere) but seismic studies reveal significant differences between the upper and lower mantle. One of the main unresolved questions of geology is whether the material in the lower mantle mixes with that in the upper.

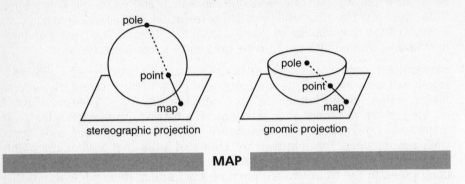

stereographic projection   gnomic projection

MAP

map (*mathematics/geography*) An image of part of the physical world, usually at a smaller size and often on a flat surface, is a map. Geographic maps often use symbols or colors to locate features that may not be physically present, such as political borders or economic activity. Such maps have been made since at least 13,000 BC.

A mathematical map is any image of 1 set of points constructed by applying some transformation to another set of points. Maps of a sphere onto a plane are important both in mathematics and geography, although such maps always distort either size or shape. A widely used shape-preserving map is stereographic projection of a sphere onto a plane; a line from 1 pole of the sphere connects each other point on the sphere to its map point. For gnomic projection the line originates at the center of the sphere and maps a hemisphere onto the plane.

maple (*plants*) Native to temperate regions of the Northern Hemisphere, maples (*Acer*) are deciduous shrubs and trees known for the brilliant red and yellow colors of their autumn foliage. In early spring they produce clusters of small flowers. The fruit, called a samara, consists of a pair of small nuts, each with a long papery wing. Maples depend on the wind both for pollination and seed dispersal. Among the tallest species is the sugar maple (*A. saccharum*) of North America, which grows to 130 ft (40 m); it produces a sweet sap that is made into maple syrup and sugar.

marble (*rocks and minerals*) Marble is metamorphic rock produced from limestone by heat or pressure. Varieties produced by heat alone are contact marbles. Contact marbles occur in the vicinity of the source of heat (usually an intrusion of granite) and grade into limestone. Marble used for buildings and sculptures (because of its hardness, fine grain, and often interesting color patterns) is likely to have been subject to pressure as well as heat. Although marble is largely calcite, some varieties contain considerable dolomite. Veins and banding in colored marble may be any one of a number of minerals.

marmoset (*animals*) Primates of the family Callithrichidae comprise 26 species of marmosets and tamarins. Some species are called titis after the sound of their mating call. They are among the smallest primates—an adult pygmy marmoset (*Cebuella pygmaea*) weighs only 3.5 oz (100 g). Marmosets have sparsely furred or naked faces, with soft, silky fur covering the rest of the body and the long tail. The front limbs are shorter than the hind. The thumb and big toe are not opposable; the big toe has a flat nail but all other digits have sharp claws. Native to tropical forests of Central and South America, marmosets and tamarins are active during the day, moving about in trees and bushes looking for insects and fruits.

marmot (*animals*)   Marmots are stocky ground-dwelling rodents native to the Northern Hemisphere. They are the largest members of the squirrel family and include the woodchuck, or groundhog (*Marmota monax*), of North America. The woodchuck grows to 20 in. (50 cm) long, plus a tail to 7 in. (18 cm).

Mars (*solar system*)   The fourth planet from the Sun, Mars is the outermost of the 4 terrestrial planets in the solar system. Its pronounced reddish color comes from iron oxide in the Martian soil. The so-called canals on Mars—later found to be optical illusions—were first observed by 19th-century astronomers. They led to the widespread belief that there was or is life on Mars, an unproven theory.

Mars is heavily cratered by meteorites as well as by once-active volcanoes. Olympus Mons is an extinct volcano 3 times as high as Earth's Mount Everest. A feature of Martian weather is planetwide dust storms, which engulf the entire planet for months at a time. Mars has carbon dioxide ice caps at both poles. Water once flowed over Mars in long channels that extend for hundreds of miles. This water probably originated from springs rather than rainfall. Some water may still be trapped as ice beneath the surface.

Mars's atmosphere thinned after volcanic activity ceased. Atmospheric pressure is now just 0.7% of that on Earth at sea level; the predominant gas is carbon dioxide. Surface temperatures vary from a high of about 70°F (20°C) to a low of about –220°F (–140°C).

marsh (*ecology*)   A marsh is a type of wetland characterized by shallow water; soil that is neither highly acidic nor alkaline; and low, herbaceous vegetation. Inland marshes contain fresh water and are fed by rivers, springs, and precipitation. Grasslike plants such as cattails and bulrushes grow partly in, partly out of the water, while water lilies and pondweeds float on the surface. Animal life is plentiful and varied. Salt or tidal marshes contain salt water and are flushed by tides twice a day. They are important nurseries for many species of marine fish and home to numerous crustaceans, mussels, birds, and mammals. Cordgrass (genus *Spartina*) and blackgrass (*Juncus*) are typical salt marsh plants.

marsupial (*animals*)   There are about 270 living species of pouched mammals, or marsupials, all native either to the Americas or to Australia and nearby islands. They include bandicoots, kangaroos, koalas, native cats (not true cats), numbats, opossums, phalangers, wombats, and the Tasmanian devil. Marsupials differ from other mammals in their mode of reproduction. The female marsupial does not develop a placenta to provide nourishment for embryos; the only available nourishment comes from the yolk of the original egg cell and from the solution—called uterine milk—that surrounds the embryo. As a result, the gestation period is very brief. The young are essentially still embryos at birth, but they manage to claw their way to a pouch, or marsupium, on the mother's belly. Here they fasten to nipples and obtain milk from mammary glands. The babies do not leave the pouch until they are well developed and able to eat and digest food other than their mother's milk.

The smallest are the marsupial mice (not true mice); they seldom reach 4 in. (10 cm) in length and weigh less than 0.18 oz (5 g). The largest are certain kangaroos nearly 7 ft (2 m) tall and weighing up to 200 lb (90 kg). Traditionally all pouched mammals were classified in Order Marsupialia. In recent years, based largely on new information from the fossil record and molecular biology research, schemes have been proposed to divide marsupials into a number of orders with, in particular, a division between American and Australian species.

maser (*particles*)   The maser is a microwave analog to the laser, invented 6 years before the first laser, in 1954 by Charles H. Townes [American: 1915–]. Townes created a population of ammonia molecules in a high-energy state. A microwave of the right frequency induces molecules to drop into a lower state, releasing a large amount of microwave photons at the exact same frequency—microwave amplification by stimulated emission of radiation (hence, the acronym maser). Original applications were in atomic clocks and sensitive amplifiers.

mass (*physics*)   Mass is often defined as the amount of matter in an object, but that same phrase could refer to the volume of a solid or to matter measured in moles. The mass of an object is better described as the inertial mass, which reflects the amount of force needed to cause an object to move or a moving object to increase in speed. Inertial mass is exactly equivalent to gravitational mass, measurable by the force between 2 bodies separated by a given distance. In 1905 Einstein showed that mass is congealed energy, related to other forms of energy by the equation $E = mc^2$, where E is ordinary energy, m is mass, and c is the speed of light in a vacuum.

mass extinction (*paleontology*)   At certain times during the history of life on Earth, large groups of organisms were wiped from the face of the planet. At least 20 such mass extinctions have been identified, though some are matters of dispute and the causes are unclear. During the largest such event, the P/Tr event, more than 95% of all species on Earth disappeared. Also well documented is the K/T event, which was notable for the disappearance of dinosaurs.

mass spectroscopy (*particles*)   In mass spectroscopy, a beam of ionized atoms or molecules passes through a strong magnetic field. The beam bends by an amount proportional to the mass of each ion. Lighter ions strike a target farther from the direction of the beam than more massive ones, separating atoms or molecules by mass just as an ordinary spectroscope separates light into colors.

mastodon (*paleontology*)   Mastodons were relatives of modern elephants that evolved in Africa and spread to Europe, Asia, and North America. The American mastodon, *Mammut americanum,* was about the size of a small African elephant. It had a long shaggy coat, long curving upper tusks, and a long trunk. It lived in wooded areas throughout North America, becoming extinct about 10,000 years ago. Remains found associated with fires suggest that early peoples may have roasted and eaten its meat.

mastoid (*anatomy*)   The 2 mastoid bones of a mammal's skull are located behind the ears; in humans and some other mammals each mastoid bone is fused with 3 other bones to form the temporal bone. A mastoid bone is honeycombed with spaces, lined with mucous membrane, that connect to the middle ear.

materials science (*sciences*)   The intersection of chemistry and physics in the study of the properties of materials such as conductivity of heat and electricity, strength, ductility, stiffness, magnetism, crystal structure, grain structure, and polymerization is known as materials science. Biomechanics, the study of such properties in tissues of organisms, is a branch of materials science.

mathematics (*mathematics*)   Mathematics is a branch of knowledge that has been described as a language, a game, a science, and a part of logic. Mathematics is a language because it almost entirely employs symbols instead of words. Mathematics is a game because mathematicians choose sets of nearly arbitrary rules and follow where the rules lead. But parts of mathematics, like a science, stem from observation and experiment; thus there is an intimate connection between mathematics and science. Ideas developed in mathematics for their own sake often find application in science, and problems from science have initiated whole new branches of mathematics. The essence of mathematics seems to be reasoning and deduction, but attempts to derive mathematics from logic have led to unresolvable paradoxes. Many mathematicians feel that their instincts, rather than pure logic, lead them.

matrix (*algebra*)   A matrix is a rectangular (including square) array of numbers or variables treated as a single mathematical entity similar to a vector or a number. Typically the array is shown in large parentheses or between pairs of double lines to indicate a matrix. Matrices can be added, multiplied, or transformed in other ways to solve systems of equations or problems involving many different variables or numbers. Matrix multiplication is not commutative.

Matrices originally were created in 1855 for use in abstract mathematics by Arthur Cayley [English: 1821–95]. In 1927 Werner Heisenberg used matrix algebra as the basis of the first successful quantum mechanics.

matter (*physics*)   Traditionally matter is defined as anything that has mass and volume. Einstein's equation $E = mc^2$ shows that mass and energy are 2 aspects of the same underlying reality, so matter is energy that has become solid. Matter consists of aggregations of subatomic particles. Particles that are fermions, including the electrons, protons, and neutrons that make up atoms, are matter. Particles that are bosons, such as photons, transmit forces and are energy—but as Einstein's equation predicts, a high-energy photon can become an electron and a positron, particles of matter.

Maunder minimum (*solar system*)   Edward Walter Maunder [English: 1851–1928], who studied the historical records of sunspots, observed that none occurred during the period from 1645 to 1715, a time now known as the Maunder minimum. This period is often cited in studies of the relation between solar activity and weather on Earth, since it coincides with the Little Ice Age in Europe.

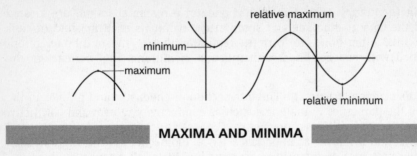

MAXIMA AND MINIMA

maxima and minima (*calculus*) The height of a curve above the horizontal axis—expressed as a negative number for points below the axis—is a maximum when all other points of the curve are lower and a minimum when the rest of the curve is higher. When a curve rises and falls more than once, relative maxima or minima (relative to points in an arbitrarily small neighborhood) occurs whenever the curve changes direction from rising to falling or vice versa.

Maxwell, James Clerk (*physicist*) [Scottish: 1831–79] Much of Maxwell's early work was based on interaction of moving particles. In 1857 he showed that Saturn's rings must consist of small particles. Three years later he determined the statistical distribution of moving molecules in gases, explaining diffusion and conduction of heat (this theory was independently derived by Ludwig Boltzmann).

Maxwell also studied color, being the first (in 1849) to show that the primary colors of light are red, green, and blue and (in 1861) demonstrating the first color photograph based on this idea.

From 1856 through 1873 Maxwell developed the laws of electromagnetism, beginning with Michael Faraday's concept of a field of lines of force. Maxwell's calculations showed that electromagnetic waves in a vacuum travel at the same speed as light; he correctly concluded that light is a form of electromagnetic wave, boldly predicting the rest of the electromagnetic spectrum.

mayfly (*animals*) Among the most ancient and primitive of insects, mayflies (Order Ephemeroptera) have a short, or ephemeral, adult stage that lasts from a few hours to a few days, depending on the species. Adult mayflies do not eat; their sole purpose is reproduction. Mayflies are delicate creatures, with very large compound eyes, large front wings, small or missing hind wings, and greatly elongated filaments on the end of the abdomen. Immature forms, called nymphs, mostly live in freshwater and feed on algae and plant matter.

McClintock, Barbara (*geneticist*) [American: 1902–92] In 1951, based on her observations of the colors of corn kernels, McClintock proposed that fragments of genetic material could "jump," or move from one area on chromosomes to another. The theory contradicted widely held beliefs and was not accepted until the late 1960s when such transposable elements were found to be common in bacteria. McClintock, whose discoveries also influenced evolution theories, received the 1983 Nobel Prize in physiology or medicine.

Mead, Margaret (*anthropologist*)   [American: 1901–78] Mead made her reputation studying adolescence, child-rearing, and sexuality in Samoa in the 1920s. During World War II she helped the U.S. military understand cultural differences between the British and Americans. After the war she concentrated on development of the family once more, both in the United States and the South Pacific.

meadow (*ecology*)   A meadow is a moist grassland community that is a stage in the ecological succession of a pond to a forest. The vegetation is comprised of grasses and herbs. Typical animals include insects, frogs, mice, garter snakes, and hawks. Gradually trees begin to replace the grasses; this in turn causes changes in animal populations. Large grazing animals help keep the area at the meadow stage.

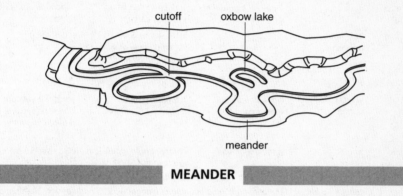

cutoff   oxbow lake

meander

MEANDER

meander (*geography*)   When a stream flows across a nearly flat region, small variations in soil or current cause the stream to develop S curves called meanders, after a river famous for such wanderings. Over time the stream cuts through such meanders, leaving curved lakes called oxbows, which lie adjacent to the stream.

measure of central tendency (*statistics*)   A measure of central tendency for a set of numbers is a single number that represents in some way a typical member. The best known measure of central tendency is the average, or arithmetic mean. Another number commonly used is the median, or "middle number" of a set. A median is the middle number for a set with an odd number of members, but if there are an even number of members there is no middle number, and the median is half the sum of the 2 numbers nearest the middle. Sometimes another measure of central tendency, called the mode, is used; this is the number in the set that occurs most frequently. Rarely a fourth measure is employed, the geometric mean, which is the nth root of the product of the n numbers in the set.

The measure that is most useful depends on the kind of numbers in the set. If a few numbers are much larger or smaller than the others, the median remains in the middle of the set, but the arithmetic mean, or average, is pulled toward the extreme numbers. This mode is used when a single value occurs much more often than the others.

Measurement

UNIT	QUANTITY	CONVERSIONS	DEFINITION	REMARKS
ampere (A)	electric current	1 A = 1 C-sec (the flow of 1 coulomb of electrons per second)	current that produces force of 0.0000007 N/m when in 2 conductors 1 m apart in vacuum	a basic SI measurement unit; current = voltage/resistance (Ohm's law)
angstrom (Å)	length	× 0.00000003937 for in.	= 10 nanometers = 0.000000001 m = 10^{-9} m	not an SI unit, but often used in measuring small cells or wavelengths of light
astronomical unit (A.U.)	distance	× 0.0000159 for light-years × 0.0000004867 for parsecs	average distance of Earth from Sun, 92,943,721.11 mi (149,578,706.91 km)	official SI abbreviation is ua, not A.U.
atomic mass unit (amu)	mass	1 amu = 1.6605402 × 10^{-27} kg	1/12 the mass of an atom of carbon-12	official SI abbreviation is u, not amu
candela (cd)	luminous intensity		intensity of 5.40 × 10^{14} Hz light source producing 1/683 W per steradian (radian measure of a solid angle)	a basic SI unit
coulomb (C)	electric charge	flow of 6.25 × 10^{18} electrons	1 ampere-sec	coulomb was a basic unit in metric system, but is derived from the ampere in SI
degree (°)	angle	π radians = 180°	central angle intercepts arc that is 1/360th of circle	also used to measure arcs
degree (°F or °C)	temperature	conversion formulas: °C = (°F − 32)/1.8 °F = 1.8°C + 32 °C = K + 273.15 °F = K + 459.7	water freezes at 32°F (0°C) and boils at 212°F (100°C)	for very high temperatures (above 10,000 degrees) Celsius and Kelvin are considered equivalent
electron volt (eV)	energy (used to express mass)	equal to 1.60217733 × 10^{-19} joule	energy of 1 electron when passed through potential difference of 1 volt	most masses for subatomic particles are expressed in MeV (1,000,000 eV) or GeV (1,000,000,000 eV)
gallon (gal)	capacity	× 0.264 for liters; × 0.25 for quarts	231 cu in.	1 gal of water weighs 8.3453 pounds
hectare (ha)	area	× 2.471 for acres	10,000 m^2	official SI unit is m^2
hertz (Hz)	frequency	period of wave = 1/Hz; or Hz = 1/period	1 cycle per sec	speed of a wave is wavelength × Hz
inch (in.)	length	× 2.54 for cm; × 0.83 for ft	0.0254 meter (m)	12 in. = 1 ft; 36 in. = 1 yd; 63,360 in. = 1 mi
joule (J)	work, energy, heat	× 4.186 for Calories; × 0.0007374 for foot-pounds	1 newton-meter (Nm)	in cm-g-sec metric system joule is replaced by erg = dyne-cm
kelvin (K)	temperature	C = K − 273.15; F = 1.8K − 459.7	1/273.16 of temperature of triple point of water	A basic SI unit; 0 K is absolute zero
kilogram (kg)	mass	× 2.205 for lb	standard mass kept in Sèvres, France	a basic SI unit; often used for weight, which is a force not a mass

UNIT	QUANTITY	CONVERSIONS	DEFINITION	REMARKS
light-year	length	$\times$ 5.88 $\times$ 10^{12} for mi; $\times$ 9.46 $\times$ 10^{12} for km	distance light travels in vacuum in 1 tropical year	tropical year, the basis of calendar, is 365.2422 days
liter (L)	volume	1 L = 1000 cm^3	1/1000 cubic meter (10^{-3} m^3)	mostly used for volumes of fluids, called capacity
meter (m)	length	$\times$ 39.37 for in.; $\times$ 3.281 for ft; $\times$ 0.0006214 for mi	in vacuum, light travels 1 m in 1/299,792,458 sec	a basic SI unit; definition makes speed of light exact
micron = micrometer (μm)	length	$\times$ 1,000,000 for m; $\times$ 0.00003937 for in.; $\times$ 1000 for Å	0.000001 m = 10^{-6} m	not a SI unit but still popular, especially in biology
mile (mi)	length	$\times$ 1.609 for km	5280 ft	km $\times$ 0.62137 for mi
mole (mol)	amount of substance	about 602,213,670, 000, 000,000,000,000 molecules; 1 mole of gas occupies 22.41410 L at 273.15 K and 101 325 Pa	amount of substance containing as many molecules (or atoms or ions) as are in 0.012 kg of carbon-12	a basic SI unit
newton (N)	force	in U.S. customary system, poundal (0.12889 N) used for force	kilogram-meters per sec per sec (kg/sec^2)	in centimeter-gram-second metric system, the newton is replaced by dyne = g-cm/sec^2
ohm (Ω)	resistance	conductance: 1 siemens (S) = amperes per volt (reciprocal of ohm)	resistance of current of 1 ampere driven by force of 1 volt	equivalent to volts per ampere; resistance = voltage/current (Ohm's law)
parsec (par)	length	$\times$ 0.306598 for light-years; $\times$ 1.91782 $\times$ 10^{13} for mi; $\times$ 3.08547 $\times$ 10^{13} for km	shift in position of 1/3600 degree as seen from points 1 A.U. apart	1 parsec is 3.2616 light-years
pascal (Pa)	pressure, stress	101,325 Pa = 1 standard atmosphere	newton per square meter (N/m^2)	meteorologists use the bar and millibar; 1 bar = 100,000 Pa
radian (rad)	angle	1 rad is about 57°; π rad = 180°	central angle that intercepts arc of circle equal in length to radius	radian measures are treated as real numbers in mathematics so they are not measurement units in ordinary sense
second (sec)	time	$\times$ 0.0002777 for hr; $\times$ 0.0000115 for days; 3600 sec in 1 hr; 86,400 sec in 1 day	duration of 9,192,631,770 vibrations of cesium-133 atom	a basic SI unit; approximates 1/86,400 of time from one sunrise to next; official SI abbreviation is s, not sec
volt (V)	electric potential		joules per coulomb (J/C) or watts per ampere (W/A)	voltage = current $\times$ resistance (Ohm's law)
watt (W)	power	$\times$ 0.00134 for horse power	joule per second (J/sec)	watts = amperes $\times$ volts

measurement (*mathematics*) Any method of assigning a number, called the measure, to an entity on the basis of multiples of a unit amount can be termed measurement. Most measures are rational or real numbers because measurement normally is applied to continuous amounts (measurement of a discrete set is called counting). To measure a line segment, for example, a unit length is placed on the segment a number of times until the unit extends beyond the segment. The unit that extends beyond is subdivided to determine the fraction that is beyond the end. The concept is generalized for measurement of any quantity that can vary in amount, such as electric current.

mechanical advantage (*mechanics*) A machine can be used to increase force at the expense of distance or vice versa. Mechanical advantage is the ratio of input force (effort) to output force (load), ignoring distance. For simple machines, such as levers, inclined planes, and compound pulleys, friction is not a big factor. Thus, theoretical mechanical advantage is near the actual multiplication of force obtained. In most other applications the actual output force is less than predicted by mechanical advantage.

mechanics (*physics*) Mechanics refers to the physical or mathematical theory of how forces on bodies interact. If the forces on solids balance, causing no motion, the subject is statics, but when unbalanced forces produce motion, the theory becomes dynamics. Liquids and gases follow laws of fluid mechanics. Subatomic particles interact via quantum mechanics.

median (*geometry*) A line segment with one endpoint on a vertex of a triangle and the other that bisects a side is called a median of the triangle. The 3 medians in a given triangle are concurrent, intersecting in what is called the median point of the triangle.

meerkat (*animals*) Native to dry grasslands of southern Africa, meerkats (*Suricata suricatta*) are small, slender members of the mongoose family of carnivorous mammals. Also called suricates, they are highly social, living in colonies and communicating through a variety of calls and barks. Active during the day, meerkats often are seen standing on their hind legs as they watch for predators.

megalith (*archaeology*) A megalith ("big stone") is a tomb or monument or temple consisting of large standing stones often with other large stones laid across them. In western Europe, starting about 4500 BC, tombs were built from several large stones capped by one laying across it. Single stones or circles of stones developed later. Still later megalithic methods were used for building observatories or temples, such as Stonehenge, built in its first form about 2200 BC and achieving the form we know today by about 1450 BC.

meiosis (*cells*) The type of cell division called meiosis results in the creation of gametes (egg and sperm cells). The first meiotic phase is much like mitosis, with the production of 2 cells each with a full complement of chromosomes. In the second meiotic phase, the 2 chromosomes of each pair separate, and the cells divide. Thus the resulting 4 cells have half the normal number of chromosomes. During sexual reproduction, an egg unites with a sperm to form a single cell that has half its chromosomes from one parent and half from the other.

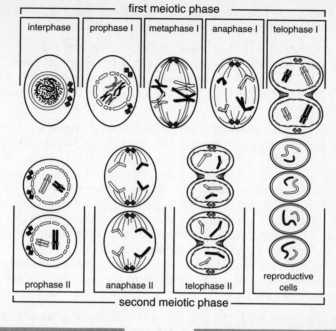

first meiotic phase

| interphase | prophase I | metaphase I | anaphase I | telophase I |

| prophase II | anaphase II | telophase II | reproductive cells |

second meiotic phase

MEIOSIS

Meitner, Lise (*physicist*)   [Austrian-Swedish: 1878–1968] Meitner collaborated with Otto Hahn [German: 1879–1968] in Germany from 1906 through 1938. Together they found the first long-lived isotope of **protactinium**, establishing it as a new element. They also produced uranium **fission**, but Hahn failed to recognize that uranium atoms were splitting. In 1938 Meitner, in exile from Germany because of her Jewish heritage, worked out the correct explanation in collaboration with her similarly exiled nephew Otto Frisch [Austrian: 1904–79], who named the process nuclear fission.

meitnerium (*elements*)   Meitnerium is a rapidly decaying radioactive metal, atomic number 109, a few atoms of which were synthesized in 1982 by the Society for Heavy Ion Research in Darmstadt, Germany. The name honors the Austrian-Swedish physicist **Lise Meitner**. Chemical symbol: Mt.

melanin (*biochemistry*)   Melanins are a group of dark brown and black pigments produced by many kinds of plant and animal cells. In humans melanins are found in hair, skin, eyes, cardiac muscle, nervous tissue, and adrenal glands. An abnormal deposition of melanin—in the skin, for example—is termed melanism. A deficiency causes **albinism**.

melting point (*physics*)   The temperature, usually at a pressure of 1 bar, at which a heated crystalline solid becomes liquid is called the melting point. Heat, in an amount specific to the solid (called heat of fusion), must be added at that temperature to produce melting. The melting point is the same temperature as the freezing point, where liquids become solid. Glasses do not have a specific melting point but gradually flow faster with heating. Most alloys melt in stages, representing their constituents. (*See also* **sublimation**.)

membrane (*biology*)   Thin coverings called membranes exist at several levels within an organism. All cells are surrounded by a cell membrane. Subcellular membranes enclose chloroplasts, nuclei, and other organelles. At the tissue level there are membranes that cover organs, such as the pericardial membrane surrounding a vertebrate heart.

Mendel, Gregor (*botanist*)   [Austrian: 1822–84] A monk who experimented with garden peas and other plants in his spare time, Mendel discovered some of the fundamental principles of genetics. In an 1865 paper he demonstrated what came to be called Mendel's laws. The law of unit characters says that characteristics of an individual are controlled by hereditary factors (now called genes) and that these factors occur in pairs. The law of dominance says that some inherited factors are dominant and can mask other, recessive factors. The law of segregation says that the 2 factors of a pair are separated during reproduction, so that only 1 goes to a particular offspring.

Sadly, the importance of Mendel's work was not recognized until 1900 when 3 botanists working independently reached similar conclusions and, in the process, discovered his paper.

mendelevium (*elements*)   Mendelevium is a radioactive actinide metal, atomic number 101 and chemical symbol Md, first synthesized in 1955 by Albert Ghiorso [American: 1915–] and coworkers. It was named for Mendeleyev, who developed the first useful periodic table. The half-life of the most stable isotope is about 2 months.

Mendeleyev, Dimitri Ivanovich (*chemist*)   [Russian: 1834–1907] In 1869 Mendeleyev published the first effective version of the periodic table of the elements. Unknown to Mendeleyev, 5 years earlier John Newlands [English: 1837–98] had proposed a similar table based on strictly increasing atomic masses forming periods of 8 elements. Mendeleyev's table used valences as well as atomic masses, enabling him to recognize periods of 18 elements later in the table. Mendeleyev was bold in predicting 3 new elements and their properties for his 1871 version of the table and in rearranging the order from that of atomic masses where needed to make properties fall in line. The 3 predicted elements were found by 1885, exactly as described, and the rearrangements were justified in 1913 by the discovery of atomic number.

meninges (*anatomy*)   Three connective tissue membranes, called meninges (singular: meninx), encase and protect the vertebrate brain and spinal cord. The outermost membrane, the dura mater, is a thick, tough layer that lines the inside of the skull and forms a loose sheath around the spinal cord. The middle membrane, the arachnoid mater, is delicate, with weblike fibers. The innermost membrane, the pia mater, contains many blood vessels and is attached directly to the brain and spinal cord.

menstrual cycle (*physiology*)   Adult females of most primates experience a recurring series of changes termed a menstrual cycle, which differs from the estrous cycle of other mammals in that there is no period of heat, or estrus. The menstrual cycle—controlled by various hormones—begins with the release of an egg from an ovary. The endometrium (lining of the uterus) thickens in preparation for implantation of a fertilized egg. If the egg is not fertilized, the endometrium breaks down and is shed as a bloody discharge called menstruation.

mercury (*elements*) At commonly encountered temperatures, mercury, atomic number 80, is a heavy, silvery liquid metal, the only metal liquid below 80°F (27°C). Mercury's chemical symbol, Hg, comes from *hydrargyrum* ("liquid silver" in Latin), similar to another name, quicksilver ("living silver"). Quicksilver's density is great compared to its surface tension and it does not wet objects, so it breaks into small spheres that roll about as if they were alive ("quick"). The formal name, mercury, is the same as that of the Roman messenger god and the swiftest planet.

Mercury occurs as a mineral in native form—not combined as part of a compound. Thus it was known to the ancients, who also learned to obtain it by heating another mineral, cinnabar (mercuric sulfide). Many metals, including silver and gold, dissolve in mercury to form alloys called amalgams, popular with alchemists and still used today in filling teeth. A constant coefficient of expansion when heated makes mercury useful in thermometers, and mercury vapor is the basis of a bright electric lamp. Although mercury compounds also have many uses (paints, pesticides, and electric batteries among them), mercury in almost all forms is poisonous. A major environmental problem is the gradual accumulation of mercury in lakes, where it becomes concentrated in fish through biological magnification.

Mercury (*solar system*) The planet closest to the Sun, Mercury is a waterless, airless world that alternately bakes and freezes. On Mercury's sunlit side temperatures reach 950°F (510°C) but plummet to -346°F (−210°C) on the dark side. These extremes result from Mercury's slow rate of rotation; one rotation takes more than 2 months. Mercury is slightly larger than the smallest planet, Pluto. Its surface is scarred with hundreds of thousands of meteorite craters, most blasted out soon after the formation of the solar system. In many regions, however, the craters have been smoothed over by ancient lava flows. The surface is crisscrossed by huge cliffs that formed as Mercury cooled and shrank. Mercury is so dense for its size that its rocky outer crust must be very thin, covering a planet that is mostly iron. Astronomers conjecture that during the early bombardment, a large object hit Mercury so hard that it blasted most of the original crust away.

Mersenne number (*numbers*) In 1644 Father Marin Mersenne [French: 1588–1648] discussed numbers of the form $2^p - 1$, where p is a prime; hence, the term Mersenne numbers. Many are prime numbers (for example $2^2 - 1 = 3$, $2^3 - 1 = 7$, $2^5 - 1 = 31$, $2^7 - 1 = 127$, $2^{13} - 1 = 8191$, but $2^{11} - 1 = 2047$, which is not prime). Studies of Mersenne primes have led to discovery of many large prime numbers. As of 1999 the largest known prime is the Mersenne prime, $2^{6972593} - 1$, a number with 2,098,960 digits.

mesa (*geography*) A mesa is a large flat-topped hill with steep sides, generally found in regions of low rainfall. Its flat top consists of hard, level rock that covers softer material. Mesas are pieces of a plateau, separated when, in wetter times, streams cut canyons. The rare rainfall of today erodes the mesa's sides, creating cliffs. Mesa tops may be less than a square mile (a butte) or many square miles.

meson (*particles*) Particles formed from 2 quarks are called mesons. The name was given to mean medium-sized particle (*mesos* means "middle" in Greek), but is no longer very descriptive. Pions are ordinary matter, but other mesons only occur at higher energies. The 6 quarks and their antiquarks can combine in 140 ways that are stable enough to count as mesons.

Mesopotamia (*archaeology*) Mesopotamia ("between the rivers"), strictly speaking, is the region in the Middle East between the Tigris and Euphrates rivers, which today is largely eastern Iraq. It often includes the region of early civilization that extends into surrounding nations and north of the rivers into southeastern Turkey.

mesosphere (*atmospherics*) The mesosphere, above the stratosphere and below the thermosphere, or from about 30 mi (50 km) to about 50 mi (80 km) high, cools with height. Most meteors light up in the mesosphere. In its coldest heights, ice crystals form thin noctilucent clouds, visible, as the name suggests, only at night, lit by a Sun that is already below the horizon.

mesquite (*plants*) Native to warm, dry regions of the Americas, mesquites (*Prosopis*) are spiny shrubs and trees that range from 3.3 ft (1 m) to 33 ft (10 m) in height. The various species and hybrids differ mainly in the appearance of their compound leaves. The small greenish flowers are borne in dense spikes and mature into yellow or red pods that are edible. The honey mesquite (*P. glandulosa*), a common shrub of deserts in the U.S. Southwest, has a taproot that can be more than 100 ft (30 m) long. It is an excellent example of a phreatophyte—a perennial with very deep roots that obtains its water from permanent subsurface water sources and therefore is not dependent on rainfall.

metabolic rate (*physiology*) The speed at which an organism carries out metabolism is its metabolic rate. The higher the metabolic rate, the greater the amount of energy produced, the higher the body temperature, the faster the heart rate—and the greater the amount of food and oxygen that must be consumed. The basal metabolic rate, usually measured by oxygen consumption, indicates the amount of energy produced by an organism at rest. The metabolic rate increases rapidly with exercise. For instance, during exercise a person consumes 15 to 20 times as much oxygen as is consumed at rest. In general, birds have higher metabolic rates than mammals, which have higher rates than the other vertebrates.

metabolism (*physiology*) Metabolism is the sum of all the chemical reactions occurring in the cells of a living organism. It includes anabolic reactions, in which energy is used to build substances used or stored by the body for growth and repair, and catabolic reactions, in which substances are broken down to release energy. Metabolic reactions typically occur in a sequence of small steps that make up a metabolic pathway.

metal (*astronomy*)   Astronomers speak of high- or low-metal stars. In stellar astronomy, metal means any element heavier than helium. Hydrogen and helium were produced by the big bang. The other elements—the metals—result from fusion or supernova explosions.

metal (*materials science*)   A metal is an element or mixture of elements in which electrons move easily from atom to atom providing the bond that holds the substance together. The atoms of metals have an outermost electron shell that has 1, 2, or 3 electrons that may easily be passed from atom to atom, accounting for the ability of metals to conduct an electric current. Mobile electrons also cause metals to be good heat conductors. Metallic bonds are not as rigid as covalent bonds, so metals bend easily and are ductile. Finally, loose electrons interact with photons of light, giving metals a characteristic luster.

metal ages (*archaeology*)   In 1816 Christian Jürgensen Thomsen [Danish: 1788–1865] started separating archaeological finds into the Stone Age, Bronze Age, and Iron Age, publishing his system 20 years later. Today archaeologists recognize that for a brief period from 500 to 1000 years about 4000 BC, copper was widely used and bronze scarcely known—the Copper Age. By 3000 BC, starting in Mesopotamia, tin was added to copper to produce harder bronze. The Bronze Age lasted until about 1500 BC, when improved methods of iron smelting were developed.

metamorphic rock (*rocks and minerals*)   Rock that has been altered by heat or pressure or, most commonly, both is called metamorphic, whether it began as an igneous rock such as granite (metamorphic form, gneiss) or a sedimentary rock such as limestone (metamorphic form, marble). Often metamorphic rock can be recognized from the folded nature of its mineral layers, giving it a wavy pattern. Sometimes a soft rock, such as shale, becomes harder from metamorphism (metamorphic form, slate).

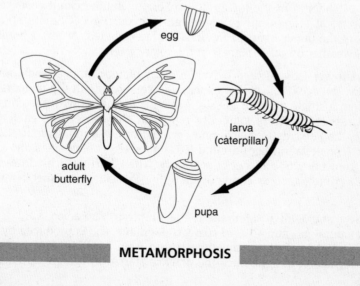

egg

larva
(caterpillar)

adult
butterfly

pupa

METAMORPHOSIS

metamorphosis (*physiology*)   Some animals go through a series of major changes in body form during their life cycle. The process, called metamorphosis, is controlled by **hormones** secreted at specific times. A frog, for example, has 3 distinct stages in its life cycle: egg, tadpole, adult. Almost all insects undergo metamorphosis, either incomplete or complete. Incomplete metamorphosis, as in grasshoppers, occurs in 3 stages: egg, **nymph**, and adult, with the nymph looking rather like a miniature, immature adult. Complete metamorphosis, as in butterflies, has 4 stages: egg, **larva**, **pupa**, and adult, with radical differences in the structure of the larva and the adult. (See art on previous page.) Each stage of metamorphosis may be accompanied by distinctive behaviors. For example, tadpoles feed on algae and adult frogs eat mainly insects.

meteor (*solar system*)   The visible streak in the night sky caused when a small fast-moving object glows with the heat of friction is a meteor or, colloquially, a shooting star. Most meteors are produced by small particles of dust, often from comets, and last less than a second. Meteors as bright as or brighter than Venus are called **fireballs**. An object large enough to survive passage through the atmosphere and collide with Earth is termed a meteorite.

meteoroid (*solar system*)   Small rocky or metallic bodies in interplanetary space are meteoroids (larger solid bodies of about 30 ft [10 m] or greater in diameter are **asteroids**). If meteoroids impact on Earth, the recovered bodies of stone, iron, or carbon are meteorites. At one time scientists did not believe that meteorites existed; however, a fall of about 200 in L'Aigle, France, in 1803 convinced most of their existence. Most meteoroids are thought to be fragments formed by asteroid collisions. In 1983 researchers learned that at least some meteorites come from the Moon or from Mars.

meteorology (*sciences*)   Meteorology is the study of weather not of meteors, but the connection is the Greek word *metoros*, meaning "high in the air." Technically, meteorologists study Earth's atmosphere, but weather consists almost entirely of changes in the lower atmosphere, including cloud cover, precipitation, electrical effects, air pressure, and air currents. Climate studies and changes in the upper atmosphere may also be included. A newer branch is the study of weather on other planets in the solar system.

meteor shower (*solar system*)   When many meteors appear each hour in the night sky, the event is called a meteor shower. Nearly all the meteors seem to come from a single point in the sky, which is identified with the constellation that locates it. Showers generally repeat at the same time each year; for example, the Perseids (constellation Perseus) occur around August 12 and the Leonids (constellation Leo) around November 18. The showers are caused by debris from comets and occur regularly when Earth's orbit intercepts the orbit of a comet, such as comet Swift-Tuttle (Perseids) or comet Temple-Tuttle (Leonids).

methane (*compounds*)   Methane is the simplest **hydrocarbon**, CH_4. It is about 90% of natural gas and 40% of coal gas. Methane also is produced by decay of organic materials, so it is sometimes piped from landfills. Methane has been burned for heat and light since the practice started in China, perhaps as early as 300 BC. Methane is occasionally used to manufacture other chemicals, such as acetylene and the electrically conducting plastic polyacetylene.

mica (*rocks and minerals*)   A number of different minerals are called mica, including greenish glauconite, pale lilac lepidolite, white muscovite, and black biotite. Chemically the micas have in common a combination of potassium, aluminum, silicon, and oxygen. The color changes reflect additional aluminum for white, iron and magnesium for black, chlorine for green, and lithium for lilac. Micas can be split into parallel-faced pieces, often very thin ones. A thin sheet of mica is sometimes used as a transparent covering (isinglass) or as electrical insulation.

Michelson-Morley experiment (*waves*)   In 1887 Albert Michelson [American: 1852–1931] and Edward Morley [American: 1838–1923] used interferometry to measure the speed of light in perpendicular directions, hoping to determine how Earth moves with respect to a presumed invisible substance called ether, thought to pervade space and transmit electromagnetic waves. They found no difference in speed between the directions, casting doubt on the existence of ether. In 1905 Einstein's theory of relativity, inspired in part by the Michelson-Morley result, explained electromagnetic waves without recourse to ether.

microbe (*biology*)   Any microorganism (microscopic organism) can also be called a microbe. However, common usage of the term often is restricted to disease-causing microorganisms. The 3 major groups of microbes are viruses, bacteria, and protists. Most microbes are harmless; many are beneficial to humans and other organisms; and a comparatively small percentage cause infectious diseases.

microbiology (*sciences*)   Microbiology is the study of living things too small to be seen by the naked eye. The term is often restricted to the study of bacteria, a science also called bacteriology. Other terms are used for the study of viruses (virology), algae (algology), animal-like protists (protozoology), etc.

The study of microorganisms is critical in the work of many scientists. Pathologists study disease-causing bacteria and viruses. Geneticists use bacteria extensively in their efforts to understand hereditary mechanisms.

microclimate (*ecology*)   The climate of a comparatively small site may differ significantly from the surrounding area. This microclimate can have major effects on organisms. In winter, for example, the sunny and shady sides of a tree have different temperatures. Stoneflies may be active on the sunny side, but they are not found on the cooler shady side.

Human activities create a variety of microclimates; those of cities are the best examples. A city's buildings and asphalt absorb and retain more heat than the surrounding countryside. A city also has a larger proportion of vehicles, factories, and other fuel-using devices that generate heat. As a result of this "heat island" effect, many cities in temperate zones have frost-free growing seasons that are a month longer than in surrounding rural areas.

microlith (*paleoanthropology*)   Microliths are small stone flakes or points shaped like triangles, crescents, or trapezoids, manufactured at the very end of the Paleolithic Age and during the Neolithic Revolution. They were designed to attach to wood, bone, or antler, often in combinations, to form arrows and many other tools, such as scythes or knives. Paleolithic peoples gradually turned from tools that used large pieces of stone with a small cutting edge, such as hand axes, to tools with long cutting edges for small amounts of stone, characteristic of microliths.

microscope (*optics*)   A lens or combination of lenses used to enlarge an image is called a microscope. The trade-off for a larger image is less brightness, so a microscope that enlarges many times also uses some method such as a mirror and bright light source to compensate. A visible light, or optical, microscope is limited in part by the wavelength of light—objects near the size of a wavelength cannot be imaged. The apparent solution to this limitation—use of shorter wavelengths, such as ultraviolet light or X rays—is hampered because such electromagnetic waves have higher energies and therefore do not refract easily. It is almost impossible to make a lens for gamma rays, which have even shorter wavelengths. The electron microscope, however, uses electrons, which have very short wavelengths but are charged particles that can be focused with magnets. Electron microscopes enlarge images about 1000 times as much as the best optical microscopes.

Some microscopes are based on moving a tiny fiber over a surface to obtain an image (scanning tunneling microscopes and atomic force microscopes). These are capable of imaging individual molecules or atoms.

microwave radiation (*waves*)   Microwave radiation is that part of the electromagnetic spectrum with a frequency between about 100,000,000 Hz and 1,000,000,000,000 Hz or a wavelength between 30 cm and 1 mm; that is, between the longer radio waves and shorter infrared radiation. Medium-length microwaves have the right energy level to speed up water molecules, so they can be used to heat food that contains water without heating a ceramic or plastic plate. They are small enough to be reflected from medium-sized objects, such as ships or planes, and therefore are the basis of radar. Both FM radio and television are broadcast on frequencies near the radio end of the microwaves. The cosmic background radiation consists of microwaves.

midden (*archaeology*)   A midden ("muck heap") is the pile of garbage that commonly develops over centuries near any human habitation. Sometimes it is a kitchen midden, consisting mainly of refuse from cooking and eating—near the sea, the kitchen midden may consist largely of nonbiodegradable shells of clams or mussels. Middens are often productive places to learn about the culture of a vanished civilization.

mid-oceanic ridge (*geology*)   The mid-oceanic ridge system is a chain of undersea mountains running north and south in the middle of the Atlantic Ocean, bending to pass below Africa into the Indian and Pacific oceans, branching in the Pacific past Australia's east coast into ridges that head separately toward Asia, North America, and South America. This globe-girdling chain was first discovered in 1925 by Germany's ship *Meteor*, whose early sonar identified the mid-Atlantic ridge. In 1956 Maurice Ewing [American: 1906–74] determined the system's worldwide extent. Soon after, geologists recognized that the ridge forms where tectonic plates move apart. As the plates separate, hot rock from the mantle rises, producing the elevated ridge.

migration (*ethology*)   Regular movements of animals from one place to another are termed migrations. Migrations occur to take advantage of seasonal food supplies, improve survival rates of offspring, and complete life cycles. Many birds, for example, move from summer nesting grounds in cool temperate and arctic areas to winter grounds in warmer climes, then return in spring to the nesting grounds. Animals generally follow familiar routes when migrating, arriving at the same destinations each time. A round trip may take an entire lifetime, as with Pacific salmon; or an individual may make many migrations in its life, as with blue whales. Both environmental and physiological factors may be involved in initiating migration. In birds, for instance, enlargement of reproductive organs in spring and their reduction in autumn stimulate migration.

mildew (*fungi*)   Mildew is a broad term, referring both to certain fungi and to their growths—usually whitish—on the surface of living plants or dead organic matter. Powdery mildews, of Phylum Ascomycota, are parasites of clovers, lilacs, roses, grapes, and many other plants; they mainly attack leaves, appearing as pale dusty patches. Downy mildews, of Phylum Zygomycota, are named for the fuzzy, or downy growth they produce. They, too, are parasites, usually of leaves. Among the most notorious is downy mildew of grape (*Plasmopara viticola*), which was accidentally introduced from North America into Europe around 1870; the common species of European grape was highly susceptible to the fungus and the French wine industry was almost wiped out.

milk (*biochemistry*)   Milk is the white fluid produced and secreted by mammary glands of female mammals for the nourishment of their young. It contains sugars (mainly lactose), proteins, fats, vitamins, and minerals—all the nutrients needed by young mammals until they are able to eat the same foods as their parents. Milk's composition varies according to species. For example, human milk averages 87.60% water, 7.00% lactose, 1.41% protein, 3.78% fat, and 0.21% other substances. Reindeer milk is 63.31% water, 2.50% lactose, 10.30% protein, 22.44% fat, and 1.45% other substances.

The term "milk" also refers to any natural liquid that resembles milk, such as the liquid in unripened coconuts, called coconut milk.

milkweed (*plants*)   The milkweed family, Asclepiadaceae, comprises some 2000 species of herbs, vines, and shrubs. All produce a milky latex in the roots, stems, and leaves. The flowers are borne in clusters called umbels. When mature, the podlike fruits split open along a seam, exposing numerous seeds that usually bear tufts of silky hair; even slight air currents will lift out and disperse the seeds. The milkweeds (*Asclepias*), perennial herbs native to Africa and the Americas, grow to 5 ft (1.5 m) tall.

Milky Way (*galaxies*)   The Milky Way is the galaxy to which the Sun and Earth belong. If you are in a place without much light pollution, you can see a faint band crossing the night sky. The ancient Greeks named the whitish band the Milky Way (galaxy in Greek). Early in the 19th century, William Herschel determined that our Sun is a star in a vast lens-shaped star system, and that the Milky Way is the part of that system we see from our vantage point inside it. The Milky Way galaxy is a medium-sized spiral galaxy about 100,000 light-years across, containing an estimated 10,000,000,000 stars.

millet (*plants*) Members of various genera in the grass family, millets can grow in semiarid areas and relatively infertile soil. They are widely cultivated as fodder and for their edible grains (fruits). Their flower clusters are long spikes, or panicles, at the end of stalks up to 10 ft (3 m) high.

millipede (*animals*) Millipedes ("thousand legs") are arthropods with 2 pairs of legs on each body segment, with some species having more than 100 segments. The body is cylindrical, from about 0.1 to 9 in. (0.25 to 23 cm) long and protected by a tough exoskeleton containing calcium salts. Millipedes live in rotting logs and other dark, humid places. Most feed on decaying plant matter and fungi.

mimicry (*zoology*) The superficial resemblance of one organism—the mimic—to another organism that lives in the same area—the model—is termed mimicry (from the Greek *mimos*, meaning "mime"). The mimic evolves colors, shapes, scents, behaviors, or other characteristics that are sufficiently like those of the model to fool a predator or to provide some other benefit to the mimic. For example, female *Photuris* fireflies can mimic the pattern of flashing lights emitted by females of other firefly species. Males of the latter species are attracted by the pattern, then seized and eaten by the *Photuris* female. The main types of mimicry are Batesian mimicry and Müllerian mimicry.

mimosa (*plants*) The genus *Mimosa* includes a variety of herbs, vines, shrubs, and trees. Best known are the sensitive plants, such as *M. pudica* of tropical America, so-called because at the slightest touch the compound leaves quickly fold. Mimosas and their close relatives acacias (some of which are commonly called mimosas) are sometimes classified in the legume family; in other schemes they are part of a separate family, Mimosaceae.

mineral (*rocks and minerals*) Minerals are chemical compounds that have a specific crystal structure. Rocks are made from one or more minerals, usually more. Minerals are identified not only by crystal form but also by density, hardness (measured with Mohs' scale), habit of breaking (called cleavage or fracture), color, luster, and other properties. Some fairly simple minerals are apatite (a hydrate of calcium fluoride phosphate), cinnabar (mercury sulfide), corundum (aluminum oxide), cuprite (copper oxide), fluorite (calcium fluoride), galena (lead sulfide), halite (sodium chloride), hematite (iron oxide), olivine (magnesium or iron silicate), and zircon (zirconium silicate). Other familiar minerals are discussed in separate entries.

mineral nutrient (*biochemistry*) Living things require a number of chemical elements, or minerals, for normal functions. These include calcium, chlorine, copper, iron, magnesium, manganese, molybdenum, phosphorus, potassium, sodium, sulfur, and zinc. Animals obtain minerals from inorganic compounds in food; plants absorb minerals from the soil. Some minerals are needed in comparatively large amounts; one such macronutrient in humans is calcium, needed for bone formation. Other minerals, while equally essential, are needed in only trace amounts, as micronutrients; in humans, iodine—a constituent of the thyroid hormone thyroxin—is an example.

mint (*plants*)   Native to northern temperate regions, mints (*Mentha*) are herbaceous perennials noted for their aromatic foliage. They grow about 3 ft (0.9 m) tall. Like almost all members of the mint family, Labiatae, they have square stems, 2-lipped flowers often colored blue or purple, and glands that secrete a fragrant oil. The family also includes basils (*Ocimum*), bergamot or bee balm (*Monarda didyma*), catnip (*Nepeta cataria*), lavenders, oreganos (*Origanum*), rosemary, sage, and thymes (*Thymus*).

Mira (*stars*)   Mira is a long-period variable star that is also a binary, although its periodic changes in brightness, from the tenth magnitude—invisible to the naked eye—to a bright magnitude 2, are almost all caused by the main star, a red giant. David Fabricus [German: 1564–1617] noted Mira in 1596; it faded away, but was back by 1603 when Johann Bayer [German: 1572–1625] incorporated it into his new naming system as Omicron Ceti. As the first periodic star to be recognized by science, it finally was named Mira ("wonderful") in the middle of the 1600s.

missing mass (*cosmology*)   Evidence points to matter that is in the universe but that is not observed except for its gravitational force—thus, missing mass. Galaxies rotate as if they were embedded in larger, invisible bodies and would fly apart unless something beyond observed matter holds stars in place. In addition, accepted theories of cosmology predict even more missing mass than the unobservable, or dark matter, surrounding galaxies. This additional mass is needed to explain the rate of expansion of the universe.

mistletoe (*plants*)   Species of 2 related genera, *Viscum* and *Phoradendron*, are partial parasites commonly called mistletoes. These mistletoes grow on the branches of deciduous trees with no connections to the ground. They have special rootlike structures called haustoria, which grow into the bark and vascular tissues of hosts to rob them of water and nutrients. Mistletoes have chlorophyll in their leaves and carry out some photosynthesis. The female plants produce clusters of small whitish berries, which are popular with thrushes and other birds, who distribute the seeds in their feces.

mite (*animals*)   Small arachnids with head, thorax, and abdomen fused into 1 unit, mites are found worldwide in both aquatic and land environments. They exhibit great diversity of mouthparts; some species feed on plants, others are predators, and still others are parasites. Human pests include the follicle mites (genus *Demodex*), which live in human hair follicles and sebaceous glands, and the itch mites (genus *Sarcoptes*), which burrow through skin.

mitochondrion (*cells*)   The site of cell respiration in virtually all plant, animal, fungal, and protist cells are organelles called mitochondria. Here, energy-rich molecules are broken down, or oxidized, releasing the energy that can power cell activities. Each mitochondrion contains its own DNA, called mitochondrial DNA, similar to the circular chromosomes of bacterial cells. It is theorized that mitochondria are descendants of bacteria that became established in host cells early in the evolution of life on Earth. Most cells contain hundreds of mitochondria, though the number can vary from one to several thousand per cell.

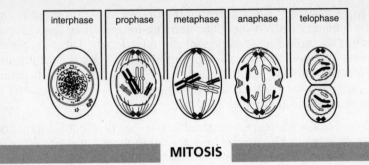

| interphase | prophase | metaphase | anaphase | telophase |

MITOSIS

mitosis (*cells*)   The type of cell division called mitosis results in the creation of 2 cells that are exactly like the original cell. The original cell is the parent cell; the 2 new cells are called daughter cells. For example, an onion root tip cell divides to form 2 onion root tip cells; an amoeba divides to form 2 amoebas. Mitosis proceeds through a series of phases during which the chromosomes are replicated so that each daughter cell has a full complement of hereditary material.

mixture (*chemistry*)   A combination of 2 or more substances that do not bond chemically is a mixture. Mixtures do not have a fixed composition. The substances can be separated using physical means such as sorting or evaporation; compounds separate only when bonds are broken.

mobbing (*ethology*)   Birds, sometimes of different species, may form groups to harass, or mob, a common enemy. For example, arctic terns (*Sterna paradisaea*) gather to mob and drive away egg-eating gulls. Mobbing may include making lots of noise, repeatedly performing displays, and swooping and striking at the enemy. It can be very effective, even against extremely large predators.

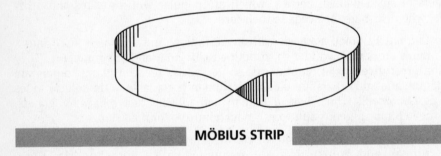

MÖBIUS STRIP

Möbius strip (*geometry*)   A Möbius strip (or band—often spelled Moebius) is formed when one end of a rectangle is given a half turn and connected to the opposite end. Ferdinand Möbius [German: 1790–1868] discovered the 1-side, 1-edge strip in 1858. An ant can crawl anywhere on the side without crossing an edge. A cut completely along the middle of its single side does not cause the strip to fall into 2 pieces.

mockingbird (*animals*) Perching birds of the mockingbird family, Mimidae, are mimics that imitate the songs of other birds—and of frogs and other animals, too. Native to the Americas, they include catbirds, mockingbirds, and thrashers. The birds are 8 to 12 in. (20 to 30 cm) long, with 10 flight feathers on each wing and a comparatively long tail. They feed on insects and plant matter.

modular arithmetic (*arithmetic*) Operations performed with a finite set of consecutive natural numbers that range from 1 to some larger number are called modular arithmetic. The most familiar example is clock arithmetic using numbers from 1 to 12. On a 12-hour clock, the sum of 7 + 8 (8 hours after 7 o'clock) is 3. In number theory the same idea is expressed by calling 2 numbers congruent if they have the same remainder when divided by a particular number, called the modulus. The clock arithmetic statement in this system is recast as 7 + 8 is congruent to 3, *modulo* 12, since 15 ÷ 12 has the remainder 3.

Mohorovicic discontinuity (Moho) (*geology*) In 1909 Andrija Mohorovicic [Croatian: 1857–1936] observed a sharp increase in the speed of certain seismic waves, which he identified as a boundary between 2 different types of rock. The boundary exists at an average depth of 30 mi (50 km) under eastern Europe. This boundary acquired the formal name Mohorovicic discontinuity, soon abbreviated to Moho. By the 1940s seismologists had observed the same change in velocity all over the world, noting that the Moho is much nearer the surface under oceans. In 1954 geologists agreed to accept the Moho as the lower boundary of Earth's crust and upper boundary of the mantle.

Mohs' scale (*rocks and minerals*) Hardness is one of the chief ways used to recognize a mineral. In 1822 Friedrich Mohs [German-Austrian: 1773–1839] created a hardness scale based on common minerals that is still used. The concept is that a harder mineral can scratch a softer one. The scale, from softest to hardest, is 1 talc, 2 gypsum, 3 calcite, 4 fluorite, 5 apatite, 6 orthoclase feldspar, 7 quartz, 8 topaz, 9 corundum, 10 diamond. Other materials may be correlated to this scale. For example, steel used in a pocketknife is about 5.5 and glass is about 6.

mold (*fungi*) Mold is a broad term applied to a variety of fungi that form woolly or fuzzy growths on living and dead organic matter. Among the best known are the black bread molds, including *Rhizopus*, many of which grow on bread, fruits, and vegetables. White at first, they turn dark as large numbers of their spores mature. Another group, the blue and green molds, also get their characteristic colors from their mature spores; *Aspergillus* and *Penicillium* are common examples. Slime molds are not fungi but protists.

mole (*animals*) Small, burrowing insectivore mammals of the family Talpidae, moles are highly adapted for digging, with powerful front limbs and long claws on spade-shaped front feet. It is not uncommon for a mole to dig a tunnel more than 300 ft (91 m) long in 1 day, pushing soil upward to form a pile on the surface. Native to North America and Eurasia, moles feed mainly on insect larvae and earthworms, eating half their weight daily.

mole (*chemistry*) A mole is the amount of a substance with the same number of molecules as there are carbon atoms in 12 g of carbon-12, an amount known as the gram molecular weight or molar mass of carbon-12. This number of molecules, called Avogadro's number, is about 602,213,670,000,000,000,000,000 molecules. Under standard conditions 1 mole of an ideal gas assumes the volume 22.41410 L. All real gases have volumes close to that of an ideal gas.

molecular weight (*chemistry*)   The sum of the atomic masses in a molecule is called the molecular weight or molecular mass. For example, a water molecule with 2 atoms of hydrogen (atomic mass 1) and 1 atom of oxygen (atomic mass 16) has a molecular weight of 18. The gram molecular weight (GMW—also called molar mass), is the amount of a substance whose mass is the same in grams as the molecular weight. The GMW is useful in determining the molecular composition of a compound.

molecule (*chemistry*)   Two or more atoms that are connected by covalent or ionic bonds form a molecule. A molecule is often defined as the smallest particle of a substance that has the properties of the substance in bulk, but this is not exactly so. Small clusters of molecules (fewer than a hundred or so) usually have special properties, while atoms of metals and some other elements form crystals without forming separate molecules. Single atoms of noble gases are sometimes called molecules.

mollusk (*animals*)   Members of Phylum Mollusca are a diverse group of mostly marine invertebrates. They have soft bodies (hence the name, derived from the Latin *mollus*, meaning "soft"). Usually the body is enclosed in a hard shell secreted by a tissue called the mantle. But the wormlike aplacophorans apparently never develop a shell; some species, such as nudibranchs, have no shell in the adult stage; and species such as squids have only an internal remnant of a shell called the pen.

Most of the more than 50,000 living species are classified in 5 groups: chitons, bivalves, tooth shells, gastropods, and cephalopods. They range in length from less than 0.4 in. (1 cm) to the largest of all invertebrates, the giant squid (*Architeuthis princeps*), which grows to 60 ft (18 m).

molting (*physiology*)   The shedding of an outer covering such as a cuticle, scales, feathers, or hair is termed molting. Young insects grow each time they molt their exoskeleton, in a process called ecdysis; before a molt, a new, soft exoskeleton grows beneath the older exoskeleton. Snakes molt their outer epidermis several times a year. Birds periodically molt to replace worn and faded feathers. In all cases molting is induced by hormones.

molybdenum (*elements*)   Molybdenum, atomic number 42, is a hard metal with a high melting point. It was first recognized in 1778 by Karl Wilhelm Scheele in a mineral thought to be lead ore, hence its name from the Greek *molybdos*, "lead." Molybdenum confers its toughness on other metals in alloys, its main use. Plants require molybdenum for growth, so it is an important trace element in soils. Chemical symbol: Mo.

moment of inertia (*mechanics*)   A force that pushes a body around a fixed point develops a quantity called moment or torque, measured as a force-unit (such as pounds) times distance. The mass does not move in the same direc-

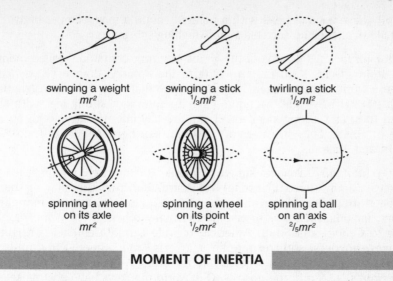

swinging a weight
mr^2

swinging a stick
$\frac{1}{3}ml^2$

twirling a stick
$\frac{1}{2}ml^2$

spinning a wheel
on its axle
mr^2

spinning a wheel
on its point
$\frac{1}{2}mr^2$

spinning a ball
on an axis
$\frac{2}{5}mr^2$

MOMENT OF INERTIA

tion as the force; instead the force is tangent to the circular path of the mass. The square of the radius of that circle times the mass is the moment of inertia or rotational inertia. Different shapes or centers of rotation result in different moments of inertia.

momentum (*mechanics*) The size of the momentum of a moving object is the product of its mass (m) and velocity (v), or mv. The conservation law for momentum is Newton's third law of motion. This law accounts for the flight of a rocket or the recoil of a shotgun. Notice that momentum is a vector quantity, which means that it has direction. When a shotgun fires, the momentum of the gun is directly opposite the direction in which the shot moves. In a collision momentum for the system does not change, so one billiard or croquet ball can impart nearly all of its momentum to another of the same mass. (*See also* **angular momentum**.)

Monera (*monera*) The most ancient of the 6 kingdoms of living things, existing at least 3,500,000,000 years ago, Monera is comprised of bacteria and blue-green algae. About 2700 living species have been identified. All monera are prokaryotes—that is, 1-celled organisms that lack a membrane-bound nucleus and membrane-bound organelles. The genetic material is DNA; it is found in the center of the cell as a circular double-stranded molecule. Most monera are heterotrophs, requiring organic materials—either living or dead—for food. Some are autotrophs, able to make food by either photosynthesis (using light as the energy source) or chemosynthesis (obtaining energy by oxidizing inorganic substances). Monera usually reproduce asexually by fission or budding.

mongoose (*animals*) Small carnivorous mammals native to Eurasia and Africa—and widely introduced elsewhere—mongooses are fierce hunters, feeding mainly on rodents and snakes. The Indian mongoose (*Herpeses edwardsi*) is famed for its quick reflexes; it is able to seize and crack the skull of a cobra before the snake can bite and inject its venom.

The mongoose family, Herpestidae, also includes meerkats. Family members generally have a small head with a pointed snout, a long, slender body, and a long tail; an anal gland secretes a foul-smelling substance.

monitor (*animals*) Lizards of the genus *Varanus*, commonly called monitors, range from species 8 in. (20 cm) long to the Komodo dragon (*V. komodoensis*)—the world's largest lizard, reaching more than 9 ft (2.7 m) in length and 200 lb (90 kg) in weight. Monitors have a strong jaw, strong legs with sharp-clawed toes, and a long slender tail. All are daytime predators with an acute sense of smell. They are native to Africa, southern Asia, the Indo-Pacific islands, and Australia.

monkey (*animals*) Primates known as monkeys comprise 2 groups. Old World monkeys (Family Cercopithecidae) are distributed from Gibraltar in southern Europe throughout Africa and Asia. They include macaques, mangabeys, baboons, langurs, proboscis monkeys, and many others. New World monkeys, native to Central and South America, include Family Cebidae, consisting of capuchins, howlers, squirrel monkeys, ukaris, and others; and Family Callithrichidae, consisting of marmosets and tamarins. A major distinction between the 2 groups is seen in the nostrils. Old World monkeys have nostrils set close together and opening forward and downward. New World monkeys have widely spaced nostrils opening to the sides of the nose.

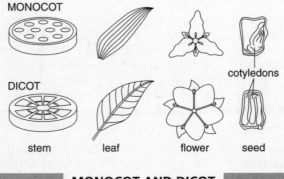

MONOCOT

DICOT

cotyledons

stem   leaf   flower   seed

MONOCOT AND DICOT

monocot and dicot (*botany*) Flowering plants are divided into 2 classes, monocots and dicots. Monocots have a single cotyledon, or seed leaf; dicots have 2 cotyledons. In monocots bundles of vascular tissue are scattered throughout the stem; in dicots the bundles are arranged in a circle. Monocots have leaves with parallel veins; dicot leaves have branching veins. Monocot flowers have parts in multiples of 3; dicot flowers in multiples of 4 or 5. Most flowering plants—from cacti to sunflowers—are dicots. Monocots include grasses, lilies, palms, and orchids.

monoculture (*environment*) The cultivation or growth of a single crop or organism is called monoculture. This agricultural practice is widespread, employed from coffee plantations to catfish farms. Environmental problems associated with monoculture include the increased likelihood of epidemics. For example, large-scale monoculture of banana plants makes it easier for banana wilt and other fungal pests to flourish and cause extensive crop damage; similarly, spruce budworms spread more easily in solid stands of balsam fir than in firs mingled among other species.

monoecious and dioecious plants (*botany*) Plant species that bear both male and female reproductive organs on the same plant are said to be monoecious (from Greek words meaning "1 house"). For example, a pine tree has small male (staminate) cones, which produce pollen, and larger female (pistillate) cones, which produce seeds. Many flowering plants with unisex flowers also are monoecious. The corn plant has a cluster of male flowers that form a plumelike tassel at the top of a stalk and female flowers that form the cob lower on the stalk.

Plant species that bear male and female reproductive organs on separate plants are dioecious ("2 houses"). For example, some junipers bear male cones, others bear female cones. Similarly, flowering plants with unisex flowers may be dioecious. Willows and holly are examples.

monogamy and polygamy (*ethology*) Some animals remain paired for years while others meet solely to copulate. Monogamy is the practice of mating with only 1 member of the opposite sex, at least through 1 mating season. Many bird species are monogamous, including swans and geese, which mate for life. Polygamy involves having more than 1 mate during a breeding season. Usually it is the male who mates with more than 1 female, a practice called polygyny. For example, a male elephant seal (*Mirounga angustirostris*) may have a harem of 20 females. In polyandry, it is the female who has several mating partners at one time. The female wattled jacana (*Jacana jacana*) pairs simultaneously with a harem of males, leaving the males to incubate the eggs and then raise the chicks.

monopole (*particles*) Ordinary magnets and charged particles have 2 poles each, usually called north and south. Theories that attempt to unify all forces and particles in a single mathematical structure almost always contend that at the high energies available during the big bang, particles with 1 magnetic pole, called monopoles, were produced. Such monopoles are thought to be stable, so they should exist today, but attempts to locate them have been unsuccessful so far.

monotreme (*animals*) Reptilelike mammals that lay eggs, monotremes comprise the platypus and echidnas. The name means "single opening," referring to the fact that monotremes, like reptiles and birds but unlike other mammals, have a cloaca.

monsoon (*meteorology*) A monsoon (from the Arabic for "season") is a large-scale wind produced by differences in warming and cooling rates of land and water. In summer land warms faster than water, so winds blow from the ocean toward the land, carrying moisture with them. In winter land cools faster than water and air above continents cools with it, sinking and flowing toward the warmer oceans. The best known example is in India, which depends on the summer monsoon to bring rain.

Small-scale winds with a similar cause are sea and land breezes near ocean shores. Sea and land breezes alternate daily instead of seasonally, with sea breezes, from the sea onto the shore, developing in daytime and land breezes appearing during the night.

Moon (*solar system*)   The Moon is Earth's only natural satellite and, at 238,900 mi (384,400 km) the brightest object in the nighttime sky. The equatorial diameter is 2160 mi (3476 km), but the Moon is slightly egg-shaped. Because of this shape. the same side of the satellite always faces Earth. Thus, the Moon rotates once during each revolution. The Moon is airless, waterless, and devoid of life. Temperatures range up to 273°F (134°C) on the bright side to –274°F (–170°C) on the unlit side. The surface is pockmarked with craters up to 56 mi (90 km) across caused by bombardment by meteoroids. The near side also has large regions (called maria, or seas) of solidified lava. Lunar regolith, a mixture of fine powder and broken rock, blankets the surface. (*See also* **Phases of Moon**.)

moose (*animals*)   The largest member of the deer family, the moose (*Alces alces*) is native to northern forests. Its common name, meaning "eater of twigs," comes from the Algonquin and refers to the animal's vegetarian diet. In Europe and Asia this animal is called the elk. The moose is a huge animal, with males attaining a shoulder height of 6.5 ft (2 m) and weight exceeding 1600 lb (725 kg). The male's head is adorned with a rack of antlers that can reach 6 ft (1.8 m) in width and 50 lb (22.7 kg) in weight. Moose have a long nose, a large shoulder hump, long legs, a 2-layered insulating coat, and excellent senses of smell and hearing.

moraine (*geology*)   All rock material transported by a glacier, ranging from boulders to sand, is called moraine. Deposits of such material in long piles are moraines. A glacier pushes moraine into strips along its edges, called lateral moraines, and builds a pile in front called a terminal moraine. Moraine frozen in the bottom of the ice is called the ground moraine. When the glacier melts, the moraines remain, often becoming hills on land or islands near the coasts of continents.

morel (*fungi*)   Members of the sac fungi, morels (genus *Morchella*) have a fleshy, hollow fruiting body 2 to 4.5 in. (5 to 10 cm) tall, with a thick stalk and a cap that is honeycombed with deep pits and raised ridges. Morels commonly appear in spring, often on burnt land. They are considered great delicacies; indeed, people in Europe used to set fire to woodlands in hopes of creating a bountiful crop of morels. False morels (genus *Gyromitra*) look somewhat similar, but with a wrinkled—not pitted—cap; they are poisonous.

Morgan, Thomas Hunt (*geneticist*)   [American: 1866–1945] Working mainly with the small fruit fly *Drosophila melanogaster*, Morgan concluded that genes are arranged in a line on chromosomes, and that chromosomes are present in the nucleus of every living cell. Morgan also noted occasional recombinations of inherited characteristics and proposed that this results from an exchange of genes between the 2 chromosomes of a pair, a process he named crossing over. Morgan drew chromosome maps with relative locations of genes. He was awarded the 1933 Nobel Prize in physiology or medicine.

morning glory (*plants*)   The morning glory family, Convolvulaceae, comprises some 1500 species. Most are weak-stemmed herbs that trail along the ground or climb by twining around other vegetation. Their flowers have 5 petals that are partially united, forming a bell or funnel. The common morning glory (*Ipomoea purpurea*), native to the American tropics, is an annual that grows to 13 ft (4 m) long and usually has purple blooms. The sweet potato (*I. batatas*), also native to the American tropics, has a large edible underground stem, or tuber. The family also includes bindweeds (*Convolvulus*).

morning star (*solar system*)   Venus and Mercury, the planets that orbit the Sun inside Earth's orbit, must always be in about the same direction from Earth as the Sun. Venus and Mercury can only be seen when the Sun is a small distance below the horizon. Before sunrise, the planets appear as "morning stars" (planets appear, like stars, as bright spots of light in the night sky, although planets usually twinkle less). Venus is so much brighter than Mercury or than any real star that it is known as *the* morning star. A morning star soon vanishes in the light of the rising Sun.

morphology (*sciences*)   Morphology is the study of the shape, appearance, and architecture of organisms. It is concerned with how organisms are built and how all the parts work together. Major divisions of this science include gross morphology, which examines structures that can be seen with the unaided eye, such as muscles and bones; histology, which looks at tissues and how they are built into organs; and cytology, which focuses on the structure and components of individual cells. The science of anatomy, which deals with the structure of organisms, can also be considered part of morphology.

Moseley, Henry Gwyn Jeffreys (*physicist*)   [English: 1887–1915] Moseley used X-ray diffraction in 1913 to show that each element has an atomic number. With this tool he was able to predict the 6 remaining gaps in the periodic table and to resolve identities of rare earths.

mosquito (*animals*)   Constituting Family Culicidae of the fly order of insects, mosquitoes have long, slender antennae; sucking mouthparts; and scales on the body and wings. Only the females have mouthparts designed for piercing, and only they suck blood. Mosquitoes are important transmitters of bird and mammal diseases, including malaria, yellow fever, dengue, and encephalitis. The larvae are aquatic and found in both fresh and salt water.

moss (*plants*)   True mosses constitute a group of small bryophyte plants that typically grow in dense carpets, usually in moist, shady areas. They frequently are pioneer species on bare rock and soil. Mosses are very sensitive to air pollution, particularly sulfur dioxide; their abundance in an area can serve as an indicator of pollution levels.

Various plants other than true mosses also are commonly called mosses. Certain red algae are called sea mosses, a large lichen is called reindeer moss, *Lycopodium* are called club mosses, and a bromeliad is called Spanish moss.

motmot (*animals*)   Beautifully colored birds of the American tropics, motmots (Family Momotidae) are named after the sounds made by the common motmot (*Momotus momota*). Motmots have a broad curved beak and a long tail that often is wagged from side to side. Motmots hunt perched on tree branches, swooping down to grab spiders, insects, and small lizards.

mountain (*geology*)　A mountain is usually thought of as a big hill with steep sides, which is true, but there are several entirely different processes that result in mountains. Volcanoes produce mountains by depositing ash or lava or both on top of existing rock. Where an ocean plate moves under a continent, it creates not only volcanoes but also batholiths that push up the overlying rock. Smaller intrusions called laccoliths can raise mountains far from plate boundaries. Pushing rock layers together can fold existing rock into mountains; pulling rock apart can result in mountain ranges caused by faulting. Fault-block mountains have steep sides along the fault, gentler slopes on the other side.

mouse (*animals*)　A variety of small rodents are called mice. They typically have a somewhat pointed head and a long tail covered with scales instead of hair. Among the best known and most widely distributed is the house mouse (*Mus musculus*), native to Europe, Asia, and Africa; albino varieties are bred as pets and laboratory animals.

Mousterian (*paleoanthropology*)　Stone points and blades flaked from a prepared core (known as Levallois technology) began to appear about 200,000 years ago in Africa, Europe, and the Near East, usually associated with Neandertals. They were part of the long-lived Mousterian tool industry that also included some hand axes and many stone tools known as scrapers. This industry disappeared about 35,000 years ago, as did the Neandertals.

mouth (*anatomy*)　Food enters an animal's body through the mouth, the anterior opening of the gastrointestinal tract. Mechanical digestion, such as chewing and grinding, begins here; in animals with salivary glands, chemical digestion also begins. Tentacles or specialized mouthparts surround the mouth of many invertebrates—jellyfish and bees, for example—to aid in food gathering. Among vertebrates the mouth typically contains teeth (birds being the major exception) and a tongue.

mouthpart (*anatomy*)　In insects and other arthropods, the mouth is surrounded by paired, jointed appendages called mouthparts that help obtain and manipulate food. There is great diversity of form and structure of mouthparts. They may be modified for chewing and grinding grass (as in the grasshopper), elongated for piercing skin and sucking blood (horsefly), and so on.

mucous membrane (*anatomy*)　Cavities and tubes in the vertebrate body that open to the outside are lined with a modified skin tissue called mucous membrane, or mucosa. The gastrointestinal tract plus respiratory, urinary, and genital passages are lined with mucosa. Specialized goblet cells within the membrane produce a lubricating fluid called mucus, which is secreted onto the membrane's surface.

mucus (*physiology*)　Specialized cells in the mucous membranes of vertebrates secrete a thick, slimy fluid called mucus, which functions as a lubricant, moisturizer, and defense against bacteria and other foreign particles. Mucus moistens inhaled air and traps smoke in the nose and windpipe, preventing it from reaching the lungs. It lubricates food and makes swallowing easier and coats the inner surface of the stomach, protecting the stomach from digestive juices. Some invertebrates and plants also secrete slimy or sticky fluids that are called mucus. For example, a snail moves over a track of mucus secreted by a gland below the mouth.

Muir, John (*naturalist*)   [American: 1838–1914] A lifelong traveler who recorded his observations of nature along the way, Muir was the first to show that the Yosemite Valley in California was formed by glacial erosion; he also discovered Glacier Bay in Alaska, including a glacier that now bears his name. Muir's most important contributions were in the area of wildlife and wilderness preservation. Through his efforts, the U.S. Congress established Yosemite and Sequoia national parks in 1890. In 1892 Muir and his supporters founded the Sierra Club, a conservation organization.

mulberry (*plants*)   The mulberry family, Moraceae, comprises about 1400 species of trees, shrubs, climbers, and herbs—including breadfruit, figs, osage orange (*Maclura pomifera*), and castilla rubber trees (*Castilla*). Family members generally have a milky sap, simple leaves, and small unisexual flowers borne in clusters. Various species of mulberry trees (*Morus*) are cultivated as ornamentals and for their edible fruits. The white mulberry (*M. alba*), a native of Asia, has been cultivated for more than 5000 years for its leaves, which are fed to silkworms.

Muller, Hermann (*geneticist*)   [American: 1890–1967] A student of Thomas Hunt Morgan, Muller was the first to identify a mutagen (mutation-inducing agent). In 1927 he showed that the rate of mutations in fruit flies (*Drosophila*) could be greatly increased by exposing the flies to X rays. Because most of the mutations he produced were harmful, he stressed the dangers of human exposure to radiation. Muller was awarded the 1946 Nobel Prize in physiology or medicine for his work.

Müllerian mimicry (*zoology*)   First described by Fritz Müller [German: 1821–97] in 1879, Müllerian mimicry involves 2 unrelated species that live in the same area, resemble one another in appearance, and are both unpalatable, thought to reinforce the effects of warning. For example, the monarch butterfly (*Danaus plexippus*) and viceroy butterfly (*Limenitis archippus*) have similar coloration that warns would-be predators that they are distasteful. It has long been known that monarchs are distasteful, but it was originally thought that viceroys were palatable and that the similar appearance was a case of Batesian mimicry. In 1991, however, it was shown that viceroys are just as unpalatable as monarchs.

mullet (*animals*)   Constituting Family Mugilidae, mullets are moderate-size fish seldom exceeding 2 ft (0.6 m) in length. They inhabit coastal salt and brackish waters in tropical and temperate seas around the world. Mullets are bottom feeders, eating mainly organic debris. They have a thick, streamlined body, no lateral line, 2 dorsal fins, the first containing stiff spines, and a forked tail.

An unrelated species native to the Mediterranean is the red mullet (*Mullus barbatus*), which belongs to the goatfish family, Mullidae. Like all goatfish, it has a pair of long fleshy barbels (feelers) under its chin, used to find small invertebrate prey on the sea bottom.

multiplication (*arithmetic*) For natural numbers, the operation of multiplication can be understood as repeated addition, with one number (the multiplicand) added as many times as a second number (the multiplier) to produce the answer, called the product: 4×3 means $4 + 4 + 4$. The multiplicand and multiplier are also known as factors. Natural number multiplication can also be interpreted as the number of ways that 2 sets can be matched; for example, the members of {a, b, c} with 3 members and {D, E,}with 2 can be matched as {aD, bD, cD, aE, bE, cE}, a set with 3×2 members. The repeated-addition model hints at how to extend multiplication to numbers expressed as fractions; 4×3 is 3 of 4 (things) so 1/2 of 4 is interpreted as $1/2 \times 4$. A careful definition of multiplication for all numbers is based on the idea that any product of irrational numbers can be approximated as closely as needed with rational numbers.

muon (*particles*) The muon is a higher energy analog to the electron—sometimes called a "fat electron." The muon and its neutrino form the second generation of leptons. Although muons are not part of ordinary matter, they are the lightest independent particles from higher energy levels, so they are easily produced in cosmic ray collisions or even by interactions of fast neutrinos. Because they are charged, they are easy to detect. Despite their role as one of the workhorses of particle physics, no one has offered a good reason for them to exist at all.

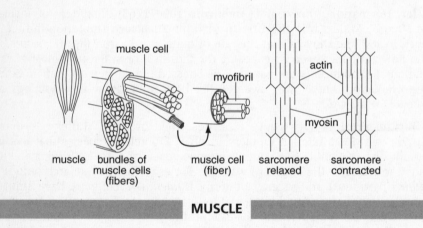

muscle / bundles of muscle cells (fibers) / muscle cell (fiber) / muscle cell / myofibril / actin / myosin / sarcomere relaxed / sarcomere contracted

MUSCLE

muscle (*anatomy*) Muscle is an animal tissue or organ that has the ability to contract. Every animal movement—running, flying, breathing, moving blood through arteries, pushing a fetus from the uterus, and so on—depends on the contraction of muscles. Humans and other vertebrates have 3 types of muscle tissue: skeletal muscle, which is attached to and moves bones; smooth muscle, which occurs mostly in the walls of internal organs; and cardiac muscle, which makes up the walls of the heart.

The basic unit of a muscle is the myofibril, which consists of a number of sarcomeres arranged end to end. Within a sarcomere are longitudinal filaments of the proteins actin and myosin. According to the generally accepted sliding-filament theory, contraction occurs as the actin and myosin filaments slide past each other.

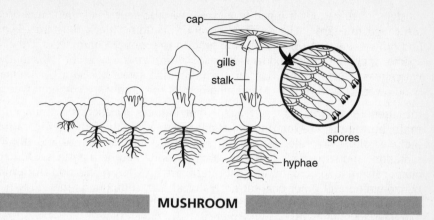

cap

gills

stalk

spores

hyphae

MUSHROOM

mushroom (*fungi*) Club fungi that produce a fruiting body with a stalk and a cap are called mushrooms or toadstools. The vegetative body (mycelium) of the mushroom is a mass of tiny threads (hyphae) growing in the soil, under leaf mold, or in tree bark. The fruiting body is the familiar mushroom, or toadstool. It forms as a result of the fusion of 2 hyphae, either from the same mycelium or from mycelia of 2 organisms of the same species. The fruiting body starts as a small button but grows rapidly and soon releases mature spores from the underside of the cap. Many species of mushrooms are edible, but others are poisonous. Particularly notorious are certain *Amanita* species, which produce toxins that often are fatal to humans.

musk (*biochemistry*) The male musk deer (*Moschus moschiferus*) of central Asia has an abdominal gland—the musk pod—that produces musk during the rutting season, a period of sexual excitement in males. Musk is a greasy, highly fragrant secretion; the odor results from the presence of the organic compound muscone. People use musk in making perfume.

muskrat (*animals*) Native to North America but introduced elsewhere, the muskrat (*Ondatra zibethica*) is a large rodent with a dense undercoat of fur covered by coarse guard hairs. It is 9 to 14 in. (23 to 36 cm) long; the hairless tail is flattened from side to side and up to 11 in. (28 cm) long. The muskrat spends most of its time in or near water. Adaptations for swimming include the partially webbed hind feet and the tail, which serves as a rudder. Musk secretions from anal glands give the animal its name.

mussel (*animals*) Members of the mollusk class of bivalves, mussels are common clamlike animals in both fresh and salt water. Freshwater mussels include the European zebra mussel (*Dreissena polymorpha*), which has become a serious pest in North America. Marine mussels often hang in masses from wharves and rocks, attached by means of byssus threads. They include edible species of genus *Mytilus*, easily recognized by blue-black shells.

mustard (*plants*) The mustard family, Cruciferae ("cross bearer") or Brassicaceae ("cabbagelike"), is comprised of more than 3300 species, most of them herbs native to temperate regions. The small flowers, usually borne in loose clusters, have petals arranged in a cross. Various species of the genus *Brassica* are called mustards. The seeds of several, including black mustard (*B. nigra*) and brown mustard (*B. juncea*), are ground to produce condiments. Other food species include cabbage, horseradish, radish (*Raphanus*), wasabi (*Wasabia*), and watercress (*Nasturtium*). Garden plants include sweet alyssum (*Alyssum*) and wallflowers (*Cheiranthus*). Many members of the family are common weeds.

mutagen (*biochemistry*)   Any agent that induces mutation is a mutagen. The main mutagens are ionizing radiation and chemicals. Mutagenic radiation includes shortwave electromagnetic radiation, such as X rays and ultraviolet radiation, and subatomic particles, such as alpha and beta particles. Mutagenic chemicals include compounds that interfere with DNA replication (for example, benzopyrene and other coal-tar derivatives) and compounds that directly alter DNA (for example, nitrous acid and formaldehyde). Many mutagens are also carcinogens (agents that cause cancer).

mutation (*genetics*)   Any change in a gene that can be passed on to daughter cells during cell division is a mutation. Mutations that occur in somatic (body) cells usually are of consequence only to the affected individual. Mutations in germ (sex) cells can be transmitted to subsequent generations. Mutations result from mutagens (mainly chemicals and radiation), and errors (such as crossing over) during replication of nucleic acids or cell division. Mutations are the source of the great variety among organisms on Earth, and the basis of evolution. Some mutations benefit organisms, making them better adapted to environmental conditions. Many mutations have no effect on organisms. And some mutations, such as those leading to cancer, are harmful or even lethal.

mutualism (*ecology*)   The symbiotic relationship in which organisms of 2 different species live together and benefit is called mutualism. The relationship may be permanent or temporary and intermittent. Wood-eating termites have a mutualistic relationship with protozoa that inhabit their intestines and produce an enzyme needed to digest cellulose. These protozoa break down enough cellulose to satisfy both their own needs and those of their hosts, who lack the enzyme. Temporary mutualism occurs between the rhinoceros and starling-sized birds called oxpeckers. The rhinos get a cleaning as the oxpeckers eat ticks and other pests that infest their hide. Furthermore, the oxpeckers act as an early warning system, uttering a shrill alarm when danger approaches.

mycorrhiza (*fungi*)   Mycorrhiza is an example of mutualism. It refers to the association that exists between certain soil fungi and members of most plant families. Hyphae (filaments) of a fungus surround a plant's roots and in some cases actually penetrate the roots. Other hyphae extend outward over a wide area of soil. In effect, the fungus acts as a second root system, absorbing water and dissolved minerals for the plant. In return, the fungus obtains sugar and other food materials from the plant.

myelin (*anatomy*)   The axons of many neurons (nerve cells) are fibers surrounded by Schwann cells, which produce a pearly white substance called myelin. Composed of protein and fat, myelin protects the axons and acts as an electrical insulator, ensuring that impulses move rapidly along.

White matter of the brain and spinal cord is made up of masses of nerve fibers covered by myelin sheaths; gray matter of the brain and spinal cord consists of unmyelinated nerve fibers.

myoglobin (*biochemistry*)   Related to hemoglobin, myoglobin is a vertebrate protein found mainly in cardiac and skeletal muscles. It gives the muscles their color. Myoglobin's function is to bind with oxygen, storing it until the muscle cells need it for respiration. Mammals such as whales and dolphins, which dive deep into the sea and remain submerged for long periods of time, have particularly abundant supplies of myoglobin. The atomic structure of myoglobin was determined in 1960 by John Kendrew [English: 1917–97] using X-ray diffraction. This led to the determination of the structure of the much larger hemoglobin molecule by Kendrew's colleague Max Perutz [Austrian: 1914–]. The two men shared the 1962 Nobel Prize in chemistry.

myrtle (*plants*)   Native mainly to the subtropics of South America and Australia, myrtles (*Myrtus*) are fragrant evergreen shrubs often grown as ornamentals. The best-known species, *M. communis*, native to Mediterranean regions, grows to 10 ft (3 m) or more in height. In Greek mythology, myrtle was sacred to Aphrodite, the goddess of love and beauty. The myrtle family, Myrtaceae, includes many shrubs and trees that produce aromatic oils, among them allspice, eucalyptus, guava, Jamaica bayberry, and the clove tree (*Syzygium aromaticum*), whose dried flower buds are called cloves.

nacre (*biochemistry*)   The "mother-of-pearl" lining of a mollusk shell is called nacre. It is formed by the mantle, a thin fold of epithelium tissue that surrounds the organs of the mollusk's body. The nacre consists of layers of calcium carbonate; the pearly or iridescent effect results from the arrangement of the tiny calcium carbonate crystals and the reflection of light from them. Pearls also are made of nacre.

nail (*anatomy*)   Four-legged terrestrial vertebrates have a thin, hard nail covering and protecting the upper tip of each finger and toe. Originating from the skin, nails are composed largely of keratin. Nails grow continuously from the base to compensate for wear. In some animals, nails are sharp and curve downward; these nails are called claws or, in eagles and other birds of prey, talons.

nanotube (*materials science*)   Structures of pure carbon formed in long hollow tubes about a nanometer (0.00000004 in. or 0.000001 m) in diameter are called nanotubes; they may be 10 to 100 nanometers long. Nanotubes are similar to buckminsterfullerene molecules that have not closed. Their unusual electrical properties may lead to important applications.

nasal cavity (*anatomy*)   Vertebrates have a pair of openings on the head called nostrils or external nares (singular: naris). These nares lead from the exterior to the nasal cavity, which contains organs of smell. Fish have only external nares. Air-breathing vertebrates also have internal nares, through which air passes between the nasal cavity and the oral cavity (and, ultimately, to and from the lungs) in respiration. The nasal cavity is lined with mucous membrane that bears numerous tiny hairs. As air moves through the nasal cavity from the outside, it is warmed and moistened; dust particles are trapped by the hairs and mucus to prevent them from entering the lungs.

natural number (*numbers*)   The natural numbers, also called counting numbers, are 1, 2, 3, . . . , with 0 sometimes included. Often n is used as a variable to represent any natural number. An expression such as $2n - 1$ can describe a whole sequence of numbers—for example, when $n = 1$, then $(2 \times 1) - 1 = 1$; when $n = 2$, then $(2 \times 2) - 1 = 3$; etc. Natural numbers can be defined either in terms of the number of members in a set or as the numbers formed by starting with 1 (or 0) and adding 1 for each succeeding number.

natural resource (*environment*)   Materials that occur naturally and can be used by humans for various purposes are considered natural resources. Renewable resources are ones that, if managed properly, can be replaced.

Food and timber are examples. Nonrenewable resources are ones that cannot be replaced and can eventually be exhausted. Metal ores and fossil fuels, such as coal and natural gas, are examples.

natural selection (*evolution*)   The major force in evolution is natural selection, a process first proposed by Darwin after he observed that farmers improved yields by artificially selecting certain cows or corn plants for breeding. Natural selection recognizes that there is great variety of inherited traits among members of a species. Organisms with the most favorable traits have a survival advantage and thus are more likely to reproduce and pass on their traits to the next generation. The process has been referred to as "survival of the fittest." Natural selection can take thousands of years to produce a significant change in a species. Or changes can happen very quickly, as witnessed by the emergence of strains of bacteria resistant to antibiotics and insects resistant to pesticides.

nautilus (*animals*)   Members of the genus *Nautilus* are mollusks with a coiled brown-striped shell. The shell's interior is separated into a series of progressively larger chambers. The animal occupies the newest, biggest chamber, protruding its tentacles into the sea to catch shrimp and other prey. The abandoned chambers are filled with gas; a tube running from those chambers to the nautilus allows the animal to control its buoyancy. Nautiluses live in deep waters of the South Pacific and Indian oceans.

navigation (*ethology*)   Many animals are able to find their way, or navigate, from one specific place to another. A petrel flies out to sea, then returns to its nest; a green turtle engages in migration between feeding areas off Brazil and breeding grounds on Ascension Island; *Anguilla* eels hatch in the Sargasso Sea, then journey to European rivers. Animals use various methods to determine position and course. Salmon rely on their sense of smell to locate their stream of origin. Bees use the Sun as a reference point. Nighttime migrators such as warblers depend on the stars. Bats use echolocation.

Nazca Plate (*geology*)   The lithosphere of the Pacific Ocean to the west of South America, called the Nazca Plate after the Nazca high desert region of Peru, plunges under the edge of the continent, raising the volcanic Andes mountains and producing very strong earthquakes. The plate is also bordered by spreading centers to the south (the Chile rise) and west (the East Pacific rise) and by the Cocos Plate to the north.

Neandertal (*paleoanthropology*)   The first early hominids to be recognized were the Neandertals, named for skeletons first found in 1856 in a cave in the Neander River Valley (*Neander Tal*) in Germany. (The traditional English spelling, Neanderthal, has been abandoned by most paleoanthropologists.) While most Neandertal fossils are from Europe, the Neandertals also inhabited the Near East and Africa, in some places side by side with archaic *Homo sapiens*. The Neandertals were the only hominid inhabitants of Europe at the height of the ice ages, flourishing some 200,000 to 35,000 years ago. Neandertals used tools similar to those of anatomically modern humans, but with their own distinct traditions. They also followed similar customs, such as burial of the dead.

Neandertals disappeared abruptly about the same time as or shortly after early modern humans, or **Cro Magnons**, arrived in Europe.

Limited DNA studies indicate that the Neandertals formed a separate species, *Homo neandertalis*, characterized in part by a strong, stocky body, a sloping forehead with a prominent brow ridge, and a brain that is on the average larger than that of *H. sapiens*. Other evidence, however, suggests that the Neandertals were a subspecies of modern humans with Ice Age adaptations or that they interbred with Cro Magnons.

nebula (*astronomy*)   Nebulas (or nebulae, meaning "clouds") are observable patches of gas and dust. Stars are points of light, even through a telescope. Galaxies and nebulas have definite shapes and sizes, but galaxies can be resolved into individual stars with good telescopes (some older sources still use the name *nebula* to refer to galaxies). Planetary nebulas are usually spheres of gas emitted from single stars. Gas nebulas emit light, fluorescing in response to radiation from stars. Some dust nebulas also glow, reflecting light from nearby stars. Other dust nebulas block out part of the sky, but are observable when glowing gas surrounds opaque dust or vice versa.

nectar (*botany*)   Many types of plants have glands called nectaries that secrete a sugary liquid—nectar—that is gathered and eaten by various insects and other animals. Some plants have nectaries on their leaves, but usually nectaries are modified parts of a flower. As an animal visits the flower and laps up the nectar, pollen grains cling to its body, to be transferred to the next flower it visits.

negative number (*numbers*)   When the real numbers are arranged in order on a line from least to greatest, all the numbers less than 0 (on the opposite side of 0 from 1) are negative numbers. Notice that imaginary numbers and complex numbers that are not also real are never negative. Although the opposite of $2i$ is $-2i$, $-2i$ is not a negative number.

neighborhood (*mathematics*)   A neighborhood is a region that contains an element of a set (such as a point or a number) and other elements that are in some sense nearby.

It is a generalization of the idea that all the points in a small circle (or other closed curve) around a point p in the plane are in the neighborhood of p. Similarly, for a number n the set of points that have a distance less than d from n are in n's neighborhood. The concept of neighborhood of a number is the basis of the definition of limit for calculus. Several different types of neighborhoods defined specifically for sets are the foundation of the theory of topological spaces, an important branch of modern mathematics.

nekton and benthos (*ecology*)   Animals that live in aquatic environments can be grouped into 3 general categories, nekton, benthos, and plankton. Nekton are strongly muscled and can swim against water currents. They include fish, whales, and squid. Benthos are bottom dwellers. They are mainly crawlers or sessile creatures, living in burrows or attached to rocks, seaweed, and other substrates. Sponges, corals, starfish, clams, snails, lobsters, and polychate worms are examples of benthic animals.

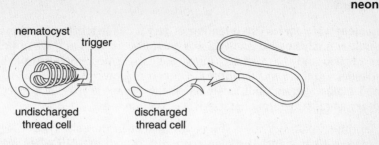

nematocyst   trigger

undischarged   discharged
thread cell   thread cell

NEMATOCYST

nematocyst (*cells*)   A structure unique to jellyfish, sea anemones, and other cnidarians is the thread cell, or cnidoblast. This cell has an organelle called a nematocyst, which is a bladder containing a coiled, slender thread. When something touches the thread cell's trigger, the thread explodes from the bladder. Some threads are sticky and entangle prey. Others have numerous tiny spines that penetrate prey. A poison may also be injected into the prey around the point of contact.

Thread cells are concentrated on a cnidarian's tentacles and are important in capturing and subduing prey. Some animals that feed on cnidarians, such as certain flatworms, are able to incorporate nematocysts into their own bodies. The nematocysts migrate to the surface of their new "owners" and function exactly as they did in the cnidarians.

neodymium (*elements*)   Neodymium, atomic number 60, is the least rare of the rare earths in the crust, more common than iodine but somewhat less abundant than lead. It was discovered in 1885 by Karl Auer (Baron von Welsbach) [Austrian: 1858–1929]. The name, from the Greek for "new twin," refers to its recognition, along with praseodymium, as 1 of 2 elements in a substance named "twin" because it appeared to contain an element very much like lanthanum. Neodymium, with its twin, is used primarily in coloring special glasses for technical applications. Chemical symbol: Nd.

Neolithic Revolution (*paleoanthropology*)   Gordon Childe [Australian: 1892–1957] named the period immediately following the Paleolithic Age the Neolithic Revolution. This time, about 11,000 years ago, at the end of the most recent Ice Age, is marked by the beginning of the Agricultural Revolution as well as many changes in tools. Large polished or ground stone tools, such as axes, began to replace microliths. Ceramics were improved and used to make pots and building materials. Permanent villages became more common and some grew into cities. The Neolithic period extended for about 3000 to 5000 years, depending on the region, and is considered to have ended when copper tools were regularly produced.

neon (*elements*)   Neon is a noble gas, atomic number 10, discovered in 1898 by Alexander Ramsay [Scottish: 1852–1916] and Morris William Travers [English: 1872–1961]. It was named from the Greek *neo*, "new." Because neon produces a strong reddish orange glow in response to an electric current, it has become the generic name for gas-discharge signs, whether using neon or not. Chemical symbol: Ne.

neoteny (*biology*)   The retention of larval characteristics in a sexually mature adult animal is called neoteny. For example, the axolotl ("water doll" in Aztec) of Mexico and the western United States is a neotenic form of the tiger salamander (*Ambystoma tigrinum*). Though it grows to lengths of 15 in. (37 cm) and mates like an adult salamander, the axolotl keeps black external gills and other larval structural characteristics its entire life.

Neptune (*solar system*)   The eighth planet in the solar system, Neptune is the outermost gas giant. The planet was discovered in 1846 when irregularities in the orbit of Uranus provided astronomers with its correct location. Neptune derives its pale bluish color from its hydrogen and helium atmosphere. Beneath the atmosphere, Neptune may have an ocean of liquid hydrogen and helium, a mantle of solidified gases and water ice, and a hot, rocky core some 15 times as massive as the planet Earth. One unresolved mystery is that Neptune radiates 2.7 times as much heat as it receives from the Sun. Like the other gas giants it has rings encircling it.

neptunium (*elements*)   The first of the radioactive actinide metals to be synthesized, neptunium, atomic number 93, was produced from uranium, atomic number 92, in 1940 by Edwin McMillan [American: 1907–91] and Philip Abelson [American: 1913–]. It was named for the planet Neptune, the first beyond Uranus. Since neutron capture occurs in uranium ores, traces of neptunium can be found in nature, although the longest-lived isotope has a half-life of about 5000 years. Chemical symbol: Np.

nerve (*anatomy*)   Located outside the central nervous system, a nerve is a bundle of nerve fibers with associated connective tissue and blood vessels, all enclosed in a sheath of connective tissue. Afferent nerves carry impulses toward the brain and spinal cord; efferent nerves carry impulses away from the brain and spinal cord; mixed nerves contain both afferent and efferent fibers. Some nerves, such as the olfactory nerve, which carries smell impulses to the brain, are comparatively short. Others, such as the sciatic nerve, which begins at the pelvis and has branches that extend down the leg to the foot, are very long.

nervous system (*anatomy*)   The purpose of an animal's nervous system is to control bodily functions—that is, to tell structures what to do and when to do it. In invertebrates such as jellyfish the nervous system is very primitive, consisting of a simple network of nerve cells. At the other extreme are the nervous systems of vertebrates, controlled by a brain that reaches its height of complexity in humans. The vertebrate nervous system has 2 basic parts: the central nervous system, made up of the brain, spinal cord, and nerves that come from them; and the peripheral nervous system, comprised of nerves that connect the central nervous system with various parts of the body.

nest (*zoology*)   A bird's nest is the place where it lays its eggs and raises its young. The nest may be a shallow depression in the ground or an elaborate structure built of mud, grasses, sticks, saliva, and other materials. Hummingbird nests may be only 1 in. (2.5 cm) across and 1 in. (2.5 cm) deep. In contrast the white stork (*Ciconia ciconia*) builds a nest up to 5 ft (1.5 m) across and 6 ft (1.8 m) deep, weighing a ton or more.

Other animals also prepare nests, which may serve as a home not only for eggs and the immature but also for adults. Gorillas, for example, build nests of leaves and branches in which to sleep at night.

nettle (*plants*)   Nettles are herbs found in temperate regions. They range in height from a few inches to about 10 ft (3 m). When a person or animal brushes against a nettle (*Urtica*), sharp leaf hairs penetrate the skin and discharge a histamine-laden juice that causes a burning sensation. Some members of the nettle family, Urticaceae, particularly ramie (*Boehmeria nivea*), are cultivated for their tough, silky fibers.

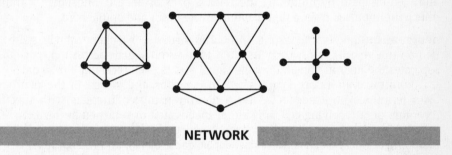

NETWORK

network (*geometry*)   Figures called networks can be formed by connecting distinct points by line segments or arcs. The kind of connection does not matter in topology, where shapes and sizes do not count. The points are called the vertices of the network while the connections are sometimes called edges, terminology inspired by networks that define polyhedrons in 3-dimensional space. In a network all vertices are connected—you can travel via the edges to any vertex from any other—and edges do not intersect except at vertices.

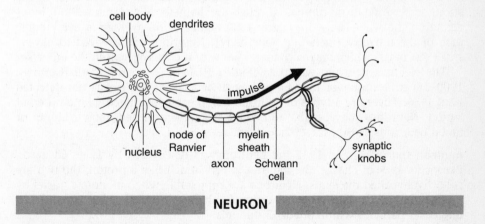

NEURON

neuron (*cells*)   The basic unit of an animal's nervous system is the neuron, or nerve cell. Its function is to transmit impulses. Neurons come in many shapes and sizes, but typically have 3 main parts: cell body, dendrites, and axon. The cell body contains the nucleus and carries out most cell metabolism. Dendrites pick up impulses from other cells and carry them toward the cell body. The axon—often enclosed in an insulating myelin sheath—conducts impulses away from the cell body. The axon does not actually touch the dendrites of other neurons; the cells are separated by a gap called a synapse.

neuropeptide (*biochemistry*) Neuropeptides are neurotransmitters that play various roles in a vertebrate's awareness of and sensitivity to pain. They include endorphins, enkephalins, and substance P. Endorphins suppress pain. They also have been connected to learning, memory, and sexual activity. Enkephalins reduce pain by inhibiting the release of substance P. Substance P transmits pain impulses from pain receptors to the brain and spinal cord.

neurotransmitter (*biochemistry*) A neurotransmitter is a chemical released by a neuron (nerve cell) at a synapse. Its job is to transmit signals across the space that separates the neuron transmitting the signal and the cell receiving the signal.

Neurotransmitters are synthesized by neurons and stored in the neurons' axon terminals until needed. After a neurotransmitter interacts with specific receptors on a receiving cell, it is either inactivated or returned to the axon terminals. Several dozen chemicals are known to function as neurotransmitters, including acetylcholine, norepinephrine, dopamine, serotonin, and neuropeptides.

neutralization (*chemistry*) When an acid and a base are combined, a chemical reaction occurs. If the product formed creates no change when an acid-base indicator is applied, the reaction is called neutralization. For neutralization to occur, the acid and base must be combined in amounts that cancel each other's activity. The result of neutralization is a salt and water.

neutrino (*particles*) Neutrinos—"little neutral ones"—are leptons that interact with other matter with the weak interaction and gravity only, which means hardly at all. Neutrinos are not subject to either the electromagnetic force or the strong force. A neutrino can pass through the diameter of a planet without interacting on the way. There are nearly 1500 neutrinos on average in every cubic inch of the universe (including your body). Neutrinos are believed to have a small rest mass, which may account for some of the missing mass in the universe.

The electron neutrino was predicted in 1930 by Wolfgang Pauli [German: 1900–58] to account for missing energy in beta decay, but it was not detected until 1956. There are 3 types of neutrino in all, each associated with a different lepton. Neutrinos may be able to change from electron neutrinos to muon or tauon neutrinos as they travel though space.

neutron (*particles*) In 1932 the neutron was discovered by James Chadwick. Except for lack of charge, the neutron is very much like a proton; the two are collectively called nucleons. Neutrons are kept stable when in atomic nuclei by interactions with other nucleons. When outside the nucleus, neutrons decay into other particles within about 15 minutes.

neutron star (*stars*) Neutron stars are stars that have collapsed in a violent explosion, such as the birth of a supernova, so that the force keeping electrons apart is overcome. All the neutrons and protons can touch, forming the equivalent of a giant atomic nucleus. A neutron star is electrically neutral because of the charge of the collapsed electrons. Such a star may be only a dozen miles in diameter but still have a mass twice that of the Sun. High rotation rates produced by conservation of angular momentum cause intense magnetic fields and make some neutron stars into pulsars.

newt (*animals*) Newts are semiaquatic salamanders of the family Salamandridae. Their eggs are laid in ponds and streams. An egg hatches into a larva, which after a month or so metamorphoses into a terrestrial phase called an eft. The eft lives on land until it is sexually mature. Then it returns to the water, where it may remain for its entire adulthood.

Newton, Isaac (*physicist/mathematician*) [English: 1642–1727] Newton, among the greatest of mathematicians, corrected views of motion, force, light, and the universe itself that had persisted since the time of Aristotle. He showed how Kepler's laws of planetary motion derived from a universal law of gravity and he clarified and made systematic Galileo's ideas about motion.

Newton first became famous for his work from 1665 through 1676 with light, showing white light is a mixture of the colors. His experiments convinced him that a lens is inherently flawed because of aberration; he invented the reflecting telescope to solve that problem. In 1704 he summarized his research in *Optics*, based on his particle theory for light.

Newton invented the differential calculus in 1665 and the integral calculus the following year. He also developed the binomial theorem, used for calculating powers of any expression with 2 terms. Although he circulated a written account of the calculus in the early 1670s, his first publication of his ideas came with the *Principia* in 1687. Newton also proved results concerning infinite series and developed methods for approximating solutions.

New World and Old World (*ecology*) Although many view these terms as passé, biologists regularly use New World and Old World when describing the origins of organisms. New World refers to the Americas; Old World pertains particularly to Europe but also to western Asia and Africa. Thus rattlesnakes are New World snakes and pythons are Old World snakes.

Another term embedded in biologists' language is neotropical, which means native to the American tropics. For example, macaws are neotropical birds.

niche (*ecology*) The niche of a species consists of its position and interactions in a community. It includes the species' nutritional needs, temperature and humidity requirements, feeding and resting times, methods of escaping detection, and pattern of reproduction. The niche includes competitors, predators, parasites, mates, and offspring. Different species may occupy the same niche in different communities but no 2 species occupy exactly the same niche in the same community. If 2 species compete for the same niche—that is, compete for all their needs—the species that is better adapted eventually crowds out the other. Niches within a community may overlap, however; for example, white-breasted nuthatches (*Sitta carolinensis*) and gray squirrels (*Sciurus carolinensis*) both live in forested regions of the eastern United States, nest in tree hollows, and eat acorns.

nickel (*elements*)   Nickel, atomic number 28, is a member of the iron triad. A transition element, it was discovered in 1751 by Axel Cronstedt [Swedish: 1722–65] in an ore called *Kupfernickel* by German miners ("Old Nick's copper"), who claimed Satan had removed the copper from the ore. In addition to its source in ores, nickel forms about 5 to 20% of iron meteorites. Nickel has become an important part of useful alloys of both iron and copper. Chemical symbol: Ni.

nictitating membrane (*anatomy*)   The nictitating membrane is a transparent piece of skin sometimes called the third eyelid. Found beneath the upper and lower eyelids of many amphibians, reptiles, birds, and mammals, the nictitating membrane cleans the eye, protects it underwater, and keeps it moist in air.

nightjar (*animals*)   Named for the jarring cries of European species, nightjars (Family Caprimulgidae) are 7 to 13 in. (18 to 33 cm) long, with a small beak, large eyes, long pointed wings, and very small feet. They exhibit cryptic coloration, with mottled brown or gray plumage. Nightjars hunt at night, scooping up insects in their wide-open mouth while in flight. A well-known species of North and Central America is the whip-poor-will (*Caprimulgus vociferus*), named after its musical call. Nightjars also are known as goatsuckers, after the mistaken belief that they suck milk from goats.

nightshade (*plants*)   The nightshade family, Solanaceae, comprises some 2800 species of herbs, vines, shrubs, and small trees. Most contain toxic alkaloids; while some parts of a plant may be edible, other parts are deadly. The potato, for example, has edible tubers but poisonous leaves and fruits. Other members of the family include eggplant (*Solanum melongena*), peppers, petunias (*Petunia*), tobaccos, and tomatoes.

niobium (*elements*)   Niobium is a shiny, white, ductile metal, atomic number 41, that occurs with tantalum, from which it is very difficult to separate because of their similar electron structures. Niobium was recognized in 1801 by Charles Hachett [English: 1765–1847], who named it columbium after a name for the United States, then a new country. After metallic niobium was made in 1864, chemists renamed columbium after Niobe, a daughter of Tantalus. Until 1950 both names were used, but that year the official name became niobium, chemical symbol Nb. Niobium is used in various steels, and its superconductivity in magnetic fields makes it useful in powerful electromagnets.

Nirenberg, Marshall (*biochemist*)   [American: 1927–] In 1961 Nirenberg built a strand of messenger RNA composed only of the nucleotide base uracil (U) and showed that the sequence UUU coded for the amino acid phenylalanine. This opened the way for Nirenberg and colleagues to crack the genetic code in 1966, demonstrating that a sequence of 3 nucleotide bases (a codon) determined each of 20 amino acids. Nirenberg shared the 1968 Nobel Prize in physiology or medicine with Robert W. Holley [American: 1922–] and Har Gobind Khorana.

nitrate and nitrite (*compounds*)   Nitrates and nitrites each contain radicals formed from nitrogen and oxygen, but the nitrate radical has 3 atoms of oxygen to 1 nitrogen (NO_3), while the nitrite has only 2. Nitrates provide nitrogen for plant growth and are widely used as fertilizers, especially ammonium nitrate (NH_4NO_3), also the basis of a powerful explosive. Nitrites are used in preserving meat, such as bacon, giving a characteristic reddish color. Lightning produces nitric acid (HNO_3) from nitrogen in air, a nitrate that plants can use. However, most nitrate in the soil must be produced by bacteria from decaying organic remains.

nitrification (*biochemistry*)   The conversion of ammonia to nitrate by certain soil bacteria is called nitrification. Some bacteria, including *Nitrosomonas* and *Nitrosococcus*, convert ammonia into nitrites. *Nitrobacter* species convert nitrites to nitrates, which can be used by green plants to produce proteins.

nitrogen (*elements*)   Nitrogen is a gas, symbol N and atomic number 7, recognized in 1772 by Daniel Rutherford [Scottish: 1749–1819]. Because 4 out of 5 molecules of air are elemental nitrogen, it is the most available element on Earth with about 4,000,000,000,000 tons (3,600,000,000,000 metric tons) floating about us. Although elemental nitrogen does not support life, nitrogen compounds, including all proteins, are essential for life. Ammonia provides nitrate fertilizer for modern agriculture. Nitric oxide (NO) is an important cellular messenger, but nitrogen-oxygen compounds (No_x) are among the main forms of air pollution, causing smog and contributing to acid rain. Nitrogen is named from the Latin for "saltpeter," a compound of nitrogen.

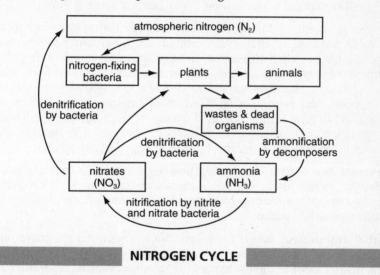

NITROGEN CYCLE

nitrogen cycle (*biochemistry*)   Nitrogen, an essential part of the proteins found in all living organisms, is continually circulated through the environment. Molecular nitrogen in air is abundant but not usable by most organisms. Thanks largely to bacteria, nitrogen is incorporated into compounds that can be used by green plants to make proteins. Animals eat the plant proteins, break them down into amino acids, and recombine the amino acids to form animal proteins.

Human activities affect the cycle in various ways. Industrial processes and the internal combustion engine convert atmospheric nitrogen to ammonia. Nitrate fertilizers added to soil increase crop yields. But when these fertilizers are carried into streams and lakes by rainwater, they can cause a rapid increase in the algae population, resulting in dense concentrations called blooms.

nitrogen fixation (*biochemistry*)   Some microscopic organisms have the ability to convert nitrogen gas into ammonia in a process called nitrogen fixation. The ammonia can then be assimilated by plants (either directly or after further conversion to nitrates) and used to make proteins. Some nitrogen fixers are free-living soil bacteria. Others are symbiotic bacteria that fix nitrogen only when living in close association with a plant. Among the most important are *Rhizobium* bacteria, which form nodules on the roots of beans, clover, and other legumes. The *Rhizobium* provide the legumes with a plentiful supply of nitrogen compounds and in return obtain food from the plants. Certain blue-green algae are important nitrogen fixers in aquatic systems.

nobelium (*elements*)   Nobelium is a radioactive actinide metal, atomic number 102, for which all the isotopes have a half life of 3 minutes or less. It was synthesized in 1958 by Albert Ghiorso [American: 1915–] and coworkers and named indirectly for Alfred Nobel [Swedish: 1833–96], inventor of dynamite. The name was given because the element was at first claimed by scientists from the Nobel Institute in Sweden a year earlier. Eventually Ghiorso's group concluded that the Swedish effort may indeed have been first. Chemical symbol: No.

noble gas (*elements*)   The noble gases, so called after the noble metals but much more resistant to chemical combinations, owe their nobility to a completely filled outer shell of electrons. They are also called inert gases and, sometimes, rare gases. Since they fail to combine naturally with other elements—although chemists have forced a few compounds—argon, helium, krypton, neon, and xenon are found in small amounts in air. Helium, discovered by spectroscopy in the Sun, also occurs in large amounts in natural gas. The remaining noble gas, radon, consists of short-lived radioactive isotopes constantly being created in decay chains of uranium and thorium.

noble metal (*elements*)   The metals that resist corrosion, notably gold, silver, and platinum, were designated noble metals by alchemists. Chemically, however, gold and silver are closer to copper in many ways, while platinum forms a triad with osmium and iridium.

nocturnal and diurnal habits (*ecology*)   Many organisms are active only during a certain part of a 24-hour day. Nocturnal organisms, such as owls and skunks, restrict most of their activities to nighttime. Diurnal organisms, such as hawks and squirrels, are active during daytime. Organisms active mainly at dawn and dusk are called crepuscular; deer and rabbits are examples. Diurnal flowers open during the day and close at night, while nocturnal blooms do the opposite. Structural and physiological adaptations support these various behavioral patterns. For example, the eyes of nocturnal mammals are much more sensitive to light than the eyes of diurnal mammals; nocturnal flowers are likely to use scent to attract pollinators while diurnal flowers often use color.

node (*botany*)   The region on a plant stem where 1 or more leaves arise is the node; the part of the stem between one node and the next is the internode.

Leaf arrangement differs depending on the number of leaves borne at a node. They are alternate if only 1 leaf occurs at a node, opposite if there are 2 leaves, and whorled if there are 3 or more.

node (*waves*)   If 2 waves of equal and opposite wavelength and amplitude pass through each other from opposite directions, they create stationary points called nodes. Waves between nodes, called standing waves, fail to transmit energy. Standing waves may induce other waves, however. For example, a plucked string vibrates with several nodes and induces sound.

Noether, Emmy (*mathematician*)   [German: 1882–1935] Between 1916 and 1933 Noether and her pupils created much of abstract algebra by studying the inner structure of any entity that obeys a few specific rules, such as the group or ring. Early in her career, Noether also created the mathematical version of Einstein's general theory of relativity. As part of this work in 1918 she demonstrated that every mathematical symmetry is equivalent to a conservation law in physics.

noise (*physics*)   Noise is any physical event that occurs at random, such as waves of random sizes or the random motions of molecules. A mixture of sound or electromagnetic waves of random amplitude and frequency is white noise (because white light consists of a mixture of frequencies). Although noise is often undesirable, since 1981 scientists have recognized that noise can also increase the amplitude of regularly occurring events. If a molecule has a feature that acts as a ratchet, noise in the form of Brownian motion can propel the molecule in a specific direction, a principal means of transport in organic cells.

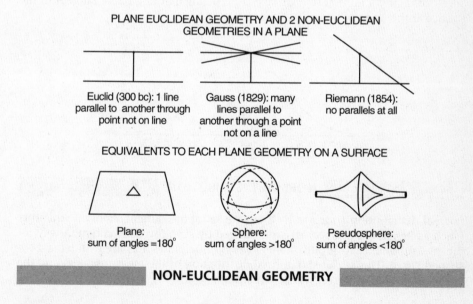

PLANE EUCLIDEAN GEOMETRY AND 2 NON-EUCLIDEAN
GEOMETRIES IN A PLANE

Euclid (300 bc): 1 line parallel to another through point not on line

Gauss (1829): many lines parallel to another through a point not on a line

Riemann (1854): no parallels at all

EQUIVALENTS TO EACH PLANE GEOMETRY ON A SURFACE

Plane: sum of angles =180°

Sphere: sum of angles >180°

Pseudosphere: sum of angles <180°

NON-EUCLIDEAN GEOMETRY

non-Euclidean geometry (*geometry*)   About 1830 a trio of mathematicians (Gauss, Janos Bolyai [Hungarian: 1802–60], and Nikolai Ivanovich Lobatchevsky) independently discovered a self-consistent system of geometry different from that logically established by Euclid about 300 BC, which had been thought to be a description of reality. The basis of non-Euclidean geometry, first recognized by Gauss, who did not publish his discovery, differs from plane Euclidean geometry in that many lines through a point can be parallel to a line not through the point; or, equivalently, in that the sum of angles in a triangle is less than 2 right angles (180°).

Euclid's geometry was based on 5 postulates, of which the fifth concerned parallel lines. The parallel postulate had been suspect since antiquity, with numerous attempts to replace or prove it, some coming very close to non-Euclidean geometry.

In 1854 Bernhard Riemann established general principles of geometry including other non-Euclidean geometries. Mathematicians proved that these are as consistent as Euclidean geometry. Einstein's general relativity theory of 1916 showed that the true geometry of space may be non-Euclidean.

norepinephrine (*biochemistry*)   Also known as noradrenaline, norepinephrine is a neurotransmitter produced by neurons (nerve cells) in parts of the vertebrate brain and spinal cord. In humans it affects mood and may be released during hunger, thirst, dreaming, and sexual behavior. Norepinephrine is also secreted by the central part of the adrenal glands into the bloodstream and functions as a hormone. It helps maintain a constant blood pressure.

normal (*geometry*)   A normal is a perpendicular line, but usually one perpendicular to a curve or a surface (although the curve can be a line and the surface a plane). For a curve the normal is the line perpendicular to the tangent to the curve. Similarly a normal to a surface is perpendicular to a plane tangent to the surface at the point of tangency.

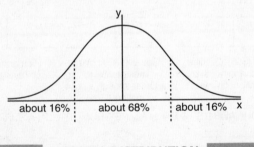

about 16%   about 68%   about 16%   X

NORMAL DISTRIBUTION

normal distribution (*statistics*)   If a large set of measurements of one quantity (such as heights or test scores) are arranged by size, the most common way for the measurements to group themselves approximates the area under a bell-shaped curve. This is called a normal distribution of the measurements and the curve is the normal curve, whose equation is

$$y = \frac{1}{\sqrt{2\pi}} e^{-\frac{x^2}{2}}$$

In a normal distribution, more than 2/3 of the measurements fall in the central region and about 1/6 of all measurements are extremely high or low.

North American Plate (*geology*)   The North American Plate stretches from Cuba and Guatemala to include not only most of the continent but also Greenland, the western half of Iceland, and the western part of the North Atlantic. The tiny part of the continent not on the plate is Baja California and the part of California west of the San Andreas Fault, a plate boundary that passes east of Los Angeles and just north of San Francisco before entering the Pacific Ocean. Because most of North America is far from the plate boundaries, the only volcanoes and regions of frequent, large earthquakes are near the west coast.

north and south pole (*geography*)   Earth rotates about an axis. The north and south poles are where the axis meets the surface. An observer on Earth perceives what appears to be movement of the Sun and stars about that same axis, so ancient sky watchers identified north and south based on the points about which the dome of the sky turns.

north star (*solar system*)   Earth maintains approximately the same orientation in space for centuries. When its axis of rotation if extended would pass through a star as seen from the Northern Hemisphere, that star is called a north star. At present, the north star is Polaris, but in about 5000 years it will be the star Alpha Cephei and in about 12,000 years it will be Vega. At present there is no south star.

not (*logic*)   Not, usually symbolized by $\sim$, is the logical connective negation, although it applies to a single sentence. A sentence preceded by not is false if the original sentence is true, and true if the original is false. In electronic devices the not gate changes 1 to 0 and 0 to 1. Not can be combined with any one of the connectives and, or, or if . . . then to produce a complete system of logic for propositions. For example, if p and q are 2 sentences, then the conditional $p \rightarrow q$ is the same as the disjunction $\sim p \vee q$.

notochord (*anatomy*)   A defining characteristic of animals in the phylum Chordata (which includes the vertebrates) is the presence of a notochord, either throughout life or at least during embryonic stages. A notochord is a firm but flexible rod that extends the length of the body, from the brain to the tail. It lies immediately below the spinal cord.

In the most primitive vertebrates, lampreys and hagfish, the notochord serves as a supportive structure and remains throughout life. In other vertebrates the notochord appears early in the development of the embryo. But it is soon supplanted by a backbone of vertebrae, which forms around the notochord and spinal cord. Most of the notochord disappears, though the disks of tissue that separate one vertebra from another are believed to derive at least in part from the notochord.

nova (*stars*)   Early astronomers noted the occasional appearance of a new (nova) star in the sky. Ancient Chinese astronomers called them guest stars. There is no actual new star; instead, a dim, existing star suddenly brightens. Before the telescope the dim stars could not be seen. There are two main types of guest stars, called novas and **supernovas**. Novas are less bright than supernovas and often appear more than once. Nova brightening occurs when material from one star in a binary pair falls on the other star, causing it suddenly to flare up.

nuclear physics (*sciences*)   Analysis of the **nucleus** of the atom, a study that began with its discovery by **Rutherford** in 1911, became a separate branch of physics after nuclear fission was discovered in 1938. The discovery led to the development of nuclear weapons. The atomic nucleus is the source of **radioactivity**, and nuclear reactions are the basis of **astrophysics**, so nuclear physics extends well beyond weaponry. The atomic nucleus is an important subatomic particle itself.

nuclear waste (*environment*)   Nuclear waste consists of unusable radioactive materials created during the production of nuclear energy. It includes large quantities of residue, called tailings, produced during mining and processing of uranium ore. High-level nuclear wastes, such as spent fuel from nuclear power plants, are highly radioactive. Low-level nuclear wastes, such as protective clothing, filters, paper towels, and cleaning rags that have come in contact with radioactive material, are slightly radioactive. Transuranic wastes consist of plutonium and other transuranic elements formed when uranium absorbs extra particles. Produced mainly at nuclear reprocessing plants, transuranic wastes are not as intensively radioactive as high-level wastes, but they take much longer to decay. Because nuclear wastes remain radioactive for thousands to millions of years, disposal has been a controversial, and largely unresolved, issue.

nucleic acid (*biochemistry*)   Nucleic acids—so-called because they were first seen in the nucleus—are organic compounds that carry the genetic material of all living matter. A nucleic acid molecule is very long and composed of repeating units called **nucleotides**. A nucleotide has 3 components: a phosphate group, a pentose (5-carbon) sugar, and a nitrogen base. There are 2 types of nucleic acids: DNA and RNA.

nucleon (*particles*)   Protons and **neutrons** are collectively termed nucleons, since they are the only constituents of an atomic nucleus. Within the nucleus protons and neutrons change back and forth to each other as an effect of the color, or strong, force.

nucleotide (*genetics*)   DNA and RNA molecules are made up of building blocks called nucleotides. Each nucleotide has 3 parts: a pentose (5-carbon) sugar unit, a phosphate unit, and a nitrogen base. In DNA the sugar is deoxyribose and there are 4 different bases: adenine (A), thymine (T), guanine (G), and cytosine (C). The shape of adenine is complementary to thymine; every adenine in 1 of the 2 DNA chains is attached by hydrogen bonds to a thymine in the other chain. Similarly guanine and cytosine are complementary. Thus if 1 chain has the sequence A-C-C-T-G-G-A then the second chain has T-G-G-A-C-C-T. RNA nucleotides are similar but they contain the sugar ribose and instead of thymine the base uracil (U) is complementary to adenine.

nucleus (*cells*)   Discovered and named in 1831 by Robert Brown [Scottish: 1773–1858], the nucleus ("little nut") is the cell structure that contains the genetic material DNA. DNA directs the operations of the cell and passes on genetic information to daughter cells during cell division. When the cell is not dividing, the DNA is dispersed in the nucleus as a tangle of thin strands. During cell division, the DNA strands condense to form chromosomes.

The nucleus is enclosed in a double membrane, with pores extending through the membranes. One or more dense bodies—each a nucleolus—are in the nucleus. They consist of proteins and ribosomal RNA, which is assembled into ribosome subunits that then enter the cytoplasm.

nucleus (*particles*)   Each atom consists of a cloud of electrons around a center of positive charge, which is called the nucleus. The nucleus is a small region that behaves somewhat like a single subatomic particle. This part of the atom consists of protons and, except for hydrogen, neutrons—similar particles that together are called nucleons. The positively charged protons tend to move away from each other. The nucleus is held together, however, by the strong force between nucleons, which is produced by the exchange of quarks in the form of virtual pions. Virtual particles appear and disappear so quickly that the pions are not counted as part of the nucleus.

nudibranch (*animals*)   Among the most colorful and bizarrely shaped of marine invertebrates, nudibranchs are small sluglike mollusks that lack shells in the adult stage. The name—"naked gills"—refers to feathery gills on the tail end of many species; some species absorb oxygen through numerous hornlike growths on the back, called cerata, and some take in oxygen through the skin. Found worldwide in coastal waters, nudibranchs seldom exceed 2 in. (5 cm) in length. They feed on sponges, sea anemones, and other small animals, locating prey mostly by touch.

number (*mathematics*)   Several related entities—the natural numbers, integers, rational numbers, real numbers, and complex numbers—are numbers. They are all different subsets of the complex numbers. However, it is common to start with the natural numbers that are used for counting and build the other entities. For example, integers are all sums and differences of natural numbers; rational numbers are all ratios of integers (except for division by 0); real numbers are the limits of sequences of rational numbers; and complex numbers are the sums of real numbers and products of real numbers with the square root of negative 1.

number line (*mathematics*)   A number line relates geometry to numbers. A segment, called the unit, is marked between 2 arbitrary points on a line. Then one endpoint (conventionally the left-hand one) is marked 0 and the other 1. Adjacent segments congruent to the first are laid off in each direction and the endpoints of the segments are numbered with the integers in order. Since each point on such a line is associated with 1 and only 1 real number, this is also called a real line or a real number line.

The axes used in Cartesian coordinates are both number lines, while the axes in an Argand diagram are a horizontal real number line and a similar vertical line in which each number is imaginary.

number theory (*mathematics*)   Number theory (sometimes called arithmetic or higher arithmetic) is the study of properties of the natural numbers, especially those connected with solution of diophantine equations, divisibility (division with

0 remainder), and **prime numbers**. Studied for its own sake since the time of Diophantus [Hellenic: c200–99] number theory in recent years has had important applications in codes used for protecting information transmitted by computer.

numeral (*numeration*)   A standard symbol for a number is a numeral. Familiar examples include Roman numerals such as XLVII (47 in Hindu-Arabic numerals). Teachers sometimes explain arithmetic operations as other names for the same number (38 + 9 is a name for 47), but such nonstandard names are usually not considered numerals.

numeration system (*numeration*)   Any set of rules used to express numbers is called a numeration system. Early systems repeated a symbol, as with tally marks. Near the dawn of civilization, Egyptians improved on tallies with separate symbols for 10, 100, 1000, etc., and people in Mesopotamia developed a more sophisticated place-value system based on 10, 60, and 360. Our present numerals use the Hindu-Arabic numeration system based on an "alphabet" of 10 digits used in a place-value decimal system.

nutation (*astronomy*)   Gravitational interactions with the Sun and Moon cause Earth to lean at slightly different angles in a period of about 18.6 years, corresponding to the Metonic cycle, after which the phases of the Moon return to same days of the calendar year. This motion was named nutation (for "nodding") by James Bradley [English: 1693–1762], who discovered it in 1748.

nutmeg (*plants*)   Native to the Moluccas (once called the Spice Islands), the nutmeg tree (*Myristica fragrans*) is a fragrant evergreen that grows to 60 ft (18 m). It has long leaves and clusters of small yellow flowers that give rise to fleshy fruits. Each fruit has a single brown seed that is the source of 2 spices: the dried kernel is nutmeg and the seed coat is mace.

nymph (*zoology*)   Insects that undergo incomplete metamorphosis hatch from eggs as nymphs. The nymph resembles the adult but lacks basic features. A grasshopper nymph, for example, lacks wings. The nymph goes through a series of molts, looking increasingly like the adult. Metamorphosis to the adult stage is complete after the final molting.

oak (*plants*)   Members of the **beech** family, oaks (*Quercus*) include about 450 species of trees and shrubs easily recognized by their fruit, the acorn, which consists of a large seed in a cuplike husk. Some species are deciduous; other species, commonly called live oaks, are evergreen. Among the most massive species is the white oak (*Q. alba*) of eastern North America, which can attain heights of 150 ft (46 m).

oat (*plants*)   Members of the **grass** family, oats (*Avena*) are widely cultivated in cool and warm temperate regions for their edible **grains** (fruits) and as fodder and hay. The slender stems—called culms—grow 8 in. to 6.5 ft (20 cm to 2 m) high, depending on the species and variety. Flowers form at the top of the stem in branching clusters called panicles.

obsidian (*rocks and minerals*)   Obsidian is a rock **glass** formed when ejected **magma** or even flowing **lava** cools very quickly. It is not transparent, but commonly black or less often brown or green, depending on chemical composition. Obsidian can be polished into mirrors or jewelry. It breaks into sharp fragments the way that manufactured glass does, making it useful for rock weapons and tools during the **Stone Ages**.

Occam's razor (*philosophy of science*)   William of Occam [or Ockham; English: c1285-c1349] suggested that "entities are not to be multiplied beyond necessity." Scientists have interpreted this to mean that the simplest theory that explains all observations should be taken as true. This rule of thumb has come to be called Occam's razor, as it shaves off unnecessary complications.

occultation (*astronomy*)   An occultation occurs when one celestial body is hidden by another moving in front of it—for example, a solar eclipse is an occultation of the Sun by the Moon. Occultations of stars or planets by the Moon help locate them exactly. Occultations of stars by planets are used to refine estimates of planetary sizes or density of a planet's atmosphere.

ocean (*oceanography*)   About 72% of Earth's surface is covered with water. If Earth were flat the water would be over 2 mi (3 km) deep). Instead most of the water is classified as belonging to large bodies of salty water called oceans. Geographers recognize 4 oceans: the Atlantic from the Americas to Europe and Africa; the Indian from Africa to the East Indies and Australia; the Pacific from Asia and Australia to the Americas; and the Arctic surrounding the north pole. Sometimes the waters surrounding Antarctica up to about 60° S latitude are called the Southern or Antarctic Ocean.

Oceans are not just very large salty lakes; they are determined by their geology. Ocean floor is mostly a thin skin of **basalt** covered with sediment. New ocean floor is created at **mid-oceanic ridges**. Old floor plunges into the mantle at **trenches**. Thus the oldest part of any ocean is only about 200,000,000 years old, while parts of the continents are 20 times older.

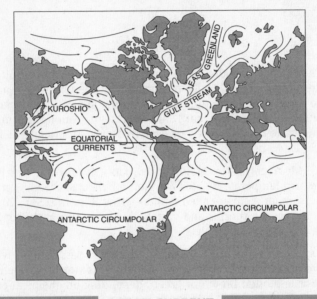

OCEAN CURRENT

ocean current (*oceanography*)   An ocean current is the undersea equivalent to wind—the large-scale movement of fluid in a single direction. Unlike wind, however, currents flow along regular paths almost all of the time. Most currents occur because denser water sinks and pushes aside less dense water, but some surface currents are propelled by prevailing winds. In general cold water is denser than warm, and salty water is denser than fresher water. Currents develop because cold polar water sinks and flows toward the equator or because salty water sinks from seas with high evaporation rates. Deep currents caused by dense water sinking produce countercurrents flowing in the opposite direction at the surface. The Gulf Stream, perhaps the most famous current, is actually a countercurrent.

oceanography (*sciences*)   Oceans cover about 70% of Earth. All aspects of the oceans are the province of oceanography, even the biology of ocean life. The core portions of oceanography are currents and waves; sedimentation, strata and plate tectonics; ocean chemistry; transmission of light and sound; and the use of ocean data to reconstruct past climates.

ocher (*rocks and minerals*)   Ocher is the name used for red hematite, the mineral form of ferric oxide (Fe_2O_3). When ground into a powder, some samples have the color of fresh blood. This is perhaps why Neandertals and early *Homo* *sapiens* buried their dead sprinkled with ocher and may have used it for body decorations. Ocher also provided the red for the painted cave art of 12,000 to 24,000 years ago. Hematite, which in bulk is more grayish black than red, is also the principal iron ore.

octahedron (*geometry*)   Although any polyhedron with 8 faces is an octahedron ("8 bases"), most of the time the octahedron is a regular figure, one of the Platonic solids. The regular octahedron's 8 faces are congruent equilateral triangles. It can be formed by cutting off each of the 8 vertices of a cube at the midpoints of the sides of the cube.

octopus (*animals*)   Found in marine waters worldwide, octopuses are mollusks characterized by 8 arms attached to the head and equipped with suckers. They have a well-developed brain and eyes similar to those of vertebrates. If attacked, an octopus secretes a dark ink from a gland in the mantle cavity; the ink clouds the water, giving the octopus a chance to escape—which it does by jet propulsion, drawing water into the mantle cavity and then expelling it with great force. (*See also* **squid**.)

Oersted, Hans Christian (*physicist*)   [Danish: 1777–1851] Oersted recognized that electricity and magnetism are probably related. In 1820 he demonstrated this by placing a magnetic compass below an electric current. The current caused the needle to move perpendicular to the wire. Oersted also did some first-class experimental work in both chemistry and physics.

Ohm's law (*electricity*)   In a given conductor, the amount of electric current is proportional to the potential, but the current for a given potential is less in a conductor that resists conduction. In 1826 Georg Simon Ohm [German: 1787–1854] developed the rule known as Ohm's law to describe how current varies with potential when resistance is taken into account. For wire of a given conducting material, resistance doubles if the length of the wire is doubled, but is halved for a given length if the diameter of the wire is doubled. Wire also shows more or less resistance depending on how good a conductor its material is. Resistance can be measured in ohms, with 1 ohm equal to a volt/ampere. Ohm's law is commonly expressed in terms of R for resistance, E for voltage (electromotive force), and I for amperage (intensity of current): $R = E/I$. Any device that uses electric current, such as a lamp or a motor, increases resistance on the circuit; such a device is often known as "the resistance" in circuit diagrams.

Olbers' paradox (*cosmology*)   Almost as soon as astronomers recognized that stars are not on a transparent sphere surrounding the universe, they wondered why the night sky is black instead of bright. Johannes Kepler, Edmond Halley, and others perceived that an infinite number of stars produces an infinite amount of starlight. Recently, the riddle of the dark sky has been known as Olbers' paradox after a discussion in 1823 by Heinrich Wilhelm Olbers [German: 1758–1840]. An acceptable explanation for the dark sky was put forward without proof by author Edgar Allan Poe [American: 1809–49], who speculated in 1848 that light from very distant stars, moving at a finite speed, has not yet reached Earth. Today we have evidence that the age of the universe is finite, so light does not have an infinite time to reach Earth. Furthermore, in an expanding universe the Doppler effect increases the wavelength of light from distant stars beyond visibility.

old growth forest (*ecology*)   Woodlands dominated by trees of multiple species, hundreds of years old and hundreds of feet tall, are known as old growth forests. In addition to healthy, growing trees, these forests contain broken trees (snags), decaying timber, and a rich layer of forest floor debris, each providing unique habitats for species that often are unable to survive anywhere else. Once widespread, old growth forests have been almost totally destroyed by logging operations. In places where the land is replanted, the forest is typically replaced by single-species tree farms or monocultures, which will never resemble the original forest.

Oldowan (*paleoanthropology*)   The earliest stone tool industry, dating from about 2,400,000 to 1,500,000 years ago, is the Oldowan industry of Africa. Originally Oldowan tools were thought to be rounded cobbles that were broken so that sharp edges appeared, but later researchers contend that the broken-off flakes were the most useful parts. The cobble tools are often called choppers or scrapers.

olive (*plants*)   Native to lands around the eastern Mediterranean, the olive (*Olea europaea*) is an evergreen tree that grows to heights of 40 ft (12 m). It produces numerous fragrant cream flowers that give rise to small edible fruits with a high oil content. Olive trees have been cultivated for their fruits since prehistoric times. The olive family, Oleaceae, comprises some 600 species, mostly shrubs and trees. It includes forsythia, privet, ash, and lilacs.

Omar Khayyam (*mathematician*)   [Persian: 1048–1123] Known in the West for his poetry, Omar also wrote an algebra in which he gave correct geometric solutions to cubic equations and produced a sophisticated attempt to prove Euclid's parallel postulate. Omar was best known in his own time as an astronomer, but most of his work in that science has been lost.

omega particle (*particles*)   The omega particle, a hyperon made from 3 strange quarks, is among the few subatomic particles that were predicted before discovery, along with the neutrino, positron, pion, and charm, bottom, and top quarks. Prediction of the omega in 1961 by Murray Gell-Mann and its subsequent discovery in 1964 led directly to the quark theory.

omnivore (*ecology*)   Animals that eat both plants and animals are called omnivores, a name that comes from Latin words meaning "all eaters." Bluebirds, foxes, rats, bears, and humans are examples. The ability to eat many kinds of food is an advantage, for if one food is scarce the animal can eat something else.

oncogene (*genetics*)   A gene that can potentially cause cancer is called an oncogene. Cellular oncogenes are mutated versions of normal genes. Viral oncogenes are found in certain viruses; they can transform normal cells into cancerous cells in the animals they infect. Most oncogenes work when cell "guardians," called tumor suppressor genes, fail to function normally. Tumor suppressor genes regulate cell division and instruct cells with DNA damage to commit suicide so they cannot divide and pass on the DNA damage to daughter cells. But if tumor suppressor genes are mutated, they may not send the proper signals to the cells and the cells may lose their ability to counteract the oncogenes.

one-to-one correspondence (*sets*)   Two sets are in one-to-one correspondence when each element in the first set is matched with exactly 1 element in the second and each element in the second is matched with exactly 1 element in the first. This idea can be used to explain the meaning of the natural numbers since each natural number can be abstracted from all the sets in one-to-one correspondence with a given set. For example, 5 could be described as the quantity abstracted from all sets in one-to-one correspondence with the fingers on 1 hand. The concept of one-to-one correspondence is vital to discussions of infinite sets and also is useful in solving certain problems in algebra.

onion (*plants*)   Many members of the genus *Allium* are cultivated as food plants, including onions (*A. cepa*), chives (*A. schoenoprasum*), garlic (*A. sativum*), leeks (*A. porrum*), and shallots (*A. ascalonicum*). These herbs share a characteristic odor, which ranges from mild to highly pungent. They have underground stems—either bulbs or rhizomes—long flat or hollow leaves, and clusters of small flowers.

onychophoran (*animals*)   Often referred to as velvetworms because of their velvety outer covering and soft wormlike bodies, members of Phylum Onychophora ("claw bearing") share features with both annelids and arthropods and may represent an evolutionary link between the 2 groups. Velvetworms have up to 43 pairs of short, stumpy legs, each ending in 2 claws. Species range in length from 0.8 to 6 in. (2 to 15 cm). Most species live in dark, moist environments—for example, under fallen logs in tropical rain forests—coming out at night to prey on other small invertebrates.

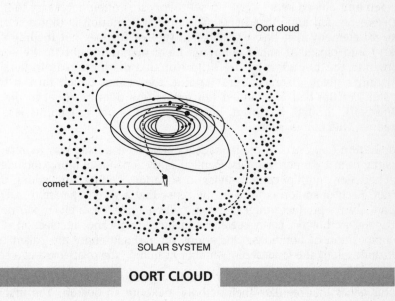

SOLAR SYSTEM

OORT CLOUD

Oort cloud (*solar system*)   In 1950 Jan H. Oort [Dutch: 1900–92] demonstrated that long-period comets have orbits that can be explained only if the comets originate at a distance of 50,000 **astronomical units** (A.U.) to 150,000 A.U. from the Sun. His calculations showed that there are about 190,000,000,000 comets in a great cloud that surrounds the solar system. The comets are so far from the Sun that the gravitational tug of passing stars disturbs the comets' orbits, sending a few toward the Sun, thus enabling observation from Earth. (*See also* **Kuiper belt**.)

ooze (*oceanography*)   Much of the ocean floor is covered with a coat of shells of small organisms such as **Foraminifera, diatoms,** and **radiolarians** mixed with fine clay. If more than 30% of the mixture is shelly material, oceanographers call it an ooze. Oozes based on Foraminifera, which cover half the ocean floor, are mainly calcium carbonate. Below about 13,000 ft (4000 m) calcium carbonate dissolves in the high-pressure seawater, so foraminiferal ooze is not found on the deepest ocean bottom. Diatoms and radiolarians, which have shells of silica, are more limited in distribution. Siliceous oozes develop away from continental sedimentation that tends to overwhelm the amount of shelly material.

opacity (*optics*)   Opacity occurs in materials that absorb light by increasing the vibrations of particles (heating the material) instead of raising the energy levels of electrons. All opaque materials, however, capture some energy in electrons and radiate them at the surface, producing color—black materials capture nearly all energy as heat. Metals have free electrons that can collect and radiate most wavelengths, which is why metals are shiny and often good reflectors.

open and closed sets (*sets*)   In set theory it is often necessary to distinguish between open and closed sets. The formal definition is that a set is open if every element in it has a **neighbor** in the set (it does not include its boundary) and closed if some elements have neighbors not in the set (it does include the boundary). The origin of this distinction is in sets of **points** within a given distance from 1 point, the center, which form a **ball**. If the "skin" of the ball is included (all points less than or equal to the radius of the ball), the ball is closed, but if the skin is not (all points less than the radius), the ball is open.

operation (*algebra*)   A mathematical operation is a fixed way to obtain a single entity from 1 or more entities. Familiar operations for numbers include addition, subtraction, multiplication, division, squaring, cubing, and taking the square root. Each operation is fixed in the sense that a given operation, such as addition, always produces the same result when applied to a given pair of numbers. Beyond arithmetic, **binary operations** include union and intersection of sets and composition of functions. Operations on a single entity are called unary and include taking the **factorial** of a number or finding the **complement** of a set.

operon (*genetics*)   A cluster of structural genes that are expressed as a group plus genes that regulate their activity make up an operon. The first such system, the lac operon, was elucidated in the early 1960s by Jacques Monod [French: 1910–76] and François Jacob [French: 1920–] in the bacterium *Escherichia coli*. The lac operon controls the production of 3 enzymes needed to change the sugar lactose into the sugars glucose and galactose. It consists of 3 structural genes (called cistrons) and 3 control genes: a promoter, an operator, and a repressor. When no lactose is present, the repressor binds with the operator and prevents transcription of the structural genes. When lactose is present it binds to the repressor, causing the repressor to lose its affinity for the operator. This allows the enzyme RNA polymerase to bind to the promoter and begin transcription.

opossum (*animals*)   Small to medium-sized marsupials native to the Americas, opossums are good climbers, with strong clawed feet and a prehensile tail for wrapping around branches. They can swim, too; the yapok (*Chironectes minimus*) has webbed hind feet and spends much of its time in water. Active mainly at dawn and dusk, opossums feed on both plant and animal matter.

opposite (*numbers*)   Two real numbers or imaginary numbers at equal distances from 0 are called opposites. The integers are often defined as the natural numbers, including 0, and their opposites.

optics (*waves*)   Optics (Greek for "visible") refers to the study of visible light and also infrared radiation to ultraviolet radiation. The main laws of light transmission, reflection, and refraction have long been known. Optics today concerns not only materials and methods used in improving lenses and mirrors but also lasers, optical versions of transistors, and transmission of signals using light in fibers (fiber optics).

or (*logic*)   The connective or, symbolized by $\vee$ and also called disjunction, produces a combined sentence that is true if either one of the sentences it connects are true. For example, "The sky is blue or the Sun is shining" is true if there is a blue sky whether or not it is sunny; and also true if it is sunny, whether or not the sky is blue. It is only false if the sky is not blue and also the Sun is not shining. Similarly, the or gate in an electronic device passes a signal if either incoming signal is 1 or both are 1, but registers 0 for a signal with 2 0s.

orangutan (*animals*)   Highly endangered, the orangutan (*Pongo pygmaeus*) is a hominoid primate that inhabits swampy forests on Borneo and Sumatra (orangutan is a Malay word meaning "man of the jungle"). Orangutans spend most of their time in treetops, using their long, powerful arms to travel from tree to tree. They mainly eat fruits. Orangutans have long reddish brown hair and weigh up to 170 lb (77 kg). Adult males have a large inflatable throat pouch.

orbit (*astronomy*)   The path that a celestial body takes as a result of gravitational attraction is called its orbit. When only 2 bodies are considered, the path is a conic section, typically an ellipse for a closed path and a hyperbola for an open one. Actual orbits vary slightly from the elliptical or hyperbolic because of the gravitational attraction of other bodies.

orbital (chemistry)   An electron in an atom viewed as a point has no exact location (because of the uncertainty principle) or, when interpreted as a wave, is spread out over a region. The region where there is a high probability of a point-electron may be located or where the density of a wave-electron is great-

PROBABLE ELECTRON LOCATIONS

| 1s orbital (centered about nucleus) | 2s orbital (centered on nucleus and also in a shell) | 1 of 3 p orbitals (lobes of each side of nucleus) | 1 of 5 d orbitals (4 with 4 lobes and 1 with 2 lobes and a torus) |

ORBITAL

est is called the electron's orbital; this concept replaces the earlier idea of a point-electron in a definite orbit about the nucleus. The single electron of a hydrogen atom occupies different orbitals at different energy levels; in an atom with more electrons, several orbitals are occupied at once. By the Pauli exclusion principle each orbital can contain at most 2 electrons with opposite spins.

orchid (*plants*) An extremely diverse group of more than 15,000 species, orchids make up the family Orchidaceae. Most species are native to moist tropical and subtropical areas, where many live as epiphytes on trees. Many, such as *Cattleya*, are popular ornamentals; *Vanilla* fruits are the source of vanilla flavoring. Orchid flowers are greatly modified, with the male and female reproductive structures fused to form a column. The flowers often are very showy and highly scented. For instance, *Ophrys* flowers are shaped like bees and other insects that are attracted by the shape and in the process pollinate the flowers; *Bulbophyllum* smells like rotten meat, which attracts blowflies. The orchid's fruit is a pod that contains thousands of seeds no bigger than specks of dust.

ordered pair (*mathematics*) Ordered pairs—2 numbers or variables with 1 identified as first and the other as second—are familiar as coordinates (x, y) of points in a plane. Ordered pairs are useful in many ways in mathematics. Relations and functions are defined as sets of ordered pairs. The integers, rational numbers, and complex numbers may also be treated this way. Ordered pairs become a part of set theory when an ordered pair (x, y) is defined as an abbreviation for the set $\{x \{y\}\}$, replacing first and second with the concept of membership in a set.

ordinal and cardinal number (*numbers*) Numbers used to describe the position of entities arranged in a specific sequence are called ordinal numbers. Examples are first, second, third, and so forth. These are in contrast to the numbers known as cardinal numbers, which are the natural numbers (0, 1, 2, 3, . . .) used to describe the number of members in a set.

Both ordinal and cardinal numbers have been extended to infinite sets. The first ordinal infinity is defined as the smallest number greater than any finite ordered set of natural numbers; the first cardinal infinity is the number of all the natural numbers.

ore (*rocks and minerals*)   An ore is any rock that contains enough of a valuable element to make it commercially worthwhile to mine the rock and extract the element. Most metals are extracted from ores, even those that also occur native such as gold, silver, or mercury. Few nonmetals are extracted from rock; naturally occurring native sulfur, diamond, graphite, corundum (aluminum oxide), borax, and gypsum are somewhat purified or treated, but are used almost as mined. Even so, commercially useful amounts are considered ores.

organ (*biology*)   In most multicellular organisms, various types of tissues are integrated into units called organs. These organs perform specific functions and are themselves generally organized into systems. Plant roots, stems, and leaves are organs, as are animal hearts, stomachs, and brains.

organelle (*cells*)   Organized structures in the cytoplasm of a cell, organelles ("little organs") are analogous to organs in a multicellular organism. Each type of organelle has a specific function. Examples include mitochondria, nuclei, plastids, ribosomes, and vacuoles.

organic (*chemistry*)   A compound is organic if its molecule contains 1 or more carbon atoms, although very simple compounds, such as silicon carbon (SiC) or carbon disulfide (CS_2), are often not considered organic. The word organic is used because all living organisms are mainly carbon compounds and because, before Friedrich Wöhler synthesized urea—$CO(NH_2)_2$—from an inorganic compound in 1828, chemists believed the chemistry of living organisms to be essentially different from that of minerals. Organic chemistry is the branch that specializes in such carbon compounds as alcohols and hydrocarbons.

organism (*biology*)   Any living creature is an organism. Some organisms, such as bacteria and certain algae, consist of a single cell. Other organisms, such as worms and willows, are multicellular. Each kind of organism is unique, yet all organisms have basic similarities that are characteristic of life.

Orion Nebula (*astronomy*)   The brightest nebula in the sky is barely visible to the unaided eye as Orion's sword. The Orion Nebula—M42 or NGC 1976—shines from ionization of gas by newly formed stars within. Behind the shining nebula is a dark one of dust, where other new stars develop. Other nebulas, including the dramatic Horsehead Nebula, are, by astronomical standards, in the same vicinity.

orogeny (*geology*)   Orogeny is a process that leads to mountain building, including the creation of faults and folds in rock. A past episode of orogeny may be named after the mountains that resulted, such as the Appalachian orogeny. Typically the region where orogeny occurs is a long, fairly narrow orogenic belt, now recognized as the result of tectonic plates meeting as they move in opposite directions.

orrery (*solar system*)   Any mechanical model of the solar system is called an orrery after a famous clockwork device made for the fourth Earl of Orrery about 1710. Earlier models, including some from ancient times, were based on geocentric concepts and did not show true motions of the solar system.

orthogonal (*geometry*)   The concept of orthogonal is a generalization of perpendicular (*ortho-* as a prefix means "right angle" in mathematics, derived from the Greek for "upright"). In geometry orthogonal curves meet so that their tangents are perpendicular. For lines orthogonal and perpendicular are synonymous. The concept has been generalized far beyond geometry; for example, functions, transformations, and vectors can be orthogonal.

orthomyxovirus (*viruses*)   Members of the virus family Orthomyxoviridae infect animals. The most virulent orthomyxovirus is the influenza A virus, which causes respiratory disease in a wide variety of birds and mammals, including humans. A typical orthomyxovirus particle is spherical with a lipid envelope and surface spikes formed by glycoproteins that act as antigens. The genetic material consists of RNA.

osmium (*elements*)   The hard metal osmium, atomic number 76, was discovered in 1803 by Smithson Tennant [English: 1761–1815] in association with platinum. He named it from the Greek *osme*, "smell"—powdered osmium combines with oxygen in air to produce osmium tetroxide, which not only smells bad but also is very toxic. Osmium is best known as the densest (heaviest for a given volume) metal, although its density is so close to that of iridium that it is not certain which is the denser. Chemical symbol: Os.

osmoregulation (*physiology*)   The process by which an animal regulates water and salts within its cells and fluids is called osmoregulation. Many animals live in environments that have very different concentrations of water or salts from those found within the animals. For example, the concentration of salt in the ocean is much greater than it is in a seabird, yet seabirds take in large amounts of salt water with their food. The salt must be excreted or the seabirds will die. Seabirds have special salt glands near the eye that remove salt from the blood and expel it through the nostrils. In mammals osmoregulation is carried out by kidneys; in addition, the skin is relatively impermeable to water so that water loss to the comparatively dry air is minimal.

osmosis (*physics*)   Osmosis ("pushing") is the process of molecules in a solution moving through a semipermeable membrane such as the wall of a cell or a thin sheet of cellophane. The less concentrated solution passes through as if being pushed, especially if dissolved molecules on the other side are too large to pass through easily. Osmotic pressure caused by sugar solutions (sap) in tree roots surrounded by water causes sap to rise in trees. Osmotic pressure also keeps cells from collapsing.

ossification (*physiology*)   During the early embryonic development of vertebrates, the skeleton consists almost entirely of cartilage. Except in sharks and other cartilaginous fish, most of the cartilage is gradually replaced in a process called ossification. The cartilage cells become bone cells called osteoblasts; these start secreting calcium phosphate and other compounds that make up the bulk of bones. In humans ossification occurs in most parts of the skeleton, beginning in the second month of development and continuing until about age 25, when bones stop growing. Some cartilage, such as that in the larynx and the outer part of the ears, is permanent.

ostrich (*animals*)   The largest living bird is the ostrich (*Struthio camelus*), which stands almost 8 ft (2.4 m) tall and weighs 300 lb (136 kg) or more. It has a small head, with large eyes and a tough beak, a long neck, small wings, and long, muscular legs. Native to dry grasslands of Africa, ostriches cannot fly but they are fast runners, reaching speeds of about 40 mi (64 km) per hour.

otter (*animals*)   Members of the weasel family, otters are long-bodied, short-legged, web-footed mammals adapted to an aquatic existence. River otters (*Lutra*), found in Eurasia, Africa, and the Americas, live in dens near the streams where they hunt, coming out at night to feed mainly on fish.

The sea otter (*Enhydra lutris*), native to coastal waters of the North Pacific, has webbed hind feet that resemble a seal's flippers. It feeds mainly on mollusks and sea urchins, diving to depths of 300 ft (90 m) or more, then carrying its food to the surface. As the sea otter floats on its back to eat, it often uses a rock to break a mollusk's shell.

outer planet (*solar system*)   The asteroid belt separates the planets of the solar system into two groups. Except for Pluto, the outer planets—Jupiter, Saturn, Uranus, and Neptune—are giant balls of cold gas surrounded by rings and several large rocky or icy satellites. Although Pluto is among the outer planets, it is more like a giant comet than it is like the gas-giant planets.

ovary (*biology*)   The female reproductive organ of plants and animals that produces eggs is the ovary. In a flower the ovary is an enlarged structure in which seeds are formed following fertilization. A vertebrate has 2 ovaries, which in addition to producing eggs that are released to oviducts generally secrete the hormones estrogen and progesterone.

oviduct (*anatomy*)   The oviduct is a tube in the female of many animal species that carries mature eggs away from an ovary. In fish, amphibians, reptiles, and birds the oviduct leads to a cloaca. In mammals an oviduct consists of 3 parts: the fallopian tube, into which the egg passes from the ovary; the uterus, where an embryo develops; and the vagina, which leads to the exterior. Glands in the oviduct wall of reptiles and birds surround eggs with materials such as albumen (egg white) and shell.

oviparity (*zoology*)   In oviparous animals eggs are laid by the mother and hatch outside her body. Among oviparous reptiles and birds the eggs are fertilized before they are laid. Among oviparous fish and amphibians the eggs are fertilized after they are laid. The eggs of oviparous animals have shells or other tough protective coats and contain yolk that nourishes the developing embryo.

ovipositor (*anatomy*)   The tip of the abdomen of many female insects is modified into an ovipositor, through which eggs are laid. Insects that place eggs on twigs usually have relatively simple ovipositors. Insects that dig depressions in soil, bore into trees, or pierce the skin of animals before laying eggs have more elaborate ovipositors. In some insects, such as bees and wasps, the ovipositor is modified into a stinger and associated with poison glands.

Oviraptor (*paleontology*)   Because the first fossil of this dinosaur was found in the 1920s near a nest of dinosaur eggs, scientists thought it fed on eggs and named it *Oviraptor* ("egg hunter"). The discovery in the 1990s of a nest of *Oviraptor* eggs—including one containing an embryo—suggests, however, that *Oviraptor* was not an egg thief but an animal that incubated or protected its eggs.

Oviraptor was about 6 ft (1.8 m) long and walked on well-developed hind legs. Its unusual skull was birdlike, with large eyes and a mouth that lacked teeth and ended in a short, hard beak.

ovoviviparity (*zoology*)   In ovoviviparous animals the embryos develop within the mother's body but rarely obtain nourishment from the mother. The embryo is separated from the mother by egg membranes. The common adder (*Vipera berus*) is an example of an ovoviviparous animal; its eggs develop fully within the mother and the young are born live or hatch immediately after the eggs are laid.

owl (*animals*)   Unlike other birds of prey, owls are nocturnal, with very large forward-looking eyes equipped to detect the faintest light rays. The only birds with external ears, owls also have exceptionally fine hearing. Their soft plumage blends into their surroundings, making owls inconspicuous during daylight hours. Their wing feathers are fringed, breaking air into tiny streams and making owls silent fliers. The smallest species is the elf owl, about 5 in. (13 cm) long with a wingspread about 15 in. (38 cm). Among the largest is the great gray owl, more than 30 in. (75 cm) long with a wingspread to 60 in. (150 cm).

oxidation and reduction (*chemistry*)   The term oxidation is applied to chemical reactions in which a substance loses electrons to another, after which the original substance is oxidized. (The element oxygen need not be involved.) The substance that gains the electrons is reduced. Reactions involving oxidation and reduction are called redox reactions. For example, when iron reacts with hydrochloric acid, the product is ferrous chloride and hydrogen gas. In ferrous chloride the iron has become oxidized, losing an electron to each of 2 chlorine atoms. The chlorine is now reduced, since each atom has captured an electron. Chlorine is a typical oxidizing agent, tending to gain electrons. Most metals act as reducing agents.

oxygen (*elements*)   In 1771 Karl Wilhelm Scheele and independently in 1774 Joseph Priestley discovered oxygen by heating oxides and collecting the gas. Atomic number 8, oxygen gas is essential to most living creatures, although many bacteria and Archaea find it toxic. Nearly half of Earth's crust (by mass) is oxygen, but oxygen is so reactive that there was no free oxygen until photosynthesis began about 500,000,000 years ago. Lavoisier named oxygen from the Greek for "acid forming," incorrectly believing that oxygen forms acids. Chemical symbol: O.

oyster (*animals*)   Members of the mollusk class of bivalves, "true" or edible oysters (Family Ostreidae) have a thick, rough shell composed of 2 valves unequal in size and shape and held together not by a hinge but by a single muscle. Oysters live in shallow marine and brackish waters, often in large colonies called beds, with each oyster permanently attached to a rock or other hard object. They lack a foot. Oysters feed on tiny organic particles, which they filter from the water. Other bivalves called oysters include the pearl oysters (Family Pteriidae) and spiny oysters (Family Spondylidae), whose shells are armed with long spines.

ozone (*element*)   Oxygen gas consists of molecules of 2 atoms each, but with extra energy supplied, molecules with 3 atoms, called ozone, form. Ozone is more reactive than oxygen, making it harmful to living things. But ozone high in the stratosphere absorbs ultraviolet radiation that would be even more harmful. In the stratosphere, short-wavelength ultraviolet radiation from sunlight forms ozone; somewhat longer ultraviolet radiation breaks it apart. The combination removes the most harmful ultraviolet rays. Various synthetic chemicals release chlorine in the stratosphere, interfering with ozone formation and permitting higher levels of energetic ultraviolet radiation to reach Earth's surface.

ozone layer (*environment*)   A concentration of ozone gas occurs in a layer of the atmosphere some 12 to 30 mi (19 to 48 km) above Earth's surface. This ozone layer protects life by absorbing damaging ultraviolet radiation from the Sun. In the 1980s scientists discovered that the ozone layer was thinning. Furthermore, the thinning was so dramatic that each spring an "ozone hole" formed over the South Pole. More recently significant thinning was also recorded elsewhere, including northern temperate areas.

Chemicals from human activities, primarily chlorofluorocarbons (CFCs), rise to the upper atmosphere and destroy the ozone. As a result increased amounts of ultraviolet radiation reach ground level, threatening the health of a broad range of organisms. Most nations have agreed to stop production of CFCs and certain other ozone destroyers.

paca (*animals*) Native to tropical America, pacas (*Cuniculus*) are burrowing nocturnal rodents closely related to agoutis. They are about 2 ft (0.6 m) long and weigh about 20 lb (9 kg). Pacas have large eyes, small ears, a short tail, and a broad nail on each toe. Their reddish brown fur is marked with rows of white spots.

pacemaker (*anatomy*) Located in the wall of a vertebrate's heart, the pacemaker is a node of specialized muscle cells that regulates the rate at which the heart beats. The pacemaker generates regular electrical impulses. This current spreads to muscle cells in the atrium (or atria, depending on the vertebrate), causing them to contract. The current also stimulates a second node, which sends impulses to ventricle muscle cells, causing them to contract milliseconds after contraction of atrium muscle cells is complete.

Pachycephalosaurus (*paleontology*) A member of the bonehead family of armored dinosaurs that lived during the late Cretaceous, *Pachycephalosaurus* ("heavy-headed lizard") was a plant eater that walked on its hind legs. Its 2-ft-(0.6-m-) long skull was topped by a dome of solid bone 8 in. (20 cm) thick. Bony knobs jutted up from its nose and the back of its dome. *Pachycephalosaurus* probably used its domed head in butting attacks.

Pacific Plate (*geology*) The largest section of Earth's lithosphere is the Pacific Plate, bordered on the west and north by deep trenches and on the south and southeast by mid-oceanic ridges. Along the coast of North America parts of the Pacific Plate reach into the continent slightly, encompassing all of Baja California and the lower coast of the state of California to the San Andreas Fault, which runs from the gulf of California to a point north of San Francisco. The plates surrounding the Pacific Plate are the Australian, Philippine, Eurasian, North American, Juan de Fuca, Cocos, Nazca, and Antarctic.

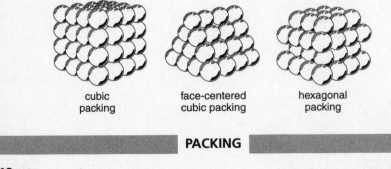

| cubic packing | face-centered cubic packing | hexagonal packing |

PACKING

packing (*geometry*)   Ways of stacking figures, usually spheres, are packings. Circles are "spheres" in 2 dimensions. Six circles touching a seventh is the tightest packing of the plane. In 1611 Kepler guessed that in 3-dimensional space the packing called a face-centered-cubic lattice, commonly used for stacking round fruits, is most efficient. In this packing each sphere touches 12 others, but there is *almost* room for a thirteenth. In 1998 Thomas C. Hales [American: 1958–] used complex computer programs to demonstrate that Kepler was right—there is no way to squeeze in a thirteenth sphere or otherwise improve the tightness of the packing.

paddlefish (*animals*)   The 2 species of paddlefish are large freshwater fish that constitute Family Polyodontidae. They are named for their long, paddle-shaped snout; equipped with taste buds, the snout probably helps the fish find plankton and other food. Paddlefish have a cartilage skeleton; the backbone extends into the tail, as in the closely related sturgeons. The Chinese paddlefish (*Psephurus gladius*) grows to 21 ft (6.4 m) long; the much smaller paddlefish of North America (*Polyodon spathula*) reaches 7 ft (2.1 m).

paedogenesis (*biology*)   The ability of larval animals to produce unfertilized eggs that develop into a new generation of larvae is called paedogenesis (from Greek words meaning "child birth"). Paedogenesis, a type of parthenogenesis, was first described in gall midges of the genus *Miastor*. The phenomenon also occurs in one of the larval stages of liver flukes.

palate (*anatomy*)   The roof of the vertebrate mouth, separating the oral cavity and the nasal cavity, is the palate. In mammals the front portion is supported by bone and is called the hard palate. The back portion, the soft palate, is composed of muscular tissue that usually hangs down into the throat. During swallowing the soft palate is raised to prevent food from entering the nasal cavity.

paleoanthropology (*sciences*)   Paleoanthropology is the study of ancestors of *Homo sapiens*, but many paleoanthropologists also study early representatives of our own species. The genera of hominoids of particular interest to paleoanthropology include *Ardipithecus, Australopithecus,* and *Homo*; earlier ancestors are studied by primatologists. Not only physical remains of species are considered, but also tool use and other cultural aspects of ancestors of humans.

Paleolithic Age (*paleoanthropology*)   The period of chipped and flaked stone tools, starting with the earliest Oldowan pebble tools (2,400,000 year ago) and ending with the Neolithic Revolution (staring about 11,000 years ago), when tools made from ground or polished stone were made, is called the Paleolithic Age. Sometimes a transitional Mesolithic Age is carved out of the last of the Paleolithic and first of the Neolithic.

paleomagnetism (*geology*)   Indications of Earth's former magnetic lines of force that are preserved when rocks form are termed paleomagnetism ("ancient magnetism"). Certain minerals in magma or lava contain atoms that align with the lines of force while the minerals are hot and maintain that alignment after cooling. Similarly, sediments that slowly settle in a lake or calm sea can form rock that retains magnetic orientation. Sensitive instruments can detect these alignments, which have been used to determine how tectonic plates have moved in the past and to date rocks by patterns of past reversals of Earth's magnetic field.

paleontology (*sciences*)   Paleontology is the study of prehistoric life as told by fossil remains. Most fossils lie in strata (layers) of sedimentary rock; as paleontologists study the distribution of fossils in various strata around the world they not only add to our understanding of the evolution and distribution of organisms but also make important contributions to geology.

The many subdivisions of paleontology include the study of fossil plants (paleobotany), ancient climates (paleoclimatology), and Earth's geography as it existed at various times in the past (paleogeography).

palladium (*elements*)   The platinum-group metal palladium, atomic number 46—discovered in 1803 by William Hyde Wollaston [English: 1766–1828]—was named for the second known asteroid, Pallas, discovered the previous year. The ability of palladium to absorb large amounts of hydrogen suggested to some that hydrogen might undergo "cold fusion" in palladium electrodes, but this has not been convincingly demonstrated. Palladium alloyed with gold is the white gold of jewelers. Chemical symbol: Pd.

palm (*plants*)   The more than 2500 species in the palm family, Palmae or Arecaceae, constitute a varied group of vines, shrubs, and trees mostly native to tropical regions. Best known are the palm trees characterized by an unbranched trunk and topped by a crown of stiff, long-stalked, feathery or fan-shaped leaves. The leaves of raffia palms (*Raphia*)—the source of raffia fiber—can exceed 65 ft (20 m) in length! Palm flowers are small and generally arranged in large clusters. The talipot palm (*Corypha umbraculifera*), which blossoms and fruits only once, when it is 30 to 75 years old, produces a cluster some 20 ft (6 m) high that contains tens of thousands of individual flowers. Palm fruits are berries, nuts, or drupes, depending on the species. Many palms are economically significant, including the coconut palm, date palm (*Phoenix dactylifera*), and oil palm (*Elaeis guineensis*).

palp (*anatomy*)   Some invertebrates have paired appendages called palps on the head or around the mouth. Palps often have a sensory function, helping the animals find food or avoid enemies. In bivalve mollusks leaflike palps generate feeding currents and help push food into the mouth.

pampas (*ecology*)   The flat treeless grassland that covers a vast portion of South America is called pampas. Like other temperate grasslands, pampas have rich, fertile soil. Native grasses and herbivores such as guanacos have been largely replaced as pampas have been developed for agriculture and ranching.

pancreas (*anatomy*)   Located near the stomach in the abdomen of vertebrates, the pancreas is a gland with both exocrine and endocrine functions. The exocrine portion produces digestive enzymes, which are carried by ducts to the duodenum portion of the small intestine. Scattered throughout the exocrine tissue are clusters of endocrine cells called islets of Langerhans. They produce the hormones insulin and glucagon, which are secreted into the blood for transport throughout the body.

panda (*animals*)   The giant panda (*Ailuropoda melanoleuca*) is native to cool bamboo forests of central and western China. Weighing up to 35 lb (160 kg) and standing as tall as 67 in. (170 cm), it is easily recognized by its coat of white and black fur. It feeds in an upright position, using its opposable thumb to grab and manipulate food, which consists primarily of bamboo.

The lesser, or red, panda (*Ailurus fulgens*), native to forests of the Himalayas, weighs up to 13 lb (6 kg) and is about 42 in. (105 cm) long. Its fur is reddish brown on the upper part of the body, black on the belly and legs. It feeds mainly on plant matter, especially bamboo.

Classification of pandas has long been a matter of much controversy. Because of DNA similarities, pandas are now placed in the bear family.

pandanus (*plants*)   The shrubs and trees comprising the genus *Pandanus* are tropical evergreens commonly called pandanus or screw pines. They have long, sword-shaped leaves that are arranged spirally and, in many species, edged with spines. The fragrant flowers, which are unisexual and lack petals, form in dense clusters at the end of branches. The large fruits, each the product of numerous flowers, resemble pineapples.

Pangaea (*geology*)   For millions of years before 180,000,000 years ago, all the continents were joined as one in the supercontinent Pangaea ("all Earth"), named by Alfred Lothar Wegener in 1912. Since that time, continental plates below the supercontinent have rifted and the movement of the plates has carried the continents to their present positions. Before Pangaea, other supercontinents existed and broke apart, a process called continental drift. About 500,000,000 years ago the major part of Pangaea, consisting of Gondwanaland, North America, and Europe, came together to form the core of Pangaea, with what is now Asia added about 300,000,000 years ago.

pangolin (*animals*)   Also called scaly anteaters, pangolins (Order Pholidota) are covered on the back and sides—from the nose to the tip of the long tail—with large overlapping scales. When attacked, pangolins roll into a ball, causing the scales to stand erect. In the toothless snout is a tongue, wormlike in appearance, that may be 10 in. (25 cm) long. Native to tropical Africa and southern Asia, pangolins are mainly nocturnal. They eat ants and termites, either on the ground or in trees. The largest, the giant pangolin (*Manis gigantea*) of Africa, grows to more than 5 ft (1.5 m) long, with the tail accounting for about half its length.

papaya (*plants*)   Native to the American tropics, the papaya (*Carica papaya*) is an evergreen tree that grows to about 20 ft (6 m). It has a stout, unbranched trunk crowned by a mass of large compound leaves. Yellow funnel-shaped flowers give rise to edible pear-shaped fruits up to 18 in. (45 cm) long. The fruits have a leathery, greenish skin; fleshy pulp; and many small black seeds. The fruits, seeds, and leaves contain papain, an enzyme that breaks down protein.

paperbark tree (*plants*)   A number of species of the genus *Melaleuca* are commonly called paperbark trees because of their bark, which consists of thin, papery layers. The trees generally prefer damp or wet soil. The genus is predominantly Australian. One species, *M. quinquenervia*, which grows to heights of 70 ft (21 m) or more and forms dense thickets, was introduced to southern Florida in the early 20th century and planted as an ornamental, soil stabilizer, and windbreak. It quickly spread to the Everglades and other wetland habitats, where it has displaced native plants and reduced biodiversity.

papovavirus (*viruses*)   The virus family Papovaviridae consists of animal viruses, including the papillomaviruses, that cause warts in birds and mammals and that have been tentatively linked to the development of certain cancers, including genital tumors. Papovavirus particles are 20-sided polyhedrons with a protein coat but no envelope; the genetic material is arranged in double-stranded circular DNA.

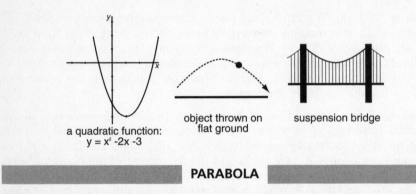

a quadratic function:
$y = x^2 - 2x - 3$

object thrown on flat ground

suspension bridge

PARABOLA

parabola (*geometry*)   One of the conic sections, the parabola is an open curve that continues to widen indefinitely away from the vertex, or turning point. It is frequently encountered as the curve of the quadratic function, the path of an object traveling in a uniform field of gravity, or as the curve formed by the cable of a suspension bridge.

paradigm (*philosophy of science*)   A paradigm is a model, but in the philosophy of science it is taken to mean the overall view of a significant part of science, encompassing not only stated theories but hidden or background assumptions. Large changes in assumptions, whether from an Earth-centered solar system to a Sun-centered one or from created species to evolved species, are called paradigm shifts.

paradox (*logic*)   A logical paradox is a declaration or argument that implies both a proposition and its opposite. The oldest may be the statement by Epimenides the Cretan [Greek: c300 BC] that "All Cretans are liars," which, if true, implies that the statement is a lie. This was put into a more precise form as "The statement I am making is false." Some paradoxes cause doubts about the foundation of mathematics, such as the 1901 "great paradox" of Bertrand Russell [English: 1872–1970]. This asks the question, "If sets that are not members of themselves are normal, is the set of normal sets itself normal?"

parallax (*astronomy*)   An apparent shift in position of an object that results from different locations of observation is called parallax. The main use of paral-

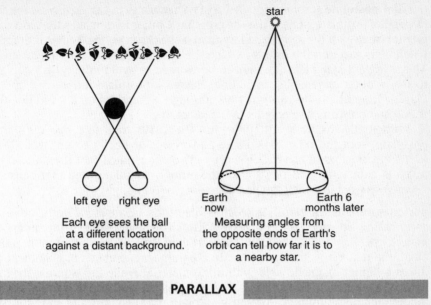

left eye   right eye

Each eye sees the ball
at a different location
against a distant background.

Earth
now

Earth 6
months later

Measuring angles from
the opposite ends of Earth's
orbit can tell how far it is to
a nearby star.

PARALLAX

lax is to determine the distance to the object: if the distance between 2 locations is known and the angle of observation is measured, the distance can be
calculated by trigonometry. The most dramatic use of parallax is in finding the
distances to nearer stars by observing shifts as seen from 2 opposite points of
Earth's revolution about the Sun. The angles involved are very small, however,
and parallax cannot be used for distant stars or other galaxies.

parallel (*geometry*)   Lines are parallel if they maintain the same distance at
every point; equivalently, in a plane parallel lines do not meet. Segments of parallel lines are also parallel. In analytic geometry, lines with the same slope are
parallel, as are other curves with the same slope for every x.

Parallel planes do not meet, but maintain the same distance at every point.
When planes are parallel, lines in different planes may be parallel or skew.

parallelepiped (*geometry*)   A parallelepiped is a polyhedron whose faces are all
parallelograms or, equivalently, a prism with parallelogram bases. The rectangular
parallelepiped, or box, has rectangles for all 6 faces. The volume (V) of a box is
the product of the 3 edges that meet at a vertex, or length (l), width (w), and
height (h): $V = lwh$. A cube is a box for which all faces are squares; if the edge
of the cube is e, then $V = e^3$.

parallelogram (*geometry*)   A parallelogram is a quadrilateral for which both
pairs of opposite sides are parallel; as defined, rectangles are parallelograms
containing a right angle, but usually parallelograms without right angles are
referred to. Opposite sides of a parallelogram are equal in length. A parallelogram that has all sides equal in length is a rhombus. The area of a parallelogram
is the product of one side (the base, b) and the altitude (or height, h): $A = bh$.

parallel postulate (*geometry*)   Of Euclid's 5 postulates in the *Elements*, the fifth, or parallel postulate, states "if a line crossing 2 other lines makes the sum of the interior angles on the same side less than a straight angle, the 2 lines if produced indefinitely meet on the side for which the sum is less than the straight angle." This awkward statement (thought to be worded to avoid infinity) is equivalent to many other statements, including Playfair's postulate introduced by John Playfair [English: 1748–1819] in 1795: through a point not on a given line there is one and only one line parallel to the given line.

   Mathematicians from antiquity until the 1830s tried to prove the parallel postulate from the other postulates and axioms of Euclid to no avail. In the 1830s the creators of non-Euclidean geometry demonstrated that the parallel postulate is completely independent of the other 4 postulates and that consistent geometries exist in which the postulate does not hold.

paramagnetism (*materials science*)   Electrons in an atom both move and carry charge, so each atom possesses a small magnetic field. If the atoms in a material pervaded by a magnetic field arrange themselves so that atomic fields align, the material is paramagnetic. In a paramagnetic material, alignment produces a higher magnetic field within the material while the external magnetic field is present, but it vanishes when the field is removed. Paramagnetic elements include aluminum, oxygen gas, platinum, sodium, and uranium. (*See also* **ferromagnetism**.)

paramecium (*protists*)   Members of the protist group called ciliates, paramecia are one-celled slipper-shaped organisms with a posterior end more pointed than the anterior end. Their bodies are covered with rows of thousands of short hairlike cilia. Coordinated beating of the cilia propels the paramecium through the water. Particularly strong cilia are located in the slitlike oral groove, where they create a flow of water inward toward the mouth. Bacteria and other minute food particles in the water are thus swept into the mouth and down the gullet.

   Depending on the species, paramecia are 0.003 (0.07 mm) to 0.012 inch (0.30 mm) long. They occur throughout the world, mainly in fresh water.

parameter (*mathematics*)   A parameter is a variable that stands for a specific number (that is, a constant) that controls some aspect of an equation. This occurs with the general equations of analytic geometry such as $y = mx + b$. The parameter m gives the slope of the graph and b is the y intercept.

   Parametric equations are pairs of equations of the form $x = at + b$, $y = ct + d$ where a, b, c, and d are parameters controlling t (usually time) to give the location of a point as (x, y).

paramyxovirus (*viruses*)   The family Paramyxoviridae consists of various animal viruses, including those that cause mumps, measles, and several respiratory infections in humans. The spherical virus particle is surrounded by a lipid envelope and has its genetic material in a single strand of RNA.

Paranthropus (*paleoanthropology*)   *Paranthropus*, with several species, represents a hominid genus that lived from about 2,500,000 to 1,000,000 years ago in eastern and southern Africa. *Paranthropus* species can be recognized by their strong teeth and jaws, adapted for chewing hard items, such as roots, instead of the softer foods favored by other australopithecines. The first *Paranthropus* fossils were found starting in 1938 in South Africa by Robert Broom [South African: 1866–1951], who named the species *P. robustus*. Louis and Mary Nicol Leakey established an East African species now called *P. boisei*, starting with a skull found in Olduvai Gorge, Tanzania, in 1959. Although once grouped with *Australopithecus* species that are thought to be human ancestors, *Paranthropus* is now viewed as an evolutionary dead end.

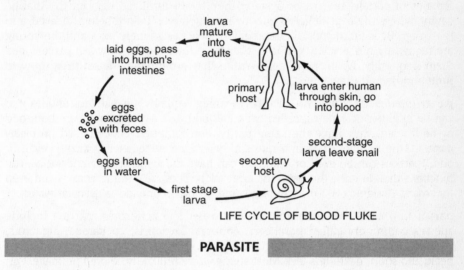

larva
mature
into
adults

laid eggs, pass
into human's
intestines

primary
host

larva enter human
through skin, go
into blood

eggs
excreted
with feces

second-stage
larva leave snail

eggs hatch
in water

secondary
host

first stage
larva

LIFE CYCLE OF BLOOD FLUKE

PARASITE

parasite (*ecology*)   A parasite is an organism that lives on or in an organism of another species—called the host—obtaining food, shelter, and other benefits at the latter's expense. Most parasites weaken but do not kill their hosts. Some, such as wheat rust, have very specific host requirements, but others, like mistletoe, parasitize a variety of species. Many parasites, such as the flukes that cause the disease schistosomiasis (also called bilharzia), alternate between 2 or more host species during their life cycle (see illustration). Most parasites produce large numbers of offspring to increase the chance that some will find new hosts.

Leeches and other parasites that live on the skin are called ectoparasites; tapeworms and other internal parasites are called endoparasites.

parasitoid (*ecology*)   Insects that are parasites during their larval stage but free-living as adults are called parasitoids. Ichneumons (Family Ichneumonidae) are examples. A female ichneumon places her eggs on or in another insect or a spider. When the eggs hatch, the larvae feed on their host. By the time the larvae metamorphose into adults, they have killed the host.

parathyroid (*anatomy*)   Located behind or within the thyroid gland of humans and many other vertebrates, parathyroids are endocrine glands that produce the hormone PTH (parathyroid hormone). Humans usually have 4 parathyroids, 2 on each thyroid lobe, each the size of a pea.

parental care (*ethology*)   The greater the amount of care provided by parents, the better the survival chances of their young. Animals that provide no parental care, such as sea urchins, produce millions of eggs and sperm, very few of which eventually develop into mature adults. Primates, which lavish extensive care on their young, generally produce 1 or 2 offspring at a time and raise them to maturity.

The simplest parental care occurs when the female chooses a protected place in which to lay her eggs. A greater degree of care is seen among the females of certain species of oysters; they retain fertilized eggs in the mantle cavity, where the eggs develop into swimming larvae before being released into the water. Bees build special chambers in their nests where eggs and the young are tended by nursery worker bees. Cats keep their kittens in warm places, feed them frequently, wash them, and—when they grow older—teach them how to hunt and defend themselves.

parity (*mathematics/physics*)   Parity (meaning "equality") is used for entities that can be in either of 2 states, called even and odd. If 2 natural numbers are both odd or both even, they have the same parity. For computer codes based on binary numerals, the number of 1s in a numeral determines whether its parity is even or odd. Functions can be even or odd, and can have the same or different parity. The functions that describe the waves for subatomic particles are odd for fermions and even for bosons. Assigning $+1$ to even parity and -1 to odd provides a quantum number.

parrot (*animals*)   Constituting the bird order Psittaciformes, parrots include species commonly called parakeets, conures, lovebirds, cockatoos, amazons, macaws, lories, and lorikeets. All have a large, rounded head; a distinctive curved beak; and short, dexterous feet. Most species have brilliantly colored plumage. Parrots are native mainly to tropical rain forests, where they live in the tree canopy, communicate via raucous calls, and feed on seeds and fruits. Almost all parrots are monogamous and form mating pairs that may endure until one of the pair dies. They are long-lived, with some larger parrots living to age 80 in captivity.

parsec (*astronomy*)   The most commonly used measure by stellar astronomers is not the light-year, but the parsec, 3.2616 light-years. A star at this distance would show a parallax shift of exactly 1 arc second (1/3600 degree) as seen from the opposite sides of Earth's orbit. Parallax shift is half the observed change in position. However, no star is quite this close, as the nearest is about a light-year farther than a parsec. The advantage of using parsecs for stars close enough to show parallax is that the distance in parsecs, d, is the reciprocal of the measured angle in arc seconds, a; that is, $d = 1/a$.

parthenocarpy (*botany*)   Some kinds of plants can produce fruit without pollination or fertilization, in a process called parthenocarpy (from Greek words meaning "virgin fruit"). Because pollination or fertilization does not occur, such fruit are seedless. Parthenocarpy is often seen among fig, banana, and pineapple plants.

parthenogenesis (*biology*)   The development of an organism from an unfertilized egg is called parthenogenesis, a term derived from Greek words meaning "virgin birth." Parthenogenesis occurs regularly in many invertebrate species, including rotifers, certain snails, and various insects. Some species of fish and lizards reproduce by parthenogenesis, as do several plants, including dandelions. In most cases these species can reproduce both sexually and parthenogenetically. The small crustaceans of the genus *Daphnia*, for example, reproduce by parthenogenesis when food is abundant, producing all female offspring. But when the environment changes—perhaps winter approaches or the pond in which they live begins to dry up—they produce both males and females and reproduce sexually. The resulting eggs have a good supply of yolk and are enclosed in a protective capsule that can withstand freezing and drying until environmental conditions improve.

particle accelerator (*particles*)   Accelerators used in particle physics crash fast-moving particles into other particles. Energetic collisions can break atomic nuclei apart, giving particle accelerators their common nickname, "atom smashers," but accelerators also turn energy into new particles and produce new elements by fusing large atomic nuclei. Particle beams can be aimed at stationary targets, although higher energies are reached by colliding 2 beams head-on.

Since 1932 accelerators have been the main tool of particle physics. That year 3 different methods of particle acceleration were introduced. Two employed high-voltage electricity to speed up protons or electrons. The third, called the cyclotron led the way for most accelerators since. It used pulsing magnetic fields to accelerate any type of charged particle with a series of pushes. Since many pushes are needed, the original design was a ring—although there are advantages to very long, straight accelerators, called linear accelerators. Giant rings several miles in diameter and linear accelerators several miles long are the main accelerators today.

particulate matter (*environment*)   Very fine particles found in air or in emissions from smokestacks, motor vehicles, and other sources of air pollution are called particulates. They include soot (carbon dust), smoke, mist, and fumes. Particulates are a component of smog and have been shown to have adverse health effects.

parvovirus (*viruses*)   Members of the virus family Parvoviridae are the smallest known viruses. Each virus particle is a 20-sided polyhedron. The genetic material is arranged in a single strand of DNA, which is surrounded by a protein coat but no envelope. Parvoviruses cause infections in a wide variety of animals; feline parvovirus and canine parvovirus can be serious problems for cats and dogs, though vaccines are available for these diseases.

Pascal, Blaise (*mathematician/physicist*)   [French: 1623–62] Pascal excelled at mathematics, contributed to physics, and invented the adding machine. Furthermore, Pascal is recognized as among the great writers of French literature. At 16 his first short paper, on what is now called Pascal's mystic hexagram, opened a subject explored by geometers ever since. In 1654 Pascal corresponded with Fermat on probability, essentially creating the subject. During this period he also investigated the pattern now known as Pascal's triangle and developed the mathematics of the cycloid. His work on the sine was so close to the calculus that reading it inspired Leibniz to invent calculus.

Pascal's law of 1648 was his main contribution to physics, but Pascal also conducted the crucial experiment that showed that a barometer measures air pressure.

Pascal's law

Pascal's law (*mechanics*)   In 1653 Pascal determined that applying pressure to a nonflowing enclosed fluid transmits the pressure equally and perpendicularly to every part of the container and every part of the fluid (Pascal's law). Pressure is force per area so a larger area of the container receives more force. For example, a pressure of 1 newton per square centimeter (1 N/cm^2) applied to the fluid results in the same pressure on each cm^2 region, so an area of 10 cm^2 feels a force of 10 N. Since the behavior of fluids is called hydraulics, devices using Pascal's principle have names such as hydraulic brakes or hydraulic press.

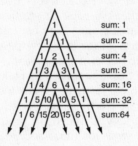

PASCAL'S TRIANGLE

Pascal's triangle (*mathematics*)   The numbers in each row of Pascal's triangle are generated by adding pairwise the numbers in the row above. The triangle was investigated by Pascal in 1654; but it was known as the Old Method in a Chinese book of 1303 and was probably also known to Omar Khayyam as early as 1100. Omar may have learned it from even earlier Indian or Chinese writers. The rows are the binomial coefficients and their sums are powers of 2. Other patterns appear on the diagonals: the second diagonal is the natural numbers and the third is the triangular numbers.

Pasteur, Louis (*chemist/microbiologist*)   [French: 1822–95] One of the founders of microbiology, Pasteur demonstrated that microorganisms are responsible for fermentation and spoilage of milk. This sounded the death knell for the theory of spontaneous generation and led to the germ theory of disease. Pasteur found that he could kill many microorganisms in wine by heating and then rapidly cooling the wine—a process now called pasteurization. In 1877 Pasteur began to study anthrax, a disease mainly of cattle and sheep. He developed a vaccine using a weakened strain of the anthrax bacterium. In 1885 he developed the first vaccination against rabies in humans.

patella (*anatomy*)   The patella (from the Latin for "little plate") is a small, flat bone located in front of the knee joint protecting this joint from injury. It is present in most mammals and some birds and reptiles. In humans it is commonly called the kneecap.

The patella is the largest and most conspicuous of the sesamoid bones. These bones develop within tendons, particularly tendons that function over bony ridges.

The knee-jerk reflex occurs when the tendon of the thigh muscle (quadriceps femoris) is tapped below the patella.

pathogen (*biology*)   Organisms that cause disease in other organisms are said to be pathogenic. Most pathogens are microscopic in size, including various species of viruses, bacteria, fungi, and protozoa. Pathogens may damage a host's tissues during growth and reproduction. For example, as ergot fungi (*Claviceps*) spread through a flower, they attack and destroy the flower's ovary. In other cases pathogens produce materials that are poisonous to the host. For example, toxins produced by staphylococcus bacteria cause many cases of human food poisoning.

To be successful, pathogens must overcome a broad variety of defenses, ranging from protective barriers—such as an animal's skin and a tree's bark—to sophisticated immune responses.

Pauli exclusion principle (*particles*)   The exclusion principle, first noted by Wolfgang Pauli [German: 1900–58] in 1925, is that no 2 electrons in an atom can share the same 4 basic quantum numbers. The first 3 quantum numbers describe the orbit of an electron in an atom; Pauli added the fourth, now known as spin. This principle explains why electrons form shells of certain sizes in atoms, the basis of the periodic table. Later theoretical developments showed that the exclusion principle applies to all fermions and accounts for the familiar idea that no 2 material objects can be in the same place at the same time.

Pauling, Linus (*chemist*)   [American: 1901–94] Pauling revolutionized the understanding of mineral structure, the formation of molecules, and the relation between molecule shape and chemical action. Early work on minerals—using atomic shapes and sizes to explain structure—presaged his better known analysis of chemical bonds. Pauling used quantum theory to analyze bonds from basic principles and applied this concept to determine characteristic structures in proteins. Such structures led to the lock-and-key theory of antibody action, provided Pauling's correct molecular explanation of sickle-cell anemia, and inspired James Watson and Francis H.C. Crick to unravel DNA. In later years, Pauling championed large doses of vitamin C to combat colds and promote long life, but this theory remains controversial.

Pavlov, Ivan (*physiologist*)   [Russian: 1849–1936] Pavlov profoundly influenced studies of animal behavior by demonstrating conditioned responses in dogs; he taught the dogs to salivate at the sound of a bell, proving that animals could be trained to respond to a neutral stimulus. Pavlov also did pioneering research on digestion, for which he was awarded the 1904 Nobel Prize in physiology or medicine.

PCB (*environment*)   Polychlorinated biphenyls (PCBs) are a group of compounds that are good heat conductors, nonconductors of electric current, and nearly insoluble in water. Beginning in the 1940s they were widely used by electric utilities for insulating purposes and by other industries in manufacturing products such as lubricants, paints, and paper coatings. Introduced into the natural environment, mainly through the disposal of wastes, PCBs proved to be extremely toxic and persistent, breaking down very slowly and remaining to be recycled through food chains for many years. Organisms that are highly susceptible to PCB poisoning include shrimp, which can be killed by water with 1 part per billion (ppb) of PCBs, and trout, which can be killed by water with 8 ppb. Sale of PCBs has been banned in the United States and elsewhere.

pea (*plants*)   Various bushy and climbing legumes produce edible seeds called peas; the plants themselves are called peas too. The green pea (*Pisum sativum*), native to western Asia, has been cultivated since prehistoric times. It is an annual vine, with pod fruits that contain as many as 10 round seeds. The seeds are smooth or wrinkled depending on the variety.

peafowl (*animals*)   Members of the pheasant family, peafowl are commonly called peacocks, a name more accurately limited to males, with females called peahens. The best-known species is the blue peafowl (*Pavo cristatus*), native to forests of India. The male has an elaborate tail that may be 5 ft (1.5 m) long; during mating season he lifts and spreads the tail, a display designed to attract females.

Peano, Giuseppe (*mathematician*)   [Italian: 1858–1932] Peano investigated the boundary between geometry and algebra. He created the Peano curve, which has continuity and yet no direction (that is, no tangent) at any point. Furthermore, the curve completely covers a given 2-dimensional region. Peano also invented a useful symbolic logic and created the most accepted axiomatic system for the natural numbers.

peanut (*plants*)   A member of the legume family native to South America, the peanut (*Arachis hypogaea*) is widely cultivated for its edible seeds. The plant, which grows to about 3 ft (90 cm) tall, has yellow flowers that last only a day. After a flower withers, it grows down into the soil and forms a pod that typically contains 2 nutlike seeds—hence the alternative name for peanuts: groundnuts.

peanut worm (*animals*)   Members of the phylum Sipuncula are unsegmented marine invertebrates. Smaller species are shaped somewhat like peanuts, hence the common name. The mouth, usually surrounded by tentacles, is at the tip of a slender proboscis that can be turned inside out. Sipunculids range in size from 0.1 in. (.25 cm) to about 24 in. (60 cm). They are bottom dwellers, found from the intertidal zone to the ocean depths.

pear (*plants*)   Members of the rose family native to Europe and Asia, pears (*Pyrus*) are deciduous trees that can grow to 60 ft (18 m) tall. Several species, particularly the common pear (*P. communis*), are cultivated for their edible, fleshy fruits; numerous varieties, or cultivars, have been developed.

pearl (*zoology*)   Certain mollusks, mainly oysters and mussels, form pearls around foreign objects that accidentally penetrate the space between the shell and the mantle (a thin tissue that surrounds the organs of the animal's body). A foreign object, such as a particle of sand, is irritating and in response the mantle secretes layers of nacre—the substance that also lines the shell—around it. The secret to culturing pearls, which are among the oldest and most valued gems, was discovered independently by 3 Japanese at the beginning of the 20th century.

peat (*soils*)   When plant remains fall into swamps or marshes, water prevents atmospheric oxygen from reaching the remains, which turn into peat. Carbon does not form gases or soluble compounds without free oxygen, so carbon remains while other plant compounds are lost. Peat is more than 50% elemental carbon, although twigs, leaves, and roots remain evident. Mosses in swamps also form peat when they die. Heat and pressure from burial in the earth turns peat into coal.

peccary (*animals*) Hoofed mammals constituting Family Tayassuidae, peccaries resemble pigs, though they have long slim legs and a thick coat of bristly hair. They stand up to 30 in. (75 cm) high at the shoulder and weigh up to 65 lb (30 kg). Native from the southwestern United States through most of Central and South America, peccaries live in groups of 3 to 100 or more. They are most active at dawn and dusk, feeding mainly on plant food that they root out with their snout. Peccaries have poor eyesight. A large musk gland on the lower back gives off an unpleasant scent that may help a group stay together.

Peking Man (*paleoanthropology*) Starting in 1927 *Homo erectus* fossils from 40 different individuals were excavated from Zhoukoudian, a large cave near Beijing (formerly Peking), China. The cave had been inhabited by humans and their ancestors off and on for at least 200,000 years. The find is still the largest single group of early hominid fossils from a single site. The fossils, collectively known as Peking Man, were lost during World War II. Although casts exist, the originals have never been found.

pelagic zone (*ecology*) The open waters of an ocean are its pelagic zone, as opposed to the bottom areas, which form the benthic zone. Pelagic animals may feed on the benthic community but most of them feed on pelagic organisms. Unicellular algae are the most important food producers of the pelagic zone. Animal plankton feed on the algae and in turn are preyed on by bigger animals.

pelican (*animals*) Easily recognized by the large expandable pouch that hangs from the lower half of the bill, pelicans (genus *Pelecanus*) live and breed in large colonies near water. They often fish together, driving prey into shallow water, then scooping up their catch. Pelicans are found in most tropical and temperate regions. The largest are more than 5 ft (1.5 m) long, with wingspans exceeding 10 ft (3 m).

pelvis (*anatomy*) The pelvis, or pelvic girdle, of a vertebrate skeleton serves to support legs or, in fish, pelvic fins. In humans and other mammals, the pelvis is shaped like a basin and consists of several fused bones. It joins the lower (sacral) vertebrae at the back and curves forward to the front of the abdomen. On each side is a large socket—the hip joint—that receives the head of the femur (thighbone).

pelycosaur (*paleontology*) Members of the order Pelycosauria were the earliest mammal-like reptiles, living about 300 million to 250 million years ago and giving rise to the therapsids. They included both plant and meat eaters and ranged in length from 2 to 13 ft (0.6 to 4 m). Some species—the carnivorous sphenacodonts such as *Dimetrodon*—had large sails on their back.

pendulum (*mechanics*) A weight (bob) suspended by a nonstretching string or light rod is a simple pendulum. Gravity pulls the weight toward the center of Earth. When the bob is pulled to one side and released, it rises nearly to the same height on the other side, losing some energy to friction. This pattern repeats over and over in approximately the same time (or period), especially for small swings. The period depends on the length of the pendulum, not on weight. The period is approximately 2π times the square root of length divided by acceleration due to gravity (increasing somewhat for a wider swing).

penguin (*animals*)   Found only in the Southern Hemisphere, penguins are stocky aquatic birds with dark backs and white breasts. They are flightless—their wings resemble stiff flippers—and clumsy on land. But they are superb swimmers and divers; emperor penguins (*Aptenodytes forsteri*) can reach depths of 870 ft (265 m) and stay submerged for as long as 9 minutes. Unlike most seabirds, penguins have a layer of blubber and a dense plumage of small, scalelike feathers that help them withstand cold weather. The emperor is the largest species, with a length of about 42 in. (115 cm); the little blue penguin (*Eudyptula minor*) is the smallest, about 16 in. (40 cm) long.

penicillin (*biochemistry*)   Discovered by Alexander Fleming in 1928, penicillin is an antibiotic produced by *Penicillium* mold. It interferes with formation of bacterial cell walls, causing the cells to burst. Since the 1940s, penicillin has been used to treat a broad range of infectious diseases. However, strains of bacteria have evolved that are resistant to penicillin. For example, many *Staphylococcus* bacteria produce penicillinase, an enzyme that destroys penicillin.

Penicillium (*fungi*)   Classified in the sac fungi phylum Ascomycota, *Penicillium* is a common genus of blue and green molds. Most species are saprophytes that grow on decaying fruits and other foods, and on a great variety of moist plant and animal products (paper, lumber, leather, etc.). As such, they are responsible for tremendous economic losses. Other species cause disease in animals, including penicillosis morneffi in humans. But a few species are economically beneficial. *P. roqueforti* and *P. camemberti* are involved in the ripening of Roquefort and Camembert cheeses; as they metabolize compounds in milk, they impart flavor to the cheeses (the blue-green "veins" in the cheeses are masses of fungal spores). In 1928 Alexander Fleming discovered that *P. notatum* produces a substance that kills bacteria. Named penicillin, it became the first antibiotic.

penis (*anatomy*)   The male organ of copulation in mammals, a few birds, and some reptiles is the penis. Its function is to deposit sperm directly into the body of a female. In mammals a tube called the urethra passes through the penis. During copulation, sperm—mixed with a fluid to form semen—are ejaculated from the penis. At other times the urethra carries urine from the urinary bladder through the penis to the outside.

pentagon (*geometry*)   Any polygon with 5 sides is a pentagon ("5 angles"), although the most commonly encountered pentagons are regular figures. A regular pentagon has 5 sides of equal length and 5 angles of 108° each.

pentagram (*geometry*)   A pentagram (sometimes called a pentacle) is a 5-pointed star that is a regular figure; it is most easily formed by connecting each of the vertices of a regular pentagon to the 2 opposite vertices. When created this way, the sides will form a second pentagon inside the first that is not part of the pentagram. The same procedure can be repeated again and again to form an endless sequence of smaller and smaller pentagrams and pentagons.

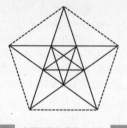

PENTAGRAM

penumbra (*solar system*)   *Umbra* is Latin for shadow, and penumbra literally means "almost shadow." Although a shadow from a single light source seems to have sharp edges, waves always bend slightly near edges, producing a lighter shadow, or penumbra, around the main shadow. For a lunar or solar eclipse, the penumbra is larger than the shadow itself. Lighter rings around sunspots are also called penumbras, although they are not shadows at all.

peony (*plants*)   Famed for their large, fragrant flowers, peonies (*Paeonia*) are shrubby perennials native mainly to Eurasia and North America. The tallest is the tree peony (*P. suffruticosa*), which grows to 7 ft (2 m). Extensive hybridization has created hundreds of varieties, or cultivars.

pepper (*plants*)   Species of 2 different plant families are commonly called peppers. Members of the genus *Capsicum*, Family Solanaceae, are native to tropical and subtropical Central and South America. These peppers bear large fleshy-walled fruits that contain a colorless irritant called capsaicin. Sweet or bell peppers have only small amounts of capsaicin while hot or chili peppers contain larger doses. Dried fruits are ground up to produce paprika and cayenne pepper.
   Peppercorns are dried berries of the vine pepper (*Piper nigrum*), Family Piperaceae, native to India. Black pepper is made from the entire berry; white pepper is made from berries with the outer skin removed.

peptide (*biochemistry*)   Two amino acids combine with one another by forming a peptide bond, which links the carboxyl group of one amino acid molecule to the amino group of the other. Any organic compound whose molecules consist of 2 or more amino acids is called a peptide. A dipeptide is formed from 2 amino acid molecules. A polypeptide is a long chain of amino acids. Proteins are made up of 1 or more polypeptide chains.

percent (*arithmetic*)   Percent is the ratio of one number to another expressed as the equivalent ratio of some number to 100. For example, the ratio of 35 to 70 is equivalent to the ratio of 50 to 100, so 35 is 50 percent, written 50%, of 70. Percents use numbers, but are not numbers themselves, and cannot be added, subtracted, multiplied, or divided. To operate with percents, convert them from ratios to numbers, usually to decimal fractions (50% is converted to 0.5 or to 1/2).

perching bird (*animals*)   The largest order of birds, Passeriformes ("sparrow-shaped"), comprise the approximately 5500 species of perching birds. The birds range from wrens a few inches long to the raven (*Corvus corax*), which averages 2 ft (0.6 m) long with a wingspread exceeding 4 ft (1.2 m). Perching birds have

4 toes on each foot, 3 pointing forward and 1 backward; together with muscles and tendons this arrangement enables the birds to easily grasp and tighten a grip on perches such as branches, leaf stems, and telephone wires.

The great majority of perching birds are classified as oscines, or songbirds. They include larks, mockingbirds, sparrows, and warblers. Though not the only birds that make vocal sounds, songbirds produce the most intricate and melodious songs.

perennial plant (*botany*)   Plants that live for more than 2 growing seasons are called perennials. In the first year or first few years, growth is exclusively vegetative, limited to roots, stems, and leaves. Thereafter, most perennial species flower annually. Herbaceous perennials, which have soft, green stems, may die back to underground roots and stems at the end of a growing season, then grow new aboveground parts at the beginning of the next growing season. Dahlias, peonies, and rhubarb are examples of herbaceous perennials. Woody perennials, including trees and shrubs, have rigid stems that grow in length and diameter every growing season.

perfect number (*numbers*)   A perfect number is a natural number for which the sum of its divisors (including 1 but not including itself) is the number itself. The number 6 qualifies because its divisors are 1, 2, and 3 for which $1 + 2 + 3 = 6$. Most numbers are deficient (the sum of the divisors is less than the number) and many are abundant (the sum is greater than the number). The next perfect number after 6 is 28 ($1 + 2 + 4 + 7 + 14 = 28$); the third is 496.

perigee (*solar system*)   The lowest point of an orbit about Earth, whether the orbiting body is the Moon or an artificial satellite, is its perigee. The suffix *–gee* means Earth. A similar point in an orbit about the Sun is a perihelion, where the suffix *-helion* means Sun.

perihelion (*solar system*)   The point in an orbit about the Sun that is nearest to the Sun is the perihelion, while the farthest is the aphelion. The perihelion is often confused with the perigee, the point in an orbit about Earth nearest the Earth.

perimeter (*geometry*)   The combined length of all sides of a polygon is the perimeter. Similarly the length of any simple closed curve is its perimeter, although the perimeter of a circle is more often called the circumference.

period (*waves*)   In mathematics the length of a wave from crest to crest is its period. For example, a sine wave has a period of 2π. Any functions that regularly repeat values also have periods, such as the tangent function, period π.

For physical waves, such as sound or light, the quantity that repeats is measured over time, not distance, and period (p) is the reciprocal of the frequency (f): $p = 1/f$. Any repeated motion, as for a pendulum or vibrating atom, similarly has a period—the amount of time for 1 repetition.

periodic function (*trigonometry*)   Trigonometric functions repeat the same values at regular intervals, making them periodic functions. The basic interval of repetition, or period, is 360° or 2π radians for the sine, cosine, secant, and cosecant; and 180° or π radians for the tangent and cotangent. All periodic functions of real numbers are variations of the 6 trigonometric functions, but some functions of complex numbers, known as elliptic functions, are periodic with 2 periods for each function.

periodic table (*chemistry*)   In 1869 Mendeleyev created the first periodic table. He arranged a list of the then-known 63 elements by atomic mass, demonstrating that similar properties occur every 8 elements (with minor adjustments). Mendeleyev correctly predicted the properties of 3 new elements to fit blanks in the table. In the 20th century the periodic table was improved by arranging the elements in order of atomic number instead of atomic mass. (See table on following page.) This clearly showed where blanks existed.

The modern form of the periodic table is more complex, but as with Mendeleyev's original, each row starts with an alkali metal—hydrogen is thought to become a metal under great pressure. Elements in a column have similar properties. As you proceed along a row, elements become less metallic. Each row ends with a noble gas. The columns from left to right each represent 1 more electron in the outer shell. By the Pauli exclusion principle the shells fill at 2, 8, 8, 18, and so forth electrons, so the last column represents atoms with filled shells.

peristalsis (*physiology*)   The walls of the digestive tract—from the esophagus to the large intestine—contain muscles that contract and relax in sequence, creating a wavelike movement called peristalsis. Peristalsis gradually pushes matter through the digestive tract. In the stomach and small intestines, peristalsis also helps to mix food with digestive juices.

peritoneum (*anatomy*)   A smooth, transparent membrane of the abdominal cavity of vertebrates, the peritoneum consists of 2 layers. One layer, the parietal peritoneum, lines the cavity; the other layer, the visceral peritoneum, covers the organs. Between the 2 layers is a space, the peritoneal cavity. Mesenteries are double layers of peritoneum that attach organs to abdominal walls and through which nerves and blood vessels connect to the organs.

permafrost (*soils*)   In arctic regions or near glaciers, water-soaked soil may become permanently frozen into a kind of soil called permafrost. Permafrost can extend as far as 1600 ft (500 m) from the surface. The top layer of permafrost may melt for a few feet in summer, causing buildings or heavy objects to sink. Because lower levels stay frozen all year, remains of animals or plants may be preserved in permafrost with soft tissues intact for thousands of years.

permutation (*mathematics*)   Any arrangement of the members of a set into a particular order is called a permutation. Counting the number of permutations for a set of a particular size is the beginning of combinatorial mathematics (combinatrics), often called permutations and combinations. The number of permutations of a set with n members is the factorial $n!$. Consider a set with 5 members, for example. Any of the 5 can be in first place, leaving 4 choices for second place, 3 for third, 2 for fourth, and 1 for last; this means there are $5 \times 4 \times 3 \times 2 \times 1 = 5!$, or 120, permutations of 5 members. Questions involving permutations of some smaller number r of n members in a set, or permutations when some of the members are identical, are more difficult, but are analyzed by the same basic approach.

perpendicular (*geometry*)   Two lines or segments of lines that meet at right angles are perpendicular to each other. A line can also be perpendicular to a plane; it must meet the plane so that it is perpendicular to both of 2 intersecting lines.

periodic table

THE PERIODIC TABLE OF THE ELEMENTS

Legend: atomic number — chemical symbol — atomic mass — name of element (example: 2 / He / 4.00 / Helium)

IA	IIA	IIIB	IVB	VB	VIB	VIIB	VIII	VIII	VIII	IB	IIB	IIIA	IVA	VA	VIA	VIIA	noble gases
1 H 1.01 Hydrogen																	**2** He 4.00 Helium
3 Li 6.94 Lithium	**4** Be 9.01 Beryllium											**5** B 10.81 Boron	**6** C 12.01 Carbon	**7** N 14.01 Nitrogen	**8** O 16.00 Oxygen	**9** F 19.00 Fluorine	**10** Ne 20.18 Neon
11 Na 22.99 Sodium	**12** Mg 24.31 Magnesium											**13** Al 26.98 Aluminum	**14** Si 28.09 Silicon	**15** P 30.97 Phosphorus	**16** S 32.07 Sulfur	**17** Cl 35.45 Chlorine	**18** Ar 39.95 Argon
19 K 39.10 Potassium	**20** Ca 40.08 Calcium	**21** Sc 44.96 Scandium	**22** Ti 47.88 Titanium	**23** V 50.94 Vanadium	**24** Cr 52.00 Chromium	**25** Mn 54.95 Manganese	**26** Fe 55.85 Iron	**27** Co 58.93 Cobalt	**28** Ni 58.70 Nickel	**29** Cu 63.55 Copper	**30** Zn 65.39 Zinc	**31** Ga 69.72 Gallium	**32** Ge 72.61 Germanium	**33** As 74.92 Arsenic	**34** Se 78.96 Selenium	**35** Br 79.90 Bromine	**36** Kr 83.80 Krypton
37 Rb 85.47 Rubidium	**38** Sr 87.62 Strontium	**39** Y 88.91 Yttrium	**40** Zr 91.22 Zirconium	**41** Nb 92.91 Niobium	**42** Mo 95.94 Molybdenum	**43** Tc (98) Technetium	**44** Ru 101.07 Ruthenium	**45** Rh 102.91 Rhodium	**46** Pd 106.4 Palladium	**47** Ag 107.87 Silver	**48** Cd 112.41 Cadmium	**49** In 114.82 Indium	**50** Sn 118.71 Tin	**51** Sb 121.74 Antimony	**52** Te 127.60 Tellurium	**53** I 126.90 Iodine	**54** Xe 131.29 Xenon
55 Cs 132.91 Cesium	**56** Ba 137.33 Barium	Lanthanide series (see below)	**72** Hf 178.49 Hafnium	**73** Ta 180.94 Tantalum	**74** W 183.85 Tungsten	**75** Re 186.21 Rhenium	**76** Os 190.2 Osmium	**77** Ir 192.22 Iridium	**78** Pt 195.08 Platinum	**79** Au 196.97 Gold	**80** Hg 200.59 Mercury	**81** Tl 204.38 Thallium	**82** Pb 207.2 Lead	**83** Bi 208.98 Bismuth	**84** Po (209) Polonium	**85** At (210) Astatine	**86** Rn (222) Radon
87 Fr (223) Francium	**88** Ra 226.03 Radium	Actinide series (see below)	**104** Rf (261) Rutherfordium	**105** Db (262) Dubnium	**106** Sg (263) Seaborgium	**107** Bh (262) Bohrium	**108** Hs (265) Hassium	**109** Mt (266) Meitnerium	**110** (269)	**111** (272)	**112** (277)		**114** (281)		**116** (289)		**118** (293)

Notes: alkali metals (IA); alkaline earth metals (IIA); transition metals (IIIB–IIB); nonmetals; noble gases.

rare earth elements – Lanthanide series

57 La 138.91 Lanthanum	**58** Ce 140.12 Cerium	**59** Pr 140.91 Praseodymium	**60** Nd 144.24 Neodymium	**61** Pm (145) Promethium	**62** Sm 150.4 Samarium	**63** Eu 151.96 Europium	**64** Gd 157.25 Gadolinium	**65** Tb 158.93 Terbium	**66** Dy 162.50 Dysprosium	**67** Ho 164.93 Holmium	**68** Er 167.26 Erbium	**69** Tm 168.93 Thulium	**70** Yb 173.04 Ytterbium	**71** Lu 174.97 Lutetium

Actinide series

89 Ac 227.03 Actinium	**90** Th 232.04 Thorium	**91** Pa 231.04 Protactinium	**92** U 238.03 Uranium	**93** Np 237.05 Neptunium	**94** Pu (244) Plutonium	**95** Am (243) Americium	**96** Cm (247) Curium	**97** Bk (247) Berkelium	**98** Cf (251) Californium	**99** Es (252) Einsteinium	**100** Fm (257) Fermium	**101** Md (258) Mendelevium	**102** No (259) Nobelium	**103** Lr (260) Lawrencium

PERIODIC TABLE

Two planes are perpendicular if each contains a line that is perpendicular to the line formed by the intersection of the planes; the planes form a right dihedral angle.

On the coordinate plane, lines are parallel if the product of their slopes is -1; that is, the slopes are negative reciprocals of each other.

perpetual motion (*thermodynamics*)   Nonexistent engines that continue to move with no outside source of energy are said to possess perpetual motion; but the name is also applied to nonexistent machines that provide more work than the energy used to operate them. The laws of thermodynamics, especially the second law of thermodynamics, ensure that neither type can be constructed. Sometimes the statement that perpetual motion is impossible is called the first law of thermodynamics.

Perseid shower (*solar system*)   In most years, the largest number of meteors that can be seen in a short period of time comes around August 12, when a "shower" of meteors—perhaps as many as 65 an hour—appears to emanate from the constellation Perseus. The Perseid shower usually begins in late July and ends by the middle of August. It is caused by Earth passing through debris left by the comet Swift-Tuttle 1852 III.

pesticide (*environment*)   People use synthetic chemicals called pesticides to kill organisms that are harmful or in some other ways considered to be pests. Major types of pesticides include fungicides, to control fungi; herbicides, for weeds; and insecticides, for insects. Although pesticides are valuable in killing disease-carrying organisms and protecting crops, they have a high environmental cost. Broad-spectrum pesticides kill beneficial as well as harmful organisms. Many pesticides do not break down quickly into harmless substances but remain in the environment for years. As a result of biological magnification they accumulate in the fatty tissues of fish, birds, and mammals, where they can cause reproductive failure, birth defects, and other health problems.

Peterson, Roger Tory (*naturalist/illustrator*)   [American: 1908–96] Millions of people identify unfamiliar birds by leafing through field guides written and illustrated by Peterson, who is credited with making birding a hobby accessible to all. His text and drawings are uncluttered, focusing on distinctive markings of similar species—the dark head of the greater scaup (*Aythya marila*) and the white face mark of the female lesser scaup (*A. affinis*), for instance.

petiole (*botany*)   The leaf stalk, or petiole, supports the leaf blade and carries materials between the blade and the vascular tissue in the stem. In some plants, petioles can bend, positioning the blade to take maximum advantage of sunlight. Leaves that lack petioles are termed sessile.

petrel (*animals*)   Members of the shearwater family, petrels are seabirds with hooked bills; long, powerful wings; and webbed feet. They feed on fish and other seafood, snatching prey in their bill as they hover at the water's surface. Petrels come ashore only to mate and raise their young, often in colonies of thousands.

petrification (*paleontology*)   A fossil formed when the tissues of a dead organism are slowly impregnated and replaced by minerals, in such a way that the form and perhaps even the cell structure of the organism is retained, is said to be petrified (from Latin words meaning "made into rock"). Petrified wood, such as the 50-million-year-old tree trunks in Petrified Forest National Park in Arizona, is the best-known example.

petroglyph (*archaeology*)   A petroglyph ("rock carving") is a line drawing that has been carved or painted on a rock, especially on a large boulder, cliff, or cave wall. Although the term petroglyph in the United States often refers to carvings and drawings in the American Southwest, petroglyphs are found nearly everywhere that erosion does not remove them, including in dry places in Europe and the deserts of Africa. The oldest known petroglyphs, in Australia, date from 45,000 years ago.

petroleum (*rocks and minerals*)   Petroleum, sometimes called crude oil, is the liquid hydrocarbon mixture that occurs in natural deposits. It may consist of 500 or more compounds, ranging from simple methane to complex aromatics. Petroleum is distilled into various useful materials, including gasoline, kerosene, diesel fuel, lubricating oils, paraffin wax, and tar, listed here roughly in ascending order of boiling point and carbon content.

pH (*chemistry*)   The strength of acid or base solutions is measured on a scale called pH (pronounced "pee aitch) that ranges from near 0 for very strong acids (for example, hydrochloric acid) to near 14 for very strong bases (for example, sodium hydroxide, or lye). A pH of 7 is neutral. The numerical values are the negatives of the common logarithms of the number of moles of hydrogen ions (protons) per liter of the solution; the name means potential (or power) of hydrogen. Since pH is a logarithm, a difference of 1 means an increase or decrease by a factor of 10 in the amount of acidity.

phagocyte (*cells*)   Animal cells called phagocytes ("eating cells") ingest bacteria, dead cells, and other debris, thereby playing an important role in protecting against disease. A phagocyte moves like an amoeba, forming pseudopods to surround and engulf a particle. The particle becomes trapped in a vacuole within the phagocyte's cytoplasm, where it can then be destroyed. Human phagocytes include macrophages and certain white blood cells.

phalanger (*animals*)   A diverse group of marsupials native to Australia and nearby islands, phalangers get their name from the fact that the second and third toes (phalanges) of their back feet are joined together. Phalangers are small, with soft fur, and adapted for life in trees, with feet designed for climbing and grasping and, in many species, a prehensile tail that wraps around branches. Phalangers are nocturnal, feeding on insects and plant matter. The group includes possums (an entirely different group from the opossums), cuscuses, and koalas.

pharynx (*anatomy*)   The pharynx (from the Greek word meaning "throat") is the part of the digestive tract between the mouth and the esophagus. It often is muscular—planarian flatworms can extend the pharynx out from the mouth to capture microscopic food. Some animals, such as squid, have teeth in the pharynx.

   In fish and aquatic amphibians, the pharynx contains gill slits and has no digestive function. In air-breathing vertebrates, the pharynx is a passageway both for food and for air moving to and from the lungs. Mammals have a flap of tissue called the epiglottis at the base of the tongue, above the opening to the larynx (voice box) at the top of the trachea—the tube leading to the lungs. Prior to swallowing, the epiglottis covers the larynx to prevent food and liquids from going into the larynx and trachea.

phase (*waves*) Phase shift for a mathematical wave, such as a sine curve, is an amount of displacement to the left or right from the origin, the same as phase angle. Physical waves can be measured from any point, so phase only has meaning for 2 or more physical waves. Identical waves are in phase when the high points coincide in space and out of phase when high points of one match low points of the other. Identical waves that are exactly out of phase cancel each other. Waves that are not identical or identical waves partly in phase show a pattern of interference as they go in and out of phase.

phase angle (*trigonometry* The amount that a sine or cosine wave is displaced from the vertical axis is called the phase angle of the wave; for example, the function $y = \sin (x - \pi/4)$ has a phase angle of $\pi/4$ because it has been shifted that far to the right of $y = \sin x$.

phase of matter (*materials science*) Traditionally matter exists in 3 states, solid, liquid, and gas, but the modern view is that there are 5 phases of matter: the 3 traditional states plus plasma and a phase that combines Bose-Einstein condensates with a similar low-temperature quantum phase called a Fermi-Dirac condensation. For most elements phases of matter change in response to heating. From cold to hot, then, phases consist of the condensations, solid, liquid, gas, and plasma. Some elements fail to exist in all phases.

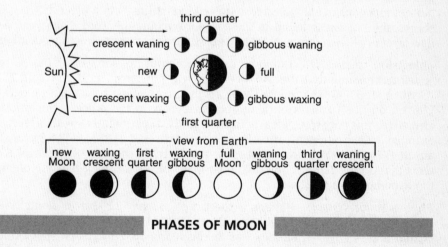

PHASES OF MOON

phases of Moon (*solar system*) At any time, half the Moon is lit by the Sun and half is dark, but we on Earth usually see part of the sunlit side only. That part goes from completely lit (full Moon) to half lit (quarter Moon waning) to a sliver (crescent Moon waning). Then the sunlit part is on the far side of the Moon, so we see no Moon at all (new Moon)—although a faint new Moon may be visible in earthshine. These changes, or phases of the Moon, then reverse from crescent Moon waxing to full Moon as the Moon moves about Earth over a period that takes 29 days for all phases. Waning refers to the 2-week period when illumination is lessening, while waxing phases are increasing.

pheasant (*animals*)   The family Phasianidae—worldwide in distribution except in polar regions—includes pheasants, partridges, jungle fowl, peafowl, and quails. Unlike other fowl, these chickenlike birds do not have feathers on their feet or hiding their nostrils. Many have combs, wattles, and other areas of bare skin on the head. In many species the males have long, elaborate tail feathers.

pheromone (*biochemistry*)   Pheromones are chemical messengers secreted and released into the environment by organisms to elicit a very specific response in other members of the same species. For example, female gypsy moths secrete a pheromone to attract male gypsy moths. Ants secrete a pheromone to mark trails leading to food. Male dogs secrete a pheromone in their urine to alert other male dogs of their presence. A queen termite secretes a pheromone that prevents the development of another queen in the colony.

Philippine Plate (*geology*)   The Philippine Plate, although named for the Philippine Islands, does not carry those or any islands of consequence. This relatively small section of Earth's lithosphere in the western Pacific is bounded by deep trenches, including the Marianas Trench, the lowest point on Earth's surface, some 35,000 ft (10,800 m) deep, on the east, the Ryukyu Trench on the northwest, and the Philippine Trench on the southwest.

phloem (*botany*)   One of the two types of conducting, or vascular tissue of plants, phloem functions mainly to carry dissolved food throughout a plant. In trees and shrubs, the phloem is found in the bark surrounding the wood. Each year, as a new layer of phloem forms, older phloem is pushed outward, into the outer bark.

phlogiston (*chemistry*)   Common experience with fire shows that the process is self-terminating and the product, ash, will not burn. In 1697 Georg Stahl [German: 1660–1734] proposed that a substance, which he named phlogiston ("to set on fire") causes burning and disappears in the process. Superficially attractive, this incorrect idea persisted. Early experimenters with gases, for example, called nitrogen (or carbon dioxide, in some cases) de-phlogistonated air. In 1774, however, Lavoisier showed that burning consists of chemical reactions that do not remove any unknown material, establishing that the phlogiston hypothesis is not needed.

Phobos (*solar system*)   Phobos is the larger—13 mi (22 km) in diameter—of the 2 irregularly shaped satellites of Mars. Named after one of the god Mars's chariot horses ("fear"), Phobos was discovered in 1877 by Asoph Hall [American: 1829–1907].

phonon (*particles*)   Vibrations of atoms in a solid give rise to small waves that correspond to ordinary sound waves. However, the waves are so small that they obey quantum laws, and therefore can also be interpreted as particles. The particles, known as phonons, are bosons, like photons of light. A well-known effect of phonons is ordinary superconductivity (high temperature superconductivity may have some other explanation). In some materials at temperatures close to absolute zero, electrons interact with phonons to form pairs that can carry a current without loss.

phosphorescence (*materials science*)   Phosphorescence is the glow caused when energy absorbed by atoms from light is released, as electrons fall back into lower energy states. An incident photon can in some materials be absorbed by an electron, lifting it temporarily to a higher state. A photon of the same energy will be released when the state returns to normal, which may be some time later. All high-energy electromagnetic waves can induce this effect, including ultraviolet light and X rays. Incident electrons can also produce phosphorescence in suitable materials; this is the basis of cathode-ray tubes used in displays such as television or computer monitors. (*See also* **bioluminescence**.)

phosphorus (*elements*)   Phosphorus was the first new element to have been discovered since ancient times. In 1669 alchemist Hennig Brand [German: c1630–c1692] obtained a glowing white nonmetal from urine. He named it from the Greek *phosphoros* ("light-bringer"). White phosphorus glows because it oxidizes in air—in effect, slowly burning. A different allotrope, red phosphorus, is less reactive, but able to start matches burning. Phosphorus, chemical symbol P, atomic number 15, is essential for life; it contributes the P to the ATP molecule and is vital for many other organic processes.

photoelectric phenomena (*particles*)   There are several interactions of substances involving photons of light and electrons. The photovoltaic effect is a flow of current between 2 electrodes when one is exposed to light; it is the basis of photovoltaic cells used to produce electricity from sunlight. The photoconductive effect occurs when light strikes certain materials, such as selenium, and a flow of current increases. This process is used in electric eyes and remote controls. The photoelectric (or photoemissive) effect occurs when electromagnetic waves, or light, strike a metal and free electrons from its surface. In 1905 Einstein explained the characteristics of this effect in terms of quanta of light, now called photons. The basis of all photoelectric phenomena is that the photon carries the electromagnetic force that affects all charged particles.

photon (*particles*)   The photon is an electromagnetic wave behaving as a particle, an effect first explained by Einstein in 1905. People usually think of the photon as the particle of light, but it is also the particle form of radio waves, X rays, and gamma rays. Photons interact well with all charged particles, but we perceive the world largely as a result of the interactions between photons and electrons. An energetic photon can suddenly turn into an electron and a positron, the reverse of the interaction that occurs when a particle and an antiparticle annihilate each other and produce energy in the form of a photon.

photoperiodism (*botany*)   Growth and developmental processes in plants often are related to the changing seasons, which plants measure by detecting the relative lengths of day and night in a response called photoperiodism. Onset of flowering is often affected by the photoperiod. Short-day plants, such as violets and most chrysanthemums, flower when the daily photoperiod is shorter than 13 to 14 hours. Long-day plants, such as clover and corn, flower when the daily photoperiod is longer than 13 to 14 hours. Flowering by day-neutral plants, such as tomatoes and dandelions, is not affected by the photoperiod.

photosphere (*solar system*)   The part of the Sun we usually see from Earth is called the photosphere. It is thought of as the "skin" of the Sun, even though the Sun is made from gas and has no solid surface. When you read that the radius of the Sun is 432,000 mi (700,000 km), that is the distance from the center of the Sun to the photosphere.

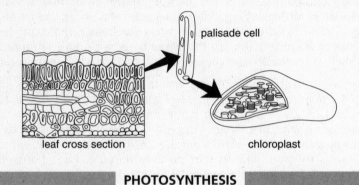

palisade cell

leaf cross section   chloroplast

PHOTOSYNTHESIS

photosynthesis (*botany*)   Without the food-making process called photosynthesis, life as we know it would not exist. In the presence of light, green plants and algae produce their own food and, in the process, release oxygen into the air. Animals eat the plants for food, either directly or indirectly, and use oxygen to obtain energy from the food.

Photosynthesis involves a complex sequence of light and dark reactions that can be summarized in a simple formula:

$$6CO_2 + 6 H_2O + \text{light energy} \rightarrow C_6H_{12}O_6 + 6O_2.$$

This says that 6 molecules of carbon dioxide plus 6 molecules of water plus light energy are converted into 1 molecule of glucose sugar plus 6 molecules of oxygen. The light energy is changed into chemical energy that holds the glucose molecule together. Plants and algae can then turn the glucose into other substances, including complex sugars, starches, proteins, and fats.

Photosynthesis takes place in special cell structures, typically chloroplasts. It requires the presence of the light-absorbing pigment chlorophyll plus various enzymes.

physics (*sciences*)   Physics is the basis of all sciences because it deals with the most elemental parts of the universe. Originally physics was concerned with matter, forces, and motion, but later discoveries led physics to incorporate electromagnetism; energy, including thermodynamics; atoms and subatomic particles; and the structure of space (relativity) and the universe itself (cosmology).

physiology (*sciences*)   The goal of physiology is to understand how molecules, organelles, cells, tissues, and organs function individually, in relation to one another, and as integrated parts of a living organism. Three premises are basic to physiology: function is dependent on structure; the chemical and physical properties of living structures are genetically determined; and maintaining a constant internal environment (homeostasis) is critical to an organism and to each of its parts.

pi (π) (*numbers*) The number π is an infinite decimal that begins 3.14159265358979 . . . , often approximated by 3.14 or by 22/7. Originally, π was the ratio of the circumference of any circle to its diameter (π is the first letter of Greek for circumference), but many other important occurrences in mathematics are based upon variations of this definition. Proof that π is an irrational number was found in 1761 by Johann Heinrich Lambert [German: 1728–77] and that π is a transcendental number in 1882 by Ferdinand Lindemann [German: 1852–1939]. Lindemann's proof demonstrates that it is impossible to "square the circle"; that is, to construct a square with a straightedge and compass that has the same area as a given circle.

picornavirus (*viruses*) Members of the virus family Picornaviridae infect and cause disease in various animals, including humans. The family includes viruses that cause the common cold (rhinovirus), poliomyelitis (poliovirus), and infectious hepatitis (hepatitis A virus). Each picornavirus particle is a 20-sided polyhedron with a protein coat but no envelope; the genetic material is arranged in a single strand of RNA.

piezoelectricity (*electricity*) When a force is applied to both sides of any of various asymmetrical crystals, including quartz, positive charge builds on 1 face and negative charge on the opposite face. This potential difference can induce a small electric current, called piezoelectricity, in a circuit. The effect also works in reverse. Applying a current changes the length of the crystal as the faces repel or attract each other.

An alternating current causes a piezoelectric crystal to vibrate rapidly, while rapid changes in pressure on such a crystal produce alternating current, making piezoelectricity the basis of microphones, loudspeakers, and other acoustic devices.

pig (*animals*) Family Suidae comprises hoofed mammals known as pigs, boars, and hogs; all are native to Eurasia and Africa. Most species have a movable snout, used to root for food, small ears and eyes, and a stocky body. The domestic pig, of which there are numerous breeds, is believed to be derived from the European wild boar (*Sus scrofa*).

pigeon (*animals*) The bird family Columbidae consists of approximately 300 species. In general larger species are called pigeons and smaller species are called doves. Most have a small head, compact body, short legs, and sturdy feet. Unusual among birds, pigeons drink by sucking up water without raising their head. The largest is the crowned pigeon (*Goura cristata*) of New Guinea, which may be 33 in. (84 cm) long. The many breeds of domestic pigeons have evolved from the rock dove (*Columba livia*). Pigeons and doves feed on seeds, fruits, berries, buds, and in some cases, insects. During nesting season both sexes produce "pigeon milk," a fat- and protein-rich secretion from the lining of the crop. This is fed by regurgitation to the young during their first days of life.

pika (*animals*)   Slightly smaller than their relatives, rabbits and hares, pikas (Family Ochotonidae) have short, rounded ears and short legs, with front and hind limbs of nearly equal length. They are native to cool regions of Asia and western North America.

pike (*animals*)   Native to fresh waters of the Northern Hemisphere, pikes are long predatory fish with voracious appetites. Their mouth, shaped like a duck's bill, is armed with sharp teeth. The fins lack spines; the dorsal and pelvic fins are located far back on the body. The most widely distributed species, the northern pike (*Esox lucius*), may grow to 7 ft (2.2 m) long. The pike family, Esocidae, also includes pickerels and the muskellunge.

pill bug (*animals*)   Also called sow bugs, pill bugs are not bugs but crustaceans of the order Isopoda, which also includes aquatic forms. The approximately 5000 known species of pill bugs are by far the most successful group of terrestrial crustaceans. They hide under stones and in other moist spots during the day and feed at night, mainly on vegetation. They are less than 0.6 in. (1.5 cm) long. Female pill bugs have a brood pouch, or marsupium, in which the eggs and young are carried.

Piltdown Man (*paleoanthropology*)   In 1912 a successful hoax was perpetrated on the paleoanthropology community by faked fossils of an early human that had been put into a gravel bed in Piltdown, England. The fossils, christened *Eoanthropus dawsoni*, and popularly known as Piltdown Man, suited English paleoanthropologists because they suggested early humans developed in England. In 1953 chemical analysis showed that the human skull of Piltdown Man was a modern one stained to match a fossil jawbone of an orangutan. The two were deliberately planted in the same site for rediscovery, probably as a practical joke that got out of hand.

pine (*plants*)   Native mainly to the Northern Hemisphere, pines (*Pinus*) are large evergreen trees with long needles that usually grow in clusters of 2 to 5. The seed-forming cones are the largest of all conifer plants. The sugar pine (*P. lambertiana*) of northwestern North America, which grows to heights of almost 250 ft (75 m), has cones up to 20 in. (50 cm) long. The bristlecone pine (*P. aristata*) of southwestern North America grows to about 45 ft (14 m) tall and often is much shorter, with a stunted appearance, but it has a life span of more than 4000 years. The pine family, Pinaceae, comprises about 150 species of conifers. In addition to pines, these include cedars, firs, hemlocks, larches, and spruce.

pineal body (*anatomy*)   The pineal body is a small organ located deep within the brain of most vertebrates; in humans it is slightly bigger than a pea. In some animals, including certain fish and amphibians and the primitive reptile tuatara, pineal cells are sensitive to light and form a so-called third eye. More properly termed the parietal eye, it cannot form images but helps the animal measure day length.

   In mammals the pineal body is a gland. It synthesizes and secretes the hormone melatonin, almost entirely at night. In seasonally breeding animals, melatonin production decreases in the spring, as days become longer; this appears to be linked to the seasonal enlargement of sex organs and the onset of mating.

pineapple (*plants*)   A member of the bromeliad family, the pineapple (*Ananas comosus*) is a tropical plant native to South America. Growing to heights of about 3 ft (0.9 m), it has 30 to 40 long stiff leaves with spiny margins arranged in a circular cluster around a thick stem. Many varieties exist—some cultivated for their fruits, others grown as ornamentals. What we call the fruit actually is a grouping of many small hexagonal fruits. It forms from a cluster of many flowers that become fleshy and fuse as they mature.

pink (*plants*)   Native to Europe, Asia, and Africa, pinks are annual or perennial herbs that have long been cultivated for their fragrant flowers. The genus name, *Dianthus*, was first given to these plants by Theophrastus about 300 BC. Well-known species—in addition to species called pinks—include the carnation (*D. caryophyllus*) and sweet William (*D. barbatus*).

   The pink family, Caryophyllaceae, comprises some 2000 species, most of them small herbs with narrow leaves and stems swollen at the nodes. The family includes campion (*Lychnis*), chickweeds (*Cerastium, Stellaria*), and baby's breath (*Gypsophila*).

pinniped (*animals*)   Seals, sea lions, and walruses—known collectively as pinnipeds ("fin-footed")—are large carnivorous marine mammals. They rely completely on the sea for food but move onto land or ice to give birth. Their teeth are designed for grasping fish and other prey rather than chewing, and most food is swallowed whole. Adaptations for an aquatic life include a streamlined body, 4 limbs modified into flippers, very small or no external ears, slitlike nostrils, a dense coat of fur, and, beneath the skin, a layer of blubber.

pion (*particles*)   Pions are the lightest mesons, each consisting of some paired combination of up or down quarks or antiquarks. Although free pions exist, the best known are virtual pions that carry the strong force binding atomic nuclei. As protons and neutrons of the nucleus exchange up and down quarks with each other, the passing quarks create virtual pions. Free pions were discovered by Cecil Powell [English: 1903–69] in 1947 in cosmic rays, where they are knocked into brief existence by impacts with nuclei but almost immediately decay into still lighter particles.

pioneer species (*ecology*)   The first organisms to colonize an area are called pioneer species. They must be hardy species that can grow on rock or bare soil or in heavily disturbed sites, such as new roadsides and land swept by fire. Lichens are common pioneers. Pioneer species dominate an area only for a while. As the pioneers change the original conditions, they are gradually replaced by species better adapted to the new conditions.

pipefish (*animals*)   Classified with sea horses in Family Syngnathidae, pipefish (*Syngnathus*) are small, primarily marine fish found mainly in shallow tropical waters; many can change color to blend in with surrounding sea grasses. Looking rather like a pipe cleaner, a pipefish has a slender, flexible body encircled by bony rings. The mouth lacks teeth; the tubular snout sucks in animal plankton and other tiny prey. Pipefish can swim vertically and horizontally. They range in length from 1 to 18 in. (2.5 to 46 cm).

pistachio (*plants*)   A member of the cashew family, the pistachio (*Pistacia vera*) is a deciduous tree that grows to a height of about 30 ft (9 m). It is native to the Mediterranean region and western Asia. The pistachio has small, unisexual flowers. A female flower gives rise to a fruit about 1 in. (2.5 cm) long; it contains an edible seed commonly called a nut.

pitch (*waves*)   The frequency of a sound wave is perceived as pitch just as the frequency of a wave of visible light is perceived as color. Higher pitches correspond to higher frequencies. The standard A of musicians is 440 Hz.

pitcher plant (*plants*)   Several genera of carnivorous plants use tubular or pitcher-shaped traps to catch insects. The insects are attracted to the traps by vivid colors or sweet nectar. Just below the entrance, however, the inner surface of the pitcher becomes slippery. Insects lose their footing and tumble downward. Downward-pointing hairs within the pitcher prevent the insects from crawling out. The insects eventually hit bottom, where they drown in digestive liquids.

pituitary (*anatomy*)   Located at the base of the brain in vertebrates, the pituitary is a small endocrine gland that secretes a number of hormones. At one time it was considered the "master gland" because its hormones regulate so many vital functions, including the activities of other endocrine glands. But the pituitary has its own master: it is under the control of a part of the brain called the hypothalamus, to which it is attached by a short stalk.

placeholder (*numeration*)   The first place-value systems did not have a method for showing a place not used in a particular numeral. Later, leaving a blank or inserting a dot were tried as a way to hold the place empty. In Southeast Asia about AD 700 a loop or egg shape was introduced as a placeholder, the ancestor of 0. Adding 0 to the digits 1 to 9 made numeration more effective. Mathematicians today recognize that 0 is not simply a placeholder, but a number essential to make number systems complete.

placenta (*anatomy*)   The females of most mammals (marsupials are the main exception) develop a structure in the uterus called the placenta to nourish developing embryos. The placenta consists of both maternal and embryonic tissue; it is firmly attached to the uterine wall and connected to an embryo by an umbilical cord. The placenta acts as the embryo's organs of respiration, digestion, and excretion. Oxygen and food are transferred—across a thin layer of cells—from the mother's blood to the embryo's blood, and carbon dioxide and excretory wastes from the embryo to the mother. Shortly after birth the placenta is expelled from the mother's body as the afterbirth.

place value (*numeration*)   Place value allows a small number of symbols to express an infinite number of ideas, also characteristic of alphabet-based writing. Both the alphabet and place-value numeration systems began in Mesopotamia near the dawn of civilization, but the system we now use originated in India some time before AD 700. Indian mathematicians used the digits 1, 2, 3, 4, 5, 6, 7, 8, and 9 in different places in numerals to indicate different values. A 5 at the far right meant 5, but 5 in the next place to the left meant 50. Later Indians introduced 0 to mark empty places in a numeral.

placoderm (*paleontology*)   Among the first fish with jaws and paired fins, the placoderms appeared in the Silurian and were abundant during the Devonian before dying out about 360 million years ago. Their name—"plated skins"— refers to the hard bony armor that protected the head and front part of the body. One of the largest and fiercest placoderms was *Dunkleosteus,* which reached lengths of more than 30 ft (9 m).

placodont (*paleontology*)   The placodonts ("flat tooths") were a group of swim-ming reptiles that lived in shallow coastal waters during the Triassic. They had paddlelike limbs and webbing between fingers and toes. Their teeth—located in both the upper and lower jaws and on the palate—were broad and flat, and designed for cracking open shellfish. Some placodonts had a turtlelike shell.

planarian (*animals*)   Looking rather like flattened arrows, planarians—includ-ing members of the genus *Planaria*—are free-living flatworms of the class Turbellaria. They range in length from 0.02 in. (0.5 mm) to 6 in. (15 cm) and typically are black, brown, or gray. Most species live in shallow water, but a few live in moist environments on land. Planarians avoid strong light and are most active at night, feeding on small organic particles. They have amazing powers of regeneration. Sexually mature planarians are hermaphrodites.

Planck's constant (*particles*)   In 1900 Max Planck [German: 1858–1947] developed a formula for the energy of packets of electromagnetic radiation, now called photons: it is the product of the frequency of the radiation and a very small constant, Planck's constant, signified by h and approximately 0.0000000000000000000000000000000006626 joule/Hertz. The constant is the size of the unit of action, and occurs throughout particle physics.

plane (*geometry*)   A plane is a geometric object in 2 dimensions that extends without bound in both directions. Any line though 2 points on the plane lies entirely within the plane (all of the points on the line are also points of the plane). A specific plane can be located in 3-dimensional space by 2 intersecting lines or by 3 points in the plane. Plane geometry is the study of figures whose points are all in the same plane. The coordinate plane is determined by the intersecting x and y axes.

planet (*astronomy*)   A planet is a large lump of ordinary matter that does not produce its own visible light. Planets ("wanderers") received their name in antiq-uity because they wandered among the "fixed" stars. After 1781, when William Herschel discovered the first previously unknown planet, astronomers searched for additional examples. When Giuseppe Piazzi [Italian: 1746–1826] discovered the asteroid Ceres in 1801, it too was called a planet—but finding 3 more small bodies in similar orbits over the next 6 years caused astronomers to reject the asteroids as planets (some call these bodies minor planets). Pluto is sometimes deemed too small and too much like large comets of the Kuiper belt to be a planet. With the discovery of extrasolar planets in the 1990s, the definition became more of an issue. Several extrasolar planets are so large that some would classify them as brown dwarfs instead of planets—meaning that they are large enough to undergo fusion. Although all definitely identified planets orbit stars or pulsars, in theory an object not in such an orbit could be termed a planet.

planetary nebula (*stars*) Although stars are normally seen as points of light, there are objects beyond the solar system that appear as spheres and other shapes that can be seen even with small telescopes. Because William Herschel thought they resembled planets, he named them planetary nebula. Each is an expanding sphere or cloud of gas, perhaps as much as a light-year across, that was emitted by a very hot star. Energy from the star causes the gas to glow. Astronomers are not sure whether most stars go through a phase of emitting gas—there are about 1000 that do so in the Milky Way—but the nebula appears to be a one-time event in the life of the stars that have them. After about 100,000 years the gas dissipates completely.

planetesimal (*solar system*) The current theory of the origin of planets in the solar system is that small bodies similar to asteroids or comets first existed as a broad ring around the Sun. Because most of these eventually became incorporated into planets as the bodies collided and were gathered by gravity into large spheres, the generic name for such bodies is planetesimal. Asteroids are thought to be planetesimals that failed to coalesce into a planet.

plane tree (*plants*) The deciduous trees that make up the genus *Plantanus* are variously known as plane trees, buttonwoods, and sycamores. They have bark that sheds in large pieces to reveal a greenish, white, or yellow inner bark. The lobed leaves resemble those of maples but occur alternately on branches (maples have opposing leaves). The tiny, unisexual flowers are borne in dense round clusters. The female flowers develop into round seed heads comprised of single-seeded nuts. The London plane (*P. acerifolia*), which may grow to more than 100 ft (30 m) tall, is widely cultivated in cities in temperate regions because of its ability to withstand air pollution.

plankton (*ecology*) Organisms that float near the surface of a body of water and are moved hither and yon by winds, waves, and currents are termed plankton (from the Greek *planktos*, meaning "wanderer"). Most plankton species are microscopic and of enormous importance, for they are at the base of almost all aquatic food chains. They typically have adaptations that help them stay afloat, such as a large surface-to-volume ratio, cilia, and elongated appendages.

There are 2 basic types: phytoplankton and zooplankton. Phytoplankton have chlorophyll and like plants can carry out photosynthesis. Indeed, phytoplankton account for more than half of all photosynthesis on Earth. In seas diatoms and dinoflagellates are the predominant phytoplankton, while in fresh water green algae or blue-green algae usually dominate. Zooplankton are consumers, feeding on phytoplankton or on other zooplankton. Some zooplankton, particularly small crustaceans such as *Daphnia*, live their entire lives as plankton. Other species, including many marine fish and most shellfish, are part of the plankton while young but then become swimmers or bottom forms.

plant (*botany*) Plants differ from the other main group of multicellular organisms, animals, in several basic ways. Most plants contain chlorophyll and can make food by photosynthesis. Their cells have cellulose walls and they commonly store carbohydrates as starch. Plants usually react slowly to stimuli. A plant's form can vary greatly from that of other members of the same species. Plants continue to grow throughout their lifetimes. They usually are attached to the ground or some other substrate for their entire lives.

There are approximately 260,000 known species of living plants. They range from mosses and ferns to cone-bearing and flowering plants. Some are almost microscopic while others are gigantic trees.

plasma (*anatomy*)   The clear, straw-colored liquid portion of blood is called plasma. Consisting mostly of water, it serves as a transport medium for blood cells, platelets, and a host of suspended and dissolved materials—amino acids, simple sugars, fatty acids, vitamins, hormones, cell wastes, etc. Hundreds of different plasma proteins are present, carrying out important functions within the blood. The major types of plasma proteins are albumins, globulins, and fibrinogens. Albumins help regulate the movement of water between the blood and the extracellular fluid that fills spaces between cells; they also transport metabolic wastes and other chemicals. Some globulins transport cholesterol, vitamins, and other substances; other globulins—particularly gamma globulins—are important in the fight against infection. Fibrinogen plays a central role in clotting. In humans plasma constitutes 55% of blood.

plasma (*materials science*)   Plasma is the phase of matter in which many atomic electrons have separated from their nuclei, resulting in matter composed largely of ions. Plasmas are caused by extremely high temperatures and are known on Earth only in fusion experiments, thermonuclear bombs, and perhaps strong lightning strokes. However, in a preponderance of stars, most matter is plasma rather than gas. Unlike gases, plasmas are excellent electrical conductors. Since the overall positive and negative charges are equal, plasma itself is electrically neutral.

plasmid (*genetics*)   In addition to a chromosome—which has all the genes needed for growth and reproduction—almost all bacteria have 1 or more small, circular DNA molecules called plasmids. A plasmid, containing 2 to 30 genes, replicates independently from the chromosome and is not essential for life. Some plasmids, however, have genes for resistance to antibiotics. As bacteria exchange plasmids, resistance to antibiotics can spread quickly, rendering ineffective medicines against diseases such as tuberculosis.

Plasmids are used extensively in genetic engineering. Typically a plasmid is cut using a restriction enzyme, and a small segment of foreign DNA is inserted. When the plasmid replicates, the foreign DNA also replicates. For example, researchers inserted the human gene that synthesizes the hormone somatostatin into plasmids, turning bacterial cells into miniature factories for the hormone.

plastic (*materials science*)   Plastics are stable, uniform organic materials, including such natural materials as casein, cellulose, or resins, as well as a host of synthetics that nearly always are polymers. The materials are called plastic because "plastic" means capable of being shaped, especially molded. Although metals, ceramics, and glasses can also be molded, they are inorganic. Some plastics are liquefied and spun into fibers or formed by pushing the material in a soft form through a hole for shaping (extrusion).

plastid (*cells*)   An organelle found only in cells of plants and some protists, the plastid is bounded by 2 membranes, and contains DNA. These features suggest that plastids may have evolved from 1-celled organisms that lived within the host cells in a symbiotic relationship.

Main types of plastids include chloroplasts, where photosynthesis occurs; chromoplasts that contain red, yellow, and orange pigments that give color to some petals and fruits (red tomatoes, for example); and leucoplasts, which store starch and fats.

plateau (*geography*) A plateau is a large, relatively flat region higher than its surroundings, but geologists also call hilly or mountainous regions plateaus if they are high—at minimum about 3000 ft (1000 m) above surrounding lands. Some, such as the Tibetan Plateau in Asia, rose when tectonic plates collided. Others, such as the Columbia and Snake River plateaus of the American West, began as giant outpourings of lava.

platelet (*cells*) Sometimes called thrombocytes, platelets ("little plates") are small colorless cell fragments found in vertebrate blood. They have no nucleus but consist solely of chemical-laden cytoplasm enclosed by a cell membrane. They break off from very large cells called megakaryocytes found in bone marrow. Platelets play a critical role in initiating blood clotting.

Plateosaurus (*paleontology*) *Plateosaurus* apparently was the most common of the prosauropods, a group of early, mainly herbivorous, dinosaurs that appeared during the late Triassic and were once believed to be the ancestors of the sauropods. *Plateosaurus* had a heavy body, very thick tail, and huge back legs that enabled it to rear up and strip leaves off tall trees. It was more than 20 ft (6 m) long and weighed about a ton.

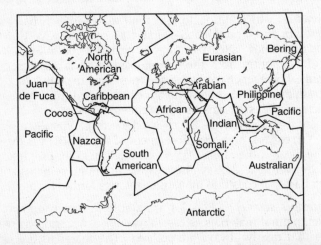

PLATE TECTONICS

plate tectonics (*geology*) In the 1950s geologists developed the theory of plate tectonics, which explains apparent continental drift, features of the oceans, and the locations of mountain ranges, volcanoes, and earthquakes. It states that Earth's solid outer layer is broken into more than a dozen large sections called plates as well as a similar number of smaller plates. The plates form at mid-oceanic ridges, where new material is added, and then move away from the ridges. When it meets other plates, a plate may be forced down into Earth by a process called subduction. In addition, plates may buckle to form mountains or one may slide past another along a great fault.

platinum (*elements*)   Platinum is a corrosion-resistant metal, atomic number 78, discovered by science in 1735 by Antonio de Ulloa [Spanish: 1716–95], but known earlier to pre-Columbian inhabitants of the Americas. It is the most common of the platinum-group metals: iridium, osmium, palladium, rhodium, and ruthenium along with platinum itself. The name comes from the Spanish for "little silver," reflecting the properties that make platinum popular for jewelry. Platinum is a vital catalyst used in many industries, but especially in cracking petroleum and in catalytic converters used in automobiles. Chemical symbol: Pt.

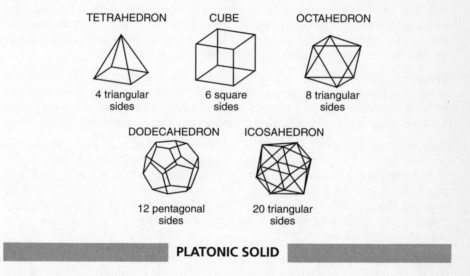

TETRAHEDRON — 4 triangular sides

CUBE — 6 square sides

OCTAHEDRON — 8 triangular sides

DODECAHEDRON — 12 pentagonal sides

ICOSAHEDRON — 20 triangular sides

PLATONIC SOLID

Platonic solid (*geometry*)   The 5 regular figures in 3 dimensions are the cube, tetrahedron, octahedron, dodecahedron, and icosahedron. They are known as the Platonic solids because Plato proposed about 350 BC that the 4 elements of fire, earth, water, and air are made from atoms shaped like tetrahedrons, cubes, icosahedrons, and octahedrons, respectively, and that the universe is a dodecahedron. The last proof in Euclid's *Elements* states that these 5 are the only regular polyhedrons possible, but Euclid overlooked the possibility of concave polyhedral angles. If these are allowed, there are 4 additional regular polyhedrons, all shaped like stars with different numbers of points.

platypus (*animals*)   The duck-billed platypus (*Ornithorhynchus anatinus*) is a monotreme mammal of eastern Australia. It has a flexible snout shaped like a duck's beak, webbed feet, and a flattened, beaverlike tail. A good swimmer, the platypus spends much of its time in water, moving along the bottom in search of shrimp, snails, worms, and other prey. Following mating, the female digs a burrow in which she lays and incubates her eggs. Her babies suck milk that oozes onto the abdomen from her mammary glands.

Pleiades (*stars*)   The Pleiades is not counted as a constellation today, although the 6 stars easily visible to the naked eye represent the Seven Sisters, daughters of Atlas who became stars. (A person with very good vision under good seeing conditions can see the seventh star.) There are some 200 fainter stars nearby, forming what astronomers call an open cluster. Unlike the old stars of a globular cluster, the stars of the Pleiades and of other open clusters are very young.

plesiosaur (*paleontology*)   The plesiosaurs were large marine reptiles that flourished during the Jurassic and Cretaceous. Unlike their contemporaries, the fishlike ichthyosaurs, plesiosaurs had broad, barrel-shaped bodies, short tails, and paddlelike limbs. The largest were up to 65 ft (20 m) long.

There were 2 main types. The long-necked plesiosaur had a long, mobile neck and small head armed with numerous needle-sharp teeth. It hunted near the surface, catching fish with quick darts of the neck. The short-necked plesiosaur had a much shorter neck and a large, powerful head with an enormous jaw and massive teeth. It probably fed on fish and squid.

plexus (*anatomy*)   A network of interlaced nerve cells or ganglia is called a plexus. For example, in vertebrates the brachial plexus is a network of nerves in the neck and front limb and is associated with limb movement; the solar plexus is located behind the stomach, with branches to various abdominal organs.

Pliny (*science writer*)   [Gaius Plinius Secundus: Roman: c23–79] Known as Pliny the Elder (his nephew, Pliny the Younger, was a well-known writer also), Pliny compiled the known science knowledge (along with much myth and misinformation) into his *Natural History*, 37 volumes published in AD 77. In 79 he observed the eruption of the volcano Vesuvius at close hand and was killed by its gases; the account of his death by Pliny the Younger is an important document in early volcanology.

plover (*animals*)   Small to medium-sized shorebirds with short bills, plovers and other members of Family Charadriidae, including lapwings and the killdeer, are strong fliers and swift runners. The black, brown, and white design of their plumage is excellent camouflage, making them inconspicuous on beaches and in fields. All use distraction displays to divert predators from eggs and young. Most migrate, often great distances and in flocks of thousands.

plumage (*zoology*)   All the feathers on a bird at any one time make up its plumage. In addition to its adaptation for flight, the plumage generally is waterproof and helps protect a bird's tender skin from injuries. Air spaces among feathers insulate the bird, helping to maintain body temperature. Plumage also may play a role in courtship or—by blending into the background—protection against predators. Plumage is distinctive to a species, and can differ according to sex and season. An example of sexual dimorphism is seen in ring-necked pheasants (*Phasianus colchicus*). The male has an iridescent green or purple head and neck and other bright colors while the hen has a mottled brown plumage. Rock ptarmigans (*Lagopus mutus*) are among the birds that exhibit seasonal dimorphism; the plumage of both sexes is white in winter, when the environment is covered with snow, and brown in summer.

plumeria (*plants*)   Also called frangipani and sambac, plumeria (*Plumeria*) is a shrub or small tree up to 23 ft (7 m) tall. Native to tropical Central America, it is widely cultivated as an ornamental. It has large dark green leaves and clusters of showy, highly fragrant flowers—in white, yellow, pink, purple, or red—that are borne in clusters at the ends of branches.

Pluto (*solar system*)   Pluto is usually considered the ninth and outermost planet of the solar system. This ball of frozen gases about the size of Earth's Moon was discovered in 1930 by Clyde Tombaugh [American: 1906–97]. Because of its size and chaotic orbit (which at times crosses inside Neptune's orbit), Pluto is sometimes thought to be one of the large comets in the Kuiper belt. Pluto's single moon Charon is about half Pluto's size; the two are sometimes called a double planet.

plutonium (*elements*)   Plutonium is often called the most dangerous element. In 1945 radioactive plutonium formed the first nuclear fission ("atomic") bomb. Tiny amounts of ingested plutonium accumulate in bone, where alpha radiation fatally destroys marrow cells. Nevertheless, plutonium, atomic number 94, is the most common artificial metal, used in nuclear bombs and some nuclear power plants. It was first synthesized in 1940 by Glenn T. Seaborg and coworkers and named for the ninth planet, Pluto (following elements named for Uranus and Neptune in the periodic table). Chemical symbol: Pu.

pocket gopher (*animals*)   Native to North and Central America, pocket gophers (Family Geomyidae) are rodents with a fur-lined pouch ("pocket") on the outside of each cheek. As it forages, a pocket gopher uses its paws to stuff seeds and other food into its pouches for later storage or eating. Pocket gophers generally are less than 9 in. (23 cm) long. They have small eyes and ears and a short tail. Large front claws are adapted for digging. The animals live solitary lives except during breeding season, and spend most of their time in burrows. They are active both day and night.

Pocket mice, kangaroo mice, and kangaroo rats (Family Heteromyidae) are small rodents that also have fur-lined cheek pouches that open next to the mouth. They are native to arid regions of the western United States south to northern South America. Unlike pocket gophers, they have weak front feet and a long tail. They are strictly nocturnal.

podzol (*soils*)   A whitish soil caused by a combination of high rainfall and evergreen conifers is called podzol. The soil is a light gray because most of the iron oxides in the soil have been dissolved by the very acid leaves of conifers and then washed away.

Poincaré, Henri (*mathematician/physicist*)   [French: 1854–1912] Poincaré is best known for having discovered parts of special relativity independently of Einstein and for his report of having established an important result in mathematics while stepping out of a bus (after trying to develop an idea in this area to no avail for several months). Psychologists have used Poincaré's experience in discovery as a keystone to understanding creativity.

point

Among mathematicians, Poincaré is recognized as founder of modern topology, and for generalizing the concept of periodic functions. He also redeveloped in modern terms the field of mathematical astronomy.

point (*geometry*)   In an axiomatic system for geometry, point is an undefined term. Often textbooks describe a point as an exact location in space, but this does not take moving points into account. In analytic geometry a point is an entity located by one set of coordinates; on a plane, for example, (2, –3) locates a point that is unique and (x, 2x + 3) locates all the points on the line $y = 2x + 3$. The identification of points with coordinates is so close that a point can be defined as a coordinate.

point (*paleoanthropology*)   The stone tools that appear to be arrowheads or spear points are called points by paleoanthropologists. In most cases, however, the type of tool is unknown—what appear to be spear points could be wide knives or axe heads, for example. Some points taper at both ends. Points were probably hafted to long shafts (arrows, spears, and harpoons) or handles (axes, sickles, and other tools).

point and nonpoint sources (*environment*)   Any identifiable source of pollution, such as a tanker ship or a factory smokestack, is a point source. Any diffuse source of pollution that does not discharge pollutants through a single, specific outlet is a nonpoint source. Farms, construction sites, mines, and urban streets are examples of nonpoint sources, with pollutants generally carried off in storm waters.

point-slope equation (*analytic geometry*)   A line in a plane is completely specified by a point and a number called the slope. In Cartesian coordinates, if the point has the coordinates (x_1, y_1) and the slope is the number m, the equation $y - y_1, = m(x - x_1)$ has the given line as its graph. The equation is called the point-slope equation.

poison (*biochemistry*)   Any substance that in relatively small amounts damages or disrupts the functioning of cells is a poison. A wide range of naturally occurring substances can be poisonous: heavy metals (lead), corrosive compounds (sulfuric acid), gases (carbon monoxide), etc. Poisonous proteins produced by disease-causing bacteria, fungi, spiders, and certain other organisms are called toxins.

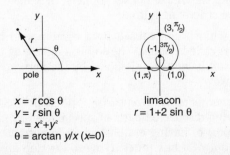

$x = r \cos \theta$
$y = r \sin \theta$
$r^2 = x^2 + y^2$
$\theta = \arctan y/x \ (x \neq 0)$

limacon
$r = 1 + 2 \sin \theta$

POLAR COORDINATES

polar coordinates (*analytic geometry*)   A coordinate system called polar coordinates locates each point as *r*, its directed distance from a given point (called the pole) and θ, the angle that the line segment between pole and point makes with a fixed ray originating at the pole. Equations based on trigonometry and the Pythagorean theorem relate polar and Cartesian coordinates. Often curves that would be complex using Cartesian coordinates, such as the limaçon, are simple in polar coordinates. The generalization of polar coordinates to 3 dimensions is called spherical coordinates.

polar form (*numeration*)   A complex number in polar form is the product of a real number with the sum of a cosine of an angle and the imaginary number *i* times the sine of the same angle. The angle is usually expressed in radians. For example, 1 + *i* in polar form is $\sqrt{2}[\cos (\pi/4) + i \sin (\pi/4)]$. Although these 2 numbers are the same, since $\cos (\pi/4) = \sin (\pi/4) = 1/\sqrt{2}$, the polar form shows that the number is $\sqrt{2}$ units from the origin of the complex plane at an angle of $\pi/4$ (45°), while the standard form shows it as 1 unit from the *iy* axis and 1 from the *x* axis.

Polaris (*stars*)   The current north star is Polaris, a binary with an abundance of names (officially Alpha Ursae Minoris, but sometimes called Cynosura). The larger star is a yellow supergiant Cepheid variable, although its magnitude varies only from 2.1 to 2.2.

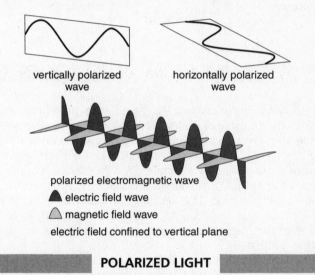

vertically polarized wave

horizontally polarized wave

polarized electromagnetic wave

▲ electric field wave

△ magnetic field wave

electric field confined to vertical plane

POLARIZED LIGHT

polarized light (*optics*)   The wave fronts of most light waves are complex, moving in all directions perpendicular to the line of travel. Such transverse waves can be filtered into a 2-dimensional structure (more like a wave along a rope) called polarized light. Light can be polarized into a plane at any angle, including vertically or horizontally. Polarization occurs when light passes through certain crystals or polymers, or when reflected from nonmetallic surfaces. If light polarized at one angle is then filtered through a material that polarizes it perpendicular to that angle, the light is completely blocked. Fishers use sunglasses polarized vertically, to block out the horizontal polarized light reflected from the water's surface.

pollen (*botany*)   As the male reproductive organs of seed plants mature, they produce vast quantities of microscopic pollen grains, particles smaller than 0.000008 in. (0.0002 mm) in diameter. The decay-resistant outer wall of most pollen grains has a characteristic shape, size, and ornamentation that can be identified at least to genus and often to species. By analyzing pollen preserved in sediments and peat, scientists can learn about vegetation and climates of the past.

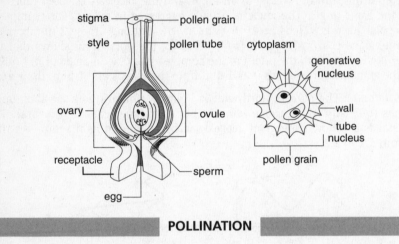

POLLINATION

pollination (*botany*)   Pollination—a basic step in reproduction of seed plants—involves the transfer of pollen from the male reproductive organ to the female reproductive organ. This transfer may take place on a single plant (self pollination) or involve 2 plants (cross pollination). In cone-bearing plants, pollen is transferred from a male cone to a female cone. In plants that bear flowers, it is transferred from the anther to the stigma. Outside agents such as wind, water, and insects are often involved. The design of flowers frequently suggests which agents are important to a species. For example, flowers with large white petals are designed to attract night-flying insects while small flowers that lack petals tend to be wind-pollinated.

If pollination is successful, the pollen forms a slender pollen tube that grows through the style and ovary until it nears the female egg in the ovule. The pollen then releases 1 or more sperm and fertilization occurs.

pollution (*environment*)   The presence of impurities in the environment is called pollution. Pollutants can be solids (household garbage, for example), liquids (runoff from mines), gases (sulfur dioxide from power plants), or energy (radiation from nuclear operations). Carried by winds and water, pollutants can travel great distances and have a variety of harmful effects. They interfere with life processes, causing disease or death in all kinds of organisms and upsetting the balance of natural communities. They also cause climatic changes, make soils infertile, foul beaches, and damage buildings and other objects. Often pollutants represent wasteful use of fuels and other natural resources. (*See also* **air pollution; water pollution.**)

polonium (*elements*)   Polonium is a rare naturally occurring radioactive element, atomic number 84, discovered in 1898 by Marie Sklodowska Curie and Pierre Curie and named by Marie Curie for her native Poland. It is a metal with properties like those of the near-metal tellurium. The isotope polonium 210 is used as a source of alpha particles. Chemical symbol: Po.

polygon (*geometry*)   A polygon is a figure completely in a plane consisting of a finite number of points and the line segments between them, provided that there are at least 3 points and that none of the line segments cross; that is, that no 2 line segments have a point in common except for endpoints. Commonly encountered polygons are triangles, various quadrilaterals, pentagons, and hexagons.

polyhedral angle (*geometry*)   The figure formed by the parts of 3 or more planes that meet at a point and that also contain each of the sides of a polygon in a different plane from the point is a polyhedral angle. The parts forming the angle are the interiors of the triangles that have the point as one vertex and the endpoints of the side of the polygon as the other vertices. Any vertex of a polyhedron is also the vertex of a polyhedral angle.

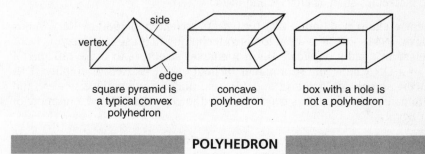

square pyramid is
a typical convex
polyhedron

concave
polyhedron

box with a hole is
not a polyhedron

POLYHEDRON

polyhedron (*geometry*)   A polyhedron is a closed surface with flat sides and straight edges that can be stretched without breaking into a sphere. "Closed" here means that the surface separates 3-dimensional space into inside and outside. The most common polyhedrons, such as pyramids, cubes, and prisms, are *convex;* if you extend any side of a convex polyhedron, the rest of the figure lies on one side of the plane only. A *concave* polyhedron can extend on both sides of such a plane.

polymer (*materials science*)   A polymer ("many parts") is a material consisting of very long molecules of indeterminate length. Typically the molecules are made from a repeating unit, called a monomer, or from a limited number of different units. Neoprene, an artificial rubber, and most plastics are simple polymers—for example, polyethylene consists of linked chains of ethylene. Rubber

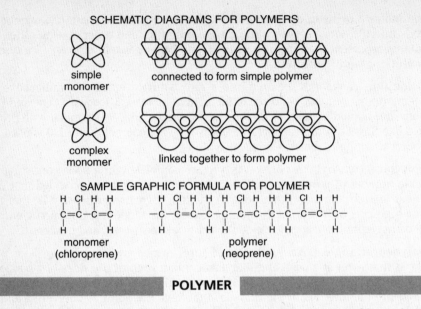

SCHEMATIC DIAGRAMS FOR POLYMERS

simple
monomer

connected to form simple polymer

complex
monomer

linked together to form polymer

SAMPLE GRAPHIC FORMULA FOR POLYMER

monomer
(chloroprene)

polymer
(neoprene)

POLYMER

is a common natural polymer. Modern chemists consider all sorts of long chain molecules polymers, including inorganic chains such as nanotubes and lengthy organic molecules such as proteins and DNA.

polymerase chain reaction (PCR) (*genetics*)   Invented in 1980 by Kary Mullis [American: 1944–], PCR is an automated technique that uses the enzyme DNA polymerase to rapidly duplicate DNA. It enables scientists to make millions of copies of a DNA molecule in a matter of hours. PCR is used to amplify DNA sequences, screen for bacteria and viruses, diagnose genetic diseases, and study human evolution. It has even been used to copy the DNA of a mummy of ancient Egypt.

polymorphism (*genetics/evolution*)   Polymorphism is the presence in a population of 2 or more distinctive forms, or morphs, of a trait. The blood types A, B, AB, and O are an example of polymorphism in humans. Some polymorphism appears to be neutral in terms of mating behavior and natural selection. In other cases nonrandom mating results. For example, there are 2 morphs of the snow goose (*Chen caervlescens*), white and blue. Members of each morph generally prefer to mate with other geese of the same morph. Under certain conditions, this can result in genetic drift.

polynomial (*algebra*)   An expression formed by adding any finite number of integral powers of a variable with numerical coefficients is a polynomial in 1 variable, for example, $x^4 + 3x^3 - 2x^2 + 5x + 8$. Polynomials in 1 variable behave like integers, including possession of closure for addition, subtraction, and multiplication. A polynomial equation is simply a polynomial in 1 variable set equal to 0; a polynomial function is similarly a function of the form $f(x) = a_0 x^n + a_1 x^{n-1} + a_2 x^{n-2} + \ldots a_n x^0$. Polynomials in several variables are the sums of the products of integral powers of the variable with numerical coefficients.

polyploidy (*genetics*) Plants and animals typically have 2 sets of chromosomes (2*n*), one inherited from the male parent, the other from the female parent. Sometimes, however, chromosomes fail to separate normally during cell division. The result is a polyploid, an organism having 3 or more sets of chromosomes. This condition is not uncommon among plants but is rare among animals. For example, Einkorn wheat is diploid (2*n*) while durum wheat is a tetraploid (4*n*). If 2 individuals with different chromosome numbers breed, their offspring may be hybrids with an intermediate chromosome number. If a polyploid's offspring can interbreed only among themselves, the offspring constitute a new species.

polysaccharide (*biochemistry*) The most abundant carbohydrates are polysaccharides, which are long chainlike or branched molecules formed by linking together many sugar molecules; a typical polysaccharide molecule consists of thousands of sugar units. Unlike sugar, polysaccharides are not sweet; nor are they soluble in water. Among the most important polysaccharides are storage materials, such as starch and glycogen, and structural materials, such as cellulose and chitin. During digestion, polysaccharides are broken down into simple sugars.

pomegranate (*plants*) Native to southern Asia, the pomegranate (*Punica granatum*) is a dense shrub or small tree that grows to heights of about 20 ft (6 m). The showy red or white funnel-shaped flowers give rise to fruits about the size of oranges, with a leathery red or yellow skin that surrounds a juicy pulp containing numerous seeds. In addition to being edible, the fruit has long been used medicinally; the Ebers papyrus of about 1500 BC mentions its effectiveness as a cure for tapeworms.

poplar (*plants*) Members of the willow family, poplars (*Populus*) are medium to tall deciduous trees native to northern temperate regions. They have broad leaves and minute flowers borne in clusters called catkins—with male and female flowers on separate trees. Prominent species include the white poplar (*P. alba*), a native of Europe and Asia that grows to about 100 ft (30 m). The genus also includes species commonly called cottonwoods and aspens.

poppy (*plants*) Native mainly to temperate and subtropical regions, members of the poppy family, Papaveraceae, are mostly herbs. Showy solitary flowers and a milky sap, usually whitish in color, characterize them. They produce a fruit called a capsule that contains many seeds. The unripe capsules of the opium poppy (*Papaver somniferum*) are the source of the drug opium and its derivatives morphine, codeine, and heroin.

population (*ecology*) All the members of a species that live in the same place, or habitat, at the same time make up a population—for example, all the wood ducks (*Aix sponsa*) that nest on Prince Edward Island or all the pink abalone (*Haliotis corrugata*) in Monterey Bay. Food availability, predator populations, and many other factors determine the size of a population and its growth potential. Variations in an environment generally lead to population fluctuations.

Populations I and II (*stars*) Although there are some gradations between, most stars can be grouped as young Population I stars that consist mostly of hydrogen and helium or as old Population II stars that contain heavier elements. Population I stars are found in open clusters, such as the Pleiades, or where gas and dust are suitable for star formation, such as the spiral arms of the Milky Way and similar galaxies. Population II stars are the main component of globular clusters and are common near the center of the galaxy.

porcupine (*animals*)   Large, stocky rodents, porcupines wear an armor of sharp, often intricately barbed, spines called quills that are loosely attached to the skin. When attacked, a porcupine shakes out quills; their tips stick into the skin of the attacker. Porcupines, primarily nocturnal, eat plant matter. Species of Family Hystricidae are native to wooded habitats of Europe, Asia, and Africa; they do not climb trees. Members of Family Erethizontidae are found in similar habitats in the Americas and are excellent climbers.

porphyrin (*biochemistry*)   Porphyrins are a group of organic substances consisting of complex ring systems with a metal ion (magnesium, for example) in the center. They form part of several important biological molecules, including pigments such as chlorophyll, cytochrome, hemoglobin, and myoglobin.

Portuguese man-of-war (*animals*)   A marine cnidarian closely related to hydras, the Portuguese man-of-war (*Physalia physalia*) looks rather like a large jellyfish, with a crested, gas-filled float and tentacles that may be more than 60 ft (18 m) long. But this creature actually is a colony of specialized polyps and medusas—some capturing or digesting food, some handling reproduction, and so on.

positive number (*numbers*)   When the real numbers are arranged in order on a line from least to greatest, all the numbers greater than 0 (on the same side of 0 from 1) are positive numbers.

positron (*particles*)   The positron, or antielectron, is the antiparticle of the electron, predicted in 1930 by Paul A.M. Dirac and discovered 2 years later by Carl Anderson. It normally appears in a rare form of beta decay and as part of an electron-positron pair created by an energetic photon. A jet of positrons flows from the galactic nucleus for unknown reasons.

postulate (*logic*)   In modern logic a statement accepted as true without proof and used as the basis for structuring a body of knowledge is called a postulate or, synonymously, an axiom. In older usage the term postulate was reserved for statements accepted as true within a given subject, such as "All right angles are equal to one another," while the term axiom applied to statements thought to be true of all entities.

potassium (*elements*)   Potassium is an alkali metal, atomic number 19, discovered in 1807 by Humphry Davy, who obtained it from potash (in Latin *kalium*), from which the chemical symbol K derives. It was the first metal found through electrolysis. Potassium is necessary for plant growth; fertilizer is rated in terms of NKP (nitrogen, potassium, phosphorus). The isotope potassium-40 is radioactive and important in isotopic dating of fossils and artifacts by paleoanthropologists.

potato (*plants*)   A member of the nightshade family, the white or Irish potato (*Solanum tuberosum*) is native to South America. It is widely grown for its edible tubers (underground stems); hundreds of varieties, or cultivars, have been developed. The plant has compound leaves, flowers of white, yellow, pink, or blue, and a berry fruit. The sweet potato (*Ipomoea batatas*), cultivated for its edible tuberlike roots, is a vine in the morning glory family.

potential (*electricity*)   An important measure used with both static electricity and electric current is potential difference, the force pushing one electron away from another (or attracting an electron to a region of positive charge). When divided by amount of charge present, the result is electrical potential, also known as voltage and as electromotive force. It is called potential because it is the electrical version of potential energy, such as that caused by separation of masses or by tension of a compressed spring. Electric current flows when there is a potential difference between 2 locations and a conductor between them—although if the potential is high enough, the conduction can be very weak and current will still flow. Potential is measured in volts, with 1 volt equal to a joule/coulomb or, equivalently, a watt/ampere.

potential energy (*mechanics*)   Potential energy is energy stored in a system as the location of a body or as structures in parts of the system. A rock balanced at the edge of a high cliff has stored the gravitational energy used to lift the rock to that place or to erode the cliff from below it. If the rock falls the potential turns to kinetic energy. A rubber band stores the work used to stretch it as potential energy. An electric battery stores potential energy in the separation of charges, releasing it when the circuit is completed. Arrangements within complex molecules in food store potential energy, released when oxygen is used to simplify the molecules.

pothole (*ecology*)   Pothole marshes, formed in depressions created by glaciers, dot the prairies of central North America. Generally less than 10 acres (4 hectares) in size, their water levels can fluctuate dramatically; during droughts they often dry out completely. Potholes are important as breeding grounds for many ducks, geese, and songbirds.

power (*mechanics*)   Power is work divided by the time it takes to do it. Depending on the situation, power is measured as horsepower, watts (the official SI measurement unit) or kilowatts, or erg/seconds.

power (*numeration*)   The power of a number is the value of a numeral indicated by an exponent; for example, 8 is the third power of 2, shown as 2^3.

poxvirus (*viruses*)   Members of the virus family Poxviridae are among the largest and most complex of all viruses. Each rod-shaped virus particle is encased in a lipid envelope and has its genetic information stored in double-stranded DNA. The group, which infects vertebrates and insects, includes the viruses that cause human smallpox, swinepox, and rabbit myxoma.

prairie (*ecology*)   Two types of natural grasslands once covered much of the North American interior. Tall-grass prairies, extending westward from Illinois, have rich, moist soil that supports species that grow 6 to 10 feet (1.8 to 3 m) tall. Farther west, in the drier Great Plains region east of the Rocky Mountains, are short-grass prairies, where grasses 1 to 3 ft (0.3 to 0.9 m) tall dominate.

Between these 2 prairies lies a transition zone, the mixed prairie, where even tall-grass species are seldom more than 4 ft (1.2 m) tall. Historically, prairies were home to herds of bison, pronghorns, and other large grazers, as well as large predators such as wolves. But little of the original prairie ecosystems remain. Like temperate grasslands elsewhere, they have been almost entirely replaced by farms and cities.

prairie dog (*animals*)   Prairie dogs (*Cynomys*) are stout, short-legged rodents native to prairies of North America. Once widespread, their range and numbers have been greatly reduced by shooting, habitat destruction, and other human actions. Prairie dogs live in underground colonies called towns that contain thousands of individuals. They feed on grass at the surface, with some of the animals keeping watch for predators. When sentinels spot danger they utter a high-pitched alarm call and all the prairie dogs retreat into their burrows. These burrows also shelter ferrets, burrowing owls, and other creatures.

praseodymium (*elements*)   A rare earth, atomic number 59, discovered in 1885 by Karl Auer (Baron von Welsbach) [Austrian: 1858–1929], praseodymium is named from the Greek for "green twin." Its oxide is green as is a spectral line that provides the name while it is a twin to the element neodymium in both its origin and properties. Its combination with neodymium is used for coloring glass in welder's masks. The oxide is extremely resistant to melting and used as a core for carbon arc lights. Chemical symbol: Pr.

Precambrian era (*geology*)   The Cambrian period, which began some 545,000,000 years ago, saw the beginning of organisms with hard body parts, such as shells, which fossilize easily. Thus Precambrian era came to be a general name for Earth's earliest history, a time of almost no fossils that stretches back about 4,000,000,000 years before the Cambrian.

precession (*mechanics*)   Precession is movement of the axis of a rotating body, such as a top, gyroscope, or planet, around the surface of a cone, so that any point on the axis except the center of the body turns in a circle. Precession occurs whenever a force slightly changes the action of rotation. Its speed is inversely related to the speed of rotation. For Earth, the gravitational pull on the slight bulge at the equator causes precession of the equinoxes.

precession of equinoxes (*solar system*)   A rapidly rotating symmetrical body maintains its orientation because of balanced inertial forces, but any instabilities cause the point where the axis of rotation touches the surface to drift in a circle, an effect called precession. Earth's precession can be seen as changes in position with respect to the stars, an occurrence noticed by ancient astronomers even though the full circle takes about 26,000 years. As the orientation of Earth with respect to its plane of rotation changes, the equinox moves through the calendar; thus, Earth's precession is called precession of the equinoxes. In addition, the star directly above a pole changes, so the north star of today will no longer be above the pole after several thousand years.

precipitate (*chemistry*)   A precipitate is a solid substance that forms from a solution when the amount of dissolved material exceeds the saturation point; or from a **colloid** for which a chemical reaction has caused particles to aggregate; or from any chemical reaction in liquids that produces solid particles.

precipitation (*meteorology*)   The use of the word precipitation to describe dew, rain, sleet, snow, and hail began in 1751 with Charles Le Roy, who was the first to recognize the **dew point**. He incorrectly believed that water dissolves in air and then **precipitates** like a solid out of a solution. The word has proved to be a handy term, however, for meteorologists, especially in the United States.

precision (*measurement*)   The precision of a measurement depends on the size of the unit used in making the measurement—for example, measuring to the nearest inch is more precise than measuring to the nearest foot. Numerically, precision can be equated with the **error**, which is half the size of the unit used in measuring (measuring to the nearest foot has an error of $\frac{1}{2}$ ft or 6 in.), or to the amount represented by the final **significant digit**.

precocial and altricial young (*zoology*)   Precocial birds and mammals are active and almost fully developed at birth. For example, a deer baby is born fully haired, with its eyes open, and able to stand and move about. Altricial birds and mammals are born naked, blind, and immobile; they are completely dependent on the parents for food and care. Examples include robins, sparrows, rabbits, and most rodents.

predator (*ecology*)   Organisms that hunt and eat other organisms are called predators. Sharks, snakes, lions, and killer whales are examples. Predators have adaptations that enable them to capture their prey. Eagles, for example, have sharp eyesight to locate prey, strong feet with sharp claws to grasp prey, and a hooked bill for tearing flesh.

preening (*ethology*)   One of the most important behaviors of a bird is caring for its feathers. Called preening, this involves drawing the feathers through the bill to remove dirt and parasites. The action also distributes oil onto the feathers from a gland located at the base of the tail. The oil keeps the feathers from drying out and makes them water repellant. Parrots, pigeons, and some other birds practice mutual preening, which is limited to the head and upper neck, areas that a bird cannot reach with its own bill.

pressure (*mechanics*)   Pressure is an amount of **force** divided by area—for example, air pressure at sea level is about 14.7 lb per sq in. ($10,132.5 \ N/m^2$). The total force is proportional to the area. For a solid, pressure concentrates force for a smaller area; hitting a big rock with the point of a small one exerts more force than hitting it with the long side, for example. For a fluid, pressure depends only on depth, an application of **Pascal's law**. Thus a diver experiences increased pressure during deeper dives.

Priestley, Joseph (*chemist*)   [English: 1733–1804] Priestley was a brilliant teacher and Unitarian minister who learned theories of electricity from **Benjamin Franklin**. Although Priestley was an amateur scientist, his experiments with electricity and later with gases contributed enormously to the field of chemistry. Before Priestley the only known gases were air, carbon dioxide, and hydrogen, but Priestley discovered 10 more, most notably oxygen.

primary color (*optics*)   Colors that in suitable mixtures can be used to create all other colors are called primary. For a mixture of light waves, the additive primary colors are red, green, and blue, which in equal proportions produce white. Black in this case is the absence of light waves. Pigments on a white paper show colors by reflecting 1 color and absorbing the others, so the subtractive primary colors are different. Each primary pigment reflects 2 central wavelengths of light: yellow pigment reflects both red and green, a pigment called cyan reflects blue and green, while magenta reflects red and blue. An equal combination absorbs all wavelengths of light and appears black, while absence of pigment reflects all and is white.

primate (*animals*)   The approximately 235 species of the mammalian order of primates range in size from pygmy lemurs and marmosets to 6-ft (1.8-m) gorillas and humans. The order is divided into 2 suborders, prosimian and simian. Primates are characterized by forward-facing eyes, flexible arms and legs with 5 digits on each hand and foot, and well-developed brains. Most species, particularly monkeys and other large forms active during daytime, are social creatures that live in groups called troops or bands.

prime number (*numbers*)   A prime number is a natural number greater than 1 that has no divisors other than itself and 1; the other natural numbers are composite. Thus, 2, 3, 5, 7, 11, 13, 17, 19 are the prime numbers less than 20 while 4, 6, 8, 9, 10, 12, 14, 15, 16, and 18 are composite (1 is neither). Prime numbers are the basis of modern codes used in commerce as well as of many important theorems of arithmetic. The fundamental theorem of arithmetic is that any natural number greater than 1 is either prime or the product of a unique set of prime numbers. For example, although 12 is 6×2 and 3×4, the prime representation as $2 \times 2 \times 3$ is unique.

primrose (*plants*)   Native to northern temperate regions, including mountainous areas, primroses (*Primula*) are low-growing perennial herbs. They have a basal rosette of leaves, tubular flowers that occur in a multitude of colors, and a capsule fruit containing numerous seeds. Many primroses are grown as ornamentals.

Evening primroses (*Oenothera*), an unrelated group native to North America, also are popular ornamentals. They are herbs that grow to heights of 6 ft (1.8 m). Most species have flowers (usually yellow, often showy) that open in the evening and close in the morning. But a few species, called sundrops, have flowers that open during the day.

prion (*biology*)   A group of degenerative neurological diseases known as spongiform encephalopathies appear to be caused by extremely tiny infectious agents that consist solely of a protein; that is, these pathogens, called prions, lack not only a cell structure but also nucleic acids. They are believed to be abnormal variants of proteins that form part of the surface of nerve cells. Prions have been linked to scrapie in sheep, bovine spongiform encephalopathy ("mad cow disease") in cattle, and Creutzfeldt-Jakob disease in humans.

prism (*geometry*)   Any polyhedron that has 2 faces that are both parallel and congruent is a prism. The parallel, congruent faces are bases, while the lateral faces are necessarily parallelograms (since the sides in the bases are parallel and equal in length). A prism is a right prism if the lateral faces are rectangles. The volume (V) of any prism is the product of the area of a base (B) and the altitude (h) as shown in the formula $V = Bh$.

prism (*optics*)   A prism is a transparent material of varying thickness, such as one whose cross section is triangular or curved, used to create dispersion of light into separate wavelengths. If a narrow band of white light passes through a prism, the result is a spectrum. Incandescence from a single element typically produces a number of monochromatic (single-wavelength) bands.

probability (*mathematics*)   The theory of probability is a branch of mathematics concerned with numbers (called probabilities) that reflect the concept of chance. The theory is an abstract structure built on the idea that an event that occurs a certain number of times (called successes) in so many tries has a probability that is the ratio of the successes to the number of tries. Thus an event with no chance of happening has a probability of 0, while one that always happens has a probability of 1. In a coin toss where heads and tails have equal probability, the probability of heads is 1/2. The probability of any of a die's 6 faces facing up is $\frac{1}{6}$.

proboscis (*anatomy*)   An elongated snout, or proboscis, is best seen in elephants, which sometimes are called proboscidians. The elephant proboscis—commonly known as the trunk—is composed of a very long nose and part of the upper lip. It may be more than 6 ft (1.8 m) long, has over 40,000 muscles, and, at the tip, contains organs of touch and smell. The elephant uses its proboscis to grasp objects, carry food and water to the mouth, and sense its surroundings.

Proconsul (*paleoanthropology*)   *Proconsul* was a genus of early apes (named after Consul, a chimpanzee at the London Zoo) that lived in East Africa as much as 20,000,000 years ago. The 3 known apes in this genus are generally recognized as the ancestors of all later hominoids.

producer (*ecology*)   Producers are organisms that can make food from inorganic materials, generally using sunlight as the source of energy. Chlorophyll-containing plants and algae are examples. Producers are the first link in every food chain—a series of organisms each dependent on the preceding one for food and, hence, energy.

projective geometry (*mathematics*)   Projective geometry grew out of perspective in painting. Transferring a scene in 3 dimensions to a plane so that parts appear to have the same relationships is one type of projection. Parallel lines, such as receding railroad tracks, converge and seem to meet behind the painting.

In mathematics projection often involves 2 planes or a sphere and a plane. Imagine a transparent plane between a viewer and a second plane angled to the first. Connect each point, P, on the second plane by a line to the eye of the viewer; the intersection, p, of that line and the transparent plane is the projection of P.

In formal projective geometry, measurement is ignored. Lines all intersect, with parallel lines meeting at a point at infinity. All such points form a line at infinity. Results in projective geometry are all dual theorems.

prokaryote (*biology*)   Unlike other organisms, bacteria and blue-green algae of Kingdom Monera are prokaryotes, made up of simple prokaryotic cells that do not have a membrane-bound nucleus; the genetic material—a single strand of DNA—is in the cytoplasm. Prokaryotic cells also lack other organelles found in the cells of eukaryotes, including mitochondria, vacuoles, plastids, and endoplasmic reticulum. The cell walls contain peptidoglycan, a substance unique to monerans. In general, prokaryotic cells are much smaller than eukaryotic cells.

promethium (*elements*)   Promethium is a radioactive rare earth, or lanthanide metal, synthesized in 1945 and named for the Greek god Prometheus, who stole fire from heaven. Promethium, atomic number 61, is the last element to be found or created with an atomic number less than 92 (the highest for a naturally occurring element on Earth)—by Charles D. Coryell [American: 1912–] and coworkers in 1945. Although not occurring on Earth, promethium has been detected spectroscopically in the star Arcturus. Chemical symbol: Pm.

prominence (*solar system*)   Large parts of the Sun's chromosphere that project far above the photosphere are called prominences. They can be observed during total solar eclipses or with special filters at other times. A prominence may extend for thousands of miles from the Sun and can last for weeks or months supported by magnetic forces.

pronghorn (*animals*)   The only living species of the family Antilocapridae, the pronghorn (*Antilocapra americana*) is a hoofed mammal native to grasslands and deserts of western North America. Incorrectly referred to as an antelope, it stands about 3.3 ft (1 m) high at the shoulder and weighs up to 155 lb (70 kg). Both males and females have a pair of slightly curved horns, each with a forward-projecting prong. The pronghorn is the continent's swiftest mammal, able to run at speeds over 40 mph (65 km/hr).

proof (*logic*)   Perhaps the greatest innovation of Greek mathematics was proof by deduction, said to have been introduced by Thales around 600 BC. Instead of asserting, for example, that vertical angles are equal as a result of measurement, Thales argued that each angle plus the single angle between them add up to a straight angle and, since all straight angles are equal, subtracting the angle between shows the vertical angles to be equal. Later such informal proofs were incorporated into formal axiomatic systems. Some proof techniques, such as indirect proof and mathematical induction, are generally used outside the framework of formal axiomatic systems.

proper motion (*stars*)   The apparent motion of a star with respect to the background of stars and galaxies is called proper motion. Stars and galaxies appear to move together through the sky because of Earth's movements; they move away from each other as a result of the expansion of the universe. Proper motion is not the same as actual motion with respect to the rest of the universe, since we see only the part of the movement that is perpendicular to our line of vision.

proportion (*arithmetic*)   A proportion is a true statement that 2 ratios are equal. It can be written in fraction form, such as $3/4 = 9/12$, but older books often use a special notation reserved for ratio and proportion in which that same proportion is written as 3 : 4 :: 9 : 12, or 3 is to 4 as 9 is to 12. In the old notation, the numbers 3 and 12 are called extremes while 4 and 9 are called means. This older language is used in rules for proportions, such as "the product of means equals the product of extremes," even when modern fraction notation is employed.

proposition (*logic*)   A proposition is a sentence that does not contain a variable and therefore can be assigned either the value true or false.

prosimian and simian (*animals*)   The primates are divided into 2 groups, or suborders. The prosimians include lemurs, galagos, lorises, and tarsiers. They are small or medium-sized furry animals. Their lower incisors and canines are fused to form a dental comb used in grooming. Simians, or anthropoids, include monkeys, apes, and humans. They are bigger than prosimians, with a larger brain, better vision, but poorer smell, and no dental comb.

prostaglandin (*biochemistry*)   A group of organic compounds first isolated in the 1930s from sheep and human prostates (hence the name), prostaglandins are unsaturated fatty acids synthesized throughout the body. A prostaglandin molecule contains 20 carbon atoms, 5 of which form a ring.

Although prostaglandins are present only in much smaller quantities than hormones, they are involved in many body processes. For example, they stimulate smooth muscle contraction, regulate fat metabolism, affect pulse rate and blood pressure, cause inflammation, control hormone secretion, and promote contractions of the uterus.

prostate (*anatomy*)   The prostate gland is part of the reproductive system of male mammals. It surrounds the urethra (urinary canal) just below the urinary bladder. The prostate produces a watery, alkaline fluid that is part of the semen. Its function is under the control of hormones called androgens.

protactinium (*elements*)   Credit for discovery of the radioactive metal protactinium, atomic number 91, is commonly assigned to Otto Hahn [German: 1879–1968] and Lise Meitner who located a long-lived isotope in 1918, about 5 years after the element was first recognized. The name comes from *proto-* ("before") plus actinium—the longest lived isotope of protactinium decays to actinium. Traces occur in uranium ores, since uranium-235 decays to protactinium. Chemical symbol: Pa.

protea (*plants*)   Native to southern Africa, proteas (*Protea*) are evergreen shrubs named after the Greek god Proteus, who could change himself into many different shapes. Proteas have flower heads composed of a tight cluster of filamentous flowers surrounded by brightly colored petal-like leaves called bracts. Heads range from less than 1 in. (2.5 cm) to more than 12 in. (30 cm) in diameter. The protea family, Proteaceae, also includes the banksias (*Banksia*) of Australia, evergreen shrubs with cylindrical, spiky flower heads.

protein (*biochemistry*)   Proteins are large, complex polypeptide molecules made up of dozens or even hundreds of amino acid molecules joined together by peptide links. The amino acids are in a specific order; any change in an amino acid or its position can make the protein nonfunctional. The protein molecule has 4 levels of organization. The primary structure is the sequence of amino acids in a chain. The secondary structure is the way that the chain is twisted around its axis or pleated into sheets. The tertiary structure describes the folding back and forth of the molecule. Quaternary structure is the configuration created by the combination of 2 or more polypeptides. (Hemoglobin, for example, is composed of 4 polypeptide chains.)

Even a simple bacterium such as *Escherichia coli* contains as many as 800 different proteins. A typical human cell has about 10,000 different proteins. Some proteins are soluble and function mainly as enzymes or carriers (hemoglobin carries oxygen, for example). Other proteins are fibrous and insoluble; they are mainly structural materials, forming cell membranes, muscles, cartilage, hair, etc.

protein synthesis (*biochemistry*) The production of a particular protein is directed by a specific gene (segment of DNA) in the cell nucleus. The gene acts as a template, or pattern, for the manufacture of mRNA (messenger RNA). The mRNA leaves the nucleus and attaches to organelles in the cytoplasm called ribosomes. Another kind of RNA, tRNA (transfer RNA), carries amino acids to the ribosomes. In a process called translation, the amino acids are linked together—in the order coded by the mRNA—to form the protein.

protist (*protists*) Members of the kingdom Protista are the simplest eukaryotes (organisms whose cells have a distinct membrane-bound nucleus). The kingdom contains an extremely diverse group of more than 50,000 species, and something of a catchall: organisms are classified as protists largely because they do not fit into any of the other kingdoms. Included in the Protista are all single-celled eukaryotes plus many simple multicellular forms. In popular terms, they often are divided into two groups: algae and protozoa.

proton (*particles*) The proton is a baryon made from 3 quarks. At least 1 proton is always found in the nucleus of every atom, usually held tightly to other protons and to neutrons by the strong force. Protons are generally considered to have been discovered by Rutherford, who in 1911 established the atomic nucleus. But it was not until 1920 that Rutherford introduced the word "proton" as the name for a heavy particle with a single unit of positive charge exactly equal to the electron's negative charge.

protoplanet (*solar system*) Early in the formation of the solar system, small planetesimals clumped together to form protoplanets (*proto* means "first"). Some protoplanets melted completely or had large chunks blasted off during continued bombardment by planetesimals. About 4,500,000,000 years ago, intense bombardment ceased and planets more like those of today emerged.

protozoan (*protists*) One-celled protists that do not contain chlorophyll and thus must take in food are referred to as protozoa. They include animal-like forms such as amoebas, paramecia, and flagellates, plus the funguslike slime molds. Some are free-living, found in fresh and salt water and in damp land habitats; others are parasites of plants and animals.

Proust, Joseph-Louis (*chemist*) [French: 1754–1826] In 1799 Proust established that elements combine so that the ratio by mass of each element in a compound to the other elements is always the same; this is known as the law of definite proportions. Four years later, Dalton recognized that Proust's law is valid because compounds are made from atoms so that, for example, sodium carbonate, Na_2CO_3, contains 2 atoms of sodium to 1 of carbon to 3 of oxygen. Proust was also the first to distinguish among the sugars.

Proxima Centauri (*stars*)   Our nearest neighbor, Proxima Centauri, was discovered in 1915 by Robert Thorburn Aytoun Innes [South African: 1861–1933]. It is the dim (magnitude 10.7) third star in the Alpha Centauri group, but currently about 0.1 light-year closer to us than the bright binary pair. It is so far from the binary that it may take as much as 1,000,000 years to complete a single orbit.

AMOEBA

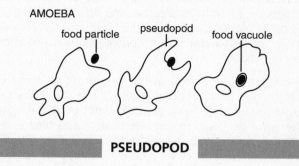

PSEUDOPOD

pseudopod (*cells*)   A pseudopod ("false foot") is a temporary protrusion from the body of certain protists, such as amoebas, and from the white blood cells of vertebrates. It is used in locomotion and to engulf particles—either food (protists) or bacteria, antigens, and dead cells (white blood cells). The pseudopod completely surrounds a particle. The cell membrane breaks and rejoins, trapping the particle in a membrane-bound region called a vacuole.

psilophyte (*plants*)   The simplest plants with vascular tissue are the psilophytes, or whisk ferns. They are not ferns, though recent evidence suggests they may have had fernlike ancestors. Psilophytes have true stems but no true roots or leaves. Their life cycle includes an alternation of generations with a dominant spore-producing phase that has green stems that carry out photosynthesis and a gamete-producing phase that consists of a tiny, subterranean plant that obtains nourishment as a saprophyte. *Psilotum* species, widespread in tropical and subtropical lands, are often grown as ornamentals.

pterosaur (*paleontology*)   Popularly referred to as pterodactyls, pterosaurs were flying reptiles that lived at approximately the same time as the dinosaurs, to whom they were closely related. Each of their two wings consisted mainly of skin membranes supported by an extremely elongated fourth finger and attached to the ankle. Early species had heavy teeth and long, bony tails; later species lost these features. About 60 genera have been identified, with species ranging from sparrow-sized animals to giants such as *Quetzalcoatlus*, which had a 33-ft (10-m) wingspan.

Ptolemaic system (*solar system*)   The visible movements of the planets and Sun against the background of fixed stars were explained at one time in terms of circular orbits around Earth. This incorrect view of the solar system is known as the Ptolemaic system, although forms of it considerably precede its description by Claudius Ptolemaeus [Hellenic: c100 – c170, better known as Ptolemy]. Ptolemy based his work on that of Hipparchus, one of the greatest of astronomers, who introduced the epicycle to make the apparent planetary motions fit circular, Earth-centered orbits.

P/Tr event (*paleontology*)   The boundary between the Permian and Triassic some 245 million years ago was marked by the largest mass extinction of all time. In the ocean, trilobites disappeared and there were precipitous declines in the variety of brachiopods, foraminifera, corals, and fish. On land, the broad-leaved seed ferns and other plants that created the great coal forests became extinct, as did most types of amphibians and reptiles. The causes of this extinction remain a matter of debate.

puberty (*physiology*)   The period at the beginning of human adolescence marked by changes that lead to sexual maturity and the ability to reproduce is called puberty. In males testes enlarge, beard and body hair grow, shoulders widen, and the voice deepens. In females breasts and hips enlarge, body fat increases, body hair forms, and the menstrual cycle begins. These changes are controlled by a variety of hormones produced mainly by the pituitary gland and reproductive organs.

puffball (*fungi*)   Members of the club fungi, puffballs are named for their spherical fruiting bodies, which become dry and leathery on maturity. Pressure from wind, rain, or animals squeezes the balls and causes puffs of spores to be expelled through an opening at the top. Puffballs grow in moist humus soil and on decaying logs and stumps in fields and wooded areas. Most puffballs are only an inch or two in diameter, but the giant puffball (*Calvatia gigantea*) can grow to more than 4 ft (1.2 m). Many species are edible when young.

pulley (*mechanics*)   The basic pulley is a freely rotating wheel and a rope that passes over the rim. Pulling one end of the rope raises the other, but the only effect of a simple pulley is to change direction of force. Archimedes is thought to have invented the compound pulley, an arrangement of 2 or more connected pulleys that reduces the distance that one end of the rope moves compared with that of the other, thus increasing the force. An encased pulley is called a block, so another name for compound pulley is block and tackle.

pulsar (*stars*)   Pulsars are neutron stars that emit electromagnetic signals from their magnetic poles in a direction that reaches Earth. All neutron stars emit signals and rotate very rapidly, at least when they are first formed; they gradually slow down. These signals form a tight beam. If the beam intersects Earth, a radio telescope observes a fast pulsing on and off of the signal. The pulses are so regular that when they were first discovered, they were thought to be the work of extraterrestrial beings.

pulse (*physiology*)   As the left ventricle of the heart contracts, a spurt of blood leaves the heart, creating a pressure wave that travels along the arteries. This wave, or pulse, can be detected in arteries close to the body surface, such as the one on the inner wrist.

puma (*animals*)   Also called the cougar or mountain lion, the puma (*Puma concolor*) is the most widespread wild cat native to the Americas, found from western Canada to the southern tip of South America. It can grow to a length of 77 in. (195 cm) and weigh more than 200 lb (90 kg). It lives in a variety of habitats, including grasslands, deserts, swamps, and forests. Unlike the so-called big cats (lions, tigers, jaguars, and leopards), the puma cannot roar; instead, it produces a piercing scream.

pumice (*rocks and minerals*)   Pumice begins as froth on lava or magma ejected from a volcano. The froth has solidified so rapidly that it has trapped bubbles of air or of gases given off by the magma. The solid part of pumice is a glass, similar to obsidian in chemical composition, but not "glassy" in appearance because of its spongelike texture. The trapped gases reduce the density of pumice enough that it often floats on water, forming great rafts in the ocean near erupting volcanoes.

punctuated equilibrium (*evolution*)   In 1972 Stephen Jay Gould [American: 1941–] and Niles Eldredge [American: 1943–] pointed out that the fossil record seldom shows a gradual transition from one species to another. Rather, there are long periods of species stability, or equilibrium, punctuated by the sudden appearance of new species. Gould and Eldredge proposed that evolution from one species to another occurs not as slow, gradual change, as Darwin thought, but rather so rapidly that it often cannot be seen in the fossil record. Initially controversial, this theory of punctuated equilibrium is now widely accepted; only among microscopic protists does gradualism seem to prevail, possibly because these organisms generally reproduce asexually.

pupa (*zoology*)   The pupa is the intermediate stage between the larva and the adult in insects that undergo complete metamorphosis. During the pupal stage, profound changes in body structure take place. Many of the larval tissues and organs break down; at the same time, new adult structures form. At the end of this stage, the full-grown adult emerges. Pupae, such as those of butterflies and moths, may be enclosed in a protective case called a cocoon.

Purkinje, Johannes (*physiologist*)   [Czech: 1787–1869] In addition to expanding our knowledge of the structure and function of the eye, Purkinje was a pioneer in the study of cells, particularly after acquiring a compound microscope in 1832. His discoveries include the nucleus in birds' eggs, sweat glands in the skin, large flask-shaped nerve cells with numerous dendrites in the cerebellum (now called Purkinje cells), and muscle fibers in the heart that conduct stimuli from the pacemaker (Purkinje fibers). He was the first to realize that fingerprints can be used for identification, the first to describe the movement of cilia, and the first to use a microtome (an instrument used to slice tissue into thin sections for microscopic examination).

PVC (polyvinyl chloride) (*compounds*)   The plastic PVC (polyvinyl chloride), is a polymer of the carcinogenic gas vinyl chloride (C_2H_3Cl). PVC is the plastic in pipes of all kinds, linings for pools, food wrap, plastic bottles, and credit cards. One virtue is that it is easy to shape. PVC by itself is stiff and brittle, but plasticizers can be added to make it more flexible.

pyramid (*geometry*)   A pyramid is a polyhedron that can have any polygon as the base, attached to faces that are triangles with a common vertex (the vertex of the pyramid). Each type of pyramid is named for the base polygon (triangular pyramids, square pyramids, and so forth). If the altitude from the vertex meets the center of the base, the pyramid is a right pyramid. The familiar pyramids of Egypt are right square pyramids. The volume of a pyramid is 1/3 the product of the area of the base and the length of the altitude ($V = 1/3\ Bh$).

pyrite (*rocks and minerals*)   Pyrite ("fire rock," since it can produce sparks when struck) is the mineral iron sulfide (FeS_2). Pyrite is widely distributed, but not used as iron ore. It is best known as "fool's gold," which suggests correctly that pyrite is bright, yellow, and metallic appearing, causing people to mistake it for gold.

Pythagoras (*mathematician*)   [Ionian: c560–480 BC] Pythagoras was among the earliest Greek philosophers. He proposed number as the basis of reality, basing his belief on his discovery that musical harmony can be explained with number (for example, doubling the length of a plucked string produces a note 1 octave lower). The Pythagorean theorem was certainly known long before Pythagoras' time, but he may have offered the first proof.

Credit for many discoveries attributed to Pythagoras probably belongs to other members of his secret society. Pythagoreans recognized that Earth is round, that the morning and evening stars are the same (the planet Venus), and that certain lengths, such as the diagonal of a square whose side is a natural number, cannot also be natural numbers or their ratios.

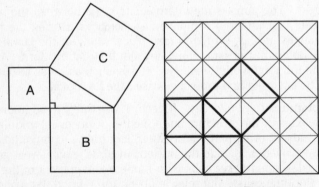

area of squares A + B
equals area of square C

tiling shows Pythagorean
theorem for isosceles
right triangle

PYTHAGOREAN THEOREM

Pythagorean theorem (*geometry*)   The Pythagorean theorem for a right triangle states that the sum of the squares of the lengths of the 2 sides of the right angle (the legs) equals the square of the length of the side opposite the right angle (the hypotenuse); in a triangle with sides *a, b,* and *c,* where *c* is the hypotenuse, $c^2 = a^2 + b^2$. This theorem, which is fundamental to much of mathematics, was known to Chinese and Babylonian scholars long before Pythagoras, although he may have provided the first of what are now several hundred proofs. The theorem generalizes in a number of ways; the version for triangles in general (not just right triangles) is known as the law of cosines.

python (*animals*)   Like their relatives the boas, pythons are nonpoisonous, heavy-bodied snakes that kill by constriction. Unlike boas, which give birth to living young, pythons lay eggs. Native to Africa, Asia, and Australia, pythons are the largest Old World snakes, with species such as the reticulated python (*Python reticulatus*) growing to 27 ft (8 m) long.

quadrant (*analytic geometry*)   The intersecting axes in Cartesian coordinates separate the plane into 4 regions called quadrants, conventionally labeled with Roman numerals. In I both x and y are positive; in II x is negative and y positive; in III both x and y are negative; and in IV x is positive and y is negative. Note that the axes themselves are not part of any quadrant.

quadratic equation (*algebra*)   An equation in 1 variable of the second degree is called quadratic. The general form is often shown as $ax^2 + bx + c = 0$ where a, b and c are numbers and x is the variable. A quadratic equation can always be solved, one way being with the quadratic formula. If the discriminant $b^2 - 4ac$ is less than 0, the solutions are not real even if a, b, and c are real numbers.

quadratic function (*analytic geometry*)   A function in one variable is called quadratic (from the Latin for "square") if the highest power of the variable is 2. When a, b, and c are numbers ($a \neq 0$), the equation of such a function is $y = ax^2 + bx + c$. The graph of a quadratic function is a parabola.

quadrilateral (*geometry*)   All polygons with 4 sides are quadrilaterals ("4 sides"), including parallelograms, rectangles, and kites. The area of any quadrilateral can be found by using a line segment from opposite sides to make 2 triangles, finding the areas of the triangles, and adding.

quantifier (*logic*)   Sentences that contain a variable such as x can be made into true or false propositions by using a quantifier, which assigns a truth value to the sentence based on the entire set of values allowed for the variable. Two quantifiers are commonly used, the universal quantifier, symbolized by $\forall$, and usually read as "for all"; and the existential quantifier, symbolized by $\exists$, usually read as "there exists." A sentence such as "x is a boojum" when quantified by $\forall$ ("For all x, x is a boojum") is false if there is a value of x not a boojum and true otherwise, while when quantified by $\exists$ ("There exists an x such that x is a boojum") is false only if there is not a value of x that is a boojum.

quantum (*particles*)   A quantum is something that can be counted—what mathematicians call discrete. In 1900 Max Planck [German: 1858–1947] resolved difficulties in explaining the spectrum of electromagnetic radiation under some circumstances with a formula that required radiation to come in small packets, which he called quanta, instead of in a continuous flow. The size of each quantum is determined by Planck's constant. Five years later Einstein borrowed Planck's idea of the quantum nature of light to explain the photoelectric effect. In 1913 Niels Bohr explained the spectrum of hydrogen in terms of discrete orbits of electrons based on Planck's constant. This led to a fundamental reinterpretation of physics based on the quantum.

quantum chromodynamics (*particles*)   The part of the standard model of particle physics that deals with quarks and gluons is called quantum chromodynamics (QCD) in analogy to quantum electrodynamics (QED). The prefix chromo- is used because the force involved in QCD is known as the color force. The force in QED is charge, or electromagnetism.

quantum electrodynamics (*particles*)   The part of quantum theory known as electrodynamics deals mainly with the interactions of the electromagnetic field with electrons and positrons. Several versions of quantum theory developed during the late 1920s and the 1930s were successful, but not entirely so. Among the main problems was that most calculations from theory dealing with electrons inappropriately gave infinity as an answer for known quantities, such as the mass of the electron. In 1947 theorists resolved such difficulties by a method known as renormalization, in which positive and negative infinites almost cancel each other, but leave a finite residue. This residue corresponds to experimental results for as many decimals places as anyone has calculated.

quantum mechanics (*particles*)   The methods of calculating the effects of moving electrons that were developed in the 1920s and 1930s are often called quantum mechanics, where the sense of *mechanics* is analysis of effects of forces on matter. In 1925 Werner Heisenberg developed matrix mechanics, an abstract method of calculating correct results for electrons based on mathematical entities called matrices. The following year Erwin Schrödinger independently created an equation with a strong physical basis that is the heart of wave mechanics. Despite seeming entirely different in all respects, both matrix mechanics and wave mechanics give the same results.

quantum number (*particles*)   Quantum theory is based on integers or multiples of 1/2 called quantum numbers. The 3 original quantum numbers for atoms specified characteristics of an electron in orbit, while the fourth was spin. Later quantum numbers for subatomic particles included charge and isotopic spin. In quark theory, the strange, charm, bottom, and top quarks each provide its own quantum number. A given combination of quantum numbers describes a particular subatomic particle.

quantum theory (*particles*)   The beginning of quantum theory was the discovery in 1900 by Max Planck [German: 1858–1947] that energy comes in small packets called quanta instead of flowing continuously. This idea was further developed by Einstein and Niels Bohr into the beginnings of quantum theory, which concerns the behavior of subatomic particles. Quantum theory states that entities have properties of both particles and waves, a duality that has a large influence on small scales. As it grew, quantum theory developed offshoots known as quantum mechanics, quantum electrodynamics, and quantum chromodynamics. The alternative explanation of force and matter, which applies to matter at a larger scale, is called classical physics.

quark (*particles*)   In 1964 Murray Gell-Mann introduced 3 particles that he named quarks. Two, known as up and down, form protons and neutrons, while the third, called strange, is combined with up, down, or itself in pairs and triplets to form the heavy particles that had been observed in cosmic rays since 1947. In 1974 new particle discoveries added the charm quark. When it became apparent in 1975 that a third generation exists, scientists quickly found the bottom quark— but it took another 20 years to complete the list of 6 quarks with the top.

quartz (*rocks and minerals*) Quartz is a hard, transparent-to-translucent mineral that occurs widely in igneous rock and metamorphic rock. The rock granite is at least 10% quartz and most sand particles are quartz, since it is hard, common, and does not dissolve easily in water. Quartz is the basic mineral of the silica group (SiO_2). Purple quartz is called amethyst.

quasar (*astronomy*) Quasars are distant sources of great energy. The name quasar is short for quasi-stellar object, since quasars appear to be about the size and have the general appearance of dim, reddish stars. At wavelengths other than visual light, however, quasars release far more energy than stars. The energy may pour from the accretion disk of a giant black hole. There is some evidence that quasars are central black holes in distant galaxies. The stars in the galaxies are too far away to see, so we observe only the bright quasar.

quaternion (*numbers*) A quaternion is the sum of a real number and a vector expressed in terms of unit vectors **i**, **j**, and **k**; for example, $4 + 2\mathbf{i} - \mathbf{k} + 5\mathbf{j}$. Quaternions were invented by Sir William Rowan Hamilton [Irish: 1805–65] in 1843 in a failed effort to expand numbers beyond the complex numbers. Quaternions found little use but were important in introducing commutative and associative properties—quaternions are associative—and in leading to development of algebras.

quetzal (*animals*) An extraordinarily beautiful bird, the quetzal (*Pharomachrus mocino*) lives in humid cloud forests of Central America. The male's upper plumage is a glittering green, his belly is crimson, and the underside of his tail is white. His most notable feature is his long green tail plumes. The bird's body is about 15 in. (38 cm) long; the plumes add another 25 in. (63 cm) or more. The female is not as brightly colored and lacks long tail plumes.

quince (*plants*) A member of the rose family, the quince (*Cydonia oblonga*) is a small deciduous tree native to western Asia often cultivated for its aromatic edible fruit. It grows to about 20 ft (6 m) tall. Flowering quinces (*Chaenomeles*), also of the rose family, are deciduous or semievergreen shrubs native to eastern Asia. They produce showy flowers in early spring and are cultivated as ornamentals.

quinoa (*plants*) Native to South America, where it has been cultivated for thousands of years, quinoa (*Chenopodium quinoa*) is an annual herb that grows about 6 ft (1.8 m) tall. It has a hollow stem and triangular leaves. Its clusters of small greenish flowers give rise to numerous large seeds that are rich in protein and can be eaten whole or ground into flour.

R

rabbit and hare (*animals*)   Members of the mammal order Lagomorpha, rabbits and hares (Family Leporidae) have long, mobile ears, short front limbs, long, powerful hind limbs, and a short tail. As adults they have 2 pairs of upper incisors that never stop growing; however, constant gnawing on leaves and grass keeps the teeth worn to an appropriate length. Rabbits and hares are native to the Americas, Africa, Europe, and Asia. In Australia, where they were introduced, they are considered pests. Nondomesticated species range in size from about 10 to 20 in. (25 to 50 cm) in length.

Rabbits are generally smaller species that nest in burrows; their young are born hairless and blind. Hares are larger and do not dig burrows but form their nests on the ground; their young are born furred and with open eyes. In popular usage, the terms are frequently misused. For instance, the white-tailed jackrabbit (*Lepus townsendii*) of North America is actually a hare.

raccoon (*animals*)   Carnivorous mammals native to the Americas, raccoons (*Procyon*) have a broad head with a pointed muzzle, a plump body covered with long fur, short legs, and a bushy tail marked with blackish rings. They are up to 3.3 ft (1 m) long, including the tail. Raccoons eat almost anything, from fish and fruits to insects and garbage.

The raccoon family, Procyonidae, also includes coatis, the cacomistle and the kinkajou. Members of the family vary widely in appearance and feeding habits. Most inhabit forests and are good climbers.

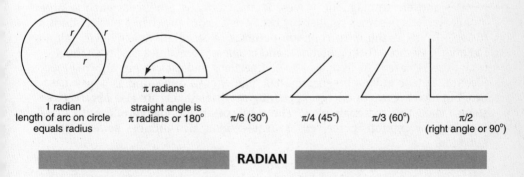

1 radian
length of arc on circle
equals radius

straight angle is
π radians or 180°

π/6 (30°)   π/4 (45°)   π/3 (60°)   π/2
(right angle or 90°)

RADIAN

radian (*geometry*)   A radian is a real number represented as an angle. An angle of 1 radian is the central angle of a circle that holds between its sides an arc whose length is the same as the radius of the circle. In degree measure, a radian is about 57° 17′ 44.6″ or 57.2958° or exactly $(180/\pi)°$ since a semicircle contains π radians. Problems in calculus or probability often must be interpreted in terms of radians.

radiation (*physics*)   Radiation is a broad term that can be defined as transmission of energy by particles, but often the particles are viewed as waves. The main form of radiation is transmission of energy by photons, which as waves form the electromagnetic spectrum from radio through visible light to gamma rays. When we say that heat is transmitted by radiation, we mean that the energy of certain electromagnetic waves sets molecules (or atoms) in motion. For most molecules, the part of the electromagnetic spectrum called infrared light is most effective at transmitting heat.

Radiation of subatomic particles produced by nuclear reactions, including high-energy photons, is called radioactivity.

radical (*chemistry*)   A group of atoms acting as a chemical unit is called a radical or a polyatomic ion. The hydroxide ion OH⁻ is an example, as are radicals whose names end in -ate or -ite (carbonate, nitrate, nitrite, phosphate, sulfite). Radicals either donate electrons (ammonium, NH_4, with a charge of $+1$, for example) or accept them (sulfate, SO_4, with a charge of -2). Although radicals are normally parts of compounds, in solution they may become free radicals that actively interact with molecules.

radical (*numeration*)   A number shown with the sign $\sqrt{\ }$ is called a radical; for example $\sqrt{2}$, the positive square root of 2, can also be called radical 2. In this notation, the number 2 is said to be inside or under the radical and such a number is called a radicand. For roots other than square roots, the same sign is used with a small numeral greater than 2 written in the $\vee$ of the radical sign. The small numeral, called an index, shows which root is indicated—3 for the cube root, 4 for the fourth root, etc. When a radical has an even index, it always means the positive root.

radioactivity (*physics*)   Radioactivity is radiation produced by decay from one element to another or by nuclear fusion or fission. It consists of both subatomic particles and electromagnetic waves. Radioactivity is produced when a large atomic nucleus becomes smaller (by decay or fission) or when 2 nuclei join (by fusion). In decay the most common particles released are helium nuclei (alpha radiation), electrons (beta radiation), and neutrinos; sometimes positrons (antielectrons) are released. In fission or fusion neutrons and protons are also released. Electromagnetic radiation generally has high energy, appearing as gamma radiation, but lower levels of energy, appearing as visible light, are also observed, so some radioactive materials glow. The recoil of all the energy released causes more atomic motion, perceived as heat—which also induces some infrared radiation.

radiocarbon dating (*paleoanthropology*)   In 1947 Willard Libby developed a technique called radiocarbon dating for determining the age of organic materials based on the ratio of radioactive carbon-14 to ordinary carbon-12. Carbon-14 is formed from nitrogen in air at a relatively constant rate by energy from the Sun and decays with a steady half life of 5730 years. When an aerobic organism is alive, the ratio of the 2 carbons stays the same, but after death the amount of carbon-14 steadily decreases, making it possible to determine the approximate date of death. This method is reliable for dates less than about 40,000 years ago.

radio galaxy (*galaxies*)   Radio galaxies can be observed in visible light but emit much brighter radio-wave signals. The most intense radio waves are nearly always produced by 2 large lobes outside the starry region of the galaxies. These lobes are the largest known continuous objects in the universe— some 20 times as wide as the galaxy of their origin. The source of the radio waves is synchrotron radiation produced by acceleration of electrons in clouds of plasma. A quasar or black hole at the galaxy's center may be the source of energy.

radiolarian (*protists*)   Radiolarians are one-celled amoeboid protists. Unlike their relatives the amoebas, however, almost all radiolarians secrete an intricate glassy skeleton on the outside of the cell membrane. Slender pseudopods project through holes in the skeleton to trap prey. Radiolarians are part of the marine plankton throughout the world. In areas where populations are especially dense, skeletons of dead radiolarians are the main constituents of ocean-floor sediments called radiolarian ooze. Fossil remains of radiolarians in Precambrian rocks testify to the great age of these organisms.

radio wave (*waves*)   Radio waves form the part of the electromagnetic spectrum below about 10,000,000 Hz or with a wavelength longer than about 0.5 m. They can be used for long-range communication because they are reflected by ionized layers in the upper mesosphere and thermosphere.

radium (*elements*)   The only alkaline earth metal radioactive in all its forms, radium, atomic number 88, was discovered in 1898 by Marie Sklodowska Curie and Pierre Curie. Its name, from the Latin *radius*, "ray," reflects the glow produced by radium and its compounds. Before people realized that radioactivity can promote cancer, radium was widely used in watch dials and as an inappropriate cure for disease. Chemical symbol: Ra.

radius (*geometry*)   A radius is either the line segment from the center of a circle or sphere to the circle or sphere or the length of that line segment. The idea has been extended to regular figures in 2 ways: the long radius goes from the center to a vertex; the short radius from the center is perpendicular to a side.

radon (*elements*)   Radon is a radioactive gas, atomic number 86, discovered in 1900 by Friedrich Ernst Dorn [German: 1848–1916], who called it radium emanation. The current name stems from "radium", with its suffix changed to *-on*, which marks all noble gases except helium. Since radon is released by common rocks, such as granite, that contain uranium, it can accumulate in buildings and become a health hazard because its decay products are long-lived radioactive metals. Chemical symbol: Rn.

radula (*anatomy*)   Except for bivalves, which are filter feeders, mollusks have a filelike organ, the radula, at the back of the mouth. The radula has thousands of tiny, rasplike teeth arranged in transverse rows (number, size, and pattern of the teeth vary greatly from one species to another). A mollusk can extend its radula from the mouth, using it to scrape algae and other food particles from underwater rocks and other surfaces.

ragweed (*plants*)   In the oldest known textbook on drugs, written by the physician Pedanius Dioscorides [Greek: fl AD 60], the name ambrosia—"food of the gods"—was given to a plant native to Africa. Dioscorides' reason for this is unknown, but the name has survived botanically for the plant and its relatives, which today are commonly called ragweeds (*Ambrosia*). In addition to several African species, the genus includes species native to North America. Herbaceous members of the composite family, ragweeds have hairy stems, divided leaves, and inconspicuous green flowers in drooping clusters. Their pollen is a primary cause of the allergy known as hay fever.

rail (*animals*)   The rail family of birds, Rallidae, includes rails, coots, crakes, gallinules, soras, and wekas. These birds are found worldwide except in polar regions. All live on the ground in wet habitats; except to migrate, they seldom if ever leave their marshes, swamps, or bogs. They mainly eat seeds and small animals (including insects, crustaceans, and fish). Family members typically have long legs, long toes, and a short tail. Their body is slender, enabling the birds to move quickly through dense vegetation. All can swim and dive. Most have cryptic coloration that blends in with reeds and grasses, but gallinules are brightly colored. Among the largest is the king rail (*Rallus elegans*) of North America, 15 to 19 in. (38 to 48 cm) long.

rain (*meteorology*)   Liquid water falling through the air in distinct drops is rain, although a light rain of small drops is drizzle and tiny droplets that do not fall or fall very slowly form mist. Rain begins when water vapor in clouds begins to condense into liquid on small bits of dust or salt crystals suspended in air. In 1911 Alfred Lothar Wegener proposed that these tiny droplets freeze and begin to collect more water vapor, forming snowflakes. The snowflakes grow in size and float in the clouds until they become too heavy. As they fall to Earth, they reach regions of higher temperature and melt, becoming raindrops. Today meteorologists believe that most rain in temperate regions forms as Wegener theorized, but rain in the tropics develops from water precipitating on salt crystals or through other mechanisms.

rainbow (*optics*)   Sunlight falling on nearly spherical raindrops is dispersed by refraction into a broad spectrum whose colors are red, orange, yellow, green, blue, indigo, and violet (often memorized as Roy G. Biv), a phenomenon called the rainbow. The observer facing away from the Sun into the rain sees red at an angle of 42° and violet at an angle of 40°. Since these angles are constant, the rainbow appears as the arc of a circle interrupted by the horizon. From an airplane or a mountain peak the whole circle can be seen.

rain forest (*ecology*)   One of Earth's major biomes is the tropical rain forest, a term coined in 1898 by Andreas F.W. Schimper [German: 1856–1901]. This biome is found in warm equatorial regions that receive at least 100 in. (250 cm) of rain annually, with the rainfall distributed evenly throughout the year. The forests are among the most complex and diverse communities on the planet, with a majority of their species probably not yet identified by scientists. Plant growth is prolific, consisting mainly of tall broadleaf trees, climbing vines, ferns, and epiphytes. Animal life is extremely diverse and includes orangutans, tamarins, and other endangered species. Extensive deforestation has destroyed about 50% of Earth's tropical rain forests. Scientists fear this deforestation will lead not only to extinction of species but to climatic changes and increased global warming.

Rain forests also occur in temperate areas. The coniferous forests of the Olympic Peninsula in Washington State and the lenga (*Nothofagus pumilio*) forests of southern Chile are examples.

Ramón y Cajal, Santiago (*histologist*)   [Spanish: 1852–1934] Improving on a technique for staining nervous tissue developed by Camillo Golgi [Italian: 1843–1926], Cajal was able to detect the structure of neurons (nerve cells) in gray matter of the central nervous system, and to trace long nerve fibers. He showed that nerve fibers do not fuse with one another, and that the endings of one neuron never actually touch endings of adjacent neurons. Cajal and Golgi shared the 1906 Nobel Prize in physiology or medicine.

random (*probability*)   In probability theory, an event is random if all possible outcomes to an experiment have the same probability. In that case, performing the experiment produces a random sample. While this seems to correspond to the ordinary meaning of random, there are significant differences. For example, computers can generate sequences of "random numbers," but even if these sequences meet the equal-probability test, they are not truly random, since computers produce digits by following a rule. For example, the sequence of digits 1234567890123456789012345678901234567890 . . . has equal probabilities for each digit, but is clearly not random. For a sequence of digits to be truly random, it must have been generated by a physical process, not a rule. There is no way to tell by looking at a sequence whether it is random, although one can recognize that a sequence is not.

random walk (*probability*)   Consider a climb up a steep hill for which there is a chance of advancing and sliding back with every step. This is an example of a random walk. The most basic random walk can be modeled as an infinite series with an equal probability for $+1$ and -1 (the series is an indicated sum such as $1 + 1 - 1 + 1 - 1 - 1 + 1 \ldots$). Whenever the random walk is terminated, there is a definite sum: stopping after 2 terms produces the sum $+2$ and after 6 terms, the sum 0. In a long random walk with equal probabilities in each direction, there is a high probability that the sum will be either positive or negative most of the time. For the hill climb, if there is an equal probability of going up and sliding down, the chance is high that the climber will either advance or slip back, rather than stay in roughly the same position. Random walks may also have different probabilities for each direction or steps of different sizes.

range (*statistics*)   The range of a set of numbers is the difference between the largest and smallest in the set. It is the simplest measure of how variable a set is from number to number, but can be heavily influenced by 1 or 2 numbers that are especially large or small.

rare earth (*elements*)   The rare earth elements form the lanthanide series of elements in a way similar to the actinide series of radioactive elements—that is, the rare earths have such similar chemical properties that they are placed in the same location on the periodic table. They include elements from atomic number 57 (lanthanum) through 71 (lutetium). Despite their name, the rare earths are relatively common metals (not earths in the chemical sense). They share properties that make them important components of electronic devices.

rat (*animals*)   A variety of rodents are called rats, including mole rats (*Spalax*), pack rats (*Neotoma*), and kangaroo rats (*Dipodomys*). But the name usually refers to members of the genus *Rattus*. Native to Europe, Asia, and Africa, many species are now distributed worldwide. For example, the Norway, or brown, rat (*R. norvegicus*)—considered by many to be the most destructive mammal on Earth—arrived in North America by ship in about 1775. Rats have been so successful because they are highly prolific and able to adapt easily to different habitats and diets.

rate (*mathematics*)   A rate, mathematically speaking, is comparison by division, like a ratio, but is treated as direct variation. Rates match such pairs as miles and hours (miles per hour) or legs and animals (legs per animal). The word "per" means that for each one of the entities named after per there are a certain number of the entities named before the word. At 55 miles per hour there are 55 miles for each hour, or for 6 legs per insect the number of legs is 6 times the number of insects.

ratio (*arithmetic*)   The amount of one entity with respect to another is a ratio. If a collection of eggs has 15 brown and 10 white, the ratio of brown eggs to white is 15 to 10 and of brown to total eggs is 15 to 25. Equivalent ratios can be found by dividing by a number such as 5 so that the ratio of brown to white is 3 to 2 and of brown to eggs is 3 to 5, also expressed as 3 out of 5 or as $^3/_5$. In older notation, the ratio 3 out of 5 is written as 3 : 5. Ratios are essentially the same as rates, which may relate 2 different types of entities. A common example is miles per gallon of gasoline. If the rate is 25 miles to the gallon, a trip of 100 miles requires 4 gallons, since 100 is to 4 as 25 is to 1.

rationalization of denominator (*numeration*)   It is considered undesirable to write a fraction with a radical showing in the denominator. For example, the fraction $1/\sqrt{2}$, common in trigonometry, is always written in the equivalent form $\sqrt{2}/2$. If the denominator contains a binomial with a radical, multiplying the numerator and denominator by the conjugate of the denominator will eliminate the radical.

rational number (*numbers*)   A rational number is any number that can be represented as the ratio of an integer to a nonzero integer. The positive and negative fractions (with integral parts) are rational numbers, but so are all the integers. As decimals, rational numbers either terminate (that is, become a string of 0s after some decimal place, such as 0.25 = 0.2500000 . . .) or repeat a finite pattern over and over (such as 0.33 . . . = $^1/_3$ or 0.142857142857142857 . . . = $^1/_7$).

ratite (*animals*) Flying birds have a sternum (breastbone) with a median keel to which flight muscles attach. The flightless penguins also have a keeled sternum. Although descended from flying birds, the flightless ratite birds have a flat sternum without a keel. Ratites include the ostrich, rheas, cassowaries, emus, kiwis, and extinct elephant bird and moas.

rattlesnake (*animals*) Native to the Americas, rattlesnakes are heavy-bodied pit vipers 2 to 8 ft (0.6 to 2.5 m) long. They typically have a series of horny segments—a rattle—at the end of their tail; each segment is a scale that covered the tip of the tail and was retained after the skin was shed. When vibrated the rattle creates a buzzing sound, thought to be a warning to keep away. Rattlesnake venom—injected into prey via 2 large, movable fangs—contains enzymes that destroy nerve and blood tissue.

Ray, John (*botanist*) [English: 1627–1705] A prolific cataloger and writer, Ray introduced the first significant system of classifying plants and animals based on anatomy and physiology (it was superseded by the scheme of Linnaeus in 1735). Ray separated flowering plants into the 2 groups, monocots and dicots, still recognized today. He also was the first to use species as the fundamental unit of classification.

ray (*animals*) Mainly bottom dwellers in tropical and temperate seas, rays are cartilaginous fish with a flat, disklike body. A huge winglike fin extends from each side. Water enters through large holes called spiracles on the top of the head and passes out through gill slits on the underside of the body. Many species, such as the stingrays, have poisonous spines. Electric rays have electric organs on each side of the head. The largest ray, the Atlantic manta (*Manta birostris*), can be more than 20 ft (6 m) across and weigh 3000 lb (1360 kg).

ray (*geometry*) A ray is the part of a line on one side of a point including the point. If the point is not included, the figure is a half line. A ray continues without bound (to infinity) in a single direction.

real number (*numbers*) The totality of rational numbers and irrational numbers comprises the real numbers (so called to contrast with imaginary numbers). The real numbers are most easily pictured as any number that can be found as a point on a line at a specific distance from 0, with positive real numbers on one side of 0 and negatives on the other. Each such point corresponds to exactly 1 real number and vice versa. A second useful way to picture the real numbers is as any number that can be represented by an infinite decimal, such as $0.142857142857142857 \ldots = 1/7$, which repeats, or $3.14159265358979 \ldots = \pi$, which does not. A minor problem with the infinite-decimal representation is that 2 different decimals can equal the same real number, as, for example, $2.0000 \ldots$ and $1.99999. \ldots$

recapitulation (*biology*) Based on observations of the similarity among embryos of different species, Ernst Haeckel [German: 1834–1919] proposed the theory of recapitulation, or biogenetic law. This theory states that during its development (ontogeny) an organism summarizes, or recapitulates, the stages passed through by its ancestors (phylogeny). The theory was discredited as

20th-century scientists showed that it is not possible to determine evolutionary lineage on the basis of embryological development.

reciprocal (*arithmetic*)   The number obtained when a given number is divided into 1 is the reciprocal of the given number; for example, the reciprocal of 5 is 1 divided by 5, or ⅕. For common fractions, the reciprocal can be obtained by interchanging numerator and denominator, so the reciprocal of ³/₇ is ⁷/₃.

recombination (*genetics*)   The formation of new gene combinations is termed recombination. It can occur naturally as a result of **crossing over**. Recombination also can result from the laboratory manipulation of DNA—for example, removing a segment of DNA from one organism and inserting it into the DNA of another organism. Thus when the human gene that directs production of the hormone insulin is inserted into the DNA of bacteria, the bacteria—and all their descendants—manufacture human insulin. The first such hybrid, or recombinant DNA, was created in 1971 by Paul Berg [American: 1926–] and marked the start of the field of **genetic engineering**.

rectangle (*geometry*)   A **polygon** that consists of 2 pairs of **parallel** line segments with segments that meet forming **right angles** is a rectangle. A square is a rectangle with equal-length sides. The **area** of a rectangle is the product of the lengths of 2 of the adjacent sides ($A = lw$); the area of the square becomes the numerical square of the length of a side ($A = s^2$).

rectum (*anatomy*)   The terminal portion of an animal's intestine, opening into the cloaca or anus, is a short muscular tube called the rectum. Feces collect and are stored here prior to defecation. In insects wastes collected from the blood by Malpighian tubules also are passed into the rectum, and some reabsorption of water, salts, and amino acids occurs from the rectum to the blood.

recycling (*environment*)   The 3 Rs of environmentalism—reduce, reuse, and recycle—slow depletion of **natural resources**, conserve energy, and help reduce **pollution**. Reducing means producing less waste; for example, using both sides of a sheet of paper reduces paper waste. Reusing means cutting down on waste by using something over and over again; for example, using a cloth shopping bag hundreds of times rather than hundreds of plastic bags once each. Recycling is the process of recovering and processing discarded materials into new resources and products; for example, crushing glass, melting it, and making new bottles.

red algae (*protists*)   Algae of the phylum Rhodophyta are branching multicellular seaweeds typically less than 3 ft (0.9 m) long. Their red color comes from pigments called phycoerythrins. These pigments enable the algae to carry out photosynthesis in deeper ocean waters than those inhabitable by other algae. In addition to cellulose, the cell walls of red algae contain large amounts of complex sugars; these are the source of agar, a substance used as a solidifying agent in foods and a culture medium in laboratories.

Coralline algae are species of red algae that are of great importance in the building of coral reefs, often contributing far more to the reefs than coral polyps. The algae concentrate lime from seawater, then secrete it as a hard shell around their cells.

red blood cell (*cells*)   The most numerous cells in vertebrate blood are the erythrocytes, or red blood cells (RBCs). The pigment hemoglobin, an iron-containing protein present in RBCs, gives the blood its characteristic red color and carries out the cells' functions: the transport throughout the body of oxygen, carbon dioxide, and nitric oxide.

After birth, RBCs are normally produced only in the bone marrow, with excess cells stored in the spleen in case of emergency. In mammals, mature RBCs are flat disks, thinner in the center than around the outside. Except for camels, mammalian RBCs lack nuclei. (RBCs of other vertebrates have nuclei.) The liver and spleen destroy worn-out RBCs, with the hemoglobin molecule reused in the formation of new RBCs. In humans, the bone marrow produces some 200,000,000,000 RBCs every day.

red dwarf (*stars*)   Stars of low luminosity and a diameter less than half that of the Sun are red dwarfs, the longest living stars on the main sequence. Because of their long life span, red dwarfs are very common.

red giant (*stars*)   Red giants are stars that have moved above the main sequence in a Hertzsprung-Russell diagram. When stars on the main sequence, such as the Sun, consume all or almost all of their hydrogen, they begin to fuse helium, which heats up the core. Heat from the core expands the outer layers of the star, increasing diameter enormously. The outmost layers cool because they are now far away from the core, causing light from the star to be reddish.

Redi, Francesco (*biologist*)   [Italian: 1626–97] In simple experiments Redi showed that flies develop only from eggs laid by other flies of the same species. This supported the theory that "life comes from life" and greatly discredited the centuries-old theory of spontaneous generation, which claimed that flies could develop from decaying meat. Redi also pioneered the scientific study of worms.

red shift (*cosmology*)   Not only does the Doppler effect increase wavelengths of waves produced by a receding source, but also the expansion of the universe stretches the length of waves that have traveled great distances. Since the longest wavelengths of visible light are red, this lengthening is called a red shift. The farther away a galaxy or quasar is, the greater its red shift. Relativity theory adds a third source of red shift, which is a strong gravitational field.

red tide (*ecology*)   The so-called red tide is a misnomer, for it has nothing to do with tides. Rather, it results when populations of *Gonyaulax* and certain other marine dinoflagellates (1-celled algae) reproduce in such great numbers that they turn the seawater red. Usually these blooms are not harmful, but in other cases the algae produce potent neurotoxins that kill large numbers of fish or accumulate in shellfish to such high levels that they cause serious illnesses and even death in people who eat the shellfish.

redwood (*plants*)   The tallest of all living trees are the redwoods (*Sequoia sempervirens*), named for their reddish brown bark and wood. They are evergreen conifers native to the coastal region of northern California and southern Oregon. Some specimens grow to heights of more than 360 ft (110 m). The lower part of the trunk has supporting buttresses. Redwoods have stiff scalelike leaves and small cones.

The dawn redwood (*Metasequoia glyptostroboides*), native to China, is a deciduous conifer that grows to heights of more than 100 ft (30 m). It has a buttressed trunk, bright green needles that turn brown before falling in autumn, and small brown cones.

reed (*plants*)   Various members of the grass family with hollow, jointed stems are commonly called reeds. One of the most widespread species is *Phragmites communis*, found in streams, swamps, and other wet habitats. It grows to 20 ft (6 m) tall and has long flat leaves and flowers borne in large clusters. It spreads mainly by means of horizontal underground stems (rhizomes).

reflection (*geometry*)   Reflection in a plane is a transformation that takes each point on one side of a line and moves it so that the distance of its image from that line is unchanged and the line segment connecting the point and its image is perpendicular to the line. Thus the part on one side of the line is a mirror image of the part on the other. In a Cartesian coordinate reflection about an axis involves replacing either x with $-x$ (reflection about the y axis) or y with $-y$ (about the x axis). Reflection is the basic transformation in a plane: reflection twice about 2 parallel lines is the same as translation, while reflection twice about 2 intersecting lines is the same as rotation.

reflection (*optics*)   All materials reradiate some of the light that falls on them, a phenomenon called reflection. Light briefly raises the energy of electrons but they quickly fall back to the lower state by emitting light. Free electrons in shiny metals reflect most of the light, but so does a transparent material in front of an opaque one. White materials reflect much of the incident light, but also spread it out. Black materials reflect almost no light.

When you see an image, called a reflection, in a shiny material, the light that you observe has taken the shortest possible path from its origin to your eyes. Along this path, light forms an angle at the point of reflection. The beam of light makes the same angle with the normal to that surface when it reaches the reflecting point as when it leaves. This is often expressed by saying that the angle of incidence is equal to the angle of reflection.

reflex (*physiology*)   An involuntary, automatic response to a stimulus—jumping when frightened, constricting of the eye's pupil on exposure to bright light—is a reflex. It involves a simple chain of neurons (nerve cells) called a reflex arc. For example, when your hand touches a hot stove, the heat stimulates a sensory receptor in your skin. An impulse travels along a sensory neuron to the spinal cord, where the neuron passes the impulse to an interneuron. The interneuron passes the impulse to a motor neuron that ends in an arm muscle, which then contracts to jerk your hand away from the stove. All this happens before a message reaches your brain and you become conscious of pain. Some reflexes are learned rather than inborn. Such learned reflexes are conditioned responses.

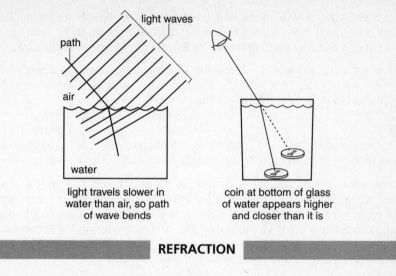

light travels slower in
water than air, so path
of wave bends

coin at bottom of glass
of water appears higher
and closer than it is

REFRACTION

refraction (*optics*)   When light passes from one material to another at an angle other than straight on (90°, called the normal), a difference in the speed of light for the 2 materials causes the light to bend toward the material with the slower speed. This bending is called refraction. The same occurs with light passing into a material from a vacuum; sunlight is refracted by the atmosphere, making the Sun appear higher in the sky than it actually is, especially near dawn and sunset. (*See also* **lens.**)

regeneration (*biology*)   Many plants and animals have the ability to regenerate, or grow back parts that have been removed. Some salamanders and lizards can grow a new tail if the original tail is lost or injured. Starfish can replace lost arms and crayfish can replace lost pincers. Planarians can regenerate entire bodies from even small body fragments.

Regeneration also describes the regrowth of tissues. Many human tissues, including liver and skin tissues, have at least some capacity to regenerate themselves after injury or disease.

regolith (*soils*)   A regolith ("blanket of rock") is weathered rock and small particles of rock, including soil, lying on the bedrock of the crust. But the word is most often used to mean the inorganic, or soil-free, mix of dust and small rocks found at the surface of terrestrial planets or other rocky bodies, such as the Moon.

regular figure (*geometry*)   A regular figure is a polygon or polyhedron that has the same kind of symmetry as the number of sides or faces. For a polygon this means that both the sides and the angles between them must be equal. For a polyhedron the faces must be congruent regular polygons and the polyhedral angles formed must all be congruent. (*See also* **Platonic solid.**)

regurgitation (*ethology*)   A variety of animals feed certain members of their species through regurgitation, casting up incompletely digested food or other substances. African wild dogs that have had a successful hunt feed kin by regurgitating meat from a kill. Pigeons regurgitate a nutritious "milk" to feed their young and storks regurgitate bits of fish for the young to gobble up.

relation (*analytic geometry*)   A relation is a way of connecting 2 sets of numbers. One way to define a relation is as any set of ordered pairs; when a statement of a relation is true, the ordered pair is in the set, but when the statement is false, the ordered pair is not in the set. Relation statements often use special symbols, such as = ("equals"), < ("is less than"), or ∈ ("is a member of"). A true statement such as 3 < 5 means the ordered pair (3, 5) is in the relation for "is less than," while the false 6 < 2 means that (6, 2) is not in that relation.

relative error (*measurement*)   The ratio of the error resulting from the precision of the instrument used in making a measurement to the amount measured is called relative error. For example, measuring a distance of 10 mi to the nearest $1/2$ mi has a relative error of 1 to 20, or 5%, while measuring the same distance to the nearest 10 ft has a relative error of 1 to 5280, or 0.01893%. The smaller the relative error, the more accuracy for the measurement.

relativity (*physics*)   Relativity refers to frames of reference in relative motion— for example, one reference frame could be Earth while the other might be a moving train. In 1905 Einstein developed laws connecting one reference frame to another moving steadily with respect to the first. Postulate (1): Physical laws remain exactly the same in either frame. Postulate (2): The speed of light is the same in each. These postulates led to the special theory of relativity. Time depends on motion in space; together they form space-time with 4 dimensions; the speed of light is the maximum possible; objects contract in the direction of motion; and energy and mass are equivalent.

   In the next decade Einstein developed the theory for acceleration of one frame of reference with respect to the other, the general theory of relativity. Postulate (3): Gravity and inertia are exactly equivalent. General relativity describes gravity as the consequence of space curving in the presence of mass.

REM sleep (*physiology*)   Recordings of the electrical impulses produced by the brain show that the sleep of humans and other primates has five distinct stages. Stages 1 through 4 progress from light to deep sleep. In the fifth stage, called rapid eye movement (REM) sleep, the eyes dart back and forth beneath closed eyelids but the rest of the body muscles are completely relaxed; it is during this stage that individuals are most likely to dream. Throughout its sleep period a primate cycles between REM and non-REM sleep. In a human this cycle repeats itself every 90 to 110 minutes. Thus a person may have 4 or 5 REM periods each night.

reovirus (*viruses*)   Members of the virus family Reoviridae infect animals, including humans. The group includes the rotaviruses, which cause infant enteritis. Reovirus particles are 20-sided polyhedrons that lack envelopes. The genetic information is stored in a double strand of RNA.

reproduction (*biology*)   One of the fundamental characteristics of living things is the ability to produce new individuals of their own kind. Unlike the other life processes, reproduction is not necessary for the life of an individual organism. But because organisms have limited life spans, reproduction is essential for the continued existence of a species; if its members do not reproduce, a species becomes extinct. There are 2 types of reproduction: asexual reproduction involves only 1 parent; sexual reproduction involves 2 parents. Plants and some other organisms have a life cycle that involves an alternation of generations, with an asexual generation followed by a sexual generation.

reproductive system (*anatomy*)   Unlike other animal systems, which are essential for the life of an individual animal, the reproductive system is needed to ensure continuation of the species. The reproductive system consists of organs that produce gametes (sex cells)—testes in males form sperm and ovaries in females form eggs—plus ducts and perhaps accessory organs designed to carry the gametes to the outside. In mammals and some other animals, the female reproductive system also is home to the developing embryo.

reptile (*animals*)   Reptiles are air-breathing vertebrates with dry, scaly skin and, usually, a 3-chambered heart. They are ectothermic (cold-blooded), warming themselves by absorbing heat from the environment and thus quite dependent on climate and temperature; hence, most species live in tropical and temperate regions. Most reptiles are 4-legged; snakes and legless lizards are descended from 4-legged ancestors that lost their limbs in the course of evolution Many snakes and lizards give birth to living young, but most reptiles lay eggs.

Reptiles were the first vertebrates totally adapted to life on land. The approximately 6000 living species are far fewer than existed in the past, particularly during the Mesozoic, when dinosaurs and other reptiles were the dominant animals on Earth. Only 4 orders have living representatives: turtles, crocodilians, tuataras, and lizards and snakes. The smallest reptile is thought to be the Virgin Gorda gecko (*Sphaerodactylus parthenopian*), at an average length of 0.7 in. (1.8 cm). The largest reptiles are the gavial (*Gavialis gangeticus*) and the saltwater crocodile (*Crocodylus porosus*), crocodilians that may exceed 20 ft (6 m) in length.

resin (*botany*)   Many trees and shrubs, particularly conifers, secrete complex organic compounds called resins. If you prune a pine tree, for example, a sticky resin flows from the wound. On exposure to air, the resin hardens, protecting the damaged area. Resins also may be produced in response to infections. Sometimes resins form a sticky covering on buds, providing protection and reducing water loss, as in horse chestnut buds.

resolution (*astronomy*)   Ability to observe details is called resolution. The term derives from resolving power, a measure of ability to discern as separate images 2 very close stars. The angular distance between stars that can be resolved is inversely proportional to a telescope's diameter. Interferometry gives 2 telescopes together a resolving power based on distance between them instead of diameter.

resonance (*particles*)   Particle resonances are high-energy states of hadrons that quickly decay. They are called resonances because a smaller particle seems to push a larger particle with just the right energy to create a high peak on a graph of particle interactions. Because resonances decay so quickly, physicists did not want to call them particles at first, but today such resonances as the delta, sigma, and xi particles are accorded full particle status.

resonance (*physics*)   Resonance refers to raising the amplitude (height) of a periodic motion, such as a wave. Any constrained system, such as the vibration of a rod or the electric current in a circuit or the swing of a pendulum or the movement of a molecule in a crystal, has a natural period. If a series of small waves or motions match that natural period, they can increase or decrease amplitude. A familiar example is using small motions of the legs and body in a swing to increase the height of the movements.

respiration (*physiology*)   In common usage the term respiration is a synonym for breathing—that is, drawing air into the lungs and then expelling it. However, respiration more properly refers to the phase of metabolism during which food molecules are broken down, releasing energy in the form of ATP. Every living cell carries out such cellular respiration—which consists of a series of many small steps, or reactions—for the cell needs a continuous supply of energy to perform life processes.

respiratory system (*anatomy*)   The function of an animal's respiratory system is to obtain oxygen from the environment and give off carbon dioxide. Gills are organs designed to extract oxygen from water; lungs remove oxygen from air. Many invertebrates have gills, as do fish. Most amphibians respire through gills when young, then metamorphose into adults with lungs. Reptiles, birds, and mammals—including aquatic mammals such as whales—breathe through lungs. Some invertebrates have structures unique to their group. Insects, for instance, have tubes called tracheae that carry air directly to all cells of the body. Spiders also have tracheae, plus gill-like structures called book lungs.

rest mass (*particles*)   Because the mass m of a subatomic particle is tiny, it is usually measured in terms of energy, E. Einstein's famous equation $E = mc^2$, where c is the velocity of light in a vacuum, shows that the energy of a particle is much larger than its mass. Particle energies are measured in units of a million electron volts, or MeV. Einstein's formula implies that a particle moving at a high speed will gain mass, but physicists calculate the energy as if the particle were stationary, a number called the rest mass.

restoration (*environment*)   Ecological restoration attempts to return a disrupted habitat to its original state. It is a costly, time-consuming process that involves the undoing of alterations (usually made by humans) to the habitat. For example, plans to restore the Everglades of Florida include removal of hundreds of miles of canals and levees, redistribution of water, and elimination of paperbark trees (genus *Melaleuca*) and other exotic species that have crowded out native plants.

restriction enzyme (*biochemistry*)   A bacterial enzyme that cuts DNA molecules at a specific sequence of nucleotides is termed a restriction enzyme or endonuclease. Several hundred endonucleases have been identified. Their presumed purpose is to destroy the DNA of invaders. Endonucleases have become important tools for manipulating DNA in genetic engineering research.

retina (*anatomy*)   The retina, or inner layer of the back of the vertebrate eye, has millions of light receptor cells called cones and rods. Cones, short and thick, are concentrated in the retina's center, at the fovea, where the lens of the eye focuses light rays. Cones contain pigments sensitive to bright light and are responsible for color vision. In contrast, rods—characteristically long and narrow—are found outside the fovea in increasing concentrations toward the periphery of the retina. Rods contain the pigment rhodopsin, which is extremely sensitive to low light levels.

retrograde motion (*solar system*)   The solar system emerged from a single rotating disk, so the initial motion of all parts reflects that rotation and corresponds to Earth's rotation from west to east and similarly directed revolution about the Sun. Although all planets revolve in the same direction, Venus and perhaps Uranus have the reverse rotation, called retrograde. Some planetary satellites show retrograde revolution (or rotation or both). Scientists believe that most retrograde motion results from collisions or gravitational capture of asteroids by planets. Comets from the Oort cloud can be launched into either prograde (opposite of retrograde) or retrograde orbits by passing stars.

retrovirus (*viruses*)   Members of the virus family Retroviridae are pathogens of vertebrates. A number of them are associated with cancerous tumors in birds and mammals. Two of the most notorious retroviruses are the human immunodeficiency virus (HIV), which causes acquired immune deficiency syndrome (AIDS) in humans, and the feline leukemia virus, the most common cause of serious illness and death in domestic cats. Retroviruses have RNA as their genetic material, but in order to reproduce within the cells of their hosts, their RNA must be copied to DNA. This is done by reverse transcriptase, an enzyme found only in retroviruses and so named because it performs the reverse of the usual transcription of RNA from DNA.

revolution (*astronomy*)   Motion in a recurring path (orbit) around another body is called revolution. The length of time for a body in orbit to travel 1 full orbit is that body's period of revolution, also known as a year.

rhabdovirus (*viruses*)   Members of the virus family Rhabdoviridae infect animals and plants. The best-known genus is *Lyssavirus,* which causes rabies. The surface of each bullet-shaped virus particle is covered with prominent glycoprotein spikes that bind to receptors on the surface of host cells. The genetic information is stored in a single strand of RNA.

rhea (*animals*)   Looking somewhat like fluffy ostriches, rheas are large flightless birds that live on grasslands in South America. The bigger of the 2 rhea species, the greater rhea (*Rhea americana*), may be 5.6 ft (1.7 m) tall and weigh 55 lb (25 kg). Rheas have long legs, can run rapidly to escape predators, and kick and claw their enemies when cornered. They travel in small groups during most of the year.

rhenium (*elements*)   Although not usually considered a part of the platinum group of metals, rhenium shares most of its attributes (at atomic number 75, it is next to the platinum metals on the periodic table). Unlike platinum-group metals, rhenium is widely distributed instead of residing most often in platinum ores. Nevertheless, rhenium was first discovered in platinum ores, in 1925 by Walter Noddack [German: 1893–1960] and coworkers, but soon found elsewhere by the same team. The name is from Latin *Rhenus*, "Rhine." Chemical symbol: Re.

rhinoceros (*animals*)   Members of the family Rhinocerotidae are large hoofed mammals with a big head. They are armed with—depending on the species—1 or 2 solid, boneless horns made of fused hair. There are 5 living species, 2 native to Africa and 3 to Asia. All are endangered species. The largest, the white rhino (*Ceratotherium simum*) of Africa, may be 6 ft (1.8 m) tall at the shoulder and weigh close to 3 tons; it is actually gray-brown in color. The more than 30 known fossil species include the even bigger *Indricotherium*.

rhizome (*botany*)   A rhizome is a specialized stem that grows horizontally at or just below the soil surface. Leaves and roots of a new plant can grow from nodes along the rhizome—a form of vegetative propagation. Rhizomes may be long and slender, as in some grasses, or thick and fleshy, as in irises.

rhodium (*elements*)   Rhodium, atomic number 45, is one of the platinum group of metals. It is produced in small quantities for hardening platinum and palladium. William Hyde Wollaston [English: 1766–1828] discovered rhodium in 1803 and named it from the Greek *rhodon*, "rose," for the color of its salts. Chemical symbol: Rh.

rhododendron (*plants*)   The shrubs and small trees that constitute the genus *Rhododendron* in the heath family are native mostly to northern temperate regions. They display great variety in size and flower color. Species with evergreen leaves and showy clusters of bell-shaped flowers are commonly called rhododendrons. Species that have deciduous leaves and tubular flowers are commonly called azaleas.

rhodopsin (*biochemistry*)   Also called visual purple, rhodopsin is a light-sensitive purple pigment in the retina of vertebrate eyes. Localized in rod cells, it is sensitive to low levels of light, enabling animals to see at night. In the presence of light, rhodopsin breaks down into the protein opsin and retinal, a derivative of vitamin A. This triggers signals in nearby cells of the optic nerve. The lack of rhodopsin causes night blindness.

rhubarb (*plants*)   Members of the buckwheat family, rhubarbs (*Rheum*) are perennial herbs native to China, where they have been cultivated for more than 2000 years. Growing to 10 ft (3 m) tall, rhubarbs have a thick underground stem, or rhizome, from which large leaves arise. Each leaf has a long edible stalk but its greatly expanded blade contains the poison oxalic acid.

ribbon worm (*animals*)   Members of the invertebrate phylum Nemertina are unsegmented worms with flat, slender bodies. They are capable of greatly elongating and contracting their body. Most ribbon worms are less than 8 in. (20 cm) long but *Lineus longissimus* can stretch to 90 ft (27 m). At the front end, above the mouth, is an opening through which a proboscis can be extended and withdrawn. The proboscis, armed with spines in some species, is used to capture small invertebrate prey. Most ribbon worms are marine, living on the bottom of the sea. They have great powers of regeneration.

ribosome (*cells*)   Ribosomes are tiny structures found in all cells. They are composed of ribosomal RNA plus various proteins. Under the direction of DNA in the cell's nucleus, ribosomes assemble amino acids into proteins. Some ribosomes are scattered in the cytoplasm, others are attached to a membrane called the endoplasmic reticulum. The free ribosomes mainly synthesize proteins used within the cell; endoplasmic reticulum ribosomes produce proteins that are secreted from the cell.

rice (*plants*)   Considered the most important of all crops because it is the principal food for more than half of humanity, common rice (*Oryza sativa*) is a perennial grass, though commonly grown as an annual. It is native to tropical Asia. Thousands of varieties suitable for a broad range of environmental conditions have been developed. But rice grows best in warm habitats and flooded soil. Rice has stems that average 5 ft (1.5 m) in height, though some varieties are more than 16 ft (5 m) tall. At the top of each stem is a head of flowers that develop into edible grains (fruits).

Wild rice (*Zizania aquatica*), native to North America, is an annual grass that grows in shallow water along the margins of lakes and ponds. It has stalks up to 10 ft (3 m) high, with flower clusters at the top. Manchurian wild rice (*Z. caducifolia*), native to northeastern Asia, is similar but smaller.

Richter scale (*geology*)   Sizes of earthquakes are measured using a system originally devised in 1935 by Charles Richter [American: 1900–85] and based on the amplitudes of seismic waves converted to numbers called magnitudes (M). The Richter scale of magnitudes uses the logarithm of the wave's height adjusted so that values range from 0 to about 8.9. The logarithmic scale means that a change in 1 of magnitude results from a change of 10 times in the wave's height. Energy levels vary even more dramatically; a difference in 1 unit of magnitude means total energy differs with a factor ranging from 30 to 60 times.

rickettsia (*monera*)   Discovered in 1909 by Howard Ricketts [American: 1871–1910], rickettsia are extremely tiny rod-shaped bacteria that live and reproduce within other cells. They live in mammals and most commonly are transmitted to other mammals by bloodsucking ticks, mites, and other arthropods (which are not affected by the rickettsia). *Rickettsia rickettsia*, transmitted by ticks, causes Rocky Mountain spotted fever; *R. prowazekii*, transmitted by lice, causes typhus.

Riemann, Bernhard (*mathematician*)   [German: 1826–66] Riemann's work made mathematics more general or abstract. He put the definite integral on a firmer basis with a generalization now called the Riemann integral. In 1851 Riemann invented a new way to show functions of complex numbers on a plane and developed fundamental ideas of topology to handle such representations. In 1854 Riemann addressed the foundations of geometry in spaces of n dimensions, suggesting a new form of non-Euclidean geometry that later became the geometry of Einstein's general theory of relativity. In 1859 Riemann put forward a conjecture about a complex function called the zeta function that remains among the main unresolved issues of mathematics.

rift valley (*geology*)   Wherever tectonic plates are pulled apart from each other, a steep-sided canyon called a rift valley develops. Such a rift valley runs completely around Earth along the mid-oceanic ridge system. On land the largest rift valley runs from the river Jordan and Dead Sea in the Middle East through the great lakes of east Africa, but it is largely inactive. The rift valleys along ocean ridges and on Iceland, however, pour out lava, forming new ocean floor or, on Iceland, extending the island.

Rigel (*stars*)   Rigel (Arabic for "foot") is a bright blue star—a blue supergiant, actually—in Orion. Picture the constellation of the hunter Orion as a parallelogram of bright stars enclosing the 3 stars of the belt with reddish Betelgeuse as the head; Rigel is the opposite vertex, the left foot as Orion faces you.

right angle (*geometry*)   If 2 lines meet and all 4 angles formed are equal in size, each angle is a right angle; from this, a right angle is half a straight angle. In degree measure, the right angle measures 90°; in radian measure, $\pi/2$. Lines or planes that meet at right angles are termed perpendicular or orthogonal.

right ascension (*astronomy*)   Location of celestial bodies is described by 2 measures, declination, or height in degrees, and right ascension. Right ascension is measured as time, with 1 hour the equivalent to 15° of longitude. Instead of measuring right ascension from an arbitrary line, such as the prime meridian, it is measured from the vernal equinox, which is the point of intersection in the northern spring of the ecliptic and the extension into space of Earth's equator. Right ascension is measured in positive hours and minutes, corresponding to east from the vernal equinox; hence, saying that Sirius has a right ascension of 6 hours, 44 minutes means that it is about 100° east of the vernal equinox. Since the equinoxes are slowly changing in the precession of equinoxes, an exact right ascension depends on the date.

ring (*algebra*)   A ring is an algebraic structure formed from a set with 2 operations, such as $+$ and $\times$. The elements and 1 operation, such as $+$, must form a group that is commutative (an abelian group). The other operation, in this case $\times$, must produce unique answers and exhibit the associative property. The 2 operations are connected by requiring $\times$ to follow a distributive property with respect to $+$, so if a, b, and c are elements of the ring, $a \times (b + c) = (a \times b) + (a \times c)$ and also $(b + c) \times a = (b \times a) + (c \times a)$.

   If $\times$ possesses the commutative property and there is an identity element for $\times$, the ring becomes an integral domain, a structure familiar from the natural numbers and integers.

ring (*solar system*)   Gas-giant planets in the solar system are encircled by rings made from millions of small particles, although Jupiter's rings are so faint they are often disregarded. The particles range from large boulders to bits of dust. Saturn has the most complex ring system, easily observed with a small telescope. Although Galileo noticed Saturn's rings in 1610, he could not determine their shape, so credit for recognition goes to Huygens in 1656. The rings around Uranus were first detected from an airplane-mounted telescope in 1977, while Neptune's partial rings were first recognized in 1984 when a star passed behind one. Astronomers believe that the ring systems are a key to understanding planet formation, but unsolved mysteries about the rings abound.

Ring of Fire (*geology*)   At the edges of tectonic plates, where 2 plates are moving apart or pushing into each other, volcanic fissures and active volcanoes are common. Because the Pacific Ocean is largely bounded by such plates, its border is called the Ring of Fire. The western part of the Ring of Fire follows long lines of volcanic islands from New Zealand through New Guinea and the Philippines up to Japan rather than along the western shore of the Pacific.

river (*geography*)   A river is a natural path created by the flow of freshwater through land. Rivers vary in size, type of flow, and origin. They may flow from lakes or glaciers or begin as small streams flowing down mountains or hills. The origin is called the river's head; the end is the river's mouth. One river can grow so that its head drains another (stream capture). When the mouth of one river reaches another river, the first is a tributary. Tributaries and the main river form a river system.

Slow-moving rivers that make meanders are sometimes called old rivers, while speedy mountain streams are called young; but these differences result from topography, not age.

RNA (*genetics*)   Ribonucleic acid, or RNA, is needed by all organisms for protein synthesis. In certain viruses RNA also acts as the hereditary material that is transmitted from one generation to the next. Like molecules of the other nucleic acid, DNA, an RNA molecule is a chain of nucleotides. Each nucleotide is composed of a phosphate group, the sugar ribose, and 1 of 4 nitrogen bases: adenine, guanine, uracil, and cytosine.

RNA occurs in several forms. In viruses, it is single- or double-stranded, depending on the species. RNA involved in protein synthesis is single-stranded and includes messenger RNA (mRNA), ribosomal RNA (rRNA), and transfer RNA (tRNA).

rock (*rocks and minerals*)   Rock is a hard substance made from one or more minerals, usually more than one. Rocks are identified by the minerals they contain; by the size of their crystals or other recognizable lumps, called grains; and by general color, ranging from dark to light. Rocks are also classified by origin as igneous rock, metamorphic rock, or sedimentary rock. For example, the igneous rock diorite, a mixture of such minerals as feldspar and hornblende, is recognized by its speckled black and white large grains, while the sedimentary rock chalk, a pure form of calcite, is white with very fine grains.

rodent (*animals*)   Chinchillas, gerbils, guinea pigs, hamsters, lemmings, marmots, mice, nutrias, pacas, porcupines, rats, squirrels, and voles are members of Order Rodentia. The order is the largest of all mammal orders, both in number of species and number of individuals. All rodents eat vegetation; many also eat animal food. Rodents lack canines but have 4 chisel-shaped incisors adapted for gnawing; these teeth grow throughout a rodent's life and thus are not worn down to stumps. Most rodents are less than 8 in. (20 cm) long, including the tail, but the capybara can attain a length of more than 4 ft (1.2 m).

Roentgen, Wilhelm (*physicist*) [German: 1845–1923] Although Roentgen was an accomplished experimental physicist, he is remembered for one giant discovery: X rays. In 1895 he observed that experiments with a cathode-ray tube produced a glow some distance away on a card coated with fluorescent paint. Since cathode rays (streams of electrons) cannot travel far through air, he recognized that the glow had to have some other cause. Roentgen correctly guessed that his X rays were short electromagnetic waves.

roller (*animals*) Found mainly in Africa but also in Europe, Asia, and Australia, rollers (Family Coraciidae) are brightly colored birds 10 to 14 in. (25 to 36 cm) long. Their name alludes to the aerial acrobatics performed during courtship—a spectacular mix of rolls, dives, twists, loops, and other maneuvers. Rollers have a powerful beak that ends in a hook, used to capture insects and small vertebrates either midair or on the ground.

roosting (*ethology*) When a bird rests or sleeps it is said to be roosting. The place on or in which a bird customarily rests and sleeps is called its roost. Birds typically roost in the same kinds of places where they nest—grosbeaks in evergreen trees, woodpeckers in tree cavities, and so on. Many species assemble in groups to roost near one another. Bobwhite quail, for example, roost on the ground in groups of up to 30, forming a tightly packed circle with their heads pointed outward.

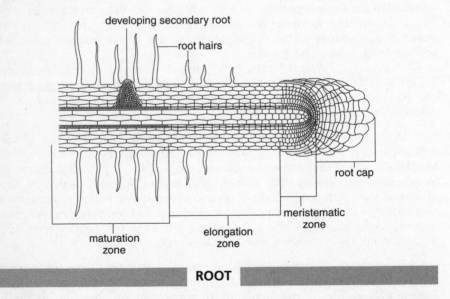

developing secondary root

root hairs

root cap

meristematic zone

maturation zone

elongation zone

ROOT

root (*botany*) The main functions of roots, which are the underground organs of most plants, are to anchor plants in the soil and to absorb water and minerals. In addition, the roots of some plants, such as beets and radishes, are important food storage organs.

Growth in length of a root occurs in the meristematic zone, which is covered by a protective root cap. Behind the meristematic zone, in the elongation zone, cells enlarge and begin to differentiate. In the maturation zone, cells reach their full size and assume their functions. Many of the surface epidermal cells in the maturation zone have cylindrical root hairs, which increase the absorbing surface of the root.

root (*numbers*) A root of a number, when multiplied by itself several times, produces a given number called the **power**. The square root is multiplied twice— $(-5) \times (-5) = 25$ so -5 is a square root of 25. The cube root must be multiplied 3 times—for example, -5 is a cube root of -125. Beyond the cube root, ordinal numbers identify how many multiplications are used. The real fourth roots of 16 are 2 and -2, while the real fifth root of -32 is -2. When complex numbers are allowed, there are as many roots as the ordinal number; for example, the 4 complex fourth roots of 64 are $+2\sqrt{2}$, $-2\sqrt{2}$, $+2i\sqrt{2}$, and $-2i\sqrt{2}$.

root system (*botany*) The first root that develops from a plant embryo is the primary root. If this root gives rise to all or most of the plant's roots, the plant has a primary root system. If all or most roots grow from stems or leaves, the plant has an **adventitious root** system. If the primary root continues to be the largest root, it is called a taproot. Carrot and dandelion plants have taproot systems. If early in the primary root's growth it is replaced by numerous slender branching roots, the plant has a fibrous root system. Grasses and African violets are examples. Adventitious root systems usually are fibrous.

rose (*plants*) The rose family, Rosaceae, comprises some 3000 species of herbs, shrubs, and trees. The stems often are armed with spines. The leaves may be simple (**pear**) or comprised of leaflets (rose). The flower typically has 5 sepals, 5 petals, numerous stamens, and numerous carpels (sections of the ovary). The fruit may be a drupe (peach), a pome (**apple**), or aggregate (raspberry). Roses (*Rosa*) have been cultivated since ancient times for their beautiful, often exquisitely scented, flowers. Other members of the family are known for their edible fruits, including cherry, medlar (*Mespilus*), quince, and strawberry. Mountain ashes (*Sorbus*), service berries (*Amelanchier*), and spiraeas (*Spiraea*) are among the species grown as ornamentals.

rosemary (*plants*) A member of the mint family native to the Mediterranean region, rosemary (*Rosmarinus officinalis*) is an aromatic evergreen shrub. Many varieties, or cultivars, exist; some grow to about 6.5 ft (2 m) tall, but others are much shorter. Rosemary has needlelike leaves, used to flavor food since ancient times, and flowers that usually are a shade of blue, pink, or purple.

rotation (*astronomy*) Turning about an imaginary line that passes through a body—the axis—is rotation. All known astronomical bodies rotate, each with its own constant speed, although a star, being fluid, can rotate so that parts spin at different speeds. The period of rotation helps determine, but is not equal to, a day.

rotation (*geometry*) Rotation is a **transformation** without **reflection** that holds 1 point in a plane unchanged (or all the points on 1 line in 3-dimensional space) while keeping all **distances** between points the same. A rotated figure is turned about the unchanged point (center of rotation) or line (**axis of rotation**) while every other point moves along a circular path. Rotations are identified by the **central angle** of the **arc** between the point and its **map** after the transformation, such as a rotation through 90°, or quarter turn. If no points move, the rotation is 0°, the identity transformation. Rotation can be replaced by reflection in intersecting lines.

rotifer (*animals*)   Typically less than 0.04 in. (1 mm) long, members of the phylum Rotifera are among the smallest of all multicellular animals. These invertebrates are sometimes called "wheel bearers" because of the distinctive crown, or corona, of cilia that encircles their mouth; the movement of the beating cilia resembles a rotating wheel. Some rotifers use the corona to propel themselves through water. Others use their corona to sweep food particles into the mouth. Rotifers are common in freshwaters worldwide; some species live in brackish water or on damp mosses and lichens.

roundworm (*animals*)   Members of the phylum Nematoda (or Nemathelminthes) are slender, unsegmented roundworms. Many of the 12,000 known species are free-living, occurring in great numbers in soil and water habitats, where they feed on bacteria, algae, fungi, and tiny animals. The species *Caenorhabditis elegans*—generally called *C. elegans*—has become an important research organism for geneticists.

Other roundworms are parasites, some of them causing much damage to cultivated plants, domestic animals, and humans. *Ascaris*, filarial worms, guinea worms, **hookworms**, and **trichina** are among approximately 50 species that infect humans. Many roundworms are tiny—less than 0.04 in. (1 mm) long—but a few parasitic species can be more than 3.3 ft (1 m) in length.

RR Lyrae stars (*stars*)   Some **variable stars** change so rapidly they are called pulsating stars. These are the very short period **Cepheid variables** called RR Lyrae stars after the first one discovered. Periods of variation are less than a day. The typical Cepheid relationship between magnitude and period makes these stars especially useful in calculating distances.

rubber (*materials science*)   Natural rubber is a **polymer** made from **latex**, usually sap from a rubber tree (*Hevea brasiliensis*), although other plant saps can serve. Among rubber's useful properties are high **elasticity** and ability to form thin, impermeable films. The name rubber was provided by **Joseph Priestley** because rubbing it on a pencil mark erases the mark. Natural rubber is somewhat unstable. In 1839 Charles Goodyear [American: 1800–60] discovered how to toughen natural rubber by cross-linking molecules (vulcanization). Synthetic rubber, made from petroleum, is widely used today.

rubber (*plants*)   The main source of natural rubber is a member of the **spurge** family, *Hevea brasiliensis*. Called the rubber tree, it is native to rain forests of Central and South America. It attains heights up to 100 ft (30 m) and has long-stalked compound leaves and clusters of tiny whitish flowers. The tree produces a milky fluid, latex, that is a mixture of water and small globules. The latex is present in a network of tiny vessels under the bark; when cuts are made in the bark the latex oozes out. On drying, the latex becomes crude rubber. Rubber also is derived from various other plants, including castilla trees (*Castilla*) of the mulberry family.

rubidium (*elements*) The alkali metal atomic number 37 was discovered in 1861 by Gustav Robert Kirchhoff and Robert Bunsen as a red line in a spectrogram and so named rubidium, from the Latin *rubidus*, "red" (chemical symbol: Rb). It was the second element to be discovered spectroscopically. It is liquid at warm room temperatures (melting at a temperature slightly higher than body temperature) and slightly radioactive.

rue (*plants*) Native to dry habitats in the Mediterranean region, common rue (*Ruta graveolens*) is a shrubby evergreen perennial that attains a height of about 2 ft (0.6 m). It has aromatic, finely cut blue-green leaves and small yellow flowers that give rise to capsule fruits. Due probably to its pungent odor, rue has been used medicinally since ancient times. Romans ate rue to protect against the evil eye; in the Middle Ages people wore it around their neck in hopes of warding off disease. The rue family, Rutaceae, also includes citruses, kumquats (*Fortunella*), and sapotes (*Casimiroa*).

Rumford, Count (Benjamin Thompson) (*physicist*) [American-English: 1753–1814] Rumford devised and supervised the experiment of boring a large cannon with a blunt tool to measure heat and change in weight, effectively proving that heat is not a form of matter. He correctly suggested that heat is a consequence of particle motion. Rumford's experiments also measured with moderate accuracy the relationship between input work and heat output.

ruminant (*zoology*) Hoofed mammals that chew their cud are called ruminants. They include cattle, sheep, goats, deer, antelopes, and giraffes. Ruminants are grazers, feeding on grass and other vegetation. The stomach is divided into 4 compartments, the first of which is the rumen. The rumen is very large, allowing the animal to eat a lot of food in a short time without chewing it thoroughly. After eating, the animal can retire to a secluded spot, bring up the food—now called a cud—and chew it at leisure.

runoff (*environment*) Water from precipitation, snow melt, and irrigation that flows off the surface of land into streams and other bodies of water is called runoff. Runoff can wash soil, fertilizers, pesticides, feedlot wastes, and other materials from the land into the water. Poor environmental practices such as clear-cutting forests and strip mining increase runoff.

rush (*plants*) Superficially similar to grasses and sedges, rushes (*Juncus*) have round, upright stems that develop from a creeping stem, or rhizome. Inconspicuous greenish flowers, borne in clusters, give rise to capsule fruits. Most species are perennials and live in cold and temperate regions, growing in wet habitats such as swamps and the margins of ponds.

rust (*fungi*) Members of the club fungi, rusts are named for their reddish-brown spores, which appear on the leaves and stems of plants they infect. They are among the most destructive of all fungi, causing fatal diseases also known as rusts. Many rust fungi have very complex life cycles. Wheat rust

(*Puccinia graminis*), for example, must alternate between 2 hosts, wheat and barberry—a practice known as heterocism ("different houses").

rust (*materials science*)   Rust is the common name for ferric oxide (Fe_2O_3), the reddish brown coating that forms on iron exposed to air and water. Rust does not occur in dry air or in water with no dissolved oxygen, so the process is not a simple chemical combination. Rust is always a hydrate rather than a simple compound, so its complete formula is $Fe_2O_3 \cdot 3H_2O$ or a similar formula with fewer or more water molecules (H_2O). Hydrogen ions play a part in the reaction, so acids speed rusting. One possible chemical equation for rusting is $4Fe + 3 O_2 + 6 H_2O = 2(Fe_2O_3 \cdot 3H_2O)$.

ruthenium (*elements*)   Ruthenium, atomic number 44, is a typical platinum-group metal—hard, heavy, catalytic, and corrosion-resistant. It was named in 1827 for Ruthenia in the Urals, where its ore was first found, by Emile Osann [German: 1787–1842], who thought he had located new elements in platinum ore. Credit, however, is usually given to Carl Claus [Estonian: 1796–1864], who isolated ruthenium in 1844, making it the last of the group to be found. Chemical symbol: Ru.

Rutherford, Ernest (*physicist*)   [English: 1871–1937] In 1911 Rutherford and his coworkers were the first to envision the atom as a positive nucleus surrounded by negative electrons. They identified 3 types of radiation, which Rutherford named alpha (helium nuclei), beta (electrons), and gamma (high-energy electromagnetic waves). Rutherford established that radioactive elements change into other elements by emitting radiation. He is credited with the discovery of the proton. In the 1920s Rutherford was the first to "smash" atoms, breaking up light atoms with alpha particles. In the 1930s he demonstrated fusion of 2 atoms of hydrogen with 1 neutron (deuterium) into tritium (hydrogen with 2 neutrons).

rutherfordium (*elements*)   The radioactive metal now called rutherfordium was simply element 104 for about 30 years. Its first synthesis was claimed both by Russian scientists at Dubna in 1964 and by a U.S. team led by Albert Ghiorso [American: 1915–] in 1969; finally, in 1997, a compromise permitted the element to be named for particle-physics pioneer Ernest Rutherford, Ghiorso's choice. Rutherfordium is the first synthetic radioactive element beyond the 15 members of the actinide group. Chemical symbol: Rf.

rye (*plants*)   A close relative of wheat in the grass family, rye (*Secale cereale*) is widely cultivated for its edible grains (fruits) and as fodder and hay. It has a cluster of unbranched stems up to 5 ft (1.5 m) tall. Flowers are borne on a long narrow spike at the top of a stem.

sabertooth (*paleontology*)   Sabertooths were members of the cat family that possessed two huge upper canines—sabers—used to stab prey. Their jaws opened extremely wide to accommodate the sabers' action. Sabertooths lived from the Oligocene to the Pleistocene. They preyed on mammoths and other large herbivores. Scientists believe they became extinct because their prey died out, possibly as a result of human hunting.

sac fungus (*fungi*)   Forming the phylum Ascomycota, sac fungi derive their name from their saclike spore-producing container, called an ascus, first described in 1837 by Joseph Léveillé [French: 1796–1870]. Sac fungi range from single-celled yeasts to *Penicillium* and morels. Many sac fungi are partners with algae in lichens.

Sagan, Carl (*astronomer*)   [American: 1934–96] Sagan promoted planetary astronomy and the search for intelligent life, becoming one of America's best-known scientists through television and popular books. He explained the run-away greenhouse effect on Venus and contributed to the theory that dust from large explosions, such as asteroid impacts, causes mass extinctions by lowering temperatures and stopping photosynthesis over all of Earth.

sage (*plants*)   Members of the mint family, sages (*Salvia*) are aromatic herbs and shrubs up to 5 ft (1.5 m) tall. The leaves of many species, particularly of common sage (*S. officinalis*), are used to season food.

sagebrush (*plants*)   Several shrubs of the genus *Artemisia*, in the composite family, are commonly called sagebrushes because the scent of their aromatic oil resembles that of sage. They are widespread in arid habitats of western North America. The genus also includes a number of Eurasian species called wormwoods, whose oil has been used since ancient times to ward off disease and insects such as moths and fleas.

salamander (*animals*)   Unlike other amphibians, salamanders have long tails. They also have a large, somewhat flattened head, a cylindrical body; and, usually, 4 legs. Salamanders live beneath rocks and logs and in other damp places. They feed mainly on insects, worms, and other small invertebrates. More than 230 of the approximately 380 known species are lungless salamanders (Family Plethodontidae), which absorb oxygen through the skin. Other salamanders include newts, mud puppies, sirens, and the axolotl.

salinization (*environment*)   The accumulation of sodium, potassium, and magnesium salts in soil or water is called salinization. In arid and semiarid areas with high evaporation rates, thin deposits of salts may actually form on the soil's surface. Salinization can be caused by improper irrigation techniques, such as spraying large quantities of water over an entire field rather than sprinkling smaller amounts of water directly onto plant roots. Salts are harmful to many plants and can turn a once flourishing habitat into a wasteland.

saliva (*physiology*)   Certain animals, including insects and land vertebrates, have salivary glands that secrete a fluid called saliva into the mouth or—in insects such as houseflies and bedbugs—into the hosts from which they suck out sap or blood. Composition of saliva varies but generally includes water, salts, mucus, and enzymes. In bloodsuckers such as the human louse (*Pediculus humanus*) and vampire bats, saliva contains chemicals to prevent blood clotting. In most mammals, including humans, saliva contains amylase, an enzyme that breaks down starch to maltose.

salmon (*animals*)   Streamlined fish native to cold waters of the Northern Hemisphere, salmon are mainly ocean dwellers but they migrate to freshwater streams to spawn. The Atlantic salmon (*Salmo salar*) grows to 5 ft (1.5 m) long. Salmon are classified with trouts, chars, and whitefish in Family Salmonidae. All have small scales, a prominent lateral line, and, between the dorsal fin and the tail, an adipose (fatty) fin. The fins do not have spines.

Salmonella (*monera*)   Members of the genus *Salmonella* are gram-negative rod-shaped bacteria. All are pathogens that cause disease in vertebrates. They are common agents of food poisoning, gastroenteritis, and enteric fevers (such as typhoid fever) in humans.

salt (*compounds*)   Salts are ionic compounds formed when an acid and a base are mixed. The hydrogen ion of the acid and the hydroxide ion from the base combine to form water (H_2O), leaving a product that is neutral. Most salts replace the hydrogen ion of acid with a metal ion. For example, when hydrochloric acid (HCl) is mixed with lye (sodium hydroxide, or NaOH), the products are table salt (sodium chloride, or NaCl) and water. Other familiar salts include calcium fluoride in toothpaste, silver bromide for photography, and calcium chloride as a drying agent.

salt dome (*geology*)   In seas near the edges of continents, sediments can accumulate over ancient regions where evaporation of ocean water has produced layers of salt. The salt can flow slowly, like a thick fluid, as a result of great pressure from the overlaying rock. Salt moves into cracks, pushing its way from depths as great as 5 mi (8 km) into tall, buried cylinders of salt. These are called salt domes because the rising salt lifts overlaying layers into domes.

samarium (*elements*)   Samarium is a rare earth, atomic number 62, discovered spectroscopically in 1879 by Paul-Emile Lecoq de Boisbaudran [French: 1838–1912] in the mineral samarskite (named for Colonel Samarski, a Russian mine official). Samarium compounds have many uses ranging from stable permanent magnets to ingredients in lasers. Chemical symbol: Sm.

sample (*statistics*) Any part of a whole used to predict properties of the whole is a sample. In statistics the properties of a whole population are assumed to vary from one sample to another, requiring special sampling techniques and mathematical tools for analysis. A random sample is one for which each member has an equal probability of being in the sample. A stratified sample consists of random samples from several subpopulations chosen to be representative of the whole. In either case the sample average and sample variance can be used as reasonable estimates for the average and variance of the whole population.

sample space (*probability*)   The basis of modern probability theory is the sample space. Probability theory requires outcomes to an experiment that can be enumerated, such as tossing a coin twice, which has 4 possible outcomes. Representing heads with H and tails with T, the only possible outcomes to that experiment (discounting the coin landing on edge) are HH, HT, TH, and TT. These 4 outcomes form the sample space for that experiment. Each of the simple events, such as HH, is also called a sample point. A compound event, such as getting 1 head and 1 tail, is represented by several sample points (in this case 2, HT and TH).

sand (*soils*)   Sand is the general name for soil particles that range in size from 5/64 in. (2 mm) on down to 1/400 in. (0.06 mm). Most of these particles are quartz or other hard minerals that rarely are worn any smaller than sand, so sand is the source of the silicon dioxide used in glass. Beach sand, worn by waves, and desert sand, broken by temperature changes and concentrated by winds, are typical. Sometimes beach sand reflects local geology—for example, black beaches of particles of lava or pink sands with high coral content.

sandalwood (*plants*)   The sandalwood family, Santalaceae, comprises some 400 species of herbs, shrubs, and trees that are widespread in tropical and temperate regions. Most and perhaps all species are partially parasitic. They make their own food but absorb water and minerals from host plants via rootlike structures called haustoria that penetrate deep into the host's tissues. White sandalwood (*Santalum album*) is an evergreen tree native to India that grows up to 60 ft (18 m) tall. A fragrant oil distilled from its bark, wood, and roots is used in perfumes.

sand dollar (*animals*)   Sand dollars are echinoderms constituting the order Clypeasteroida, with a round, flat disk-shaped body that is densely covered with short movable spines used for locomotion. Five rows of tube feet radiate from the axis and are used for respiration. Sand dollars live in shallow coastal waters, mostly buried in sand, sometimes in great clusters.

sandpiper (*animals*)   The sandpiper family, Scolopacidae, comprises a varied group of ground-dwelling birds that live mainly on seacoasts and the shores of lakes and rivers. They include sandpipers, curlews, dowitchers, godwits, phalaropes, snipes, woodcocks, and yellowlegs. These birds have a slender body with a long slender bill that is either straight or curved; long, rather pointed wings; and long legs. The plumage, typically mottled brown or gray on the upper body and pale on the belly, blends with the surroundings. Most nest on the Arctic tundra. The birds migrate south for the winter—often as far as southern South America—and fly back north again in spring. They range from no bigger than the sparrow to the long-billed curlew (*Numenius americanus*) of North America, which grows to 2 ft (0.6 m) long.

sandstone (*rocks and minerals*)   Sandstones are rocks formed when sand deposits become glued together by silica, calcite, or iron oxides. Consequently they tend to be predominately quartz. Sand that accumulated in the desert tends to produce stone with more rounded grains and redder color than sand deposited underwater. Sandstones are usually permeable and can store water in underground aquifers when located above a layer of impermeable rock, such as slate.

saprophyte (*ecology*)   A plant, fungus, or bacterium that obtains its food in solution from dead and decaying organic matter is a saprophyte or saprobe. Saprophytes secrete digestive enzymes onto the food to break it down into simple molecules that they can absorb. Saprophytes are an extremely important part of an ecological community; in addition to ridding the community of wastes, their digestion releases nitrates and other inorganic materials that can be used by green plants to make food.

Some organisms can live as either parasites or saprophytes, depending on circumstances. For example, the bluish-green molds often seen on oranges feed on living tissues of a host until the host dies, then feed on the dead tissue.

Sargasso Sea (*oceanography*)   The currents in the North Atlantic—Gulf Stream, North Atlantic Drift, Canary Current, and North Equatorial Current—form a giant circle moving counterclockwise, called a gyre. Within the gyre, ocean water is warmer, saltier, and lacking in nutrients compared with water in the currents or near to shore. This region, the Sargasso Sea, is named after the sargasso algae (*Sargassum natans* and *S. fluitans*) that float there, forming the basis of the sea's ecosystem. In addition to other animals that live in the sea, both American and European eels breed there and marine turtle hatchlings live there during early stages of growth.

satellite (*astronomy*)   Any body in orbit about another can be called a satellite, but the term originally meant moons orbiting planets. Kepler first used the term, which originally meant courtiers attending nobility, to describe the 4 large moons of Jupiter. Devices put into orbit about planets or other members of the solar system are called artificial satellites. Some asteroids are orbited by other asteroids, which are then called satellites.

saturation (*chemistry*)   In solutions, saturation refers to the amount of dissolved material (solute) normally possible at a specific temperature. For water at 68°F (20°C), 36 mass units of sodium chloride (table salt) dissolved in 100 mass units of water makes a saturated solution—added salt fails to dissolve. More salt will dissolve if the temperature of the solution is raised. Carefully lowering the temperature of a saturated solution can result in a higher concentration than the solution normally holds; the solution is supersaturated. Disturbing a supersaturated solution, especially by adding a crystal of solute, causes excess solute to precipitate.

Hydrocarbons, such as fatty acids, are saturated when their carbon atoms are linked by single bonds and unsaturated when double bonds are present. Fats that are solid at room temperature usually consist of saturated fatty acids.

Saturn (*solar system*)   Saturn is the farthest planet that can be identified easily in Earth's nighttime sky with the unaided eye. It is the sixth planet of the solar system and the second largest, after Jupiter. Saturn's ring system, more extensive than that of any other planet, is visible through a telescope. Saturn has far more than a thousand narrow rings. One of the brightest rings is under 500 ft (152 m) thick. The planet is composed of densely compacted hydrogen, helium, and other gases. Liquid or metallic hydrogen probably exists underneath the planet's thick atmosphere, above a solid core of rock about twice the size of the planet Earth. Storms thousands of miles across rage on Saturn, along with a wide band of extremely high winds—up to 1000 mi (1600 km) per hour—at the equator. Saturn ties Uranus for the most moons of any planet—18—but, except for Titan, they are relatively small.

sauropod (*paleontology*)   The sauropods made up one of the major groups of lizard-hipped dinosaurs. They were plant eaters, with flattened teeth. They typically had a tiny head, long neck, massive body, and long tail. Their front legs were smaller than their back legs but sufficiently well-developed to support the front end of the body and enable the animal to walk on all fours. Well-known sauropods include *Apatosaurus*, *Camarasaurus*, and *Diplodocus*. *Sauroposeidon*, discovered in 1999, stood about 60 ft (18 m) tall—with individual neck bones up to 4 ft (1.2 m) long.

savanna (*ecology*)   A tropical grassland with scattered trees, savanna is a major biome in South America, Africa, western India, and northern Australia. The African savanna, with its herds of elephants, giraffes, wildebeests, and antelopes—followed closely by lions and other predators—is particularly well known. Temperatures in a savanna fluctuate very little during the year. Annual precipitation averages 35 to 60 in. (89 to 150 cm), with marked wet and dry seasons. The grasses have dense networks of roots, which absorb lots of water during the rainy season and which can send up new shoots after prolonged droughts or after fire sweeps through the area.

saxifrage (*plants*)   Most species of the genus *Saxifraga* are low-growing perennial herbs native to temperate and arctic regions. They have leaves in basal rosettes and flowers produced in clusters on long stalks. Saxifrages inhabit rocky cliffs and crevices. This may explain the name, which means "rock breaker." Or the name may refer to the fact that saxifrages were once considered a cure for gallstones.

scalar (*algebra*)   The word scalar originally indicated that multiplication by a real number, called a scalar in this context, changes the length but not the direction of a vector—that is, the operation changes scale. When the components of a vector are treated as numbers, each is also called a scalar, which broadens the meaning to allow complex numbers to be scalars. Multiplication of a vector by a scalar consists of multiplying each component of the vector by the scalar to form a new vector.

scale insect (*animals*)   Classified in the same order as cicadas, scale insects are small and inconspicuous. The females are wingless; males normally have 1 pair of flying wings but lack mouthparts and do not feed as adults. All secrete some kind of material; for example, lac insects (genus *Laccifer*) secrete a thick wax and mealy bugs (genus *Pseudococcus*) secrete a powdery substance. Scale insects feed on plant juices and often cause great harm to cultivated plants.

scallop (*animals*)   Members of the mollusk class of bivalves, scallops have a 2-piece, fan-shaped shell. There is a winglike projection at each end of the hinge and ridges radiating outward from the hinge. Scallops are marine animals that live on sandy or muddy bottoms. Most swim by opening and closing their shells, thereby expelling jets of water that propel them through water.

scandium (*elements*)   Scandium, the metal with atomic number 21, was discovered in 1876 by Lars Fredrik Nilson [Swedish: 1840–99] and named for Scandinavia. Scandium was later recognized as the "eka-boron" predicted by Mendeleyev on the basis of a gap in his periodic table. Scandium is rare and difficult to isolate; the first nearly pure pound was made in 1960. Chemical symbol: Sc.

scarp (*geology*)   A scarp (or escarpment) is a long, steep slope or cliff on Earth or another rocky body, such as the Moon or an inner planet. Sometimes geologists reserve the term for steep slopes (called fault scarps) formed by faulting or by differential erosion of 2 types of rock. Long and high scarps on Mercury are lines of cliffs formed as the crust of the planet shrunk.

scattering (*particles*)   Whenever waves encounter small particles or rough surfaces, they are broken into waves of different frequencies; this is the original meaning in physics of scattering. The sky is blue because air molecules scatter light, for example. Early in the 20th century, Rutherford and coworkers found that particles were scattered by atoms as they passed through thin sheets—in this case, the particle paths bent as they passed near atomic nuclei, establishing the existence of nuclei. About the same time Charles G. Barkla [English: 1877–1944] used scattering of X rays by electrons in atoms to establish their positions outside the nucleus.

scavenger (*ecology*)   Scavengers are animals that feed on carrion—the meat of animals that die natural deaths or are killed by predators. Vultures, hyenas, crabs, and carrion beetles are examples. Some animals, such as albatrosses and coyotes, are both scavengers and predators. By eliminating dead animals, scavengers help keep their environments clean.

Scheele, Karl Wilhelm (*chemist*)   [Swedish: 1742–86] Scheele discovered chlorine in 1774 but failed to recognize it as an element. He discovered oxygen in 1771, but publication was delayed. Thus English chemists are credited with both discoveries. But Scheele indisputably was the first to compound 10 of the most familiar acids, including tartaric, oxalic, and lactic acids, as well as the 3 gases hydrogen fluoride, hydrogen sulfide, and hydrogen cyanide. He was also the first to recognize the effect of light on silver compounds, the basis of photography.

schist (*rocks and minerals*)   Schist (rhymes with mist) is a general term for a family of metamorphic rocks that are all coarse grained and that flake easily. A schist has a coarser structure than a slate, but is not separated into layers the way gneiss is. Many schists contain large amounts of mica or hornblende.

Schleiden, Matthias (*botanist*) [German: 1804–81] In 1838 Schleiden proposed that all plants are composed of cells and growth results from forming new cells; together with his friend Theodor Schwann he formulated the cell theory of life. Schleiden observed various cell structures and activities such as protoplasmic streaming. He also showed that fungi are associated with the roots of some plants in what was later called a mycorrhizal relationship.

Schliemann, Heinrich (*archaeologist*) [German: 1822–90] Inspired by reading Homer's *Iliad,* Schliemann determined that he would find the site of the book, the city-state of Troy, which many considered mythical. He succeeded in 1871, finding the correct location, although misidentifying the incarnation of the city that was Homer's Troy. He continued to search for and find rich remains of the Mycenaean civilization, which he identified (probably incorrectly also) as Agamemnon's tomb.

Schmidt, Ernst Johannes (*biologist*) [Danish: 1877–1933] In the early 1900s Schmidt located the spawning grounds of cod and several other North Atlantic fish. He then investigated freshwater eels, which as adults migrate downstream to the sea to spawn and die. By collecting young eels at various locations in the Atlantic and noting their size, Schmidt determined that freshwater eels spawn in the Sargasso Sea and that ocean currents then carry their offspring back to the mouths of rivers.

Schrödinger, Erwin (*physicist*) [Austrian: 1887–1961] Schrödinger is known primarily as the creator of the wave equation for quantum mechanics, which he developed in 1926, just after Werner Heisenberg and coworkers had solved the same problems with matrix mechanics. Schrödinger's equation allows physicists to compute energy levels for electrons and remains an important tool in particle physics. After World War II Schrödinger considered fundamental problems in biology, expressing his views in the philosophical book *What Is Life?*

Schwann, Theodor (*biologist*) [German: 1810–82] In 1839 Schwann proposed that all organisms are composed of cells; together with Matthias Schleiden he formulated the cell theory of life. Schwann also discovered the sheath of cells (since named Schwann cells) that surrounds nerve axons and conducted experiments that helped disprove the theory of spontaneous generation.

Schwarzschild radius (*astronomy*) In 1916 Karl Schwarzschild [German: 1873–1916] calculated the mathematical implications of Einstein's newly published general theory of relativity. Part of the solution to Einstein's equations implies that if a mass is concentrated so that it is entirely within a distance from its center less than twice the gravitational constant divided by the square of the speed of light, it will collapse into a black hole. This distance is known as the Schwarzschild radius. It varies with mass—for a mass the size of the Sun it is about 2 mi (3 km); for a mass the size of Earth it is only about 0.4 in. (1 cm).

science (*philosophy of science*) Science is both a branch of human knowledge and a way of deriving conclusions. The conclusions of science concern physical truth instead of cultural or spiritual truth. A sociological view is that science

consists of the reports produced by processes of directed investigation conducted by members of the scientific community. Science requires observation
and may also use logical deduction, mathematics, measurement, and special
tools to enhance observation or create particular physical situations (for example, temperatures near absolute 0).

scientific method (*philosophy of science*)   Traditionally the scientific method
consists of observation; making a testable hypothesis to explain the observation; performing an experiment to test the hypothesis; accepting or rejecting
the hypothesis on the basis of the experiment; and building a theory that
encompasses hypotheses concerning related situations. The recognition that
science is a community activity has led to adding to this list recording data,
reporting results, and repetition of experiments by other scientists. The report
of observations or an experiment is usually not considered "official" until it has
undergone peer review by scientists working on similar problems and approved
for publication in a journal dedicated to such problems.

scientific notation (*numeration*)   Scientific notation is a system of writing
numbers as the product of a number between 1 and 10 and a power of 10. A
number such as 2.398×10^9 takes less space and, if you know the system, is
easier to read than 239,800,000. Negative exponents provide scientific notation
for small numbers also: 0.00000087 in scientific notation is 8.7×10^{-7}. This system is also used to indicate the precision of a measurement; the number
2.39800×10^9 is 100 times as precise (measured to the nearest 1000) as $2.398 \times
10^9$ (which is to the nearest 100,000).

Scientific Revolution (*philosophy of science*)   The Scientific Revolution refers
to the period from 1543 through at least the death of Galileo and birth of Newton
in 1642 (making a convenient 100 years) or perhaps to the 1687 publication of
Newton's *Principia*. The year 1543 saw the publication of the great works of
Copernicus and Vesalius that instituted astronomy on a rational basis and
anatomy based on observation. During the next 100 years such tools of science
as the thermometer, telescope, and microscope were invented while ground
was laid for the barometer and pendulum clock. The first scientific society
started in 1560 and the Royal Society in 1660, with the first scientific journals
beginning publication in 1665.

scorpion (*animals*)   Referred to as living fossils because they have undergone
very little change since evolving some 420,000,000 years ago, scorpions are
medium to large arachnids characterized by a pair of massive lobsterlike pincers
and—at the end of a long, upturned abdomen—a stinger attached to poison
glands. Of the approximately 1500 known species, about 25 produce a neuro-
toxin capable of causing human death. Scorpions generally inhabit warm, dry
areas. They spend their lives in or near their burrows. Nocturnal hunters, they
prey mainly on insects and spiders.

screw (*mechanics*)   One of the basic machines is the ramp known as an
inclined plane. When an inclined plane is wrapped around an axis, the machine
becomes a screw. The most familiar screws are those used to pull themselves
into material as they turn, a common kind of fastener. The Archimedian screw

is a screw in a tube that can lift a fluid when it turns. Screws are also used to lift heavy objects; one common type of automobile jack is an example.

sea anemone (*animals*)   Named for their resemblance to flowers, sea anemones are marine cnidarians that have only a polyp stage. They are common in tide pools, living attached to rocks or burrowing into sand. Sea anemones have a stout, muscular body and a mouth on the top end circled by several rows of petal-like tentacles. When disturbed or during low tide, a sea anemone pulls its mouth and tentacles inside, then contracts a muscle just below the mouth, rather like pulling a drawstring tight on a laundry bag.

Seaborg, Glenn T. (*physicist*)   [American: 1912–99] In 1940 Seaborg started working with elements with higher atomic numbers than uranium's 92 (transuranic elements). That year he was the second to produce an isotope of neptunium (93) and the first to produce plutonium (94). In 1944 his team added americium (95) and curium (96), followed in 1949 by californium (97) and berkelium (98). Seaborg recognized that thorium, uranium, and transuranic elements through element 103 form the actinides, a series with very similar properties.

seaborgium (*elements*)   Synthesis of the radioactive metal with atomic number 106 was claimed by Russian scientists at Dubna and by a U.S. team from Lawrence Berkeley and Livermore Laboratories, both in 1974. After 22 years of bickering between the 2 teams over its name, in 1997 the metal, a short-lived analog of tungsten, was christened seaborgium for the American physicist Glenn T. Seaborg, who had been instrumental in synthesizing plutonium and other elements. Chemical symbol: Sg.

sea cucumber (*animals*)   Unlike other echinoderms, sea cucumbers have an elongated, soft body. Their mouth is at the front end surrounded by tentacles; at the rear end is the anus. Five rows of tube feet run from the mouth to the anus. Like other echinoderms, sea cucumbers have amazing powers of regeneration. If disturbed, a sea cucumber can disgorge its digestive system. Feeding on this satisfies a predator—and in a few weeks the sea cucumber has regenerated the lost organs. Sea cucumbers live on the ocean floor, often at great depths. Species range in length from less than 1 in. (2.5 cm) to more than 3 ft (0.9 m).

seafloor spreading (*geology*)   In 1963 Harry Hess [American: 1906–69] proposed that new ocean floor rises and hardens at mid-oceanic ridges, pushing the older seafloor on the edges of the ocean into deep trenches, a process called seafloor spreading. The following year Frederick Vine [English: 1939–] and Drummond Matthews [English: 1931–] discovered strong evidence favoring Hess's idea in magnetic fields trapped in ocean basalts. Today many geologists believe that cool ocean floor sinking in ocean trenches powers the process, but the name seafloor spreading is still used.

sea horse (*animals*)   In 1758 Linnaeus chose the genus name *Hippocampus* ("bent horse") for the small marine fish that have come to be called sea horses. These creatures, found in coastal waters worldwide, have a snout that sucks up prey, a head bent at right angles to the body, an armor of bony plates, and a prehensile tail that can curl around kelp and eelgrass. Sea horses are weak swimmers, with only 1 small propelling fin, and swim upright. The male possesses a pouch on his abdomen into which the female deposits eggs (fertilized as they enter the pouch). The male incubates the eggs until they hatch; the pouch then opens and tiny sea horses come out.

seal (*animals*) Members of 2 families of pinniped mammals are called seals. All are powerful swimmers and divers. Family Otariidae comprises eared seals, fur seals, and sea lions. These seals have small external ears and can turn their hind flippers forward to support the body and aid in walking on land. Family Phocidae comprises hair and elephant seals. These lack external ears and cannot turn their hind flippers forward; on land they move in a hunching fashion. The largest are the elephant seals (*Mirounga*), up to 20 ft (6 m) long and weighing almost 4 tons.

seamount (*oceanography*) Away from the mid-oceanic ridge system, many isolated volcanoes, active or extinct, break through the thin ocean crust to form mountains. Where there is a hot spot beneath a moving plate, the volcanoes form a chain. If volcanic mountains reach the surface, they are islands, but those below sea level are called seamounts. Many seamounts are former islands that have since sunk below the waves. These, called guyots, can be recognized by their flat tops, produced by wave erosion.

sea snake (*animals*) Inhabiting tropical waters, mainly in the Indian and Pacific oceans and adjacent gulfs and bays, sea snakes are highly poisonous, using their venom to subdue the fish on which they feed. Their compressed body and wide, paddlelike tail enables them to swim quickly. Like all reptiles, sea snakes breathe through lungs and thus must rise to the water's surface for air. Most sea snakes are 1.6 to 3.3 ft (0.5 to 1 m) long, but Stokes' sea snake (*Astrotia stokesii*) may attain lengths of almost 6.6 ft (2 m).

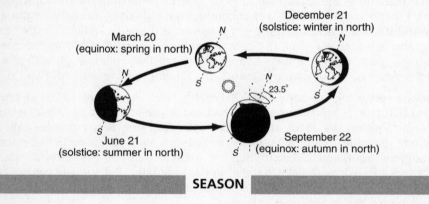

March 20
(equinox: spring in north)

December 21
(solstice: winter in north)

June 21
(solstice: summer in north)

September 22
(equinox: autumn in north)

23.5°

SEASON

season (*solar system*) Earth's axis of rotation points in the same direction for thousands of years at a time, always tilting at an angle of about 23°. The Northern and Southern Hemispheres gradually move each year from leaning toward the Sun to leaning away from it—when the north leans toward the Sun the south faces away and vice versa. Leaning toward the Sun allows for more hours of daylight each day and a shorter path for sunlight through the atmosphere, raising temperatures, especially in middle latitudes. Leaning away produces shorter hours of daylight and oblique rays of sunlight. The seasons officially begin when the tilt toward the Sun is greatest (summer solstice), zero (autumnal equinox), least (winter solstice), and zero again (spring equinox).

sea urchin (*animals*)   Sea urchins are echinoderms with a rounded body enclosed in a spherical shell of firmly joined plates. They are covered with numerous movable spines. Five rows of tube feet radiate from the mouth on the lower surface to the anus on the upper surface. Sea urchins are bottom dwellers that live at various depths in the ocean, feeding on both plant and animal matter. The largest have shells about 12 in. (30 cm) across.

seaweed (*protists*)   Multicellular marine algae that have leaflike structures and in other ways look like plants are commonly called seaweeds. They include various species of green, red, and brown algae. Seaweeds are important producers at the base of marine food chains and their beds provide homes for many animals.

sebaceous gland (*anatomy*)   The skin of mammals has sebaceous glands, which secrete an oily substance called sebum, generally into the hair follicles. The secretion lubricates the surface of the skin, prevents hair from drying out and becoming brittle, and helps waterproof the hair.

secant (*geometry*)   Any line that intersects a circle or another curve in 2 places is a secant. When 2 secants, or a secant and a tangent, to the same circle form an angle whose vertex is outside the circle, they define 2 arcs inside the angle, called intercepted arcs; the size of the angle is half the difference of the intercepted arcs.

secant (*trigonometry*)   One of the 6 functions of trigonometry, the secant (sec) is defined as the reciprocal of the cosine; that is, $\sec \theta = 1/\cos \theta$. In a right triangle $\sec \theta$ is the ratio of the length of the hypotenuse (side opposite the right angle) to the length of the angle adjacent to θ.

secondary sex characteristic (*anatomy*)   Physical characteristics that distinguish males from females but do not function directly in reproduction are called secondary sex characteristics. The long comb and wattles of a male fowl, the mane of a male lion, the egg-laying organ (ovipositor) of a female grasshopper, and the breasts of a female human are examples. Development of these characteristics is determined largely by hormones: testosterone in males and estrogen in females.

second law of thermodynamics (*thermodynamics*)   Without outside help, heat always flows from a hot region to a cold region. This rule is the second law of thermodynamics. Energy must be added to a system to reverse the direction of heat flow, which is why refrigerators and air conditioners require power supplies. The second law can also be stated as entropy, or disorder, in a closed system always increases. For example, as heat flows into cooler regions and equalizes temperatures, the system becomes more uniform. The entropy rule is more general, applying to all sorts of systems that become more uniform when left alone (for example, mixtures of 2 fluids). (*See also* **thermodynamics**.)

secretary bird (*animals*)   Native to grasslands south of the Sahara, the secretary bird (*Sagittarius serpentarius*) is named for its crest of long black feathers reminiscent of the quill pens worn behind the ears of 19th-century clerks. The bird stands 4 ft (1.2 m) tall. It has a long neck, large wings, long legs, and a long narrow tail. It preys on snakes, lizards, rodents, young birds, and insects. The secretary bird is a strong flier but prefers to run or walk rapidly away from danger.

section (*geometry*)   The intersection of a plane and a solid (often including the points inside the solid) is a section, or plane section. If the plane is perpendicular to the lines used to define a cylinder or to the faces of a prism, the section is a right section. Note that the conic sections are sections of a cone that do not include the region inside the cone.

sedge (*plants*)   Members of the sedge family, Cyperaceae, superficially resemble grasses but generally can be easily distinguished by their 3-sided stems. Sedges have grasslike leaves and clusters of tiny flowers. Found worldwide, they are most common in swamps, along the edges of ponds, and in other moist habitats. Particularly well known is the papyrus (*Cyperus papyrus*), which grows to 15 ft (4.6 m) tall and has stalks as thick as an adult human's wrist. It was used in ancient Egypt almost 5000 years ago to make the first paper.

sedimentary rock (*rocks and minerals*)   A rock formed by the deposit of small pieces of other rock, by chemicals that have precipitated, or by layers of the remains of organisms is called sedimentary. When sedimentary rock is formed from small pieces, such as sandstone from sand or limestone from the skeletons or shells of animals, it may be glued together by silica or another mineral; or the sediment may have dissolved and recrystallized leaving little trace of the parent material. Sedimentary rocks such as gypsum rock (whose main mineral is gypsum) that form by precipitation are called evaporites. Sedimentary rock tends to form in deep beds. Fossils are normally found in sedimentary rock or in metamorphic rock that originated as sedimentary rock.

seed (*botany*)   The result of sexual reproduction in gymnosperms (cone-bearing plants) and angiosperms (flowering plants) is a seed. The seed characteristically has 3 parts: an embryonic plant, stored food, and a protective seed coat. Angiosperm seeds are further protected by developing within fruits, whereas gymnosperm seeds are exposed to the air, a difference first noted by Robert Brown [Scottish: 1773–1858] in 1825. Some seeds grow, or go through germination as soon as they are mature. Others pass through a dormant, or inactive, period—the length of which varies with the species—before germinating.

Seeds vary markedly in shape and size. Orchid seeds are no bigger than dust particles; a million may weigh less than 1 oz (28 g). In contrast, the huge heart-shaped seed produced by coco-de-mer palms (*Lodoicea maldivica*) may weigh 40 lb (18 kg).

seed dispersal (*botany*)   Plants have evolved a variety of methods to ensure that their seeds are dispersed over wide areas. Lightweight milkweed seeds have parachutelike tufts of fine hairs and depend on wind for dispersal. Coconut seeds contain air, making them light enough to float on water. Burdock seeds have spines that catch in the fur of passing animals.

Many plants produce sweet, fleshy fruits that are eaten by animals; the animals usually eat the indigestible seeds, too, which later pass out of their digestive tracts without being damaged. Other plants have dry fruits that explode on maturity, shooting out seeds over great distances. Tumbleweeds, such as Russian thistle, scatter seeds as they are pushed about by the wind.

seed fern (*paleontology*)   The seed ferns, or pteridosperms, lived about 300,000,000 years ago, during the time when much of Earth's coal was formed. Some species grew to heights of 30 ft (9 m) or more, with a stout trunk topped by a crown of fernlike leaves. Other species were vines that grew up tree trunks. "Seed ferns" is a misnomer; these were not ferns, which reproduce by spores, but seed plants that produced tiny seeds attached to the leaves.

seed plant (*botany*)   Any plant that reproduces by seeds is a seed plant. There are 2 groups. **Gymnosperms**, including cycads, ginkgoes, and conifers, produce seeds in cones. **Angiosperms**, or flowering plants, produce seeds by flowers. Angiosperms, with some 230,000 known species, are far more numerous than gymnosperms, with 600 to 700 species.

segment (*geometry*)   The part of a line between 2 points is a segment. The 2 points are called the endpoints of the segment. If both endpoints are part of the segment, it is called a line segment. When neither endpoint is included, it is an open segment. A segment missing 1 endpoint is only half open.

Similar parts of other curves (between 2 points) are also called segments of a curve. For other geometric figures, especially circles and spheres, regions bordered by lines or planes are called segments.

Segrè, Emilio (*physicist*)   [Italian-American: 1905–89] In 1937 Segrè was the first to create an element that had never existed on Earth (technetium). Similarly, in 1955 he became the first to create and identify a subatomic particle not previously found on Earth (the antiproton). He was also a member of the teams that in 1940 created the artificial elements astatine and plutonium.

seismic wave (*geology*)   The movement of rock that is the origin of an earthquake initiates waves that travel in all directions through the rock. Some of these, known as **P** waves (primary waves, since they travel faster than other waves and therefore arrive first) are pressure waves. **P** waves are longitudinal waves, essentially the same as sound waves. These are followed by **S** waves (secondary waves or shear waves) that are transverse waves like electromagnetic waves. The difference in time of arrival between **P** and **S** waves is used to determine the distance to an earthquake.

Much of the damage from earthquakes comes from surface waves generated by **P** and **S** waves, including up-and-down movements of Rayleigh waves, analyzed in 1885 by Baron John Strutt Rayleigh [English: 1842–1919], and side-to-side undulating Love waves, recognized in 1911 by Augustus E.H. Love [English: 1863–1940].

seismogram (*geology*)   The record of earthquake waves produced by a device called a seismometer is a seismogram. A marker is held nearly motionless with respect to Earth's center while a moving strip of paper is on a device attached to the ground. Earthquake waves shake the device holding the paper up and down; this is recorded by the movement of the paper rolling past the marker, making the seismogram. Normally 2 seismograms, 1 perpendicular to the other, are needed to collect both **P** and **S** seismic waves. In modern seismometers, magnetic fields and electronic devices are used to increase sensitivity.

selenium (*elements*)   Selenium is a red, gray, or black nonmetal, atomic number 34, discovered in 1818 by Jöns Jakob Berzelius in association with its neighbor on the periodic table, tellurium. This connection led to the name, from the Greek *selene* ("Moon"). Selenium displays photoelectric phenomena, converting light into electric current and decreasing in electrical resistance in response to light. It is the key to xerography. Although the element is nontoxic, some of its compounds are poisons; soil containing high levels of selenium compounds is unsuitable for agriculture. Chemical symbol: Se.

self pollination (*botany*)   The transfer of pollen from a male reproductive organ of a seed plant to a female reproductive organ on the same plant is called self pollination. This is accomplished in various ways. For example, at low temperatures some violet flowers do not open enough for pollen to leave and these flowers pollinate themselves. However, many species have mechanisms that limit self pollination and encourage cross pollination; these include different maturation times of male and female organs or male and female flowers on separate plants, increasing the likelihood of new gene combinations.

semen (*physiology*)   The mixture of fluids and sperm produced by the male reproductive organs of animals, particularly those with internal fertilization, is called semen, or seminal fluid. Semen leaves the body in a process called ejaculation. About 1 teaspoon (5 ml) of semen, containing some 400,000,000 sperm, is released in an average human ejaculation.

semicircular canal (*anatomy*)   The inner ear of vertebrates contains a series of fluid-filled semicircular canals that detect changes in the position of the head, information that is critical for maintaining balance. Movement of the head causes the fluid to move and hit receptors called hair cells, triggering impulses that are sent to the brain.

semiconductor (*materials science*)   A semiconductor responds to electric current in a manner partway between conducting and insulating. The best known semiconductors are silicon and germanium, both used in computer chips and other electronic applications. Semiconductor applications in electronics require inserting atoms of other elements into the crystal structure of the semiconductor, a process known as doping. Depending on the choice of atom, the doped semiconductor can be made into a conductor or insulator.

For molecular solids the range from insulation to conductivity depends on how tightly electrons are bound to atoms. In a semiconductor such as boron, the electrons become less tightly bound as the material is heated, making it more of a conductor. Other materials change from insulation to conducting in the presence of light, which also frees up electrons. Xerographic copying depends on this property of the semiconductor selenium.

semipermeable membrane (*materials science*)   A semipermeable membrane has pores that are large enough to permit small molecules, such as dissolved salts, to pass through, but not large enough to permit passage of large colloid molecules, such as proteins. Each living cell is contained in a semipermeable membrane that also uses active methods to transport large molecules. Dialysis is a process using a semipermeable membrane to concentrate molecules or to filter out molecules. Cellophane, parchment, and pig's bladders are used as semipermeable membranes in chemistry or medicine.

sense organ (*anatomy*)   Sense organs are receptors designed to report changes in an animal's body and environment. In humans they include organs of touch (in the skin), taste (in the tongue), smell (in the nasal cavity), vision (in the eyes), hearing (in the ears), and balance (in the semicircular canals). A change detected by a sense organ initiates an impulse along associated nerve pathways to a control center—typically the brain—which processes the information and determines a response.

sentence (*logic*)   The concept of sentence, or proposition, in logic usually corresponds to a declarative sentence, such as "The sky is blue," in English and is any entity that can have the truth value of true or false. In some branches of logic, however, the word "sentence" refers to entities that contain variables that do not have a truth value unless the variable is assigned a particular value (for example, "The sky is x" is a sentence whose truth or falsity is not assigned until a particular value for x is chosen). Sentences with variables can also be assigned truth values when preceded by quantifiers such as "for all x."

sequence (*calculus*)   A sequence is a set of members (called terms) arranged in a definite order. Any sequence can be put into one-to-one correspondence with part or all of the natural numbers; that is, there is a first term, a second term, a third term, and so on. A sequence in one-to-one correspondence with the whole set of natural numbers is infinite. Typically a sequence has a rule (the general term) expressed as a function of n, where n is the natural number corresponding to a term. For example, the sequence of odd numbers $(1, 3, 5, 7, \ldots)$ has the general term $2n - 1$, while the sequence $1, 3, 6, 10, 15, 21, \ldots$ has the general term $\frac{1}{2} n(n + 1)$, so the seventh term is $\frac{1}{2} \times 7 \times 8 = 28$.

sequoia (*plants*)   The biggest of all living organisms are the sequoias (*Sequoiadendron giganteum*), evergreen conifers native to the Sierra Nevada of California. Sequoias are not quite as tall as their close relatives, the redwoods, though some specimens are more than 300 ft (90 m). But sequoias have a more massive trunk, which can weigh over 4,000,000 lb (1,800,000 kg). Sequoias can live more than 3000 years.

series (*calculus*)   A series is the indicated sum of a sequence, such as $1 + 3 + 5 + 7$ (a finite series with 4 terms) or $1 + \frac{1}{2} + \frac{1}{4} + \frac{1}{8} + \frac{1}{16} + \ldots$ (an infinite series with the general term $\frac{1}{2}^{n-1}$). The sum of an infinite series, if one exists, is the limit of the sequence formed from the sums of the first n terms of the series as n increases without bound. For example, the sum of $1 + \frac{1}{2} + \frac{1}{4} + \frac{1}{8} + \frac{1}{16} + \ldots + \frac{1}{2}^{n-1} + \ldots$ is the limit of the sequence $1, 1\frac{1}{2}, 1\frac{3}{4},$

$1^7/_8$, $1^{15}/_{16}$, . . . , $1 + (2^{n-1} - 1)/2^{n-1}$, ... This sequence has the limit 2, so 2 is also the sum of the series.

serotonin (*biochemistry*)   Serotonin is a substance that performs a variety of functions in the vertebrate body. It is a **neurotransmitter** produced by cells in the brain and spinal cord, and is thought to play an inhibitory role in sleep and other states of consciousness. Serotonin also is manufactured by specialized cells in the lining of the digestive tract; here it acts as a hormone, inhibiting gastric secretion and stimulating muscles in the intestinal wall. Serotonin is also found in blood platelets, which release the chemical at the site of bleeding, where it constricts blood vessels, thus limiting loss of blood.

serum (*physiology*)   The term serum has several meanings. Blood serum is the liquid portion of the blood minus the clotting protein fibrinogen. Immune serum, or antiserum, is blood serum from an immunized animal that contains antibodies for a specific germ; it can protect against that germ when injected into another animal. Finally, the **peritoneum** and other serous membranes that line body cavities in vertebrates secrete a clear, watery fluid called serum.

set (*sets*)   A set is any collection of entities sufficiently well defined so as to always be certain whether or not an entity is in the collection. The entities are called elements or members of the set. The fundamental relationship of **set theory** is that an element a belongs to a set A, often symbolized as $a \in A$. Two sets are equal when they have the same members. Sets can be indicated by naming their elements in curly braces; the set of colors of the rainbow is {red, orange, yellow, green, blue, indigo, violet}, although the order of naming the elements does not matter. Note that a set and an element are different. The set {red} is not the same as the element red.

set theory (*mathematics*)   In mathematics, the more inclusive the category, the more fundamental. Set theory, which deals with simple operations on any kind of recognizable collection, has become the underlying framework of mathematics; all other branches are developed implicitly or explicitly on **sets** and the **axioms** for set theory. (Logic, which underlies all **axiomatic systems**, is not usually considered part of mathematics.)

Various highly technical axiomatic systems have been developed for set theory, but all have a few ideas in common. A set is something that is made from elements and for which you can tell whether or not a given element is part of the set. Sets with the same elements are identical—that is, there is really only 1 set with 2 or more names. There is an **empty set** with no elements. Every set but the empty set contains other sets.

Sets of points are the basis of geometry, while all numbers can be defined in terms of sets. The structures of algebra and other branches of mathematics consist of a set and operations on its elements.

sex chromosome (*genetics*)   In animals as diverse as fruit flies and elephants a pair of chromosomes are involved in determining sex and **sex-linked characteristics**. In mammals the sex chromosomes are labeled X and Y; a female has 2 X chromosomes, a male has 1 X and 1 Y. The X chromosome is larger than the Y chromosome and carries much more genetic information. Among birds the male has 2 similar Z chromosomes while the female has a Z and a W chromosome.

sex-linked characteristic (*genetics*)   Traits determined by genes located in the sex chromosomes were discovered by Thomas Hunt Morgan, who found that the gene for eye color in the fruit fly *Drosophila melanogaster* is carried on the X chromosome. Since a female fruit fly—like a female human—has 2 X chromosomes, she must have 2 recessive alleles (one from each parent) for a recessive trait to be expressed. But a male has only 1 X chromosome (from his mother); any allele on his X chromosome will be expressed. In humans color blindness and hemophilia are sex-linked characteristics; they are much more common in males than females.

sexual dimorphism (*biology*)   Differences in appearance, morphology, and behavior of adult males and females of a species is termed sexual dimorphism. Male deer have antlers but, except in caribou, females do not. Female gray kangaroos average about 2/3 the size of males; in many species of deep-sea anglerfish, females are 10 times larger than males. Sexual dimorphism is particularly common among birds.

Some sexual dimorphism is seasonal. Male frogs have loud distinctive calls during breeding season but seldom are heard at other times of the year. Male bobolinks resemble females in their fall and winter plumage but grow a markedly different coat of feathers as spring begins.

sexual reproduction (*biology*)   Sexual reproduction typically involves 2 parents and results in offspring that are not identical genetically to either parent. Unlike asexual reproduction, sexual reproduction promotes variability, with some new genetic combinations perhaps improving individuals' ability to survive—and reproduce. In sexual reproduction each parent forms special sex cells, or gametes, by the process of meiosis. When the gametes from the 2 parents combine, fertilization occurs and a new individual begins to grow. In cases of hermaphrodites, self-fertilization may occur, but even in those cases there is some genetic variation in the offspring since the gametes that fused formed in separate organs.

sexual selection (*ethology*)   Among many animals there is a tendency for one sex (usually females) to favor mates that express certain genetically determined characteristics to a pronounced degree—for example, the greatest size, brightest colors, or longest tail feathers. Darwin proposed that sexual selection leads to sexual dimorphism, especially in species in which a small percentage of males monopolize the females. Those males have greater reproductive success, passing along their genes to numerous offspring.

Seyfert galaxy (*galaxies*)   Disk-shaped galaxies, including spirals, that have very bright centers and a spectrum indicating violent activity are called Seyfert galaxies after Carl Seyfert [American: 1911–60], who first recognized that they form a group. Most are powerful emitters in all parts of the electromagnetic spectrum. A quasar is thought to be the bright heart of each Seyfert galaxy.

shadow (*optics*)   Shadows, places where light is blocked by an opaque object, are familiar but more complicated than they appear. A shadow of a small object blocking the Sun or produced by an object near to a surface has sharp edges and a dark interior. This shadow is called an umbra. When the source of light is larger than the object blocking it or when the object casting the shadow is far from the shadow, light that has bent past the object forms a light, fuzzy shadow called a **penumbra**, most familiar as the region outside the total phase of a lunar eclipse. For a small object and a distant source, overlapping parts of the penumbra can cause the shadow to disappear completely. (*See also* **diffraction**.)

shale (*rocks and minerals*) As clay gradually hardens into rock over time, it first becomes mudstone and then shale. Shale is a common rock of relatively young geological formations. It is soft and easily breaks into layers. When subject to heat and pressure, it becomes the metamorphic rock slate.

shaman (*anthropology*) In many societies a part-time occupation of a few people in each band or village consists of using magic to heal (or inflict) illnesses, settle disputes, and predict the future. In anthropology such magician-priests are called shamans (the name used among peoples of northeastern Asia). The belief in good and evil spirits, who can be summoned, placated, or dispatched on missions, is called shamanism. Most shamans use tobacco, mushrooms, or other drugs to induce a trance state that they believe frees the spirit from the body. Shamans are called witches or witch doctors in some societies.

shark (*animals*) Cartilaginous fish with 5 to 7 gill slits on each side of the head, sharks are basically marine, though a few species enter freshwater. All are predators. Sharks typically have a protruding snout, several rows of sharp teeth, a long body, and a backbone that extends into the tail. The whale shark (*Rhincodon typus*), found in warm seas around the world, is the largest species, attaining a length of least 40 ft (12 m) and a weight of 20 tons or more.

shear (*materials science*) Shear is deformation of a solid caused by forces acting tangent to the surface of the solid. Shear can cause twisting or cause one side of a body to deform parallel to the opposite side. When forces act on a solid body, they often have 2 components—tension or compression and shear.

shearwater (*animals*) Named for their habit of flying close to the water's surface, shearwaters are large seabirds with a hooked bill, long, pointed wings, and webbed feet. They come ashore only to mate and raise their young.

The shearwater family, Procellariidae, also includes petrels, fulmars, and prions. The biggest member of the family is the giant petrel (*Macronectes giganteus*) of the Southern Hemisphere; it is up to 39 in. (99 cm) long, with a wingspan as much as 81 in. (205 cm). Like their relatives the albatrosses, Procellariidae are called tubenoses; their nostrils extend as 1 or 2 tubes on the sides or top of the upper bill.

sheep (*animals*) Members of the bovid family of ruminant mammals, sheep (*Ovis*) are closely related to goats but have scent glands on the face, lack a beard in males, and generally are stockier. Wild species inhabit mountainous areas of the Northern Hemisphere; they include the bighorn and Dall's sheep of western North America, the mouflons of Europe and Asia, and the urial and argali of Asia. Sheep are believed to have first been domesticated more than 8000 years ago, in southwestern Asia, probably from mouflons.

shell (*anatomy*)   The most obvious structure of most mollusks and brachiopods is their hard outer shell. Composed mainly of calcium carbonate, the shell provides protection and support. In some animals, such as snails, the shell is a single piece. In others, such as clams and brachiopods, the shell consists of 2 pieces hinged together. Among squid the shell exists only as an internal plate called the pen. The vast variety of shells makes them valuable for classification of species.

The largest shell belongs to the giant clam (*Tridacna gigas*), which lives among coral reefs in the Indo-Pacific region. Measuring up to 4 ft (1.2 m) in diameter, a giant clam can weigh 600 lb (272 kg), of which only 20 to 25 lb (9 to 11 kg) is the clam itself.

shield (*geology*)   A shield is the original part of a continent's crust, always changed by the forces of time and plate tectonics into metamorphic rock. A shield is sometimes called a craton, although the craton usually includes stable sedimentary rock surrounding the actual shield. Any continental shield dates from the Precambrian era and has been neither overlaid with layers of sedimentary rock nor raised by episodes of recent mountain building. In North America the Canadian Shield is a large region north of the Great Lakes centered around Hudson Bay but also covering Greenland.

shield volcano (*geology*)   A volcano that produces thin lava that flows easily does not form a high cone-shaped mountain. Instead a broad mountain called a shield volcano is created, mostly from lava rather than ash. Shield volcanoes can be very high—Mauna Loa on Hawaii rises 13,678 ft (4169 m) above sea level and 6 mi (10 km) from the ocean floor—but their great breadth makes them look different from classic volcanic cones.

shoebill (*animals*)   Native to swamps of tropical Africa, the shoebill (*Balaeniceps rex*)—also called whale-head or bogbird—is a storklike bird up to 4 ft (1.2 m) tall. It has a massive bill, with the upper mandible hooked at the tip. The shoebill uses its bill to probe the muddy bottom for fish and other prey. It is a shy, secretive bird, and usually feeds at dusk. Like storks, it rattles its bill when excited.

shrew (*animals*)   Among the smallest mammals, shrews are insectivores that look rather like mice, with pointed noses and soft velvety fur. They have varied habits and habitats, though most live on the ground near moist places and feed mainly on insects and worms. Shrews have an incredibly high rate of metabolism, with the heart beating up to 1000 times a minute. Maintaining this rate necessitates almost constant eating. Elephant shrews and tree shrews, once classified as insectivores, have been found to have little in common with true shrews and are now classified in their own orders.

shrike (*animals*)   Named for their shrieking songs, shrikes (Family Laniidae) are aggressive perching birds native mainly to the Old World. They have a hooked bill and strong feet with sharp claws. Unlike other songbirds, they are predators, hunting lizards, small birds, and rodents—which they often impale on thorns for later eating.

shrimp (*animals*)   Shrimp belong to a group of crustaceans known as Caridea. They have a carapace covering the thorax, 2 pairs of pincers (the front ones often greatly enlarged), and a long, curved, muscular abdomen. Most shrimp are marine, living either close to the bottom or swimming in the plankton. Some species inhabit freshwater.

Prawns are physically similar to shrimp but show differences in the abdomen. Also, female shrimp brood their eggs while female prawns shed their eggs into the water.

shrub (*botany*)   Perennial plants that typically have several woody stems of more or less equal size arising above the soil are called shrubs. Although there often is no clear distinction between shrubs and trees, shrubs generally are shorter and have more stems than trees; the tallest shrubs seldom grow to more than 20 ft (6 m). Well-known shrubs include the azalea, camellia, hydrangea, rhododendron, and rose.

sidereal period (*solar system*)   Astronomers use several ways to measure the time between celestial events because timing is affected by Earth's motion. The sidereal period between events is measured with respect to the stellar background. The Moon, for example, appears to move among the stars with a period of 27⅓ days, its sidereal period.

significant digit (*measurement*)   The total number of units in measuring a specific amount, expressed as a decimal fraction or a whole number, is the number of significant digits for that measurement. A measurement reported as 23.6 implies measurement to the nearest tenth (236 tenths in all) and shows 3 significant digits. The same quantity measured in hundredths might be 23.60 to the nearest hundredth (2360 hundredths), showing 4 significant digits. For whole numbers, a 0 at the end of the numeral may or may not be significant—23,600 may have been measured in hundreds (236 hundreds) and have 3 significant digits, or in tens (2360 tens) and have 4 or even in ones, with 5. Scientific notation clarifies this since all digits in scientific notation are significant—23,600 is written as 2.36×10^4 to mean 3 significant digits and as 2.360×10^4 for 4 or 2.3600×10^4 for 5.

silica (*rocks and minerals*)   Silica is the name of the mineral form of silicon dioxide (SiO_2), but it appears in several different guises—as quartz (including amethyst) and chalcedony (including the gems carnelian, agate, onyx, and jasper). Hydrated silica is known as opal. The most transparent and largest crystals of silica are the "crystals" of mystics.

silicate (*rocks and minerals*)   Many of the most common minerals are silicates, based on a tetrahedron that consists of a silicon atom surrounded by 4 oxygen atoms. Silicates easily accommodate various metal ions, including sodium, potassium, calcium, magnesium, and iron. The tetrahedrons can be arranged in chains (pyroxine) or rings (beryl), but one of the most common forms, olivine, consists of isolated tetrahedrons linked by metal ions. Olivine is thought to be the main mineral of Earth's mantle.

silicon (*elements*)   Silicon is, after oxygen, the most abundant element on Earth—silicon dioxide, or silica, is found as quartz or flint and is the main compound in sand (the name derives from Latin *silex* or *silicis*, "flint"). For at least 3000 years silicon has been the basis of useful materials from glass to computer chips. A nonmetal, atomic number 14, silicon resembles carbon chemically. Silicon, chemical symbol Si, was discovered in 1824 by Jöns Jakob Berzelius.

silk (*biochemistry*) A strong, elastic protein produced by spiders, caterpillars, and certain other arthropods, silk derives its name from the Chinese empress Shi Ling-chi, who according to legend discovered how to unravel a silkworm's cocoon more than 4000 years ago. Silk is secreted as a liquid by specialized glands within the animal; it solidifies into a thread as it leaves the body and comes in contact with air.

Spiders have large abdominal glands that produce and secrete silk. The silk passes out through hundreds of tiny spigots on the spinning organs, or spinnerets. Spiders use silk to spin webs, form cocoons for eggs, make draglines to prevent falls, and move from place to place by ballooning.

In caterpillars, the liquid silk is produced by modified salivary glands and secreted through an opening in the lower lip. It is used to make cocoons, tents, and other kinds of shelter, and in some species is used for ballooning.

silkworm (*animals*) The common name silkworm usually refers to caterpillars of the moth *Bombyx mori*, although it also is given to any caterpillar that spins cocoons of silk that can be used to make cloth.

Native to Asia, *B. mori* has a short heavy body and white wings decorated with brown stripes. A newly hatched caterpillar is 0.25 in. (6.4 mm) long. It eats continuously and grows rapidly, reaching a length of 3 in. (76 mm) in about 6 weeks. It then stops feeding and spins a cocoon around itself. A single *B. mori* cocoon may contain more than 2 mi (3.2 km) of silk fiber.

sill (*geology*) An intrusion of magma that spreads apart layers of existing rock and then cools, forming floors that are parallel to the existing layers, is called a sill. Like a dike, a sill is usually basalt or diabase.

silt (*soils*) Technically any soil particles larger than clay particles or smaller than sand grains—that is, between 1/400 in. (0.06 mm) and 1/6400 in. (0.004 mm)—are called silt. Deposits of silt are usually produced when such particles have been transported by water and then fall to the bottom when water speed diminishes. Thus river banks and deltas as well as parts of the ocean floor are covered with silt.

silver (*elements*) Like copper and gold, silver can be found in near pure, or native, form, but more often it occurs in compounds as part of ores. Its atomic number, 47, places silver just above gold in the periodic table and, like gold, a principal use is in jewelry. Silver compounds react to visible light, making silver essential for photography, but pure, untarnished silver is the best reflector of visible light. Silver is also the best conductor of heat and electricity, but cheaper copper is more often used when good conduction is required. The chemical symbol, Ag, is from its Latin name, *argentum*.

silverfish (*animals*) Silverfish are members of the bristletail order of insects, Thysanura, a group characterized by long antennae, no wings, and 3 bristlelike appendages at the rear of the abdomen. Silverfish have a flat, silvery body. They are common household pests, living in dark places and feeding on bookbinding, wallpaper paste, and stored food.

Prefixes used in SI Measurement

FACTOR	SCIENTIFIC NOTATION	PREFIX	SYMBOL
1/1,000,000,000,000,000,000	10^{-18}	atto-	a
1/1,000,000,000,000,000	10^{-15}	femto-	f
1/1,000,000,000,000	10^{-12}	pico-	p
1/1,000,000,000	10^{-9}	nano-	n
1/1,000,000	10^{-6}	micro-	μ
1/1,000	10^{-3}	milli-	m
1/100	10^{-2}	centi-	c
1/10	10^{-1}	deci-	d
10	10^{1}	deka-	da
100	10^{2}	hecto-	h
1,000	10^{3}	kilo-	k
1,000,000	10^{6}	mega-	M
1,000,000,000	10^{9}	giga-	G
1,000,000,000,000	10^{12}	tera-	T
1,000,000,000,000,000	10^{15}	peta-	P
1,000,000,000,000,000,000	10^{18}	exa-	E
1,000,000,000,000,000,000,000	10^{21}	zetta-	Z
1,000,000,000,000,000,000,000,000	10^{24}	yotta-	Y

SI measurement (*measurement*)   The system of measurement used by nearly all scientists is called SI measurement, where SI stands for *Système International d'Unités* (International System of Units). SI, instituted in 1960 by an international conference, refines and extends the metric system. SI starts with 7 basic units, for length, mass, time, electric current, temperature, amount of substance, and luminous intensity. These are combined with a set of prefixes that establish the power of 10 used as a coefficient for the basic unit. For example, the unit of length is the meter, the prefix kilo- means 10^3, so a kilometer is 1000 meters. Other SI units are derived from the basic ones—force, for example, is the product of mass and length divided by the square of time.

similar figure (*geometry*)   Figures are similar if they have the same shape. For polygons, this means that corresponding sides must all be in the same ratio, while corresponding angles must be equal. Two parallelograms in which sides of one are double the other, for example, are similar only if the angle between a

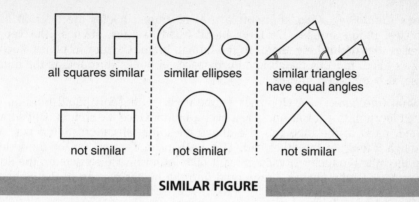

all squares similar | similar ellipses | similar triangles have equal angles

not similar | not similar | not similar

SIMILAR FIGURE

pair of corresponding sides is equal in each. Triangles are unusual in that they are similar if 2 angles of one triangle are equal to 2 of the other. Triangles are also similar if all 3 sides have the same ratio. All circles, like all squares, are similar, but for other curves it is necessary to specify which parts must be in the same ratio to produce similar curves. Ellipses are similar if the longest line segment with both endpoints on the curve and shortest segment of this type in one ellipse are in the same ratio with the corresponding segments of the other.

sine (*trigonometry*)   The sine (abbreviated sin) derives its name from the chord of a circle since it was originally the length of the chord intercepted by a central angle, a length used in early astronomy. Later Indian astronomers adopted half this chord for the same purpose and it passed into modern mathematics as the half chord. Thus, in terms of a unit circle definition of trigonometric functions, sin θ equals the *y* coordinate of the point where the terminal side of θ meets the circle. The sine is the most basic trigonometric function—all others can be defined in terms of the sine and its complement the cosine. In a right triangle sin θ is the ratio of the length of the side opposite θ to the length of the hypotenuse (side opposite the right angle).

singularity (*physics*)   Sometimes an equation predicts a physical measure, such as gravity, mass, or charge, that has infinity as its apparent value. When that happens, physicists call the impossible prediction of a physical infinity a singularity. Some singularities, such as black holes, exist physically, but have unusual attributes (although infinity is not actually present). The universe before the big bang appears as such a singularity. Other singularities, such as calculations predicting infinite mass for electrons, are thought to be incorrect answers and are removed with various techniques. Such techniques for eliminating singularities are the basis of quantum electrodynamics.

sinkhole (*geology*)   In a region with Karst topography, water with dissolved carbon dioxide acts as a weak acid and dissolves the limestone that underlies the soil. When support for the soil becomes weak enough, the soil suddenly falls into the underground pit, creating a large depression known as a sinkhole. The sinkhole may be part of the roof of a cave or cavern that has collapsed. In some cases the sinkhole fills with water, forming a deep pond called a cenote.

sinus (*anatomy*)   Two kinds of cavities in animal bodies are referred to as sinuses. Blood sinuses are large blood-filled cavities, such as the venous sinuses that drain the veins from the human brain. Cavities in bones, such as the air-filled, mucous membrane-lined spaces in bones surrounding the human nose, also are called sinuses.

siphon (*mechanics*)   A siphon is a tube used to transport liquid from an elevated location to a lower one, even over a hump. Once the siphon is filled with liquid, parts of the tube can rise above the initial elevation; liquid will flow through it and out the other end. The mechanism depends on the liquid leaving the tube producing a vacuum as it falls. Atmospheric pressure on the liquid entering the tube pushes more liquid into the higher portion of the siphon in response to the vacuum.

Sirius (*stars*)   The brightest star as seen from Earth is Sirius, whose name means "glowing." It is a binary star, first recognized from the wobble caused by its unseen companion. The companion, observed telescopically since 1862, is a white dwarf with the mass of the Sun, but much smaller in diameter. Sirius, in the constellation Canis Major, has long been called the dog star and its companion is sometimes called the Pup.

Ancient Egyptians learned that when Sirius rises closest to where the Sun rises, the Nile's annual flood would begin. They used this information to develop a calendar based on a 365-day year.

skate (*animals*)   Bottom dwellers found in tropical and temperate seas, skates are cartilaginous fish closely related and similar in appearance to rays. Species range in length from less than 2 ft (0.6 m) to about 8 ft (2.4 m). A skate has a flat disklike body; a huge winglike fin extends from each side. On top of the head are 2 eyes and respiratory holes called spiracles, through which water enters. The mouth, equipped with rows of small teeth, is on the underside of the head. The tail typically is short, with 2 dorsal fins.

skeleton (*anatomy*)   All vertebrates have an internal skeleton. (Some invertebrates have an exoskeleton.) A skeleton has several functions. It supports the body; provides surfaces for the attachment of muscles that move its parts at a variety of joints, thereby allowing locomotion; protects delicate organs, such as the brain and heart; and stores calcium and phosphorus. Bone marrow, protected by skeletal bone, is the site of blood cell production.

The skeleton consists of 2 kinds of connective tissue, bone and cartilage. The skeleton of cartilaginous fish is composed entirely of cartilage. The early embryonic skeleton of other vertebrates also is made of cartilage, most of which is replaced by bone in the process of ossification.

skew (*geometry*)   Lines in a space of 3 dimensions that are neither parallel nor intersecting are skew. A quadrilateral is skew if it is formed by joining 4 points for which 1 point lies in a different plane from the other 3; each point is joined to 2 and only 2 others.

skin (*anatomy*)   The skin that covers the body of a vertebrate protects underlying tissues, acts as a barrier against disease organisms, prevents excessive

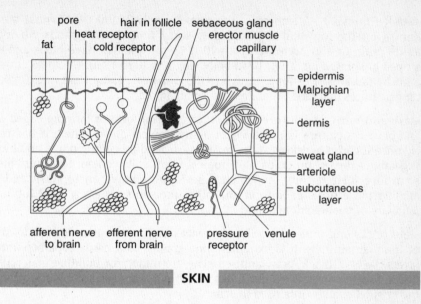

fat · pore · heat receptor · cold receptor · hair in follicle · sebaceous gland · erector muscle · capillary

epidermis
Malpighian layer
dermis
sweat gland
arteriole
subcutaneous layer

afferent nerve to brain · efferent nerve from brain · pressure receptor · venule

SKIN

evaporation, and contains sense organs and nerve endings that detect touch, pressure, pain, and heat. Glands produce various secretions, such as sweat and oil in mammals and mucus in fish. In fish and amphibians some respiration takes place through the skin. In birds and mammals the skin helps regulate body temperature through sweating, contraction and expansion of blood capillaries, and insulation, such as feathers and hair.

The skin has 2 layers, the epidermis and dermis. Cells of the innermost part of the epidermis, the Malpighian layer, continually divide, replacing surface cells that die and are worn away. Below the dermis is the subcutaneous layer, which contains numerous fatty cells that act as insulation.

skink (*animals*) Lizards constituting the family Scincidae, skinks typically are slim and agile, with a forked tongue, smooth, similarly sized scales, and small limbs that may be reduced or absent in burrowing species. Skinks are found worldwide in a variety of temperate and tropical habitats.

skua (*animals*) Powerful, fast-flying seabirds that resemble gulls, skuas and their relatives, jaegers, live along coasts and at sea at higher latitudes of both the Northern and Southern hemispheres. They feed on rodents, ship refuse, the eggs and young of other birds, and fish that they steal from other seabirds. The great skua (*Catharacta skua*) is about 21 in. (53 cm) long.

skull (*anatomy*) The skull is the part of a vertebrate's skeleton located in the head. It consists of the cranium ("brain box"), which encloses and protects the brain; capsules for sense organs (eyes, ears, nose); and paired arches that form jaws, support for the tongue, and support for gills. In sharks and rays the skull is composed of cartilage throughout life. In bony fish and other vertebrates, bone tissue replaces the cartilage of young embryo skulls.

skunk (*animals*) Members of the weasel family, skunks (*Mephitis*) are best known for the foul-smelling fluid they spray from 2 anal glands when menaced—with an aim accurate up to 10 ft (3 m). Skunks have black-and-white striped or spotted fur. Their head-body length is 10 to 15 in. (25 to 36 cm), with a bushy tail almost as long. Native to the Americas, skunks are nocturnal, solitary creatures that feed on plants and small animals.

slash-and-burn agriculture (*environment*) The form of farming called slash-and-burn agriculture is practiced in many tropical areas, where it has been a significant cause of deforestation. People clear land by cutting down and burning vegetation, then plant crops. But tropical forest soils contain few nutrients and soon lose their fertility, at which time farmers move on to slash and burn another piece of land. Erosion and other problems generally limit future productivity of the abandoned land.

slate (*rocks and minerals*) Slate is a metamorphic rock often derived from shale or from certain tuffs. It is extremely fine grained, but easily separates into flat layers. Beds of black, gray, and red slate are mined for building materials, the flat surfaces forming walks or floors or roofing. Large slabs of black slate were the original chalkboards used in schools.

sleep (*physiology*) Most vertebrates appear to have a daily pattern of rest and activity. The average total sleep time varies according to species. For example, a bat sleeps an average of 19.9 hours per 24-hour day, a hamster 15 hours, a cat 12.1 hours, a human 8 hours, a cow 3.9 hours, and a giraffe 2 hours. Total sleep time also varies according to age, with young animals needing more sleep than their elders. Sleep has a restorative function, helping the body recover from all its waking activities; stimulates growth; and builds immunity to viral infection. But the periodic occurrence of REM sleep in at least some vertebrates suggests that sleep also serves yet-unknown purposes.

sleet (*meteorology*) When rain forms it is usually first ice or snow, which melts as it falls. Raindrops falling into a temperature inversion in winter encounter a colder region of the atmosphere and can refreeze, becoming sleet. Since sleet is frozen rain, it does not have the layers of ice of hail or the crystal structure of snow.

slime mold (*protists*) Slime molds are organisms that resemble amoebas in their feeding stage but at other times produce spores like fungi. Hence, there is disagreement among taxonomists on how to classify them.

During most of its life cycle, a true, or acellular, slime mold exists as a plasmodium—a jellylike mass of cytoplasm containing many nuclei and surrounded by a cell membrane. Becoming as large as 3.3 ft (1 m) in diameter, a plasmodium creeps along a forest floor, engulfing bacteria, yeast, and other food matter. But if environmental conditions become unfavorable, the plasmodium may send up fruiting bodies called sporangia. Each sporangium produces many spores, which develop into flagellated swarm cells. Two swarm cells may mate, or fuse, to form a new plasmodium or a single swarm cell may become a plasmodium without fusion.

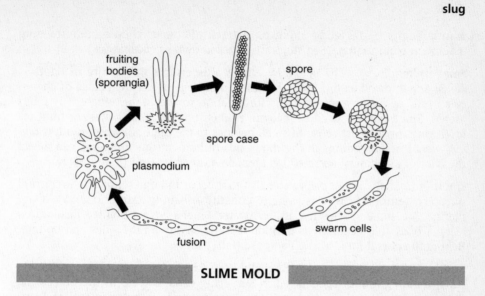

SLIME MOLD

slope (*analytic geometry*)   The steepness of a curve is measured as a number called the slope. For a line through points (x_1, y_1) and (x_2, y_2), the slope is the ratio of the directed distance between the y values to the directed distance between the x values, or $(y_2 - y_1)/(x_2 - x_1)$. The slope is 0 for a horizontal line, +1 for a line making a 45° angle with the positive x axis (measured counterclockwise), –1 for a line making a 135° angle with the positive x axis, and undefined for a vertical line. For a curve other than a line, the slope is the slope of the tangent to the curve (in calculus, this is the same as the first derivative at that point).

slope-intercept equation (*analytic geometry*)   One of the commonest forms of equation for a line is based on knowing the line's slope, expressed as m, and the y intercept, expressed as b. The slope-intercept form is $y = mx + b$.

sloth (*animals*)   Slow-moving mammals that live in trees of Central and South American rain forests, sloths (Family Bradypodidae) spend most of their life hanging upside down; females even give birth in this position! There are 2 genera: 2-toed sloths (*Choloepus*) and 3-toed sloths (*Bradypus*). All have a slender body and long limbs covered with coarse brown-gray fur; algae often grows on the fur, giving the sloths a greenish tint. Sloths are about 20 to 28 in. (50 to 70 cm) long and weigh 10 to 20 lb (4.5 to 9 kg). They feed on leaves at night.

slowworm (*animals*)   A very slender lizard with no legs, the slowworm (*Anguis fragilis*) lives in grassy habitats in Europe, western Asia, and northern Africa, where it feeds on small invertebrates. The extremely fragile tail—often as long as the head and body together—breaks off if seized by an enemy, allowing the slowworm to escape. Slowworms are sometimes called blindworms, a misnomer since slowworms are not blind.

slug (*animals*)   Land mollusks closely related to snails, slugs lack the external shell of their ancestors. They breathe air through a lunglike modification of the mantle cavity. Slugs glide along on a track of mucus they secrete, feeding mainly on plant matter. (Sea slugs are better known as nudibranchs.)

smell (*physiology*)   The sense by which animals become aware of odors is called smell, or olfaction. Specialized receptor cells detect molecules of chemicals in air or water and send this information to the brain, where it is interpreted. Smell may be used to recognize other animals, find mates, or hunt for food. Perhaps the best sense of smell belongs to the male silkworm moth (*Bombyx mori*); nerve endings in the thousands of hairs on his antennae can detect the odor of a chemical secreted by a female more than 2 mi (3 km) away.

smelt (*animals*)   Small cold-water fish that occur in large schools in arctic and northern temperate waters, smelts constitute Family Osmeridae. Most are marine, but some species live in freshwater. Smelts have a pointed head and a slender body that generally is dark on top and silvery on the sides and bottom. Behind the dorsal fin is a small adipose (fatty) fin.

Smith, William (*geologist*)   [English: 1789–1839] In a time of considerable canal building, Smith specialized in surveying canal routes. In the process he discovered that fossils can be used to recognize strata found many miles apart. He proposed the principle of superimposition, stating that lower strata are older than those above them, and published the first geological maps of Great Britain, starting in 1815.

smog (*environment*)   A mixture of smoke and fog called smog is a form of air pollution caused mainly by emissions from motor vehicles and commercial and industrial sources. The action of sunlight on hydrocarbons and nitrogen oxides produces photochemical smog, of which ozone (a highly reactive form of oxygen) is the main component. Ozone tends to break down tissues and cells, thereby causing leaf damage in plants and respiratory ailments or other health problems in animals.

smut (*fungi*)   Members of the club fungi, smuts are named for the powdery black masses of spores they produce on barley, oats, wheat, and other flowering plants they attack. Many species are parasitic on a single host species—corn smut (*Ustilago maydis*), for example, causes large tumors on the ears and other fast-growing parts of the corn plant. Despite its alarming appearance, corn smut is a delicacy in Mexican cooking under the name *huitlacoche*. When a tumor bursts open, numerous tiny spores are released into the air.

snail (*animals*)   An extremely diverse group of mollusks, snails have a 1-piece, unchambered spiral shell, a broad flat foot, and a prominent head with tentacles. The foot and head can be withdrawn into the shell; in many species a horny plate, the operculum, blocks the opening when the animal is inside the shell. Snails inhabit both aquatic and land environments. Most feed on plant matter but some are carnivores, scavengers, or parasites.

snake (*animals*)   Composing the reptile suborder Serpentes, snakes have greatly elongated bodies and no limbs. Their body may be cylindrical or, as in sea snakes, flattened. Supporting the body is a backbone of 100 to more than 300 vertebrae, all except the first 2 bearing a pair of ribs. Snakes are covered by horny scales either placed close together or overlapping (the scale arrangement is important in classifying snakes); atop the scales is a thin, transparent layer of skin. This outer skin and the scales are shed periodically and replaced from underneath by new skin and scales. Snakes move by coiling and bending the body, using movements characteristic of their species. The narrow, forked tongue, which regularly darts out of the mouth, is extremely sensitive to smell and touch. Snakes are predators, with a flexible jaw and elastic throat, enabling them to swallow prey wider than themselves.

snow (*meteorology*)   Water vapor high in a cloud is often supercooled, remaining a gas at temperatures below freezing for water. When small particles, such as dust, salt crystals, or ice, are present, the supercooled water begins to freeze onto the particles. Because of the structure of the water molecule, the water freezes into 6-sided crystals. As these tiny crystals fall slowly through the cloud, they attract more supercooled water vapor, which tends to create projections from each side of the original crystal. Soon a complex pattern with sixfold symmetry is made from ice—a snowflake. If the air is cool enough between the cloud and the ground, millions of snowflakes, each different from almost all the others, will fall as snow.

SNP (*genetics*)   A single nucleotide polymorphism (SNP)—pronounced "snip"—is a common alteration in one nucleotide base in a sequence of DNA occurring in at least 1% of the population. It's rather like a misspelled word ("bread" instead of "broad") where a change of 1 letter changes the word. The vast majority of SNPs are probably of no significance, but some contribute to disease. For example, sickle cell anemia in humans is the result of SNP (a thymine replaced by an adenine) in the gene that specifies how to make hemoglobin in red blood cells.

sociobiology (*sciences*)   Sociobiology is the study of the social behaviors of animals—bee swarms, fish schools, bird flocks, wolf packs, etc.—particularly from an evolutionary perspective. Sociobiologists contend that many social behaviors exist because, like positive physical characteristics, they increase the likelihood of propagation by individuals or survival by a population or species. For example, many social birds have alarm calls that indicate the presence of a predator, signaling other birds to take heed. When threatened by wolves, a herd of musk oxen forms a compact circle with young animals inside, presenting the wolves with an ominous display of horns. Female bats of some species hibernate together in large clusters, thereby helping conserve heat.

Sociobiology, first defined by Edward O. Wilson in 1975, overlaps the discipline of ethology in terms of its concepts, theories, and goals. Attempts to apply sociobiology to human behavior have been controversial.

sodium (*elements*)   Sodium is an alkali metal, atomic number 11, discovered in 1807 by Humphry Davy using electrolysis of molten lye (sodium hydroxide), also known as caustic soda, from which the name sodium is derived. The symbol Na is from a Latin name for lye, *natrium*. Sodium is lighter than water and so reactive that it ignites in the presence of water even though it is the least reactive of the alkali metals.

sodium chloride (*compounds*)   Table salt—chemically sodium chloride (NaCl)—is the major salt in ocean water and an important nutrient, providing sodium ions used in many physiological processes (buffering, osmosis, and transmission of nerve signals among them). Chlorine, in addition to having similar roles as those of sodium, is part of digestive hydrochloric acid. Sodium chloride has a vast number of uses, ranging from curing hams to glazing one form of pottery. The mineral sodium chloride, called halite, is found in large deposits where shallow seas evaporated.

soil (*rocks and minerals*)   Soil is a mixture of rock particles and organic remains that can sustain the growth of plants with roots. This common definition expands to include such complexities as frozen soils in which plants can grow if the soil thaws or buried soils that sustain growth when brought to the surface. The restriction "plants with roots" is needed because some mosses—as well as algae and lichens—can grow on bare rock, not counted as soil.

Residual soil derives from the bedrock beneath it, while transported soil comes from another locality. Residual soil typically appears in 3 layers, called the soil's profile. The top layer, a mixture of clay, sand, and humus called topsoil, contains so many living organisms that some call topsoil itself alive. Topsoil often lacks soluble minerals. The subsoil lacks the humus but often contains the missing minerals. Below subsoil, loose rock mixed with clay sits on top of bedrock.

solar constant (*meteorology*)   Solar constant is the name meteorologists have given to the energy that reaches Earth from the Sun. The word constant in this instance is misleading since the amount varies 1% or so from year to year. Numerically the solar constant is defined as the number of calories caused by direct sunlight striking 1 cm^2 (about 0.15 sq in.) of the top of the atmosphere in 1 minute when the Sun is at its average distance. The solar constant currently measures about 2 cal/min-cm^2 (22 w/s-m^2). This number was defined and first calculated by Charles Greeley Abbot [American: 1872–1973].

solar cycle (*solar system*)   The Sun reverses the orientation of its magnetic field every 11 years, creating a 22-year solar cycle. Since sunspots, prominences, and flares are all largely magnetic phenomena, they all reflect this cycle—for example, the number of sunspots rises and falls in an 11-year pattern (which some think influences weather on Earth). The maximum number of sunspots typically occurs 2 or 3 years before the pole reversal.

solar system (*solar system*)   The Sun—*sol* in Latin—and all the bodies that orbit the Sun form the solar system, sometimes including the solar wind and magnetic field. When the Sun was born through the collapse of a cloud of gas, some gas spread into a disk about the Sun. Freezing, collisions, and gravitational accretion resulted in numerous small bodies in the disk. These bodies swept up gases and stuck to each other as they collided. The solar wind pushed light gases away from the Sun. The largest bodies near the Sun became rocky terrestrial planets, heavily bombarded by remaining smaller bodies. Many smaller bodies are still in orbit, most beyond the terrestrial planets in the asteroid belt. Farther from the Sun, light gases collected to form four giant planets. Beyond the giant planets, collisions are less frequent, and small balls of gas and dust freeze into clouds of comets.

Planets of the Solar System

PLANET	DIAMETER	DISTANCE FROM SUN	ROTATION TIME	TIME FOR 1 ORBIT	MASS	ESCAPE VELOCITY
Mercury	3032 mi (4880 km)	35,500,000 mi (57,900,000 km)	59 days	88 days	5.53% of Earth	2.7 mi/sec 4.3 km/sec
Venus	7521 mi (12,104 km)	67,200,000 mi (108,200,000 km)	243.7 days (retrograde)	224.7 days	81.5% of Earth	6.4 mi/sec 19.4 km/sec
Earth	7927 mi (12,756 km)	92.9600,000 mi (149,600,000 km)	23.9 hours	365.256 days	100% of Earth (5970 trillion metric tons)	7.0 mi/sec 11.2 km/sec
Mars	4222 mi (6794 km)	141.600,000 mi (227.900,000 km)	24.6 hours	686.98 days	10.7% of Earth	3.1 mi/sec 5.0 km/sec
Jupiter	88,850 mi (142,984 km)	483,600,000 mi (778,300,000 km)	10.0 hours	4332.59 days, or 11.86 years	31,783% of Earth, or 317.83 × mass of Earth	37.0 mi/sec 59.5 km/sec
Saturn	74,901 mi (120,536 km)	887,700,000 mi (1,427,000,000 km)	10.5 hours	10,759.22 days, or 29.45 years	9516.2% of Earth, or 95.162 × mass of Earth	22.1 mi/sec 35.5 km/sec
Uranus	31,765 mi (51,118 km)	1,783,000,000 mi (2,869,600000 km)	17.2 hours	30,865.4 days, or 84.5 years	1453.6% of Earth, or 14.536 x mass of Earth	13.2 mi/sec 21.3 km/sec
Neptune	30,779 mi (49,532 km)	2,794,000,000 mi (4,496,600,000 km)	16.1 hours	60,189 days, or 165 years	1741.7% of Earth, or 17.417 × mass of Earth	14.6 mi/sec 23.5 km/sec
Pluto	1413 mi (2274 km)	3,674,500,000 mi (5,913,500,000 km)	153 hours, or 6.4 days	90,456.0 days, or 274.6 years	0.21% of Earth	0.7 mi/sec 1.1 km/sec

solar wind (*solar system*)   The Sun produces electrons and ions that are driven away in all directions at about 500 mi per sec (750 km/sec); these form the solar wind. The solar wind points comets' tails away from the Sun, pushes magnetic fields of planets into giant teardrops, and causes auroras on Earth. It reaches well beyond Pluto's orbit to about 100 astronomical units from the Sun, a boundary called the heliopause.

solid (*geometry*)   The name solid is commonly used for polyhedrons and other closed figures in a Euclidean space of 3 dimensions, such as cones, cylinders, and spheres, all figures defined as surfaces. This usage in geometry is somewhat different from that in ordinary language, where the adjective solid indicates that an object is not hollow. Thus the interior of a geometric solid (and its "skin," the actual figure) is often meant outside of strict mathematical usage, especially for applications.

solid (*materials science*)   A solid is a substance whose molecules, atoms, or ions vibrate but do not move easily from one part of the substance to another. Thus a solid keeps its shape and does not take the shape of its container. For most solids, atoms or ions arrange themselves into small geometric forms that repeat over and over, forming crystals. For an amorphous substance, such as a glass, atoms or molecules remain fixed (except for vibration) with respect to each other, but do not form crystals.

solstice (*solar system*)   The point on the horizon where the Sun rises or sets each day gradually moves south for part of the year and then turns north. The day on which the direction changes is the solstice, meaning "Sun stops." The Sun reaches its northernmost position about June 21, the summer solstice and the official beginning of summer in the Northern Hemisphere. The winter solstice is about December 23. The change in day length is caused by the tilt of Earth's axis with respect to the plane of revolution. The Northern Hemisphere tilts about 23° toward the Sun at its summer solstice and the same amount away at its winter solstice.

solution (*algebra*)   Numbers from the domain of a variable that make an equation true when substituted for the variable are solutions to the equation. The equation $x + 1 = 2$, where x is a natural number, has the solution 1, often written as $x = 1$ or as the set $\{1\}$. Frequently there is more than a single solution. For example, the equation $x^2 - 3x + 2 = 0$ has the solutions 1 and 2. Another name sometimes used for the solution is root of the equation.

solution (*chemistry*)   A solution is a mixture for which the ratios of the 2 or more substances mixed are the same for any small sample (unlike a mixture of sand and salt, which may in some samples have more sand and in others more salt); thus any homogenous mixture is a solution. Because gases mix uniformly with each other, all mixtures of gases, including air, are solutions. Gases, liquids, and solids all can be dissolved (form solutions) in liquids; examples include carbonated water (carbon dioxide gas in water); antifreeze (liquid ethylene glycol in water), and sugar syrup (solid sucrose in water). Solid metals form solutions called alloys, although the solutions are usually formed when the metals are molten. Carbon, silicon, and other nonmetals can also form alloys with metals.

Solutrean (*paleoanthropology*)   The Solutrean tool industry introduced pressure flaking, characterized by pushing flakes off a core with a bone tool instead of striking them off with another stone. This very advanced stone technology produced long, thin blades called laurel leaves that may have been ceremonial, since they seem too thin for practical applications. The Solutrean culture was characterized by early cave art featuring engravings and bas reliefs of animals.

solvent (*chemistry*)   A solvent is the substance in a solution (except for solutions of gases) present in the greatest amount. Often a solid or small amount of liquid forms a solution with a large amount of liquid. The most abundant liquid is the solvent. The solid or small amount of liquid, called the solute, is said to dissolve in the solvent.

Somali Plate (*geology*)   The Somali Plate is identified as the part of the lithosphere that is breaking away from the African Plate along Africa's major rift valley. It carries the east coast of Africa, the island of Madagascar, and the surrounding part of the Indian Ocean to the east.

sorghum (*plants*)   Members of the grass family native to tropical Africa and Asia, sorghums (*Sorghum*) are widely cultivated for their edible grains (fruits) and as fodder. Broomcorn sorghums are grown for their stalks, which are made into brooms. Sweet sorghums are grown for a syrup extracted from the stalks. Sorghums grow to 10 ft (3 m) tall. The flowers and, later, the grains are borne on heads at the top of the stalks.

sound (*waves*)   Vibration of a physical object in air produces longitudinal waves of alternating higher and lower pressure that humans perceive as sound, provided the vibrations are faster than about 20 Hz and slower than about 20,000 Hz (animal ranges differ). Sound waves need a medium through which to travel—in air they travel about 0.2 mph (340 m/s), in water about 1 mph (1530 m/s), and in steel about 3.7 mph (5960 m/s). Any longitudinal pressure waves, such as some seismic waves, may be called sound waves also.

soursop (*plants*)   The soursop family, Annonaceae, is comprised of shrubs and trees native mainly to tropical regions. Several are cultivated for their edible fruits, including the soursop or custard apple (*Annona muricata*), cherimoya (*Annona cherimola*), and papaw (*Asimina triloba*).

South American Plate (*geology*)   Although the border in the Atlantic Ocean between the North American Plate and the South American Plate is not well defined, the different organisms found in North and South America indicate that the 2 continents were separated in the past. They became joined when the movement of the Cocos Plate into the Caribbean Plate raised Central America. In addition to the continent, the South American Plate includes the South Atlantic Ocean to the mid-oceanic ridge and extends south to the Antarctic Plate.

space (*astronomy*)   The region between bodies of matter in the universe, including planets, stars, nebulas, and galaxies, is called space. In common language, the locations of those bodies is termed outer space—roughly any part of the universe beyond that part of Earth's atmosphere dense enough to support an airplane. Astronomers separate space into interplanetary space (within the solar system), interstellar space, and intergalactic space. Although all of space is a near vacuum, interplanetary space is filled with the solar wind and magnetic fields of planets. Interstellar space outside of dust and gas clouds contains fast-moving particles; when these reach Earth they are called cosmic rays. Various pieces of evidence show that intergalactic space, with the exception of some intergalactic clouds, contains fewer than 1 atom per 13,000 cu yd (10,000 ^{3}), making it nearly empty.

space (*geometry*)   Although there are many named types of space in mathematics, all can be thought of as consisting of either points, vectors, or sets. The familiar Euclidean spaces are the archetypes of point spaces. These are also examples of metric spaces, a common type in which distance has a meaning. Vector spaces are algebraic structures defined by a field of scalars (used also as the **components** of the vectors) and a procedure for creating the whole space from a finite number of its members. Spaces based on sets, such as topological spaces, are very general and usually discussed only in specialized texts.

space-time (*physics*)   A geometrical formulation of **Einstein's** 1905 special theory of **relativity** by Hermann Minkowski [Russian-German: 1864–1909], published in 1907, showed that relativity is best described in terms of a universe of 4 dimensions, the 3 dimensions of space plus time. The basic stuff of the universe in this system is called space-time, with **coordinates** (x, y, z, t). The 4 coordinates are not symmetrical—although x, y, and z could be interchanged, t maintains a separate status. Einstein, a former pupil of Minkowski, adopted space-time as a concept and used it in later work.

Spallanzani, Lazzaro (*biologist*)   [Italian: 1729–99] Spallanzani conducted important experiments in many areas. Working with microorganisms, he helped discredit the theory of **spontaneous generation**. He also showed that boiling water is a better sterilizing agent than hot air, and that some microorganisms can live for days in a vacuum. Spallanzani studied plant and animal reproduction and carried out experiments on the artificial insemination of frogs.

sparrow (*animals*)   Most **perching birds** called sparrows are members of the finch family, but several—including the common house, or English, sparrow (*Passer domesticus*)—are in the weaverbird family. Sparrows are small songbirds. They are gregarious and often found in large flocks. The diet varies with the species; some sparrows are seedeaters while others feed mainly on insects.

spear-thrower (*paleoanthropology*)   A device for extending the human arm to produce a higher velocity when casting a spear or harpoon is called a spear-thrower (or throwing stick or atlatl—the Aztec name). The spear-thrower was invented in Magdalenian times and was sometimes carved from reindeer bone and engraved with animal pictures. Similar devices, usually of wood, have been used by Australian and North American natives.

speciation (*evolution*)   The formation of a new species from a preexisting species is termed speciation. Geographic isolation is one circumstance that can lead to speciation—when 2 populations are separated by a natural barrier such as a mountain or river, they can no longer interbreed; over time, each group evolves a different **gene pool**. Other ways that speciation can occur include adaptive radiation and **polyploidy**.

species (*taxonomy*)   Species is the basic—and usually the smallest—unit in the **classification** of organisms. Members of a species are closely related genetically and are nearly identical in structure and behavior. They can interbreed, producing fertile offspring. Each species has its own unique **binomial name**. The typhus bacterium (*Rickettsia prowazekii*), poinsettia (*Euphorbia pulcherrima*), and screech owl (*Otus asio*) are examples of species.

specific heat (*materials science*)   Specific heat is the amount of heat needed to raise the temperature of a particular substance, which varies considerably from one material to another. Measurement of specific heat is linked to water, for which 1 g rises 1 K (1°C) when 4.187 joules (1 Calorie) of heat is absorbed.

Iron, for example, has a specific heat about ¹⁄₁₀ that of water, while that of hydrogen gas is more than 3 times that of water. Specific heat varies slightly at different temperatures and 15°C is considered standard.

spectral type (*stars*)   In the 1890s astronomers at Harvard College Observatory in Cambridge began classifying stars by the strength of the hydrogen line in the spectrum. The types originally were intended to run from A through O, but later measurements eliminated some and caused others to be transposed. Today every astronomer learns the sequence of types as OBAFGKM, for which the mnemonic is Oh Be A Fine Girl (Guy), Kiss Me!

   The types indicate temperature from about 50,000°F (30,000°C) for type O down to about 4000°F (2000°C) for type M. Colors range from the blue of type O through blue-white (B), white (A), yellow-white (F), yellow (G), orange (K), down to red (M). Types of some well known stars are Rigel (B), Sirius (A), Procyon (F), Sun (G), Arcturus (K), and Betelgeuse (M).

spectrum (*optics*)   The result of passing light through a prism or separating it by wavelength through diffraction is called a spectrum (from the Latin for "image"). Each wavelength of visible light falls at a different place, from red at one end to violet at the other with infrared and ultraviolet regions beyond the ends. The spectrum of a glowing element shows separated regions, each representing a particular change in energy level of an electron. Atomic hydrogen, with just a single electron, has one of the simplest spectrums with about 8 individual lines, but hydrogen gas consists of 2-atom molecules and the spectrum contains dozens of individual lines that run together in a wide band in the violet.

spectrum (*physics*)   The original spectrum ("image") is the rainbow produced by refraction of white light. About 1855 Gustav Kirchhoff and Robert Bunsen discovered that refracted light from each hot, glowing chemical element produces a different pattern of single lines of color, the bright-line spectrum. The bright-line spectrum was soon used to detect new elements. Kirchhoff also recognized that dark Fraunhofer lines in the Sun's spectrum match the colored lines of bright-line spectrums. A refracting camera called a spectroscope reveals not only the elements in stars through their Fraunhofer lines but also, through the Doppler effect, their motions.

   The electromagnetic spectrum is all forms of light, from radio waves to gamma radiation. Any method of spreading radiation to reveal composition is also called a spectrum. The mass spectrometer, invented in 1919, uses a magnet to separate a beam of ions by mass, producing a spectrum that reveals the elements in the beam.

speed (*mechanics*)   Speed is the rate of change of distance with respect to time, ignoring the direction traveled. It can be compared to velocity, a vector quantity that takes direction into account; speed is the absolute value of velocity. Speed is sometimes called rate (r), as in the distance (d) formula $d = rt$ where t is time.

speed of light (*waves*)   The speed of light in a vacuum, c (from *celeritas*, "velocity"), is exactly 299,292,458 m/s since the meter is defined in terms of this speed (about 186,250 mi per sec). James Clerk Maxwell recognized in 1873 that c is always exactly the same. Einstein considered observers moving relative to each other without acceleration for whom *c* would measure the same. He determined that the speed of light is the maximum possible for any object or even any signal, the cornerstone of the special theory of relativity.

Light moving in a substance other than a vacuum travels more slowly than *c*. In water it is about 140,000 mi per sec (225,000 km/s), in glass about 124,000 mi per sec (200,000 km/s), and in diamond about 77,000 mi per sec (124,000 km/s). The ratio of the speed in a vacuum to that in a substance is the index of refraction for the substance (1.33 for water, 1.5 for glass, and 2.42 for diamond).

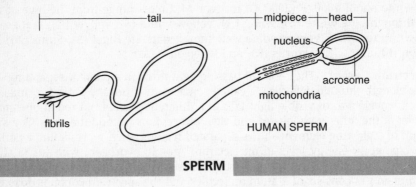

HUMAN SPERM

SPERM

sperm (*cells*) Male sex cells, produced in testes of male organisms, are called sperm, or spermatozoa. A sperm's function is to unite with, or fertilize, an egg from a female. Sperm are usually microscopic in size—typically a tiny fraction of the egg in volume—and usually able to swim in fluid. The human sperm, for example, is about 0.002 in. (0.05 mm) long. It has 3 main parts: the head, a midpiece, and a long tail, or flagellum. Atop the head is an acrosome, which contains enzymes that dissolve a section of the egg's membrane, allowing fertilization to occur.

In mammals, the formation of sperm, called spermatogenesis, occurs continuously throughout adult life. However, it is cyclical in mammals with specific breeding seasons; in these mammals, spermatogenesis ceases during the nonbreeding season.

spermatophore (*zoology*) In certain species of animals with internal fertilization, the male does not deposit sperm directly into the female's body. Rather, he deposits his sperm in a packet called a spermatophore. Depending on the species, the female either takes the packet into her reproductive system or admits its tip, so that the sperm can enter her body. For example, most male salamanders and newts deposit a spermatophore on the ground and perform a courtship dance. The female picks up the spermatophore and inserts it into her cloaca, where it is stored until her eggs are ready to be fertilized. Many snails, mites, and scorpions also produce spermatophores.

Sphagnum (*plants*) Plants of the genus *Sphagnum* are mosses sometimes called peat moss. They grow in bogs, marshes, and other wet places, where they may form dense mats of vegetation. A *Sphagnum* plant may grow to 12 in. (30 cm) tall. It has a branching shoot covered with many leaflike structures composed of 2 types of cells: narrow chlorophyll-containing cells and wide porous cells that can absorb and store large amounts of water.

sphere (*geometry*) The fundamental meaning refers to a sphere in 3 dimensions, where the sphere is the set of points equidistant from a given point (called the center). Mathematicians extend the same idea to "spheres" in any dimension; for example, a sphere in 2 dimensions is a circle. Spheres are often used as neighborhoods; a point in the neighborhood of a given point is within a sphere that surrounds the given point. The volume of a 3-dimensional sphere of radius r is $\frac{4}{3}\pi r^3$ and its surface area is $4\pi r^2$. (*See also* **ball**.)

spider (*animals*) Spiders differ from their relatives the insects in important ways. Spiders have a 2-part body: a head-thorax region called the cephalothorax and an abdomen, with a narrow waist connecting the 2 regions. They have no antennae but most have 8 simple eyes. Spiders have 4 pairs of legs and no wings. In front of a spider's mouth is a pair of jaws called chelicerae; these have movable fangs used to stab prey. In almost all species, poison glands are connected to the fangs. A spider has silk glands in its abdomen that secrete silk, which may be used for ballooning or to weave spider webs. On its legs are sensory hairs that can detect even the slightest vibrations.

Some 35,000 different species of spiders have been identified. They make up the arthropod order Araneae in the class Arachnida (named after Arachne, a peasant girl who, according to Greek legend, challenged the goddess Athena to a spinning contest). Spiders range in size from about 0.02 in. (0.5 mm) to large tarantulas of South America with leg spans exceeding 10 in. (25 cm).

spider web (*ethology*) Not all spider species spin webs. Those that do use their own unique design, adapted to the environment and needs of the species. Orb weavers make webs that resemble pinwheels, with many spiral turns crossed by "spokes" that meet in the center. Sheet-web spiders weave broad 2-dimensional webs, often in grass or close to the ground. Tangled-web spiders weave messy looking webs that, once abandoned, become dusty "cobwebs." Purse-web spiders weave tubelike nests. And ogre-faced spiders weave tiny rectangular webs that they hold with front legs until they come upon an insect; the spiders then stretch the web and toss it over the insect.

spin (*particles*) Spin, as proposed by Wolfgang Pauli [German: 1900–58] in 1924, is a quantum number that describes electrons in an atom. It gives each charged particle a north and south pole, related to the poles of a magnet, and is only apparent in a magnetic field. A spinning charged particle sets up its own magnetic field according to classical electrodynamics, so the electron is pictured as rotating. The quantum number spin has only 2 values, often called up and down, corresponding to spinning clockwise or counterclockwise as viewed from the north pole.

Spin is also a fundamental property of all particles, including atoms. This intrinsic spin acts like the speed of rotation and has the same dimensional units as angular momentum. Electrons and all other fermions have a spin of $\frac{1}{2}$, while bosons have integral spin.

Physicists also use isotopic spin (sometimes called isospin and, rarely, isobaric spin), which has nothing to do with isotopes or ordinary spin. Isotopic spin is a quantum number whose properties in calculations are similar to those of spin. It accounts for the different types of hadrons and was an important part of the steps leading to the discovery of quarks.

spinal cord (*anatomy*)   The spinal cord is a hollow cylinder of nerve tissue lodged within and protected by the backbone. It extends down the back from the brain and is connected to muscles, glands, and other body parts by 31 pairs of spinal nerves. Each spinal nerve is attached to the spinal cord by 2 roots: one that carries impulses from body organs to the spinal cord and a second that carries impulses from the spinal cord to the organs. A major function of the spinal cord is to transmit impulses between the brain and the body. The spinal cord also relays automatic responses called reflexes.

spine (*biology*)   Hard, sharp spines, present on a variety of plants and animals, are generally defensive structures, designed to protect against hungry animals. Many plant spines, such as those of cacti, are modified leaves; they should not be confused with thorns, which are modified stems. Plant spines also can grow on fruits, such as that of the durian tree, and on leaves, such as those of thistles. In the animal kingdom, spines are found on both invertebrates (sea urchins, starfish) and vertebrates (stickleback fish, porcupines).

spinneret (*anatomy*)   Near the tip of a spider's abdomen are up to 3 pairs of fingerlike appendages called spinning organs, or spinnerets, that form threads of silk. Each spinneret has hundreds of microscopic tubes through which liquid silk flows to the outside, hardening on exposure to air.

spiral (*geometry*)   Several plane curves are called spirals. Their common characteristic is that they can be pictured as the path of a point moving in an ever-widening curve; that is, any point is always farther from the initial point than the points before it. The different spirals—Archimedian, hyperbolic, logarithmic, and parabolic—are usually defined with equations in polar coordinates that describe the relationship of a line segment from the initial point to a point on the spiral and the angle that segment makes with some other line. In the logarithmic, or equiangular, spiral the angle between the segment to a point and the tangent at that point has a constant measure.

spirillum (*monera*)   Coiled and spiral-shaped bacteria are called spirilla. They occur singly or in chains—*Vibrio,* for example, are tiny, comma-shaped spirilla that often are linked in long spirals. All spirilla have flagella for movement. Most spirilla are decomposers but others are parasitic. *Bdellovibrio* species are unique in being parasitic on other bacteria.

spirochete (*monera*)   Spirochetes are spiral-shaped bacteria with special flagella called periplasmic flagella. These flagella do not extend outward from the cells as do ordinary flagella but rather wrap around the cells. The periplasmic flagella enable spirochetes to move with a unique twisting motion. Some species of spirochetes are saprophytes that live in water and mud. Others, such as *Treponema pallidum*, which causes syphilis, are parasites of animals, including humans.

spleen (*anatomy*) Present in the abdominal cavity of most vertebrates, the spleen is a spongy, reddish-purple organ made of cells similar to those found in lymph nodes. However, the spleen is filled with blood rather than lymph. (In humans the entire blood supply passes through the spleen about once every 90 minutes.) The spleen produces some kinds of white blood cells, serves as an emergency reservoir of red blood cells, and removes from the blood and breaks down defective and worn-out red blood cells.

sponge (*animals*) The simplest, most primitive of all animals, sponges are aquatic invertebrates that live attached to the bottom, mainly in marine waters. The more than 3000 species, ranging from less than 0.4 in. (1 cm) to more than 6 ft (180 cm) in diameter, constitute the phylum Porifera ("bore bearers"). A typical sponge has many tiny holes, called incurrent pores, on its sides. Water enters the sponge through these pores. Specialized collar cells capture food particles in the water. The water is then expelled through a large opening at the top. The pores and internal passageways are kept open by an internal skeleton of either hard, needlelike spines called spicules or a more flexible protein, spongin, depending on the species.

spontaneous combustion (*chemistry*) Spontaneous combustion is any instance of a substance initiating fire without heat from an outside source. Reactive elements such as sodium or phosphorus burst into flame when exposed to oxygen in air. Slower reactions that produce heat can lead to spontaneous combustion when heat is trapped by insulation. Green hay in large piles accumulates heat caused by bacterial action (fermentation) in central parts of the stack, but fire does not occur because oxygen cannot penetrate. If the stack is opened, however, spontaneous combustion occurs.

spontaneous generation (*biology*) It was long believed that living things arose spontaneously from nonliving materials. Aristotle taught that flies arose from decaying matter. Some people believed that snakes formed from horsehairs that fell into water. And some said that toads formed from the mud on the bottom of ponds. Work in the 17th and 18th centuries by Francesco Redi, Lazzaro Spallanzani, and others discounted the theory of spontaneous generation of life, or abiogenesis. It wasn't until the 1860s, however, when Pasteur demonstrated that microorganisms could arise only from other microorganisms, that the theory was finally overthrown.

sporangium (*biology*) An asexual reproductive structure found in algae, fungi, and plants, in which spores are produced and stored, is called a sporangium. The round, black cases that appear on bread molds such as *Rhizopus* are an example; each of these sporangia produces about 50,000 spores. The brown dots on the underside of fern fronds are clusters of sporangia called sori.

spore (*biology*) Fungi, ferns, and many other organisms reproduce by means of small, usually 1-celled, structures called spores. Spores often are produced in great numbers—millions of spores are discharged from the underside of a mushroom cap, for example. Spores of many species have thick, resistant coats that protect against unfavorable conditions, such as dryness or extreme temperatures. When conditions become favorable, perhaps after the passage of months or years, a spore germinates and gives rise to a new organism.

sporophyte and gametophyte (*botany*)   During its life cycle, a plant has 2 different stages. This alternation of generations involves an asexual sporophyte generation and a sexual gametophyte generation. The sporophyte produces spores, which develop into gametophytes. The gametophytes produce eggs and sperm, which following fertilization grow into sporophytes. In the simplest plants, such as mosses, the gametophyte is the dominant generation. In most other plants, the sporophyte is dominant. In a flowering plant, for example, the male gametophyte is the pollen tube, which contains the sperm, and the female gametophyte is the embryo sac, which contains the egg; all the rest of the plant—roots, stem, leaves, etc.—is the sporophyte.

sporozoan (*protists*)   Sporozoans are parasitic protozoans that produce spores at some stage in their often complex life cycles. They lack cilia and flagella and depend on their hosts for transportation. Best known are *Plasmodium* species that cause malaria in many kinds of birds and mammals. Transmitted by female *Anopheles* mosquitoes, the *Plasmodium* spend part of their life cycle in red blood cells feeding on hemoglobin.

spring (*geology*)   Any place where water naturally flows out of soil or rock is called a spring. Some springs occur because land is below the water table, and water finds its way through an opening. Where water flows or seeps below a layer of underground rock, pressure from the weight of the overlying rock can force water upward through a crack, forming a spring far above the water table. In a region with Karst topography, an underground stream may surface as a large spring. Hot springs occur when water flows from great depths where rock is warm or in volcanic regions where magma is near the surface.

springtail (*animals*)   Insects of the order Collembola are named for their unique abdominal appendage, the furcula. Pushing the furcula forcefully against the ground causes the insect to catapult into the air. Another unique abdominal appendage, the tubelike collophore, helps a springtail cling to surfaces. Springtails rarely exceed 0.2 in. (0.5 cm) in length. Most live in the soil but others are adapted to different environments, including the snow-flea (*Achorutes nivicola*), which is sometimes abundant on snow in early spring.

sprite (*meteorology*)   Sprites are very short, mostly red, flashes that appear directly above thunderstorms during lightning strokes. They are usually brightest around 40 to 45 mi (65 to 75 km) above the ground. Although reported occasionally earlier, their existence was not accepted until a sprite was accidentally caught in a photograph in 1989. Sprites occur above only about 1 in 100 lightning strokes, appearing and disappearing so fast that they are difficult for humans to observe; however, they can be captured on film or video.

spruce (*plants*)   Members of the pine family, spruces (*Picea*) are tall evergreen trees native to cooler regions of the Northern Hemisphere. The trees have a pyramidal shape with branches arranged in tiers. Several species, including the Norway spruce (*P. abies*) of Europe, can grow to 200 ft (60 m). Spruces typically have 4-sided needles, which are arranged in spirals around twigs. The seed-forming cones hang downward, usually at the tips of branches.

spurge (*plants*)   Most members of the spurge genus, *Euphorbia*, are succulent cactuslike plants native to warm temperate and tropical regions. The best-known species is the poinsettia (*E. pulcherrima*), an erect shrub native to tropical America. The poinsettia's white, pink, or red "petals" actually are modified leaves called bracts, which surround clusters of small yellowish or greenish flowers.

The spurge family, Euphorbiaceae, comprises more than 7000 species of herbs, shrubs, and trees of greatly varied form and habitat. Some live in deserts, others float on water. Some are minute forms or clinging vines; others are trees that attain heights of 200 ft (60 m). Almost all produce a milky sap (poisonous in some species) and have tiny unisexual flowers that lack petals and, frequently, sepals. The family includes cassavas (*Manihot*), crotons (*Croton*), the rubber tree, the candlenut tree (*Aleurites moluccana*) of Asia, and the sand box tree (*Hura crepitans*) of Central and South America.

square and square root (*numbers*)   Multiplying a number by itself produces a square of that number (so called because it is a figurate number), which can be indicated with the exponent ^{2}. For real numbers the square is always positive; not only does $3^2 = 9$ but also $(-3)^2 = 9$, for example.

For a given positive number p, the square root, r, is a number that, when multiplied by itself, has the product p, or $r^2 = p$. In that case, $r = \pm \sqrt{p}$. For example, the 2 square roots of 9 are $+3$ and -3.

squid (*animals*)   Found in marine waters worldwide, squids are cephalopod mollusks with 8 sucker-bearing arms and 2 longer tentacles around the mouth. The tentacles, which can contract, are used to seize prey and draw it toward the mouth, where it is held by the arms while torn apart by strong jaws. Squids have a streamlined body and are the fastest invertebrate swimmers. They move through the water by jet propulsion, drawing water into the mantle cavity and then expelling it with great force. When attacked, squids secrete a dark ink that clouds the water, giving them a chance to escape.

squirrel (*animals*)   Inhabiting all continents except Antarctica, squirrels are rodents of Family Sciuridae. Tree squirrels, such as the red squirrel (*Sciurus vulgaris*) of Europe and Asia and the gray squirrel (*S. carolinensis*) of North America, spend much of their time in trees. They typically have a long, richly furred tail that is held curled over the back. Ground squirrels include chipmunks, marmots, and prairie dogs. They make their homes in underground burrows.

Tree squirrels and ground squirrels are active during the day. In contrast, flying squirrels are nocturnal. They have a loose membrane of skin extending along each side of the body, between the front and hind limbs; when stretched out, this membrane enables a squirrel to glide from tree to tree.

stalactite and stalagmite (*geology*)   Water seeping through limestone into underground caves and caverns forms icicle-shaped formations from calcite that precipitates from the solution. The rock formed this way is called dripstone. Hanging from the roof are dripstone stalactites, while similar elongated cones, called stalagmites, form point-up on the floor.

standard condition (*measurement*) In SI measurement standard conditions are a temperature of 273.15 K and a pressure of 101,325 pascals, equivalent to the more familiar temperature of 32°F (0°C) and pressure of 1 atmosphere (about 1.013 bar).

standard deviation (*statistics*) The standard deviation of a set of numbers is the positive square root of the variance. The standard deviation, symbolized as σ, is especially useful for measurements that follow a normal distribution since about 68% of all the measurements will be within a distance of 1 standard deviation from the average (arithmetic mean), about 95% will be within 2 standard deviations, and virtually all will be within 3 standard deviations.

standard model (*particles*) The theory of particle physics established by a century of experiments is called the standard model. While continually challenged by new experiments and by theories designed to extend it, the standard model remains the basis for understanding high-energy interactions such as those in cosmic rays, thermonuclear reactions in stars or bombs, the origin of the universe, and events inside particle accelerators. The standard model is based on a branch of mathematics known as group theory that is used to order elementary particles into 3 generations of 2 leptons and 2 quarks each, with these particles connected to each other by color and electroweak force mediated by 4 particles. The main recognized flaw is that the standard model does not include gravity.

standard time (*geography*) When travel was slow, local time was kept by height of the Sun or passage of the stars. With the advent of faster transportation, notably railroads, adjustments to local time became complicated. Standard time was introduced.

Earth turns about 15° each hour. Boundaries for time zones were set approximately every 15° of longitude, and time considered the same within each zone. Clocks are set forward or back by exactly 1 hour leaving one zone and entering the next.

Staphylococcus (*monera*) Members of the genus *Staphylococcus* are gram-positive spherical bacteria (cocci) that often occur in clusters (*staphylo* is derived from the Greek for "bunch of grapes"). They are tiny, lack flagella, and are nonmotile. Many cause diseases of humans and other animals. *S. aureus* is a normal inhabitant of human skin and mucous membranes but can cause pneumonia, septicemia (blood poisoning), abscesses, and other ailments in hospital patients and others whose immunity is lowered.

star (*stars*) A star is a body whose mass is so great that atoms in its interior are forced together so closely that nuclei fuse. Fusion releases energy as heat and as subatomic particles, such as neutrons and high-energy photons (gamma rays), which produce further heat. Atoms above the fusion zone become ionized and also form convection currents, producing strong magnetic fields. The fusion energy eventually reaches the surface of the body, dissipating in part as heat and light. The light from some stars is so intense that we can see it with the naked eye even though it is thousands of light-years away.

Ordinary stars range in size from brown dwarfs 10 to 80 times the size of Jupiter to supergiants a thousand times the size of the Sun. Neutron stars can be much smaller, perhaps 10 mi (16 km) in diameter.

25 Brightest Stars as Seen from Earth

COMMON NAME	ASTRONOMICAL NAME	ABBREVIATION	APPARENT MAGNITUDE	ABSOLUTE MAGNITUDE	DISTANCE IN LIGHT-YEARS	SPECTRAL TYPE
1. Sirius	Alpha Canis Majoris	α Cma	−1.46	1.42	8.6	A
2. Canopus	Alpha Carinae	α Car	−0.72	−5	74	F
3. Rigil Kentaurus	Alpha Centauri (binary)	α Cen A α Cen B	−0.01 1.33	4.37 5.71	4.3	G
4. Arcturus	Alpha Boötis	α Boo	−0.04	−0.10	34	K
5. Vega	Alpha Lyrae	α Lyr	0.03	0.65	25.3	A
6. Capella	Alpha Aurigae (binary)	α Aur A	0.08	−0.40	41	G
7. Rigel	Beta Orionis	β Ori	0.12	−7	815	B
8. Procyon	Beta Canis Minoris	β CMi	0.38	2.71	11.4	F
9. Achernar	Alpha Eridani	α Eri	0.46	−1.7	69	M
10. Betelgeuse	Alpha Orionis	α Ori	0.50 (variable)	−7	425	M
11. Hadar	Beta Centauri	β Cen	0.61	−4.3	320	A
12. Altair	Alpha Aquilae	α Aql	0.77	2.30	16.8	A
13. Aldebaran	Alpha Tauri	α Tau	0.85	−0.49	60	K
14. Spica	Alpha Virginis	α Vir	0.98	−3.2	230	B
15. Antares	Alpha Scorpii (binary)	α Sco	0.96	−5.4	220	M
16. Pollux	Beta Geminorum	β Gem	1.14	1.00	40	K
17. Fomalhaut	Alpha Piscis Austrinus	α PsA	1.16	2.02	22	A
18. Deneb	Alpha Cygni	α Cyg	1.25	−7.2	1630	A
19. Mimosa	Beta Crucis	β Cru	1.25	−4.6	460	B
20. Regulus	Alpha Leonis (binary)	α Leo	1.35	−0.38	69	B
21. Adhara	Epsilon Canis Majoris	ε CMa	1.50	−4.9	570	B
22. Acrux	Alpha Crucis (binary)	α Cru	1.58	−3.8	510	B
23. Castor	Alpha Geminorum	α Gem	1.58	0.72	46	A
24. Gacrux	Gamma Crucis	γ Cru	1.63	−2.5	120	M
25. Shaula	Lambda Scorpii	λ Sco	1.63	−3.3	325	B

starch (*biochemistry*)   The main food storage compound in plants, starch is a polysaccharide formed from hundreds to many thousands of glucose units. Most starch molecules have 2 types of components: amylose, which is a straight chain of glucose units, and amylopectin, which consists of branched chains of glucose units. Starch is most abundant in storage tissues such as roots (cassava), tubers (potatoes), rhizomes (arrowroot), and grains of cereal crops (wheat). Enzymes can rapidly break down starch to yield glucose and energy. In animals surplus sugar is stored as glycogen, sometimes called animal starch.

starfish (*animals*)   Echinoderms known as starfish or sea stars have a flattened body with a central disk from which radiate 5 or more tapering arms. Numerous tube feet protrude on the underside of the arms. Each tube foot has a sucker at its outer end. The tube feet are used for locomotion, to hold onto surfaces, and to capture prey. Using its tube feet, a starfish pulls on the shell of a clam or other mollusk until the shell opens. Then the starfish turns its stomach inside out and releases digestive juices that dissolve the prey. It then slurps up the digested material, retracts its stomach, and leaves behind an empty shell. The largest known starfish is the deep-sea *Midgardia xandaros*, which can exceed 4 ft (1.2 m) in diameter.

Starling, Ernest (*physiologist*)   [English: 1866–1927] Together with William Bayliss [English: 1860–1924], Starling discovered that the intestinal lining produces a substance that stimulates the pancreas to secrete fluid. Starling and Bayliss named the substance secretin and coined the term hormone–from the Greek word meaning to set in motion—for a substance produced in one part of the body that regulates another tissue or organ. Starling's pioneering work on the heart led him to conclude that the energy expended during contractions of the heart depends on the length of muscle fibers in the heart walls.

starling (*animals*)   Perching birds of the starling family, Sturnidae, are native to temperate and tropical regions of the Old World. They range in length from 7 to 17 in. (18 to 43 cm). The plumage typically is a glossy black, with 10 flight feathers on each wing. The diet consists of insects and plant matter. Most starlings are gregarious and form large, noisy flocks. Some species, including the mynahs, can imitate the songs of other birds.

static electricity (*electricity*)   Static electricity, or electrostatics, refers to electrical phenomena caused when an object has a large excess or deficit of electrons, resulting in the object having either a negative or positive charge. The charge sets up an electric field that affects other charged bodies. Objects with the same charge repel each other, while those with opposite charges attract.

Electrons can be removed from 1 body and placed on another by rubbing; rubber stroked with fur collects electrons from the fur, while glass polished with silk loses electrons to the silk. Electrons from a body charged by rubbing can be transferred to a conductor by bringing the charged body near it or touching it (a process known as electric induction, which is not the same as electromagnetic induction). Static electricity can reach high levels in a Van de Graaff generator, a device in which electric current leaks electrons onto a moving belt that transfers them to a metal sphere.

statistics (*mathematics*)   Statistics consists of the branch of mathematics that corresponds to applied probability combined with methods for taking a sample and for analyzing and presenting data. Because of the emphasis on obtaining and interpreting data, statistics extends beyond the borders of mathematics. The application of probability to a sample, which is pure mathematics, is sometimes labeled inferential statistics.

steady-state theory (*cosmology*)   A small group of astronomers, led mainly by Sir Fred Hoyle [English: 1915–], have argued against the big bang theory of the universe. Hoyle and different collaborators have proposed various versions of a steady-state theory in which the universe has no beginning and expansion is caused by the creation of new matter.

steam (*materials science*)   The steam seen above a boiling liquid consists of small drops of water that have condensed from water vapor arising from the hot liquid; it could be called hot fog. The steam that powers a steam engine is water vapor formed by boiling mixed with heated air. At the boiling point water vapor occupies 1700 times the space of an equal mass of water, providing most of the expansion of gases that causes a steam engine to function.

steel (*materials science*)   Steel is iron alloyed with less than 1.7% carbon (often less than 1%), making steel both stronger and harder than cast or forged iron. Since the 19th century steel has been the metal of choice for applications requiring toughness and strength.

Steel was first made by lucky accident, leading to heuristic forging processes. In the 1600s forcing carbon into iron to produce an outer layer of steel, called blister steel, became the first step in steelmaking. In the 18th century René de Réaumur [French: 1683–1757] identified the role of carbon in steelmaking. Improved manufacturing processes have slowly developed since.

Stegosaurus (*paleontology*)   The best-known member of the plated dinosaurs, *Stegosaurus* had large bony plates aligned along the midline of its back and several pairs of long spikes on its tail. The function of the bony plates is uncertain. Like the spikes, they may have been used in defense. Or they may have had an extensive network of blood vessels and been used to regulate body temperature. *Stegosaurus* grew to about 20 ft (6 m) in length and weighed about 4 tons—but its brain was no bigger than a walnut. It lived about 140 million years ago.

stellar evolution (*stars*)   A star forms when a gas cloud condenses enough to initiate fusion. The energy of fusion pushing outward keeps the star from collapsing beyond a certain radius. When all the elements suitable for fusion have been used as fuel, however, the star again collapses. Sometimes the energy overpowers gravity and the star expands or even explodes.

Stellar evolution is determined by its mass. A star about the size of the Sun begins to shine even before fusion starts. When hydrogen begins to fuse, the star enters a nearly stable main sequence phase that could last 10,000,000,000 years, but eventually, as helium builds up at the core, the star expands to a red giant. When the helium begins to fuse into carbon, the star contracts again, but as the carbon builds up, the star again expands. If it becomes so great that the escape velocity at its surface becomes 0, gas begins to flow away, forming a planetary nebula. The part remaining shrinks again to a white dwarf, which will burn itself gradually into a cool cinder over millions of years.

A star 10 or more times as massive as the Sun can continue beyond fusing carbon in a sequence that ends with the fusion of iron. Iron is the end of the line for fusion caused by gravity: the star collapses and explodes at the same time, forming a supernova or perhaps a black hole.

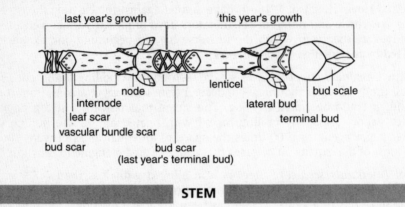

last year's growth · this year's growth

node · lenticel · bud scale · internode · lateral bud · leaf scar · terminal bud · vascular bundle scar · bud scar · bud scar (last year's terminal bud)

STEM

stem (*botany*) When a seed germinates, the part of the plant that emerges above the soil is the shoot; this grows into the stem. The major functions of stems are to support leaves and reproductive structures and to carry materials such as water from roots to leaves.

There are 2 main types of stems: herbaceous and woody. Herbaceous stems are soft and green, with comparatively little growth in diameter; the stems of annual plants and many biennial plants are herbaceous. Woody stems are rigid and indicate growth—both lengthwise and in diameter, and including the formation of wood—over a period of more than 2 years. The surface features of twigs (small woody stems of trees and shrubs)—including buds, lenticels, and nodes—are distinctive and often can be used to identify species.

In addition, there exists a variety of specialized stems. Tubers and other underground stems are often organs of food storage and vegetative propagation. Other modified stems include tendrils, thorns, and rhizomes.

stem cell (*cells*) Cells that can generate various types of cells are called stem cells. For example, embryonic stem cells, found in embryonic tissue, can differentiate into all types of cells. Hematopoietic stem cells, produced in bone marrow, can differentiate into blood cells. Mesenchymal stem cells, also produced in bone marrow, can differentiate into muscle, bone, cartilage, and tendon cells.

step function (*analytic geometry*) A step function maintains the same value for a continuous part of its domain, jumps to a different value for the next continuous part, and repeats this pattern throughout its domain. A familiar example is postal rates, which are the same for any weight less than 1 ounce, jump to a different amount between 1 and 2 ounces, and continue to rise in several steps.

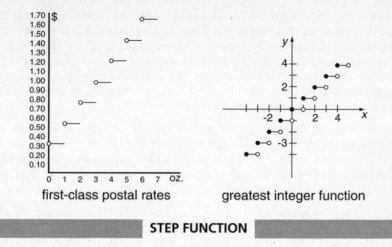

first-class postal rates   greatest integer function

STEP FUNCTION

The greatest integer function (defined as the greatest integer less than a given integer) is a slightly different type of step function.

steppe (*ecology*) The temperate grassland of Eurasia is called steppe—a term that sometimes is also applied to the short-grass prairies of North America, the pampas of South America, and the veld of southern Africa. The Eurasian steppe exhibits zonation; it includes a wooded steppe with deciduous trees and rainfall of more than 16 in. (41 cm) annually; tillable steppe, with rich black soil and rainfall of 10 to 16 in. (25 to 41 cm) annually; and semiarid steppe on the fringes of deserts, with less than 10 in. (25 cm) of rain annually.

steroid (*biochemistry*) A large family of lipids, steroids include many compounds essential for life. A number of hormones, including those secreted by the adrenal cortex and sex glands, are steroids. Bile salts, which emulsify fats in the small intestine, are steroids. Fatty alcohols, or sterols, of which cholesterol is the best known, also are commonly classed as steroids.

A steroid molecule differs from other lipid molecules in consisting of 4 attached carbon rings; in addition, the molecule may have a hydrocarbon side chain, or "tail."

stimulus-response (*physiology*) Any change in the environment that can be detected by an organism is called a stimulus; any reaction to the stimulus is called a response. The strength and speed of the response generally depends on the strength of the stimulus and the nature of the organism. For instance, plants usually respond much more slowly than animals. Also, a specific stimulus typically causes a specific response: food in a person's mouth causes the secretion of saliva, while a sudden loud noise causes the person to clap hands over ears.

stinkbug (*animals*) Named for the foul-smelling secretion of their stink glands, stinkbugs (Family Pentatomidae) are primarily plant feeders. Some are serious pests, including the harlequin bug (*Murgantia histrionica*), which attacks plants of the cabbage family.

stochastic process (*probability*)   The word stochastic, deriving from a Greek word for guessing, means random, so a stochastic process is a sequence of random events. The simplest examples of stochastic processes are Markov chains, sequences of events in which the probability of one event depends on the outcome of the immediately preceding event. An example is the location after each step of a random walk for which the probability of stepping in one direction is different from that of the other direction.

stolon (*botany*)   A stolon, or runner, is a specialized stem that grows along the surface of the ground. When the tip or a node of the stolon touches the ground it roots and forms a daughter plant, which eventually produces stolons of its own. Strawberry plants and saxifrage are examples of plants that spread by this type of vegetative propagation.

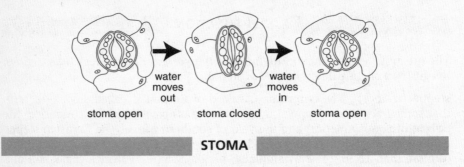

water
moves
out

water
moves
in

stoma open   stoma closed   stoma open

STOMA

stoma (*botany*)   Water vapor and other gases move between the environment and the inside of a leaf through pores called stomata. Surrounding each stoma are 2 guard cells. As the guard cells absorb water, they become curved, opening the stoma. As they lose water, they straighten, closing the stoma. Typically, stomata are numerous on the lower surface of leaves and sparse on the upper surface. A single sunflower leaf may have 2,000,000 stomata on its lower surface! Generally, stomata are open during the day, when photosynthesis takes place, and closed at night to limit water loss through transpiration.

stomach (*anatomy*)   A muscular organ of the gastrointestinal tract, following the esophagus and leading into the intestines, the stomach has important digestive functions: it serves as a storage space for food, and contraction of muscles in its walls crush and mix food with digestive juices that chemically break down some materials.

The design and secretions of stomachs differ greatly among animals. The human stomach is a saclike organ shaped like a gourd (the shape changes as a person stands or lies down). A ruminant's stomach is divided into 4 compartments: in the first, bacteria digest plant cellulose; in the second and third, both fermentation and digestion take place; in the fourth, further digestion occurs. A bird's stomach has 2 compartments: a glandular stomach (proventriculus) where digestive juices function and a muscular stomach (gizzard) where food is ground, often aided by sand or pebbles.

Stone Ages (*paleoanthropology*)   In 1816 Christian Jurgensen Thomsen [Danish: 1788–1865] began to classify early artifacts as belonging to the Stone, Bronze, or Iron Age, terms that have been used since, although with considerable refinement and modification. The Stone Age has been subdivided, initially in 1865 by Sir John Lubbock [English: 1834–1913] into an Old Stone Age, or Paleolithic Age before about 9000 BC, and a New Stone Age, or Neolithic Age, that predates copper smelting, which started about 4500 BC. Some paleoanthropologists insert a Middle Stone Age, or Mesolithic, that corresponds to the transition between the Paleolithic and the Neolithic Revolution.

stonefly (*animals*)   Members of the insect order Plecoptera are commonly called stoneflies because the immature nymphs typically live clinging to stones in well-oxygenated streams. The adults remain near their breeding grounds. Small and drably colored, with 4 membranous wings, most stoneflies are poor fliers. They feed on algae, lichens, and plants.

stork (*animals*)   Native to tropical and warm temperate regions worldwide, storks are big birds with a long pointed bill, long neck, long broad wings, and long legs with partly webbed toes. The largest species are about 4 ft (1.2 m) tall, with a wingspan up to 8.5 ft (2.5 m). Storks feed on fish, frogs, insects, and other small animals. Some species, such as the white stork (*Ciconia ciconia*) of Eurasia—legendary as the deliverer of babies—migrate in flocks.

straight angle (*geometry*)   A straight angle is recognized by identifying the 2 parts of 1 line on each side of a point as rays. Alternatively, rotating a ray about its endpoint until its terminal position forms a line with the initial position produces a straight angle. Degree measure is defined by setting the straight angle equal to 180°, so 1° is 180th of a straight angle. In radian measure, the definition of radian implies that the straight angle equals π radians.

strain (*mechanics*)   Strain is a measure of the change in length (or another measure, including volume) of a solid produced by applying stress. Strain is the ratio of the original measurement to the measurement after stress has been applied, so it is always a pure number without a unit of measure attached. Shear results in strain that compares amount of twisting to the original shape.

strangeness (*particles*)   Strangeness is a property conveyed to hadrons by possession of the strange quark or antiquark. When new particles were discovered in cosmic rays in the 1940s and 1950s, most decayed in ways that were so different from those in theory that they were called strange particles. In 1953 Murray Gell-Mann described a new quantum number called strangeness whose properties account for the odd decay rate. In 1964 Gell-Mann identified the third quark as strange; the presence of a strange quark conveys strangeness on a subatomic particle.

strange quark (*particles*)   The original third quark in the 1964 theory of Murray Gell-Mann is called strange because Gell-Mann himself had in 1953 introduced a concept he called strangeness. Strangeness explains why heavy particles found in cosmic rays last longer before decaying into familiar particles than previous theories predicted. In the quark theory Gell-Mann accounted for strangeness with a heavy quark that decays into the lighter up and down quarks.

stratigraphy (*geology*)   Stratigraphy is the use of the layered nature of rock (rock layers are called strata) to understand the past. In the 18th and 19th centuries geologists developed a number of tools based on strata. All strata are originally horizontal. If undisturbed, lower strata are older than those above them. Strata that contain the same set of fossils have the same age. When strata are disturbed so that they cross or cut each other, working backward to an original state reveals age relationships. In the 20th century new stratigraphic tools have been developed, including isotopic dating, paleomagnetism, and plate tectonics.

stratosphere (*atmospherics*)   A steady decline in temperature with height (the lapse rate) characterizes the lower atmosphere. In 1902 Léon Philippe Teisserenc de Bort [French: 1855–1913] showed that this decline ceases about 10 mi (15 km) above Earth's surface. He named the lower part of the atmosphere the troposphere ("sphere of change") and the part above the stratosphere ("sphere of layers") on the false assumption that air at that height would separate by density. The stratosphere actually becomes warmer with height because of the formation of ozone by ultraviolet radiation, a process that captures radiation. The heating halts at about 30 mi (50 km) when air becomes too thin for ozone production.

strawberry (*plants*)   Members of the rose family, strawberries (genus *Fragaria*) are low-growing perennial herbs. The compound leaves have 3 toothed leaflets. Each small white flower develops into an edible red "berry" more properly called an accessory fruit; it consists of an enlarged flower receptacle with tiny 1-seeded fruits embedded in its surface. Most strawberry plants produce long stems called runners that grow along the surface of the soil and root at nodes to create new plants.

Streptococcus (*monera*)   Members of the genus *Streptococcus* are gram-positive spherical bacteria (cocci) that often occur in chains. Most species are parasitic; many are serious pathogens of humans and other mammals. *S. pneumoniae* causes pneumonia, meningitis, and ear infections. *S. mutans* causes tooth decay. But *S. lactis* is used in the dairy industry to make cheese.

stress (*mechanics*)   Stress is the same as pressure when a force is pushing parts of a solid together; it is measured in the same units even when pulling a solid apart. Thus stress, like pressure, is applied to every part of a body and measured in lb per sq in. (N/m^2 or Pa). A difference between stress in a solid and pressure in a fluid is that pressure in a fluid is exerted in every direction, but the direction of stress varies with the situation. Strain, the effect of stress, is often confused with stress.

stridulation (*ethology*)   The rubbing together of 2 body parts is termed stridulation. It is the process by which the noisiest insects produce sound. Grasshoppers rub together the edges of their wings, causing resonating areas in the wings to vibrate and produce sounds. Crickets hold their front wings above the back and rub vigorously; by altering the space between the wings and the back the crickets can change the volume and pitch of their songs.

string theory (*particles*)   String theory replaces the particles of the standard model with very short curves, or strings, having 1 dimension, length. Some strings have ends while others form loops. Strings are as short compared to an atom as an atom is to the solar system. Strings exist in 10 dimensions, including the 3 familiar space dimensions of length, width, and height, a time dimension, and 6 other space dimensions that are wound up too tightly to be observed. Strings interact by joining or splitting, described by a 2-dimensional surface in space-time called the *worldsheet* of the string. One reason for the popularity of string theory is that it is mathematically simpler than particle theory. Its main flaw is that no known experiment can establish that strings exist.

strip mining (*environment*)   The use of huge power machinery to scrape away vegetation, soil, and rock to reach coal and other mineral deposits is called surface or strip mining. Stripping destroys wildlife habitats, makes the land vulnerable to erosion and flooding, and pollutes waters. Restoration of stripped land is now required in some places.

stromatolite (*monera*)   Evidence that photosynthesizing organisms lived some 3,500,000,000 years ago comes from rocklike formations called stromatolites. These flat or moundlike formations consist of layer upon layer of hardened sediment loaded with fossils of bacteria, mainly cyanobacteria (blue-green algae). The ancient stromatolites resemble those being formed by bacteria today in Shark Bay, Australia, and several other shallow aquatic environments with high temperatures or high salt concentrations. The cyanobacteria grow as mats of sticky intertwined filaments. Sediment and other debris is trapped in the mats. This kills all but the uppermost cyanobacteria, which grow on top of each other at a rate of 0.04 in. (1 mm) a year. Gradually the deeper layers of sediment harden and become rigid.

strong force (*particles*)   The strong force is the force that holds positively charged protons and neutral neutrons close to each other in an atomic nucleus. In 1935 Hideki Yukawa theorized that this force is produced by the exchange of particles. The particles Yukawa described, now called pions, were observed in 1947, but when the quark theory was developed, physicists recognized that the strong force is a manifestation of the color force between quarks. Protons and neutrons exchange quarks instead of pions, but as the quarks pass each other they form virtual pions.

strontium (*elements*)   Strontium, atomic number 38, was discovered in 1808 by Humphry Davy. It is named for Strontian, a town in Scotland. Nuclear explosions produce a radioactive isotope, strontium-90, that is dangerous because it easily replaces the chemically similar calcium in the body. Strontium can also substitute for other alkaline earth metals, but is rarer and more expensive. Its unique use comes from the bright red color of its flames when substituted for magnesium in fireworks. Chemical symbol: Sr.

Struthiomimus (*paleontology*)   Its small head, long neck, short body, and long legs seemed ostrichlike to the scientists who named this theropod dinosaur *Struthiomimus* ("ostrich mimic"). Standing about 8 ft (2.4 m) high, it grew to lengths of 12 ft (3.7 m) or more, including its long tail. *Struthiomimus* had no teeth, and may have used the strong, curved claws on its hands to tear open eggs or insect nests. It lived during the late Cretaceous.

strychnine (*plants*) The deadly poison strychnine comes from seeds of *Strychnos nux-vomica*. Native to tropical Asia, this evergreen tree grows to about 65 ft (20 m). It produces clusters of fragrant greenish white flowers that give rise to fruits that look like miniature oranges. Related species, native to Central and South America, are the source of the poison curare.

sturgeon (*animals*) Constituting Family Acipenseridae, native to the Northern Hemisphere, sturgeons are primitive fish up to 30 ft (9 m) long. Large bony plates cover much of their body. The skeleton consists of cartilage and the backbone extends into the long upper lobe of the tail. Four sensory barbels hang under the toothless snout. A sturgeon is a bottom-feeder, sucking up small organisms as it moves along. Some species live in freshwater, others are marine but enter freshwater to reproduce.

subatomic particle (*particles*) Although ancient Greeks and 19th-century chemists defined the atom as the smallest particle of matter, since 1897 it has been apparent that there are smaller particles, known as subatomic. The first to be found was the electron, followed by the photon in 1905, the proton in 1911, predictions of the neutrino in 1930, discovery of the neutron in 1932, prediction of the pion in 1935, and recognition of the muon in 1947. These were followed soon after by the discovery of large numbers of subatomic particles. (See table on page 553.) Nearly 200 subatomic particles are known today, a number that doubles in theories of supersymmetry.

subduction (*geology*) When lithospheric plates collide, one of them plunges below the other, a process called subduction of the plate that passes into the mantle. Subduction causes deep ocean trenches, raises mountain ranges, and produces earthquakes and volcanoes.

It appears that the denser plate is pushed under the lighter one. A different idea is that the denser plate is sinking at the plate boundary because it has cooled. Warm plate edges form when new lithosphere is created at a spreading center, such as a mid-oceanic ridge. As part of a plate moves away from the spreading center, it gradually cools and therefore becomes denser. Eventually the edge of the plate is denser than the asthenosphere and mantle below it, so it sinks (subducts). In this view, the edge that is subducting pulls the whole plate along, opening up the spreading center at the other edge of the plate.

sublimation (*materials science*) Some materials change directly from a solid to a gas without first becoming a liquid, a process called sublimation. This is most familiar from frozen carbon dioxide, known as dry ice because it sublimes. Water ice also sublimes slowly at temperatures below 32°F (0°C).

submarine canyon (*oceanography*) Many of the largest canyons on Earth are beneath the oceans along the continental shelf and slope. These submarine canyons originate when turbidity currents scour them along the edges of continents. Some, such as the Hudson River Canyon in the Atlantic, begin where river mouths have been buried by rising seas, but others are not associated with rivers. At the base of a submarine canyon, the sediment from the turbidity current forms an alluvial feature called a fan that spreads out to help form an abyssal plain.

Subatomic Particles

PARTICLE	SYMBOL(S)		CHARGE	MASS (MEV)	AVERAGE LIFE (SECONDS)	SPIN
First Family						
electron	e		−1	0.511	stable	1/2
positron (antielectron)	e̱		+1	0.511	stable	1/2
electron neutrino	v_e		0	less than 0.000015	stable	1/2
up quark	u		+2/3	.04	stable	1/2
down quark	d		−1/3	8	stable	1/2
Second Family						
muon	μ		−1	105.6	0.0000022	1/2
mu neutrino	v_μ		0	less than 0.17	stable	1/2
strange quark	s		−1/3	150	unstable	1/2
charm quark	c		+2/3	1500	unstable	1/2
Third Family						
tauon	τ		−1	1784	0.0000029	1/2
tau neutrino	v_τ		0	less than 20	stable	1/2
bottom quark	b		−1/3	4700	less than 5×10^{-12}	1/2
top quark	t		+2/3	175,000	c 10^{-18}	1/2
Mesons (samples)						
pions	π	u$\bar{\text{d}}$	+1	140	2.6×10^{-8}	0
		u̱d	−1	189.6	2.6×10^{-8}	0
		mix	0	135	8×10^{-17}	0
kaons	K	u̱s	+1	493.7	1.2×10^{-8}	0
		u̱s	−1	493.7	1.2×10^{-8}	0
	K^0_s	mix	0	497.7	9×10^{-11}	0
	K^0_L	mix	0	497.7	5.2×10^{-8}	0
B meson	B_0	d$\bar{\text{b}}$	0	5279	1.65×10^{-12}	0
Baryons (samples)						
proton	p	uud	+1	938.3	stable(?)	1/2
antiproton	p̱	$\overline{\text{uud}}$	−1	938.3	stable (?)	1/2
neutron	n	udd	0	939.6	0.0918	1/2
lambda	λ	uds	0	1116	3×10^{-10}	1/2
charmed lambda	λ^+_c	udc	+1	2285	2.06×10^{-11}	1/2
omega minus	Ω^-	sss	−1	1672	1.3×10^{-10}	1/2
Bosons (samples)						
gluons	g		0	0	stable	1
photon	γ		0	0	stable	1
W bosons	W^+		+1	80,410	10^{-20}	1
	W^-		−1	10^{-20}	1	
Z boson	Z^0		0	91,160	10^{-20}	1

subscript (*mathematics*)   A subscript is a small numeral or variable written partly below the line and to the left or right (most often the right) of a full-size symbol or word. A numerical subscript with a variable indicates a specific number (constant), such as x_1 for a particular value of x. The subscript i indicates subscripts that run through the natural numbers, as in a_i. With C (for combinations), as in $_nC_r$, the subscripts n and r mean "n things r at a time." Subscripts are also used in chemistry: to the right to show numbers of atoms in a molecule (as in H_2O with 2 atoms of hydrogen, H); or to the left for atomic number, such as in $_{17}Cl$.

subset (*sets*)   If all elements in set A are also elements in set B, then A is a subset of B. This definition includes the case where A and B are equal (each set is a subset of itself) and also the case where A or B or both have no members (the empty set is a subset of every set). Proper subsets are all subsets except the set itself.

subspecies (*taxonomy*)   When populations of a species become isolated from one another, perhaps due to geographical barriers, they may gradually develop differences that are significant enough for the populations to be formally recognized as subspecies. For example, chimpanzees are classified as *Pan troglodytes* but scientists recognize three main subspecies that differ in size and other characteristics: the upper Guinean chimpanzee (*Pan troglodytes verus*), the lower Guinean chimpanzee (*Pan troglodytes troglodytes*), and the eastern long-haired chimpanzee (*Pan troglodytes schweinfurthii*).

subtraction (*arithmetic*)   Subtraction of whole numbers can be viewed as removing a part of a set ("take-away" subtraction) or as finding an amount needed to complete a set (subtraction as the inverse of addition). In take-away subtraction the question is "how many are left?" while in subtraction as the inverse of addition the question is "how many more are needed?" For numbers other than natural numbers, subtraction can be formally defined as the inverse of addition, although the take-away idea is still important. For example, if you pour 3.549 gallons of water from a barrel holding 18.7 gallons, the subtraction concept is still take-away.

succession (*ecology*)   The series of changes in an area that results in one community of organisms gradually replacing another is called ecological succession. The first stage encompasses a community of pioneer species able to live on rocks and bare soil. These species change the surface both physically and chemically, making it suitable for other species. The process is repeated over and over again until, eventually, a relatively stable climax community is established. The climax community persists because it does not modify the area in ways that would favor the invasion of other species. An oak forest is a climax community but a pond is not, for as sediment accumulates on the pond bottom, the plants along the edges encroach on the pond.

   If fire, flooding, or other upheaval disrupts an established community, secondary succession occurs. It usually proceeds more quickly than primary succession since the area retains some organic matter, seeds, and other remnants of the previous community.

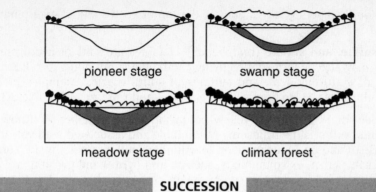

pioneer stage swamp stage

meadow stage climax forest

SUCCESSION

succulent (*botany*) Plants adapted for storing water in swollen leaves or stems are called succulents. *Sedum* and *Aloe* have fleshy leaves in which they accumulate water; *Opuntia* and other cacti have fleshy stems. Succulents are most common in deserts and other habitats where water is in short supply, and in habitats with very salty water, such as salt marshes. Some succulents also conserve water by opening their stomata only at night; this cuts down on water loss by transpiration but it also restricts photosynthesis (food making).

sucker (*animals*) The more than 60 species of the fish family Catostomidae are named suckers for their habit of sucking up mud containing small crustaceans and other invertebrates from the bottom of lakes and rivers. Their mouth is small but has large fleshy lips. Most suckers are native to North America; one species also occurs in Siberia; another lives in China. Most species are less than 24 in. (60 cm) long.

sucker (*biology*) The term sucker refers to a variety of plant and animal structures. Many suckers are designed to attach an organism to either a support or a host; for example, a sheep liver fluke (*Fasciola hepatica*) uses its ventral sucker to attach to a sheep's bile duct. Other suckers, such as the sucker mouth of the sea lamprey (*Petromyzon marinus*), are used both to attach to a victim and to obtain food. The roots or stems of some plants produce sprouts called suckers that give rise to new plants.

sugar (*biochemistry*) Simple sugars, or monosaccharides, are carbohydrates with 3 to 8 carbon atoms in a molecule. They are grouped according to the number of carbon atoms. For example, hexoses—such as glucose, fructose, and galactose—have 6 carbon atoms and the formula $C_6H_{12}O_6$ (the atoms are somewhat differently arranged in each sugar).

Double sugars, or disaccharides, are formed in a reaction that joins together 2 monosaccharides and produces a disaccharide molecule and a molecule of water. The most important disaccharides are sucrose and maltose in plants and lactose in animals; all have the formula $C_{12}H_{22}O_{11}$.

sugarcane (*plants*) A member of the grass family native to Asia, sugarcane (*Saccharum officinarum*) is a tropical perennial that grows 8 to 24 ft (2.4 to 7.3 m) tall. It is cultivated for its stems, or canes, which contain a sugary pulp. The plant has long, tapering leaves and, at the top of each stem, a large plume of flowers.

sulfide, sulfite, and sulfate (*compounds*)   Sulfides are salts of hydrogen sulfide; they may also be formed directly by combining sulfur and a metal. Sulfites are salts of sulfurous acid, and sulfates of sulfuric acid. Examples include sodium sulfide (Na_2S), sodium sulfite (Na_2SO_3), and sodium sulfate (Na_2SO_4).

sulfur (*elements*)   Sulfur is a yellow nonmetal, atomic number 16. Known to the ancients, sulfur is the brimstone of the Bible and associated with volcanism both on Earth and on Jupiter's moon Io. Although sulfur is not a major component of Earth's crust, its compounds, such as iron pyrites and gypsum, are relatively common, and many are important in commercial processes. It is also found associated with coal and petroleum; as a result, sulfur dioxide is a common air pollutant and a main source of acid rain. Chemical symbol: S.

sumac (*plants*)   Members of 2 genera in the cashew family are commonly called sumacs. They have a resinous sap that in many cases contains skin irritants. The genus *Rhus*, native to temperate and subtropical regions, consists of deciduous and evergreen shrubs, trees, and vines. Common North American species include the smooth sumac (*R. glabra*) and staghorn sumac (*R. typhina*); in autumn their large compound leaves turn a brilliant red and the female plants bear dense clusters of scarlet fruits.

All parts of sumacs of the genus *Toxicodendron* contain the potent resin urushiol, which causes severe skin rashes and blisters. These plants include poison sumac (*T. vernix*), poison ivy (*T. radicans*), and poison oak (*T. diversiloba*), all native to North America.

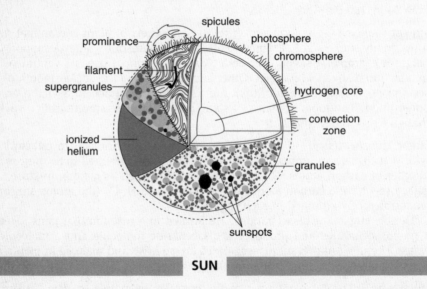

SUN

Sun (*solar system*)   The Sun is the star about which Earth orbits, as do all other bodies in the solar system (even satellites that circle a planet orbit the

Sun). The Sun is mainly hydrogen with some helium. Since it is all gas, it does not rotate as a whole; its equator rotates once in 25 days, but regions near the poles take about 30 days. The Sun fuses hydrogen to helium in its core. The energy from this nuclear fusion flows through the upper layers of the Sun to emerge as electromagnetic radiation (which we perceive as light and heat), as the kinetic energy propelling the solar wind, and as neutrinos. Movements within the layers account for the Sun's complex magnetic field, which produces sunspots, prominences, and flares.

sunspot (*solar system*) In 1611 four different astronomers noticed that dark spots move across the bright face of the Sun. Sunspots vary greatly in number and size over an 11-year solar cycle and sometimes disappear completely (as during the Maunder minimum). This probably accounts for the simultaneous discovery and the shortage of earlier records. Various odd theories were propounded to account for sunspots before 1908, when George Ellery Hale demonstrated that they are magnetic storms caused by the complex solar magnetic field.

supercluster (*galaxies*) Galaxies that are much closer to each other than they are to other galaxies form groups called clusters—our Local Group is an example. The Local Group is a member of a supercluster, the Local Supercluster, which contains about 100 clusters. Although most members of the Local Group are gravitationally bound to each other, clusters in a supercluster are too far apart from each other—some superclusters are 300,000,000 light-years across—for gravity to be important; instead, the supercluster probably developed from a region of slightly higher density just after the big bang.

superconductivity (*materials science*) In 1911 Heike Kamerlingh Onnes discovered that materials such as mercury or lead, when cooled by liquid helium to temperatures near absolute zero, can conduct an electric current with 0 resistance, a phenomenon called superconductivity. A current, once started in a superconducting ring, continues without diminishing. It also rejects completely all magnetic field lines from the interior of the superconductor.

In 1957 John Bardeen, Leon Cooper [American: 1930–], and Bob Schrieffer [American: 1931–] explained superconductivity. Their BCS theory is that pairs of electrons become so bound that individuals in the pair cannot change energy. Thus the pair moves through the material without interacting with the crystal lattice. (See also **high temperature superconductivity**.)

supercooled (*materials science*) A fluid that maintains its phase of matter at a temperature below which it would freeze or liquefy is called supercooled. Supercooling is best understood for liquids. The molecules in a liquid are disordered. When a liquid is cooled below freezing and neither disturbed nor provided with a crystal pattern, randomly arranged molecules slide past each other without chancing upon the crystal arrangement. Stirring or seeding with a small crystal causes sudden freezing. Glasses are often considered supercooled liquids that have become solid. Supercooled water vapor is a factor in precipitation; water vapor requires solid particles to initiate condensation.

superfluid (*materials science*) A superfluid has a viscosity that is nearly 0, enabling it to pass through solids, flow uphill, and maintain vortices eternally once started. When helium-4 (ordinary helium with 2 protons and 2 neutrons) is cooled below −463.12°F (275.34°C), its atoms are slowed enough that their identity as bosons, masked by motion at higher temperatures, emerges and the liquid becomes a superfluid; this effect was discovered in 1938. At even lower temperatures, helium-3 (with 1 neutron) also becomes a superfluid, for reasons analogous to a metal becoming superconducting. Because of the different mechanisms, the behavior of the 2 superfluids is not exactly the same. For example, superfluid helium-3 is affected by magnetic fields.

supernova (*stars*)   Supernovas occur when large stars explode. They are rare in any one galaxy—the last observed in the Milky Way was in 1604—but common in the universe. Since they can easily be observed from great distances (the supernova in AD 1054 was visible on Earth in daytime), supernovas are used to find distances to faraway galaxies.

Nuclear fusion in stars combines light elements into heavier ones, but the process bogs down at several stages and stops completely when iron is reached. The great energy of a supernova forces atoms together to form heavier elements and also blasts the new elements into space. Since there are elements in the solar system heavier than iron, supernovas must have at one time exploded in the neighborhood of the Sun.

superposition (*geometry*)   Superposition is moving one geometric figure to the location of another so that corresponding parts of both figures occupy exactly the same points. If superposition is possible, the figures are congruent. Although superposition was used as the basis of congruence in Euclid's *Elements*, modern geometry generally replaces it with axioms that do not involve moving geometric figures.

superposition law (*geology*)   The superposition law is a general rule that in an undisturbed sequence of rock layers, lower layers formed before layers above them. This superposition law was formulated in 1669 by Nicolaus Steno (Nils Stensen) [Danish: 1638–86] and has become one of the foundations of stratigraphy.

supersonic (*waves*)   Motion faster than the speed of sound in air (Mach 1) is supersonic. Mach 1 varies with temperature, humidity, and density of air. At 68°F (20°C), relative humidity of 50%, and pressure of 1 bar, sound travels about 1129 ft per sec (344 m/s). At a height of 6 mi (10 km) the speed falls to about 980 ft per sec (300 m/s), or about 670 mph. Twice the speed of sound is termed Mach 2 and so forth. Rocket-powered aircraft have reached speeds of about Mach 7.

supersymmetry (*particles*)   Supersymmetry is an attempt to extend the standard model of particle physics so that it includes gravity. The symmetry requirement is that for every particle in the standard model, there is also a symmetric particle that so far has not been detected. Fermions such as the electron and proton have partners named with an added initial s, as in selectron and sproton. Boson partners are shown with the suffix −ino, so the partner of the photon is the photino while the W particle has the wino as its partner. Supersymmetry can also be applied to string theory in a combination known as superstring theory.

supplementary angles (*geometry*)   Two angles are supplementary when the sum of their measures is a straight angle. Each angle is a supplement to the other. If 2 angles are each supplementary to the same angle, the 2 angles are equal in measure, or congruent.

surface (*geometry*)   A surface is a 2-dimensional "skin" in a space of 3 dimensions. A surface of infinite extent is a plane or a figure formed by bending or otherwise deforming a plane (no points are allowed to pass through each other); for example, a paraboloid, formed by rotating a parabola about its axis, is such a surface. The simplest closed surfaces are spheres or figures formed by deforming spheres, such as cylinders and tetrahedrons. More complicated closed surfaces have holes in them, such as the 1-hole torus or the cup with 2 handles.

In analytic geometry a surface is any graph of an equation in 3 variables that does not represent a point or a line.

surface tension (*materials science*)   Molecules in a liquid are free to move, but attract each other. At the boundary of a liquid there are attractive molecules only on one side. The unbalanced attraction creates a force, called surface tension. The force, proportional to length, causes the surface of a liquid, such as water, meeting a gas, such as air, to appear as a thin elastic membrane on the liquid. Surface tension makes raindrops round and causes a thin coat of water to form beads. Surface tension is a major force for tiny creatures, such as whirligig beetles; it also affects larger organisms since it is the basis of capillary action.

suspension (*materials science*)   A suspension is a mixture of finely divided particles and a fluid that is homogenous at first; however, the particles eventually settle to the bottom of the fluid. The particles are larger than colloids, which are small enough in relation to the molecules of a liquid that they do not settle. A colloidal mixture can be turned into a suspension by adding a chemical that causes colloids to condense into larger particles.

sustained yield (*environment*)   Maintaining the productivity of a woodland, wheat field, or other biological resource year after year without damaging the environment is the objective of sustainability. For example, the technique of paddock grazing divides large pastures into smaller units that are used in sequence. This prevents overgrazing of some areas and undergrazing of others, and gives the grass in each paddock time to recover completely before cattle or other livestock are returned to that area.

swallow (*animals*)   Perching birds of the swallow family, Hirundinidae, include swallows and martins. They are 4 to 9 in. (10 to 23 cm) long and slender. They have bristles on the face, a very short bill, long pointed wings with 9 flight feathers, and 12 tail feathers. Swallows are superb fliers, catching insects in the air as well as feeding on ground insects. Migrating species travel in large flocks.

Swammerdam, Jan (*anatomist*)   [Dutch: 1637–80] One of the founders of comparative anatomy, Swammerdam made detailed microscopic studies of a wide variety of organisms. He did much to refute ancient beliefs that insects have no internal organs and that they originate by spontaneous generation. He also showed that during metamorphosis an insect changes its form and structure rather than developing suddenly from a totally different kind of organism.

swamp (*ecology*)   A swamp is a type of wetland characterized by the presence of trees and shrubs that grow on small hillocks surrounded by stagnant or slow-flowing water. There is great diversity among swamps, from freshwater cottonwood swamps in desert canyons to dense, dark mangrove swamps in tidal areas. Many swamps are associated with lakes and rivers and may be saturated during only part of the year. Plant and animal life in swamps tends to be plentiful and diverse.

sweat gland (*anatomy*)   The skin of mammals contains small sweat glands that produce a watery secretion called sweat or perspiration. Sweat is excreted through a duct that leads from the gland to a pore in the skin's surface. The sweat contains sodium chloride (salt), urea, and other wastes. Evaporation of sweat helps cool the body and regulate body temperature in warm environments. Distribution of sweat glands varies greatly from species to species—for example, cats have sweat glands only between their toes and cattle have them only on the muzzle. A human has about 3 million sweat glands.

sweetener (*compounds*)   Common natural sweeteners are forms of sugar, such as sucrose ($C_{12}H_{22}O_{11}$); fructose ($C_6H_{12}O_6$), the sweetest of the nutrient sugars; and the isomer of fructose, glucose ($C_6H_{12}O_6$), which is less sweet than sucrose. Some poisons are sweet, including lead acetate and beryllium salts. In 1879 saccharin ($C_7H_5O_3NS$) was discovered. It is 300 times as sweet as sucrose, has 0 calories of nutritional value, and a bitter aftertaste. A sweetener without the calories of sucrose and no bitterness is aspartame ($C_{14}H_{18}O_5N_2$), discovered in 1965. Aspartame is a protein but in the amounts used in food has almost no calories.

swift (*animals*)   Gregarious birds named for their high-speed flight, swifts (Family Apodidae) do almost everything while aloft: catch insects, drink water, sleep, even mate. They have a streamlined body, a short bill, long pointed wings that curve backward, and very short legs with strong claws. Their plumage is mainly brown or gray. Swifts are found worldwide except in polar regions. They build nests by using saliva to glue together twigs and other materials (salivary glands enlarge during nesting season).

swim bladder (*anatomy*)   Many bony fish have a swim bladder (less accurately called an air bladder) filled with gases. The swim bladder enables a fish to float at a certain depth without having to use energy to swim in place. It also adjusts overall density as the fish moves from one depth to another. Movement of gas into the bladder makes the fish lighter, allowing it to rise in the water; expelling gas from the bladder makes the fish heavier, allowing it to sink.

swordfish (*animals*)   The only species in Family Xiphiidae, the swordfish (*Xiphias gladius*) lives in tropical and temperate oceans around the world. It grows to 15 ft (4.5 m) and more than 1300 lb (590 kg). The back of the body is dark, the belly pale. The upper jaw is prolonged and flattened into a formidable swordlike weapon. The high dorsal fin is shaped like a sickle.

syllogism (*logic*)   The verbal version of formal rules of deduction for logic consists of rules called syllogisms. Each syllogism consists of a statement called a major premise ("All men are mortal"), a minor premise ("Socrates is a man") and a conclusion derived by a rule of logic ("Socrates is mortal"). Common syllogisms include *modus ponens* (if *p*, then *q*; *p*; therefore, *q*) and *modus tollens* (if *p*, then *q*; not *q*; therefore, not *p*).

symbiosis (*ecology*)   Two organisms of different species that live together in a close relationship are said to have a symbiotic ("living together") relationship. The 3 main types of symbiosis are commensalism, mutualism, and parasitism.

symbolic logic (*logic*)   Logic in which variables such as *p* or *q* represent propositions, which are combined by operations such as not or and that have been assigned such symbols as ~ or &, is called symbolic logic. Symbolic logic is generally established as an axiomatic system so that theorems can be proved or disproved. Since the basic ideas behind arithmetic can be expressed as the consequences of symbolic logic, this form of logic is sometimes taken as a branch of knowledge that includes mathematics.

symmetry (*biology*)   Many living things exhibit at least an outward symmetry of form. Humans and most other animals show bilateral symmetry, with a right and a left side; if cut from head to tail, the right and left sides would be approximate mirror images of one another. Starfish and jellyfish have radial symmetry, with parts arranged like spokes around the hub of a wheel. Radiolarians, *Volvox* colonies, and some other protists have spherical symmetry, with parts arranged around a central point.

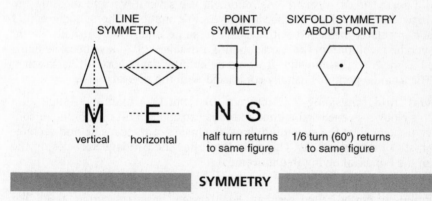

LINE SYMMETRY	POINT SYMMETRY	SIXFOLD SYMMETRY ABOUT POINT
vertical horizontal	half turn returns to same figure	1/6 turn (60°) returns to same figure

SYMMETRY

symmetry (*mathematics*)   Geometric symmetry involves one part looking exactly like another. Line symmetry exists when a figure possesses an axis for which the part on one side of the axis is the mirror image of the other. If folded along the axis, the 2 halves match. Each letter in A B C D E has line symmetry, but F and G do not. In 3 dimensions a plane of symmetry divides a symmetric figure into mirror images. The letters N, S, and Z show point symmetry; if you rotate each about its center a half turn, the parts match up to the original.

Other figures show similar symmetry for a smaller part of a turn. A **pentagon** has fivefold symmetry since ⅕ rotation makes parts match.

Symmetry in algebra uses the same idea. An expression or equation in x and y is symmetric if interchanging x and y keeps the meaning the same. Symmetry about the y axis exists if substituting $-x$ for x in the equation does not change the curve. Similarly symmetry about the x axis can be tested by substituting $-y$ for y.

Any **relation** • is symmetric if when $a \cdot b$ is true, then $b \cdot a$ is true. For example, = is a symmetric relation while > is not.

synapse (*anatomy*)   The point where an end of the axon of one **neuron** (nerve cell) meets the end of a dendrite of a neighboring neuron is called a synapse. In most mammalian synapses, the cells' ends are separated by a tiny gap, the synaptic cleft. Chemical **neurotransmitters** synthesized and released by the neurons carry signals across the gap. The neurotransmitters are then absorbed or broken down, permitting neurons to convey new information.

synchrotron radiation (*particles*)   When charged particles are accelerated in a magnetic field, they produce electromagnetic waves called synchrotron radiation; the particles lose energy to this radiation and accelerate more slowly. This effect was discovered in 1947 when the first of a then-new type of particle **accelerator** called the synchrotron was introduced. In a synchrotron, charged particles move in a circle, a motion that produces continual acceleration. Although first viewed as a problem because the radiation reduces the speed that charged particles, especially low-mass electrons, can reach, some accelerators are now built for the purpose of creating high-energy synchrotron radiation, an excellent source of bright X rays and gamma rays.

synodic period (*solar system*)   Astronomers use several ways to measure the time between celestial events because of Earth's motion. The synodic period between events is measured with respect to some other body, usually Earth. One synodic revolution of the Moon about Earth takes 29½ days because Earth as well as the Moon is moving—it would be about 2 days shorter if Earth stood still. The synodic period is usually contrasted with the **sidereal period**.

synovial fluid (*physiology*)   Each movable joint in a mammal—such as a human's elbow—is enclosed within a fibrous capsule. A thin membrane of connective tissue, called the synovium, lines the inside of the capsule and secretes a clear, sticky synovial fluid. This fluid lubricates the cartilage surfaces on the ends of the bones, allowing free movement.

system (*biology*)   The bodies of most multicellular organisms are organized into groups of **organs** called systems. Each system performs certain tasks. For example, the vascular system of a rose bush conducts food, water, and other materials throughout the bush—much as the circulatory system carries materials throughout your body.

system of equations (*algebra*)   If 2 or more equations are intended to be solved together, the set is called a system of equations. Typically such a system has as many **variables** as equations, although this is not required. Equations in the system are independent if the set of **solutions** to each equation is not the same as the set for another in the system; if 2 equations have exactly the same set of solutions, the equations are dependent. When equations have sets of solutions that include no common solution, the equations are inconsistent. The whole system is consistent if and only if there is at least 1 solution that works for all the equations. For most problems, the system is consistent and comprised of independent equations. A typical small system is

$$2x + 3y = 5 \text{ and}$$

$$3x - y = 13,$$

which has the single solution $(x, y) = (4, -1)$. Many problems, however, require systems of hundreds of equations with the same number of variables.

systole and diastole (*physiology*)   The beat of a mammal's heart has 2 phases. In the first phase, systole, the ventricles (lower chambers) contract, pushing blood into the arteries. In the second phase, diastole, the ventricles relax and blood flows into them from the atria (upper chambers). A blood pressure measurement consists of the systolic pressure followed by the diastolic pressure; a typical reading for a human adult is 120/80, meaning that the systolic pressure pushes up a column of mercury to 120 mm, and the diastolic pressure to a height of 80 mm.

Szent-Györgi, Albert von (*biochemistry*)   [Hungarian-American: 1893–1986] Szent-Györgi's first major contributions occurred during the late 1920s, when he isolated a substance from adrenal glands that he named hexuronic acid. His suspicion that this substance was vitamin C—which had been discovered in 1907—was confirmed by the work of Charles G. King [American: 1896–1986], who determined its formula, $C_6H_8O_6$. (Vitamin C later came to be called ascorbic acid.) Szent-Györgi also studied the proteins involved in muscle contractions, laying the foundation for elucidation of the Krebs cycle. Szent-Györgi was awarded the 1937 Nobel Prize for physiology or medicine.

tahr (*animals*) Native to forested hills and mountains of the Himalayas, southern India, and Oman, tahrs (*Hemitragus*) are goatlike hoofed mammals in the bovid family. They grow to about 3.3 ft (1 m) high at the shoulder and weigh up to 200 lb (90 kg).

taiga (*ecology*) Stretching across northern North America, Europe, and Asia are vast expanses of coniferous forests called taiga or boreal forests. To the north lies tundra. To the south are temperate deciduous forests and grasslands. Taiga is characterized by long, cold winters and short summers during which temperatures are high enough to completely thaw the ground. Precipitation averages about 20 in. (50 cm) annually. Streams, lakes, ponds, and bogs are numerous. Trees such as spruce, fir, and pine dominate the plant life, though birch, willow, and other deciduous trees are abundant in disturbed areas and along the edges of bodies of water. Small herbivores such as squirrels and hares live here year round; larger herbivores migrate into the taiga—moose and caribou, which winter here and summer on the tundra, and deer and elk, which summer here and move south into deciduous forests for the winter. Predators include bears, wolves, martens, and lynx. Migratory birds are abundant during summer, as are biting insects. There are few amphibians and reptiles.

talc (*rocks and minerals*) Talc is the softest mineral, 1 on the Mohs scale. It can be scratched easily with a fingernail. Soapstone, easily carved, is made from talc. Talc ground up fine is well known as talcum powder, soothing to the skin and also used for industrial purposes.

talus (*geology*) The boulders, rock, and gravel that accumulate at the base of a cliff, or even a very steep mountain or hill, are called talus. The pile of talus forms a steep talus slope against the base of the cliff.

tamarind (*plants*) The tamarind (*Tamarindus indica*) is a legume native to tropical Africa. It is a long-lived evergreen tree that attains heights up to 80 ft (24 m). It has dense, feathery foliage and hanging clusters of fragrant, cream-colored flowers. The leathery brown pods contain 1 to 12 glossy brown seeds embedded in an edible pulp.

tamarisk (*plants*) Native to Europe and Asia, tamarisks (*Tamarix*)—also known as salt cedars—are deciduous shrubs and small trees with tiny gray-green leaves and small pinkish flowers borne in long clusters. They grow fast and can tolerate a wide range of environmental conditions, including salty soils and prolonged droughts. Species introduced to North America, mainly as ornamentals and windbreaks, have become serious pests.

tanager (*animals*) Native chiefly to forests of tropical America, tanagers (Family Thraupidae) are perching birds 4 to 12 in. (10 to 30 cm) long. They have bristles around the bill, 9 flight feathers on each wing, and a short tail. Males have brightly colored plumage; female plumage is dull olive or yellow to blend with surrounding leaves. The diet consists mainly of insects and small fruits.

CIRCLES AND LINES

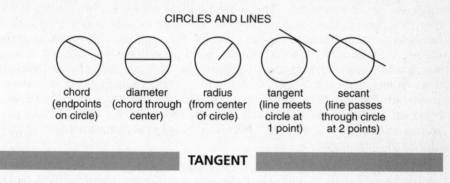

chord	diameter	radius	tangent	secant
(endpoints on circle)	(chord through center)	(from center of circle)	(line meets circle at 1 point)	(line passes through circle at 2 points)

TANGENT

tangent (*geometry*) A line that intersects a circle at 1 point only is a tangent, and always perpendicular to the circle's radius at the point. The tangent to a curve in general also touches it at 1 point, but also must be the limiting position of the curve's secant as the 2 points where the secant crosses the curve converge. For a curve in some surface other than a plane, a line tangent to the curve is also tangent to the surface. A plane is tangent to a surface at a point provided that every line in the plane through that point is tangent to the surface.

tangent (*trigonometry*) The tangent function (tan) is the ratio of the sine to the cosine: $\tan x = \sin x/\cos x$. In a right triangle the tangent is the ratio of the side opposite an angle to the side adjacent. A unit circle definition of the tangent is y/x where (x, y) is a point on the circle. A ratio of the form y/x is also the slope of a line, suggesting the relationship of this function to the slope of the tangent to a curve.

tannin (*botany*) Tannins, or tannic acids, are complex organic compounds found in some tissues, such as wood, bark, leaves, and unripe fruit, in certain plants. Their bitter, astringent taste may deter plant-eating animals from feeding and may protect against disease-causing organisms.

Tansley, Arthur G. (*ecologist*) [English: 1871–1955] A pioneer in the field of plant ecology and instrumental in the creation of organizations devoted to conservation, Tansley is best remembered for coining the term "ecosystem" in 1935. He noted that while "organisms may claim our primary interest . . . we cannot separate them from their specific environment, with which they form one physical system."

tantalum (*elements*) The dense, gray, hard metal tantalum, atomic number 73, was discovered in 1802 by Anders Ekeberg [Swedish: 1767–1813]. It was named for its ore, tantalite, in turn named after the mythical king Tantalus, who thirstily spends eternity immersed in water he cannot drink. Tantalite, and the metal tantalum, can be immersed in most acids and remain unaffected. Tantalum's low reactivity at ordinary temperatures has made tantalum useful in surgical instruments and implants. Chemical symbol: Ta.

tapeworm (*animals*) Ribbonlike flatworms of the class Cestoda, some of them growing to 20 ft (6 m) or more in length, tapeworms are parasites that live in the intestines of vertebrates. They have no digestive system but simply absorb nutrients through their body wall. Their knoblike head is armed with suckers or hooks for attachment to the host. Behind the head is a short neck, or growing region, from which body segments called proglottids are produced. A tapeworm may have several thousand proglottids. Each proglottid has 2 reproductive systems, male and female, with sperm typically fertilizing eggs of the same proglottid. Proglottids at the tapeworm's rear end—filled with fertilized eggs— break off and leave the host's body with the feces. The life cycle of most tapeworms also involves a secondary host, in which larval tapeworms form cysts in muscle or other tissue.

taphonomy (*paleontology*) When organisms die, they may be destroyed or preserved as fossils. Taphonomy—from the Greek for tomb, *taphos*—is the study of the biological, chemical, and physical processes that lead to the disintegration of an organism or to its burial and fossilization. For example, experiments may indicate the speed of bacterial decay under certain environmental conditions, while careful study of a fossil may show teeth marks indicating a deadly attack by a predator.

tapir (*animals*) Constituting the family Tapiridae, tapirs are hoofed mammals that grow to 8 ft (2.4 m) long, with weights exceeding 600 lb (270 kg). Related to horses and rhinoceroses, they look somewhat like pigs, with a rounded back, short legs, and a stumpy tail. The projecting jaws and nose form a trunklike snout. Tapirs live in wooded tropical regions of Central and South America and Southeast Asia. They are herbivores.

tarantula (*animals*) Large, hairy spiders often are given the common name tarantula. The name originated with a European wolf spider, *Lycosa tarantula*, and was then applied by explorers to giant spiders encountered elsewhere. For scientists the term refers to members of the family Theraphosidae, of which there are about 800 known species. The largest spiders of all are the South American tarantulas, also called birdeaters because they are big enough to catch birds; the goliath birdeater (*Theraphosa blondi*) has a leg span of more than 10 in. (25 cm).

tar pit (*paleontology*) At some places on Earth, oil seeps to the surface and forms pools of tar that can become exceedingly sticky. Animals that arrive to drink water lying atop the tar can become trapped in the tar. Predators that attack the trapped animals suffer the same fate. In time, their remains sink into the tar and become preserved as fossils. One of the richest sources of fossils from 10,000 to 40,000 years ago is the Rancho La Brea tar pits in Los Angeles, California. More than 565 species, ranging from microscopic one-celled organisms to sabertooths, camels, bison, mastodons, and giant ground sloths, have been recovered from these pits.

tarsier (*animals*)   Small primates native to islands in Southeast Asia, tarsiers (Family Tarsiidae) have enormous eyes and the ability to turn their head nearly 360°. They are nocturnal, feeding on insects and small vertebrates. Their body ranges up to 6.5 in. (17 cm) in length, the tail to almost 11 in. (28 cm). The front limbs are shorter than the hind, with the latter designed for leaping. The big toe is opposable but the thumb is not. All digits have flattened nails except the second and third toes, which have claws used for grooming.

Tartaglia, Niccolò (*mathematician*)   [Italian: c1501–57] Tartaglia is known mainly as the first to solve the general equation of a polynomial of degree 3 (the cubic). He also developed the rule that an elevation of 45° for a cannon provides the maximum-distance shot; was the first to publish Pascal's triangle in Europe; and translated Euclid into Italian and some of Archimedes into Latin.

Tasmanian devil (*animals*)   The size of a small dog, the Tasmanian devil (*Sarcophilus laniarius*) is a carnivorous marsupial found on the island of Tasmania though it once ranged throughout Australia. Black or blackish brown in color, it is heavily built, with short legs and a strong jaw. It preys on wombats and other small marsupials and also scavenges for food.

taste (*physiology*)   The sense by which animals recognize and distinguish among different flavors is called taste. Specialized receptor cells detect molecules of chemicals in solution (hence the importance of mixing dry food with saliva); this information is then sent to the brain, where it is interpreted. There are believed to be only 5 basic tastes: sweet, sour, bitter, salty, and umami. (Umami is the taste of glutamate, as in monosodium glutamate). All taste sensations are combinations of these, strongly influenced by smell in many animals.

tauon (*particles*)   The tauon, discovered by Martin Perl [American: 1926-] in 1975, is the massive analog to the electron that is the heavy lepton in the third generation of matter, corresponding to the muon in the second generation.

tautology (*logic*)   Originally a tautology was a statement repeating the same idea, such as "Sir Walter Scott is the author of *Waverley*" (Scott and the author of *Waverley* are one person). Today a tautology is a complex statement (one with 2 or more parts) that is true no matter whether the parts are true—for example, if p is a proposition, "p or not p" is a tautology. The axioms of symbolic logic are all tautologies, and all theorems derived from those axioms are also tautologies.

taxis (*biology*)   Movement of an organism from one place to another in response to an external stimulus is a taxis. The movement may be toward or away from light (phototaxis), heat (thermotaxis), chemicals (chemotaxis), or air or water currents (rheotaxis). For example, a bacterium exhibits positive chemotaxis as it swims toward higher concentrations of food and negative chemotaxis as it moves away from toxic substances.

taxon (*taxonomy*)   Any particular group of organisms in a hierarchical classification scheme is called a taxon (pl.: taxa)—from the Greek word *taxis*, meaning to arrange or put in order. For example, insects, which are all classified in Class Insecta, form a taxon at the level of class; tiger beetles, members of the family

Cicindelidae, form a taxon at the level of family. Members of different taxa at the same level have different characteristics. Hence, the tiger beetle family is one taxon, the stag beetle family another. The various taxa of the white pine, the human, and the wolf are shown in the table.

Some Taxonomic Hierarchies			
	WHITE PINE	HUMAN	WOLF
Kingdom	Plantae	Animalia	Animalia
Phylum	Coniferophyta	Chordata	Chordata
Class	Coniferopsida	Mammalia	Mammalia
Order	Coniferales	Primate	Carnivora
Family	Pinaceae	Hominidae	Canidae
Genus	*Pinus*	*Homo*	*Canis*
Species	*strobus*	*sapiens*	*lupus*

taxonomic hierarchy (*taxonomy*)   The classification system used to classify all known organisms, both living and extinct, is hierarchical, with large groups divided into smaller and smaller groups. The largest groups are kingdoms. Each kingdom is divided into phyla, each phylum into classes, each class into orders, each order into families, each family into genera, each genus into species. Additional categories often are used. For example, some classes within a phylum may be subdivided into subclasses or grouped together into a superclass.

taxonomy (*sciences*)   Taxonomy is the scientific classification of organisms into groups according to their similarities and differences. Taxonomists use anatomical, biochemical, embryological, genetic, and evolutionary characteristics to provide each organism with a scientific binomial name and to determine that organism's relationship to other organisms. (See table on pages 569–71.) Two main approaches to classification are used today: cladistics and evolutionary taxonomy.

tea (*plants*)   Native to warm regions of China, the tea plant (*Camellia sinensis*) is an evergreen shrub or small tree up to 20 ft (6 m) tall. Its aromatic, tannin-rich evergreen leaves are treated in various ways to obtain green tea, black tea, and other products. The genus also includes several species with showy flowers, such as *C. japonica*, that are grown as ornamentals.

teak (*plants*)   The teak (*Tectona grandis*) is a tropical deciduous tree that is native from India through Thailand and south into Indonesia. It grows to 100 ft (30 m) in height and has large, broad leaves and clusters of tiny white flowers that develop into small fruits containing oily seeds. The wood is extremely hard and decay-resistant. Teak belongs to the vervain family, Verbenaceae, which also includes herbs and shrubs of the genera *Verbena*, grown for medicinal and ornamental purposes, and *Lantana*, grown as ornamentals.

Major Groups of Living Organisms, with Representative Species

Kingdom Archaea One-celled organisms that look like bacteria but are genetically very different. They often live in hot springs and other high-temperature environments. Instead of using oxygen, they metabolize chemicals such as methane or sulfur.

Kingdom Monera One-celled, filamentous or colonial organisms. The genetic material is a single strand of DNA that is not separated from the rest of the cell by a membrane.

Phylum Eubacteria—bacteria
Phylum Cyanobacteria—blue-green algae

Kingdom Protista Very diverse group of 1-celled and colonial organisms, with a membrane around the genetic material. Some can produce their own food, others cannot.

Phylum Mastigophora—flagellates (*Trypanosoma*)
Phylum Ciliophora—ciliates (*Paramecium, Vorticella*)
Phylum Sporozoa—amoeboid parasites (*Plasmodium*)
Phylum Sarcodina—amoeboids (*Amoeba,* radiolarians, foraminifera)
Phylum Euglenophyta—flagellate algae (*Euglena, Volvox*)
Phylum Chrysophyta—golden algae, diatoms
Phylum Pyrrhophyta—fire algae, dinoflagellates (*Gonyaulax, Ceratium*)
Phylum Chlorophyta—green algae (*Spirogyra,* sea lettuce)
Phylum Phaeophyta—brown algae (kelp)
Phylum Rhodophyta—red algae (*Porphyra*)
Phylum Myxomycota—acellular slime molds (*Physarum*)

Kingdom Fungi One-celled and multicellular organisms with cell walls made of chitin. They cannot produce their own food. They reproduce asexually by spores and sexually by conjugation.

Phylum Zygomycota—bread molds (*Rhizopus*)
Phylum Ascomycota—sac fungi (yeasts, morels, truffles)
Phylum Basidiomycota—club fungi (mushrooms, puffballs, bracket fungi, rusts, stinkhorns)

Kingdom Plantae Multicellular organisms that carry out photosynthesis. The cell walls are made of cellulose.

Phylum Bryophyta—mosses, liverworts, hornworts
Phylum Lycopodiophyta—club mosses, spike mosses, quillworts
Phylum Equisetophyta—horsetails
Phylum Pterophyta—ferns
Phylum Cycadophyta—cycads
Phylum Ginkgophyta—ginkgo
Phylum Coniferophyta—conifers (pines, firs, cypresses)
Phylum Magnoliophyta—flowering plants
  Class Monocotyledones—monocots (onions, lilies, grasses, orchids, palms)
  Class Dicotyledoneae—dicots (apples, asters, legumes, maples, roses, cacti)

Kingdom Animalia Multicellular organisms that obtain food by eating other organisms or decayed organic matter. The cells do not have cell walls.

Phylum Porifera—sponges
Phylum Cnidaria—coelenterates
  Class Hydrozoa—hydroids (*Hydra*)
  Class Scyphozoa—jellyfish
  Class Anthozoa—sea anemones, corals
Phylum Ctenophora—comb jellies
Phylum Platyhelminthes—flatworms
  Class Turbellaria—turbellarians (planarians)
  Class Trematoda—flukes
  Class Cestoda—tapeworms
Phylum Nematoda—roundworms (*Trichinella, Ascaris*)
Phylum Rotifera—rotifers
Phylum Bryozoa—moss animals
Phylum Brachiopoda—lampshells
Phylum Nemertina—proboscis worms, ribbon worms

Major Groups of Living Organisms, with Representative Species
(Continued)

Phylum Phoronida—horseshoe worms
Phylum Annelida—segmented worms
  Class Polychaeta—bristle worms
  Class Oligochaeta—earthworms
  Class Hirudinea—leeches
Phylum Onychophora—velvetworms
Phylum Mollusca—mollusks
  Class Polyplacophora—chitons
  Class Bivalvia—bivalves (clams, oysters, mussels, scallops)
  Class Scaphopoda—tooth shells
  Class Gastropoda—snails, slugs, nudibranchs
  Class Cephalopoda—octopuses, squids, nautiluses
Phylum Arthropoda—arthropods
  Class Merostomata—horseshoe crabs
  Class Crustacea—lobsters, crabs, shrimp, barnacles
  Class Arachnida—scorpions, spiders, mites, ticks
  Class Chilopoda—centipedes
  Class Diplopoda—millipedes
  Class Insecta—insects (beetles, bugs, butterflies, bees, ants, flies, grasshoppers)
Phylum Tardigrada—water bears
Phylum Chaetognatha—arrowworms
Phylum Echinodermata—echinoderms
  Class Crinoidea—sea lilies, feather stars
  Class Asteroidea—sea stars (starfish)
  Class Ophiuroidea—brittle stars, basket stars
  Class Echinoidea—sea urchins, heart urchins, sand dollars
  Class Holothuroidea—sea cucumbers
Phylum Hemichordata—acorn worms
Phylum Chordata
  Subphylum Urochordata—tunicates (sea squirts)
  Subphylum Cephalochordata—lancelets
  Subphylum Vertebrata—vertebrates
    Class Agnatha—jawless fish (lampreys, hagfish)
    Class Chondrichthyes—cartilaginous fish (sharks, skates, rays)
    Class Osteichthyes—bony fish (bass, salmon, trout, seahorses)
    Class Amphibia—amphibians
      Order Gymnophiona—caecilians
      Order Caudata—salamanders, newts, waterdogs
      Order Anura—frogs, toads
    Class Reptilia—reptiles
      Order Chelonia—turtles, tortoises
      Order Crocodylia—caimans, alligators, crocodiles
      Order Rhynchocephalia—tuataras
      Order Squamata—lizards, snakes
    Class Aves—birds
      Order Struthioniformes—ostriches
      Order Rheiformes—rheas
      Order Casuariiformes—emus, cassowaries
      Order Apterygiformes—kiwis
      Order Tinamiformes—tinamous
      Order Sphenisciformes—penguins
      Order Gaviiformes—loons, divers
      Order Podicipediformes—grebes
      Order Procellariiformes—albatrosses, fulmars, shearwaters, petrels
      Order Pelecaniformes—pelicans, cormorants, gannets, boobies, anhingas, frigatebirds

Major Groups of Living Organisms, with Representative Species (Continued)

Phylum Chordata
  Subphylum Vertebrata—vertebrates
    Class Aves—birds
      Order Ciconiiformes—herons, bitterns, storks, ibises, spoonbills
      Order Phoenicopteriformes—flamingos
      Order Anseriformes—ducks, geese, swans, screamers
      Order Falconiformes—birds of prey (condors, hawks, vultures, ospreys, falcons)
      Order Galliformes—fowl (megapodes, curassows, grouse, quail, pheasants, turkeys)
      Order Gruiformes—hemipodes, cranes, rails, coots, bustards
      Order Charadriiformes—gulls, terns, jacanas, stilts, avocets, plovers, auks, puffins, murres
      Order Columbiformes—sandgrouse, doves, pigeons
      Order Psittaciformes—lories, cockatoos, parrots, parakeets
      Order Cuculiformes—cuckoos, roadrunners, turacos
      Order Strigiformes—owls
      Order Caprimulgiformes—nightjars, nighthawks, poorwills, frogmouths, goatsuckers, potoos
      Order Apodiformes—swifts, hummingbirds
      Order Coliiformes—colies, mousebirds
      Order Trogoniformes—trogons, quetzals
      Order Coraciiformes—kingfishers, hoopoes, hornbills, motmots, bee-eaters, rollers
      Order Piciformes—woodpeckers, honeyguides, toucans, barbets, jacamars, puffbirds
      Order Passeriformes—perching birds (shrikes, gnateaters, wrens, mockingbirds, thrushes, warblers, chickadees, tits, nuthatches, sparrows, blackbirds, swallows, crows, finches)
    Class Mammalia—mammals
      Order Monotremata—echidnas, platypuses
      Order Marsupialia—marsupials (opossums, bandicoots, koalas, wombats, wallabies, kangaroos)
      Order Insectivora—insectivores (shrews, moles, tenrecs, solenodons)
      Order Dermoptera—flying lemurs
      Order Chiroptera—bats
      Order Primates—primates (lemurs, marmosets, tarsiers, monkeys, gorillas, chimpanzees, humans)
      Order Edentata—edentates (anteaters, sloths, armadillos)
      Order Pholidota—pangolins
      Order Lagomorpha—rabbits, hares, pikas
      Order Rodentia—rodents (squirrels, prairie dogs, beavers, mice, rats, hamsters, muskrats, porcupines, chinchillas, capybara)
      Order Cetacea—dolphins, porpoises, whales
      Order Carnivora—carnivores (foxes, dogs, bears, raccoons, otters, weasels, skunks, mongooses, hyenas, cats)
      Order Pinnipedia—seals, sea lions, walruses
      Order Tubulidentata—aardvarks
      Order Proboscidea—elephants
      Order Hyracoidea—hyraxes
      Order Sirenia—manatees, dugongs
      Order Perissodactyla—odd-toed hoofed mammals (horses, zebras, tapirs, rhinoceroses)
      Order Artiodactyla—even-toed hoofed mammals (pigs, hippopotamuses, camels, deer, giraffes, cattle, sheep, antelopes, buffalo)

technetium (*elements*) The radioactive metal technetium was synthesized in 1937 by Emilio Segrè. Its name, from the Greek *technetos*, "artificial," signifies that it was the first artificial element. Technetium has the lowest atomic number, 43, of any element all of whose isotopes are radioactive. None are found naturally on Earth. Chemical symbol: Tc.

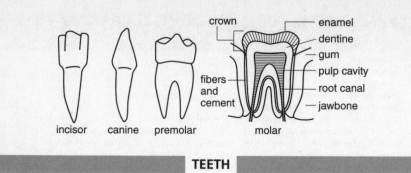

incisor canine premolar molar

crown — enamel — dentine — gum — pulp cavity — fibers and cement — root canal — jawbone

TEETH

teeth (*anatomy*) Hard bony structures borne on the jaws of vertebrates, teeth are designed to bite, tear, crush, and grind food. For many animals, teeth also are weapons of offense and defense. In fish, amphibians, and reptiles, all the teeth are similar in structure and size. Modern birds lack teeth but some early ancestors, such as *Archaeopteryx*, had teeth. Most mammals have 4 kinds of teeth. Starting at the front of the jaws these are incisors, typically designed for cutting or gnawing; canines, for tearing; and premolars and molars, for grinding. Some teeth—the enlarged canines, or tusks, of a walrus, for example—may be highly specialized. Collectively, an animal's teeth make up its dentition.

The crown forms the part of the mammalian tooth above the gum; the root is the part beneath the surface, covered with a substance called cementum and embedded in the jawbone. The tooth has a very hard enamel covering. Beneath the enamel, forming the bulk of the tooth, is the somewhat softer dentine. The pulp cavity contains blood vessels and nerve fibers.

tektite (*rocks and minerals*) A tektite is a glassy rock form, almost always very small, thought to have been produced on various occasions when a large meteorite struck Earth. The heat from the collision melted both rock and meteorite. The melt cooled quickly into glass as drops of molten rock were tossed large distances through the air. The largest concentration of tektites has been found in the Philippine Islands and Australia. A thick deposit of tektites in Haiti is considered to be evidence that the K/T event was caused by a large object striking Earth.

teleost (*animals*) The most highly evolved fish are teleosts—they include the most common and "ordinary" bony fish. Some, including herring and salmon, have fins supported only by soft rays; their swim bladder opens into the esophagus. The much larger group of teleosts, including sea bass and perch, have fins supported by both spines and soft rays; their swim bladder lacks a duct to the esophagus.

telescope (*astronomy*) About 1600 several inventors, mostly Dutch, combined 2 lenses to make distant objects appear closer. Their invention was the telescope. Galileo built telescopes in 1609 as soon as he learned of the devices. Galileo's telescopes used a convex lens to focus light and a concave lens to spread the image. This arrangement left the image right-side up and is still used in binoculars and field glasses. Kepler soon improved on Galileo's design by using 2 convex lenses.

A half-century's experience with lens-based refracting telescopes showed that the images were flawed by aberration. About 1670 several scientists, notably

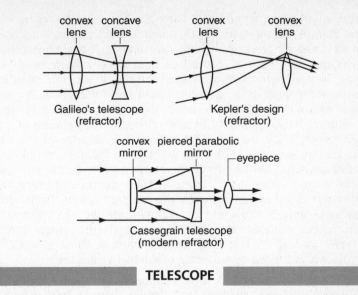

convex   concave   convex   convex
lens   lens   lens   lens

Galileo's telescope   Kepler's design
(refractor)   (refractor)

convex   pierced parabolic
mirror   mirror   eyepiece

Cassegrain telescope
(modern refractor)

TELESCOPE

Newton, designed and built reflecting telescopes (using mirrors instead of lenses), eliminating chromatic aberration and, when properly ground, eliminating spherical aberration also. Very large reflectors still make the best telescopes.

tellurium (*elements*)   Tellurium, from the Latin *tellus* ("the Earth"), at atomic number 52, lies on the boundary between metal and nonmetal. It is silvery like a metal, but brittle. Because tellurium is a semiconductor with greater conductivity in some directions than others, it has promise for electronics applications, but most commercial tellurium is used in alloys of iron, copper, and lead to improve specific properties. It was discovered in 1782 by Franz Joseph Müller [Austrian: 1740–1825]. Chemical symbol: Te.

telomere (*genetics*)   Chromosomes of humans and other eukaryotes are long DNA molecules. At each end of a DNA molecule is a specialized structure called a telomere. It consists of hundreds of copies of the same short sequence of base pairs repeated over and over. Each time a cell divides, its telomeres get shorter. It is thought that this shortening is an important feature of cell aging. When its telomeres get too short a cell dies. Cancer cells seem to overturn the aging process by activating the enzyme telomerase, which is involved in replicating telomere sequences.

temperature (*thermodynamics*)   Temperature is a measure of the average energy of motion in the particles of a substance. It should not be confused with heat, the amount of energy transferred from a high temperature to a low one; or with internal energy, the sum of the kinetic energies of the particles. Temperature determines the direction of heat flow—heat always transfers from a high temperature to a lower one. Temperature scales can be arbitrary so long as they correctly describe the direction of heat flow, but the Kelvin scale corresponds as much as possible to the actual average kinetic energy of molecules of a gas that would obey the ideal gas laws.

temperature regulation (*physiology*)   There are 2 main processes by which animals warm themselves. Birds, mammals, and other endothermic animals produce heat through metabolic processes, mainly oxidation of glucose molecules. Reptiles and other ectothermic animals absorb heat from the environment—for example, by exposing themselves to the warm rays of the sun. In general, endotherms can survive in a wider range of temperatures than can ectotherms. But producing internal heat means that endotherms require much more food than ectotherms.

Many animals have the ability to maintain body temperature at a more or less constant level. They are called homeotherms or, in popular and somewhat inaccurate terminology, warm-blooded animals. In contrast, poikilotherms, or cold-blooded animals, have body temperatures similar to the temperature of their surrounding environment. Heterotherms regulate body temperature at some times but at other times match the temperature of the environment. Owlet moths (Family Noctuidae), for example, are the same temperature as their environment while at rest. But in order to fly the temperature of their thoracic region (where the wings are located) must be about 86°F (30°C). The necessary heat is produced by shivering of the flight muscles.

tendon (*anatomy*)   A muscle is most commonly attached to bone by white cords of connective tissue called tendons. Tendons are continuous with sheaths of connective tissue that extend throughout the entire mass of the muscle. They consist largely of tough collagen fibers that absorb stress. The largest tendon in the human body is the Achilles tendon, which attaches the chief muscle of the leg calf (gastrocnemius) to the heel bone (calcaneus).

tendril (*botany*)   Many climbing plants have structures called tendrils, which coil around objects on which the plants are growing. Some tendrils, such as those of Boston ivy, are modified stems. Other tendrils, such as those of pea plants, are modified leaves or parts of leaves.

tensegrity (*sciences*)   A common architectural system in nature (and in man-made structures), tensegrity is characterized by counteracting forces of tension and compression that balance one another to provide stability. For example, an animal's bones are stabilized into a skeleton of a certain form by the continuous pull of muscles, tendons, and ligaments. An increase in tension at one point along the skeleton is balanced by an increase in compression elsewhere. Geodesic forms—frameworks of struts connected into triangles, pentagons, or hexagons—are among the most common tensegrity structures, defining the shape of buckminsterfullerenes; certain viruses, pollen grains, and cells; and many other natural objects.

tentacle (*anatomy*)   Found in a wide variety of animals, tentacles are long, flexible appendages, usually near the mouth, that are important in food-gathering. Cnidarians, such as sea anemones, have tentacles armed with stinging cells (nematocysts) that help capture prey. Squid and other cephalopod mollusks have suckered tentacles. Hagfish use the short tentacles around the mouth to attach to the body of prey.

terbium (*elements*)   In 1843 Carl Gustav Mosander [Swedish: 1797–1858] separated and named 3 elements for Ytterby, a village in Sweden—yttrium, erbium, and terbium. Terbium (atomic number 65, chemical symbol Tb), like erbium, is a rare earth. Mosander also found a fourth substance that was not itself an element, although it was later broken into the elements ytterbium and lutetium.

teredo (*animals*)   The shipworm, or teredo (*Teredo navalis*), is actually a clam. Its shell is only about 0.5 in. (1.3 cm) long, but its slender worm-shaped body can grow to more than 2 ft (60 cm). When the teredo is no bigger than a pinhead it bores a tiny hole into submerged wood, often of wharves or ships. Thereafter the teredo never leaves its wooden home. The bored opening never gets any larger, but as the teredo grows it creates an extensive network of hidden tunnels that weaken the wood.

term (*algebra*)   A term is a part of an algebraic expression that is treated as a single addend. Often a term is the result of multiplication, division, and exponentiation alone. Thus in $2x^3 - 3x + 1/x$ the 3 terms are $2x^3$, $-3x$, and $1/x$. In some cases other operations, such as taking roots or addition and subtraction grouped into a single entity, occur, as in $\sqrt{x^2+1}+\dfrac{2x+3}{4}-\dfrac{5}{1-x}$ where the terms are $\sqrt{x^2+1}$, $\dfrac{2x+3}{4}$, and $-\dfrac{5}{1-x}$.

termite (*animals*)   Termites are social insects that constitute Order Isoptera ("equal wings"). They live in colonies, with a caste system that includes a king and queen, workers, and soldiers. Both winged and wingless forms are present in a colony; those with wings have 2 pairs of similar long wings that lie on the back when at rest. Most of the approximately 1800 known species are tropical or subtropical. Their nests are dark and moist; some build nests that extend 20 ft (6 m) into the air. Termites feed on plant matter and are able to digest cellulose thanks to bacteria and protozoa that live symbiotically in their digestive tract. The best-known species are the wood-eaters, which cause extensive damage to buildings and other wooden objects.

tern (*animals*)   Members of the gull family, terns are seabirds generally found in coastal habitats, where they prey on fish and small invertebrates. They are strong fliers and may migrate long distances. The champion of all animal migrators is the arctic tern (*Sterna paradisaea*), which makes a yearly round trip of more than 22,000 mi (35,000 km) between the Arctic and Antarctic.

terrestrial planet (*solar system*)   The 4 inner planets of the solar system, including Earth, are alike in having rocky surfaces, so they are termed terrestrial ("like Earth"). Underneath their surfaces, however, and above it as well, they are all different. Mercury and Earth have powerful magnetic fields, while Venus and Mars have very weak ones. Atmospheres range from almost none (Mercury) to thin (Mars) to crushingly dense (Venus). It is far too hot for humans on Venus, almost too cold on Mars, and a bit of both on Mercury. Even so, the other 3 terrestrial planets are much more like Earth than like the other 5 planets.

territory (*ethology*)   Many animals, including most bird species, establish an area that they defend against other individuals, usually of the same species. This area, or territory, may have various purposes: it may provide food, nesting sites, access to mates, and so on. It may be large or small and may persist year-round or only temporarily (during breeding season, for example). It may belong to an individual, a pair, a family, or a large group. Generally the fittest animals get the best territories, further enhancing their chances of survival and reproductive success. Animals patrol their territories and mark them with scents (dogs), song (wood warblers), or threat displays (fiddler crabs). If this is not sufficient, the defenders fight the intruders.

Tesla, Nikola (*physicist*) [Croatian-American: 1859–1943] Tesla's successes came in the 1880s and 1890s with his invention of an electric motor that runs on alternating current and transformers that make long-distance transmission of electric power possible. In the 20th century he worked on problems of transmitting electric power without wires, an idea that failed to have any significant impact.

testis (*anatomy*) A male animal has 2 gonads, or reproductive organs, called testes. They have 2 major functions: male gametes, called sperm, are produced in coiled tubules; male sex hormones, particularly testosterone, are produced by interstitial cells lying outside the tubules.

In vertebrates the testes form in the abdomen near the kidneys. In fish, amphibians, reptiles, and birds, the testes remain in the abdomen. In mammals they descend into a pouch of skin called the scrotum.

Tethys (*geology*) The ocean that developed between Gondwanaland and Laurasia some 180,000,000 years ago is known at the Tethys Ocean or Tethys Sea, named by Alexander Du Toit [South African: 1878–1948] in 1937. As Africa and Eurasia came together over the past 45,000,000 years, some of the Tethys became trapped as the Mediterranean; the rest is part of the Indian Ocean.

tetrahedron (*geometry*) A tetrahedron ("4 bases") is any figure in 3 dimensions with 4 faces. However, in general, the regular tetrahedron, or Platonic solid, with each face an equilateral triangle, is meant. Any tetrahedron is also a triangular pyramid, so its volume V is 1/3 the area of one base, B, times the altitude from the vertex to that base, h: $V = 1/3\ Bh$.

thalamus (*anatomy*) The thalamus ("inner room") is the part of the vertebrate brain that sits atop the brain stem. It sorts and processes sensory information reaching the brain from the eyes, ears, and other sense organs. It then relays this information to the outer surface of the cerebrum.

Thales (*mathematician/scientist*) [Ionian: c625–c550 BC] Thales is credited as the first to have proved geometric theorems, such as "a diameter bisects a circle" or "base angles of an isosceles triangle are equal." He is also said to have used similar figures formed by shadows to measure the height of a pyramid and to have predicted at least 1 solar eclipse. There is no firsthand information about Thales, but many anecdotes about his wisdom were repeated by later Greek writers, who considered him the first to seek natural explanations of phenomena.

thallium (*elements*) A toxic metal, atomic number 81 was discovered in 1861 by Sir William Crookes, who first located it as a bright green line in a spectrum and named it from the Greek *thallos* ("green twig"). Thallium sulfate is a common pesticide, while the sulfide exhibits photoelectric phenomena and is employed in photoelectric cells. Chemical symbol: Tl.

thecodont (*paleontology*)   It is believed that dinosaurs evolved from a primitive group of reptiles called thecodonts, which lived in the late Permian and the Triassic. The name, from the Greek for "case teeth," refers to the fact that the animals' teeth were implanted in sockets. The earliest and most primitive thecodonts were lizardlike, with their limbs projecting sideways from the body. In later species the limbs were drawn downward and inward, as in modern crocodilians; this raised the body off the ground, allowing for faster movement.

Theophrastus (*botanist*)   [Greek: c370–c285 BC] A student and friend of Aristotle and his successor as head of the Peripatetic School at Athens, Theophrastus has been called the father of botany. One of his books deals with plant structure and physiology. Another describes and classifies some 500 kinds of plants, based mainly on habit of growth (trees, shrubs, etc.) but also on type of inflorescence, position of the ovary, and other characteristics.

theorem (*logic*)   A theorem is any proposition established through a logical proof. In an axiomatic system the theorem must be proved from the axioms and definitions. A theorem that can be proved in a very few steps from an established theorem is a corollary, while a theorem that is proved as a preliminary to a more important goal is a lemma.

theory (*philosophy of science*)   A scientific theory is a broad explanation of a set of observations that is accepted so long as no series of observations contradicts the explanation. The word "theory," however, continues to be used for explanations that have been superseded; for example, modern quantum theory replaces Niels Bohr's theory of the hydrogen atom. Useful theories predict specific results not previously observed, as Mendel's theory of heredity predicts dominant and recessive traits; or explain known observations not covered by earlier theories, as Einstein's general relativity theory correctly accounts for observed deviations in the orbit of Mercury from the predictions of Newton's gravitational theory.

therapsid (*paleontology*)   Arising from the pelycosaurs in the mid-Permian, reptiles of the order Therapsida ("mammal arch") are believed to have been ancestors of mammals. They had specialized teeth (incisors, canines, premolars, and molars) and skulls much like those of mammals. Their legs were underneath the body rather than sprawled to the side, as in more primitive reptiles. The most mammal-like therapsids were the theriodonts ("mammal toothed"), including the dicynodonts.

thermal (*meteorology*)   A rising current of air is called a thermal. The rise is powered by cooler air moving below air heated by a warm patch of Earth's surface. Thermals are important to soaring birds and gliders and are also the source of cumulus clouds. As a parcel of air rises, it reaches regions of lower air pressure; this produces expansion that causes the parcel to cool. When cooling brings the air below the dew point, it becomes a cloud.

thermal pollution (*environment*)   The discharge of heated water from industrial processes into cool rivers and lakes is called thermal pollution. By increasing temperature, the heat can affect life processes of aquatic plants and animals. For example, as temperature increases, water's ability to hold oxygen decreases. Organisms that require high concentrations of oxygen, such as trout, can no longer survive in the habitat.

thermionic emission (*thermodynamics*)   A heated conductor, such as a metal, releases from its surface some of its loosely bound electrons, especially into a vacuum, a phenomenon called thermionic emission (also known as the Edison effect).

thermodynamics (*sciences*)   The study of the movement of heat is called thermodynamics. Since everyone observes that hot matter left alone soon cools, it might seem that there is little to study. But understanding of heat is fundamental to physics. Conservation laws show that heat is a form of energy, explained by the kinetic theory of heat as average particle momentum. Among the contributions of thermodynamics are explanations of how engines work, the wide-ranging notion of entropy, and the important mathematical technique of analysis based on the Fourier series. The second law of thermodynamics, or law of entropy, is joined with the first law (conservation of energy) and the rules of absolute zero, the third law, to form the basis of thermodynamics.

thermoluminescence (*materials science*)   Thermoluminescence is light produced by heating solids that have been exposed to radioactivity or to any strong radiation. Thermoluminescent light is beyond the normal light produced when any substance reaches a high temperature. Exposure to radiation knocks some electrons into defects in the solid's structure; there they stay trapped until released by the agitation caused by heating. As they exit the traps, the electrons radiate visible light. Thermoluminescent materials, such as lithium fluoride, are used to measure accumulated doses of radioactivity, since thermoluminescent light is proportional to exposure. Archaeologists date fired ceramics with thermoluminescence since trapped electrons are eliminated by firing; however, more electrons gradually become trapped, usually in response to radioactive isotopes in the ceramic. Thus, the amount of thermoluminescence reflects the date of firing.

thermometer (*thermodynamics*)   Most substances expand in volume at higher temperatures since an increase in the average kinetic energy of the molecules causes the molecules to move farther apart. This expansion is the basis of nearly all thermometers. Thermometers are designed to reach thermal equilibrium quickly with the environment. The expansion of a substance in the thermometer is measured and reported as the temperature.

thermosphere (*atmospherics*)   The thermosphere is the part of the outer atmosphere beyond 50 mi (80 km) above the Earth, a region with so few atoms and molecules that most people consider it outer space. Its name comes from the high average speed of its few particles, which is expressed as a high temperature, although there are so few particles that heat content is low. Space suits used by astronauts in the thermosphere need to be heated.

theropod (*paleontology*)   The theropods made up one of the major groups of lizard-hipped dinosaurs. They typically had a large head with strong jaws that contained numerous bladelike teeth suitable for a carnivorous diet. Their front legs were small and had sharp, curved claws for grasping prey. Theropods stood and walked on their strong, massive hind legs. They had long tails that acted as a counterbalance. Well-known theropods include *Allosaurus, Tyrannosaurus,* and *Velociraptor*.

thistle (*plants*)   Certain species of composite plants native to the Northern Hemisphere are commonly called thistles because of their spiny stems and leaves. Spiny bracts may surround the flower head, which is composed of many small flowers that usually are pinkish or purplish in color. Some thistles, such as the bull thistle (*Cirsium vulgare*), native to Europe but introduced to North America, are troublesome weeds. Thistles of the genus *Cynara*, particularly *C. scolymus*, are cultivated for the fleshy edible bracts of their giant flower buds, known as globe artichokes.

Thomson, J. J. (Joseph John) (*physicist*)   [English: 1856–1940] In 1897 Thomson proved that a cathode ray consists of equal-mass particles that respond to both electric and magnetic fields. The particle mass is that of the smallest mass change during ionization, much smaller than a hydrogen atom. The tiny charged particle Thomson had identified was the electron.

Thomson also made the first version of a mass spectrometer (a device that separates ions by mass) and used it to find 2 isotopes of neon, physical proof that isotopes, then only suspected in radioactive decay, exist.

thorax (*anatomy*)   In vertebrates and many arthropods, the thorax is the part of the body between the head and the abdomen. In mammals the thorax is enclosed by the ribs and contains the heart, lungs, and esophagus; it is separated from the abdomen by the muscular diaphragm. In crustaceans and arachnids (spiders, etc.), the thorax is joined with the head to form a single unit called the cephalothorax.

thorium (*elements*)   The metal thorium, atomic number 90, was the second radioactive element discovered (in 1829 by Jöns Jakob Berzelius). Radioactivity was unrecognized before 1896, so thorium's name, for the lightning-bolt tossing Norse god Thor, is coincidental. Abundant thorium in common rock contributes to heating Earth and to radon gas that may accumulate in basements. Thorium oxide has a high melting point, making it useful in mantles for gas lamps and other high-temperature applications. Chemical symbol: Th.

thorn (*botany*)   Acacias, locusts, hawthorns, and other plants have stems modified as hard, sharp structures called thorns. The thorns protect the plants against plant-eating animals. Thorns should not be confused with spines, which generally are modified leaves.

threat display (*ethology*)   Fighting is dangerous and best avoided. Threat displays are designed to prevent or end disputes by scaring off opponents. For example, a male baboon stands, erects his ruff, and opens his jaws to display his canines; a penguin lowers its bill and arches its neck; a hognose snake flattens its head, inflates its body with air, and hisses.

threatened species (*environment*)   Any species whose numbers have declined to the point where it is likely to be in danger of extinction unless remedial steps are taken is considered threatened. The species may face problems in only part of its range; a migrating bird species, for example, may have plentiful wintering grounds but may find its nesting grounds being replaced by urban sprawl. (*See also* **endangered species**.)

thrips (*animals*)   Thrips (Order Thysanoptera—"fringe wings") are small insects usually less than 0.12 in. (3 mm) long. Some species have 4 long, narrow wings fringed with long hairs; others are wingless. Most thrips are plant feeders, with piercing-sucking mouthparts for drawing out juices. Species such as the onion thrips (*Thrips tabaci*) and citrus thrips (*Scirtothrips citri*) are major crop pests.

thrush (*animals*)   Widely distributed around the world, the approximately 330 species of the thrush family, Turdidae (sometimes considered a subfamily of Muscicapidae), include blackbirds, bluebirds, nightingales, robins, thrushes, and other familiar perching birds, many famed for their lovely singing. The birds range in length from 5 to 13 in. (13 to 33 cm) and have 10 flight feathers on each wing. Their diet consists mainly of insects and fruit.

thulium (*elements*)   Although thulium is the rarest—least abundant—rare earth, it is still 100 times as common in Earth's crust as gold. Thulium, atomic number 69, chemical symbol Tm, was discovered in 1879 by Per Teodor Cleve [Swedish: 1840–1905], who named the metal after *Thule*, Greek for northernmost inhabited land, which could be taken to be Sweden.

thunderstorm (*meteorology*)   A thunderstorm is heavy rain and wind accompanied by lightning; thunder from lightning gives the storm its name. The origin of a thunderstorm is a tall cumulonimbus cloud that begins with thermals that lift moisture into the upper air. When the cloud top reaches the base of the stratosphere, the inversion causes the top to flatten and spread. In the higher reaches of the cloud, water vapor freezes and ice falls, starting downward motions of air (downdrafts) among the upward flowing thermals. Convection cells form within the cloud. The downdrafts now carry both rain and electric charge toward the bottom of the cloud, creating all 3 major effects of the storm—rain, wind, and lightning.

thymus (*anatomy*)   Found in all vertebrates, the thymus is a small organ named for its resemblance to a thyme leaf. In humans and other mammals, it is located under the breastbone, just above the heart. Precursors of white blood cells called T lymphocytes (the T is for thymus) form in bone marrow, then migrate to the thymus, where they mature into T lymphocytes. Although in humans the thymus gradually decreases in size after puberty, it continues to make new T lymphocytes throughout adulthood.

thyroid (*anatomy*)   Located in the neck region of vertebrates, adjacent to the thyroid cartilage of the larynx (voice box), the thyroid is an endocrine gland that synthesizes several hormones, of which the most important are thyroxin and calcitonin.

ti (*plants*)   Common throughout Polynesia, where it traditionally has had many economic uses, ti (*Cordyline terminalis*) is an evergreen shrub or small tree with a slender stem and strong branches; the ends of the branches are decorated with a whorl of elongated glossy leaves. Ti is a popular ornamental and numerous cultivars have been developed, with leaves ranging from green to maroon to variegated. Related species also are cultivated as ornamentals, including the cabbage tree (*C. australis*), which in the wild grows to about 40 ft (12 m).

tick (*animals*) Ticks are large, blood-sucking mites with a tough, leathery exoskeleton and 4 pairs of clawed legs. They are parasites of reptiles, birds, and mammals. Some species transmit human diseases, such as Rocky Mountain spotted fever, tularemia, and Lyme disease. A tick spends most of its life on vegetation or the ground waiting for an appropriate host. When it senses a host, it climbs aboard, pierces the skin with its mouthparts, and feeds, increasing greatly in size as its digestive tract fills with blood.

tidal bore (*oceanography*) A very high tide entering an estuary can form a rapidly advancing wave called a tidal bore. The best known example is in the Bay of Fundy in eastern Canada, where a combination of suitable bottom topography and a natural period of oscillation close to that of the tides combine to produce a high tide as much as 40 ft (12 m) above low tide. A tidal bore is an example of a soliton, a single wave traveling alone.

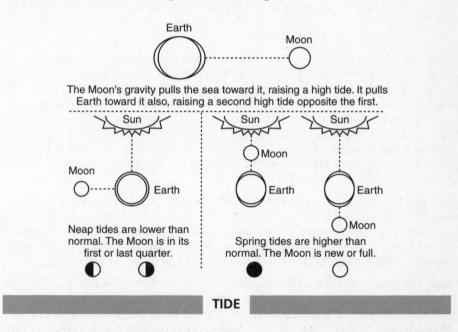

Earth

Moon

The Moon's gravity pulls the sea toward it, raising a high tide. It pulls Earth toward it also, raising a second high tide opposite the first.

Sun   Sun   Sun

Moon

Moon   Earth   Earth   Earth

Moon

Neap tides are lower than normal. The Moon is in its first or last quarter.

Spring tides are higher than normal. The Moon is new or full.

TIDE

tide (*oceanography*) The pull of gravity from the Moon and to a lesser extent the Sun raises fluids on Earth, an effect that is most noticeable in the oceans, where it is called the tide. As Earth turns below the Moon, the part of the ocean facing the Moon is elevated, producing an increase in sea level called high tide. A second high tide forms on the opposite side of Earth because the Moon's gravity pulls the solid Earth as a unit based on its center, bringing the solid Earth closer but having less effect on the more distant far-side ocean. Because the Moon is moving around the Earth while Earth rotates, high tide occurs somewhat more than every 12 hours (half a rotation)—about 12 hours 25 minutes.

When the Moon is full or new, it is lined up with the Sun, producing a higher tide than usual, called a spring tide. Similarly at the first and second quarters, the lower tides caused by the Sun and Moon being at right angles are neap tides.

tiger (*animals*)   Competing with the lion for the title of largest member of the cat family, the tiger (*Panthera tigris*) is an endangered species that lives in forests and on steppes of Asia. Tigers depend mainly on their acute hearing to stalk prey. They hunt on the ground, mostly at night. After a kill, a tiger eats its fill, then hides the carcass, returning to the carcass again and again until all the meat has been eaten.

tilapia (*animals*)   Plant-eating fish of the cichlid family, tilapias (*Tilapia*) inhabit most tropical and subtropical waters. They live mainly in freshwater but also occur in brackish and salt water. They average less than 1 ft (0.3 m) in length and 2 lb (0.9 kg) in weight. Tilapias were among the first fish raised by people; a bas-relief in an Egyptian tomb dating from 2000 BC shows tilapias being harvested in an artificial pond.

tiling (*geometry*)   Covering a portion of a plane with congruent figures (or 2 or more sets of figures that are congruent within a set) without gaps or overlaps is called tiling; the figures are tiles. Triangles, rectangles and regular hexagons are commonly used as tiles, but it is impossible to tile with regular pentagons by themselves because the pentagon's angles of 108° cannot be combined to reach 360° around a common vertex. In 1973 Roger Penrose [English: 1931–] discovered that there are small sets of tiles (2 types of congruent tiles used together at minimum) that tile a plane in such a way that the pattern fails to repeat at any scale. These are now called Penrose tilings.

till, glacial (*geology*)   Two types of drift are left by glaciers. Drift that forms where the glacier melts is called glacial till; it is recognized as a jumble of rock dust and rocks of all sizes. Drift that has been carried away from the glacier, called outwash, has rocks and dust that lie in layers produced because small pieces remain suspended in water longer than large ones.

time (*physics*)   Time is a continuous, 1-way flow of events—the quantity that clocks measure. Consider an event $S1$ that changes state to $S2$ (different from $S1$) and returns to a new version of its original state ($S3 = S1$), repeating this sequence many times. If sliding the whole sequence so that $S1$ becomes $S3$ produces no observable change in the sequence, the sequence is a clock. The official clock is the atom of cesium-133, which passes from a state $S1$ to $S3$ exactly 9,192,631,770 times in an amount of time defined as 1 second. This official second is close in value to the earlier second based on the day and equal to 1/86,400 of Earth's rotation as 1 second.

Time is a coordinate combined with 3 space coordinates in relativity theory. Each space coordinate is measured by the speed of light, but time uses a clock.

When a frame of reference is moving with respect to an observer, space coordinates shrink in the direction of motion and the time coordinate slows—the clock repeats more slowly. Time differs in each frame of reference.

Time moves in the same direction for all observers (from past toward future). The second law of thermodynamics is that entropy in a closed system increases with time. In particle physics some particles, such as kaons, show asymmetry for certain properties (matter predominates over antimatter, for example), implying mathematically that time cannot be symmetrical and therefore has a preferred direction.

tin (*elements*)   Tin, atomic number 50, was known to prehistoric smelters who added it to copper to produce bronze, the most useful metal before iron. Tin's English name has been used since Anglo-Saxon times, when mines in Cornwall were already ancient. Its symbol, Sn, is from Latin *stannum*. Tin is among the most widely used metals in alloys, including solder, pewter, and metals used for bearings, type, and dies. Tin is plated over steel, for example in "tin" cans, to reduce corrosion.

Tinbergen, Nikolaas (*ethologist*)   [Dutch: 1907–88] Among the first to study animal behavior under natural conditions, Tinbergen is best known for his investigations of the territorial behavior of stickleback fish and herring gulls, in both cases emphasizing the importance of stimulus-response processes. For example, during mating season a male stickleback is stimulated by the red belly of an intruding male to respond with an attack. Tinbergen shared the 1973 Nobel Prize for physiology or medicine with Konrad Lorenz and Karl von Frisch [Austrian: 1886–1982].

tissue (*biology*)   In multicellular organisms, cells are organized into groups called tissues, which work together to carry out a specific function. Humans and other mammals have 4 main types of tissues. The cells of a tissue may form a solid or a fluid. Some tissues consist of only 1 type of cell while others are made up of 2 or more types of cells. In most multicellular organisms, tissues are aggregated into organs.

HUMAN TISSUES	MAJOR FUNCTIONS	EXAMPLES
connective	holds together and supports other kinds of tissues; stores fat; forms scar tissue	bone, cartilage, tendons, ligaments, blood, lymph, adipose (fat-storage)
epithelial	forms a protective covering over the body, its individual organs, and internal passageways and cavities; forms glands that secrete digestive juices, hormones, and other fluids	outer layer of skin; linings of digestive tract, uterus, and blood vessels; salivary, sweat, and thyroid glands
muscle	contracts to provide movement	skeletal (attached to skeleton), cardiac (wall of heart), smooth (walls of internal organs)
nerve	conducts nerve impulses	brain, spinal cord, nerves

Titan (*solar system*)   Titan, Saturn's largest moon—3000 mi (4800 km) in diameter—is the only moon in the solar system known to have an atmosphere composed mostly of nitrogen, like that of Earth, with only a small percentage of methane and carbon monoxide. Atmospheric pressure is at least 1.5 times that on Earth and temperatures range around -294°F (-181°C). Titan appears to be a frozen version of Earth before life evolved, so it is conceivable that some precursors of life have formed there.

Titania (*solar system*)   Titania is the largest of the 18 moons of Uranus, although with a diameter of only 980 mi (1578 km) it is only slightly larger than Oberon, Umbriel, and Ariel and about 3 times the size of Miranda. Most of Uranus's moons are named for characters from Shakespeare, but Umbriel is from *The Rape of the Lock* by Alexander Pope.

titanium (*elements*)   Although titanium, atomic number 22, was discovered by William Gregor [English: 1761–1817] in 1791, it did not become commercially available until 1946, when an inexpensive process for extracting this common metal (ninth most abundant in Earth's crust) was developed. Today titanium metal, strong and light, is important in alloys and on its own, while titanium dioxide provides a bright white pigment used in paints. Named for Titans of classical mythology, its chemical symbol is Ti.

titmouse (*animals*)   Perching birds of the titmouse family, Paridae, include chickadees and titmice—called tits in Great Britain. Typically less than 6 in. (15 cm) long, they have a compact body covered with soft plumage, mostly in browns and grays. The birds have short rounded wings, each with 10 flight feathers, and often a very long tail. They have a short pointed bill and feed mainly on insects and seeds.

titration (*chemistry*)   Titration is a method of finding the concentration of an acid or base in a solution by adding an indicator a drop at a time. The name comes from titer, originally the amount of gold or silver in an alloy, but now the ratio expressing concentration of a solution.

tobacco (*plants*)   Members of the nightshade family native to the Americas, tobaccos (*Nicotiana*) are notorious for their toxicity. The most active toxin is the alkaloid nicotine, which is present in all parts of a tobacco plant. The most widely cultivated species, *N. tabacum*, grows up to 10 ft (3 m) tall. It has large, broad leaves that are harvested, dried, and used to make cigarettes and other products.

tobacco mosaic virus (*viruses*)   An infectious disease of tobacco plants known as tobacco mosaic disease, because of the pattern of bleached spots that appear on the leaves, is caused by a virus of the genus *Tobamovirus*. The virus particle is rod-shaped and consists of a protein coat surrounding a single strand of RNA. It was the first virus to be isolated, in 1935 by Wendell M. Stanley [American: 1904–71] (for which he shared the Nobel Prize in chemistry in 1946).

togavirus (*viruses*)   Many members of the virus family Togaviridae are transmitted by insects and cause human disease, including yellow fever, dengue fever, and several types of encephalitis. The rubella (German measles) virus also is a member of this family. Togavirus particles are spherical and encased in a lipid envelope. The genetic information is stored in a single strand of RNA.

tomato (*plants*) A member of the nightshade family native to South America, the tomato (*Lycopersicum esculentum*) is a short-lived perennial, though it usually is grown as an annual. Numerous varieties, or cultivars, have been developed. Tomato plants grow up to 6.5 ft (2 m) tall. They have toxic compound leaves, and clusters of small yellow flowers. The edible fruit is a berry, consisting of a fleshy, juicy ovary in which are embedded numerous seeds.

Tomonaga, Shin'ichiro (*physicist*) [Japanese: 1906–79] Tomonaga worked out details of the way that particles of the same type interact (through exchanging a particle of a different type). His theory, called quantum electrodynamics (QED), links quantum mechanics with special relativity. It is equivalent to theories of QED reached independently and simultaneously by Richard Feynman and Julian Schwinger [American: 1918–94].

tongue (*anatomy*) A fleshy, muscular organ attached at one end to the floor of the mouth, the tongue is covered with tiny sense organs, buds, that perceive taste. In many amphibians, reptiles, and birds, the tongue helps in capturing food—for example, in less than ⅓ second a chameleon can extend its tongue forward to lengths greater than its body and trap an insect on the sticky tip. In mammals the tongue is important for manipulating food during chewing.

tonsil (*anatomy*) Near the back of the throat of mammals and some other vertebrates, under the mucous membrane, are small bodies called tonsils. Part of the lymphatic system, tonsils help control infection by filtering out bacteria and viruses from lymph.

tool industry (*paleoanthropology*) A set of objects manufactured by a single culture is called its tool industry. Probably the earliest hominid tool industries used wood and other plant materials that did not survive. Stone tools produced by hominids first appeared about 2,400,000 years ago, surviving long passages of time with little physical change. Early cultures can be identified by the stone tools that were characteristic, while later cultures also have bone, horn, or even wooden tools preserved.

Tool use is common in primates, especially among chimpanzees and, of course, humans. Chimpanzee cultures are characterized in part by the types of tools they make and their use. One chimpanzee culture may use a long stick to collect ants from a hill and wipe the ants off the stick with a hand before eating, while another may use a short stick that is placed directly in the mouth. Tools from hominids that are extinct can be studied for differences in tool types or methods of manufacture, but the patterns of use must be inferred. Thus, although many hominids made tools that paleoanthropologists call choppers, scrapers, or cleavers, the way these tools were actually used is unknown.

tooth shell (*animals*) Also called tusk shells because of their resemblance to elephants' tusks, tooth shells are marine mollusks that burrow front-end first into sand or mud. The posterior projects up into the water. A constant flow of water through the posterior opening of the shell supplies the animal with food (microscopic organisms) and oxygen.

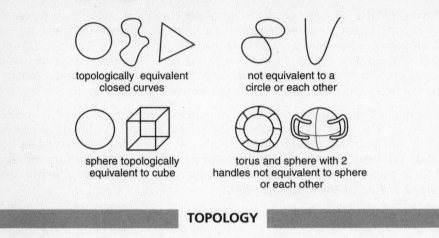

topologically equivalent
closed curves

not equivalent to a
circle or each other

sphere topologically
equivalent to cube

torus and sphere with 2
handles not equivalent to sphere
or each other

TOPOLOGY

topology (*mathematics*)   Topology began as the study of complicated surfaces, such as the Möbius strip, and invariant properties of curves. It is sometimes called rubber-sheet geometry because continuous deformations (similar to stretching a drawing on rubber) do not affect topological properties, such as connections. Two figures connected the same way are topologically equivalent. As geometry, topology includes knot theory. But modern topology, even the part concerned with sets of points, often seems far from geometry. Topological spaces are defined in terms of how neighborhoods of elements behave or the kinds of measurement employed.

top quark (*particles*)   The top quark (once known as "truth") was observed definitely for the first time at the Fermilab particle accelerator in Illinois in 1995, completing the third generation of quarks. It was the last predicted fermion from the standard model to be found, a task made difficult by the high mass of the particle.

tornado (*meteorology*)   A tornado is a whirling wind that originates in a thunderstorm when cooled air begins to hang below the main cloud and some force—thought to be winds caused by the meeting of 2 air masses—starts the hanging cloud spinning. As the cloud spins, it extends downward and shrinks in diameter, increasing wind speed as a result of conservation of angular momentum. Wind speeds can reach 300 mph (500 km/hr); within the tornado air pressure becomes very low. The tornado moves near the surface at about 30 mph (50 km/hr) over a path that may be 10 to 20 mi (15 to 30 km) long, causing great destruction when it reaches down to ground level.

torque (*mechanics*)   Torque (or moment of force) is the result of a force exerted on a body that causes rotation about some point or axis. For example, applying a force to a handle or wheel attached to an axle has an increased effect when the applied force is farther from the axle. The distance from the axis in such a case is also the radius of a circle, and torque is measured as force multiplied by that radius.

Torricelli, Evangelista (*physicist*)   [Italian: 1608–47] Galileo suggested that Torricelli investigate why vacuum pumps fail to raise water higher than about 33 ft (10 m). In 1643 Torricelli filled a long glass tube, closed at one end, with mercury, which is much denser than water, and placed the open end in a dish of mercury. When the tube was vertical, the mercury fell to about 30 in. (76 cm), forming a vacuum in the top of the tube. This was the first barometer. Torricelli correctly concluded that air pressure pushes on the mercury, lifting it. The vacuum prevents an equal push from the opposite direction. Torricelli also noted the small changes in height of the mercury related to changes in weather.

torus (*geometry*)   A torus is a geometric figure shaped like a doughnut, complete with hole. Formally the torus is a solid formed by rotation in 3 dimensions of a circle about a line in the plane of the circle whose distance, k, from the center of the circle is greater than the circle's radius, r. The volume of such a torus is $2\pi^2 kr^2$ and the area of its surface is $4\pi^2 kr$.

toucan (*animals*)   Constituting the bird family Ramphastidae, toucans are easily identified by their very large beak, which looks somewhat like a lobster claw. The beak and plumage are usually brilliantly colored. Like their relatives the woodpeckers, toucans are skillful climbers. They live in tropical forests of Central and South America, eating fruits and small animals. They are gregarious and communicate with one another in squawking cries.

touch (*physiology*)   The sense of touch depends on specialized receptor cells that "fire" and alert the central nervous system whenever they detect pressure. A human, for example, has an estimated 640,000 touch receptors, distributed unevenly and at different depths in the skin. Some of these receptors detect light pressure, others detect deep pressure, still others detect the movement of hairs. Some animals have organs of touch that protrude from the skin; the whiskers of cats and many other mammals are an example.

toxin (*biochemistry*)   Any poisonous protein produced by an organism that interferes with some vital function in another type of organisms is a toxin. For example, the cause of botulism, botulin, produced by the bacterium *Clostridium botulinum*, blocks transmission of nerve impulses; ricin, found in seeds of the castor bean plant (*Ricinus communis*), causes clumping of red blood cells.

Toxins are found among bacteria, protists, fungi, plants, and animals (those produced by animals are called venoms). An antitoxin is an antibody formed in the blood that combines with and neutralizes a toxin. Each kind of toxin stimulates formation of a specific antitoxin.

trachea (*anatomy*)   In air-breathing vertebrates, the windpipe, or trachea, leads from the pharynx in the back part of the mouth down toward the lungs. In the thoracic (chest) cavity, the trachea divides into 2 branches, the right and left bronchi. Each bronchus enters a lung and subdivides into smaller and smaller branches that terminate in the alveoli.

In most insects and some other arthropods, trachea are hollow tubes that carry respiratory gases between the body cells and the outside, via openings on the thorax called spiracles.

transcendental number (*numbers*)   A real number that is not also an algebraic number is called transcendental. Examples include π (= 3.14159265358979 . . .), which is the ratio of the circumference of a circle to its diameter, and e (= 2.718281828459045 . . .), the base of an exponential function whose rate of increase is the same as the value of the function. Although the transcendental numbers appear at first to be rare, Georg Cantor showed that there are more transcendental numbers than any other kind of real number.

transcription (*genetics*)   An early step in protein synthesis is the formation of messenger RNA (mRNA) using DNA as a template, or pattern. Called transcription, the process begins as part of the DNA molecule unwinds. A certain sequence of bases on the DNA—called a promoter—signals the enzyme RNA polymerase to begin RNA synthesis. RNA nucleotides bond to complementary bases on one strand of the DNA and are linked together to form a strand of mRNA. Each group of 3 consecutive bases on the mRNA is a codon for a specific amino acid. The order of codons indicates the order of amino acids needed to make the protein. The RNA polymerase continues to link nucleotides together until another sequence of bases on the DNA—called the terminator—signals the end of transcription. The new mRNA then separates from the DNA and moves out of the nucleus into the cytoplasm.

transduction (*genetics*)   The transfer of genetic material from one bacterium to another by bacteriophages (viruses that infect bacteria) is termed transduction. The new host bacterium incorporates the transferred genes into its own DNA and then passes them on to its descendants. Transduction occurs fairly often and has probably contributed to the great genetic diversity among bacteria.

transfinite (*numbers*)   Ordinal and cardinal numbers that are not integers are transfinite. The transfinite cardinal numbers count the members in sets with infinite numbers of members—for example, the number of natural numbers or of points in space. Transfinite ordinal numbers begin with the set of natural numbers arranged in order of size—one possible ordering—called ω (Greek omega). A different transfinite ordinal, ω × 2, is based on ordering the natural numbers with the odds followed by the evens: 1, 3, 5, . . . , 2, 4, 6, ...

transformation (*geometry*)   A general definition of geometric transformation is any rule that sets up a one-to-one correspondence between the points in one figure and another (with perhaps some points transformed into themselves). It helps to picture transformations as moving or stretching or shrinking the first figure to get the second. In elementary school common transformations are given names such as slides, flips, and turns, which correspond to the transformations known to mathematicians as translation, reflection, and rotation. These transformations all maintain congruence between the original and the transformed figure. Other transformations, such as dilatation, result in similar figures that are not necessarily congruent, while still others result in figures that do not resemble the original. In the identity transformation the transformed figure is the same as the original.

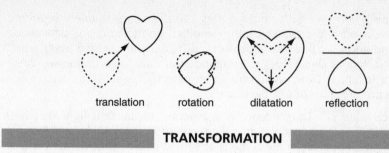

translation   rotation   dilatation   reflection

TRANSFORMATION

transit (*solar system*) When one heavenly body passes in front of another that is much larger, the event is called a transit of the smaller body; for example, the transit of Venus occurs when Venus is seen directly between Earth and the Sun. If the larger body is covered completely or nearly, however, the event is an eclipse and not a transit. From Earth both Mercury and Venus can transit the Sun. We can also observe the transits of satellites of the gas giants across their planets.

transition element (*elements*) Transition element is a category of metals in which an inner shell fills with electrons while the outer shell remains the same. Both the rare earths (lanthanide series) and actinide series of radioactive metals fit this definition. But the name transition element or transition metal refers to the hard metals from scandium (atomic number 22) through copper (29) or zinc (30) with all the atomic numbers between; from yttrium (39) through silver (47) or cadmium (48); and from hafnium (72) to gold (79) or mercury (80). Opinions differ as to whether zinc, cadmium, and mercury, for which the next-to-outer shell is completely filled, should be included.

The nearness of the outer shells to each other means that most ions display color, which also appears in compounds (for example, the blue of cobalt compounds). Except for the zinc-cadmium-mercury group, transition metals each have more than 1 valence number. Atoms are relatively small for their masses, making transition metals comparatively dense. The "heavy metals" of environmentalists include the transition metals heavier than iron as well as lead and sometimes tin or antimony.

translation (*genetics*) After information is transferred from DNA to messenger RNA (mRNA) in the process of transcription, mRNA moves into the cytoplasm and binds to ribosomes, where protein synthesis takes place. The information in the mRNA is translated into the specific sequence of amino acids that make up a particular protein. Each amino acid is represented on the mRNA by a sequence of 3 nucleotides called a codon. Transfer RNA (tRNA) picks up amino acids and pairs them with their complementary mRNA codons.

translation (*geometry*) Translation is the transformation that can be pictured as sliding a figure in a single direction. Every point in the figure is moved. In analytic geometry every point (x, y) is moved to $(x - h, y - k)$ where h and k are constants; subtraction appears because the transformation is pictured as moving the axes h and k units, which shifts the coordinates of a point $-h$ and $-k$ units. Translation can also be defined as reflection twice across 2 parallel lines.

translocation (*genetics*)   Sometimes during cell division a segment of a chromosome breaks off and attaches itself to another, nonhomologous chromosome. This phenomenon, called translocation, differs from crossing over, which involves homologous chromosomes. Translocation can have immense consequences. For instance, Down syndrome in humans may result from a translocation in the chromosomes of one parent.

translucence (*optics*)   Translucence is a reaction to incident light that is in between transparency and opacity. Some light energy is absorbed by whole molecules, raising their speeds, while some is absorbed and emitted by atomic electrons and passes through the material. When any material is a thin enough sheet, it becomes translucent because there are too few molecules to collect the energy and convert it to heat.

transparency (*optics*)   A transparent material consists of molecules that are unresponsive to a particular frequency of light, far apart, or held tightly in a crystal structure. These molecules are not pushed by the light to collide with each other, heating the material. Instead the light's energy is absorbed by atomic electrons and quickly emitted at the same wavelength. In this way the light wave or photon passes from molecule to molecule without changing frequency—although slowed by the process. We see through materials that exhibit transparency to visible light instead of opacity. Material that is transparent at one band of wavelengths is usually opaque at others. Window glass is transparent to visible light, but opaque to both infrared and ultraviolet wavelengths, for example.

transpiration (*botany*)   The evaporation of water from the inside of a plant to the outside environment is called transpiration. Almost all transpiration takes place through stomata (openings) in the surface of leaves. Many factors influence the rate of transpiration. For instance, high temperatures, high air pressure, low humidity, and wind favor high transpiration. A plant can regulate transpiration by opening and closing the stomata.

transposable element (*genetics*)   Popularly though misleadingly referred to as jumping genes, transposable elements are segments of DNA that either move to and insert themselves into a different location on the same or another chromosome or copy themselves and cause these copies to be inserted at another location on the same or different chromosome. First described in 1951 by Barbara McClintock, transposable elements may consist solely of insertion sequences or may also contain one or more genes. Transposable elements have been found in many kinds of organisms, including viruses, bacteria, fungi, plants, and insects.

transverse wave (*waves*)   The characteristic defining a transverse wave is that entities forming the wave move perpendicular to the direction in which the wave moves. A rope that has one end tethered and the other lifted up and down develops transverse waves as each particle of rope in turn moves up or down. A higher region moves from one end of the rope to the other, although any part of the rope simply moves up and down. Water waves are transverse waves spreading in 2 dimensions but moving up and down in 3 dimensions. Light and other parts of the electromagnetic spectrum are 3-dimensional transverse waves that spread in all directions.

trapezoid (*geometry*)   A trapezoid is a quadrilateral with 2 parallel sides, usually a figure for which the other 2 sides are not parallel. If nonparallel sides are equal in length, the trapezoid is isosceles. The area of a trapezoid with parallel sides B and b (called bases) and altitude h (often called height) is half the product of the altitude with the sum of the bases, or $A = \frac{1}{2}(B + b)h$.

tree (*botany*)   A tree is a woody perennial plant that usually has 1 main stem, or trunk, from which arise numerous branches that can be quite big. Some tree species attain maximum heights of less than 20 ft (6 m), but others reach heights of more than 330 ft (100 m). The sequoias of California are believed to be the heaviest organisms ever to have lived on Earth. One specimen, the General Sherman tree, is more than 272 ft (83 m) tall and has a base 36 ft (11 m) across; it is estimated to weigh more than 6000 tons.

tree line (*ecology*)   A more or less well-defined line on mountains and in polar regions beyond which trees will not grow is called the tree line. At a slightly lower altitude or latitude than the tree line is the timberline. It indicates the point beyond which normal tree growth does not occur. In the area between the timberline and the tree line, trees tend to be dwarfed and misshapen, a characteristic called krummholz.

tree shrew (*animals*)   Misleadingly named, for they are not shrews, tree shrews constitute their own order, Scandentia. An alternate common name is tupaia, derived from the sound of the animals' call. Tree shrews are small mammals about the size and shape of squirrels. Most species even have long, bushy tails. But tree shrews lack the whiskers found on squirrels and have 5 toes on each front foot (squirrels have 4). Tree shrews inhabit tropical rain forests of central and southeastern Asia. Most forage in daytime, feeding on insects and fruits.

trench (*oceanography*)   The places where subduction is occurring, where one tectonic plate is moving under another, are the lowest regions of Earth's crust, called trenches. Trenches are all at the edges of oceans or near island arcs in oceans, since subduction produces both trenches and island arcs. The deepest is the Marianas Trench, 37,800 ft (11,516 m) below sea level.

triad (*elements*)   Columns in the periodic table contain elements with similar properties, but in the middle of the table are short rows of 3 elements called triads whose members are more similar to each other than to others in their columns. The triads are iron/cobalt/nickel, ruthenium/rhodium/palladium, and osmium/iridium/platinum.

triangle (*geometry*)   A triangle, the polygon with 3 sides, is the simplest closed curve in the plane with line segments for sides. It is the only polygon completely determined by the lengths of the sides. Since the 3 angles are always the same for a given set of sides, the triangle's shape is stable, making it the basis of

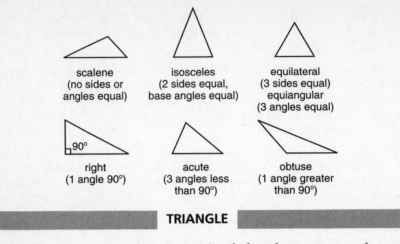

scalene
(no sides or
angles equal)

isosceles
(2 sides equal,
base angles equal)

equilateral
(3 sides equal)
equiangular
(3 angles equal)

90°

right
(1 angle 90°)

acute
(3 angles less
than 90°)

obtuse
(1 angle greater
than 90°)

TRIANGLE

rigid, load-bearing construction. Every triangle has the same sum of measures for the 3 angles, a straight angle (180°). Triangles are classified according to lengths of sides or measures of angles. The area A is half the base b times the altitude h: $A = \frac{1}{2} bh$.

tribe (*anthropology*)   Tribe is the level just above band for hunter-gatherers, but it may also refer to settled farmers or nomads. The members of a tribe generally include all humans who share a common language and culture and trace their ancestry to a legendary common ancestor. Tribes often do not form a single state, but organize themselves on the basis of local units such as villages or bands.

tribology (*materials science*)   The study of friction, abrasion or wear, and lubrication is called tribology (from Greek *tribos*, "rubbing"). Lubrication takes place when fluids or soft materials are confined between 2 surfaces moving with respect to each other, reducing friction.

Triceratops (*paleontology*)   *Triceratops* ("three-horn face") was a plant-eating horned dinosaur that lived in the late Cretaceous. Its name refers to the large horns on its head—one on the snout and one above each eye—which probably were used as weapons as *Triceratops* charged enemies. A bony shield, or frill, extended back from its head, over the shoulders. *Triceratops* grew to lengths of 30 ft (9 m) and weights of more than 6 tons.

trichina (*animals*)   The parasitic roundworms *Trichinella spiralis* live as larvae in the skeletal muscles of various mammals, including pigs, cats, dogs, and humans. If infected flesh is eaten raw or undercooked by a susceptible mammal, the larvae become adults and mate in the new host's small intestine. The new generation of larvae travels via the bloodstream to other parts of the body; larvae that reach muscles coil up and form cysts, which may survive for decades.

trichocyst (*cells*)   *Paramecium* and other ciliated protozoa have numerous threadlike structures called trichocysts located in tiny pits on the surface of their 1-celled bodies. Upon stimulation, as during an attack by a predator, a protozoan discharges its trichocysts. The protozoan may also use its trichocysts to attach to an object in its watery home.

triggerfish (*animals*)   Members of the family Balistidae are shallow-water marine fish most common in tropical seas. Among the best-known is the 8-in.-(20-cm-) long humuhumunukunukuapuaa (*Rhinecanthus aculeatus*), the state fish of Hawaii. A triggerfish has a unique protective device in its dorsal (top) fin. When the large front spine in the fin is erected, a smaller back spine moves forward to hold the taller spine erect. A triggerfish that erects its spine as it wedges itself in a narrow crevice is almost impossible to dislodge.

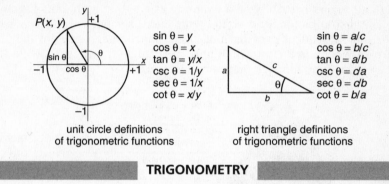

unit circle definitions
of trigonometric functions

right triangle definitions
of trigonometric functions

TRIGONOMETRY

trigonometry (*mathematics*)   Trigonometry is a small, but essential, corner of mathematics with applications to surveying, astronomy, navigation, and engineering. It concerns 6 functions that originated with the sine ("half chord"), the measure of half the chord of a circle produced by a given central angle, a relationship needed in observational astronomy. For angles less than 90°, the ratios of the sides of a right triangle give all 6 trigonometric functions, but many applications in engineering require periodic functions based on a definition using a circle and directed distances. Sines are positive in quadrants I and II, and cosines are negative in II and III, for example. Because they are periodic functions, the trigonometric functions also appear whenever waves or repeated motions are represented; they emerge unexpectedly in the integral calculus; and they form the basis of Fourier series.

trilobite (*paleontology*)   Marine arthropods of the class Trilobita were among the most common animals on Earth during the Paleozoic. Many species apparently lived on the floor in coastal waters, but some swam in the open sea. Most species were 1 to 4 in. (2.5 to 10 cm) long, but giant forms reached lengths of 2 ft (60 cm) or more. The name ("three lobed") refers to the longitudinal division of the exoskeleton into a median ridge and two flatter side areas. The oval body had three segmented regions: head, thorax, and tail. Numerous jointed legs on the underside of the thorax were used for crawling and swimming.

tritium (*elements*)   Tritium ("third substance") is the isotope hydrogen-3 that has 2 neutrons and 1 proton forming the nucleus of each atom. Tritium is radioactive with a half-life of 12.5 years. It is useful in nuclear fusion devices, such as the "hydrogen bomb," for which tritium atoms are created in particle accelerators.

Triton (*solar system*)   Neptune has 8 known moons—Triton and Nereid, discovered from Earth, and 6 others discovered by *Voyager 2* in 1989. Triton is by

Trojan asteroid

far the largest, with a diameter of 1681 mi (2706 km), and has an atmosphere. Triton travels in a direction opposite that of Neptune's rotation, suggesting that it may be a captured asteroid.

Trojan asteroid (*solar system*) In 1788 the mathematician Joseph-Louis Lagrange calculated the gravitational effect one body in orbit has on another. He found that there are 2 stable points in each planet's orbit, now called Lagrangian points, produced by interactions with the Sun, one each 60° in front of and behind the planet. Giant Jupiter has collected a swarm of asteroids at each of its Lagrangian points. The first to be found, over 100 years after Lagrange's work, was named Achilles, and subsequent ones have all been named after heroes of the Trojan War. Trojan asteroids in front of Jupiter are named for Greeks, while those behind are from Troy, although there are exceptions.

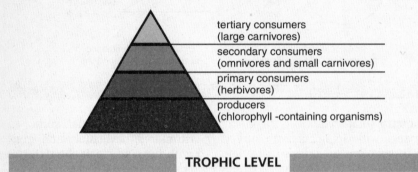

TROPHIC LEVEL

trophic level (*ecology*) A community contains several feeding, or trophic, levels, which can be illustrated using a pyramid. At the broad base is the large number of producers (plants and algae); at the apex is the small number of top predators (eagles, lions, etc.). At each step of the pyramid, energy is transferred from the food to the feeder. Some of this energy is stored but much of it is used in respiration to fuel life processes. For example, studies indicate that only about 10% of the food energy produced by plants actually becomes available for primary consumers. It is no wonder, then, that a community's producers far outweigh its consumers.

tropical storm (*meteorology*) A tropical storm begins as a warm air mass over oceans in the tropics after surface water has reached more than 76°F (25°C). The rising air creates a low-pressure area called a tropical depression, home to thunderstorms. When the thunderstorms come together to form a single body and winds reach 39 mph (63 km/hr) or more, the region is called a tropical storm. Tropical storms often strengthen to become hurricanes or cyclones—winds exceeding 75 mph (120 km/hr).

Tropic of Cancer and Capricorn (*geography*) The axis about which Earth rotates maintains a 23.5° angle to the stars throughout Earth's orbit. When one end of Earth's axis is pointed farthest away from the Sun, the region directly beneath the Sun at noon is at 23.5° of latitude. This occurs twice each year, each time at a solstice. Latitude 23.5°N is called the Tropic of Cancer ("turning of Cancer") and 23.5°S is the Tropic of Capricorn, both named from the zodiac constellations in which the Sun begins to reverse its apparent motion after the solstice.

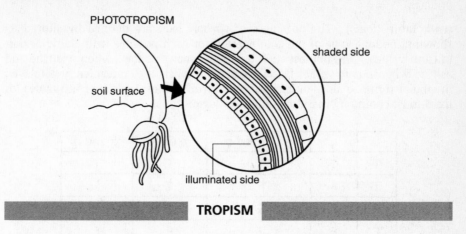

PHOTOTROPISM

soil surface

shaded side

illuminated side

TROPISM

tropism (*botany*) Plant movements that involve growth toward or away from an external stimulus are called tropisms. A positive tropism involves growth toward a stimulus; a negative tropism involves growth away from a stimulus. Positive phototropism, or growth toward light, is often seen in shoots and stems. The growth regulator auxin moves from the lighted side to the shaded side of a stem. This causes the cells on the shaded side to elongate more rapidly than those on the bright side, bending the stem toward the light.

Geotropism is growth in response to gravity. Downward growth of roots is an example of positive geotropism, while the upward growth of stems is negative geotropism. Other examples of tropisms include growth of a pollen tube down a flower's pistil (positive chemotropism), growth of roots into cracks in water pipes (positive hydrotropism), and coiling of a tendril around a support (positive thigmotropism, or response to touch).

troposphere (*atmospherics*) The atmosphere up to about 10 mi (15 km) is where all weather occurs. Nearly all clouds gather there and all precipitation begins in this layer, the troposphere. Temperature in this lowest layer of the atmosphere steadily decreases with height, reaching as low as −80°F (−60°C) just below the stratosphere, in an effect known as the lapse rate.

trout (*animals*) Fish native to the Northern Hemisphere and closely related to salmon, trouts live mainly in cool lakes and streams; some populations migrate to the ocean for part of their life cycle. Most species are 1 to 3 ft (0.3 to 0.9 m) long. Trouts have strong jaw teeth and primarily eat larva of winged insects, other small invertebrates, and small fish, including trout.

truffle (*fungi*) Forming the family Tuberaceae in the sac fungi phylum, truffles are the most highly esteemed of all edible fungi. They grow underground, on or near roots of oaks and other trees, and often are hunted by trained pigs and dogs, which locate them by their pungent odor. Truffles are round or potato shaped and range in diameter from 1 to 4 in. (2.5 to 10 cm). Their firm fleshy interior is covered by a smooth or knobby exterior that usually is white, brown, or black.

truth table (*logic*)   The operations of symbolic logic are defined by short lists showing the truth value of the combination for each possible truth value or pair of truth values. The simplest such list defines not *p* as false when *p* is true and as true when *p* is false. For binary operations the lists are conveniently displayed in tabular form, so all such lists are called truth tables. Here are truth tables for the 3 most common binary operations of logic:

p	*q*	if *p*, then *q*	*p* or *q*	*p* and *q*
T	T	T	T	T
T	F	F	T	F
F	T	T	T	F
F	F	T	F	F

Any proposition formed by combining these operations can be analyzed by using truth tables, making it always possible in this kind of logic to determine whether a proposition is true or false.

truth value (*logic*)   In symbolic logic each proposition has 1 of 2 truth values, usually called true and false, although any 2 nonoverlapping entities can be used—on and off or 1 and 0 are common. Truth tables differentiate between true and false. For example, the truth table for if *p*, then *q* specifies that it is true for 3 of the 4 possible assignments of true and false to *p* and *q*, so true differs from false. In 1921 Emil Post [American: 1896–1954] generalized this idea to logics with more than 2 truth values.

tsunami (*oceanography*)   A large mass of water impelled across the ocean by the sudden shift of part of the ocean floor during an earthquake or the collapse of an oceanic volcano is called a tsunami (also known, inappropriately, as a tidal wave). The Japanese name describes a characteristic of the waves: they rise up out of the sea as they approach land. While at sea, a tsunami's long wavelength and great speed of 440 to 500 mi (700 to 800 km) per hour results in a hardly noticeable wave only a few feet high. As the wave approaches land, however, the wave rises to 30 to 60 ft (10 to 20 m), causing great destruction along low-lying coastal regions.

tuatara (*animals*)   Rare natives of New Zealand, tuataras (genus *Sphenodon*) are the only surviving members of an ancient group of reptiles called beak-heads. Looking somewhat like a lizard, a tuatara has a spiny crest along its back that can be raised to warn or scare enemies. Males grow to about 25 in. (65 cm) long; females are somewhat smaller. Tuataras live on small rocky islands. During the day they rest in burrows. At night they hunt on the ground for insects, spiders, and other food.

tuber (*botany*)   A swollen underground stem adapted for food storage but also capable of **vegetative propagation** is the tuber, of which the potato is the best-known example. Buds (a potato's "eyes") on the tuber give rise to new plants.

Tuberous roots are thickened food storage organs that resemble tubers, but they are roots, not stems. Dahlias have tuberous roots.

tube worm (*animals*)   Various worms live on the sea bottom in protective tubes they secrete. They include beardworms (Phylum Pogonophora), horseshoe worms (Phylum Phoronida), and certain **bristle worms** (Phylum Annelida). Most use their exposed tentacles to obtain food. But giant tube worms (Pogonophora: *Riftia pachyptila*) up to 5 ft (1.5 m) long that live around hydrothermal vents depend on bacteria within their body. The worms take in oxygen and hydrogen sulfide from the environment. The bacteria oxidize the hydrogen sulfide, using energy released from this process to create organic compounds. Some of these compounds are secreted by the bacteria and used by their hosts.

tuff (*rocks and minerals*)   Volcanic **ash** mixed with larger pieces of rock accumulates in regions around an erupting **volcano**, forming layers of rock that geologists call tuff. If the ash and rock is still hot when it lands, the particles may weld together immediately. Tuffs are often easier to date than other kinds of **sedimentary rock**, so they are important to paleontologists and paleoanthropologists in regions, such as East Africa, that have a long history of volcanism.

tulip (*plants*)   Members of the **lily** family native to Europe, Asia, and northern Africa, tulips (*Tulipa*) are perennial herbs that grow from underground bulbs. Many varieties, or cultivars, exist. Tulips have a basal cluster of leaves and erect flower stalks, each bearing a single, showy flower. The flower's 3 petals and 3 sepals are similar in color and shape. The fruit is a 3-chambered capsule.

tumbleweed (*plants*)   Certain plants are known as tumbleweeds because of their distinctive method of **seed dispersal**. After flowering, the plants dry into a rounded shape and break off at ground level. As wind tumbles them along the ground, they drop their seeds. Common tumbleweeds include the Russian thistle (*Salsola kali*) and an **amaranth** (*Amaranthus albus*).

tumor (*biology*)   A tumor is an abnormal tissue mass, formed as cells in an area reproduce at an increased rate. Some tumors are benign, or harmless. Other tumors, called **cancers**, are malignant, invading surrounding tissues and spreading to other parts of the body.

tuna (*animals*)   Members of the **mackerel** family, tunas are large marine fish found worldwide in tropical and temperate waters. Unlike other fish, tunas can maintain body temperatures significantly higher than the surrounding water. Small species weigh about 18 lb (8 kg) but the bluefin tuna (*Thunnus thynnus*) reaches 1500 lb (680 kg).

tundra (*ecology*)   Circling the Arctic Ocean north of the **taiga** (coniferous forests) and south of the permanent snow and ice of the polar cap is a treeless biome called tundra (from a Finnish word meaning "barren land"). Tundra is characterized by long, bitterly cold winters and very short, cool summers. In mid-winter it is a land of darkness while in summer the sun shines 24 hours a day. The soil is frozen about 9 months of the year and is underlain by permanently frozen subsoil called permafrost. Lichens, mosses, heaths, sedges, and grasses dominate the vegetation, seldom growing taller than 12 in. (30 cm). Animals include arctic foxes, lemmings, voles, musk-oxen, and, in summer, caribou.

tungsten (*elements*)   Tungsten is best known as the filament in electric lamps, used for this purpose because it has the highest melting point of the metals. Tungsten, atomic number 74, was discovered in 1783 by Fausto d'Elhuyar [Spanish: 1755–1833], working with his older brother Juan José. The name is from the Swedish *tung sten*, "heavy stone," but the chemical symbol, W, is from German *Wolfram*, a miner's term for the "wolf" ore that "eats" tin.

Tunguska event (*solar system*)   On July 30, 1908, a collision between Earth and an asteroid near the Tunguska River in Siberia caused widespread devastation but no loss of human life because of the remote location. Calculations show that the Tunguska event can best be explained by the explosion of a stony meteorite about 100 yd (100 m) in diameter striking the atmosphere at 45 times the speed of sound and exploding in a cloud of superhot gas about 5 mi (8 km) above Earth's surface.

tunicate (*animals*)   Primitive chordates, tunicates share important features with vertebrates; they have no backbone but a notochord and spinal cord are present during the larval stage. Adult tunicates have an outer covering, the tunic, which provides protection and external support. Tunicates live in seas worldwide; most of the 2000 known species are sessile as adults, living attached to rocks, algae, and other substrates. The best-known tunicates are the sea squirts (Class Ascidiacea), which squirt jets of water from siphons when disturbed.

tunneling (*particles*)   One of the strange phenomena predicted by quantum theory, tunneling is the passage of an electron or other particle through an impenetrable barrier. Tunneling occurs because the location of a particle depends on probability and because particles also act as waves. The location must be fuzzy enough so that there is some probability that the particle will be on the far side of the barrier. In that case, after some statistically predictable interval, the particle will disappear on one side of the barrier to reappear on the other. Despite its basis in probability, tunneling can be counted on to occur and is the operating mechanism of many useful electronic devices.

turbidity current (*oceanography*)   Water in which sediment is suspended is denser than water with no sediment. Where sediment-laden water flows down a steep slope, as at the drop-off at the edge of the continental shelf, it picks up more sediment and speed as it travels, soon lifting pebbles and then large rocks. The water has become a turbidity current, the agent that scours out submarine canyons. Turbidity currents slow as they reach abyssal plains and the load of sediment falls to the floor of the ocean, sorting itself by weight into distinct layers.

turbulence (*mechanics*)   Turbulence in a fluid is flow that changes rapidly with time, expressed as irregular internal waves or vortices that quickly grow or dissipate. It is the opposite of smooth, or laminar ("in thin layers"), flow, which tends to settle into a pattern. Turbulence can occur in any flowing fluid but it is commonly observed when a solid, such as a ship or airplane, passes through. Streamlining is designing a solid body to minimize turbulence, since turbulence greatly increases friction on the body. A fluid can pass from laminar flow to turbulence almost instantly. Predicting turbulence is among the goals of chaos theory.

turgor (*botany*)   As water enters a plant cell, pressure builds up inside the cell. The pressure exerted by the cell's contents against its cell wall is turgor pressure. When the cell has taken in enough water, pressure against the cell wall is high, and the cell is said to be turgid. If water loss exceeds absorption, the cell's contents shrink away from the wall.

Turgidity maintains rigidity and support for most herbaceous (nonwoody) stems. If a plant such as a daisy does not absorb enough water, or if it loses a lot of water through transpiration, turgor pressure in the stem decreases and the stem droops.

turgor movement (*botany*)   Turgor movements are those that depend on changes in turgor pressure. The leaf of a mimosa offers an excellent example. This leaf consists of numerous tiny leaflets on a central rib. When something touches the leaf, an electrical stimulus travels to cells at the base of each leaflet, causing those cells to speedily lose water and become limp. This in turn causes the leaflets to droop. If there is no further disturbance, the cells begin taking in water in 10 to 20 minutes, causing the leaflets to return to an open position.

Turing, Alan (*mathematician*)   [English: 1912–54] Before World War II Turing showed that an idealized machine (a universal Turing machine) using a few simple operations can calculate the values of all functions. Despite this, he proved that some mathematical results remain beyond calculation.

During the war Turing was instrumental in building a machine that decoded enemy messages, a prototype for electronic computers. After the war Turing developed such computers. His "Turing test" is a game used to determine whether a computer has intelligence beyond computational ability.

Turing's 1952 study of the mathematics of fluid interactions has become important in showing how organisms develop and also in analyzing fluids with periodic properties.

turkey (*animals*)   Turkeys are large fowl native to the Americas. The wild turkey (*Meleagris gallopavo*) of North American woodlands is the ancestor of domestic turkeys. It stands up to 4 ft (1.2 m) tall and weighs as much as 20 lb (9 kg). A wattle and other bare fleshy projections appear on the head, neck, and throat. The turkey runs swiftly but can make only short flights. During mating season, a male spreads his tail and produces a gobbling sound that can be heard up to 1 mi (1.6 km) away.

turtle (*animals*)   Unlike other reptiles, turtles have a rigid shell and horny beak instead of teeth. Their shell has 2 main parts: an upper carapace and a lower plastron, joined between the legs on each side by a bridge. Some species can pull their entire bodies into the shell; others can do this only partially. The ribs are fused to the carapace. Other air-breathing vertebrates expand and contract the rib cage as they breathe but turtles depend mainly on 2 sets of abdominal muscles for respiration.

Most turtles are omnivores, eating both plant and animal matter. The approximately 250 species range from less than 4.5 in. (11 cm) to more than 8 ft (2.4 m) in length. People sometimes use the word turtle to indicate water-dwelling species and refer to land-dwellers as tortoises.

tusk (*anatomy*)   Elephants, hippopotamuses, and several other mammals have elongated, pointed teeth called tusks that extend outside of the mouth. A tusk of a large male elephant may be 10 ft (3 m) long and weigh 150 lbs (68 kg) or more. Tusks usually occur in pairs but the narwhal has a single tusk, which may be 8 ft (2.5 m) long. Animals use their tusks for various purposes—elephants for digging and stripping bark off trees, warthogs for defense, and so on.

two-point equation (*analytic geometry*)   Since 2 points determine a line, there exists an equation that gives the line through any 2 points, such as (x_1, y_1) and (x_2, y_2). The 2-point equation is often written as $(y - y_1)/(y_2 - y_1) = (x - x_1)/(x_2 - x_1)$. Its origin is clearer in the form $(y - y_1)/(x - x_1) = (y_2 - y_1)/(x_2 - x_1)$, which shows that the equation derives from setting the slope of the whole line equal to the slope between the 2 given points.

Tycho Brahe (*astronomer*)   [Danish: 1546–1601] Like Galileo, Tycho is generally known by his first name. His reputation began in 1572 when he described a supernova and used parallax to prove it to lie beyond the solar system, where, according to Aristotle, the stars are unchanging. In 1577 Tycho again used parallax to determine that a comet was farther from Earth than the Moon is, establishing comets as celestial and not atmospheric phenomena. Denmark's king built the first true astronomical observatory for Tycho, from which he and his assistants, including Kepler, produced the most accurate observations ever made without a telescope.

Tyrannosaurus (*paleontology*)   Among the largest and fiercest of the meat-eating dinosaurs, *Tyrannosaurus* ("tyrant lizard") lived in the late Cretaceous. It grew to 50 ft (15 m) long and 20 ft (6 m) high, and weighed 12,000 lb (5500 kg) or more. *Tyrannosaurus* walked upright on its massive hind legs and used its long, heavy tail for balance. Its huge head had powerful jaws armed with curved, serrated teeth up to 8 in. (20 cm) long.

U

ultrasound (*waves*) Physicians and others use sound waves of low intensity but higher frequency than humans can hear, called ultrasound, to image a developing fetus or other hidden structure. The advantage of ultrasound over lower frequency sound is that a higher frequency means a shorter wavelength, enabling relatively small structures to be observed. Sonar, or underwater echolocation, also uses ultrasound, but at a higher intensity. Very high-intensity ultrasound is used to clean surfaces or to break up kidney stones.

ultraviolet radiation (*waves*) The part of the electromagnetic spectrum with energy higher than visible light and lower than X rays—roughly from 1,000,000,000,000,000 Hz to 10,000,000,000,000,000 Hz, or from 400 nanometers to about 100 nanometers—is ultraviolet radiation (UV), also called ultraviolet light. UV has enough energy to cause biological changes, such as tanning, skin cancer, and production of vitamin D in humans and death of bacteria. Much solar UV is intercepted by ozone in the upper atmosphere. Thinning of the ozone layer by chlorofluorocarbons, which causes more harmful UV to reach Earth's surface, has been a major environmental problem.

umbilical cord (*anatomy*) In placental mammals, a soft, flexible umbilical cord connects the embryo to the placenta. The cord contains branches of the embryo's circulatory system. Umbilical arteries carry blood from the embryo to the placenta, where an exchange of food, gases, and other materials takes place between the embryo's blood and maternal blood. An umbilical vein returns the blood to the embryo. The umbilical cord ceases to function at birth and either breaks or is severed by the parent.

uncertainty principle (*particles*) In 1927 Werner Heisenberg startled physicists by showing that quantum mechanics implies that the position and momentum of a particle cannot be simultaneously measured as closely as one likes. The product of the uncertainty in each measurement must always be greater than Planck's constant divided by 2π. This uncertainty principle also applies to certain other pairs of measurements, such as energy measured and time needed for measurement. Experimental evidence strongly supports the idea that uncertainty is fundamental and impossible to evade, although the small size of Planck's constant makes it noticeable only for subatomic particles.

unconformity (*geology*) Geologists call an apparent gap between 2 layers of sedimentary rock an unconformity. An unconformity indicates that time has passed between deposits of sediment; during this interval, some of the lower layer has eroded away. Often an unconformity is angular—the lower layer is folded or partly uplifted, making the boundary between layers a break between different orientations. Where each layer has a different suite of fossils, unconformities separate geologic ages.

undefined term (*logic*)   In an axiomatic system the basic elements described by the system cannot be defined within the system. For example, modern axiomatic treatments of geometry do not specify meanings or even descriptions for point or line, leaving these as undefined terms. In part this stems from necessity, since it is impossible to define every word without employing circular reasoning. Undefined terms also make axiomatic systems more useful, since any conclusion proved within the axiomatic system can be transferred from one set of objects to another by changing the interpretation of the undefined terms.

ungulate (*animals*)   Mammals with hooves instead of nails on their toes are called ungulates. Hooves are designed for running and for protecting the feet on rough surfaces. There are 4 orders of ungulates: elephants, hyraxes, odd-toed ungulates such as horses, and even-toed ungulates such as deer.

uniformitarianism (*geology*)   The theory that the rocks, seas, mountains, and other geological features of Earth are the result of processes acting throughout history much as they can be observed today is called uniformitarianism. Erosion, volcanism, uplift of land by earthquakes in small stages, petrifaction (rock formation) of organic sediments, and similar events are the agents of change. This theory, propounded by James Hutton in 1785 and promoted by Sir Charles Lyell throughout the 19th century, replaced catastrophism and dominated geological thinking completely until 1980. In recent years strict uniformitarianism has been tempered by acceptance of a few catastrophes thought to have changed the course of evolution.

union of sets (*sets*)   The set consisting of all the elements that are in either of 2 sets is called the union of the 2 sets. On a Venn diagram this set consists of all the points in the 2 overlapping disks. Finding the union is a binary operation symbolized by the sign $\cup$. If A is all the even natural numbers and B is all the odd ones, then $A \cup B$ is the whole set of natural numbers. Note that union corresponds to the logical operation or.

universal set (*sets*)   Sets are often chosen from a particular type of entity, such as sets of numbers or sets of letters. The set of all elements of the type being considered is called the universal set or sometimes the universe of discourse.

universe (*cosmology*)   The present concept of the universe is of an expanding space that includes all matter and energy with nothing outside that space. Most astronomers think that the universe exploded in an event called the big bang about 15,000,000,000 years ago; recent evidence suggests that the expansion will continue, making distances between galaxies and even between stars so great that the night sky will be empty.

In various theories of how the universe was created as well as in one interpretation of quantum theory, this universe is not alone; at various points in history alternative universes are created and proceed with no connection to the one we inhabit.

up quark (*particles*)   The 2 light quarks that are the basis of ordinary matter are simply called up and down, names based on characteristics of a type of spin. The up quark with a charge of $+\frac{2}{3}$ occurs twice in a proton, for example, which produces a charge of $+\frac{4}{3}$, so a down quark with a charge of $-\frac{1}{3}$ is needed to give the proton an integral charge. The antiup quark has a charge of $-\frac{2}{3}$.

upwelling (*oceanography*)   One kind of upwelling is a nearly vertical current caused when prevailing winds push ocean waters away from coasts. As the warmer top layers are pushed out to sea, cool water from near the ocean floor rises to take its place. A similar effect with a different cause occurs in polar seas, where surface water cools so much that it sinks, displacing water from the bottom, which rises to the top. Water from lower levels contains many nutrients that algae and other organisms need. Consequently regions where upwelling occurs are among the most biologically productive in the ocean.

uranium (*elements*)   Uranium, the first radioactive element discovered (in 1789 by Martin Klaproth [German: 1743–1817]), was the first recognized as radioactive in 1896. Klaproth named the metal for the planet Uranus, discovered in 1781. Uranium's atomic number, 92, is greatest for any element occurring naturally in significant amounts, so people refer to "the 92 elements;" but at least 2 elements below atomic number 92 are not found in nature. Most uranium is isotope 238 with a half-life of 4,510,000,000 years, useful in dating the oldest rocks. Uranium-235, 0.71% of natural uranium, undergoes nuclear fission easily and is employed in "atomic" bombs and nuclear power plants. "Depleted" uranium, with most uranium-235 removed, is barely radioactive and useful where high density is needed, as in tank armor. Chemical symbol: U.

Uranus (*solar system*)   Uranus is barely visible with the naked eye and went undiscovered until William Herschel spotted it in 1781. The seventh planet from the Sun in the solar system, Uranus is a gas giant, but with only about 5% of Jupiter's mass. Uranus is faintly greenish, perhaps because its atmosphere contains methane. The planet's axis of rotation is tipped so far that one pole sometimes points almost directly at the Sun. In 1977 a system of 9 faint rings was discovered. The atmosphere is hydrogen, helium, and methane and very cold (–355°F, or –215°C). At lower depths these gases compress into a liquid or slushy surface "crust." Underneath there is a mantle of solidified methane, ammonia, and water, and a rock and iron core about 15 times as massive as Earth. There are 18 known moons.

urea (*biochemistry*)   In mammals and amphibians, the breakdown, or metabolism, of amino acids produces ammonia, which reacts with carbon dioxide to form the nitrogenous waste urea plus water:

$$2NH_3 + CO_2 \rightarrow CO(NH_2)_2 + H_2O$$

The urea is carried by the blood to the kidneys, where it is removed for excretion in urine. In mammals, some urea also is excreted by the sweat glands.

Urey, Harold (*chemist*)   [American: 1893–1981] Although Urey worked in chemistry, his specialty of isotopes led to his having an important role in physics, astronomy, geology, and even biology. He was the first to isolate deuterium (heavy hydrogen), in 1932, the isotope of hydrogen containing a neutron as well as a proton. Urey developed ways to use isotopes to measure temperatures of past climates that remain the mainstays of climatology. His investigations led Urey to a theory of Earth's early history, now believed to be correct, that Earth formed by accretion of planetesimals. He proposed the first experimental studies of how life may have originated.

uric acid (*biochemistry*) Uric acid ($C_5H_4N_4O_3$) is an insoluble waste product produced in many animals during the breakdown, or metabolism, of amino acids. It is the main excretory product of birds, reptiles, and insects. It can be excreted in a very concentrated form, as a thick paste or dry pellets. In birds, for example, this reduces weight, since no urinary bladder is needed for temporary storage of urine.

In humans and other primates, uric acid is produced during the metabolism of nucleic acids (DNA and RNA) and is excreted in urine. High levels of uric acid can contribute to the development of gout and kidney stones.

urine (*physiology*) The kidneys of mammals produce a fluid called urine, which consists of water and various wastes. The chief waste is an end product of protein metabolism, urea. Other wastes include additional organic materials, such as creatinine and uric acid, and ions (chloride, sodium, potassium, etc.). The volume and composition of urine is regulated by hormones such as aldosterone and ADH (antidiuretic hormone) to ensure that the organism's water, salt, and acid-base balance are kept within healthy limits.

uterus (*anatomy*) After an egg is released by a female mammal's ovary, it travels through a fallopian tube to the uterus (a Latin word meaning "womb"). If the egg has been fertilized, the resulting embryo implants and develops in the uterine wall. The uterus is pear-shaped and muscular, capable of changing shape and dilating as the embryo grows. Contractions of the uterine muscles expel the embryo and its placenta at the time of birth.

V

vaccination (*physiology*)   When a person has a disease such as mumps or smallpox, the body produces antibodies to react with disease antigens and render them harmless or inactive. After the person recovers the antibodies remain in the body, providing immunity against future attacks by the disease organisms. Vaccination is a technique that stimulates the production of antibodies without producing the disease. A person is inoculated with a vaccine consisting of weakened or dead bacteria or viruses. The vaccine is seldom potent enough to cause disease but it does create immunity. Since Edward Jenner performed the first known vaccination in 1796, vaccines have been developed against a series of once-fatal diseases of humans and other animals.

vacuole (*cells*)   Vacuoles are fluid-filled organelles within the cytoplasm of various cells. Vacuoles in plant cells are filled with cell sap—a solution of salts and other substances. A young plant cell usually contains many vacuoles; as the cell matures these vacuoles merge to form one large vacuole that may fill most of the cell. In amoebas and other 1-celled protists, food vacuoles are the site of digestion. Many protists also have excretory organelles called contractile vacuoles.

vacuum (*physics*)   A vacuum is a region of space that contains no matter. Philosophers starting in antiquity debated the existence of the vacuum. In 1643 Evangelista Torricelli used mercury to create a vacuum free from all but wisps of mercury vapor. Within a few years, Otto von Guericke [German: 1602–86] invented an air pump that produced near perfect vacuums.

After the wave nature of light was established at the beginning of the 1800s, scientists believed that the vacuum must be pervaded by something, which they called ether. But by 1905 Einstein's work had eliminated the need for ether, causing scientists to believe that a real vacuum, with nothing in it, can exist. The uncertainty principle, however, permits subatomic particles to appear out of nothing and disappear before violating any physical laws. In 1947 Willis Lamb, Jr., [American: 1913–] showed experimentally that such virtual particles always appear in vacuums. Thus any vacuum contains matter. In one version of cosmology, the big bang originated as a vacuum filled with "negative energy", called a false vacuum.

vagina (*anatomy*)   The portion of the reproductive tract of female mammals that leads from the uterus to the outside of the body is the vagina. It is a muscular organ that can expand greatly during copulation (insertion of the penis into the vagina) or birth of the young.

valence (*chemistry*)   The valence of an element or radical is an integer or integers that express the kinds of chemical bonds that are possible for an atom of that element or a single radical. Originally expressed in terms of the number of hydrogen atoms that would bond with the atom or radical, the concept has largely been replaced by the more sophisticated idea that the outer shell of an atom, known as the valence shell, forms stable compounds when the shell has the "magic number" of electrons needed to make it complete. Thus sodium, valence +1, has 1 electron in its valence shell and needs 7 more to complete it, while chlorine, valence –1, has 7 electrons in the valence shell and needs 1. In stable compounds the valence numbers add up to 0, as is the case with sodium chloride. Most compounds are more complex than sodium chloride, however, since many elements can combine in several ways and so have several valences (iron has valences of +2 and +3, for example). The valence number is usually the same as an integer called the oxidation number, which provides a slightly different way of calculating how valence-shell electrons form bonds.

vanadium (*elements*)   Vanadium is a metal, atomic number 23, discovered in 1801 by Andrès del Rio [Spanish-Mexican: 1764–1849]. Del Rio was persuaded that he had not identified a new element, however, and the element's status was not resolved until 1830. The metal's rediscoverer, Nils Sefström [Swedish: 1787–1845], named it for the Scandinavian goddess Vanadis. The chemical symbol for vanadium is V.

Van Allen radiation belts (*geology*)   Earth's magnetic field captures charged particles in the solar wind and injects them into paths that spiral around field lines, each particle going back and forth from near one magnetic pole to near the other in about 1 minute. Two levels of such particle-rich regions were found in 1958 by the first U.S. Earth-orbiting satellites in an experiment designed by James Van Allen [American: 1914–]. The lower begins some 600 mi (1000 km) above the equator and the upper starts at about 12,000 mi (20,000 km); both sink lower near the poles. As the particles spiral around field lines, they also rotates around Earth in response to changes in intensity of the magnetic field at different altitudes. A particle takes about 30 minutes per rotation—electrons drifting east and protons west—as it continues to bounce back and forth between poles.

Van der Waals force (*chemistry*)   Very weak attraction between some atoms or molecules occurs because a somewhat unevenly distributed electromagnetic field in one particle sets up the opposite field in the nearby part of a second particle. This force, called the Van der Waals force after Johannes Van der Waals [Dutch: 1837–1923], whose work brought it to light, is 1 of the 2 factors preventing real gases from behaving like the ideal gases of the gas laws (the other is the size of the molecules). At very low temperatures some organic compounds as well as many gases form solids held together only by Van der Waals forces.

variable (*algebra*)   A symbol (usually a letter, such as x) that càn represent any member of a set is called a variable. A variable in an equation is sometimes called the unknown. By custom the variable n is used for natural numbers or integers; x is for real numbers; and z is for complex numbers. In graphs the variable x represents numbers on the horizontal axis, y numbers on the vertical axis. Constants, usually letters near the beginning of the alphabet, are variables understood to have fixed values in a given situation. For example, the slope-intercept equation of a line, $y = mx + b$, is interpreted with y and x ranging over all real numbers, while m and b are constants.

variable star (*stars*) Any star that periodically changes brightness is called a variable (a nova changes brightness but not at regular intervals). The period varies with the cause of the change and the individual star. Some variables, called eclipsing binaries, are part of a binary system in which one star periodically passes in front of the other. Other kinds of variables are called Mira and Cepheid variables, after the first stars known of each type. It is not clear what causes the brightness to vary in these latter cases.

variance (*statistics*) The average of the squares of the deviations for a set is its variance. The variance also equals the average of the squares of the original numbers minus the square of the average of those numbers. For example, the small set 2, 5, 7, 8 has a mean of 5.5. To calculate the variance in the ordinary way, find the deviation of each member from 5.5, square that deviation, and take the average, which is 5.25. It is easier to square the numbers first, obtaining the set 4, 25, 49, 64 with a mean of 35.5. Subtracting $35.5 - (5.5)^2$ gives 5.25, the variance.

variation (*evolution*) In species that reproduce sexually, individuals vary in size, proportion, coloration, physiology, etc. Some variations are acquired characteristics produced by differences in environmental factors and are not inherited by offspring. Other variations are genetic and during reproduction can be passed from one generation to the next. Genetic variations play a major role in natural selection and evolution.

varve (*geology*) Distinct light and dark layers of sediment that form at the bottom of lakes or ponds, especially those caused by glaciers, are varves, Swedish for "layers." A varve is the sediment that settled during 1 year. In cold regions, summer sediment contains runoff from glaciers, streams, and surrounding land. When winter comes, surface water freezes and the only sediment is produced by fine suspended clay particles. A light sandy layer and a darker clay layer forms 1 varve. Counting varves or measuring their age with isotopic dating can be used to date objects embedded in sediment or to learn when a glacier formed a lake.

vascular tissue (*botany*) Except for mosses and their relatives, all plants are vascular plants, with fluid-conducting vascular tissue and true roots, stems, and leaves. The vascular tissue forms a network throughout the plant, much as the circulatory system forms a network throughout the human body. Xylem is vascular tissue that carries water and minerals. Phloem is vascular tissue that carries dissolved foods. In dicot plants, such as sunflowers and oaks, the xylem and phloem are separated by a thin layer of vascular cambium, which is responsible for secondary growth, or growth in diameter of roots and stems.

vector

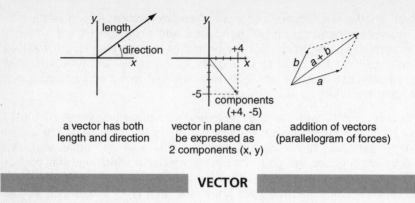

a vector has both
length and direction

vector in plane can
be expressed as
2 components (x, y)

addition of vectors
(parallelogram of forces)

VECTOR

vector (*algebra/mechanics*)   Vectors originated in physics as arrows indicating size and direction of a force or of other directed quantities such as velocity or acceleration. Such vectors in spaces of 2 or 3 dimensions remain among the most useful parts of applied mathematics. These vectors are described in terms of length combined with direction measured by angles that the arrow makes with coordinate-system axes. A vector in 3-space can also be indicated with 3 numbers, an ordered triple equivalent to the coordinates of the point of the arrow (assuming the arrow starts at the origin). More complex physical situations are described as objects in phase space where several different numbers in ordered n-tuples (n numbers listed in a given order) form a vector. Such ordered n-tuples are written in parentheses with numbers separated by commas; the expression $(1, -3, 0.25, 0, 2)$ is an ordered 5-tuple and represents a vector in 5 dimensions.

Similar ideas in mathematics had several origins—in attempts to generalize numbers beyond complex numbers, in extensions of coordinate systems to greater numbers of dimensions, in treating systems of equations, and in other developments of algebra. In each case the entities can be represented, as vectors are, as ordered n-tuples. The concept can even be extended to infinite sequences.

Two vectors **a** and **b** are equal if the n is the same for both and if the n numbers, called components, are equal at each place in the order. Adding such vectors can be accomplished by adding the components in each place, although in physical situations it is more useful to operate by combining arrows. Multiplication is more complicated, with 3 different types (multiplication by a scalar, the dot product, and the cross product) that have different applications.

vector (*biology*)   An organism that transmits parasitic organisms from one host to another is called a vector. For example, the *Anopheles* mosquito is the vector of the *Plasmodium* protozoan that causes malaria. An uninfected mosquito picks up *Plasmodium* when it feeds on the blood of an infected person (host). The *Plasmodium* live and reproduce in the mosquito (thus the mosquito is a host as well as a vector). When the mosquito bites an uninfected person, *Plasmodium* pass from the mosquito's saliva into the person's blood.

Vega (*stars*)   Vega is a star only about 3 times as wide as the Sun, and therefore its bright appearance is caused largely by its relative nearness. Because of precession, Vega will replace Polaris as the north star in about 12,000 years.

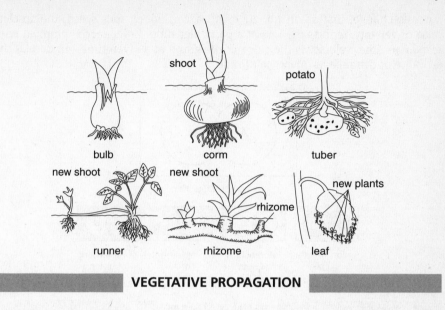

VEGETATIVE PROPAGATION

vegetative propagation (*botany*)   In addition to reproducing sexually, with the formation of seeds, many plants reproduce asexually by growing specialized organs that develop into new individuals. Onion plants develop bulbs, potato plants grow tubers, strawberry plants send out runners, etc. The new individuals are clones; that is, they are genetically identical to their parents. People artificially propagate plants by taking cuttings; willow, forsythia, and coleus are examples of garden plants that are easily propagated by cuttings.

vein (*biology*)   A vein is a tubelike structure. In animal circulatory systems, veins are vessels that carry blood toward the heart. In insect wings, veins are thickenings that enclose respiratory tubes called tracheae. In plant leaves, veins are vascular tissue responsible for conducting food and water to and from vascular tissue of the stem.

vein (*geology*)   Minerals often occur in veins, narrow deposits within a rock mass that follow a path through the rock. The paths followed are cracks in the parent rock. A very hot fluid, such as superheated water, dissolves large amounts of minerals as it flows. After it enters the crack, the fluid cools and releases the minerals from solution, forming the vein.

Velociraptor (*paleontology*)   A small, meat-eating dinosaur made famous by the film *Jurassic Park*, *Velociraptor* ("swift plunderer") hunted in packs. It was armed with a long, curved claw on the second toe of each foot. These claws—kept raised off the ground as *Velociraptor* ran—were efficient killing weapons, as were the claws on the long fingers and the sharp teeth in the enormous mouth. About 6 ft (1.8 m) long, *Velociraptor* lived in the late Cretaceous.

velocity (*mechanics*)   Velocity is the rate of change of distance when direction is taken into account, so velocity is a vector quantity. Velocity concerns instantaneous change, just as an odometer on a car, which tells speed, the absolute value of velocity, reports the rate at a particular time. If distance is graphed with respect to time, velocity is the slope of the graph at a given time—in calculus the derivative of distance as a function of time.

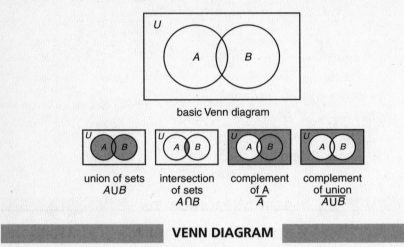

basic Venn diagram

union of sets	intersection of sets	complement of A	complement of union
$A \cup B$	$A \cap B$	$\overline{A}$	$\overline{A \cup B}$

VENN DIAGRAM

Venn diagram (*logic*)   Venn diagrams are one of several ways to represent graphically relationships among propositions or, equivalently, among operations in set theory. As developed in 1876 by John Venn [English: 1834–83] for Boolean algebra, each diagram is a pair of intersecting circles within a rectangle. Shading indicates relationships or operations. If the intersection of the circles is the only part shaded, the relationship is and for propositions as well as intersection for sets; shading both circles is or as well as union of sets. The rectangle represents all the possibilities (called the universe) under consideration, so shading outside a circle is not or the complement. Similar methods using several overlapping circles were introduced by Euler and others.

venom (*biochemistry*)   A wide variety of animals produce poisons, or venoms, either for defensive purposes or to kill or immobilize prey. Among the best-known examples are poisonous snakes, which produce venom in specialized glands located in the head. Ducts lead from these glands to teeth—fangs—that are hollow or grooved and conduct the toxin into the snakes' victims. Depending on the species of snake, the venom either interferes with the functioning of the victim's nervous system, particularly the centers that control breathing and heartbeat, or it destroys the walls of blood vessels, causing hemorrhaging.

Venus (*solar system*)   Venus is second only to the Moon in brightness in the night sky. It is the second planet from the Sun in the solar system. The Venusian sky is thick with sulfuric-acid clouds that hide the planet's surface, clouds that float in an atmosphere of about 96.5% carbon dioxide, 3.5% nitrogen, and small amounts of other substances. Pressure at the surface is 90 times that of Earth's. The lower atmosphere of Venus traps heat in an extreme greenhouse effect responsible for high surface temperatures of about 870°F (460°C). There are 2 major highland areas: one about half the size of Africa in the equatorial region and the other about the size of Australia, to the north. The highest mountain on Venus—Maxwell Montes—is higher than Earth's Mount Everest. Volcanic activity dominates Venus; the planet is covered with volcanic domes and lava channels. Venus has no moons.

Venus's-flytrap (*plants*)   Venus's-flytrap (*Dionaea muscipula*) is a carnivorous plant that uses a spring trap to catch insects. The leaf is edged with sharp hairs and hinged along its midrib; its upper surface has small, extremely sensitive trigger hairs. When an insect touches the trigger hairs, electrical signals pass from the hairs to cells in the hinge. In approximately 30 seconds the 2 halves of the leaf move together, the sharp hairs on their margins meshing to form a trap. Digestive juices flow from glands in the leaf. In 5 to 10 days the insect is digested and nutrients absorbed. The trap then reopens.

vertebrate (*zoology*)   Any animal with vertebrae—bones that surround the spinal column and make up a backbone—is a vertebrate. Vertebrates are the most highly evolved animals, with an internal skeleton and a well-developed brain enclosed in a skull. Some 44,000 living species of vertebrates have been identified. They include fish, amphibians, reptiles, birds, and mammals.

vertical angles (*geometry*)   The angles formed by 2 intersecting lines are called vertical when they have no sides in common; that is, they are directly across from each other. Extending the sides of one angle through their mutual vertex produces the other. Since intersecting lines determine a plane, vertical angles are always in the same plane. Vertical angles have the same measure since each is always a supplementary angle to the same one of the other angles formed by the pair of lines.

Vesalius, Andreas (*anatomist*)   [Belgian: 1514–64] Considered the father of modern anatomy, Vesalius made careful dissections of human cadavers. In 1543 he published *On the Structure of the Human Body*, a 7-volume text that contained the first accurate illustrations of internal human anatomy. His rejection of many of the teachings of the Ancient Greek physician Galen was controversial, and forced him to leave teaching—but this led to an appointment as physician to Holy Roman Emperor Charles V.

vestigial organ (*anatomy*)   Many organisms have certain useless structures, often of reduced size, called vestigial organs. In the organisms' ancestors, however, these organs were useful. Thus pythons and whales inherited traces of a pelvic girdle and hind limbs from their legged ancestors. Many cave-dwelling fish and insects have reduced eyes, vestiges of the eyes of ancestors that lived in light-filled habitats. Vestigial organs in humans include the appendix and caudal vertebrae (tailbone).

vetch (*plants*)   Members of the legume family, vetches (*Vicia*) are climbing and trailing herbs with compound leaves that end in single or branched tendrils. The pealike flowers—in white, yellow, blue, or purple—are borne singly or in clusters. The fruit is a flat pod. Common vetch (*V. sativa*) and several other species have long been cultivated as fodder plants.

Crown vetches (*Coronilla*), also legumes, are trailing plants grown as ornamentals and to fight erosion. They have clusters of pinkish white flowers. In autumn in cold habitats, the plants die back to the crown (the soil-level part where roots and stem join); new stems and leaves are produced the following spring.

vine (*botany*)   A plant with a flexible stem that climbs or twines around an object or creeps along a surface is called a vine. Vines of tropical rain forests, particularly those with woody, ropelike stems, are referred to as lianas.

Instead of using energy to build supporting wood cells in their stems, vines grow specialized structures that enable them to attach to surfaces that lead upward toward sunlight. Scramblers—such as many members of the rose family—have prickles or thorns on their leaves and stems that catch on vegetation and prevent slipping. Ivies produce **adventitious roots** that penetrate supports while pea plants produce **tendrils** that coil around supports. Other vines, such as morning glories (*Ipomoea hederacea*) and hops (*Humulus lupulus*), coil their main stems around supports.

violet (*plants*)   A widespread group comprising more than 400 species, violets (*Viola*) are mostly low-growing perennial herbs. The unique flower, usually violet or white, has a lower petal that extends to form a spur—a convenient ledge for insects who arrive to probe for nectar and who act as cross-pollinators. Some species produce petalless flowers that do not open and that self-pollinate. The fruit is a capsule; as it splits it shoots seeds a distance of several feet. The best-known species is the sweet violet (*V. odorata*), native to Europe and the source of perfume oils.

viper (*animals*)   Snakes called vipers have folding fangs attached to the upper jawbones. When a viper opens its mouth, the jawbones rotate, swinging the fangs forward into biting position. As the fangs strike prey—typically small mammals, lizards, and amphibians—contracting muscles squeeze venom through the fangs into the victims.

True vipers (Family Viperidae) are found in Europe, Asia, and Africa. They include asps and adders. Pit vipers (Family Crotalidae) live mainly in the Americas; they include rattlesnakes, copperheads, the cottonmouth, and the fer-de-lance. Pit vipers have a unique sense organ that opens by means of a pit between the nostril and the eye; it detects body heat given off by potential prey.

Virchow, Rudolf (*pathologist*)   [German: 1821–1902] Virchow's many discoveries include cells in bone and connective tissue; substances, such as myelin; and pathologies, such as embolism and leukemia. In 1855 he published his now-famous aphorism *omnis cellula e cellula* ("every cell stems from another cell"). He also stated that all diseases involve changes in normal cells—that is, all pathology ultimately is cellular pathology.

viroid (*biology*)   Viroids are extremely tiny infectious particles that resemble incomplete viruses. To date they have been found only in plants; they are known to cause diseases of apples, avocados, chrysanthemums, coconuts, cucumbers, peaches, potatoes, and tomatoes. A viroid consists solely of a single-stranded, circular RNA molecule of fewer than 400 nucleotides. It is apparently replicated by enzymes within the plant host's cells. Exactly how a viroid causes disease is unknown.

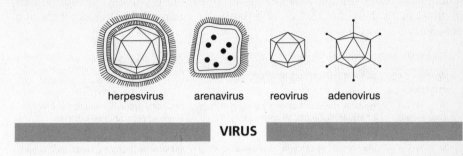

herpesvirus   arenavirus   reovirus   adenovirus

VIRUS

virus (*viruses*)   Because they cannot carry on life processes unless they are within living cells, viruses generally are not considered to be living organisms. Much smaller than the smallest bacteria, viruses can multiply only after they invade the cells of a host. They do not carry out metabolism themselves, but they are able to control the metabolism of the infected cells. They parasitize all types of organisms and many, though not all, cause disease.

A virus particle, or virion, consists of an inner core of DNA or RNA surrounded by a protective protein coat, called a capsid. Exterior to the capsid in some types of viruses is a lipid membrane or envelope. Virions are highly symmetrical; most are either rod-shaped or icosahedrons (20-sided polyhedrons).

viscosity (*materials science*)   Viscosity is the measure of the main attribute of a fluid, its ability to flow; a higher viscosity means a slower flow. The cause of viscosity is internal friction. Viscosity of liquids decreases with increasing temperature, but gases are the opposite—a hotter gas is more viscous because its molecules bump into each other more. At room temperature, glycerol is about 15 times as viscous as water and ether is about 1/5 as viscous. A glass is sometimes said to be so viscous that it seems solid.

vision (*physiology*)   Eyes are the organs of vision, the sense by which animals perceive light, color, and the appearance of objects. Eyes vary greatly in the amount of visual data they absorb and process. The eyes of a scallop measure changes in light and dark sufficient to detect moving shadows. In contrast, the eyes of humans and other primates, coupled with a large visual center in the brain, produce detailed 3-dimensional color images. There are 2 kinds of light receptor cells (photoreceptors) in the retina of primate and other vertebrate eyes: rods, responsible for night vision, and cones, for color vision.

vitamin (*biochemistry*)   For normal growth and functioning, almost all organisms require very small amounts of 13 organic compounds called vitamins. Plants can synthesize vitamins. Animals can synthesize some vitamins, and may depend on intestinal bacteria for some synthesis, but require other vitamins in their diet. Dietary requirements vary from one species to another. For example, rats can synthesize vitamin C but humans must obtain it in food, such as citrus fruits.

Excess amounts of fat-soluble vitamins can be stored in body fat; thus, an animal does not need to consume these vitamins every day. In contrast, excess amounts of water-soluble vitamins are excreted; these vitamins should be part of an animal's daily diet. A lack of a particular vitamin leads to symptoms of a deficiency disease.

VITAMIN	MAJOR FUNCTIONS IN VERTEBRATES	SIGNS OF DEFICIENCY IN VERTEBRATES
Fat-soluble		
A carotenoids	growth, differentiation of cells, synthesis of visual pigments, strengthening of immunity	retarded growth, impaired vision, lowered resistance to infection
D calciferols	absorption of calcium, maintenance of blood calcium levels, growth and maintenance of bones	skeletal deformities, weakened bones, soft teeth
E tocopherol	antioxidant; maintenance of cell membranes	reproductive problems, weakened muscles, fragile red blood cells
K quinones	synthesis of proteins involved in blood clotting	blood-clotting problems, bleeding
Water-soluble		
C ascorbic acid	synthesis of collagen, other intercellular materials, and neurotransmitters; absorption of iron; metabolism of amino acids; antioxidant; strengthening of immunity	weakening of connective tissues, with hemorrhaging of capillaries; slow healing of wounds; poor bone growth; lowered resistance to infection
B_1 thiamine	carbohydrate metabolism, conduction of nerve impulses	weakened muscles, digestive tract disorders, nerve inflammations
B_2 riboflavin	metabolism of carbohydrates, fats, and respiratory proteins; maintenance of mucous membranes	poor growth, skin disorders, sensitivity to light
niacin nicotinic acid	cell respiration	weakened muscles, fatigue, skin and gastrointestinal disorders, mental disorders
B_6 pyridoxine	metabolism of amino acids, use of fats, formation of red blood cells	skin disorders, anemia, convulsions, kidney and adrenal disorders
biotin	metabolism of carbohydrates, fats, and proteins	skin disorders, hair loss, muscle paralysis, reproductive problems
folic acid	metabolism of amino acids, synthesis of nucleic acids, maturation of red blood cells	impaired cell division, poor growth, anemia
B_{12} cobalamin	synthesis of proteins, red blood cells, and nerve sheath myelin	anemia, neurological disorders, loss of lining of intestinal tract
pantothenic acid	metabolism of carbohydrates, fats, and proteins; synthesis of sterols and other vital compounds	poor growth, skin disorders, kidney and adrenal disorders, degeneration of nervous tissue

viviparity (*zoology*)   In viviparous animals, the developing embryo receives its nourishment directly from the mother and then is born live. Among placental mammals, embryos obtain food via a **placenta**. Certain species of reptiles, amphibians, and invertebrates also are viviparous. Unlike the eggs of birds and other oviparous animals, the viviparous eggs lack a tough protective covering and contain very little yolk.

volatility (*materials science*) Volatility describes how quickly a liquid evaporates under normal conditions. It is measured by the pressure at which the vapor (gas) from the liquid reaches equilibrium with the liquid, a process called saturation—the same amount of liquid condenses as evaporates. Vapor pressure is independent of other gases, so the vapor pressure of water at room temperature, about ¾ in. (2 cm) of mercury, indicates 100% relative humidity. More volatile benzene has a vapor pressure at room temperature of about 4 in. (10 cm) of mercury. Vapor pressure increases with temperature, so perfume warmed by the human body has a higher vapor pressure than perfume in the bottle.

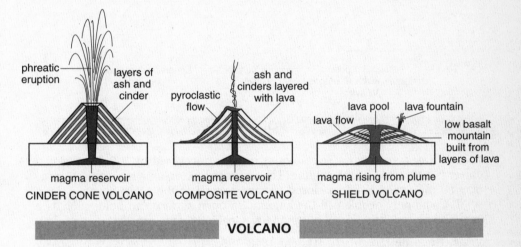

VOLCANO

volcano (*geology*) A single opening, or vent, in Earth's crust where hot or melted rock and hot gases have poured forth from a magma reservoir is called a volcano (a long crack is a fissure), as is the mountain produced by material from the vent. Some volcanoes release small amounts of gas and volcanic ash almost continually, but others undergo periods of violent release, or eruptions. A volcano is considered active so long as the magma reservoir is present— although an active volcano may sometimes be dormant, or not erupt, for thousands of years. When all magma has cooled to rock or been expelled, the volcano is extinct.

vole (*animals*) Often called field mice, voles are small rodents found throughout the Northern Hemisphere. They have small beady eyes, small ears, a stout body, and a comparatively short tail. Voles live on the ground in a variety of habitats, including meadows, sagebrush, and evergreen forests.

Volta, Alessandro (*physicist*)   [Italian: 1745–1827] Volta is remembered for his invention in 1799 of the first electric battery, chemically producing an electric current with plates of 2 different metals soaking in brine. About 25 years earlier, Volta invented the electrophorus, a device for creating large amounts of static electricity still used today. And in 1778 he was the first to isolate the gas methane (CH_4).

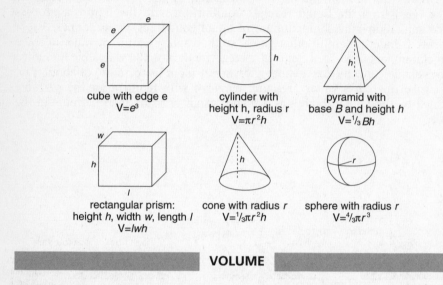

cube with edge e
$V=e^3$

cylinder with
height h, radius r
$V=\pi r^2 h$

pyramid with
base *B* and height *h*
$V=\frac{1}{3}Bh$

rectangular prism:
height *h*, width *w*, length *l*
$V=lwh$

cone with radius *r*
$V=\frac{1}{3}\pi r^2 h$

sphere with radius *r*
$V=\frac{4}{3}\pi r^3$

VOLUME

volume (*measurement*)   The amount of space occupied by a 3-dimensional object is a measure called volume. The unit used for volume is a cube. Volume may be the region inside a closed surface as well as the space occupied by a solid. Commonly people call the amount an open surface can hold, such as a box with no top, its volume, but properly speaking that is capacity.

Vries, Hugo de (*botanist*)   [Dutch: 1848–1935] In 1901, after years of closely studying a field of evening primroses (*Oenothera lamarckiana*), Vries proposed a theory to explain the phenomenon of evolution. He noted that occasionally individuals with new characteristics appear and that they breed true; that is, they pass the new characteristics to successive generations. Vries concluded that such changes in traits result from sudden changes in hereditary material. He called these changes mutations.

vulture (*animals*)   Large birds of prey that hunt during the day, vultures have a naked (featherless) head, hooked beak, and sharp eyesight. They can soar for hours on end, watching the ground for carcasses of dead animals, their main food. Vultures fulfill an important ecological role by removing this carrion from the environment. There are 2 groups. American vultures, which include the condors, constitute Family Cathartidae; they lack a syrinx (voice box). Eurasian and African vultures are part of the hawk family; they have a syrinx.

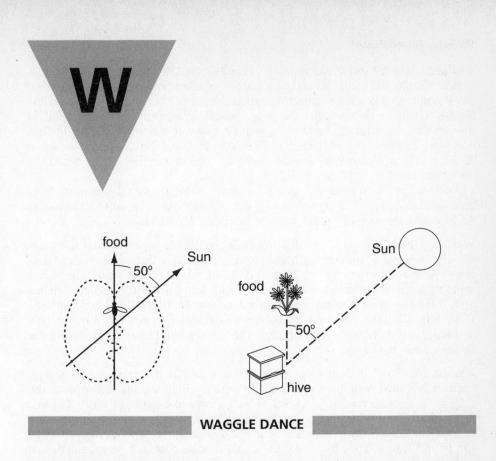

WAGGLE DANCE

waggle dance (*ethology*)   Research in the 1920s by Karl von Frisch [Austrian: 1886–1982] showed that when foraging honeybees (genus *Apis*) return to their hive, they use dances to convey information about new food sources to other bees. Round dances indicate that a food source is close to the hive. Tail-waggling dances are performed for more distant food. A bee runs along the combs of the hive, waggling its tail from side to side. It circles to the left, makes another waggling run, then circles to the right. The slower the dance, the farther away the food. The direction of the waggling run indicates the direction of the food. For example, if the bee runs straight upward, the food is toward the Sun; if the bee runs at a 30° angle to the right of the vertical, the food is 30° to the right of the Sun.

Wald, George (*biochemist*)   [American: 1906–97] Wald's contributions to our understanding of vision include the discovery that vitamin A is essential for maintaining vision and that the lens filters out ultraviolet light. He and a colleague identified the photosensitive pigments in the retina that detect light—one for absorbing light in the green sector of the spectrum, one for red, and one for blue—and showed that color blindness results from the absence of one of these pigments. Wald shared the 1967 Nobel Prize in physiology or medicine with Ragnar Grant (Swedish: 1900–91) and Haldan Keffer Hartline [American: 1903–83], who also did research into the physiology of vision.

Wallace, Alfred Russel (*naturalist*) [English: 1823–1913] Alfred Wallace is best known as the codiscoverer, with Darwin, of evolution by natural selection. In 1854 Wallace set out on a collecting expedition to the Malay Archipelago. During his travels he decided that the geographical distribution of species resulted from evolutionary forces. In 1858 he sent an essay containing his ideas to Darwin who, unknown to Wallace, had been developing a similar theory for some 20 years. Darwin presented a joint paper on their theory before the Linnaean Society in London, England, on July 1, 1858.

Wallace's work in zoogeography (the study of the geographic distribution of animals) included the observation of 2 distinct zoological regions in the Malay Archipelago separated by what came to be called Wallace's Line.

Wallace's line (*ecology*) In 1860 Alfred Russel Wallace reported that the eastern and western islands of the Malay Archipelago have 2 distinct types of animals: Australian fauna in the east and Asian fauna in the west. The water separating these 2 regions—between Bali and Lombok, and between Celebes and Borneo—was named Wallace's line by Thomas Huxley [English: 1825–95]. Wallace believed that natural selection was responsible for this boundary, but later evidence indicated that geological disturbances during the Tertiary were the cause.

walnut (*plants*) Walnut trees (*Juglans*) are known for their meaty seeds, commonly called walnuts. Commercially grown edible walnuts come from the English or Persian walnut (*J. regia*), which grows to heights of 100 ft (30 m). The walnut family, Juglandaceae, also includes hickories and the wingnuts (*Pterocarya*) of Asia. Family members are deciduous shrubs and trees with large compound leaves. They have tiny unisexual flowers that lack petals and sepals and are borne in clusters. The seed is enclosed in a husk.

walrus (*animals*) A pinniped mammal, the walrus (*Odobenus rosmarus*) is known for its huge canine teeth, or tusks, which it uses for defense and to dig in sand for clams, crabs, and worms. The male walrus grows up to 11 ft (3.4 m) long, and may weigh 3000 lb (1400 kg); the female is slightly smaller. Walruses have wrinkled skin, a thick body, a broad muzzle with stiff bristles, and no external ears. Walruses live in Arctic waters, congregating on shores or ice floes in herds of up to 1000 animals.

warbler (*animals*) Two unrelated groups of perching birds are commonly called warblers. Wood warblers (Family Parulidae), including chats, redstarts, and the ovenbird, are native to the Americas. Most are brightly colored. All have 9 flight feathers on each wing and a short pointed bill. These warblers feed mainly on insects and are not particularly good singers.

Old World warblers (Family Sylviidae, sometimes considered a subfamily of Muscicapidae) include American gnatcatchers and kinglets. All family members have 10 flight feathers per wing and a long pointed bill. They eat insects and are lovely singers.

warning coloration (*zoology*) Conspicuous colors and patterns generally are found on animals that are inedible or poisonous; they serve to warn predators to stay away. For example, frogs of the family Dendrobatidae have bright coloration—flame red in one species, black with yellow and white markings in another—that alerts predators that their skins secrete a powerful poison.

wart (*viruses*) When papillomaviruses infect the skin or mucous membranes of birds and mammals, they cause abnormal growth of cells and overproduction of fibrous proteins called keratins and result in a horny region called a wart, or papilloma. Although most types of warts are harmless, there is some evidence that genital warts may predispose female humans to cervical cancer. (Wartlike structures on a toad's skin are actually glands.)

warthog (*animals*) Named after the large warts around the eyes, warthogs (*Phacochoerus*) are pigs native to sub-Saharan Africa, where they are most common in savannas and open woodlands. They grow to about 33 in. (84 cm) tall at the shoulder and weigh up to 330 lb (150 kg). The massive head bears 2 pairs of large tusks that turn upward; the tusks on the upper jaw are longer than the pair on the lower jaw.

wasp (*animals*) Wasps are classified in a number of families in the same order of insects as ants. The second segment of their abdomen is constricted to form a "waist." In predatory species, the female's ovipositor is modified to form a stinger. Some species are solitary. Others—including paper wasps, hornets, and yellow jackets—are social insects that live in colonies with a caste system; they build paperlike nests from chewed plant matter mixed with saliva.

water (*compounds*) Most living organisms exist at temperatures where water (H_2O) is a liquid. At a pressure of air at sea level and above 212°F (100°C), water becomes the gas water vapor. Below 32°F (0°C), water that is not supercooled becomes ice. Liquid water or ice covers about ¾ of Earth's surface; water vapor floats in the air; and water flows slowly underground. Many minerals in the crust are hydrates, containing water loosely bound to other compounds.

Water has special properties that make life possible. Liquid water breaks into hydrogen ions (H^+) and hydroxide ions (OH^-) in electrical balance, making it easy for most materials to dissolve in water. Water expands slightly as it freezes, so ice floats on liquid water, protecting liquid water below.

water bear (*animals*) The microscopic water bears (Phylum Tardigrada), generally less than 0.04 in. (1 mm) long, are invertebrates with a soft body, 4 pairs of stumpy legs with claws, and an outer skin that is periodically shed, or molted. Found in a wide variety of aquatic and moist land environments, they feed on plants.

water cycle (*geology*) The travels of a drop of water are famous—from ocean to cloud, from cloud to snow on a mountain, then melting and joining a stream, perhaps used by a plant or animal and released as vapor, and finally raining back on the sea. The water cycle is the complex path that most water takes from ocean to clouds and back. Some water, however, leaves the cycle

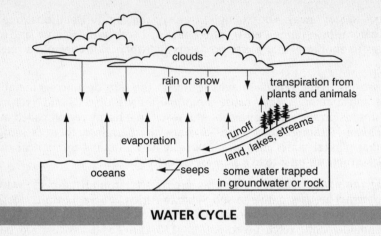

clouds

rain or snow

transpiration from
plants and animals

runoff

land, lakes, streams

evaporation

oceans

seeps

some water trapped
in groundwater or rock

WATER CYCLE

for long periods, sinking into the ground or becoming chemically entangled with rock.

waterfall (*geology*)   Most waterfalls result from a stream passing over the conjunction of a layer of hard rock with softer rock. Where the softer rock lies below or downstream, the water level immediately downstream of the boundary lowers as soft rock wears away. Water cascades down the boundary. The energy gained in falling erodes the softer rock even faster. Water may drop hundreds of feet after a few millennia. The upper part of Angel Falls in Venezuela drops 2648 ft (807 m), greatest for major waterfalls. Other waterfalls arise at hanging troughs where water tumbles into a steep-walled valley formed by a glacier.

waterfowl (*animals*)   Ducks, geese, and swans are commonly referred to as waterfowl in North America, wildfowl in Great Britain. They have a comparatively long neck, webbed feet, and a flattened bill with serrated edges. Most nest, feed, and migrate in large flocks. The largest is the trumpeter swan (*Cygnus buccinator*), up to 6 ft (1.8 m) long with a wingspan of 8 ft (2.4 m). Some waterfowl are vegetarians; others eat insects, fish, and other prey. Closely related to waterfowl are the screamers of South America, turkey-sized birds with raucous calls. Waterfowl and screamers constitute Order Anseriformes. All are swimmers or semiaquatic, with a thick coat of feathers and a preen-oil gland for keeping the plumage water-repellent. The creatures have 11 primary feathers, the first of which is always small. Their chicks, born covered with down, are quickly able to leave the nest.

water lily (*plants*)   Water lilies (*Nymphaea*) are aquatic perennials native to the Northern Hemisphere, Australia, and southern Africa. They often are cultivated as ornamentals and many hybrids have been developed. Their stems—either rhizomes or tubers—in the mud send up long-stalked leaves and many-petaled flowers. Tropical water lilies usually hold their flowers high above the water's surface. Hardy water lilies, which can grow in cooler areas, do not hold their flowers above the water; some produce flowers that change color during their 3- to 4-day blooming period. The other members of the water lily family, Nymphaeaceae, also are aquatic herbs with floating leaves and showy flowers. The Victoria water lily (*Victoria amazonica*), native to tropical South America, has leaves up to 8 ft (2.4 m) across!

water pollution (*environment*)   The discharge of sewage, feedlot wastes, pesticides, oil, heavy metals, and other harmful or objectionable materials into water is called water pollution. Most pollutants result from human activities, but some come from natural sources; for example, when groundwater moves through rocks rich in uranium, it picks up uranium and radon, which can cause cancer. Bacteria and other microorganisms in excrement can kill coral; chemicals in fertilizers can cause eutrophication; and hazardous chemicals can cause reproductive defects in fish and aquatic mammals. Thermal pollution is another form of water pollution.

waterspout (*atmospherics*)   Small-scale rotating winds form for poorly understood reasons. Over water such winds, like miniature tornadoes, pick up water vapor and swirl it high into the sky, causing it to condense into droplets as the air temperature lowers. This visible manifestation of the wind is called a waterspout. More powerful waterspouts are full-scale tornadoes that form over water. In a tornado-waterspout, not only are winds higher, but precipitation from the tornado adds to the amount of water being carried.

water table (*geology*)   Rainwater that does not evaporate or flow away sinks through soil and many kinds of rock. Eventually sinking water encounters clay or rock that stops all or most of it. Empty spaces (pores) in the region above this layer, big and small, fill with water. When the pores are full, the rock or soil is saturated. The top boundary of the saturated layer is the water table. Often the saturated layer itself is called the water table, but it is more accurately an aquifer. A shaft that penetrates the water table collects water, becoming a well. A depression deeper than the water table becomes a pond or lake.

water vapor (*materials science*)   The gas phase of water is called water vapor. Water vapor is an invisible, odorless, tasteless gas. Visible steam, fog, and clouds consist of liquid water droplets in suspension in air. (Water vapor used to power engines is also called steam.) Unlike the other gases in air, water vapor is constantly entering and leaving the mixture because of local conditions, so although it is one of air's main gases (after nitrogen and oxygen, the most abundant where humidity is high), its concentration varies greatly from place to place.

Watson, James (*biochemist*)   [American: 1928–] X-ray diffraction studies by Maurice H.F. Wilkins [English: 1916–] and Rosalind Franklin [English: 1920–58] in the late 1940s and early 1950s suggested that the DNA molecule might be a helix (spiral). Using this and other evidence, Watson and Francis H.C. Crick concluded that DNA is a double helix, consisting of 2 complementary chains wound around each other and attached by their nucleotide bases. Wilkins then verified that this Watson-Crick model is the only possible explanation for DNA diffraction patterns. Watson, Crick, and Wilkins shared the 1961 Nobel Prize in physiology or medicine.

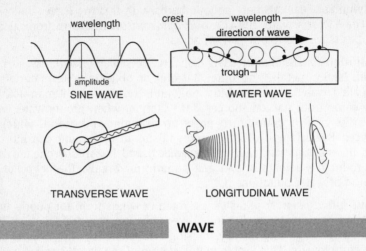

SINE WAVE

WATER WAVE

TRANSVERSE WAVE

LONGITUDINAL WAVE

WAVE

wave (*oceanography*)   The waves observed on the shores of oceans and other large bodies of water are the tops of circular movements of fluid, which originate from and are propelled by the strong winds of storms over water. As waves pushed by the storms move away, they merge, forming a regular pattern called swell. Although each drop of water moves only a short distance, swell can travel as waves for hundreds of miles. When swell reaches shallow water, each wave begins to slow, but the waves behind come in at the same rate, causing the front waves to rise up, often so high that they crash as breakers.

wave (*physics*)   A simple pattern of increase and decrease that moves through space is called a wave. Waves are able to transmit energy from place to place without actual or permanent movement of a body. Some waves must have a medium in which to exist, such as the **longitudinal waves** of sound or the **transverse waves** of water. **Light** waves and the other transverse waves of the **electromagnetic spectrum** do not need a medium since they create a pattern of moving fields in space alternating between the electric field and the magnetic field.

Quantum theory tells that every particle of matter can be viewed as a wave and furthermore every wave also acts as a particle, but this correspondence is difficult to observe except for very small particles or very short waves (the distance between repetitions of the pattern is the wavelength). In addition to length, waves are characterized by **amplitude, frequency, period,** and **phase.**

wave-particle duality (*particles*)   All entities that carry a force, whether one marble striking another or sunlight moving molecules of air, act sometimes as particles and sometimes as waves. For an object as large as a marble, the associated wave is too weak to detect, but small entities usually thought of as particles, such as electrons and atoms, can be observed as waves in certain experiments. Sunlight, which we think of as waves of electromagnetism, acts as particles when it pushes electrons from the surface of a metal. At the level at which we live, called the macroscopic level by particle physicists, this wave-particle duality is not apparent, but it is a constant presence when dealing with subatomic particles.

waxwing (*animals*)   Native to temperate regions of the Northern Hemisphere, waxwings (Family Bombycillidae) are small perching birds named for the red waxy material that forms on the tips of some feathers. They are 6 to 8 in. (15 to 20 cm) long, with a head crest, silky plumage, and 10 flight feathers per wing. Their thick bills are slightly hooked; their diet consists of insects and small fruits. Waxwings are primarily tree-dwellers. They are gregarious and often form large flocks.

weak interaction (*particles*)   The weak interaction, sometimes called the weak force, is the "force" holding some composite particles together until they decay; it is also involved in the decay of more massive leptons into lighter ones. The weak interaction is short range because the W and Z bosons that transmit it are massive. The force is called "weak" because it is about 1/100,000 as great as the electromagnetic force, although at short ranges the weak interaction is more powerful than gravity.

weasel (*animals*)   Mammals of the weasel family, Mustelidae, are generally small and medium-sized animals with long, lithe bodies and short, stocky legs. They are predators with sharp, curved claws and generally very aggressive. True weasels include the ermine, ferret, marten, mink, polecat, and wolverine. They range in size from the least weasel (*Mustela nivalis rixosa*), which weighs less than 2.5 oz (70 g), to the wolverine (*Gulo gulo*), which can weigh 70 lb (32 kg). The Mustelidae family also includes badgers, otters, and skunks.

weather (*meteorology*)   Weather is the combination of temperature, air movement, and moisture at a particular time and place (a combination strongly affected by changes in air pressure) along with the amount of sunlight or cloud cover. Thus weather can be categorized roughly as hot or cold, windy or still, damp or dry, or cloudy or sunny; for example, as 40°F (4°C) with 30 mph (50 km/hr) winds and periods of light rain beneath a sky with 90% cloud cover.

weathering (*geology*)   Weathering is rock breaking into smaller pieces or chemically changing, mainly as a result of the action of weather—heat, cold, rain, wind, and ice—but also through biological activity. Mechanical weathering breaks rock without changing its nature and includes splitting or cracking caused by ice, water, roots of plants, or animals digging holes. Chemical weathering includes oxidation of minerals by air or water, hydration (water combining with rock), and dissolution by weak natural acids.

weed (*botany*)   Any plant growing where it is not wanted is considered a weed. Many weeds are adapted to flourish in a variety of habitats. They typically have characteristics that enable them to out-compete and replace other plants—for example, deep root systems to absorb water and nutrients, rapid growth, and production of large quantities of seeds. Some host disease-causing organisms; the destructive wheat rust fungus (*Puccinia graminis*), for example, spends part of its life cycle on common barberry (*Berberis vulgaris*)—an ornamental plant in many gardens but a weed near wheat fields.

weevil (*animals*)   Certain beetles that have an elongated head with a more or less distinct snout with chewing mouthparts at the tip are called weevils. Many are serious pests of plants and stored food, both as larvae and adults. Among the most notorious are the cotton boll weevil (*Anthonomus grandis*), rice weevil (*Sitophilus oryzae*), and granary weevil (*Calendra granaria*).

Wegener, Alfred Lothar (*meteorologist/geologist*)   [German: 1880–1930] Trained as an astronomer, but working mostly as a meteorologist, Wegener is primarily remembered for his contribution to geology, the theory of continental drift. He named and described Pangaea and provided cogent arguments as to how this supercontinent broke into today's separate continents about 200,000,000 years ago. Although Wegener was not the first to recognize the breakup and separation of continents, he was the most effective advocate for the idea.

weight (*measurement*)   Weight is the force exerted by gravity on an object by the object's environment. In most environments, such as the surface of a planet, weight is proportional to mass. If 1 object has twice the mass of another and both are near each other on Earth, the first object will weigh twice as much as the second. Objects in an orbiting spaceship, however, are weightless no matter what their masses. Weight is measured as pounds or newtons, since it is a force, but because weight on Earth's surface reflects mass, people also use kilograms (which measure mass) to express weight.

weightlessness (*mechanics*)   When the force of gravity is exactly balanced by acceleration, weight—the effect of gravity on mass—vanishes, a condition called weightlessness. This can occur when a body is moving unimpeded toward the source of gravitational attraction, called free fall. Weightlessness can also be achieved when a body is on a curving path that accelerates the body away from gravitational force by an equal force.

An orbiting satellite is not beyond Earth's gravity, which extends infinitely into space, but is freely falling toward Earth at the same time as it moves forward. The combination of motions produces a satellite's elliptical orbit.

Weismann, August (*biologist*)   [German: 1834–1914] In a famous series of experiments in which he cut off the tails of mice for 22 generations, Weismann disproved the theory that acquired characteristics could be inherited.. A strong proponent of Darwinian evolution, Weismann also proposed the germ plasm theory. This theory suggested that while the body, which Weismann called somatoplasm, lives for only 1 generation, hereditary material, which he called germ plasm, is immortal, passed from generation to generation without change.

wetland (*ecology*)   Swamps, marshes, bogs, and other areas that are regularly saturated by water and that contain vegetation adapted for life in saturated soil are called wetlands. Inland wetlands contain fresh water; coastal wetlands contain salty or brackish water. Wetlands are among Earth's most valuable habitats. They are home for an enormous variety of plants and animals, serving as breeding and feeding grounds for many fish, most amphibians, and various birds and mammals. Wetlands limit flooding by acting as sponges after heavy rains and ice melts and filter contaminants from water.

The loss of wetlands, mainly due to drainage for farmland, is a serious environmental issue. Since the 1700s, for example, the contiguous United States has lost 53% of its wetlands.

whale (*animals*)   Whales are large, streamlined cetaceans found in all oceans. They include the blue whale, the largest animal that has ever lived on Earth. There are 2 types: toothed and baleen. Toothed whales, such as sperm and bottle-nosed whales, have teeth and a single blowhole. Baleen whales, such as blue and humpback whales, have sheets of baleen instead of teeth, and 2 blowholes.

wheat (*plants*)   Members of the grass family native to temperate Europe and Asia, wheats (*Triticum*) have been cultivated since prehistoric times for their edible grains (fruits) and as fodder. Today most cultivated wheats are varieties of bread wheat (*T. aestivum*) and pasta wheat (*T. durum*). Wheats typically have hollow unbranched stems up to 5.6 ft (1.7 m) high. The flowers are borne at the top of the stems, in heads of 20 to 100.

whelk (*animals*)   Large marine snails, whelks have a pointed, usually thick-walled, shell with only a few whorls, which may be smooth or decorated with knobs, ridges, and the like. The shell's opening is an oval through which the head and large foot can be completely withdrawn. Whelks are found in coastal waters worldwide, moving along the bottom in search of clams and other invertebrate prey.

white blood cell (*cells*)   Leukocytes, or white blood cells (WBCs), are colorless, comparatively large blood cells that contain nuclei. Their function is to fight infection. Some WBCs are phagocytes that engulf and destroy bacteria and other foreign particles. Other WBCs produce proteins called antibodies, which attack foreign substances. WBCs can move from the blood through the walls of capillaries into the surrounding tissues. They then return to the blood system via the lymph.

WBCs originate in the bone marrow. They can be divided into 3 groups. Granulocytes have a lobed nucleus and contain many granules in the cytoplasm; they include basophils, eosinophils, and neutrophils. Monocytes have a large horseshoe-shaped nucleus and can become phagocytic macrophages. Lymphocytes have a single large round nucleus; some are killer cells while others are important in the immune response.

white dwarf (*stars*)   White dwarfs are stars that have collapsed after fusion stops. The diameter of the white dwarf is less than that of the planet Jupiter, but usually larger than that of Earth. Since most of the star's mass remains, a single teaspoonful of the matter in a white dwarf weighs about 5 tons.

A star collapses because it has used all its lighter elements for fusion, which before being exhausted had provided energy to overcome gravity. The collapse is finally halted by the Pauli exclusion principle—the electrons cannot occupy the same space at the same time. After the collapse the star retains the heat energy from its larger size, so the surface is much hotter than it was before, accounting for the whiteness. The interior, however, is cooler than it was during fusion. Conservation of angular momentum spins the star rapidly, producing an intense magnetic field. But since there is no internal energy source, the star eventually cools to a black dwarf that is no longer visible from Earth.

whole number (*numbers*)   In elementary school arithmetic the whole numbers are the natural numbers including 0, but in later mathematics the term is often synonymous with integers.

wild type (*genetics*)   Originally the term wild type referred to the phenotype (physical characteristics) that is most common in wild (natural) populations of a species. Today the term also refers to the normal phenotype present in a laboratory population. Deviants from the wild type are called mutants. For example, wild-type *Drosophila melanogaster* fruit flies have red eyes; flies with white eyes are mutants.

willow (*plants*)   Willows (*Salix*) derive their scientific name from Latin words meaning "near water," an apt description of the usual habitat of many species. Willows range from dwarf arctic species only 1 in. (2.5 cm) tall to forms that exceed 100 ft (30 m) in height. They have narrow, tapering leaves and often colorful branches.

The willow family, Salicaceae, comprises some 300 species of willows and poplars. These are fast-growing deciduous shrubs and trees widely distributed in the Northern Hemisphere. The minute male and female flowers are borne on separate plants in clusters called catkins. The fruits are capsules containing many seeds, each with silky hairs that aid in wind dispersal.

Wilson, Edward O. (*biologist*)   [American: 1929–] Wilson first distinguished himself by becoming the world's leading authority on ants. His discoveries included the finding that ants communicate primarily through pheromones. Wilson then became the principal proponent of sociobiology; his belief that there is a biological basis for the social behavior of all animals, including humans, has aroused much controversy. Wilson also has been a major voice in efforts to maintain Earth's biodiversity.

wind (*atmospherics*)   The movement of air parallel to or nearly parallel to Earth's surface is called wind, but wind is always part of one or more rotary motions. Wind is part of a complex pattern called circulation of atmosphere. Air rises and falls in circular patterns called cells. Winds themselves turn right or left as a result of the Coriolis effect. The main cause of all this motion is unequal heating of different parts of Earth. Since Earth is almost spherical, different parts receive different amounts of sunlight, some of which must travel farther through the atmosphere to reach the surface.

That surface may be land, which heats up and cools down relatively quickly, or water, which changes temperature less rapidly. On a small scale this difference causes winds known as sea or land breezes and on a large scale causes monsoons.

Most winds are named for the direction from which they come (for example, the north wind), but strong, warm winds that flow from mountains onto plains have such regional names as Chinook (northern Rockies), foehn (Alps), Santa Ana (Sierras and Rockies), and sirocco (northern Africa).

wind-chill factor (*meteorology*)   Movement of air across a source of moisture or heat usually withdraws humid or warm air and replaces it with drier or cooler air, encouraging faster cooling in either case (since evaporation produces cooling). The amount of cooling produced by wind can be calculated for a given wind speed and temperature and is available in tables. A moderate wind of 10 mph (16 km/hr) can cause 0°F (–18°C) to have the cooling effect of –22°F (–30°C). The effect is more pronounced for lower temperatures and higher winds.

wing (*anatomy*)   Several types of animals have paired, movable appendages called wings that are specialized for flight. In birds, the 2 wings are modified forelimbs and are covered with feathers. The 2 wings of a bat also are modified forelimbs. Insect wings are thin, double-layered outgrowths of the body wall of the thorax, with longitudinal thickenings called veins (useful in species identification). Most insects possess 2 pairs of wings but some species have 1 pair and some are wingless.

Wöhler, Friedrich (*chemist*)   [German: 1800–82] Wöhler's fame is based on his synthesis of urea, understood in 1828 as the first creation of an organic compound from an inorganic one. Since then science historians have found an earlier example and have termed Wöhler's precursor chemical partly organic. Wöhler was an outstanding chemist in many ways, and, with Justus Liebig [German: 1803–73], discovered that the same radical can appear in several related compounds. Wöhler was also the first to isolate aluminum and beryllium.

wolf (*animals*)   Native to the Northern Hemisphere, wolves are members of the genus *Canis* in the dog family. Looking like large dogs with a long, hanging tail, they live in a variety of habitats, though their range has been greatly restricted by human activities. Wolves are social animals, with a highly complex pack structure and a clear dominance hierarchy in which mating occurs only between the most dominant wolves, called alpha male and alpha female. Wolves prey on small animals and eat berries and carrion; as a pack they will attack deer and other large mammals.

wombat (*animals*)   Burrowing marsupials native to Australia, wombats spend the day in underground dens, emerging at night to feed on plant matter. Looking like small bears, with short limbs and a thick body, they are about 3 ft (0.9 m) long and can weigh as much as 100 lb (45 kg).

wood (*botany*)   Each year a woody stem grows in diameter as the vascular cambium produces new xylem and phloem cells. Xylem formed in recent years is called sapwood, because it transports a lot of watery sap. As new sapwood forms each year, its innermost layers, or annual rings, are squeezed and become clogged with resins and oils. Called heartwood, this region is darker than sapwood; it can no longer conduct water, but it continues to provide support to the tree.

Rows of thin-walled cells, called rays, extend from the stem's center; these cells carry food from the phloem into the wood. The stem's age can be deter-

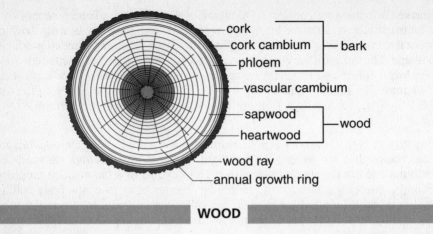

cork
cork cambium — bark
phloem

vascular cambium

sapwood
heartwood — wood

wood ray
annual growth ring

WOOD

mined by counting the annual rings in the xylem, but no such distinct regions form in the phloem.

woodpecker (*animals*)   Constituting the bird family Picidae, woodpeckers are adapted for life in trees. They have short legs with long, strong feet and stiff tail feathers that can be used as a brace against trees. The straight bill can drill and gouge into trees in search of insect prey or excavate a hole for nesting or roosting. Woodpeckers are found in most parts of the world. The approximately 200 described species, including flickers and sapsuckers, range in length from 3.5 in. (8.9 cm) to 23 in. (58 cm).

The woodpecker order, Piciformes, also includes jacamars, puffbirds, honeyguides, toucans, and barbets. All inhabit woodlands and forests and have feet designed for climbing, with 2 toes pointing forward and 2 projecting backward (called zygodactylous feet).

work (*thermodynamics*)   In physics a force that causes an object to move produces work, an entity that can be measured as the product of the force and the distance moved. Work is measured in foot-pounds or newton-meters (joules), the same units used to measure energy. If the force does not move the object, then no work is produced. Thus the definition in physics is different from the ordinary definition of work.

worm (*animals*)   A great variety of small, elongated invertebrates are commonly called worms, including the structurally dissimilar flatworms, ribbon worms, and earthworms. Certain insect larvae are also called worms—the cabbageworm of the cabbage butterfly, for example. Among vertebrates, the slowworm is a lizard.

W particle (*particles*)   The bosons that produce such weak interactions as beta decay are called the positive and negative W particles. At high energies, the weak force merges with the electromagnetic force, so the photon and W particles are to some extent analogs. W interactions are unusual in that they interact only with particles that have a left-handed spin. They were first observed in 1983 by Carlo Rubbia [Italian: 1934–].

wrasse (*animals*) Constituting Family Labridae, wrasses are marine fish—often brightly colored—found in tropical and warm temperate waters, generally in coral reefs and shallow coastal habitats. Wrasses show considerable variety in shape and range in length from less than 6 in. (15 cm) to almost 10 ft (3 m). The dorsal fin has 8 to 14 spines. Wrasses have strong jaw teeth plus teeth in the throat; most are carnivorous.

wren (*animals*) Perching birds native to the Americas, wrens (Family Troglodytidae) are energetic, quick-moving creatures with spotted and striped brown or gray plumage. They are 4 to 6 in. (10 to 15 cm) long, with a pointed bill, short wings, and a short tail typically carried straight up. Wrens generally live on or near the ground, feeding mostly on insects.

wryneck (*animals*) A relative of woodpeckers, the wryneck (*Jynx torquilla*) is native to Europe, Asia, and Africa. It is about 6.5 in. (17 cm) long, with speckled grayish-brown plumage. Like woodpeckers, the wryneck has a long tongue and zygodactylous feet (2 toes pointing forward and 2 projecting backward). However, it has a slender, sharply pointed bill and lacks the stiff tail feathers seen on woodpeckers. It feeds on ants and insect larva. The common name comes from the bird's habit, during courtship and defense, of twisting its neck and jerking its head.

wulfenite (*rocks and minerals*) Wulfenite is a lead-molybdenum ore that usually forms near lead deposits. Its crystals, often square and either transparent or translucent, may be a brilliant yellow or orange, making the ore popular with collectors.

X

X and Y chromosomes (*genetics*)   In mammals the sex chromosomes, which determine whether an organism is male or female, are called X and Y. A female mammal has 2 X chromosomes; a male has 1 X and 1 Y. X and Y chromosomes have different genes. For instance, an important gene on the Y chromosome is the testis-determining gene, which is responsible for development of a testis (the organ that produces sperm).

xanthophyll (*botany*)   Xanthophylls are pale yellow pigments synthesized in the chloroplasts of plants and algae that aid in photosynthesis. Their color usually is masked by the green pigment chlorophyll, but they are among the pigments responsible for the striking colors of autumn leaves. Xanthophylls also are common in flower petals and fruits.

xenon (*elements*)   Xenon, atomic number 54, was discovered in 1898 by Alexander Ramsay [Scottish: 1852–1916] and Morris William Travers [English: 1872–1961] and named from the Greek *xenon*, "stranger." In 1962 xenon became the first noble gas to be combined with another element (fluorine) to form a true compound. Chemical symbol: Xe.

xerophyte (*botany*)   Plants adapted to living in deserts and other habitats where water is scarce are xerophytes ("dry plants"). They have various adaptations that reduce water loss by transpiration, including sunken stomata (pores), stomata that open only at night, thick waxy cuticles on leaves, permanently rolled leaves, and leaves reduced to spines. Some xerophytes are succulents with water-storage organs. Others are ephemerals, completing their life cycle during a short wet season.

X ray (*waves*)   X rays are an energetic, short-wavelength portion of the electromagnetic spectrum between ultraviolet radiation and gamma radiation—roughly from 10,000,000,000,000,000 Hz to 1,000,000,000,000,000,000 Hz. They are best known for passing through materials opaque to visible light. Their penetrating power makes them useful in cancer therapy, since they destroy cells. X rays are also produced in high-energy stellar and galactic processes, which are studied via X-ray telescopes on satellites since X rays from space cannot penetrate the atmosphere to reach Earth's surface.

xylem (*botany*)   One of the two types of conducting, or vascular tissue of plants, xylem functions mainly to carry water from the roots to other parts of a plant. In trees and shrubs, xylem makes up the greatest portion of the stem—the wood—and its strong fibers support the plant body.

yak (*animals*)   A ruminant mammal closely related to cattle, the yak (*Bos grunniens*) is native to high plateaus and mountains of Central Asia. It has long horns that curve outward and upward, a hump at the shoulders, and a thick coat of very long hair. Wild yaks, considered an endangered species, grow to 6.5 ft (2 m) tall at the shoulder and weigh up to a ton. Domestic yaks are smaller.

Yalow, Rosalind (*medical physicist*)   [American: 1921–] Together with Solomon Berson [American: 1918–72], Yalow developed radioimmunoassay (RIA), a highly sensitive laboratory technique that uses radioactive isotopes to measure minute concentrations of hormones, enzymes, viruses, and other biological substances. Yalow shared the 1977 Nobel Prize in physiology or medicine with Roger Guillemin [American: 1924–] and Andrew Schally [American: 1926–], who used RIA to isolate and identify brain hormones.

yam (*plants*)   Several species of the genus *Dioscorea* are cultivated for their large edible tubers (underground stems) called yams. These species are twining vines native to tropical and warm temperate areas. The leaves are usually heart-shaped. The small flowers grow in clusters and produce 3-angled capsules filled with winged seeds. Yams are classified in the family Dioscoreaceae. They are not the same as sweet potatoes (*Ipomoea batatas*), which belong to the morning glory family.

yarrow (*plants*)   A hardy, aromatic perennial of the composite family, yarrow (*Achillea millifolium*) is native to Eurasia. Yarrow acquired its generic name from the Greek warrior Achilles, who reputedly used it to treat soldiers' wounds. The plant has since been shown to have antiinflammatory properties and the ability to stop hemorrhaging. Yarrow has erect stems, finely divided compound leaves, and flat flower heads of white, yellow, pink, or purple. It grows to about 39 in. (1 m) tall.

year (*solar system*)   A year is the time of one revolution of Earth about the Sun, but that can be defined in several ways. The year used for the calendar is measured from spring equinox to spring equinox and is called the tropical or solar year (365.2422 days). Because of the precession of the equinoxes this is somewhat shorter than the sidereal year (365.2564 days), which is measured against the background of the stars. Tropical years are becoming shorter and so scientists compare these with an exact year based on 365 days times 86,400 seconds (366 days in leap years), or 31,536,000 seconds, and add a second every few years to keep the calendar from slipping.

yeast (*fungi*)   Yeasts are microscopic, single-celled sac fungi; they do not have hyphae or fruiting bodies, but they do produce reproductive sacs (asci) that contain spores. However, yeasts most commonly reproduce by budding, forming tiny swellings ("buds") that grow into new individuals. The offspring may separate from the parents or remain attached as part of chainlike colonies. Most species are saprophytes. Of particular economic importance are various strains of *Saccharomyces cerevisiae*, which produce carbon dioxide and alcohol during fermentation and are used in baking and making alcoholic beverages.

yellowlegs (*animals*)   Shorebirds in the sandpiper family, yellowlegs (genus *Tringa*) are named for their bright yellow legs. They are up to 15 in. (38 cm) long, including the bill, and have a wingspan up to 26 in. (66 cm). Yellowlegs nest on the ground in Alaska and Canada during the summer. They migrate south for winter, some flying as far as Chile or Argentina. The diet consists mainly of small fish, insects, and crustaceans.

yew (*plants*)   Conifers of the genus *Taxus*, commonly called yews, are evergreen shrubs and trees native to temperate and subtropical areas around the world. The leaves are dark green needles. The fruit is a fleshy red cup that surrounds a small hard seed. Both the leaves and the seeds contain highly toxic alkaloids. The English yew (*T. baccata*), native to Europe, northern Africa, and western Asia, is an important ornamental. The bark of the Pacific yew (*T. brevifolia*), native to western North America, is the original source of the drug taxol, used for treating certain cancers.

yolk (*physiology*)   The eggs of certain animals, particularly reptiles and birds, contains a yellowish material called yolk. A mixture of carbohydrates, proteins, and fats, yolk provides nourishment for a developing embryo.

Young, Thomas (*physicist/physician*)   [English: 1773–1829] Young is famous for his work on light and vision. He showed that changes in curvature of the eye's lens focus light, found the cause of astigmatism, and proposed the correct 3-color explanation of color vision. Young's experiments with interference of light waves and diffraction of shadows established the wave nature of light. Young was the first to recognize that light travels as transverse waves.

In other areas, Young was the first to define energy in the modern sense. He improved on Hooke's law by providing a measure—Young's modulus—showing the constant of proportionality between force and stretching for different substances. He even helped establish the basis for translation of Egyptian hieroglyphics.

ytterbium (*elements*)   Discovery of the rare earth ytterbium, atomic number 70, is credited to Georges Urbain [French: 1872–1938] in 1907, although it was also found that year by Karl Auer (Baron von Welsbach) [Austrian: 1858–1929]. It is named for Ytterby, a village in Sweden, where its ore was found in combination with erbium, lutetium, and yttrium. The pure metal was not isolated until 1953. Chemical symbol: Yb.

yttrium (*elements*)   Yttrium, chemical symbol Y, is a silvery, somewhat reactive metal similar to the rare earths, or lanthanide metals. Its atomic number, 39, shows that this resemblance occurs because yttrium is above lanthanum in the periodic table. It was found in a quarry near the village of Ytterby in Sweden near Vauxhaul along with several rare earths (erbium, lutetium, terbium, and ytterbium). Carl Gustav Mosander [Swedish: 1797–1858], who isolated it in 1843, named it after Ytterby. Yttrium compounds and alloys have many important uses, providing red for color television, filtering microwaves, catalyzing plastics, and contributing to high-temperature superconducting ceramics.

yucca (*plants*) Species of the genus *Yucca* are shrubs or trees with long, sword-shaped leaves. Huge clusters of creamy, lantern-shaped flowers are borne on vertical stalks that may be 10 ft (3 m) tall. Yuccas are native to drier areas of North America, with the greatest variety of species occurring in Mexico and the southwestern United States. Spanish bayonet and the Joshua tree are well-known species. In an interesting example of coevolution, all yuccas are pollinated by yucca moths (genus *Tegeticula*).

Yukawa, Hideki (*physicist*) [Japanese: 1907–81] Yukawa proposed in 1934 that the force that holds the nucleus of an atom together, which has to be strong enough to overcome repulsion of positive protons, is produced by the exchange of previously unrecognized particles. He calculated that a particle's mass is about 200 times the electron's. The particles, pions, were found in 1947 in cosmic rays. Yukawa also correctly predicted that under certain circumstances an atomic nucleus can absorb an inner electron, changing 1 proton to a neutron.

zebra (*animals*) Easily recognized by their striped coat, zebras (genus *Equus*) are hoofed mammals in the horse family, Equidae. They are about 7.5 ft (2.3 m) long and weigh up to 770 lb (350 kg). There are 3 living species. All are native to savannas of eastern and southern Africa, where they form herds, often with other grazing animals. Another species, the quagga (*E. quagga*), became extinct in the 19th century due to overhunting. It had stripes only on its head, neck, and chest.

Zeeman effect (*physics*) Two types of splitting of the lines of the spectrum occur when the material emitting light is placed in a magnetic field. The Zeeman effect, established by Pieter Zeeman [Dutch: 1865–1943] occurs because atoms aligned with the field have slightly different energy levels in electron orbits from those not aligned. A smaller splitting, called the anomalous Zeeman effect, occurs because each electron has an up or down magnetic field that is similar to Earth's north and south poles. Discovery of the anomalous Zeeman effect led to the concept of spin and the Pauli exclusion principle.

zenith (*astronomy*) The zenith is often defined as the point in the sky directly above an observer, a definition that pictures the sky as a crystal sphere. A modern definition has the zenith as the extension above an observer of the diameter of Earth's sphere that has 1 end at the observer. The extension of the other end of this diameter is the nadir.

Zeno of Elea (*mathematician*) [Greek: c450 BC] Zeno was a critic of the idea of infinity, expressing his views in the form of paradoxes. For example, he noted that an arrow in flight has a definite location at any instant and hence cannot move. His best-known paradox concerns a slow tortoise racing a speedy Achilles. Suppose the tortoise is given a head start. Achilles must reach each point where the tortoise has been, but by that time the tortoise will have moved on—so the tortoise always wins.

zeolite (*rocks and minerals*) Zeolites are natural or artificial minerals that have many tiny pores, used to filter materials at a molecular level. Natural zeolites are typically hydrates of aluminum silicates, but the water is removed before applications. Zeolites may have catalysts inserted within pores, making it possible to expose more molecules to catalysis as they pass through the materials.

zero (*numbers*) Zero (0) is the identity element for addition, meaning that a number and its additive inverse add to 0. Zero is used as the starting place for measurements; describes the size of a set with no members; and is the placeholder needed in any place-value numeration system.

zinc (*elements*)   Zinc, atomic number 30, is among the most useful metals; it is the corrosion-resistant coating on galvanized metal, an important alloy metal, and a source of industrial compounds, especially zinc oxide. Its discovery and the origin of the name are prehistoric. The earliest use was in brass, an alloy with copper widely smelted long before zinc was recognized as a separate substance. Other alloys include German silver, typewriter metal, and solder. Zinc is an essential trace element for all animals. Chemical symbol: Zn.

zirconium (*elements*)   Zirconium is a metal, atomic number 40, discovered in 1789 by Martin Klaproth [German: 1743–1817]. It is corrosion-resistant and does not absorb neutrons easily, making it useful in nuclear reactors as well as for machine parts used in the chemical industry. The name is from zircon (zirconium silicate), thought to come from Persian for "gold color," which describes the gem. Zircon is among the oldest mineral inclusions in rocks, unchanged since early in Earth's history. Chemical symbol: Zr.

zodiac (*stars*)   As the Sun, Moon, and planets move through the sky, they pass through a group of 12 constellations called the zodiac. The zodiac constellations are those that lie within a few degrees of the plane of the ecliptic. Chaldean astronomers about 1500 BC believed that these constellations influence happenings on Earth, thus laying the groundwork for astrology. Because of precession the traditional 12 constellations are related differently to the Sun, Moon, and planets today, but astrologers base their studies on where the constellations used to be. Today the Sun even passes through a thirteenth constellation (Ophiuchus), but astrologers ignore it.

zodiacal light (*solar system*)   A faint cone of light seen after sunset or before sunrise along the plane of the ecliptic, caused by the reflection of sunlight on tiny dust particles in space, is called zodiacal light. Opposite the Sun's position is an even fainter form of zodiacal light known as gegenshein.

zonation (*ecology*)   The distribution of natural communities in distinct bands is called zonation. For example, as altitude increases on a mountainside, temperate deciduous forests give way to coniferous forests, which give way to alpine tundra. Similar zonation occurs with increases in latitude. Zonation occurs in comparatively small areas, too. On a beach, for example, different species are found in low tide, high tide, and splash zones.

zoology (*sciences*)   Zoology is the branch of biology that deals with the study of animals. It can be divided into many overlapping fields. Some zoologists specialize in fields such as animal physiology, anatomy, or behavior. Others focus on specific types of animals, such as insects (entomology), fish (ichthyology), birds (ornithology), or parasites (parasitology).

Z particle (*particles*)   The neutral boson that produces the weak interaction leading to decays into neutrino-antineutrino or quark-antiquark pairs is called the Z particle. At high energies, the weak interaction merges with the electromagnetic force, so the photon and Z are to some extent analogs. They were first observed in 1983 by Carlo Rubbia [Italian: 1934–].

Index

Boldface type indicates a main entry for the subject and on what page it can be found.

Index

Index

Index

Index

Index

Index

Index

Index

Index

Index

Index

Index

Index